Reagents for Organic Synthesis

Fieser and Fieser's

Reagents for Organic Synthesis

VOLUME ELEVEN

Mary Fieser

Harvard University

A WILEY-INTERSCIENCE PUBLICATION

JOHN WILEY & SONS

NEW YORK • CHICHESTER • BRISBANE •
TORONTO • SINGAPORE

Library of Congress Cataloging in Publication Data:

(Revised for vol. 11)

Fieser, Louis Frederick, 1899–
Reagents for organic synthesis.

Authors' names in reverse order in v. 2–7.
Vol. 8, by Mary Fieser.
Vol. 9 by Mary Fieser, Rick L. Danheiser, William Roush.
Vol. 8– has title: Fieser and Fieser's Reagents for organic synthesis.
Vol. 2–3 have imprint: New York, Wiley-Interscience.
Vol. 4– A Wiley-Interscience publication.
Includes bibliographical references and indexes.
1. Chemistry, Organic—Synthesis. 2. Chemical tests and reagents. I. Fieser, Mary A. Peters, 1909– joint author. II. Title. III. Title: Fieser and Fieser's Reagents for organic synthesis.
QD262.F5 547'.2 66-27894

ISBN 0-471-88628-9 (v. 11)

Printed in the United States of America

10 9 8 7 6 5 4 3 2 1

PREFACE

This volume of reagents includes references to papers published during the period from mid 1981 through 1982. I am indebted to many colleagues for advice and help and am particularly grateful to Alan E. Barton, Paul B. Hopkins, William Seibel, and Andrew Myers, who read large portions of the manuscript. I am indebted again for invaluable proofreading by Professor Dale Boger, Professor Paul B. Hopkins, Dr. James Heck, Dr. Steven Freilich, Thomas Eckridge, John C. Schmidhauser, Daniel F. Lieberman, Jay W. Ponder, Marifaith Hackett, Alan E. Barton, Dr. Lewis Manring, Joseph Tino, Theodore S. Widlanski, Robert W. Hahl, Ernie-Paul Barrette, Stephen Wright, Katharine Brighty, Karl Fisher, Nathan C. Ihle, Dr. Janusz Kulagowski, Regan G. Shea and Jeffrey N. Fitzner. The proofreading was again coordinated by Peter Lansbury, Jr. The picture of some members of the group was taken by Alan Barton.

MARY FIESER

Cambridge, Massachusetts
November 1983

CONTENTS

Reagents for Organic Synthesis

A

Acetic anhydride, 1, 3; **2,** 7–10; **5,** 3–4; **6,** 1–2; **7,** 1; **8,** 1–2; **9,** 1.

Pummerer cyclization.[1] Pummerer rearrangement of 4-(phenylsulfinyl)butyric acids (**1**) catalyzed by TsOH results in 4-phenylthio-4-butanolides (**2**). Oxidation followed by thermolysis results in either **3** or **4**, depending on the substitution pattern.

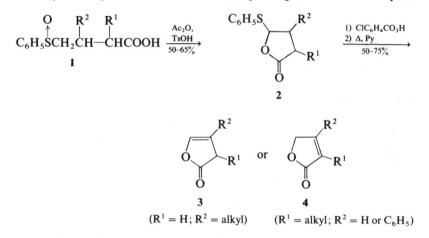

$$C_6H_5\overset{O}{\overset{\uparrow}{S}}CH_2\overset{R^2}{\overset{|}{C}}H{-}\overset{R^1}{\overset{|}{C}}HCOOH \xrightarrow[50\text{–}65\%]{Ac_2O,\ TsOH} \mathbf{2}\ (C_6H_5S,\ R^2,\ R^1) \xrightarrow[50\text{–}75\%]{1)\ ClC_6H_4CO_3H\ \ 2)\ \Delta,\ Py} \mathbf{3}\ \text{or}\ \mathbf{4}$$

1 **2**

3 (R^1 = H; R^2 = alkyl) or **4** (R^1 = alkyl; R^2 = H or C_6H_5)

[1] M. Watanabe, S. Nakamori, H. Hasegawa, K. Shirai, and T. Kumamoto, *Bull. Chem. Soc. Japan*, **54**, 817 (1981).

Acetic anhydride–Boron trifluoride etherate.

Selective acetylation.[1] This combination is useful for fairly selective acetylation of primary or secondary hydroxyl groups in the presence of a phenolic group.

[1] Y. Nagao, E. Fujita, T. Kohno, and M. Yagi, *Chem. Pharm. Bull.*, **29**, 3202 (1981).

Acetone cyanohydrin, 1, 5.

Aromatic aldehydes. Acetone cyanohydrin can be used in place of hydrogen cyanide in the Gattermann reaction for formylation of arenes.[1]

Example:

$$\text{1,3,5-}(CH_3)_3C_6H_3 + (CH_3)_2C(OH)CN \xrightarrow[48\%]{AlCl_3,\ ClCH_2CH_2Cl,\ 83^\circ} \text{2,4,6-}(CH_3)_3C_6H_2CHO$$

[1] A. Rahm, R. Guilhemat, and M. Pereyre, *Syn. Comm.*, **12**, 485 (1982).

Acetonedicarboxylic acid (3-Ketoglutaric acid), $HOOCCH_2COCH_2COOH$ (**1**). Mol. wt. 146.10, m.p. 133° (dec.). Supplier: Aldrich. Preparation.[1]

1,3-Annelation of imines.[2] A key step in a recent synthesis of lycopodine (**4**) is the reaction of **1** with the imine **2** to yield a tricyclic ketone **3**.

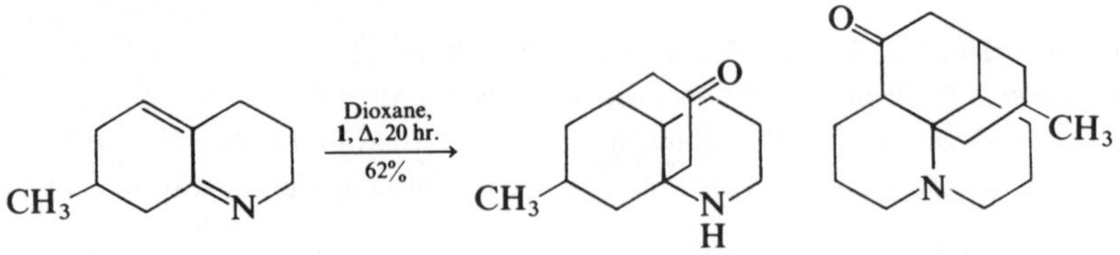

Dioxane, 1, Δ, 20 hr. 62%

2 **3** **4**

[1] T. W. Abbott, R. T. Arnold, and R. B. Thompson, *Org. Syn. Coll. Vol.*, **2**, 10 (1943).
[2] D. Schumann, H.-J. Muller, and A. Naumann, *Ann.*, 1700 (1982).

2-Acetoxy-1-methoxy-3-trimethylsilyloxy-1,3-butadiene (1).
Preparation:

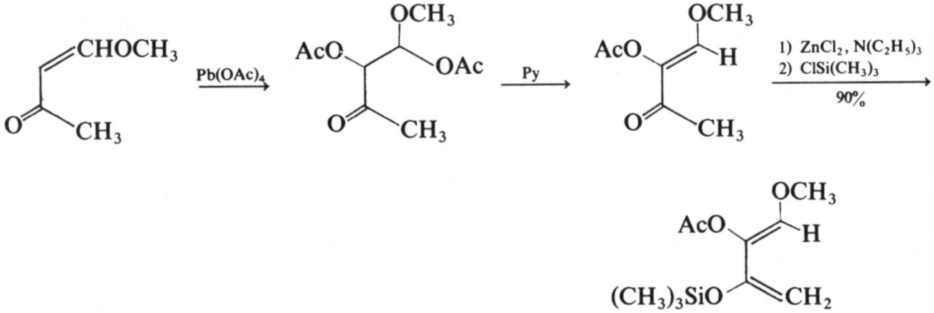

Pb(OAc)₄ → Py → 1) $ZnCl_2$, $N(C_2H_5)_3$ 2) $ClSi(CH_3)_3$ 90%

1 (b.p. 62–64°/2 mm)

Diels-Alder reactions.[1] This highly substituted silyloxydiene is comparable to 1-methoxy-3-trimethylsilyloxy-1,3-butadiene (**6**, 370; **9**, 303–304) in reactivity in Diels–Alder reactions.

Examples:

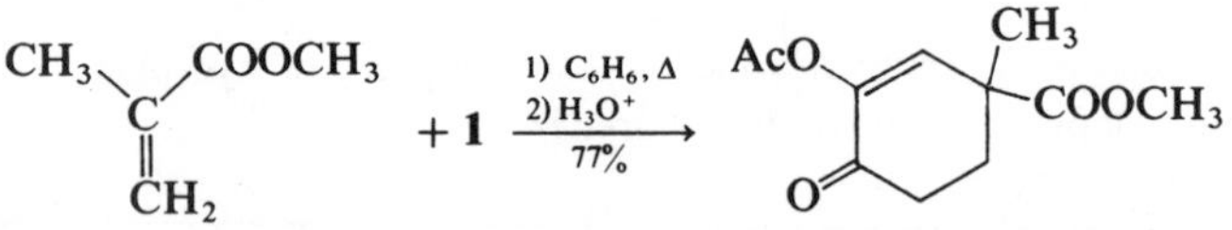

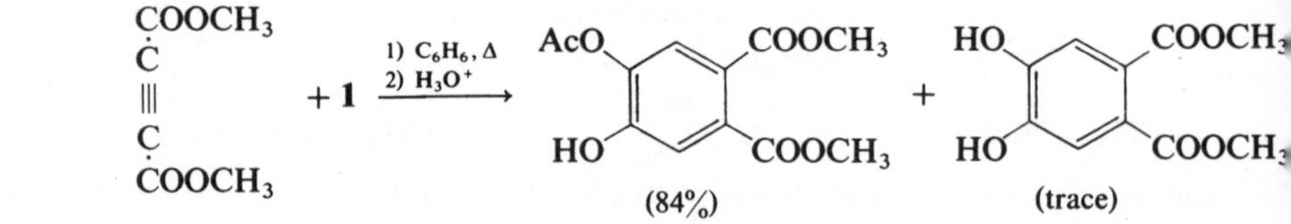

+ **1** 1) C_6H_6, Δ 2) H_3O^+ 77%

+ **1** 1) C_6H_6, Δ 2) H_3O^+ (84%) + (trace)

Cycloaddition of **1** with chiral **2**, derived from L-glutamic acid, provides optically pure L-Dopa (**3**).

1 + [structure **2**] $\xrightarrow[48\%]{120°}$ [spiro adduct] $\xrightarrow[58\%]{OH^-}$

[structure] (α_D + 1.2°) ⟶ **3**

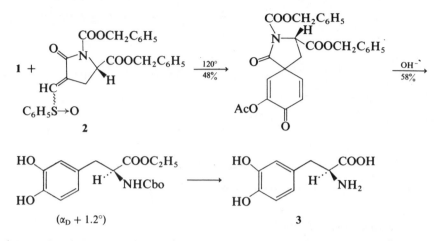

[1] S. Danishefsky and T. A. Craig, *Tetrahedron*, **37**, 4081 (1981).

(1E,3E)-4-Acetoxy-1-trimethylsilyl-1,3-butadiene (1).

Preparation:

$$(CH_3)_3SiCH_2CH=CH_2 \xrightarrow[60\%]{\text{1) } sec\text{-BuLi, TMEDA; 2) } HCON(CH_3)_2\text{; 3) } Ac_2O} (CH_3)_3Si\text{–CH=CH–CH=CH–OAc} + \text{(1E, 3Z)-isomer}$$

1 (b.p. 54°/0.2 mm)

Diels-Alder reactions. This diene can serve as a precursor to the highly oxygenated cyclohexane derivative shikimic acid (**3**), as shown in Scheme **1**. Oxidative desilylation of the Diels-Alder adduct **2** could not be effected with peracids, but was effected by *cis*-dihydroxylation (Upjohn procedure, **7**, 256–257) followed by β-elimination of $(CH_3)_3SiOH$ with TsOH. Introduction of the 4α,5β-diol system was effected indirectly from the 4α,5α-epoxide in several steps, since direct hydrolysis of the epoxide resulted in a mixture of three triols.[1]

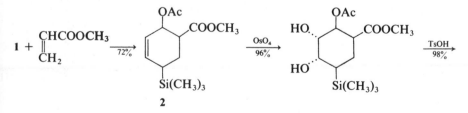

1 + $CH_2{=}CHCOOCH_3$ $\xrightarrow{72\%}$ **2** $\xrightarrow[96\%]{OsO_4}$ [diol] $\xrightarrow[98\%]{TsOH}$

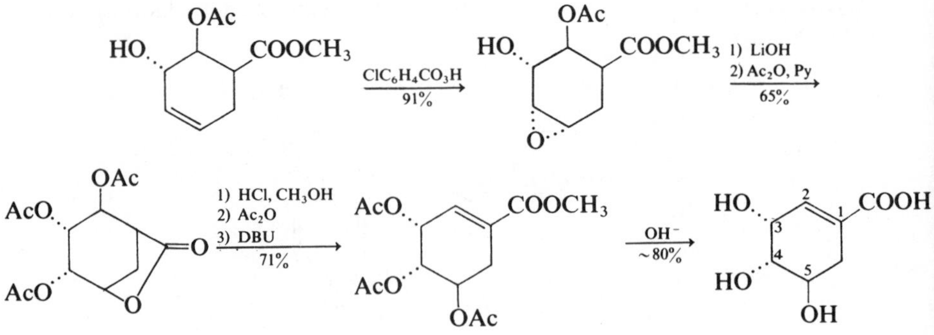

Scheme (I)

[1] M. Koruda and M. A. Ciufolini, *Am. Soc.*, **104**, 2308 (1982).

Acetyl chloride–2-Trimethylsilylethanol.

α-Methylenecyclopentenones. A small group of antibiotics has the general structure shown in A. Although these compounds are generally unstable, their synthesis has been achieved by Smith *et al.*[1,2] by retrolactonization of precursors of structure B with acetyl chloride and 2-trimethylsilylethanol.

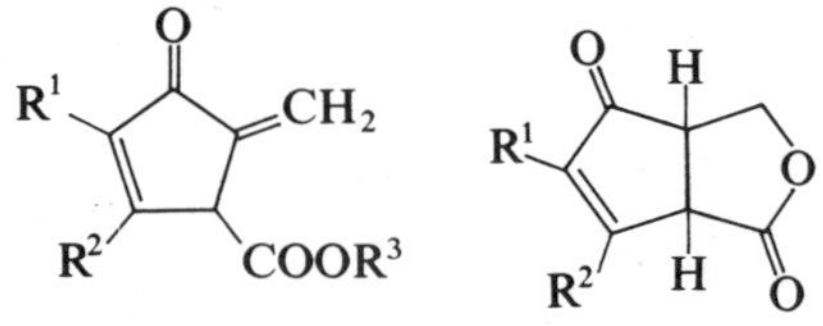

A B

An example is the synthesis of (±)-desepoxy-4,5-didehydromethylenomycin (**2**) from the bicyclic ketone **1**.

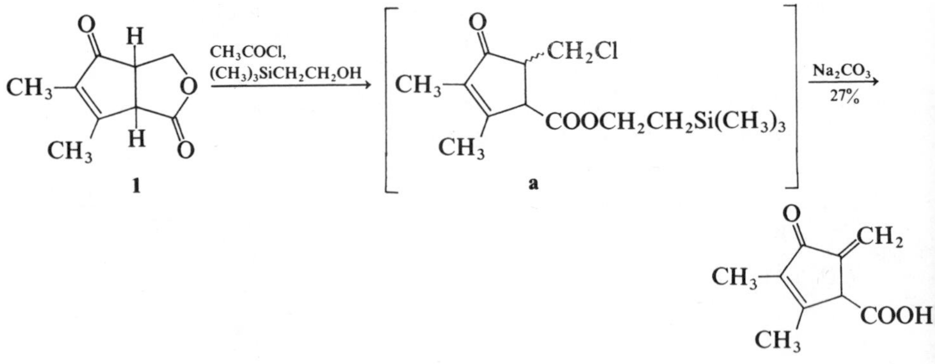

[1] D. Boschelli, R. M. Scarborough, Jr., and A. B. Smith, III, *Tetrahedron Letters*, **22**, 19 (1981).
[2] D. Boschelli and A. B. Smith, III, *ibid.*, **22**, 3733 (1981).

Acetyl hypofluorite, CH_3COOF (**1**), **10**, 1.

Fluorination. This reagent can fluorinate strongly activated arenes, but alkylbenzenes, nitrobenzene, and phenyl acetate react only slowly.[1]

Example:

$$C_6H_5NHAc \xrightarrow[55\%]{1} \text{o-}FC_6H_4NHAc + \text{p-}FC_6H_4NHAc \quad (7:1)$$

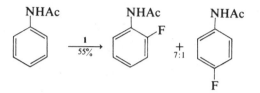

[1] O. Lerman, Y. Tor, and S. Rozen, *J. Org.*, **46**, 4629 (1981).

Acetylmethylene(triphenyl)arsorane, $(C_6H_5)_3As{=}CHCOCH_3$ (**1**). Mol. wt. 362.28, m.p. 160–162° (dec.).

Preparation:

$$As(C_6H_5)_3 + BrCH_2COCH_3 \xrightarrow[84\%]{C_6H_6,\ \Delta} (C_6H_5)_3\overset{+}{As}{-}CH_2COCH_3Br^- \xrightarrow[93\%]{NaOCH_3,\ CH_3OH} \mathbf{1}$$

Wittig-type reactions.[1] In contrast to the corresponding phosphorane reagent, **1** reacts with ketones as well as aldehydes to form α,β-unsaturated methyl ketones. Two typical products and the yields are formulated.

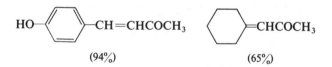

$$HO{-}C_6H_4{-}CH{=}CHCOCH_3 \ (94\%) \qquad C_6H_{10}{=}CHCOCH_3 \ (65\%)$$

The reagent reacts with α,β-enones and α,β-unsaturated esters to form *trans*-cyclopropanes, such as **2** and **3**.

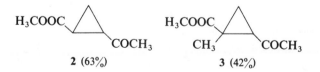

$$H_3COOC{-}C_3H_4{-}COCH_3 \ \ \mathbf{2}\ (63\%) \qquad H_3COOC(CH_3)C_3H_3{-}COCH_3 \ \ \mathbf{3}\ (42\%)$$

[1] Y. Huang, Y. Xu, and Z. Li, *Org. Prep. Proc. Int.*, **14**, 373 (1982).

Acetyl tetrafluoroborate, CH_3COBF_4. Mol. wt. 129.85, b.p. 20°/745 mm. The reagent is prepared in >98% yield by reaction of acetyl fluoride with BF_3.[1]

1,3-*Diketones.* Silyl enol ethers are C-acylated by acetyl tetrafluoroborate to give 1,3-diketones. Yields are generally higher if the reagent is generated *in situ* from CH_3COF and BF_3.[2]

Examples:

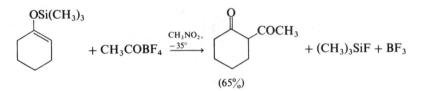

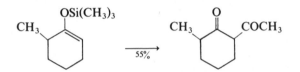

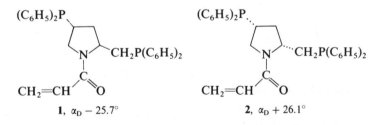

1-(Trimethylsilyloxy)cyclohexene + CH_3COBF_4 $\xrightarrow[-35°]{CH_3NO_2,}$ 2-acetylcyclohexanone (65%) + $(CH_3)_3SiF + BF_3$

$(CH_3)_2CHC(OSi(CH_3)_3)=C(CH_3)_2 \xrightarrow[84\%]{CH_3COF,\ BF_3;\ CH_3NO_2,\ -35°} (CH_3)_2CHC(=O)-C(CH_3)_2-COCH_3$

6-Methyl-1-(trimethylsilyloxy)cyclohexene $\xrightarrow{55\%}$ 2-acetyl-6-methylcyclohexanone

[1] J. H. Clark and J. Emsley, *J.C.S. Dalton Trans.*, 2129 (1975).
[2] I. Kopka and M. W. Rathke, *J. Org.*, **46**, 3771 (1981).

(2R,4R)- and (2S,4S)-N-Acryloyl-4-(diphenylphosphine)-2-[(diphenylphosphine)-methyl]pyrrolidine, 1 and **2.**

$(C_6H_5)_2P$ / $CH_2P(C_6H_5)_2$ / N–C(=O)–CH=CH_2

1, $\alpha_D - 25.7°$ **2**, $\alpha_D + 26.1°$

Asymmetric hydrogenation of N-acyl-α-aminocinnamic acids.[1] Rh(I) complexes with either **1** or **2** attached to polymers with suitable swelling characteristics are very effective for asymmetric hydrogenation of dehydroamino acids. Optical yields of about 90% are possible. As expected, polymer-bound Rh(I)-**1** results in (R)-amino acid derivatives, whereas polymer-bound Rh(I)-**2** results in (S)-amino acid derivatives.

The polymers are recovered by filtration with no loss of selectivity.

[1] G. L. Baker, S. J. Fritschel, J. R. Stille, and J. K. Stille, *J. Org.*, **46**, 2954 (1981).

Alkylaluminum halides, 6, 251; **7**, 146; **10**, 177–181. Snider and co-workers[1] have reviewed catalytic uses of these Brønsted bases, which are also Lewis acids. Although it is very hard to prepare anhydrous $AlCl_3$, BF_3, etc., alkylaluminum halides are readily obtained anhydrous, and also can scavenge undesired water. They are soluble in most organic solvents and are commercially available in heptane solutions. They can be transferred by syringe without elaborate precautions. They cover a wide range of Lewis acidity, ranging from $C_2H_5AlCl_2$, only slightly less acidic than $AlCl_3$, to $(CH_3)_3Al$, a very weak Lewis acid. $(CH_3)_2AlCl$ is available from Texas Alkyls; CH_3AlCl_2 is prepared by reaction of $(CH_3)_2AlCl$ and $AlCl_3$ in refluxing heptane. $(C_2H_5)_2AlCl$ is much cheaper than $(CH_3)_2AlCl$.

The review covers uses of these reagents as catalysts for ene reactions, [2 + 4] and [2 + 2] cycloadditions. A few years ago $(C_2H_5)_2AlCl$ was reported to catalyze also the Claisen rearrangement (yields, 89–96%).[2]

Ene reaction with methyl acrylate.[3] Methyl acrylate undergoes an ene reaction with (Z)-ethylidene steroids (**1**) in the presence of one equiv. of $C_2H_5AlCl_2$ to afford

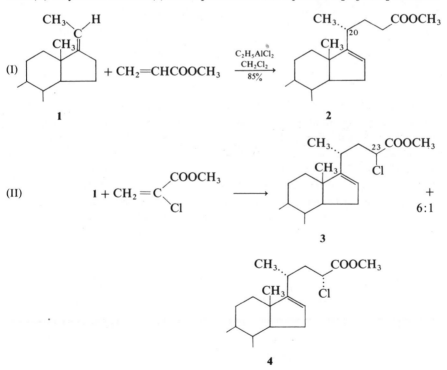

(I) **1** + $CH_2{=}CHCOOCH_3$ $\xrightarrow[85\%]{C_2H_5AlCl_2,\ CH_2Cl_2}$ **2**

(II) **1** + $CH_2{=}C(Cl)COOCH_3$ $\longrightarrow$ **3** + **4** (6:1)

products with the natural steroid configuration at C_{20} (equation I). Actually, $AlCl_3$ is also an effective catalyst if used in conjunction with a proton scavenger (pyridine). Use of this system results in a faster reaction, but in somewhat lower yield.

An additional chiral center can be introduced at C_{23} by use of methyl 2-chloroacrylate.

Allylic sulfoxides. Alkenes can be converted directly into allylic sulfoxides by an ene reaction with *p*-toluenesulfinyl chloride catalyzed by ethylaluminum dichloride in ether at 25°. In this reaction $C_2H_5AlCl_2$ serves as a Lewis acid and as a proton scavenger by reaction with the HCl liberated to form $AlCl_3$ and ethane.[4]

Example:

$$\text{methylenecyclohexane} + p\text{-}CH_3C_6H_4SOCl \xrightarrow[-C_2H_6]{C_2H_5AlCl_2,\ \text{ether}} \left[\text{1-cyclohexenyl-}CH_2S(=O\cdot AlCl_3)C_6H_4CH_3\text{-}p \right] \xrightarrow[65\%]{H_2O} \text{1-cyclohexenyl-}CH_2S(=O)C_6H_4CH_3\text{-}p$$

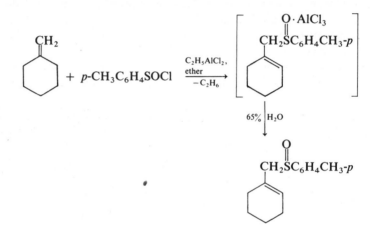

Prins reaction (*cf.* **10**, 186–187). Dimethylaluminum chloride is an effective catalyst for the ene addition of formaldehyde (as trioxane or paraformaldehyde) to mono- and 1,2-disubstituted alkenes.[5] When 1.5–2.0 equiv. of the Lewis acid is used, homoallylic alcohols are obtained, usually in high yield. γ-Chloro alcohols, formed by *cis*-addition of –Cl and $-CH_2OH$ to the double bond, are sometimes also observed when only 1 equiv. of the Lewis acid is present. The advantage of this reaction over the Prins reaction (using HCl) is that *m*-dioxanes are not formed as by-products, because formaldehyde no longer functions as a nucleophile when complexed to the Lewis acid.

Example:

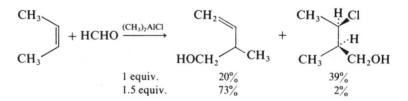

$$CH_3CH{=}CHCH_3 + HCHO \xrightarrow{(CH_3)_2AlCl} CH_2{=}CHCH(CH_3)CH_2OH + CH_3CH(Cl)CH(CH_3)CH_2OH$$

$(CH_3)_2AlCl$	homoallylic alcohol	chloro alcohol
1 equiv.	20%	39%
1.5 equiv.	73%	2%

Other aldehydes also undergo this ene reaction, particularly with 1,1-disubstituted alkenes, such as methylenecyclohexane (equation I). The absence of a strongly acidic

(I) methylenecyclohexane $+ C_6H_5CHO \xrightarrow[69\%]{(CH_3)_2AlCl}$ cyclohexenyl-$CH_2CH(OH)C_6H_5$

catalyst permits use of this reaction for synthesis of ipsenol (**1**), the pheromone of the bark beetle, albeit in low yield.

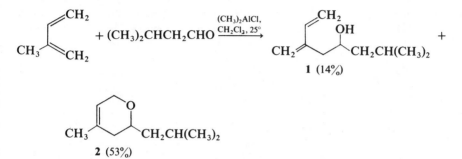

$CH_2{=}CH{-}C(CH_3){=}CH_2 + (CH_3)_2CHCH_2CHO \xrightarrow{(CH_3)_2AlCl,\ CH_2Cl_2,\ 25°}$ $CH_2{=}CH{-}C({=}CH_2){-}CH_2CH(OH)CH_2CH(CH_3)_2$ **1** (14%) + dihydropyran (CH_3, $CH_2CH(CH_3)_2$) **2** (53%)

Dihydropyrane synthesis.[6] A novel synthesis of a dihydropyrane intermediate (**2**) for a synthesis of pseudomonic acid (**3**) involves an ene reaction of formaldehyde with a 1,4-diene, followed by a quasi intramolecular Diels-Alder reaction with formaldehyde as dienophile, both reactions being catalyzed by ethylaluminum dichloride. Thus, reaction of **1** with CH_2O (3 equiv.) and $C_2H_5AlCl_2$ (4.5 equiv.) gives **2** in 35–40% yield via a number of intermediates.

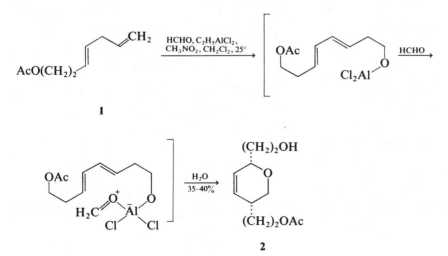

$AcO(CH_2)_2$–diene–CH_2 (**1**) $\xrightarrow{HCHO,\ C_2H_5AlCl_2,\ CH_3NO_2,\ CH_2Cl_2,\ 25°}$ [OAc … O–$AlCl_2$] $\xrightarrow{HCHO}$ [OAc … O–Al(Cl)(Cl)–$\overset{+}{O}{=}CH_2$] $\xrightarrow[35–40\%]{H_2O}$ dihydropyran with $(CH_2)_2OH$ and $(CH_2)_2OAc$ (**2**)

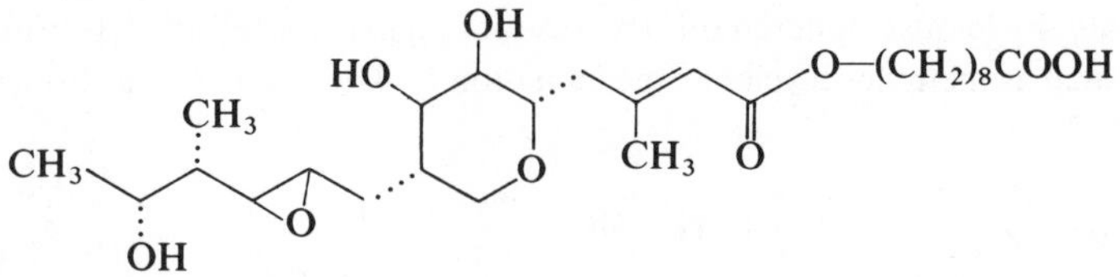

3

Cyclization of unsaturated aldehydes. Reaction of 2,6-dimethyl-$\triangle^5$-heptenal (**1**) with $(CH_3)_2AlCl$ (1 equiv.) gives essentially only the ene product **2**. However, treatment with 2 equiv. of $(CH_3)_2AlCl$ results in **3** by an ene reaction. Treatment with the stronger Lewis acid CH_3AlCl_2 results mainly in the cyclopentanone **4**. Treatment with $C_2H_5AlCl_2$ (2 equiv.) results in reductive cyclization to **5**. The ketone **4** is not an intermediate in this reaction.[7]

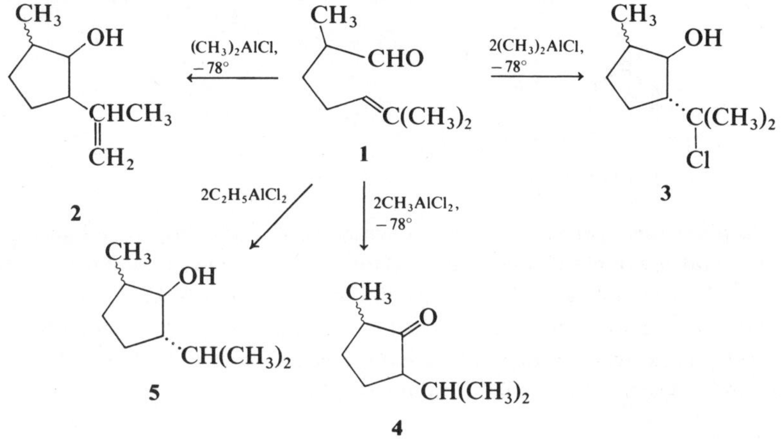

Cyclohexanol annelation.[8] In the presence of dimethylaluminum chloride, α,β-unsaturated aldehydes or ketones undergo two consecutive ene reactions with alkylidenecycloalkanes to form bicyclic alcohols. Thus, if the reaction of methylenecyclohexane and methyl vinyl ketone is conducted at $-20°$, the product (**1**) of the initial ene reaction can be isolated in 39% yield together with traces of **2**, which becomes the main product from reactions conducted at 25°. The intermediate is usually not isolable in reactions of α,β-enals. Thus, reaction of ethylidenecyclopentane and acrolein at 0° gives **3** in 72% yield. An attractive feature of this annelation is that

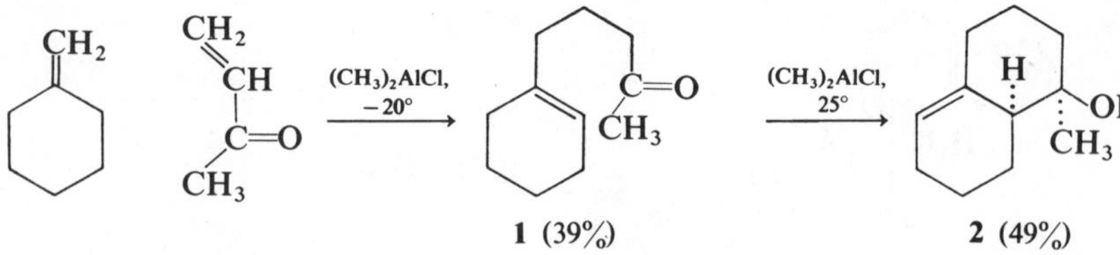

1 (39%) **2** (49%)

only one isomer is formed, that in which both substituents have the equatorial configuration.

$$\text{(ethylidenecyclopentane)} + CH_2{=}CHCHO \xrightarrow[72\%]{(CH_3)_2AlCl,\ 0^\circ} \mathbf{3}$$

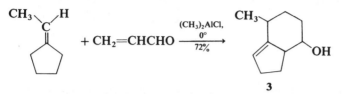

3

Friedel-Crafts acylation of alkenes.[9] $C_2H_5AlCl_2$ is an effective catalyst for acylation of alkenes with acyl chlorides or anhydrides in CH_2Cl_2. The reaction proceeds in higher yield than that previously reported with other Lewis acid catalysts (such as $ZnCl_2$). The reaction provides a useful route to β,γ-enones.

Examples:

$$(CH_3)_2C{=}CHCH_3 + CH_3COCl \xrightarrow{C_2H_5AlCl_2,\ CH_2Cl_2,\ 25^\circ} CH_2{=}C(CH_3)CH(CH_3)COCH_3\ (53\%) + (CH_3)_2CClCH(CH_3)COCH_3\ (13\%)$$

$$\text{1-methylcyclohexene} + (CH_3CO)_2O \xrightarrow{55\%} \text{2-acetyl-1-methyl-}\Delta\text{-cyclohexene}$$

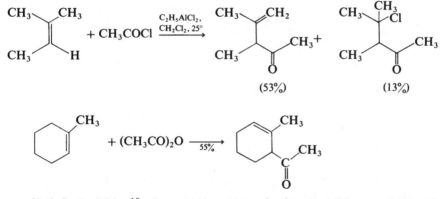

[1,4]-*Cycloaddition.*[10] Isocyanides undergo [1,4]-cycloaddition to cisoid α,β-unsaturated ketones in the presence of diethylaluminum chloride or ethylaluminum dichloride (1 equiv.). The unsaturated iminolactones formed are convertible into γ-butyrolactones.

Examples:

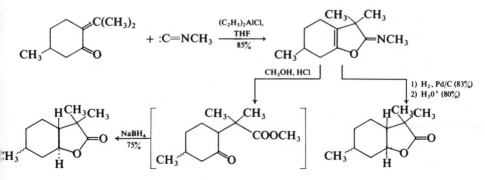

$$\text{pulegone} + {:}C{=}NCH_3 \xrightarrow[85\%]{(C_2H_5)_2AlCl,\ THF} \text{iminolactone}$$

Iminolactone $\xrightarrow{CH_3OH,\ HCl}$ [keto ester with $C(CH_3)_2COOCH_3$] $\xrightarrow[75\%]{NaBH_4}$ lactone

Iminolactone $\xrightarrow{1)\ H_2,\ Pd/C\ (83\%);\ 2)\ H_3O^+\ (80\%)}$ lactone

$$\text{2-acetyl-1-methylcyclohexene} + :C{=}NCH_3 \xrightarrow{87\%} \text{(imino lactone)} \xrightarrow[80\%]{1)\ H_2,\ Pd/C;\ 2)\ H_3O^+} \text{(lactone)}$$

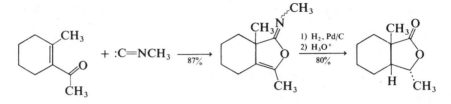

***(E)* or *(Z)*-1,3-Dienes.**[11] Trimethylsilyl allylic carbanions (**1**, *t*-BuLi, HMPT, THF, $-78°$) react with aldehydes in the presence of ethylaluminum dichloride or dicyclopentylboron chloride exclusively at the α-position to give *threo*-adducts **3**. On treatment with KH, *threo*-**3** is converted into the (Z)-1,3-diene (**4**), whereas use of BF_3 results in the (E)-1,3-diene (**6**, 199–200).

$$(CH_3)_3Si\text{-CH(Li}^+)\text{-CH=CH}_2\ (\mathbf{1}) + RCHO \xrightarrow[80\text{–}100\%]{C_2H_5AlCl_2} R\text{-CH(OH)-CH(Si(CH}_3)_3)\text{-CH=CH}_2\ (\textit{threo-}\mathbf{3})$$

$$(E)\text{-}\mathbf{4} \xleftarrow{BF_3} \textit{threo-}\mathbf{3} \xrightarrow{KH} (Z)\text{-}\mathbf{4}$$

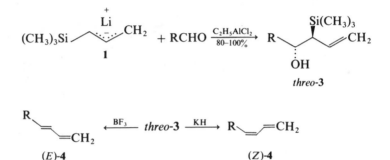

Use of $Bu_3SnCl–BF_3$ instead of $C_2H_5AlCl_2$ reverses the stereocontrol in the reaction, and *erythro*-**3** is formed exclusively (**10**, 451).

Detritylation. Diethylaluminum chloride or diisobutylaluminum chloride effects selective quantitative removal of trityl groups of deoxynucleosides, deoxynucleotides, or oligodeoxynucleotides in methylene chloride at 25° in 2–10 minutes.[12]

[1] B. B. Snider, D. I. Rodini, M. Karras, T. C. Kirk, E. A. Deutsch, R. Cordova, and R. T. Price, *Tetrahedron*, **37**, 3927 (1981).
[2] F. M. Sonnenberg, *J. Org.*, **35**, 3165 (1970).
[3] A. D. Batcho, D. E. Berger, S. G. Davoust, P. M. Wovkulich, and M. R. Uskoković, *Helv.*, **64**, 1682 (1981).
[4] B. B. Snider, *J. Org.*, **46**, 3155 (1981).
[5] B. B. Snider, D. J. Rodini, T. C. Kirk, and R. Cordova, *Am. Soc.*, **104**, 555 (1982).
[6] B. B. Snider and G. B. Phillips, *ibid.*, **104**, 1113 (1982).
[7] B. B. Snider, M. Karras, R. T. Price, and D. J. Rodini, *J. Org.*, **47**, 4538 (1982).
[8] B. B. Snider and E. A. Deutsch, *ibid.*, **47**, 745 (1982).
[9] B. B. Snider and A. C. Jackson, *ibid.*, **47**, 5393 (1982).
[10] Y. Ito, H. Kato, and T. Saegusa, *ibid.*, **47**, 741 (1982).
[11] Y. Yamamoto, Y. Saito, and K. Maruyama, *J.C.S. Chem. Comm.*, 1326 (1982).
[12] H. Köster and N. D. Sinha, *Tetrahedron Letters*, **23**, 2641 (1982).

Alkylcopper reagents.

Butatrienes. Secondary or tertiary RCu reagents react with 3-bromo-3-alkene-1-ynes (**1**), prepared as shown, to give butatrienes (**2**) in 85–95% yield. The (E)/(Z) ratio of the products is nearly equal to the (Z)/(E) ratio of the starting material (equation I).[1]

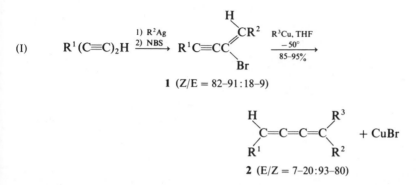

(I) $R^1(C{\equiv}C)_2H \xrightarrow[2)\ NBS]{1)\ R^2Ag} R^1C{\equiv}CC(Br){=}CHR^2 \xrightarrow[85\text{–}95\%]{R^3Cu,\ THF,\ -50^\circ}$

1 (Z/E = 82–91:18–9)

$HR^1C{=}C{=}C{=}CR^3R^2$ + CuBr

2 (E/Z = 7–20:93–80)

[1] H. Kleijn, M. Tigchelaar, R. J. Bullee, G. J. Elsvier, J. Meijer, and P. Vermeer, *J. Organometal. Chem.*, **240**, 329 (1982).

Alkyllithium reagents, 9, 5–8; **10**, 3–5.

Quinols.[1] Alkyllithium reagents react selectively at the more electrophilic carbonyl carbon of unsymmetrical quinones. In contrast, Grignard reagents react selectively with the less hindered carbonyl group.

Examples:

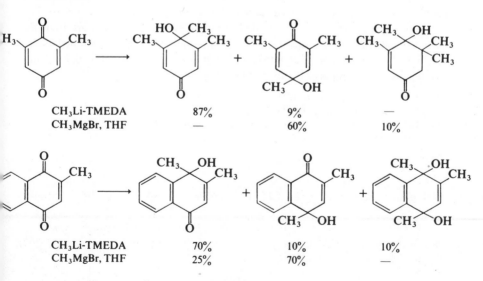

CH₃Li-TMEDA	87%	9%	—
CH₃MgBr, THF	—	60%	10%

CH₃Li-TMEDA	70%	10%	10%
CH₃MgBr, THF	25%	70%	—

Cyclopentenones. Reaction of the Fittig bislactone **2** with methyllithium results in the substituted cyclopentenone **3** in moderate yield. This product on oxidative decarboxylation gives methylenomycin B (**4**).[2]

$$(C_2H_5CO)_2O + \mathbf{1}\ \xrightarrow[53\%]{Py,\ \Delta,\ -CO_2}\ \mathbf{2}\ \xrightarrow[50\%]{1)\ CH_3Li,\ THF,\ C_6H_5CH_3;\ 2)\ NaOH}$$

COOH, COOH, COOH, C_2H_5

1 **2**

$$\mathbf{3}\ \xrightarrow[31\%]{Pb(OAc)_4,\ Cu(OAc)_2,\ Py}\ \mathbf{4}$$

CH_3, CH_3, CH_2COOH; CH_3, CH_3, CH_2

3 **4**

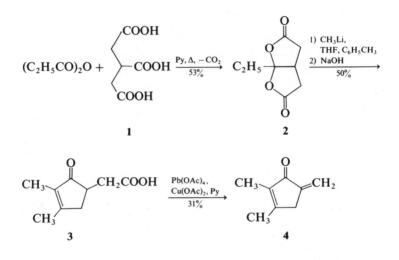

Metallation of allenes.[3] Allene is metallated at C_1 by *n*-butyllithium in THF at $-78°$. In the absence of HMPT, the lithio derivative can be alkylated to give a 1-alkylallene (86–93% yield). Under the same conditions, a 1,1-dialkylallene is convertible into a pure 1,1,3-trialkylallene, but a 1-alkylallene is metallated at C_1 or C_3, depending on the size of the alkyl group. When R contains more than four carbons, metallation is largely at C_3. Metallation of 1,1,3-trisubstituted allenes requires *t*-butyllithium or *n*-butyllithium + 1 equiv. of HMPT; 1,1-dimethyl-3,3-di-*n*-butylallene can be prepared in this way in 86% yield.

[1] D. Liotta, M. Saindane, and C. Barnum, *J. Org.*, **46**, 3369 (1981).
[2] G. M. Strung and G. S. Lal, *Can. J. Chem.*, **60**, 2528 (1982).
[3] D. Michelot, J.-C. Clinet, and G. Linstrumelle, *Syn. Comm.*, **12**, 739 (1982).

Allyl phenyl ether, $C_6H_5OCH_2CH{=}CH_2$. Mol. wt. 134.18, b.p. 192°. Supplier: Aldrich.

1-*Alkenes*; 1,5-*dienes*.[1] The anion of allyl phenyl ether (*n*-butyllithium, THF, $-78°$) reacts with a trialkylborane to form an allylic borane (**1**) (*cf.* **9**, 361–362). Protonolysis results in a 1-alkene.

Example:

$$(C_7H_{15})_3B + C_6H_5O\bar{C}HCH{=}CH_2 \;(Li^+) \longrightarrow \begin{matrix} C_7H_{15} \\ (C_7H_{15})_2B \end{matrix} \rangle CHCH{=}CH_2 \quad \mathbf{1a}$$

$$\mathbf{1a} \rightleftharpoons C_7H_{15}CH{=}CHCH_2B(C_7H_{15})_2 \;\mathbf{1b}$$

$$C_7H_{15}CH_2CH{=}CH_2 \xleftarrow[60\%]{1)\ MgCl_2,\ 2)\ H^+} C_7H_{15}CH{=}CHCH_2B(C_7H_{15})_2 \;\mathbf{1b}$$

In the presence of CuI, **1** reacts with allyl bromide to give two alkenes. The major product (**2**) is a straight-chain 1,5-diene.

Example:

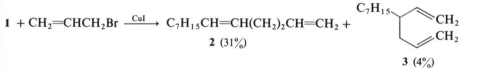

$$\mathbf{1} + CH_2{=}CHCH_2Br \xrightarrow{CuI} C_7H_{15}CH{=}CH(CH_2)_2CH{=}CH_2 + \ldots$$

2 (31%)

[branched diene bearing C_7H_{15} and two $=CH_2$ groups] **3** (4%)

[1] S. Hara, S. Imai, T. Hara, and A. Suzuki, *Syn. Comm.*, **12**, 813 (1982).

Allyltri-*n*-butyltin (1). Mol. wt. 321.02, b.p. 94–95°/0.55 mm.

Preparation[1]:

$$(C_6H_5)_3SnCH_2CH{=}CH_2 \xrightarrow[-(C_6H_5)_4Sn]{C_6H_5Li} CH_2{=}CHCH_2Li \xrightarrow[69\%]{(n\text{-}C_4H_9)_3SnCl} (n\text{-}C_4H_9)_3SnCH_2CH{=}CH_2 \quad \mathbf{1}$$

Allylation of organic halides. Two laboratories[2] have reported briefly that in the presence of a radical initiator organic halides undergo allylic substitution reactions with allyltrialkyltin compounds in moderate yield. This reaction was used in a recent synthesis of the neurotoxin (±)-perhydrohistrionicotoxin (**7**) to introduce the *n*-butyl side chain.[3] AIBN-catalyzed reaction of the bromide **2** with **1** proceeds in unexpectedly high yield and with complete stereocontrol to give a single product **3**. It is the undesired isomer, but the desired stereochemistry is obtained by epimerization of the intermediate ketone **5**. The hydroxy lactam (**6**) had previously been used for the synthesis of **7**.

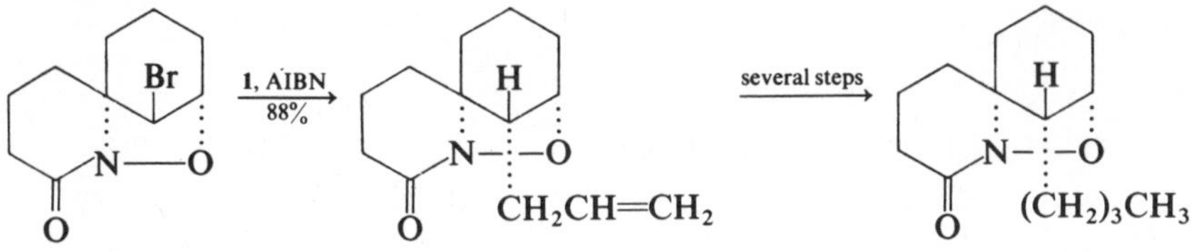

Although stereoselectivity would not be expected from a radical reaction, high stereoselectivity can be attained in other cases.[4]

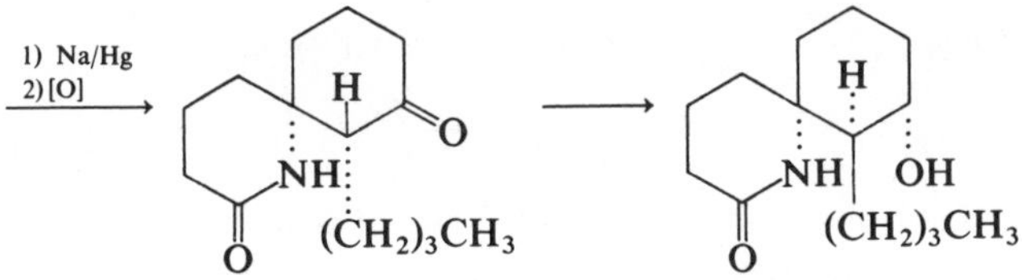

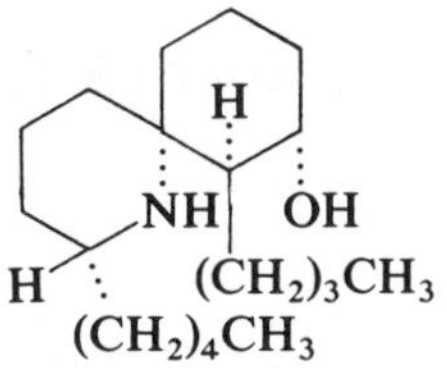

[1] D. Seyferth and M. A. Weiner, *J. Org.*, **26**, 4797 (1961).
[2] M. Kosugi, K. Kurino, K. Takayama, and T. Migita, *J. Organomet. Chem.*, **56**, C11 (1973); J. Grignan and M. Pereyre, *ibid.*, **61**, C33 (1973).
[3] G. E. Keck and J. B. Yates, *J. Org.*, **47**, 3590 (1982).
[4] G. E. Keck and J. B. Yates, *Am Soc.*, **104**, 5829 (1982).

Allyltrimethylsilane, 10, 6–8.

Amidoalkylation. In the presence of BF_3 etherate or $SnCl_4$ (1 equiv.) allyltrimethylsilane alkylates alkoxylactams, including the azetidinone **1**. The product (**2**) was used for the synthesis of the carbapenem **4**.[1]

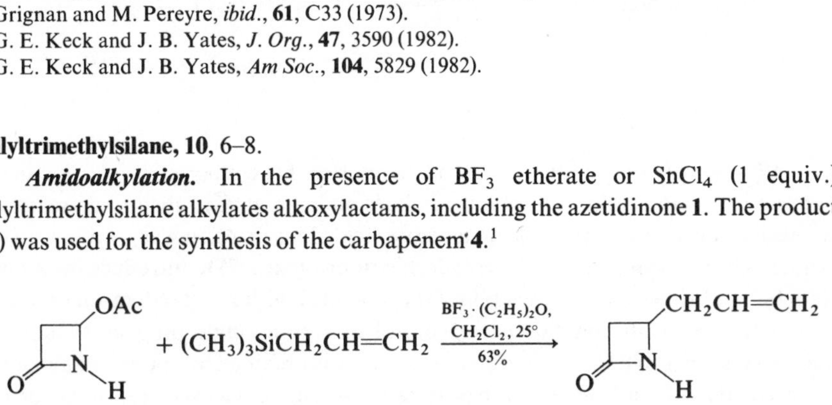

$$\mathbf{1} + (CH_3)_3SiCH_2CH{=}CH_2 \xrightarrow[63\%]{BF_3\cdot(C_2H_5)_2O,\ CH_2Cl_2,\ 25^\circ} \mathbf{2}$$

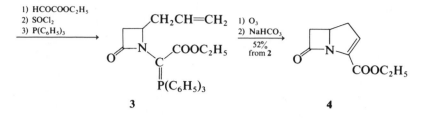

1) $HCOCOOC_2H_5$
2) $SOCl_2$
3) $P(C_6H_5)_3$
→

$CH_2CH{=}CH_2$; O; N; $COOC_2H_5$; C; $P(C_6H_5)_3$

3

1) O_3
2) $NaHCO_3$
52% from **2**
→

O; N; $COOC_2H_5$

4

Aluminum ate complexes, $(CH_3)_3SiCH{=}CHCH_2\bar{A}l(C_2H_5)_3Li^+$ (**1**). The ate complex is prepared by reaction of the lithium anion of allyltrimethylsilane with $Al(C_2H_5)_3$. In contrast with the anion of allyltrimethylsilane, which reacts with carbonyl compounds mainly at the γ-position, **1** reacts selectively at the α-position (equation I).

$$\text{(I)}\ \ RCHO \xrightarrow{\mathbf{1}} \underset{\substack{|\\ OH}}{RCH}-\overset{\substack{Si(CH_3)_3\\ |}}{C}HCH{=}CH_2 + \gamma\text{-isomer}$$

(65–100%) (minor)

The ate complex (**2**) of allyl phenyl selenide reacts somewhat less regioselectively than **1**. Even so, it was useful for a synthesis of the pheromone (**3**) of *Diparopsis castanea* (equation II).[2]

$$\text{(II)}\ \ AcO(CH_2)_8CHO + C_6H_5SeCH{=}CHCH_2\bar{A}l(C_2H_5)_3Li^+ \xrightarrow[70\%]{}$$

2

C_6H_5Se; CH_2; HO; $(CH_2)_8OAc$ $\xrightarrow[77\%]{H_3O^+}$ H_2C; $(CH_2)_8OAc$

3

C-Glycosides. Reaction of allyltrimethylsilane (**1**) and BF_3 etherate with 2,3,4,6-tetrabenzylglucopyranose (**2a**) results in a 10:1 mixture of the α- and β-allylglucopyranes **3** and **4** in 55% yield. The chemical yield can be improved by use of the α-*p*-nitrobenzoate ester **2b**. The β-C-glycoside **4** is prepared with high stereoselectivity by reaction of allylmagnesium bromide with the lactone (**5**) corresponding to **2a**, followed by reduction of the resulting hemiketal.

OBzl BzlO OBzl RO O OBzl

$CH_2{=}CHCH_2Si(CH_3)_3$, $BF_3 \cdot (C_2H_5)_2O$

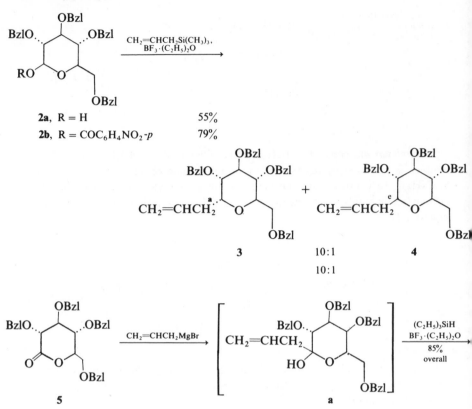

2a, R = H 55%

2b, R = $COC_6H_4NO_2$-*p* 79%

3 + **4** 10:1

10:1

OBzl BzlO OBzl O OBzl

5

$CH_2{=}CHCH_2MgBr$

a

$(C_2H_5)_3SiH$, $BF_3 \cdot (C_2H_5)_2O$, 85% overall

The same preference for substitution from the α (axial) side with **1** was found with galactose and mannose derivatives. A similar selectivity was also observed in the reaction of 2,3,4-tribenzyl-1,6-anhydroglucopyranose (**6**) with **1**.[3]

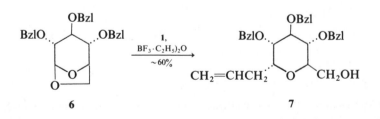

6 → **1**, $BF_3 \cdot C_2H_5)_2O$, ~60% → **7**

C-Allylglycosides are also obtained by reaction of **1** with glycal derivatives in the presence of $TiCl_4$ (1 equiv.). In all cases, the principal C_1-glycoside is the one produced by apparant axial attack at C_1 (with concomitant shift of the double bond to C_2–C_3).[4]

Examples:

OAc ...OAc OAc O $\xrightarrow[85\%]{1,\ TiCl_4}$ $CH_2{=}CHCH_2$ O ...OAc OAc 16:1 + $CH_2{=}CHCH_2$ O ...OAc OAc

OAc O OBzl $\xrightarrow[99\%]{}$ $CH_2{=}CHCH_2$ O OBzl

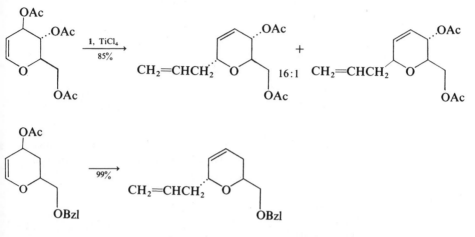

However, the reaction of glycosyl esters with allyltrimethylsilane, when catalyzed by zinc bromide, shows only slight evidence of anchimeric assistance in the four cases investigated.[5]

Optically active allylsilanes. Optically active allysilanes have now been obtained by coupling α-(trimethylsilyl)benzylmagnesium bromide with vinyl bromides, effected with $PdCl_2$ complexed with the ferrocenylphosphine (R)-(S)-PPFA (this volume, **1a**). The (R)-isomer is formed preferentially, usually in high optical purity.

(I) $R^1R^2C{=}CHBr + (CH_3)_3SiCH(C_6H_5)MgBr \xrightarrow{PdCl_2\cdot \mathbf{1a}} R^1R^2C{=}CH\text{–}C^{R}H(Si(CH_3)_3)C_6H_5$

2a, $R^1{=}R^2{=}H$	42%	**3a** (95%ee)
(Z)-**2b**, $R^1{=}H$, $R^2{=}CH_3$	38%	(Z)-**3b** (24%ee)
(E)-**2b**, $R^1{=}CH_3$, $R^2{=}H$	77%	(E)-**3b** (85%ee)

The chiral allylsilanes (**3b**) were used to determine the stereochemistry of the electrophilic substitution reactions (S_E1). Typical results are shown in equations (II) and (III). The (Z)-allylsilanes give products of (R)-configuration and the (E)-allylsilanes give the (S)-isomers. In each case, the electrophile attacks the double bond *anti* to the leaving group.[6]

The (E)- and (Z)-(R)-allylsilanes (**3b**) react with aldehydes to give optically active (E)-homoallylic alcohols **5** or **6**. In reactions with (E)-**3b**, essentially only the *erythro*-diastereoisomer **5** is formed, regardless of the aldehyde used. In reactions with (Z)-**3b**, both *erythro* and *threo*-diastereoisomers (**6**) are obtained, the ratio depending on the structure of the aldehyde; but the configuration at C_3 is identical in both products and

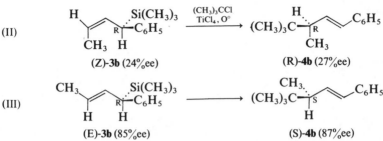

(II) (Z)-3b (24%ee) $\xrightarrow{(CH_3)_3CCl,\ TiCl_4,\ 0°}$ (R)-4b (27%ee)

(III) (E)-3b (85%ee) $\longrightarrow$ (S)-4b (87%ee)

opposite to that of **5**. The configuration at C_3 in **5** and **6** is consistent with attack of the aldehyde *anti* to the leaving silyl group. The configuration at C_4 is determined by steric interactions between R and the methyl group in the transition state.[7]

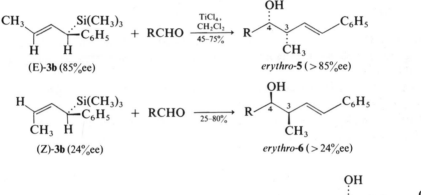

(E)-3b (85%ee) + RCHO $\xrightarrow[45–75\%]{TiCl_4,\ CH_2Cl_2}$ *erythro*-**5** (>85%ee)

(Z)-3b (24%ee) + RCHO $\xrightarrow[25–80\%]{}$ *erythro*-**6** (>24%ee) + *threo*-**6** (>24%ee)

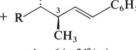

[1] G. A. Kraus and K. Neuenschwander, *J.C.S. Chem. Comm.*, 134 (1982).
[2] Y. Yamamoto, Y. Saito, and K. Maruyama, *Tetrahedron Letters*, **23**, 4597 (1982).
[3] M. D. Lewis, J. K. Cha, and Y. Kishi, *Am. Soc.*, **104**, 4976 (1982).
[4] S. Danishefsky and J. F. Kerwin, Jr., *J. Org.*, **47**, 3803 (1982).
[5] A. P. Kozikowski and K. L. Sorgi, *Tetrahedron Letters*, **23**, 2281 (1982).
[6] T. Hayashi, M. Konishi, H. Ito, and M. Kumada, *Am. Soc.*, **104**, 4962 (1982).
[7] T. Hayashi, M. Konishi, and M. Kumada, *ibid.*, **104**, 4963 (1982).

Allyltrimethyltin, $(CH_3)_3SnCH_2CH{=}CH_2$. Mol. wt. 204.86, b.p. 125.5/642 mm. The reagent can be prepared by reaction of allylmagnesium bromide with chlorotrimethyltin (85% yield).

The first step in a recent synthesis[2] of eleutherin (**7**) and isoeleutherin (**8**) involved conjugate addition of **1** to the quinone **2**. The alcohol obtained by reduction of **4** undergoes intramolecular cyclization with either mercury acetate or benzeneselenenyl bromide. The first method gives **5** and **6** in equal amounts, but phenylselenoetherification favors formation of the *cis* isomer **5**. In this reaction, the total yield of **5** and **6** is only 56%.

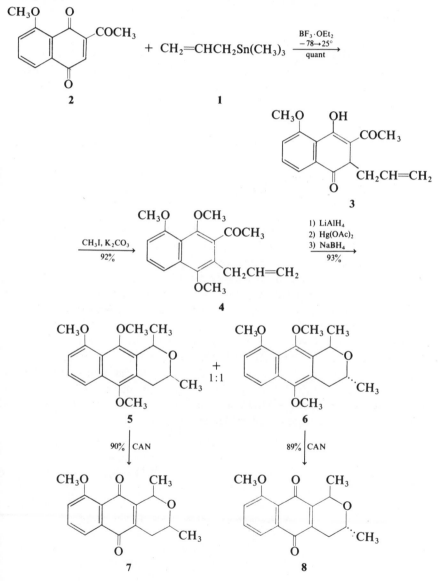

CH_3O O COCH$_3$ O **2** + $CH_2{=}CHCH_2Sn(CH_3)_3$ **1** $\xrightarrow[\text{quant}]{BF_3\cdot OEt_2,\ -78\rightarrow 25°}$

CH_3O OH COCH$_3$ $CH_2CH{=}CH_2$ O **3**

$\xrightarrow[92\%]{CH_3I,\ K_2CO_3}$ CH_3O OCH_3 COCH$_3$ $CH_2CH{=}CH_2$ OCH_3 **4**

1) LiAlH$_4$
2) Hg(OAc)$_2$
3) NaBH$_4$
93%

CH_3O OCH_3 CH_3 O CH_3 OCH_3 **5** + (1:1) CH_3O OCH_3 CH_3 O CH_3 OCH_3 **6**

5 → 90% CAN → **7**; **6** → 89% CAN → **8**

CH_3O O CH_3 O CH_3 O **7**

CH_3O O CH_3 O CH_3 O **8**

[1] E. W. Abel and R. J. Rowley, *J. Organometal. Chem.*, **84**, 149 (1975).
[2] Y. Naruta, H. Uno, and K. Maruyama, *J.C.S. Chem. Comm.*, 1277 (1981).

Alumina, 1, 19–20; **2**, 17; **3**, 6; **4**, 8; **6**, 16–17; **7**, 5–7; **8**, 9–13; **9**, 8–11; **10**, 8–9.

1-*Acetylcycloalkenes.* Acylation of 1-alkylcycloalkenes results in an equilibrium mixture of three products, depending on the Lewis acid used. Alumina is a useful reagent for conversion of the mixture to the conjugated enone.[1]

Example:

CH_3 (1-methylcyclopentene) $\xrightarrow{Ac_2O,\ ZnI_2}$ $COCH_3$, CH_3, OAc (10%) + $COCH_3$, CH_3 (70%) + $COCH_3$, CH_3 (20%)

78% $Al_2O, 20°$

$COCH_3$, CH_3

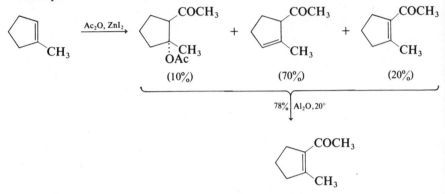

Acetylation. Primary alcohols are converted into acetates when stirred efficiently with ethyl acetate at 25–75° in the presence of Woelm-200-N alumina. Yields are generally high, even for unsaturated alcohols or chlorohydrins.[2] This reaction is possible even in the presence of secondary alcohols.[3,4]

Alcohols can be acetylated by ketene supported on alumina (or SiO_2, Celite) without other catalysts at room temperature. The method is applicable even to tertiary alcohols.[5]

Diels-Alder reactions.[6] The Diels-Alder reaction of cyclopentadiene with dienophiles of the general type $CH_2{=}CHR$ results in *endo/exo* adducts in the ratio of *ca.* 3:1, and this regioselectivity is not affected by catalysts. The ratio is affected when the reaction is conducted heterogeneously with the dienophile adsorbed on nonactivated surfaces such as Al_2O_3, SiO_2, and cellulose. The most striking effect is shown by methyl acrylate adsorbed on alumina (equation I).

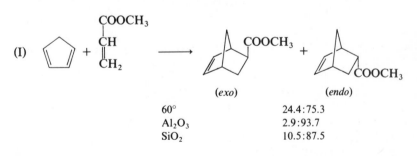

(I) cyclopentadiene + $CH_2{=}CH{-}COOCH_3$ ⟶ $COOCH_3$ (*exo*) + $COOCH_3$ (*endo*)

60°	24.4:75.3
Al_2O_3	2.9:93.7
SiO_2	10.5:87.5

(2E,4Z)-Dienoates.[7] β-Allenic esters rearrange to (2E,4Z)-dienoic esters when heated at 80–138° in benzene or xylene with slightly basic alumina. The stereoselectivity is 91–100%, but the yields are only moderate to good.

Example:

$$C_2H_5(H)C{=}C{=}C(H)CH_2COOCH_3 \xrightarrow[82\%]{Al_2O_3} C_2H_5CH{=}CH{-}CH{=}CH{-}COOCH_3 + \text{(2E, 4E)-isomer} \quad 96:4$$

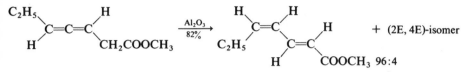

Conjugate addition of amines to α,β-enones (**10**, 9).[8] This reaction had been used to effect addition of aziridine to withaferin A (**1**). The product (**2**) has significant antitumor activity.

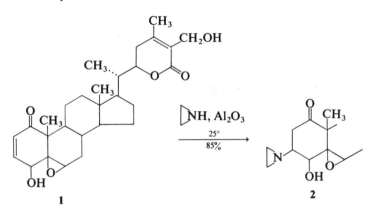

$$\mathbf{1} \xrightarrow[25^\circ,\ 85\%]{\text{aziridine (NH)},\ Al_2O_3} \mathbf{2}$$

Knoevenagel condensation.[9] Alumina can be used as catalyst for the Knoevenagel condensation, particularly with aldehydes. Only the (E)-product is formed. This catalyzed condensation fails with diaryl ketones.

Examples:

$$C_6H_5CHO + CH_2(CN)Y \xrightarrow[85\text{–}100\%]{Al_2O_3} C_6H_5(H)C{=}C(CN)Y + H_2O$$

(Y = CN, $COOCH_3$, $CONH_2$)

$$(CH_3)_2C{=}O + CH_2(CN)COOCH_3 \xrightarrow{53\%} (CH_3)_2C{=}C(CN)COOCH_3$$

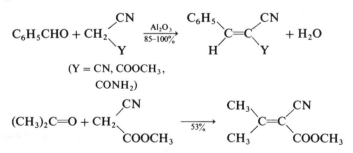

[1] T. Hudlicky and T. Srnak, *Tetrahedron Letters*, **22**, 3351 (1981).
[2] G. A. Posner, S. S. Okada, K. A. Babiak, K. Miura, and R. K. Rose, *Synthesis*, 789 (1981).

[3] G. H. Posner and M. Oda, *Tetrahedron Letters*, **22**, 5003 (1981).
[4] S. S. Rana, J. J. Barlow, and K. L. Matta, *ibid.*, **22**, 5007 (1981).
[5] T. Chihara, Y. Takagi, S. Teratani, and H. Ogawa, *Chem. Letters*, 1451 (1982).
[6] H. Parlar and R. Baumann, *Angew. Chem. Int. Ed.*, **20**, 1014 (1981).
[7] S. Tsuboi, T. Masuda, H. Makino, and A. Takeda, *Tetrahedron Letters*, **23**, 209 (1982); *J. Org.*, **47**, 4478 (1982).
[8] S. W. Pelletier, G. Gebeyehu, and N. V. Mody, *Heterocycles*, **19**, 235 (1982).
[9] F. Texier-Boullet and A. Foucaud, *Tetrahedron Letters*, **23**, 4927 (1982).

Aluminum amalgam, 1, 20–21; **3,** 7; **5,** 9.

Reduction of acylphosphoranes. Acylphosphoranes are reduced to β-keto esters by treatment with excess Al-Hg in wet THF at 15–25° with periodic addition of a proton donor such as TFA or HCl (equation I).

$$\text{(I)}\ \mathrm{RCOCl} \xrightarrow{(C_6H_5)_3P{=}CHCOOC_2H_5} \mathrm{R\overset{O}{\overset{\|}{C}}{-}\underset{\underset{P(C_6H_5)_3}{\|}}{C}COOC_2H_5} \xrightarrow[80\text{–}95\%]{\text{Al/Hg},\ H^+} \mathrm{R\overset{O}{\overset{\|}{C}}CH_2COOC_2H_5}$$

$$\text{(II)}\ \mathrm{CH_2{=}CH\overset{O}{\overset{\|}{C}}{-}\underset{\underset{P(C_6H_5)_3}{\|}}{C}COOC_2H_5} \xrightarrow[71\%]{1)\ C_2H_5CH{=}C(CH_3)Li;\ 2)\ CH_3I} \mathrm{C_2H_5CH{=}C(CH_3)CH_2CH(CH_3)\overset{O}{\overset{\|}{C}}{-}\underset{\underset{P(C_6H_5)_3}{\|}}{C}COOC_2H_5}$$

$$\xrightarrow[84\%]{\text{Al/Hg}} \mathrm{C_2H_5CH{=}C(CH_3)CH_2CH(CH_3)\overset{O}{\overset{\|}{C}}CH_2COOC_2H_5} \xrightarrow[85\%]{1)\ NaH;\ 2)\ CH_3I}$$

$$\mathrm{C_2H_5CH{=}C(CH_3)CH_2CH(CH_3)\overset{O}{\overset{\|}{C}}CH(CH_3)COOC_2H_5} \xrightarrow[91\%]{Ba(OH)_2,\ H_2O,\ C_2H_5OH} \underset{\mathbf{1}}{\mathrm{C_2H_5CH{=}C(CH_3)CH_2CH(CH_3)\overset{O}{\overset{\|}{C}}CH_2CH_3}}$$

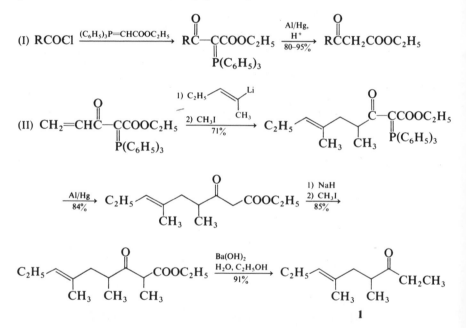

The utility of this reaction is shown by a stereoselective synthesis of 4,6-dimethyl-(E)-6-nonene-3-one (**1**), the defense substance of *L. longipes* (equation II).[1]

Allenyl alcohols. Organometallic reagents containing the trimethylsilylpropargyl group, $(CH_3)_3SiC{\equiv}CCH_2M$, can react in the propargylic or allenic form with electrophiles. In reactions with aldehydes or ketones, highest yields of propargyl alcohols are obtained under Reformatsky conditions with zinc activated with $HgCl_2$. In marked contrast, the organoaluminum species obtained with aluminum amalgam reacts almost entirely to form allenic alcohols resulting from coupling α to the trimethylsilyl group.[2]

Examples:

$$(CH_3)_2C{=}O + (CH_3)_3SiC{\equiv}CCH_2ZnBr \xrightarrow[61\%]{THF} (CH_3)_2C(OH){-}CH_2C{\equiv}CSi(CH_3)_3$$

$$(CH_3)_2C{=}O + (CH_3)_3SiC{\equiv}CCH_2Al_{2/3}Br \xrightarrow[69\%]{THF} (CH_3)_2C(OH){-}C({Si(CH_3)_3}){=}C{=}CH_2$$

[1] M. P. Cooke, Jr., *J. Org.*, **47**, 4963 (1982).
[2] R. G. Daniels and L. A. Paquette, *Tetrahedron Letters*, **22**, 1579 (1981).

Aluminum chloride, 1, 24–34; **2**, 21–23; **3**, 7–9; **4**, 10–15; **5**, 10–13; **6**, 17–19; **7**, 7–9; **8**, 13–15; **9**, 11–13; **10**, 9–11.

Friedel-Crafts reactions.[1] Ethyl cyclopropanecarboxylate undergoes Friedel-Crafts reactions with arenes in the presence of $AlCl_3$ to form indanones. The proposed reaction mechanism is formulated in the example.

Example:

$$\triangleright\!-COOC_2H_5 \xrightarrow[HCl]{AlCl_3,} \left[ClCH_2CH(CH_3)COOC_2H_5 \xrightarrow{C_6H_6,\ AlCl_3} C_6H_5CH_2CH(CH_3)COOC_2H_5 \right] \xrightarrow[93\%]{AlCl_3,\ \Delta} \text{2-methyl-1-indanone}$$

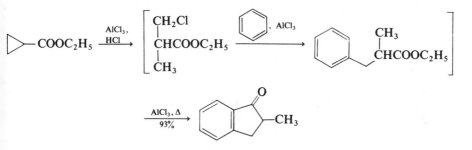

Acylation of alkynes; cyclopentenones. Acylation of alkynes with α,α-disubstituted-*β*, *γ*-unsaturated acid chlorides, catalyzed with $AlCl_3$, results in 5,5-disubstituted-2-cyclopentenones by an unexpected intramolecular cyclization-rearrangement.[2]

Example:

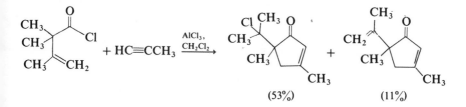

$$CH_2{=}C(CH_3)C(CH_3)_2COCl + HC{\equiv}CCH_3 \xrightarrow[CH_2Cl_2]{AlCl_3,} \text{5-(2-chloro-2-propyl)-3,5-dimethyl-2-cyclopentenone } (53\%) + \text{5-isopropenyl-3,5-dimethyl-2-cyclopentenone } (11\%)$$

Ring contraction is observed only in the absence of an α-hydrogen; thus the reaction of **1** and propyne results in a phenol (**2**).

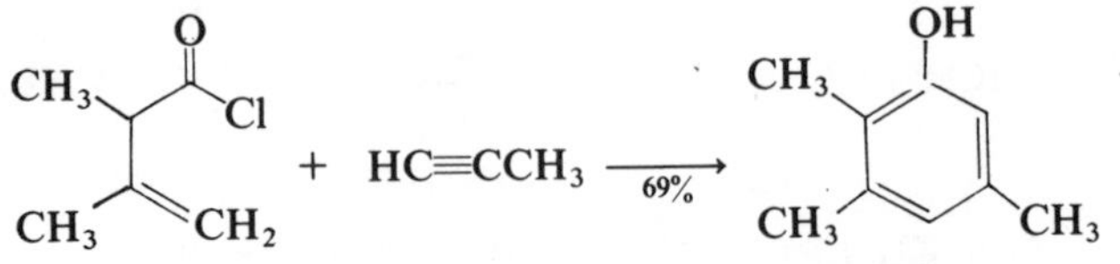

$HC{\equiv}CCH_3$ 69%

1 2

Dihydropyranes. In the presence of $AlCl_3$, α,β-unsaturated acyl nitriles (**1**)[3] react with simple olefins to give dihydropyranes.[4]

Examples:

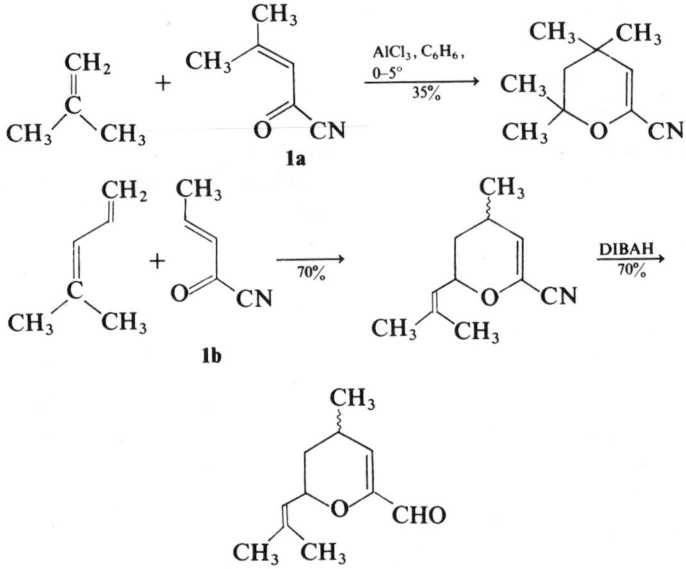

$AlCl_3$, C_6H_6, 0–5° 35%

1a

70%

1b

DIBAH 70%

Diels-Alder reactions are observed with more reactive dienes (equation I).

(I) + **1a** $(C_2H_5)_2AlCl$, C_6H_6, 25° 90%

$(C_2H_5)_2AlCl$

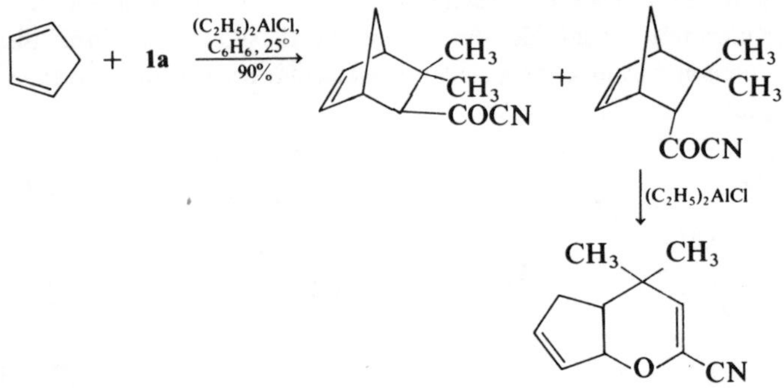

Diels-Alder catalyst. $AlCl_3$ is useful for activation of allenic esters in Diels-Alder reactions and for increasing the *endo*-selectivity.[5]

Example:

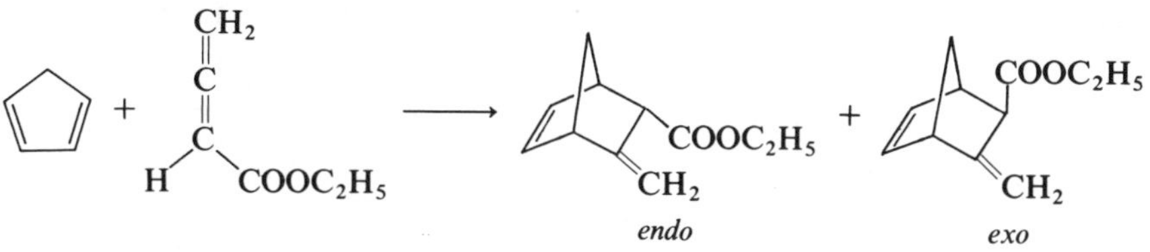

Conditions	Yield	endo : exo
C_6H_6, 80°	90%	64:36
$AlCl_3$, C_6H_6, 25°	95%	86:14

Intramolecular* [2+2]*cycloaddition.[6] A recent synthesis of coronafacic acid (**4**) proceeds through a cyclobutene (**2**) to obtain a trienone (**3**), the immediate precursor of **4**. Under the most satisfactory conditions, $AlCl_3$ in CH_2Cl_2 at 25°, the yield of this intramolacular cycloaddition is only 16%. However, since the remaining steps to **4** can be conducted without purification of the intermediates, the overall yield for a rather lengthy synthesis is reasonable.

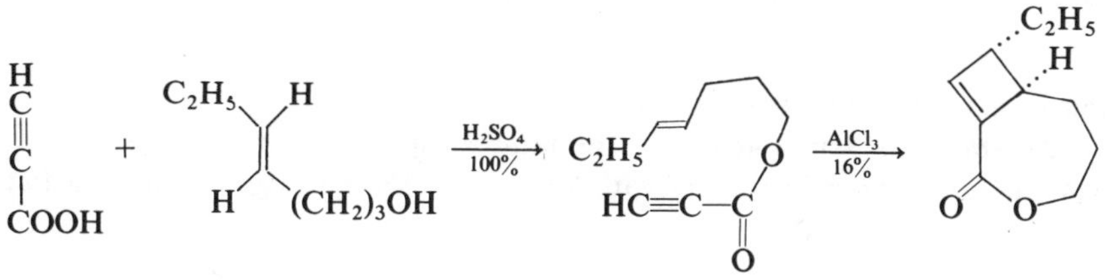

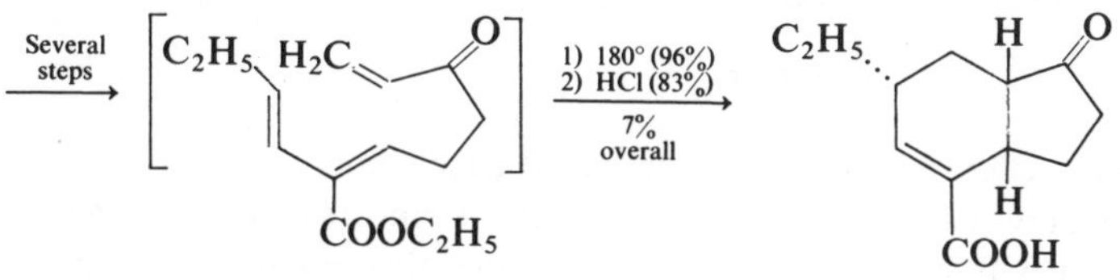

Addition of allylic sulfides to methyl propiolate. In the presence of $AlCl_3$ (1 equiv.), allylic sulfides add in Michael fashion to methyl propiolate followed by a [3.3]sigmatropic rearrangement.[7]

Example:

$$C_6H_5SCH_2CH{=}CH_2 + HC{\equiv}C{-}COOCH_3 \xrightarrow{AlCl_3} \left[\text{cyclic } C_6H_5S^+ \text{ ylide } {-}CH{=}C(OCH_3){-}O^- \cdot AlCl_3 \right] \longrightarrow H_2C{=}CH{-}CH_2{-}C(COOCH_3){=}CH{-}SC_6H_5$$

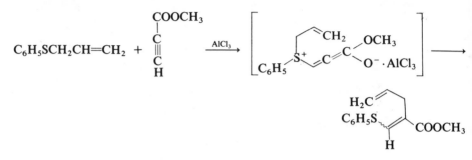

Thioacetals and -ketals. $AlCl_3$ is an effective catalyst for the condensation of aldehydes and ketones with thiols or dithiols. The reaction of thiols with carbonyl groups with an α-proton results in low yields; but the reaction with dithiols proceeds in high yield.[8]

[1] H. W. Pinnick, S. P. Brown, E. A. McLean, L. W. Zoller, III, *J. Org.*, **46**, 3760 (1981).
[2] M. Karpf, *Tetrahedron Letters*, **23**, 4923 (1982).
[3] Preparation: K. Haase and H. M. R. Hoffmann, *Angew. Chem. Int. Ed.*, **21**, 83 (1982).
[4] Z. M. Ismail and H. M. R. Hoffmann, *ibid.*, **21**, 859 (1982).
[5] Z. M. Ismail and H. M. R. Hoffmann, *J. Org.*, **46**, 3549 (1981).
[6] M. E. Jung and K. M. Halweg, *Tetrahedron Letters*, **22**, 2735 (1981).
[7] K. Hayakawa, Y. Kamikawaji, and K. Kanematsu, *ibid.*, **23**, 2171 (1982).
[8] B. S. Ong, *ibid.*, **21**, 4225 (1981); *Org. Syn.*, submitted (1981).

Aluminum chloride-Ethanethiol, 9, 13; **10**, 11.

Reaction with α-nitro ketones. The nitro group of a primary α-nitro ketone is displaced by hydrogen by reaction with $AlCl_3$–C_2H_5SH. The suggested intermediates are shown in equation (I).[1]

$$\text{(I)} \quad RCCH_2NO_2\,(C{=}O) \xrightarrow[C_2H_5SH]{AlCl_3} \left[R{-}C(SC_2H_5)_2{-}CH_2NO_2 \longrightarrow R{-}C(SC_2H_5){=}CH_2 \right] \xrightarrow[55\text{–}78\%]{C_2H_5SH} R{-}C(SC_2H_5)_2{-}CH_3$$

The reagent can effect cleavage of the activated double bond of 1-nitroalkenes to give a dithioacetal in moderate yield (equation II).[2] A similar cleavage is the main reaction observed with α-nitro ketones in which the nitro group is secondary or tertiary (equation III).

$$\text{(II)} \quad R^1CH{=}C(NO_2)R^2 \longrightarrow \left[R{-}CH{=}\overset{+}{S}C_2H_5 \right] \longrightarrow RCH(SC_2H_5)_2$$

(II) $CH_3(CH_2)_5C(O)CH(NO_2)CH_3 \xrightarrow{57\%} CH_3(CH_2)_5C(SC_2H_5)_3 \xleftarrow{40\%} CH_3(CH_2)_5C(O)C(NO_2)(CH_3)_2$

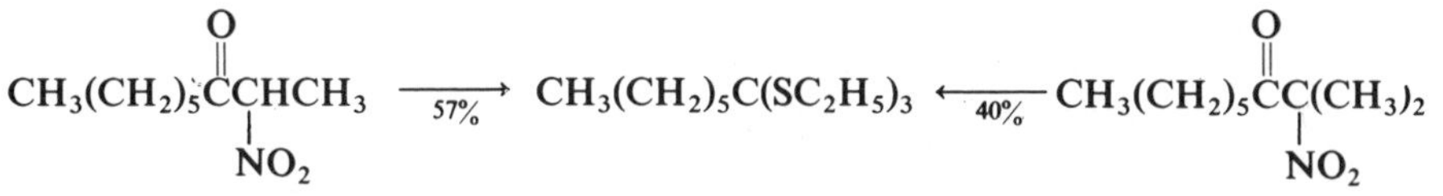

[1] M. Node, T. Kawabata, M. Ueda, M. Fujimoto, K. Fuji, and E. Fujita, *Tetrahedron Letters*, **23**, 4047 (1982).
[2] K. Fuji, T. Kawabata, M. Node, and E. Fujita, *ibid.*, **22**, 875 (1981).

Aluminum isopropoxide, 1, 35–37; **3**, 10; **4**, 15–16; **5**, 14; **6**, 19; **9**, 14–15.

***Rearrangement of α-pinene oxide* (1).**[1] α-Pinene oxide (**1**) rearranges to pinocarveol (**2**) in the presence of 1 mole % of aluminum isopropoxide at 100–120° for 1 hr. (*cf.* **9**, 14–15). The oxide **1** rearranges to pinanone (**3**) in the presence of 5 mole % of the alkoxide at 140–170° for 2 hr. Aluminum isopropoxide has been used to rearrange **2** to **3** (200°, 3 hr., 80% yield).[2]

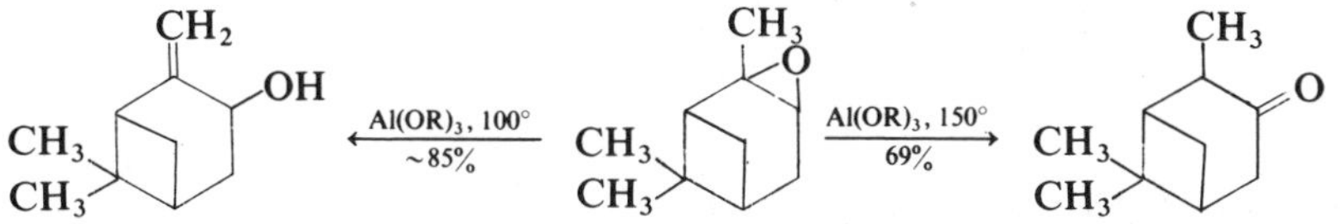

2 $\xleftarrow[\sim 85\%]{Al(OR)_3,\ 100°}$ **1** $\xrightarrow[69\%]{Al(OR)_3,\ 150°}$ **3**

[1] F. Scheidl, *Synthesis*, 728 (1982).
[2] H. Schmidt, *Ber.*, **62**, 104 (1929).

(S)-15-Aminomethyl-14-hydroxy-5,5-dimethyl-2,8-dithia[9](2,5)pyridinophane (1). Preparation.[1]

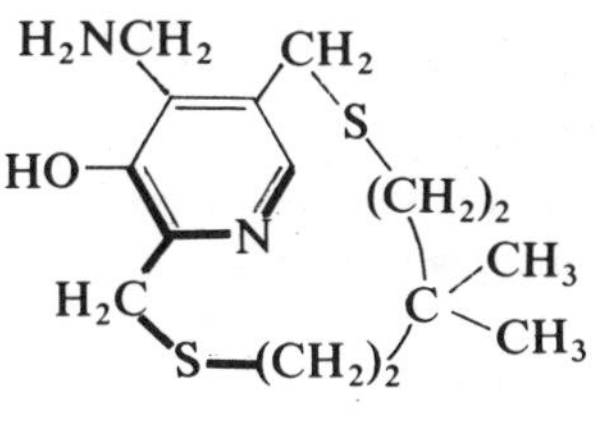

(S)-**1**

Asymmetric transamination.[2] This planar chiral pyridoxamine analog in the presence of $Zn(ClO_4)_2$ ($1/Zn(ClO_4)_2 = 1:0.5$) converts α-keto acids into (R)-amino acids in 60–96%ee. Use of (R)-**1** in place of (S)-**1** produces (S)-amino acids with the same efficiency. Chemical yields range from 50–75%. The preferred solvent is methanol. The pyridoxal-type analog is recovered in 75–85% yield. The transamination is considered to involve kinetically controlled stereoselective protonation of an octahedral Zn^{2+} chelate intermediate.

[1] T. Sakurai, H. Kuzuhara, and S. Emoto, *Acta Cryst.*, **B35**, 2984 (1979); H. Kuzuhara, T. Komatsu, and S. Emoto, *Tetrahedron Letters*, 3563 (1978).
[2] *Idem, Chem. Letters*, 1765, 1769 (1982).

(S)-2-Aminopropyl benzyl ether, polymer-supported reagent. The polymer is prepared by reaction of the N-phthaloyl derivative of (S)-2-aminopropanol with the Merrifield polymer in the presence of KH and 18-crown-6 in THF/HMPT followed by hydrazinolysis.

$$C_6H_5CH_2OCH_2\text{—}C(H)(CH_3)\text{—}NH_2 \quad \mathbf{(1)}$$

Asymmetric α-methylation of cyclohexanone.[1] The imine **2a** formed from Ⓟ **-la** and cyclohexanone is converted into (S)-2-methylcyclohexanone in 92% ee. Similar enantioselectivity is observed with the imine **2b**, but the chemical yield is much higher.

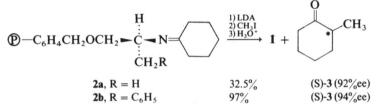

$$\text{Ⓟ—}C_6H_4CH_2OCH_2\text{—}C(H)(CH_2R)\text{—}N{=}\text{cyclohexylidene} \xrightarrow{1)\ LDA;\ 2)\ CH_3I;\ 3)\ H_3O^+} \mathbf{1} + \text{2-methylcyclohexanone}^*$$

2a, R = H — 32.5% — (S)-**3** (92%ee)
2b, R = C_6H_5 — 97% — (S)-**3** (94%ee)

Use of the unsupported **1** results in enantioselectivity of ~50%.

Asymmetric protonations. Deprotonation (LDA) of the imine **(2b)** obtained from racemic 2-methylcyclohexanone and **1b** followed by protonation with ethanol and hydrolysis gives (S)-2-methylcyclohexane **(3)** in 90% ee. The enantioselectivity depends in part on the R group; when R = H **(2a)**, (R)-**3** is formed in 22% ee. When *t*-butyl alcohol is used as the proton source, completely inactive **3** is obtained by the same sequence. No enantioselectivity is observed in protonation of the lithioenamine of 2-methylcyclohexanone and nonsupported **1a.**[2]

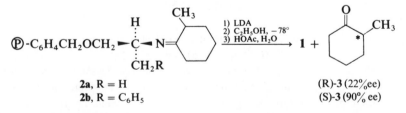

$$\text{Ⓟ-}C_6H_4CH_2OCH_2\text{—}C(H)(CH_2R)\text{—}N{=}\text{(2-}CH_3\text{-cyclohexylidene)} \xrightarrow{1)\ LDA;\ 2)\ C_2H_5OH,\ -78^\circ;\ 3)\ HOAc,\ H_2O} \mathbf{1} + \text{2-methylcyclohexanone}^*$$

2a, R = H — (R)-**3** (22%ee)
2b, R = C_6H_5 — (S)-**3** (90% ee)

[1] C. R. McArthur, P. M. Worster, J.-L. Jiang, and C. C. Leznoff, *Can. J. Chem.*, **60**, 1836 (1982).
[2] C. R. McArthur, J.-L. Jiang, and C. C. Leznoff, *ibid.*, **60**, 2984 (1982).

(R)-(+)-*o*-Anisylcyclohexylmethylphosphine (CAMP),

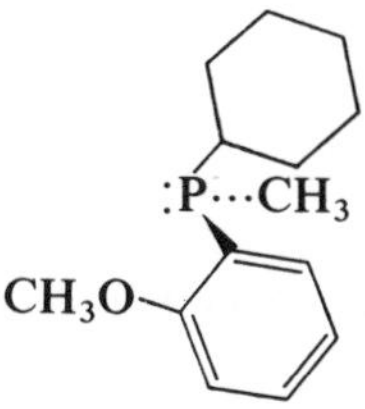

:P···CH_3 (**1**), α_D + 98.5°
CH_3O

Preparation.[1]

Asymmetric intramolecular Wittig reaction. Trost and Curran[2] have examined eight readily available optically active phosphines in a stereoselective synthesis of the diketone (**2**), a useful intermediate to several natural products. Of these, CAMP is clearly the most efficient phosphine for this purpose (equation I).

(I) [structure: CH_3, O, Br] $\xrightarrow[\text{2) } K_2CO_3]{\text{1) CAMP}}$ (S)-**2** (74%ee) + (R)-**2**
87:13

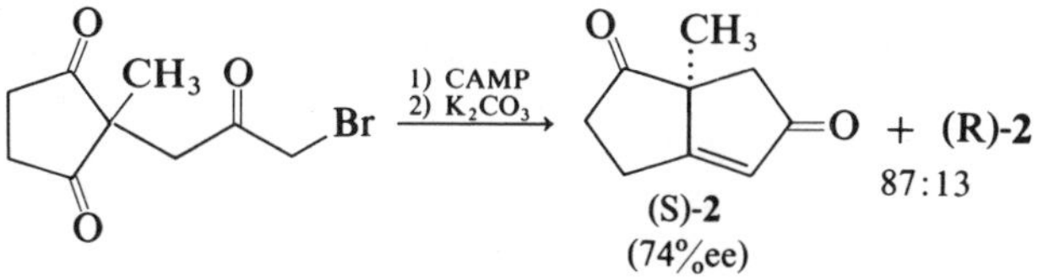

The important features for high induction are chirality at phosphorus rather than carbon, only one aryl group, and two sterically different alkyl groups. The next most efficient phosphine was (+)-DIOP, which resulted in (S)-**2** and (R)-**2** in the ratio 76:24 (52% ee).

[1] W. S. Knowles, M. J. Sabacky, and B. D. Vineyard, *Adv. Chem. Ser.*, **132**, 274 (1974).
[2] B. M. Trost and D. P. Curran, *Tetrahedron Letters*, **22**, 4929 (1981).

Aryllead triacetates, $ArPb(OCOCH_3)_3$. These reagents can be obtained by reaction of an aryl compound with lead tetraacetate in dichloroacetic acid or trichloroacetic acid (the reaction is slow in acetic acid).[1] Another general route is reaction of a diarylmercury with lead tetraacetate.[2,3]

C-Arylation.[1,4] These aryllead tricarboxylates are useful for arylation of β-diketones and β-keto esters. Yields are high with substrates containing only one acidic hydrogen.

Example:

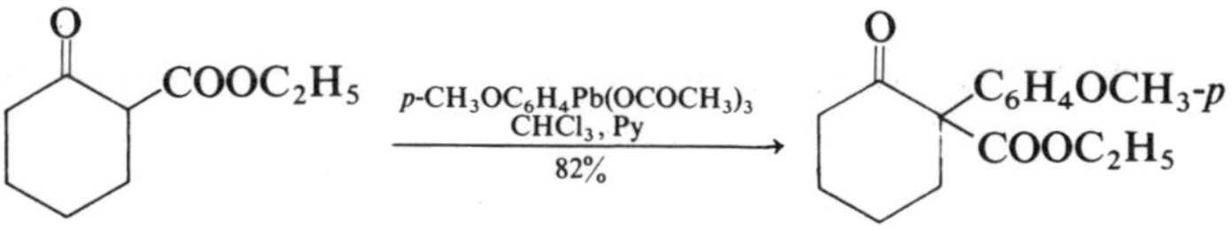

[structure: O, $COOC_2H_5$] $\xrightarrow[\text{82\%}]{p\text{-}CH_3OC_6H_4Pb(OCOCH_3)_3,\ CHCl_3,\ Py}$ [structure: O, $C_6H_4OCH_3$-*p*, $COOC_2H_5$]

Arylation of nitroalkanes. An aryllead triacetate arylates nitroalkanes and the nitronate salts at the α-position in DMSO in 60–75% yield.[5]

[1] H. C. Bell, J. R. Kalman, J. T. Pinhey, and S. Sternhall. *Aust. J. Chem.*, **32**, 1521 (1979); R. P. Kozyrod and J. T. Pinhey, *Org. Syn.*, submitted (1981).
[2] R. Criegee, P. Dimroth, and R. Schempf, *Ber.*, **90**, 1337 (1957).
[3] R. P. Kozyrod and J. T. Pinhey, *Tetrahedron Letters*, **23**, 5365 (1982).
[4] J. T. Pinhey and B. A. Rowe, *Aust. J. Chem.*, **32**, 1561 (1979); **33**, 113 (1980); *idem, Tetrahedron Letters*, **21**, 965 (1980).
[5] J. T. Pinhey and B. A. Rowe, *ibid.*, **22**, 783 (1981).

Azidotrimethylsilane, **1**, 1236; **3**, 316; **4**, 542; **5**, 719–720; **6**, 632; **9**, 21–22; **10**, 14–15.

Azides. Benzyl, allyl, and aliphatic halides react with $N_3Si(CH_3)_3$ in HMPT at 60° to form azides in 65–85% yield.[1]

[1] K. Nishiyama and H. Karigomi, *Chem. Letters*, 1477 (1982).

B

Benzeneselenenyl bromide–Silver nitrite.

1-Alkynylphenyl selenides. Benzeneselenenyl bromide and silver nitrite (1 equiv.) react to form a reagent (presumably $C_6H_5SeNO_2$) that converts 1-alkynes into 1-alkynyl phenyl selenides (equation I).[1] The same reaction has been effected in generally lower yields with phenyl selenocyanate catalyzed by Cu(I).[2] The products are useful intermediates for regio- and stereoselective synthesis of vinyl selenides.

$$\text{(I)}\quad RC{\equiv}CH \xrightarrow[65\text{–}85\%]{C_6H_5SeBr,\ AgNO_2,\ CH_3CN,\ CH_2Cl_2} RC{\equiv}CSeC_6H_5$$

Nitroselenenylation of alkenes.[3] Reaction of these reagents with alkenes results in nitroselenenylation as well as hydroxyselenenylation. The latter reaction can be significantly or completely suppressed by addition of $HgCl_2$ (C_6H_5SeBr : alkene : $HgCl_2$: $AgNO_2$ = 1.0 : 1.0 : 1.3 : 1.0). The adducts are converted into 1-nitroalkenes on oxidation in ~ 80% yield.

Examples:

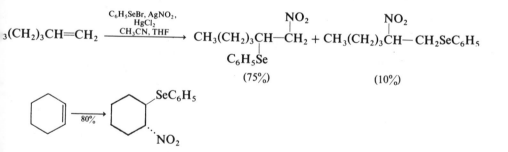

$$_3(CH_2)_3CH{=}CH_2 \xrightarrow[CH_3CN,\ THF]{C_6H_5SeBr,\ AgNO_2,\ HgCl_2} \underset{(75\%)}{CH_3(CH_2)_3\underset{C_6H_5Se}{CH}{-}\overset{NO_2}{CH_2}} + \underset{(10\%)}{CH_3(CH_2)_3\overset{NO_2}{CH}{-}CH_2SeC_6H_5}$$

$$\text{cyclohexene} \xrightarrow{80\%} \text{1-}SeC_6H_5\text{-2-}NO_2\text{-cyclohexane}$$

[1] T. Hayama, S. Tomoda, Y. Takeuchi, and Y. Nomura, *Chem. Letters*, 1249 (1982).
[2] S. Tomoda, Y. Takeuchi, and Y. Nomura, *ibid.*, 253 (1982).
[3] T. Hayama, S. Tomoda, Y. Takeuchi, and Y. Nomura, *Tetrahedron Letters*, **23**, 4733 (1982).

Benzeneselenenyl chloride–Aluminum chloride.

Cyclization of an unsaturated β-keto ester. The unsaturated β-keto ester **1** reacts with C_6H_5SeCl in HOAc to give only an addition product. However, in the presence of $AlCl_3$ this reactions results in cyclization to **2** (two diastereomers).[1]

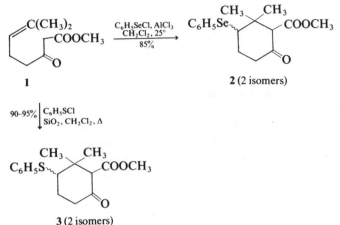

$\xrightarrow[85\%]{C_6H_5SeCl,\ AlCl_3,\ CH_2Cl_2,\ 25°}$

1 2 (2 isomers)

90–95% C_6H_5SCl, SiO_2, CH_2Cl_2, Δ

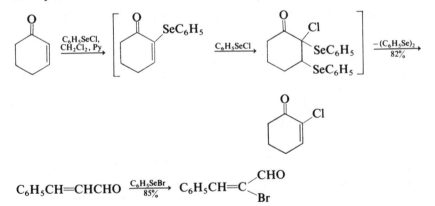

3 (2 isomers)

Use of C_6H_5SCl and $AlCl_3$ does not convert **1** into a cyclic product. In this case, the preferred reagent for cyclization to **3** is C_6H_5SCl–SiO_2.

[1] M. Alderidge and L. Weiler, *Can. J. Chem.*, **59**, 2239 (1981).

Benzeneselenenyl halides. **5**, 518–522; **6**, 459–460; **7**, 286–287; **9**, 25–32; **10**, 16–21.

α-Halo-α, β-unsaturated carbonyl compounds. Reaction of enones or enals with 2.5–3.0 equiv. of C_6H_5SeBr or C_6H_5SeCl results in formation of α-halo enones or enals via 2-phenylseleno-α,β-unsaturated carbonyl compounds.[1]

Examples:

$\xrightarrow{C_6H_5SeCl,\ CH_2Cl_2,\ Py}$ [SeC_6H_5 $\xrightarrow{C_6H_5SeCl}$ Cl, SeC_6H_5, SeC_6H_5] $\xrightarrow[82\%]{-(C_6H_5Se)_2}$ Cl

$$C_6H_5CH{=}CHCHO \xrightarrow[85\%]{C_6H_5SeBr} C_6H_5CH{=}C(CHO)Br$$

Unsaturated β-dicarbonyl compounds. This system can be prepared by reaction of enolizable β-dicarbonyl compounds with C_6H_5SeCl–pyridine (1:1, **9**, 28–29) followed by selenoxide elimination. The rate of selenenylation increases with the ease of enolization of the substrate.[2]

Examples:

$$\text{2-(hydroxymethylene)cyclohexanone (CHOH)} \xrightarrow[(1:1)]{C_6H_5SeCl\text{-}Py} \left[\text{2-}SeC_6H_5\text{-2-CHO-cyclohexanone}\right] \xrightarrow[\substack{85\% \\ \text{overall}}]{H_2O_2,\ CH_2Cl_2} \text{2-CHO-cyclohex-2-enone}$$

$$\text{keto lactone} \xrightarrow{85\%} \text{unsaturated keto lactone}$$

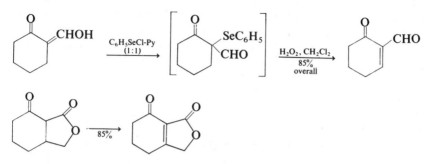

Allylic alcohols from allylsilanes.[3] The adducts of benzeneselenenyl chloride to allylsilanes on treatment with $SnCl_2$ or Florisil undergo dechlorosilylation and rearrangement to give the less substituted allyl selenide. When oxidized, the allyl selenides are converted into the allylic alcohol in which the hydroxyl group occupies the more substituted site (**6**, 338).

Example:

$$(CH_3)_3Si\text{—}CH_2CH{=}CH\text{—}R \xrightarrow{C_6H_5SeCl}$$

$$\left[(CH_3)_3SiCH_2CH(Cl)CH(SeC_6H_5)R \xrightarrow{-ClSi(CH_3)_3} H_2C{=}CHCH(SeC_6H_5)R\right] \xrightleftharpoons{80\text{–}95\%}$$

$$C_6H_5Se\text{—}CH_2CH{=}CH\text{—}R \xrightarrow[\substack{45\text{–}90\% \\ \text{overall}}]{H_2O_2,\ Py,\ CH_2Cl_2} H_2C{=}CHCH(OH)R$$

Aromatization of cyclohexenones.[4] This reaction is possible by selenenylation of the lithium enolate of the cyclohexenone, followed by oxidation of the resulting selenide. To obtain satisfactory yields of the phenol, an aromatic amine is added in the oxidation step to react selectively with the benzeneselenenic acid formed. For this purpose, 3,5-dimethoxyaniline is the most satisfactory amine.

Example:

$$\text{tricyclic enone (}CH_3\text{, H, }CH_3\text{, }CH_3\text{, O)} \xrightarrow{\substack{1)\ LDA \\ 2)\ C_6H_5SeCl\ (76\%) \\ 3)\ ClC_6H_4CO_3H,\ ArNH_2(\sim 70\%)}} \text{phenol (OH)}$$

The final step in a recent synthesis of cannabichromene (2) is the aromatization of the cyclohexenone ring of 1. Reagents used for this purpose also attack the double bond in the side chain, but the desired reaction was effected by treatment of the lithium enolate of 1 with benzeneselenenyl chloride followed by selenoxide elimination in the presence of 3,5-dimethoxyaniline.[5]

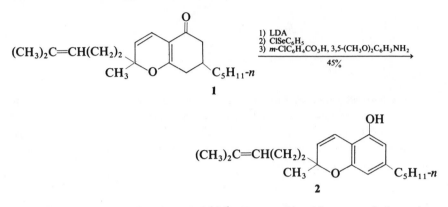

$(CH_3)_2C{=}CH(CH_2)_2$, CH_3, O, O, C_5H_{11}-*n*

1

1) LDA
2) $ClSeC_6H_5$
3) *m*-$ClC_6H_4CO_3H$, 3,5-$(CH_3O)_2C_6H_3NH_2$

45%

OH, $(CH_3)_2C{=}CH(CH_2)_2$, CH_3, O, C_5H_{11}-*n*

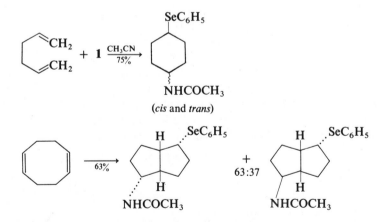

2

Benzeneselenenyl iodide, C_6H_5SeI (**1**).[6] Presumably, this reagent is formed on treatment of diphenyl diselenide with I_2 in CH_3CN. As expected, when **1** is generated in the presence of a 1,5- or 1,6-diene, selenium-substituted carbocycles are formed, by way of an intermediate that is trapped by acetonitrile.

Examples:

CH_2, CH_2 + **1** $\xrightarrow[75\%]{CH_3CN}$ SeC_6H_5, $NHCOCH_3$

(*cis* and *trans*)

$\xrightarrow{63\%}$ H, SeC_6H_5, H, $NHCOCH_3$ + H, SeC_6H_5, H, $NHCOCH_3$

63:37

[1] S. V. Ley and B. J. Whittle, *Tetrahedron Letters*, **22**, 3301 (1981).

[2] D. Liotta, C. Barnum, R. Puleo, G. Zima, C. Bayer, and H. S. Kezar, III, *J. Org.*, **46**, 2920 (1981).

[3] H. Nishiyama, K. Itagaki, K. Sakuta, and K. Itoh, *Tetrahedron Letters*, **22**, 5285 (1981); H. Nishiyama, S. Narimatsu, and K. Itoh, *ibid.*, **22**, 5289 (1981).

[4] L.-F. Tietze, G. v. Kiedrowski, and B. Berger. *ibid.*, **23**, 51 (1982).
[5] *Idem, Synthesis*, 683 (1982).
[6] A. Toshimoto, S. Uemura, and M. Okano, *J. C. S. Chem. Comm.*, 87 (1982).

Benzeneseleninic anhydride, **6**, 240–241; **7**, 139; **8**, 29–32; **9**, 32–34; **10**, 22–29.

The anhydride can be conveniently generated *in situ* by oxygen transfer from $C_6H_5IO_2$ (**1**, 511) or *m*-$O_2IC_6H_4COOH$[1] to diphenyl diselenide.[2]

Dehydrogenation of 3-ketosteroids.[2] By use of this method for generation of the reagent, dehydrogenation of 3-keto-steroids to $\Delta^{1,4}$-diene-3-ones can be a catalytic process with respect to the diselenide. Use of *m*-iodosylbenzoic acid is experimentally convenient, since the *m*-IC_6H_4COOH formed is easily separated from the ketonic product.

Examples:

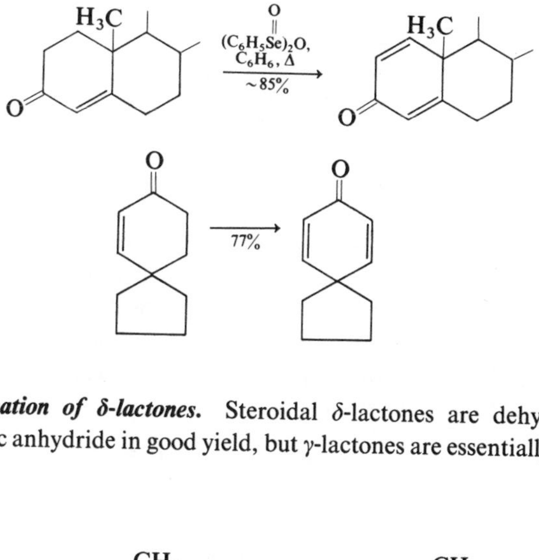

H₃C — $(C_6H_5Se)_2O$, C_6H_6, Δ; ~85% — H₃C

77%

Dehydrogenation of δ-lactones. Steroidal δ-lactones are dehydrogenated by benzeneseleninic anhydride in good yield, but γ-lactones are essentially unreactive.[3]

Example:

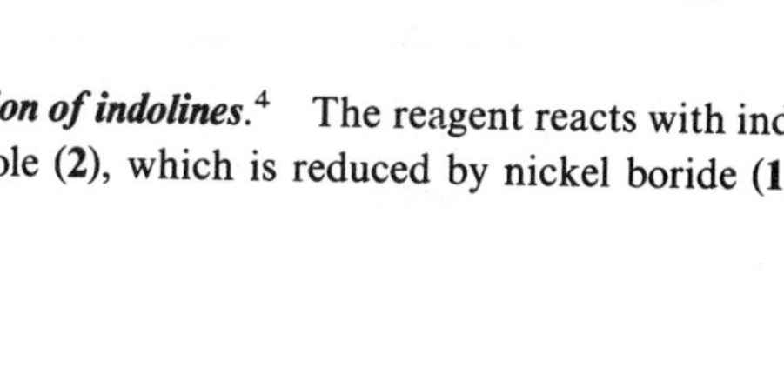

CH_3 — $(C_6H_5SeO)_2O$, C_6H_5Cl, 100°; 90% — CH_3

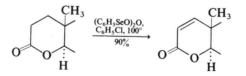

Dehydrogenation of indolines.[4] The reagent reacts with indoline (**1**) to form the β-phenylselenoindole (**2**), which is reduced by nickel boride (**1**, 472; **5**, 472) to the parent indole (**3**).

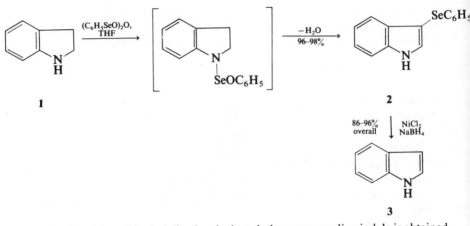

$(C_6H_5SeO)_2O$, THF; $-H_2O$, 96–98%; SeC_6H_5; $SeOC_6H_5$

1 2 3

86–96% overall; $NiCl_2$, $NaBH_4$

If the β-position of the indoline is substituted, the corresponding indole is obtained directly.

Example:

CH_3 $(C_6H_5SeO)_2O$, THF, 53% CH_3

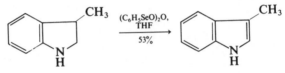

Tetrahydroisoquinoline is dehydrogenated under these conditions to 1,2-hydroisoquinoline (51% yield).

α-Aminoestrone derivatives (**8**, 29–30). Reaction of estrone with benzeneseleninic anhydride and hexamethyldisilazane (**8**, 29–30) results predominantly in the 4-phenylselenoimine **2** with lesser amounts of the 2-substituted derivative **3**. The 2-amino derivative (**6**) of estrone is best obtained by reaction of 4-bromoestrone (**5**) with the same reagent, followed by hydrogenolysis. These two reactions are probably the most efficient routes to α-aminoestrones.[5]

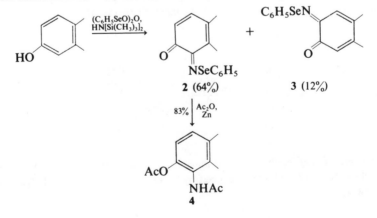

HO; $(C_6H_5SeO)_2O$, $HN[Si(CH_3)_3]_2$; O; $NSeC_6H_5$; + C_6H_5SeN; O

2 (64%) **3** (12%)

83%; Ac_2O, Zn

AcO; NHAc

4

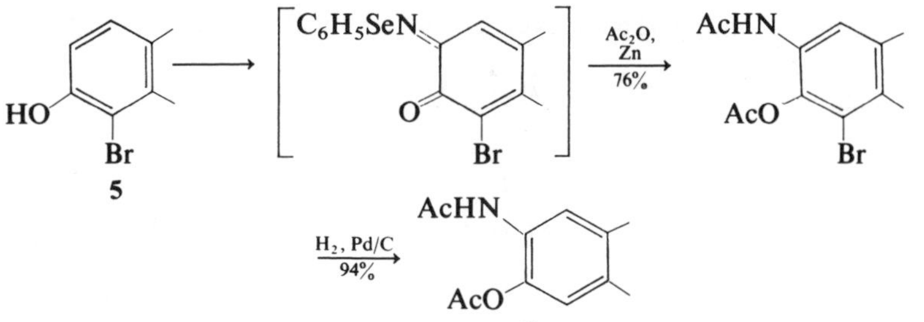

[1] $ArIO_2$ reagents can be prepared in high yield by oxidation of ArI with sodium hypochlorite (**10**, 365).

[2] D. H. R. Barton, J. W. Morzycki, W. B. Motherwell, and S. V. Ley, *J. C. S. Chem. Comm.*, 1044 (1981).

[3] D. H. R. Barton, R. A. H. F. Hui, S. V. Ley, and D. J. Williams, *J. C. S. Perkin I*, 1919 (1982).

[4] D. H. R. Barton, X. Lusinchi, and P. Milliet, *Tetrahedron Letters*, **23**, 4949 (1982).

[5] J. S. E. Holker, E. O'Brien, and B. K. Park, *J. C. S. Perkin I*, 1915 (1982).

Benzenesulfenyl chloride, **5**, 523–524; **6**, 30–32; **8**, 32–34; **9**, 35–38; **10**, 24.

1-*Allenyl dienes.* Reaction of C_6H_5SCl with the propargyl alcohol **1** results in the drimatriene derivative **2** (as a diastereomeric mixture), formed by spontaneous cyclization of the allenyldiene **a**. The triene **2** undergoes an unusual 1,6-reduction when treated with excess lithium aluminum hydride to give a *trans*-decalin (**3**) in 70% yield.[1]

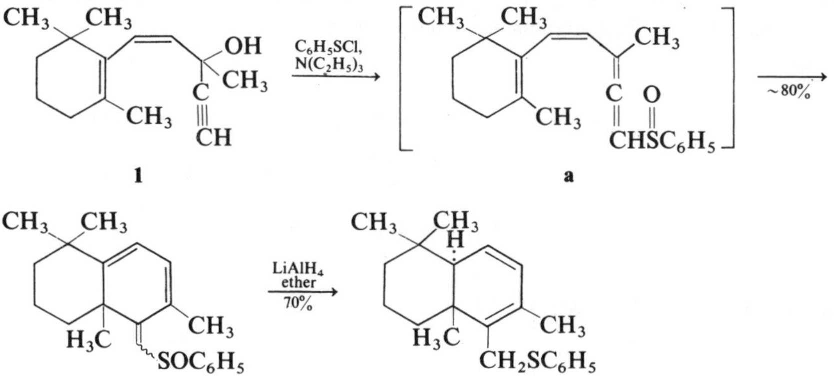

β-Lactams; azetidinones. C_6H_5SCl adds to α,β-unsaturated amides to form a kinetically controlled adduct (**1**), which gradually is converted into the more stable isomer (**2**) on standing. The adducts cyclize on treatment with base in the presence of $(Bu)_4NBr$ to azetidinones.[2]

Example:

$$CH_3C(=CHCH_3)CONHCH_2C_6H_5 \xrightarrow{C_6H_5SCl} \underset{\mathbf{1}}{C_6H_5S(CH_3)C(CONHBzl)\text{—}CH(Cl)CH_3} \xrightarrow[\text{85–90\% overall}]{25°} \underset{\mathbf{2}}{Cl(CH_3)C(CONHBzl)\text{—}CH(SC_6H_5)CH_3}$$

$$\mathbf{2} \xrightarrow[\text{45\% overall}]{KOH,\ (Bu)_4NBr} \mathbf{3}\ (\text{3-}C_6H_5S\text{-3,4-dimethyl-1-Bzl-azetidin-2-one}) \xrightarrow{\text{RaneyNi},\ (CH_3)_2C{=}O} \mathbf{4}\ (\text{3,4-dimethyl-1-Bzl-azetidin-2-one})$$

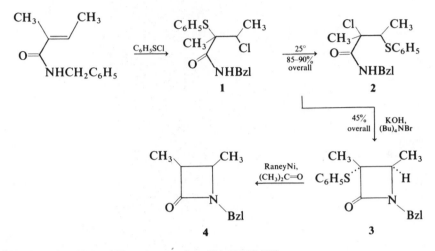

[1] W. Reischl and W. H. Okamura, *Am. Soc.*, **104**, 6115 (1982).
[2] M. Ihara and K. Fukumoto, *Heterocycles*, **19**, 1435 (1982).

Benzenesulfonyl fluoride, $C_6H_5SO_2F$ (**1**).

Enol sulfonates. Sulfonylation of enolate anions with **1** is markedly affected by the gegenion. Lithium enolates undergo mainly C-sulfonylation. Cesium or quaternary ammonium enolates undergo regioselective O-sulfonylation. The same behavior is observed with nonafluorobutanesulfonyl fluoride.[1]

Example:

$$CH_2{=}C(OSi(CH_3)_3)C(CH_3)_3 \xrightarrow[\text{90\% (isolated)}]{1)\ C_6H_5CH_2(CH_3)_3NF\quad 2)\ \mathbf{1}} CH_2{=}C(OSO_2C_6H_5)C(CH_3)_3$$

[1] E. Hirsch, S. Hünig, and H.-U. Reissig, *Ber.*, **115**, 2687 (1982).

Benzenesulfonylnitrile oxide (1) 9, 39.

Improved generation *in situ*[1]:

$$C_6H_5SO_2CH_2NO_2 \xrightarrow{CH_2N_2} C_6H_5SO_2CH{=}NO_2CH_3 \xrightarrow{Na_2SiO_3,\ H_2O\text{—}CH_2Cl_2} \underset{\mathbf{1}}{C_6H_5SO_2C{\equiv}N{\rightarrow}O}$$

$$C_6H_5SO_2CBr{=}NOH \xrightarrow{AgNO_3} \mathbf{1}$$

The second method generally provides higher yields of cycloadducts of **1** with alkenes.

[1] P. A. Wade and M. K. Pillay, *J. Org.*, **46**, 5425 (1981).

1-Benzenesulfonyl-2-trimethylsilylethane (1). Mol. wt. 242.41, m.p. 52°.

Preparation:

$$(CH_3)_3SiCH{=}CH_2 + HSC_6H_5 \xrightarrow[99.2\%]{AIBN} (CH_3)_3Si(CH_2)_2SC_6H_5 \xrightarrow[99.2\%]{H_2O_2} \underset{\mathbf{1}}{(CH_3)_3Si(CH_2)_2SO_2C_6H_5}$$

Allylsilanes.[1] A general route to allylsilanes involves addition of the anion of **1** to aldehydes and ketones to form an adduct (**2**). Mesylation *in situ* of the adducts followed by sodium amalgam reduction, results in allylsilanes (**4**) in 85–95% overall yield, based on the carbonyl compound.

$$\mathbf{1} \xrightarrow{\text{1) } n\text{-BuLi, ether; 2) } R^1COR^2} \left[\underset{\mathbf{2}}{(CH_3)_3SiCH_2CH(SO_2C_6H_5)-C(R^1)(R^2)-OLi} \xrightarrow{CH_3SO_2Cl} \underset{\mathbf{3}}{(CH_3)_3SiCH_2CH(SO_2C_6H_5)-C(R^1)(R^2)-OSO_2CH_3} \right]$$

$$\xrightarrow[\text{85–95\% overall}]{Na/Hg} \underset{\mathbf{4}}{(CH_3)_3SiCH_2CH{=}C(R^1)(R^2)}$$

The reagent has also been used to prepare terminal alkenes by alkylation of the anion followed by β-silylsulfone elimination induced by fluoride ion (**9**, 444–445).

Vinyl sulfones.[2] α-Substituted vinyl sulfones can be prepared from **1** by the sequence outlined in equation (I).

$$\text{(I)}\quad \mathbf{1} \xrightarrow[90\text{–}95\%]{\text{1) } n\text{-BuLi; 2) RX}} (CH_3)_3SiCH_2CH(R)SO_2C_6H_5 \xrightarrow[65\text{–}95\%]{\text{1) } n\text{-BuLi; 2) } t\text{-BuOCl}} (CH_3)_3SiCH_2C(Cl)(R)SO_2C_6H_5$$

$$\xrightarrow[-ClSi(CH_3)_3]{(n\text{-Bu})_4NF} \underset{(>99\%)}{H_2C{=}C(R)SO_2C_6H_5}$$

An attractive alternative is conversion of **1** into the 1-chloro derivative (**2**) followed by alkylation and elimination of chlorotrimethylsilane. The anion of **2** also adds to various aldehydes to give adducts that are convertible into 2-(benzenesulfonyl)allyl alcohols (equation II).

$$\text{(II)}\quad (CH_3)_3SiCH_2\underset{\underset{\mathbf{2}}{Cl}}{C}HSO_2C_6H_5 \xrightarrow[\text{2) RCHO}]{\text{1) } n\text{-BuLi}} (CH_3)_3SiCH_2\overset{RCHOH}{\underset{Cl}{C}}SO_2C_6H_5 \xrightarrow[\substack{90\text{–}95\% \\ \text{overall}}]{F^-} H_2C{=}C(RCHOH)(SO_2C_6H_5)$$

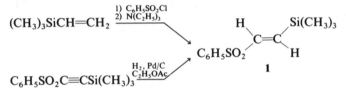

[1] C.-N. Hsiao and H. Shechter, *Tetrahedron Letters*, **23**, 1963 (1982).
[2] *Idem, ibid*, **23**, 3455 (1982).

(E)-1-Benzenesulfonyl-2-(trimethylsilyl)ethylene (1).

Preparation:

$$(CH_3)_3SiCH{=}CH_2 \xrightarrow[\text{2) } N(C_2H_5)_3]{\text{1) } C_6H_5SO_2Cl} \mathbf{1}$$

$$C_6H_5SO_2C{\equiv}CSi(CH_3)_3 \xrightarrow[C_2H_5OAc]{H_2,\ Pd/C} \mathbf{1}$$

1: H(C_6H_5SO_2)C=C(H)Si(CH_3)_3, i.e. $C_6H_5SO_2CH{=}CHSi(CH_3)_3$ (E)

Dienophile equivalent of HC≡CH or HC≡CR.[1] The use of **1** as an equivalent of acetylene in Diels–Alder reactions is illustrated in equation (I).

$$\text{(I)}\quad \text{anthracene} \xrightarrow[98\%]{160^\circ} \text{adduct } [(CH_3)_3Si,\ SO_2C_6H_5] \xrightarrow[84\%]{Bu_4\overset{+}{N}F^-} \text{dibenzobarrelene}$$

$$\text{adduct} \xrightarrow[\text{2) } CH_3I]{\text{1) } n\text{-BuLi}} \text{adduct } [(CH_3)_3Si,\ CH_3,\ SO_2C_6H_5] \xrightarrow[\substack{58\% \\ \text{overall}}]{Bu_4\overset{+}{N}F^-} \text{methyl-dibenzobarrelene } [CH_3]$$

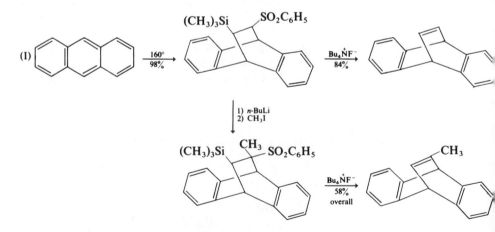

[1] L. A. Paquette and R. V. Williams, *Tetrahedron Letters*, **22**, 4643 (1981).

Benzyl *trans*-1, 3-butadiene-1-carbamate (1), 8, 34–35.
Preparation.[1]

Dendrobatid toxins. The first step in a synthesis of (±)-perhydrogephyrotoxin (**5**, a poison-frog alkaloid) involves a Diels–Alder reaction of **1** and a suitable α,β-unsaturated aldehyde, which proceeds selectively to give the *endo*-adduct **2**. Another key step is the selective reduction of a bicyclic imine (**3**) with $LiAlH_4$ in about 9:1 ratio from the sterically more hindered α-face.[2]

C_5H_{11}–CH=CH–CHO + CH_2=CH–CH=CH–$NHCO_2Bzl$ (**1**) $\xrightarrow{110°}$ **2** (cyclohexene bearing C_5H_{11}, CHO, $NHCO_2Bzl$) $\xrightarrow{\text{Several steps}}$ **3** (bicyclic imine, $(CH_2)_4$–dioxolane) $\xrightarrow{LiAlH_4}$ **4** → **5** ($HO(CH_2)_2$)

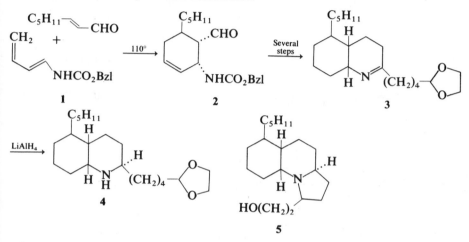

[1] P. J. Jessup, C. B. Petty, J. Roos, and L. E. Overman, *Org. Syn.*, **59**, 1 (1979).
[2] L. E. Overman and R. L. Freerks, *J. Org.*, **46**, 2833 (1981).

Benzylsulfonyldiazomethane, $C_6H_5CH_2SO_2CHN_2$ (1). Mol. wt. 196.23, m.p. 98–99°, nonexplosive. Preparation.[1]

Ring enlargement of cycloalkanones.[2] A novel method for homologation of cycloalkanones (**1**) to α-benzylsulfonylcycloalkanones (**2**) is shown in equation (I). The products can be converted into 1-benzylsulfonyl-1-cycloalkenes (**4**) by ring contraction.

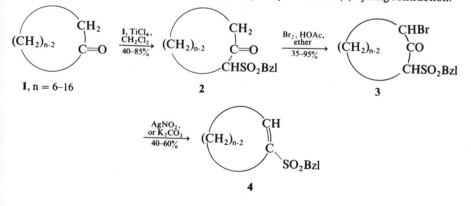

1, n = 6–16 $\xrightarrow[40\text{–}85\%]{1,\ TiCl_4,\ CH_2Cl_2}$ **2** $\xrightarrow[35\text{–}95\%]{Br_2,\ HOAc,\ \text{ether}}$ **3** $\xrightarrow[40\text{–}60\%]{AgNO_2,\ \text{or}\ K_2CO_3}$ **4**

(**1**: $(CH_2)_{n-2}$, CH_2, C=O; **2**: $(CH_2)_{n-2}$, CH_2, C=O, $CHSO_2Bzl$; **3**: $(CH_2)_{n-2}$, CHBr, CO, $CHSO_2Bzl$; **4**: $(CH_2)_{n-2}$, CH=C, SO_2Bzl)

[1] A. M. van Leusen and J. Strating, *Rec. trav.*, **84**, 151 (1965).
[2] S. Toyama, T. Aoyama, and T. Shioriri, *Chem. Pharm. Bull.*, **30**, 3032 (1982).

Benzyl p-toluenesulfonate, $CH_3C_6H_4SO_3CH_2C_6H_5$ **(1)**. Mol. wt. 265.35, m.p. 58°, moderately stable at room temperature.

Protection of phenols.[1] This highly O-selective reagent is preferred to benzyl chloride for preparation of benzyl ethers of phenols containing substituents prone to C-alkylation. Thus **2** is converted into the dibenzyl ether **3** in high yield using **1**, whereas the yield of the ether is only 18% when benzyl chloride is used.

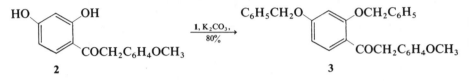

HO, OH, $COCH_2C_6H_4OCH_3$ (**2**) $\xrightarrow[80\%]{1,\ K_2CO_3,}$ $C_6H_5CH_2O$, $OCH_2C_6H_5$, $COCH_2C_6H_4OCH_3$ (**3**)

[1] P. M. Dewick, *Syn. Comm.*, **11**, 853 (1981).

Benzyl trichloroacetimidate (1). Mol. wt. 252.5, b.p. 106°/0.5 mm, m.p. 3°.

Preparation[1]:

$$CCl_3C{\equiv}N + C_6H_5CH_2OH \xrightarrow[91\%]{K_2CO_3} \underset{\mathbf{1}}{C_6H_5CH_2O(HN{=})C{-}CCl_3}$$

Benzyl ethers.[2] Benzyl ethers of sugar alcohols can be obtained in 82–98% isolated yield by reaction with **1** and catalytic amounts of triflic acid; under these conditions isopropylidene and benzylidene acetal groups are stable, as well as ester and imide groups. The reagent thus is comparable to benzyl triflate (**6**, 44).

[1] F. Cramer, K. Pawelzik, and H. J. Baldauf, *Ber.*, **91**, 1049 (1958).
[2] T. Iversen and D. R. Bundle, *J. C. S. Chem. Comm.*, 1240 (1981).

Benzyltriethylammonium permanganate, 9, 43.

Sulfones.[1] The oxidant is useful for oxidation of sulfides and sulfoxides to sulfones in CH_2Cl_2–CH_3COOH at − 10°.

[1] D. Scholz, *Monats.*, **112**, 241 (1981).

2,2′-Bipyridinium chlorochromate, 10, 30. Suppliers: Aldrich, Alfa.

Bipy · $HCrO_3Cl$ was used successfully to oxidize the allylic alcohol (**1**) to the α,β-unsaturated aldehyde **2**, a natural degradation product of linolenic acid.[1]

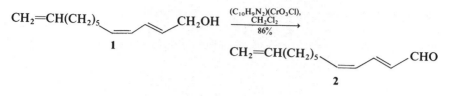

$$CH_2{=}CH(CH_2)_5\text{-}CH{=}CH\text{-}CH{=}CH\text{-}CH_2OH \ (\mathbf{1}) \xrightarrow[86\%]{(C_{10}H_9N_2)(CrO_3Cl),\ CH_2Cl_2} CH_2{=}CH(CH_2)_5\text{-}CH{=}CH\text{-}CH{=}CH\text{-}CHO \ (\mathbf{2})$$

[1] F. Bohlmann and W. Rotard, *Ann.*, 1216 (1982).

Bis(acetonitrile)chloronitropalladium(II), $Pd(CH_3CN)_2ClNO_2$ (**1**). The complex is obtained by reaction of $Pd(CH_3CN)_2Cl_2$ with 1 equiv. of $AgNO_2$ in acetonitrile.

Oxidation of alkenes to ketones. This palladium complex is an effective catalyst for air oxidation of 1-alkenes to methyl ketones in high yield via a nitro–nitrosyl redox couple.[1]

$$Pd(CH_3CN)_2ClNO_2 + RCH{=}CH_2 \longrightarrow [PdCl(NO)]_n + RCOCH_3 \quad ([PdCl(NO)]_n \xrightarrow{O_2} Pd(CH_3CN)_2ClNO_2)$$

Epoxidation of cycloalkenes.[2] Cycloalkenes, particularly strained bicyclic alkenes, can be epoxidized by treatment with 1 equiv. of **1** under N_2. In the case of norbornene (**2**) the reaction involves formation of the metallacycle **3**, which decomposes slowly to *exo*-epoxynorbornene (**4**). In fact, epoxidation is favoured over oxidation of a vinyl system in the air oxidation of **5** catalyzed with **1**.[3]

$$\mathbf{2} \xrightarrow[\text{quant.}]{1,\ N_2} \mathbf{3} \xrightarrow{>90\%} \mathbf{4} + [PdClNO]_n$$

$$\mathbf{5} \xrightarrow[30\text{–}40\%]{1,\ O_2} \mathbf{6}$$

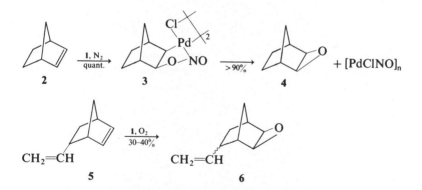

Epoxidation of simple cycloalkenes with **1** has limited practical value, because cycloalkanones, 2-cycloalkenols, and 2-cycloalkenones are formed as well, often as the major products.

[1] M. A. Andrews and K. P. Kelly, *Am. Soc.*, **103**, 2894 (1981).
[2] M. A. Andrews and C.-W. F. Cheng, *ibid.*, **104**, 4266 (1982).
[3] A. Heumann, F. Chauvet, and B. Waegell, *Tetrahedron Letters*, **23**, 2767 (1982).

Bis(acetonitrile)dichloropalladium, 7, 21–22; **8,** 39; **9,** 44; **10,** 30–31.

Cope rearrangement (**10,** 31). The Pd(II)-catalyzed Cope rearrangement of chiral 1,5-dienes occurs with virtually complete 1,4-transfer of chirality.[1]

Example:

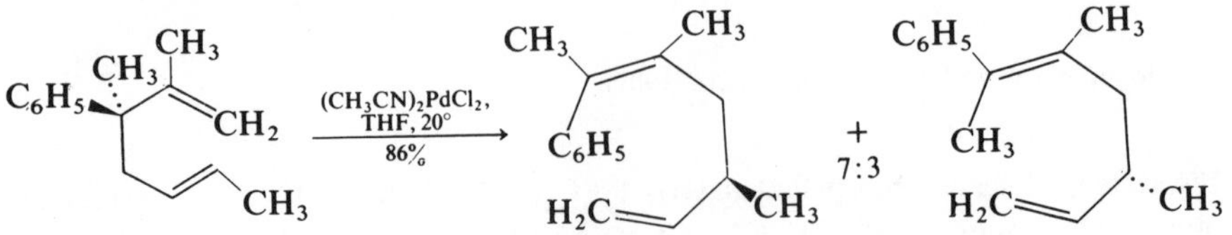

C_6H_5, CH_3, CH_3, CH_2, CH_3 $\xrightarrow[86\%]{(CH_3CN)_2PdCl_2,\ THF,\ 20°}$ (2Z, 5R), 97%ee + (2E, 5S), 96%ee (7:3)

The same induction is observed in the thermal Cope rearrangement.

Coupling of vinyl iodides with methyl vinyl ketones. Dienones can be obtained in high yield from this coupling, catalyzed by $PdCl_2(CH_3CN)_2$ prereduced with formic acid. The major dienone formed always possesses the stereochemistry of the vinyl iodide (equation I).

(I) 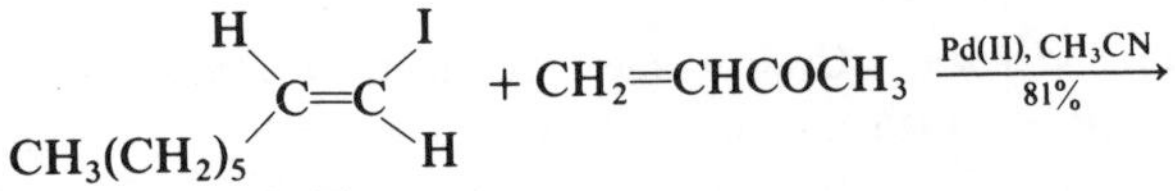 $CH_3(CH_2)_5CH{=}CHI$ (E) $+ CH_2{=}CHCOCH_3 \xrightarrow[81\%]{Pd(II),\ CH_3CN}$

$CH_3(CH_2)_5CH{=}CHCH{=}CHCOCH_3$

(E, E/E, Z > 20:1)

The coupling was extended to the synthesis of the macrocyclic lactone **1** containing a dienone group. In this case a stoichiometric amount of Pd(II) is necessary to effect ring closure.[2]

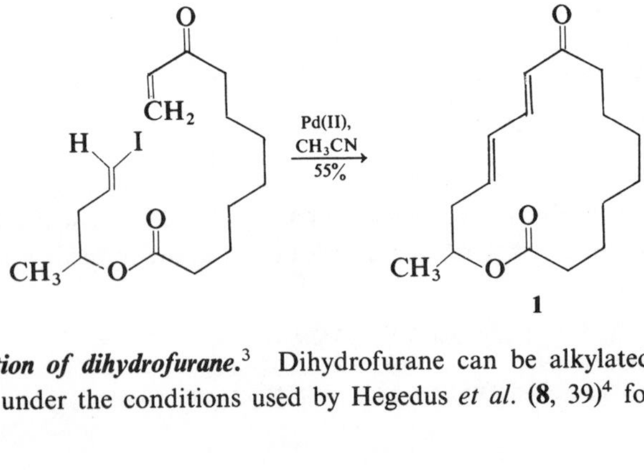

$\xrightarrow[55\%]{Pd(II),\ CH_3CN}$ **1**

Alkylation of dihydrofurane.[3] Dihydrofurane can be alkylated by stabilized carbanions under the conditions used by Hegedus *et al.* (**8,** 39)[4] for alkylation of alkenes via a Pd(II) complex. Substitution occurs exclusively at the carbon atom adjacent to oxygen with migration of the double bond into conjugation with oxygen (endo- or exocyclic).

Examples:

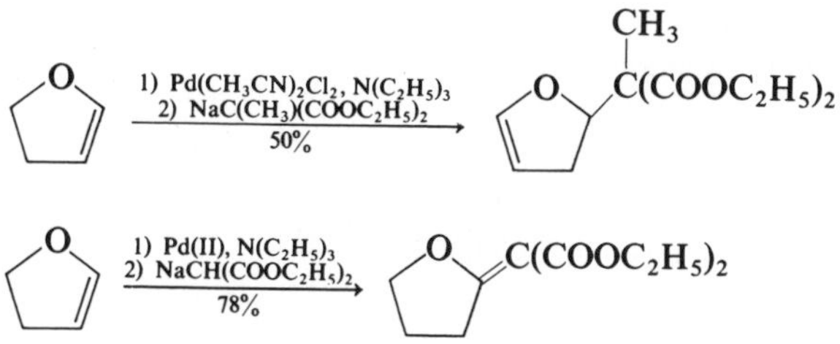

1) $Pd(CH_3CN)_2Cl_2$, $N(C_2H_5)_3$ 2) $NaC(CH_3)(COOC_2H_5)_2$, 50%; product: CH_3, $C(COOC_2H_5)_2$

1) Pd(II), $N(C_2H_5)_3$ 2) $NaCH(COOC_2H_5)_2$, 78%; product: $C(COOC_2H_5)_2$

[1] L. E. Overman and E. J. Jacobsen, *Am. Soc.*, **104**, 7225 (1982).
[2] F. E. Ziegler, U. R. Chakraborty, and R. B. Weisenfeld, *Tetrahedron*, **37**, 4035 (1981).
[3] L. V. Dunkerton and A. J. Serino, *J. Org.*, **47**, 2812 (1982).
[4] L. S. Hegedus, R. E. Williams, M. A. McGuire, and T. Hayashi, *Am. Soc.*, **102**, 4973 (1980); L. S. Hegedus and W. H. Darlington, *ibid.*, **102**, 4980 (1980).

Bis(acetylacetonate)oxovanadium (Vanadyl acetylacetonate), $VO(C_5H_7O_2)_2$. Mol. wt. 265.16. Supplier: Alfa.

Oxygenation catalyst. This vanadium catalyst is a particularly active catalyst for oxygenation of 3,5-di-*t*-butylpyrocatechol (**1**) to the muconic acid anhydride **2**, the 2-pyrone **3**, and the *o*-quinone **4**, which cannot be oxidized to **2** under these conditions.[1]

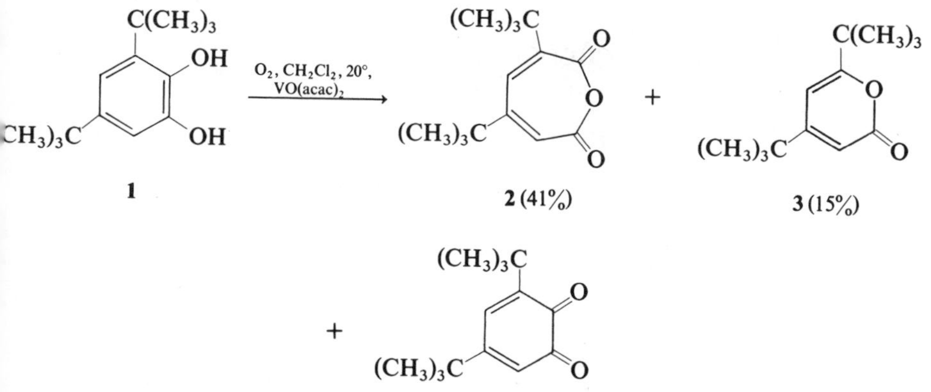

$C(CH_3)_3$, OH, OH, $(CH_3)_3C$ — **1** — O_2, CH_2Cl_2, 20°, $VO(acac)_2$ → **2** (41%) + **3** (15%) + **4** (27%)

Oxygenation of **1** with $Cl_2Ru[P(C_6H_5)_3]_3$ as catalyst affords only the anhydride **2** (26% yield) and the 2-pyrone **3** (64% yield).[2]

[1] Y. Tatsuno, M. Tatsuda, and S. Otsuka, *J. C. S. Chem. Comm.*, 1100 (1982).
[2] M. Matsumoto and K. Kuroda, *Am. Soc.*, **104**, 1433 (1982).

Bis(acrylonitrile)nickel(0), $Ni(CH_2{=}CHCN)_2$ (**1**). Mol. wt. 164.83.

Reaction with strained hydrocarbons. Bicyclo[2.1.0]pentane (**2**), when heated (160°) in the presence of an electron-deficient alkene, forms a number of monocyclic adducts.[1] In contrast, in the presence of **1** the central bond of **2** is cleaved at 70°, and cycloadducts are formed with an electron-poor alkene in useful yields. A typical reaction is shown in equation (I).[2]

(I) **2** + $CH_2{=}CHCOOCH_3$ $\xrightarrow{\text{1, 70°}}$ COOCH$_3$ (33%) + COOCH$_3$ (33%) + $(CH_2)_2COOCH_3$ (22%)

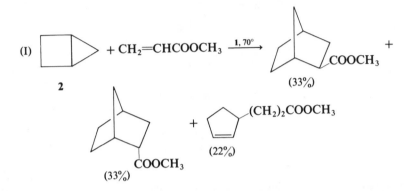

Thermolysis (200°) of bicyclo[1.1.0]butane (**3**) severs two bonds to produce 1,3-butadiene, but in the presence of **1**, **3** undergoes cleavage of the central σ-bond and one peripheral bond to generate an allyl carbenoid intermediate, which reacts with an electron-deficient alkene to give an allylcyclopropane (equation II).[3]

(II) **3** + $CH_2{=}CHCOOCH_3$ $\xrightarrow[95\%]{\text{1, 60°}}$ $CH_2{=}CHCH_2$ COOCH$_3$ + $CH_2{=}CHCH_2$ COOCH$_3$ 63:35

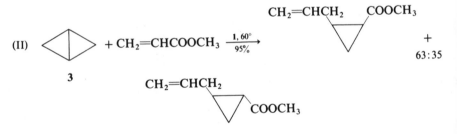

[1] G. N. Schrauzer, *Ber.*, **94**, 642 (1961).
[2] H. Takaya, T. Suzuki, Y. Kumagai, M. Yamakawa, and R. Noyori, *J. Org.*, **46**, 2846 (1981).
[3] *Idem, ibid*, **46**, 2845 (1981).

Bis(benzonitrile)dichloropalladium(II), 5, 31–32; **6**, 45–47; **9**, 44–45; **10**, 31–32.

Steroidal π-allyl palladium complexes. These complexes of unsaturated steroids are prepared conveniently by reaction in chloroform or acetone with this Pd(II) complex. Ergosterol (**1**) is converted in this way into the dimeric complex **2**, with loss of the hydroxyl group. Two interesting new reactions of the complex **2** are reported: reduction to a triene (**3**) and oxidation to a trienol (**4**).

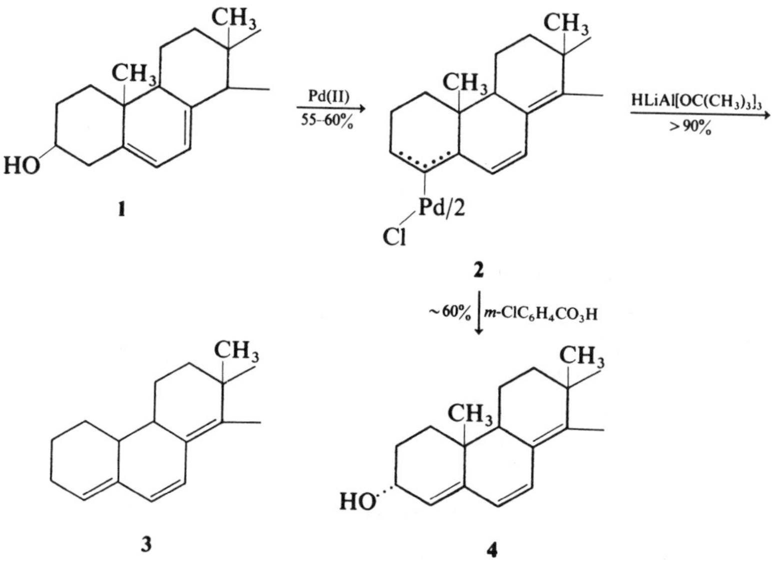

CH3 CH3 HO 1 Pd(II) 55–60% CH3 CH3 Pd/2 Cl 2 HLiAl[OC(CH3)3]3 >90% ~60% m-ClC6H4CO3H CH3 3 CH3 CH3 HO 4

Chloroallylation. In the presence of this Pd(II) complex, allylic chlorides react with alkynylsilanes to form (E)-4-trimethylsilyl-5-chloro-1,4-dienes in yields of 80–95%.[2]

Example:

$$n\text{-}C_4H_9C{\equiv}CSi(CH_3)_3 + ClCH_2CH{=}CH_2 \xrightarrow[79\%]{Cl_2Pd(C_6H_5CN)_2}$$

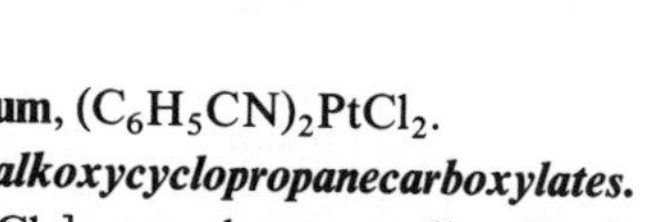

n-C4H9 Si(CH3)3 C=C Cl CH2CH=CH2

[1] C. Mahé, H. Patin, M.-T. Van Hulle, and D. H. R. Barton, *J. C. S. Perkin I*, 2504 (1981).
[2] R. Yamaguchi, H. Kawasaki, T. Yoshitome, and M. Kawanisi, *Chem. Letters*, 1485 (1982).

Bis(benzonitrile)dichloroplatinum, $(C_6H_5CN)_2PtCl_2$.

Rearrangement of alkyl β-alkoxycyclopropanecarboxylates. This Pt(II) complex, $[Rh(CO)_2Cl]_2$, and $[Ru(CO)_3Cl_2]_2$ are about equally effective as catalysts for the rearrangement of substrates **1** to vinyl ethers **2**. Both the alkoxy group and the carboethoxy group are essential for the rearrangement. Copper bronze and CuCl also catalyze this rearrangement, although at higher temperatures (170°).[1]

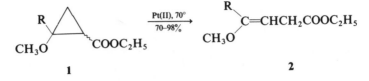

$$\text{1 (cyclopropane: R, CH}_3\text{O, COOC}_2\text{H}_5) \xrightarrow[70\text{–}98\%]{\text{Pt(II), }70^\circ} \text{2: R(CH}_3\text{O)C=CHCH}_2\text{COOC}_2\text{H}_5$$

[1] M. P. Doyle and D. Van Leusen, *Am. Soc.*, **103**, 5917 (1981).

Bis(benzyltriethylammonium) dichromate, $[C_6H_5CH_2\overset{+}{N}(C_2H_5)_3]_2Cr_2O_7^{2-}$ (**1**). Mol. wt. 600.6, orange crystals melting at 89°, fairly soluble in common organic solvents. The reagent is prepared by reaction of benzyltriethylammonium chloride with CrO_3 (1 equiv.) in 3*N* HCl.

The reagent oxidizes benzyl and allyl alcohols to carbonyl compounds in high yield. Yields are low for primary alkanols and cycloalkanols (~ 30%).[1]

[1] X. Huang and C.-C. Chan, *Synthesis*, 109 (1982).

1,3-Bis(*t*-butyldimethylsilyloxy)-2-aza-1,3-diene, Mol. wt. 315.59, b.p. 74°/0.06 mm.

Preparation:

$$HC(=O)N[Si(CH_3)_3]_2 \xrightarrow{CH_3COCl} \left[OHCN(Si(CH_3)_3)COCH_3 \right] \xrightarrow{69\%} CH_3C(=O)NHCHO \xrightarrow[86\%]{t\text{-BuMe}_2\text{SiOTf},\ N(C_2H_5)_3} t\text{-BuMe}_2\text{SiOC(=CH}_2\text{)N=CHOSiMe}_2\text{Bu-}t \quad (\mathbf{1}).$$

The corresponding trimethylsilyl ethers are thermally unstable.

α-Pyridones; α-piperidones.[1] Diels–Alder reactions of this azadiene with an acetylenic dienophile followed by desilylation provide a simple synthesis of α-pyridones (equation I). Reaction with methyl acrylate results in piperidones (equation II). A bicyclic piperidone derivative is obtained in high yield from the reaction with maleic anhydride followed by monodesilylation (equation III). The azadiene also undergoes cycloaddition with 1,4-benzoquinone to provide an isoquinoline derivative.

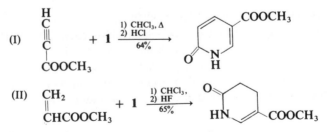

(I) $$HC{\equiv}CCOOCH_3 + \mathbf{1} \xrightarrow[64\%]{1)\ CHCl_3,\ \Delta;\ 2)\ HCl} \text{5-(methoxycarbonyl)-2-pyridone (COOCH}_3\text{, O, NH)}$$

(II) $$CH_2{=}CHCOOCH_3 + \mathbf{1} \xrightarrow[65\%]{1)\ CHCl_3,\ 2)\ HF} \text{piperidone (O, HN, COOCH}_3\text{)}$$

(III) 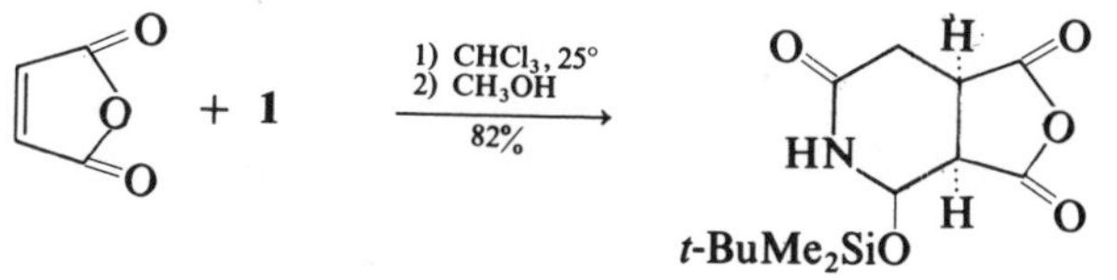 + **1** $\xrightarrow[82\%]{1)\ CHCl_3,\ 25°\quad 2)\ CH_3OH}$ t-BuMe$_2$SiO

[1] F. Sainte, B. Serckx-Poncin, A.-M. Hesbain-Frisque, and L. Ghosez, *Am. Soc.*, **104**, 1428 (1982).

Bis(1,5-cyclooctadiene)nickel(0), 4, 33–35; **5,** 34–35; **7,** 428–429; **9,** 45–60; **10,** 33.

Ullmann reaction (**4,** 33–34). Semmelhack *et al.*[1] recorded details of the coupling of aryl and vinyl halides with Ni(0) complexes, $(COD)_2Ni$ or $Ni[P(C_6H_5)_3]_4$. This reaction was used for the first synthesis of almusone (**3**), an antileukemic lignan obtained from the wood of *Alnus japonica* Steud. In the case of *ortho*-substituted aryl iodides such as **1**, reduction becomes a competing reaction and the yields are only moderate. In fact *o,o*-disubstituted aryl halides cannot be coupled under these conditions. Deliberate addition of a proton source increases formation of the reduction product. Acetic or trifluoroacetic acid are useful for this purpose. Added triphenylphosphine also promotes reduction.

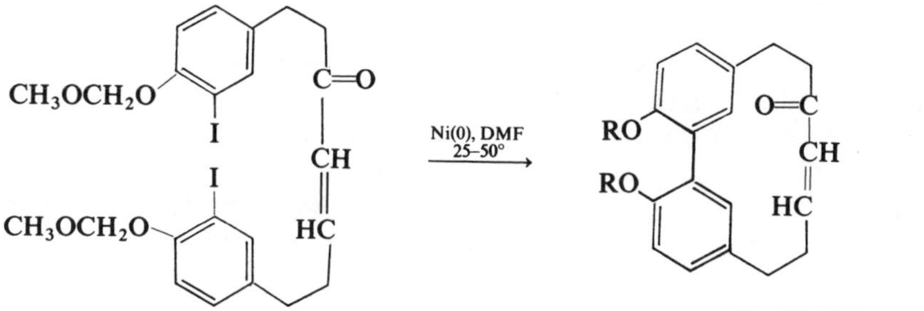

1 $\xrightarrow[25–50°]{Ni(0),\ DMF}$ **2** ($R = CH_2OCH_3$), 46%; **3** (R = H), 72%

[1] M. F. Semmelhack, P. Helquist, L. D. Jones, L. Keller, L. Mendelson, L. S. Ryono, J. G. Smith, and R. D. Stauffer, *Am. Soc.*, **103**, 6460 (1981).

Bis(cyclopentadienyl)dichlorozirconium (II), 10, 131–133.

threo-*Selective aldol condensation.*[1] The crotylzirconium derivatives **1** or **2**, prepared as shown, react with aldehydes to afford the *threo*-aldol selectively (equations I and II).

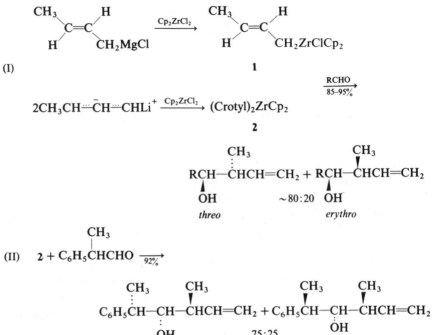

(I)

$$CH_3CH{=}CHCH_2MgCl \xrightarrow{Cp_2ZrCl_2} CH_3CH{=}CHCH_2ZrClCp_2 \ (\mathbf{1})$$

$$2CH_3CH \cdots \bar{C}H \cdots CHLi^+ \xrightarrow{Cp_2ZrCl_2} (Crotyl)_2ZrCp_2 \ (\mathbf{2})$$

$$\xrightarrow[85\text{–}95\%]{RCHO} RCH(OH)\text{–}CH(CH_3)CH{=}CH_2 \ (\textit{threo}) + RCH(OH)\text{–}CH(CH_3)CH{=}CH_2 \ (\textit{erythro}) \quad \sim 80:20$$

(II)

$$\mathbf{2} + C_6H_5CH(CH_3)CHO \xrightarrow[92\%]{}$$

$$C_6H_5CH(CH_3)\text{–}CH(OH)\text{–}CH(CH_3)CH{=}CH_2 + C_6H_5CH(CH_3)\text{–}CH(OH)\text{–}CH(CH_3)CH{=}CH_2 \quad 75:25$$

[1] Y. Yamamoto and K. Maruyama, *Tetrahedron Letters*, **22**, 2895 (1981).

Bis(cyclopentadienyl)titanium-μ-chloro-μ-methylidenedimethylaluminum, $Cp_2TiCH_2AlCl(CH_3)_2$ (**1**), **8**, 83–84; **10**, 87–88.

1,5-*Dienes*.[1] Allylic esters such as **2** can be converted into a 1,5-diene (**5**) by methylenation with **1** to give an allylic vinyl ether (**3**). Claisen rearrangement of **3** results in a γ,δ-enone **4**, which is converted to the desired diene (**5**) by a second methylenation with **1**. Actually, this three-step process can be carried out without

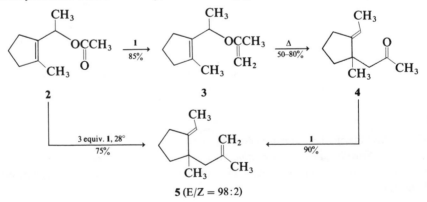

$$\mathbf{2} \xrightarrow[85\%]{\mathbf{1}} \mathbf{3} \xrightarrow[50\text{–}80\%]{\Delta} \mathbf{4} \xrightarrow[90\%]{\mathbf{1}} \mathbf{5}$$

$$\mathbf{2} \xrightarrow[75\%]{3\ \text{equiv.}\ \mathbf{1},\ 28^\circ} \mathbf{5}\ (E/Z = 98:2)$$

isolation of intermediates by treatment of **2** with 3 equiv. of **1** at room temperature for 12 hours. Presumably $(CH_3)_2AlCl$ is liberated from **1** during methylenation and this Lewis acid catalyzes the Claisen rearrangement. Five other examples of this transformation are cited.

[1] J. W. S. Stevenson and T. A. Bryson, *Tetrahedron Letters*, **23**, 3143 (1982).

Bis(dibenzylideneacetone)palladium, Pd(dba)$_2$, 10, 34.

Allylic alkylation. In the presence of this Pd(0) complex and bis(diphenylphosphine)ethane, allylic acetates can be alkylated by lithium enolates of cyclohexanone, 3-pentanone, acetophenone, and mesityl oxide in 40–80% yield. The reaction was shown to occur with overall retention of configuration in the case of the lithium enolate of acetone (equation I).

(I) OAc, C_6H_5 + $LiCH_2\overset{O}{\overset{\|}{C}}CH_3$ $\xrightarrow[\text{dppe, THF}]{Pd(dba)_2,}$ CH_2COCH_3, C_6H_5

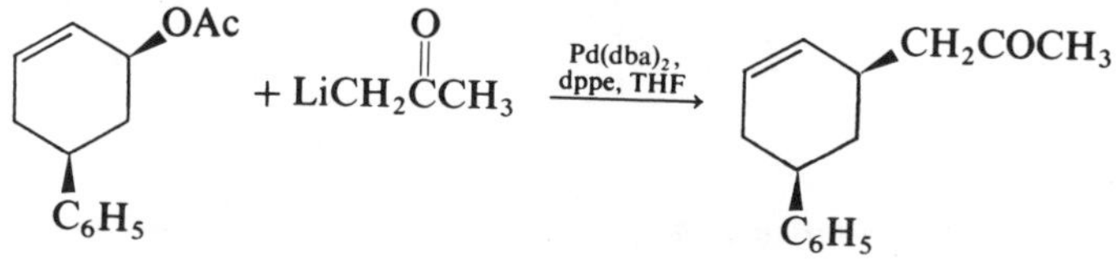

[1] J.-C. Fiaud and J.-L. Malleron, *J. C. S. Chem. Comm.*, 1159 (1981).

2,2′-Bis(diphenylphosphine)-1,1′-binaphthyl (binap), **10**, 36.

Asymmetric isomerization of allylamines.[1] The enantiomeric rhodium(I) complexes with the ligands cycloocta-1,5-diene (cod) and (+)- or (−)-binap, Rh (+)-

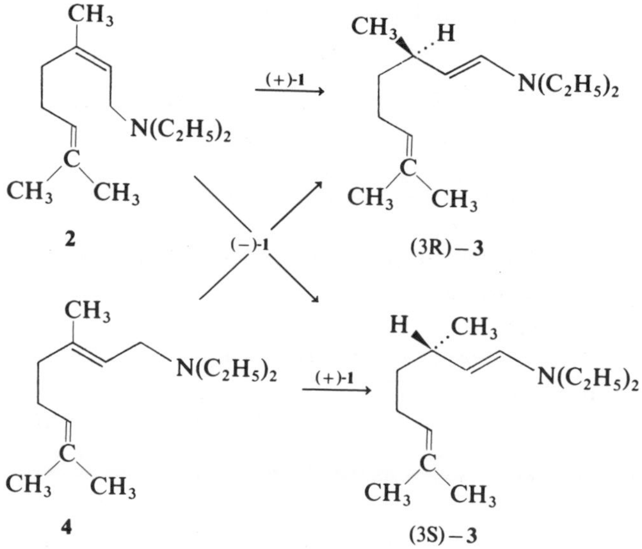

CH_3, $N(C_2H_5)_2$, C, CH_3 CH_3 **2** $\xrightarrow{(+)\text{-}\mathbf{1}}$ CH_3 H, $N(C_2H_5)_2$, C, CH_3 CH_3 (3R)−**3**

(−)-**1**

CH_3, $N(C_2H_5)_2$, C, CH_3 CH_3 **4** $\xrightarrow{(+)\text{-}\mathbf{1}}$ H CH_3, $N(C_2H_5)_2$, C, CH_3 CH_3 (3S)−**3**

Chart **1**

binap(cod)ClO_4, (+)-**1**, and Rh (−)-binap(cod)ClO_4, (−)-**1**, isomerize allylic amines to optically active enamines with virtually 100% selectivity. Thus, N,N-diethylnerylamine (**2**) is isomerized by (+)-**1** to N,N-diethylcitronellal-(E)-enamine, (3R)-**3**, with 94% ee and by (−)-**1** to (3S)-**3** in comparable optical yield. N,N-Diethylgeranylamine (**4**) is also isomerized by these rhodium catalysts to **3**, but with enantioselectivity opposite to that observed in isomerization of **2** (Chart I).

[1] K. Tani, T. Yamagata, S. Otsuka, S. Akutagawa, H. Kumobayashi, T. Taketomi, H. Takaya, A. Miyashita, and R. Noyori, *J. C. S. Chem. Comm.*, 600 (1982).

Bis[2-(diphenylphosphinyl)ethyl]amine hydrochloride (1). Mol. wt. 477.94. *Caution*: The amine hydrochloride is carcinogenic.

Preparation:

$$Cl^- H_2\overset{+}{N}(CH_2CH_2Cl)_2 \xrightarrow[90\%]{1)\ 2(C_6H_5)_2PH;\ 2)\ KOC(CH_3)_3;\ 3)\ HCl} Cl^- H_2\overset{+}{N}[CH_2CH_2P(C_6H_5)_2]_2$$

1

Water-soluble chelating diphosphines. This amine hydrochloride has been used to prepare water-soluble ligands for transition metals, particularly Rh(I). Thus, the complex Rh(I)·**2** is a highly active catalyst for homogeneous hydrogenation, and Rh(I)·**3** combines with the glycoprotein avidin to form an effective asymmetric catalyst for hydrogenation of α-acetamidoacrylic acid.

HOOC, HOOC (3,4-disubstituted benzene) — $\overset{O}{\overset{\|}{C}}N[CH_2CH_2P(C_6H_5)_2]_2$

2

HN—C(=O)—NH (bicyclic, S ring) ···$(CH_2)_4\overset{O}{\overset{\|}{C}}N[CH_2CH_2P(C_6H_5)_2]_2$

3

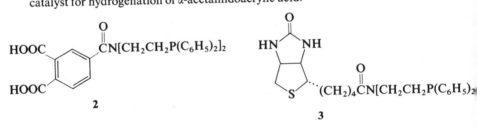

[1] R. G. Nuzzo, S. L. Haynie, M. E. Wilson, and G. M. Whitesides, *J. Org.*, **46**, 2861 (1981).

2,4-Bis(4-methoxyphenyl)-1,3-dithia-2,4-diphosphetane-2,4-disulfide (Lawesson's reagent), **8**, 327; **9**, 49–50; **10**, 39.

3,4-*Di-t-butyldithiete* (3). This thiet has been obtained by treatment of the thioxo ketone **2** with Lawesson's reagent (**1**). The reaction is not successful with the diketone corresponding to **2**.[1]

$$(CH_3)_3C\text{—}C(=O)\text{—}C(=S)\text{—}C(CH_3)_3 \xrightarrow[45\%]{\mathbf{1},\ C_6H_5CH_3,\ \Delta} (CH_3)_3C\text{—}C{=}C\text{—}C(CH_3)_3\ (S\text{—}S\ \text{ring})$$

2 → **3**

Endothiopeptides. Endothiopeptides contain one or more –C(S)NH– groups in the backbone. Endothiodipeptide esters, $H_2NCH_2C(S)NHCH_2COOR$, can be prepared without racemization by reaction of BOC derivatives of dipeptide esters with **1**, followed by removal of the protecting group with HBr–HOAc. The thionation step occurs in 91–98% yield (5 examples).[2]

Peptide synthesis.[3] Salts of N-protected amino acids react with **1** (0.5 equiv.) to form a mixed anhydride (**2**). On addition of triethylamine (2 equiv.) and the salt of an amino acid ester, the protected dipeptide (**3**) is formed in ~80–95% yield. Racemization is slight (< 0.7%), if any.

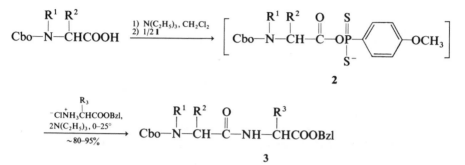

$$\text{Cbo—N(R}^1\text{)—CH(R}^2\text{)COOH} \xrightarrow[\text{2) 1/2 1}]{\text{1) N(C}_2\text{H}_5)_3,\ \text{CH}_2\text{Cl}_2} \left[\text{Cbo—N(R}^1\text{)—CH(R}^2\text{)—C(=O)—OP(=S)(S}^-\text{)—C}_6\text{H}_4\text{—OCH}_3\right]$$

2

$$\xrightarrow[\sim 80\text{–}95\%]{^-\text{Cl}\overset{+}{\text{N}}\text{H}_3\text{CH(R}_3\text{)COOBzl},\ 2\text{N(C}_2\text{H}_5)_3,\ 0\text{–}25^\circ} \text{Cbo—N(R}^1\text{)—CH(R}^2\text{)—C(=O)—NH—CH(R}^3\text{)COOBzl}$$

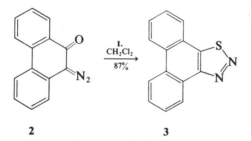

3

Aromatic thiadiazoles.[4] The phenanthro [1,2,3] thiadiazole **3** has been made by reaction of the α-diazoketone **2** with Lawesson's reagent (**1**).

$$\mathbf{2} \xrightarrow[87\%]{\mathbf{1},\ \text{CH}_2\text{Cl}_2} \mathbf{3}$$

2 **3**

[1] B. Köpke and J. Voss, *J. Chem. Research* (S), 314 (1982).
[2] K. Clausen, M. Thorsen, and S.-O. Lawesson, *Tetrahedron*, **37**, 3635 (1981).
[3] U. Pederson, M. Thorsen, E.-E. A. M. El-Khrisy, K. Clausen, and S.-O. Lawesson, *ibid.*, **38**, 3267 (1982).
[4] M. I. Levinson and M. P. Cava, *Heterocycles*, **19**, 241 (1982).

Bis(p-methoxyphenyl)telluroxide, $O{=}Te(C_6H_4{-}OCH_3\text{-}p)_2$, 9, 50.

Aldol catalyst.[1] This telluroxide (**1**) can function as a catalyst for several aldol condensations. For example, chalcone is obtained with this catalyst from benzaldehyde and acetophenone in 94% yield. A particularly interesting reaction is the cyclization of **1** to **2**.

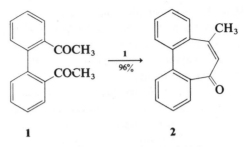

1 → (1, 96%) → 2

[1] L. Engman and M. P. Cava, *Tetrahedron Letters*, **22**, 5251 (1981).

2,4-Bis(methylthio)-2,4-dithioxocyclodiphosphathiane,

$CH_3S{-}P(=S)(\mu\text{-}S)_2P(=S){-}SCH_3$ **(1).** Mol. wt. 284.21, m.p. 112°.

The reagent is prepared by reaction of P_4S_{10} with CH_3SH under pressure or in 1,2,4-trichlorobenzene at 180°;[1] it can also be prepared by reaction of P_4S_{10} with CH_3OH.[2]

Dithioesters.[2] Carboxylic acids are converted directly into dithioesters, $RC(=S)SCH_3$, by reaction with **1** in 1,2,4-trichlorobenzene at 130°. Yields are 55–65%.

[1] C. B. Scott, A. Menefee, and D. O. Alford, *J. Org.*, **22**, 789 (1957).
[2] H. Davy, *J. C. S. Chem. Comm.*, 457 (1982).

Bis-(*p*-nitrophenethyl)phosphoromonochloridate (1).

Preparation:

$(C_6H_5O)_2P(O)H \xrightarrow{O_2N\text{-}C_6H_4\text{-}CH_2CH_2OH} (O_2N\text{-}C_6H_4\text{-}CH_2CH_2O)_2P(O)H \xrightarrow[\text{quant.}]{SO_2Cl_2} (O_2N\text{-}C_6H_4\text{-}CH_2CH_2O)_2P(O)Cl$

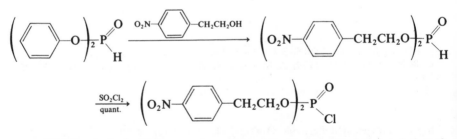

Phosphorylation.[1] This reagent can be used for 3′-and/or 5′-phosphorylation of nucleosides in the phosphotriester approach to oligonucleotides. Deblocking by β-elimination proceeds cleanly with DBU in pyridine at 25° in 24 hours.

[1] F. Himmelsbach and W. Pfeiderer, *Tetrahedron Letters*, **23**, 4793 (1982).

N,N-Bis(2-oxo-3-oxazolidinyl)phosphorodiamidic chloride (1), 10, 41.

Acid anhydrides.[1] In the presence of a tertiary base, the reagent converts carboxylic acids into the anhydride in yields >90%.

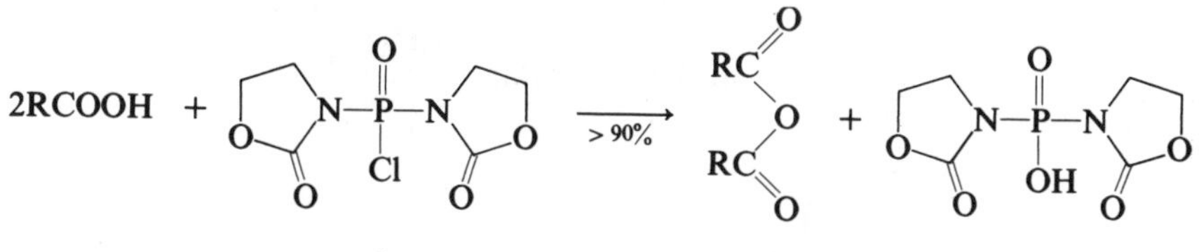

1

Macrocycles. This reagent was used twice in an assembly of aplasomycin (**5**), a boron-containing macrolide with antibiotic properties. It was used to effect esterification of **1** with **2** in 98% yield and again for cyclization of **3** to the macrocycle **4** in 71% yield under conditions of only moderate dilution. After minor adjustments of the functional groups, the metal was introduced by reaction with trimethylborate to give **5**.

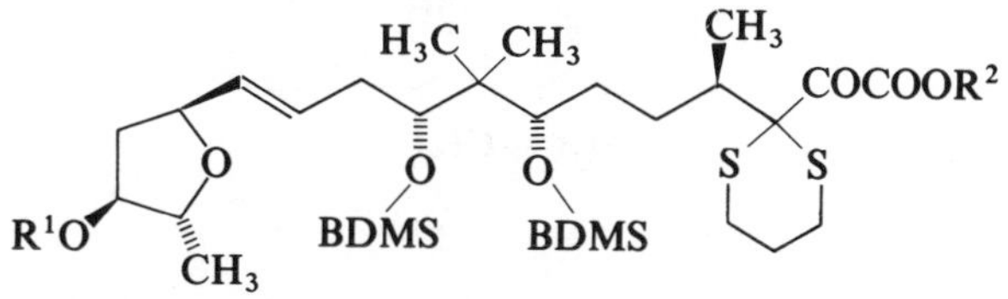

1, $R^1 = i\text{-}Pr_3Si$, $R^2 = H$
2, $R^1 = H$, $R^2 = CH_3$

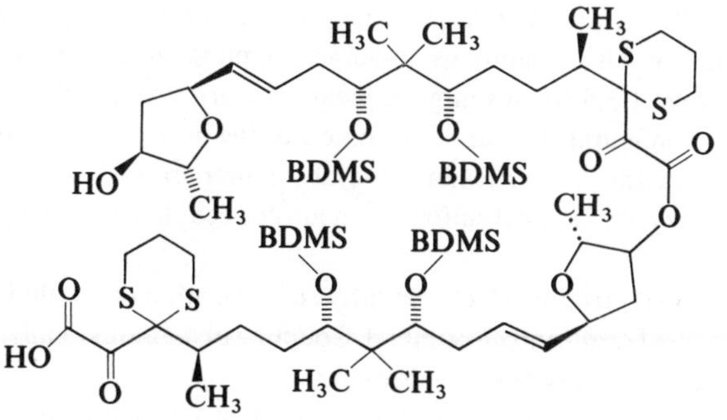

3

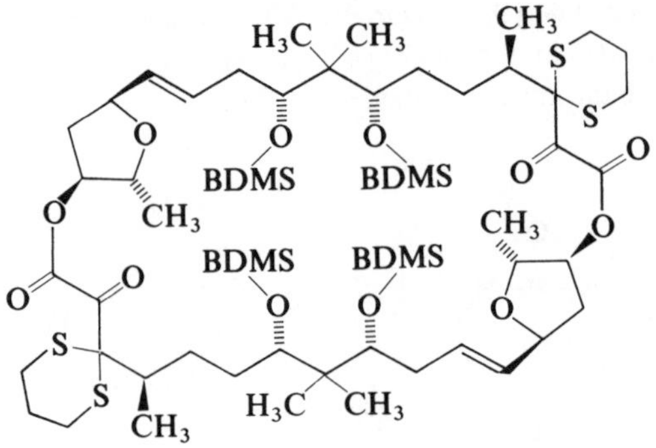

4

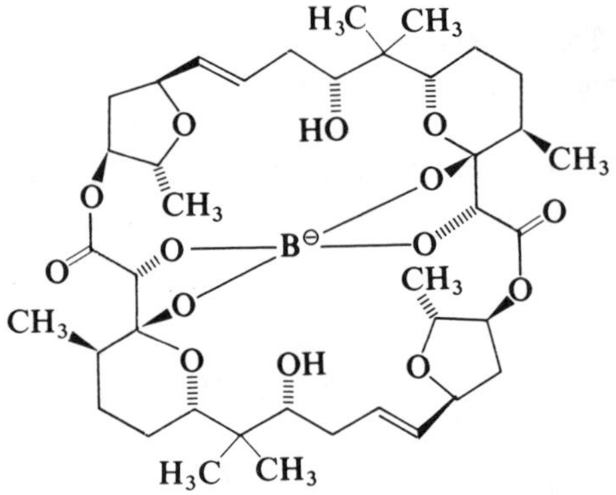

5

[1] J. Cabré, A. Palomo-Coll, and A. L. Palomo-Coll, *Synthesis*, 616 (1981).
[2] E. J. Corey, D. H. Hua, B.-C. Pan, and S. P. Seitz, *Am. Soc.*, **104**, 6818 (1982).

Bis(2,4-pentanedionato)nickel, Ni(acac)$_2$, **5**, 471, **6**, 417, **7**, 250, **9**, 51–52, **10**, 42.

Coupling of sulfones to alkenes. In the presence of a catalytic amount of Ni(acac)$_2$, α-anions of allylic sulfones undergo coupling at the α-position with desulfonylation to give 1,3,5-trienes in good yield. The original double bonds of the sulfone retain their configuration, but both isomers of the newly formed double bond are obtained. The reaction is particularly useful for preparation of C_{20}- and C_{30}-polyisoprenoids. Benzyl and *n*-alkyl sulfones can also be coupled to form symmetrical alkenes in this way.[1]

In contrast, copper (II) triflate effects oxidative dimerization of α-anions of *n*-alkyl sulfones to α-disulfones (*meso*- and *dl*-isomers). Coupling of α-anions of allylic sulfones with this catalyst occurs mainly at the γ-position.[2]

Allylsilanes. A regio- and stereoselective synthesis of allylsilanes involves nickel-catalyzed coupling of enol phosphates with trimethylsilylmethylmagnesium chloride in ether (15 hr at 25°). $Ni(acac)_2$ and $NiBr_2$ are about equally effective as catalysts. $Pd[P(C_6H_5]_4$ is somewhat less active; $PdCl_2$ is inactive.[3]

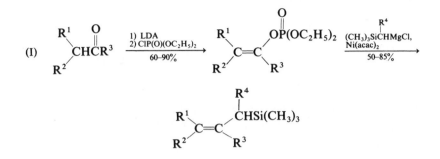

(I) $R^1R^2CHC(O)R^3$ —[1) LDA; 2) $ClP(O)(OC_2H_5)_2$; 60–90%]→ $R^1R^2C{=}C(R^3)OP(O)(OC_2H_5)_2$ —[$(CH_3)_3SiCH(R^4)MgCl$, $Ni(acac)_2$; 50–85%]→ $R^1R^2C{=}C(R^3)CH(R^4)Si(CH_3)_3$

Coupling of aryl phosphates with organometallic reagents. $Ni(acac)_2$ is the most efficient nickel catalyst for coupling of aryl phosphates with Grignard and organoaluminum reagents. Various Pd, Co, Fe, and copper salts are inactive.

Examples:

$$C_6H_5OH + ClP(O)(OC_2H_5)_2 \xrightarrow{NaH} C_6H_5OP(O)(OC_2H_5)_2 \xrightarrow[91\%]{CH_3C_6H_4MgBr,\ Ni(acac)_2} C_6H_5C_6H_4CH_3$$

1-Naphthyl-$OP(O)(OC_2H_5)_2$ —[$Al(C_2H_5)_3$, $Ni(acac)_2$; 95%]→ 1-C_2H_5-naphthalene

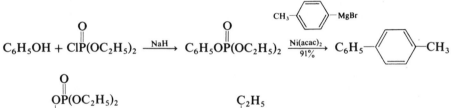

Aryl tosylates, methyl ethers, and silyl ethers do not undergo this coupling.

[1] M. Julia and J.-N. Verpeaux, *Tetrahedron Letters*, **23**, 2457 (1982).
[2] M. Julia, G. LeThuillier, C. Rolando, and L. Saussine, *ibid.*, **23**, 2453 (1982).
[3] T. Hayashi, T. Fijiwa, Y. Okamoto, Y. Katsuro, and M. Kumada, *Synthesis*, 1001 (1981).
[4] T. Hayashi, Y. Katsuro, Y. Okamoto, and M. Kumada, *Tetrahedron Letters*, **22**, 4449 (1981).

1,3-Bis(phenylseleno)propene (1).

Preparation:

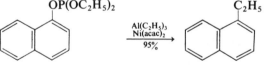

$$ClCH_2CH{=}CHCH_2Cl \xrightarrow[90\%]{2\ C_6H_5SeNa,\ KOH,\ C_2H_5OH} C_6H_5SeCH_2CH_2CH{=}CHSeC_6H_5$$

1

Enals; vinyl silyl ketones.[1] The anion of **1** reacts smoothly with various electrophiles (alkyl halides, epoxides, carbonyl compounds). The products are converted to (E)-enals by oxidation with 30% H_2O_2.

Example:

$$\mathbf{1} \xrightarrow[]{\text{1) LDA; 2) } (C_2H_5)_2C=O} C_6H_5Se\text{-}CH(C(C_2H_5)_2OH)\text{-}CH=CH\text{-}SeC_6H_5 \xrightarrow[\text{77\% overall}]{H_2O_2} (C_2H_5)_2C(OH)\text{-}CH=CH\text{-}CHO$$

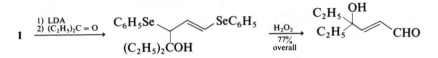

The reagent can also be converted into 3-trimethylsilyl-1,3-bis-(phenylseleno)propene (**2**), a useful reagent for preparation of silyl enones.

Example:

$$\mathbf{1} \xrightarrow[98\%]{\text{1) LDA; 2) } ClSi(CH_3)_3} C_6H_5Se\text{-}CH=CH\text{-}CH(Si(CH_3)_3)\text{-}SeC_6H_5\ (\mathbf{2}) \xrightarrow[65\%]{\text{1) } LiN(C_2H_5)_2\text{; 2) } (CH_3)_2CHCH_2Br}$$

$$C_6H_5Se\text{-}CH(CH_2CH(CH_3)_2)\text{-}CH=CH\text{-}C(Si(CH_3)_3)\text{-}SeC_6H_5 \xrightarrow[65\%]{\text{1) } CH_3CO_3H\text{; 2) } (C_2H_5)_2NH} (CH_3)_2CHCH_2\text{-}CH=CH\text{-}CH_2\text{-}C(=O)\text{-}Si(CH_3)_3$$

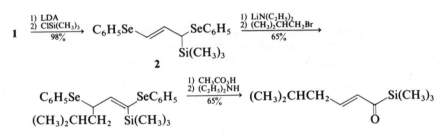

[1] H. J. Reich, M. C. Clark, and W. W. Willis, Jr., *J. Org.*, **47**, 1618 (1982).

(Z)-1,2-Bis(phenylsulfonyl)ethene,

Mol. wt. 308.37, m.p. 150°.

$$(Z)\text{-}C_6H_5O_2S\text{-}CH=CH\text{-}SO_2C_6H_5 \quad \mathbf{(1).}$$

Preparation[1]:

$$BrCH{=}CHBr + C_6H_5SH \xrightarrow[59\%]{KOH} (Z)\text{-}C_6H_5S(H)C{=}C(H)SC_6H_5 \xrightarrow[47\%]{CH_3CO_3H} \mathbf{1}$$

Diels-Alder reactions.[2] This ethylene derivative undergoes [4 + 2] cycloaddition even with relatively unreactive cyclic dienes. The adducts undergo reductive elimination of the sulfonyl groups on treatment with 2% sodium amalgam to provide bicyclodienes. The ethene therefore can function as an equivalent of acetylene in Diels-Alder reactions.

Example:

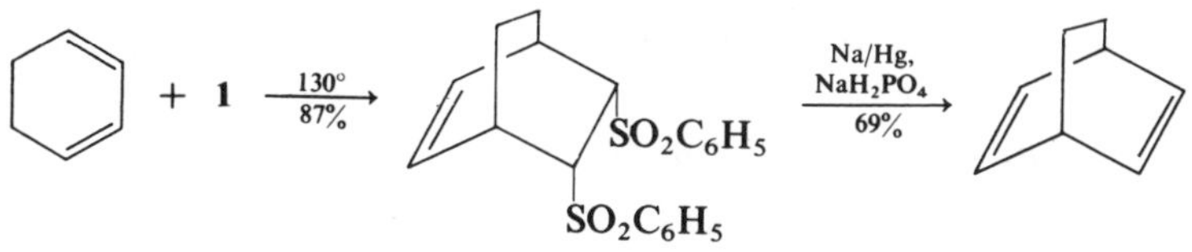

1,3-Cyclohexadiene + 1 →(130°, 87%) bicyclic adduct with two $SO_2C_6H_5$ groups →(Na/Hg, NaH_2PO_4, 69%) bicyclo[2.2.2]octadiene

[1] H. R. Snyder and D. P. Hallada, *Am. Soc.*, **74**, 5595 (1952).
[2] O. DeLucci and G. Modena, *J.C.S. Chem. Comm.*, 914 (1982).

Bis-(N-*n*-propylsalicylideneaminate)cobalt(II),

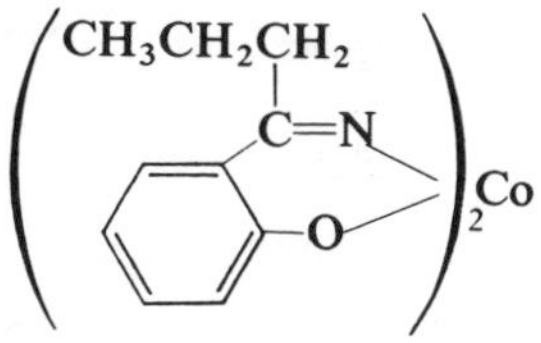

$(CH_3CH_2CH_2C{=}N\text{–}C_6H_4\text{–}O)_2Co$ **(1).**

For preparation, see **5**, 41–42.

Oxidative phenolic coupling.[1] This cobalt complex and the related salcomine (**2**, 360; **3**, 245; **6**, 507), both of which bind oxygen reversibly ($1O_2{:}2Co$), catalyze the oxygenation of the phenol **1** to give carpanone **2** in 90–94% yield. Related complexes of Fe(II) and Mn(II) are less effective. $PdCl_2$ effects this reaction in 46% yield (**4**, 370); singlet oxygen is less efficient (29% yield).

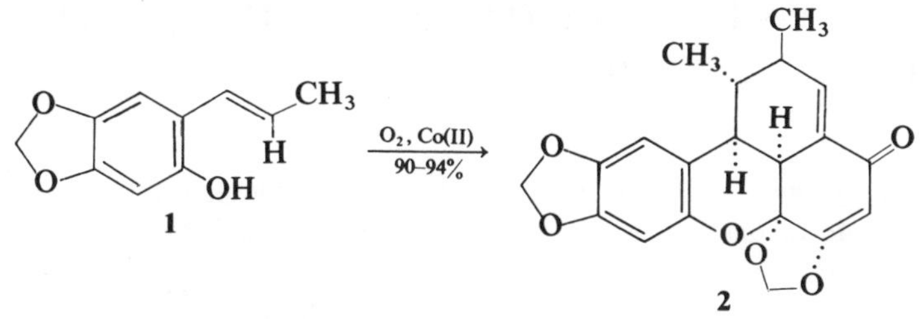

1 →(O_2, Co(II), 90–94%) **2**

[1] M. Matsumoto and K. Kuroda, *Tetrahedron Letters*, **22**, 4437 (1981).

Bispyridinesilver permanganate, $[Ag(C_5H_5N)_2]MnO_4$. Mol. wt. 385.01, m.p. 105° (dec.), purple crystals, soluble in most polar solvents, stable to impact, but ignites at 80°.

Oxidation. The reagent is particularly effective for oxidation of benzyl alcohols to carbonyl compounds and of arylamines to azines.[1]

Examples:

$$(C_6H_5)_2CHOH \xrightarrow[98\%]{C_6H_6} (C_6H_5)_2C{=}O$$

$$C_6H_4(CH_2OH)_2 \xrightarrow{95\%} C_6H_4(CHO)_2$$

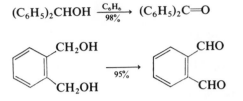

[1] H. Firouzabadi, B. Vessal, and M. Naderi, *Tetrahedron Letters*, **23**, 1847 (1982).

Bis(tri-*n*-butyl)tin oxide, $[(C_4H_9)_3Sn]_2O$ **(1)**, **6**, 56–57; **7**, 26, 27; **8**, 43–44; **9**, 53–59.

Dehydrosulfuration.[1] Primary thioamides are converted into nitriles by reaction with **1** (or di-*n*-butyltin oxide) in refluxing benzene (75–98% yield).

$$\text{(I)}\quad RC(=S)NH_2 + \underset{\mathbf{1}}{(Bu_3Sn)_2O} \longrightarrow RC(=NH)SSnBu_3 \xrightarrow[75\text{–}98\%]{-H_2O} RCN + (Bu_3Sn)_2S$$

[1] M.-I. Lim, W.-Y. Ren, and R. S. Klein, *J. Org.*, **47**, 4594 (1982).

Bis(2,2,2-trichloro-*t*-butyl) monochlorophosphate (1). Mol. wt. 467.35, m.p. 81°.

Preparation:

$$CH_3C(CCl_3)(OH)CH_3 \xrightarrow[71\%]{PCl_3,\ Py} \left(Cl_3CC(CH_3)_2{-}O\right)_2PCl \xrightarrow[65\%]{C_6H_5\overset{+}{I}OAc(OAc^-)} \underset{\mathbf{1}}{\left(Cl_3CC(CH_3)_2{-}O\right)_2P(=O)Cl}$$

Phosphorylation.[1] This reagent selectively phosphorylates the terminal 5′-hydroxyl group of nucleosides, even in the presence of a 3′-hydroxyl group. The

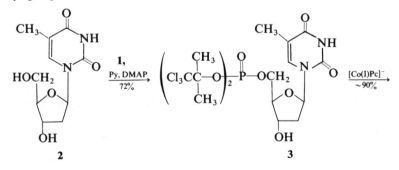

$$\underset{\mathbf{2}}{\text{thymidine}} \xrightarrow[72\%]{\mathbf{1},\ Py,\ DMAP} \underset{\mathbf{3}}{\left(Cl_3CC(CH_3)_2{-}O\right)_2P(=O){-}OCH_2\text{-thymidine (3'-OH)}} \xrightarrow[\sim 90\%]{[Co(I)Pc]^-}$$

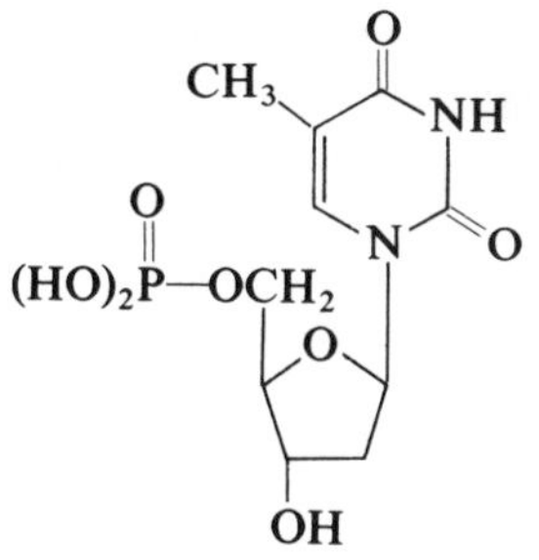

$$\text{(HO)}_2\overset{\overset{\large O}{\|}}{P}\text{—OCH}_2\text{-thymidine (5-methyl-2,4-dioxopyrimidin-1-yl deoxyribose, 3'-OH)}$$

4

trichloro-*t*-butyl group is selectively cleaved by cobalt(I)-phthalocyanine anion (**10**, 102–103).

[1] H. A. Kellner, R. G. K. Schneiderwind, H. Eckert, and I. K. Ugi, *Angew. Chem. Int. Ed.*, **20**, 577 (1981).

Bis(tricyclohexyltin) sulfide–Boron trichloride. The sulfide (**1**), m.p. 132°, is prepared by reaction of tricyclohexyltin chloride with $Na_2S \cdot 9H_2O$ in refluxing ethanol.

Sulfuration of carbonyl compounds.[1] These two reagents react to form boron trisulfide (equation I), which, when generated in this way, converts added carbonyl compounds into the corresponding thiocarbonyl compound in high yield. The effectiveness of the *in situ* reagent probably results from the solubility of a low molecular weight form generated. This method is effective for preparation of thioaldehydes, thioketones, thiolactones, and thiolactams, even from hindered carbonyl compounds.

$$\text{(I)}\quad [(C_6H_{11})_3Sn]_2S + \tfrac{2}{3}BCl_3 \xrightarrow{C_6H_5CH_3} 2(C_6H_{11})_3SnCl + 3B_2S_3$$

The selenide $[(C_6H_{11})_3Sn]_2Se$, prepared by reaction of $(C_6H_{11})_3SnCl$ with Na_2Se, can be used in this way for selenation of carbonyl groups.

[1] K. Steliou and M. Mrani, *Am. Soc.*, **104**, 3104 (1982).

1,3-Bis(triisopropylsilyl)propyne, $[(CH_3)_2CH]_3SiC{\equiv}CCH_2Si[CH(CH_3)_2]_3$. B.p. 130–135°/0.08 mm. The reagent is prepared by reaction of 3-triisopropylsilylpropynyllithium (this volume) with triisopropylsilyl triflate.[1]

(Z) or (E)-1,3-Enynes. The lithium anion of 1,3-bis(trimethylsilyl) propyne (**1a**) has recently been reported to react with aliphatic aldehydes and ketones to form a mixture of (Z)- and (E)-1,3-enynes in the approximate ratio 3:1 (equation I). Use of the anion of 3-(*t*-butyldimethylsilyl)-1-(trimethylsilyl)propyne results in a (Z)/(E) ratio of ~ 8:1.[2]

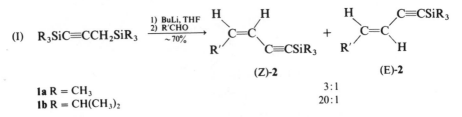

$$\text{(I)}\quad R_3SiC{\equiv}CCH_2SiR_3 \xrightarrow[\sim 70\%]{\text{1) BuLi, THF; 2) R'CHO}} \underset{(Z)\text{-}\mathbf{2}}{(H)(R')C{=}C(H)(C{\equiv}CSiR_3)} + \underset{(E)\text{-}\mathbf{2}}{(H)(R')C{=}C(C{\equiv}CSiR_3)(H)}$$

1a R = CH_3 3:1

1b R = $CH(CH_3)_2$ 20:1

Far greater (Z)-selectivity is observed in reactions of the anion of **1b**, in which the (Z)/(E) ratio is ~ 20:1. Moreover, the selectivity of this reaction is almost completely reversed by addition of HMPT. The effect of HMPT is exerted on the carbonyl addition step and not on the subsequent elimination to form the double bond. Corey and Rücker[1] suggest that lithiated **1b** reacts mainly as a lithio allene in THF, but mainly as the propargylic anion in THF–HMPT.

Benzaldehyde differs from aliphatic aldehydes in reactions with anions of **1b** in that formation of (Z)-isomers is favored only by a factor of 2–3, but again the selectivity can be reversed by the addition of HMPT.

[1] E. J. Corey and C. Rücker, *Tetrahedron Letters*, **23**, 719 (1982).
[2] Y. Yamakado, M. Ishiguro, N. Ikeda, and H. Yamamoto, *Am. Soc.*, **103**, 5568 (1981).

1,4-Bistrimethylsilyl-2-butyne (1).
Preparation:

$$(CH_3)_3SiCH_2C{\equiv}CH \xrightarrow[60\%]{\text{1) } n\text{-BuLi; 2) } ICH_2Si(CH_3)_3,\ 50^\circ} \underset{\mathbf{1}}{(CH_3)_3SiCH_2C{\equiv}CCH_2Si(CH_3)_3}$$

***Allenylsilanes*; *1,3-dienes*.** The reagent, in the presence of $TiCl_4$, reacts with acetals to form allenylsilanes (**2**) or 1,3-dienes (**3**), depending upon the stoichiometry.[1]

$$\mathbf{1} + R^1CH(OR^2)_2 \xrightarrow{TiCl_4} \underset{\mathbf{2}}{CH_2{=}C{=}C(CHR^1OR^2)CH_2Si(CH_3)_3} + \underset{\mathbf{3}}{R^1CH(OR^2){-}C({=}CH_2){-}C({=}CH_2){-}CH(OR^2)R^1}$$

[1] J. Pornet and N. Kolani, *Tetrahedron Letters*, **22**, 3609 (1981).

2,3-Bis[(trimethylsilyl)methyl]-1,3-butadiene (1). Mol. wt. 226.50, b.p. 70–75°/3 mm.
The diene is prepared by oxidative dimerization of the cuprate of (2-bromoallyl)trimethylsilane (equation I).

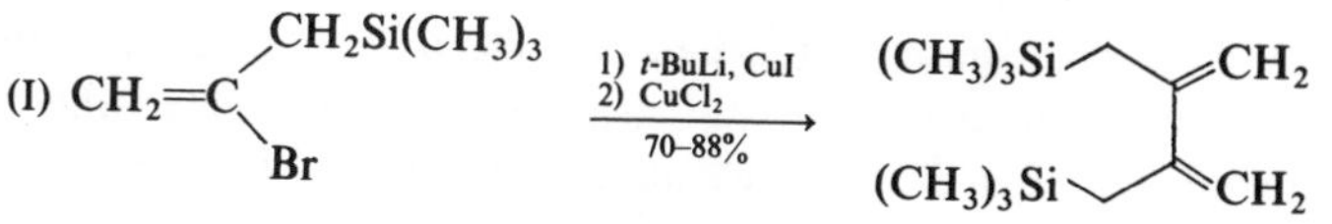

(I) $CH_2{=}C(Br)CH_2Si(CH_3)_3$ —1) *t*-BuLi, CuI; 2) $CuCl_2$; 70–88%→ $[(CH_3)_3SiCH_2]_2C_2(=CH_2)_2$ (**1**)

Tandem Diels-Alder reactions.[1] This diene, as expected, undergoes Diels-Alder reactions with the usual dienophiles. The adducts undergo oxidative bisdesilylation when treated with NBS in the presence of propylene oxide as an HBr scavenger to give dimethylenecyclohexanes, which can undergo a second Diels-Alder reaction. Because of the lack of regiocontrol in the second cycloaddition, this tandem cycloannelation is only useful when one of the dienophiles is a symmetrically 1,2-disubstituted alkene.

Examples:

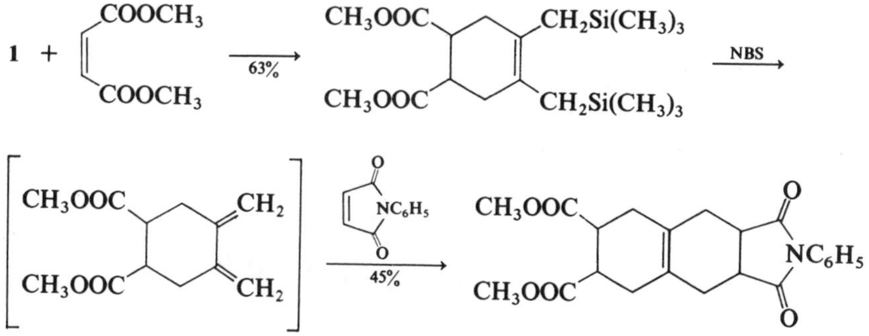

1 + CH_3OOC–CH=CH–$COOCH_3$ —63%→ CH_3OOC, CH_3OOC, $CH_2Si(CH_3)_3$, $CH_2Si(CH_3)_3$ —NBS→

[CH_3OOC, CH_3OOC, CH_2, CH_2] —N-phenylmaleimide (NC_6H_5), 45%→ CH_3OOC, CH_3OOC, NC_6H_5 (stereoisomeric mixture)

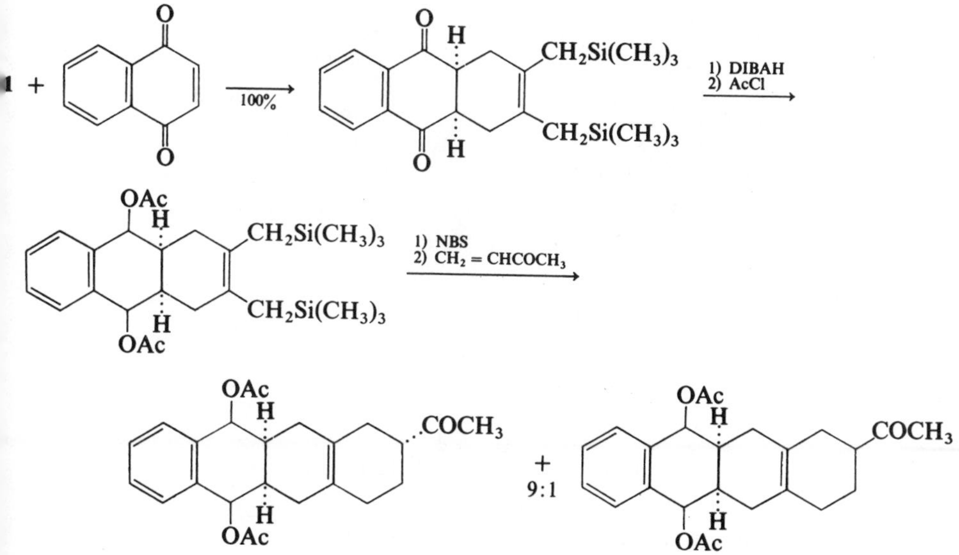

1 + naphthoquinone —100%→ H, H, $CH_2Si(CH_3)_3$, $CH_2Si(CH_3)_3$ —1) DIBAH; 2) AcCl→

OAc, H, OAc, H, $CH_2Si(CH_3)_3$, $CH_2Si(CH_3)_3$ —1) NBS; 2) $CH_2=CHCOCH_3$→

OAc, H, H, OAc, $COCH_3$ + OAc, H, H, OAc, $COCH_3$ (9:1)

The second example shows that some stereoselectivity is possible in the second Diels-Alder reaction; in general, overall stereoselectivity is greater when the more bulky dienophile is used in the first cycloaddition rather than the second.

[1] B. M. Trost and M. Shimizu, *Am. Soc.*, **104**, 4299 (1982).

1,3-Bis(trimethylsilyloxy)-1-methoxy-1,3-butadiene (1), **9**, 54–55; **10**, 46–47.

Resorcinols. Condensation of this diene with the ketal of a β-keto acid derivative results in a resorcinol with complete regiocontrol. Thus, the $TiCl_4$-catalyzed reaction of **1** with ketal ester **2** results in the resorcinol **3** in 72% yield. However, use of the acid chloride **4** corresponding to **2** in the same reaction results in the isomeric resorcinol, methyl olivetolate (**5**). The regiocontrol is based on the reactivity order acid chloride > ketal > ester. The resorcinol **5** was used in a biomimetic synthesis of the chromene Δ^1-tetrahydrocannabinol (**6**), a component of marijuana.

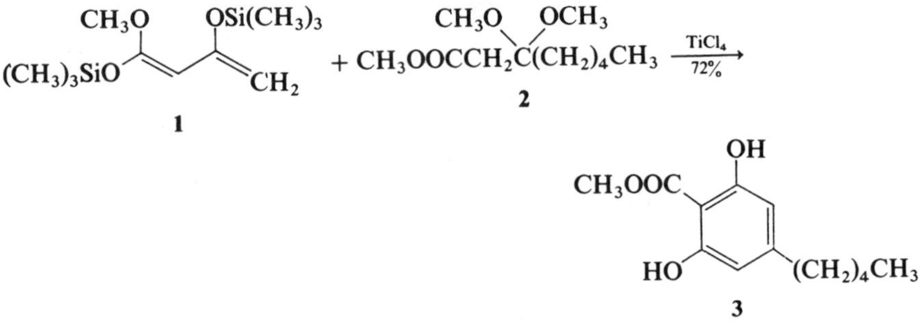

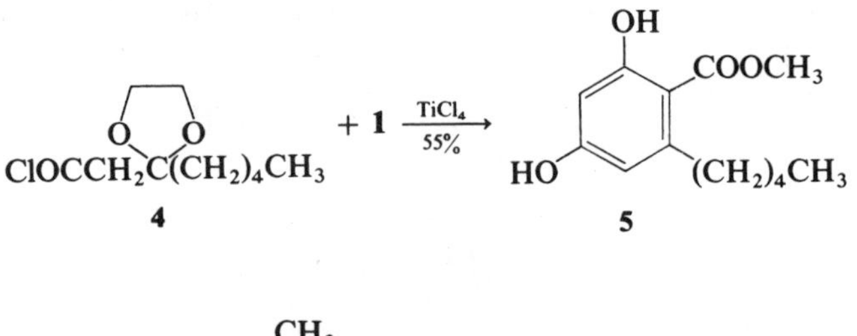

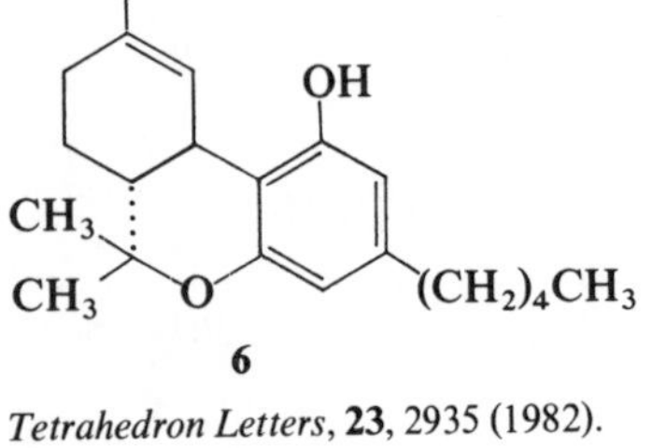

CH_3O, $OSi(CH_3)_3$, $(CH_3)_3SiO$, CH_2 **1** + $CH_3OOCCH_2C(OCH_3)_2(CH_2)_4CH_3$ **2** $\xrightarrow[72\%]{TiCl_4}$ **3** (OH, CH_3OOC, HO, $(CH_2)_4CH_3$)

$ClOCCH_2C(CH_2)_4CH_3$ (ethylene ketal) **4** + **1** $\xrightarrow[55\%]{TiCl_4}$ **5** (OH, $COOCH_3$, HO, $(CH_2)_4CH_3$)

6 (CH_3, OH, CH_3, CH_3, O, $(CH_2)_4CH_3$)

[1] T. H. Chan and T. Chaly, *Tetrahedron Letters*, **23**, 2935 (1982).

Bis(trimethylsilyl) peroxide, $[(CH_3)_3SiO]_2$ (**1**). Mol. wt. 168.37, b.p. 42°/30 mm., reasonably stable to heat, soluble in aprotic organic solvents. The peroxide is obtained most conveniently by reaction of chlorotrimethylsilane with the 1:2 complex of H_2O_2 with DABCO.[1] This complex, m.p. 112° (dec.), is readily obtained by reaction of the amine with 30% H_2O_2 and serves as a reasonably stable source of anhydrous H_2O_2.[2]

Baeyer-Villiger oxidations.[3] This peroxide in combination with trimethylsilyl triflate (1 equiv.) is highly effective for selective Baeyer-Villiger oxidation of acyclic and cyclic ketones, but not of aromatic ketones, without epoxidation of carbon–carbon double bonds.

Examples:

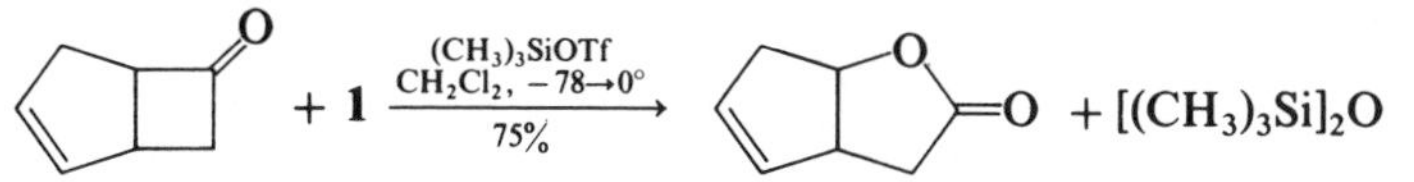

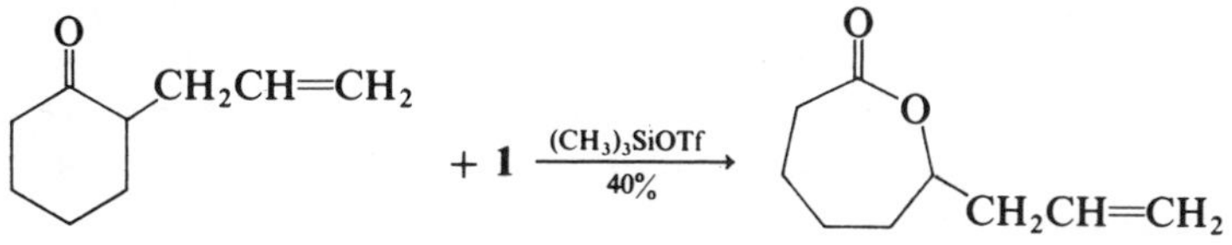

$$\text{bicyclic cyclobutanone} + \mathbf{1} \xrightarrow[75\%]{(CH_3)_3SiOTf,\ CH_2Cl_2,\ -78\rightarrow 0^\circ} \text{bicyclic }\gamma\text{-lactone} + [(CH_3)_3Si]_2O$$

$$\text{2-}(CH_2CH{=}CH_2)\text{cyclohexanone} + \mathbf{1} \xrightarrow[40\%]{(CH_3)_3SiOTf} \text{7-}(CH_2CH{=}CH_2)\text{-}\varepsilon\text{-caprolactone}$$

Oxidation of sulfur and phosphorus compounds. The peroxide has been used to oxidize the sulfide (**2**) to the corresponding sulfone in 83% yield without attack of the double bond. Geranyl phenyl sulfide is oxidized to the corresponding sulfone in similar yield.[4] Trialkylphosphines and phosphites are oxidized to the corresponding oxides in high yield.[5]

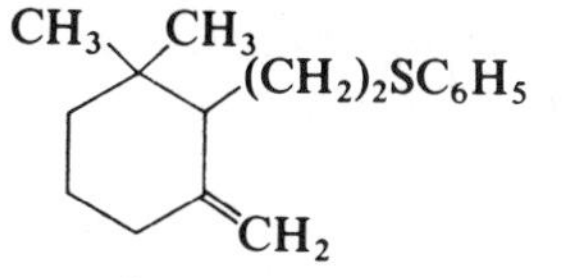

Structure **2**: 2,2-dimethyl-6-methylenecyclohexyl with $(CH_2)_2SC_6H_5$ substituent (CH_3, CH_3, $(CH_2)_2SC_6H_5$, $=CH_2$)

2

[1] P. G. Cookson, A. G. Davies, and N. Fazal, *J. Organometal. Chem.*, **99**, C31 (1975).
[2] A. A. Oswald and D. L. Guertine, *J. Org.*, **28**, 651 (1963).
[3] M. Suzuki, H. Takada, and R. Noyori, *J. Org.*, **47**, 902 (1982).
[4] P. Kocienski and M. Todd, *J.C.S. Chem. Comm.*, 1078 (1982).
[5] D. Brandes and A. Blaschette, *ibid.*, **73**, 217 (1974).

Bis(trimethylsilyl) sulfate, $[(CH_3)_3SiO]_2SO_2$ (**1**). Mol. wt. 242.4, m.p. 55–57°. Preparation.[1]

Rearrangement of vinylcyclopropanecarboxylic esters. When **1** is heated in 1,2-dichloroethane with this Lewis acid the γ-lactone **2** is obtained in 98% yield. The *gem*-dicarbonyl substitution is necessary for this transformation.

Examples:

CH=CH$_2$ / COOC$_2$H$_5$ / COOC$_2$H$_5$ (1) $\xrightarrow[98\%]{[(CH_3)_3SiO]_2SO_2,\ ClCH_2CH_2Cl}$ CH$_2$=CH … O … O / COOC$_2$H$_5$ (2)

O / CH$_3$ / COOCH$_3$ $\xrightarrow{71\%}$ O H O / O / H CH$_3$ + (63:35) O H O / O / H CH$_3$

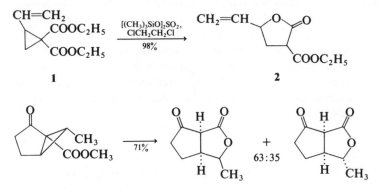

Tetrahydropyranylation.[3] The reaction of an alcohol (even tertiary) with dihydropyran is catalyzed efficiently by $[(CH_3)_3SiO]_2SO_2$. The resulting ether is isolated in 90–100% yield by addition of pyridine and concentration of the solvent. The sulfate also catalyzes alcoholysis of the derivative at 25° (10–90 minutes). It is also useful in some transesterifications.

[1] M. Schmidt and H. Schmidbaur, *Ber.*, **94**, 2446 (1961).
[2] Y. Morizawa, T. Hiyama, and H. Nozaki, *Tetrahedron Letters*, **22**, 2297 (1981).
[3] Y. Morizawa, I. Mori, T. Hiyama, H. Nozaki, *Synthesis*, 899 (1981).

Bis(trimethylsilyl)urea, $O{=}C[NHSi(CH_3)_3]_2$. The reagent is prepared from hexamethyldisilazane and urea.[1]

Silylation of alcohols and carboxylic acids.[2] The reagent has seen only limited use for silylation (**9**, 113), but is actually valuable for this purpose. It does not require acid or base catalysis and the urea formed is insoluble in CH_2Cl_2 (the solvent used). Yields of silyl ethers or carboxylates are usually > 85%.

[1] U. Wannagat, H. Burger, C. Krüger, and J. Pump, *Z. Anorg. Allg. Chem.*, **321**, 208 (1963).
[2] V. Verboom, C. W. Visser, and D. N. Reinhoudt, *Synthesis*, 807 (1981).

9-Borabicyclo[3.3.1]nonane (9-BBN), 2, 31; **3,** 24–29; **4,** 41; **5,** 46–47; **6,** 62–64; **7,** 29–31; **8,** 47–49; **9,** 57–58; **10,** 48–49.

Simplified preparation. 9-BBN (dimeric) can be prepared in excellent yield (88%) and high purity by cyclic hydroboration of 1,5-octadiene with $BH_3 \cdot S(CH_3)_2$ in monoglyme (1,2-dimethoxyethane) as the solvent, in which dimeric 9-BBN has very low solubility. One crystallization from this solvent gives large needles melting at 153–155°.

[1] J. A. Sonderquist and H. C. Brown, *J. Org.*, **46**, 4599 (1981).

Borane-Dimethyl sulfide, 4, 124, 191; **5,** 47; **8,** 49–50; **10,** 49–50.

Reduction of esters, nitriles, and amides.[1] These groups are rapidly reduced by borane–dimethyl sulfide in refluxing THF (b.p. 67°) if the dimethyl sulfide (b.p. 38°) is removed as liberated. Under these conditions, the reagent is comparable to uncomplexed diborane. Reduction of secondary and tertiary amides is best effected in the presence of boron trifluoride etherate; otherwise, excess reagent is utilized for formation of complexes with the products.

The reagent effects selective reduction in the presence of nitro, halogen, alkoxy, and sulfone groups, but carbonyl groups undergo competitive reduction and unsaturated substrates undergo hydroboration.

[1] H. C. Brown, Y. M. Choi, and S. Narasimhan, *J. Org.*, **47**, 3153 (1982).

Borane–Pyridine, 1, 963–964; **8,** 50–51; **9,** 59.

Sulfides. Aldehydes and ketones react with thiols in TFA in the presence of BH_3/pyridine to form sulfides (equation I).[1]

$$\text{(I)}\quad R^1R^2C{=}O + R^3SH \xrightarrow[52\text{–}75\%]{Py/BH_3,\ CF_3COOH} R^1R^2C(H)SR^3$$

[1] Y. Kikugawa, *Chem. Letters*, 1157 (1981).

Borane–Tetrahydrofurane (a solution of B_2H_6 in THF), **1,** 199–207; **2,** 106–108; **3,** 76–77; **4,** 124–126; **5,** 184–186; **6,** 161–162; **7,** 89; **8,** 141–143; **9,** 136–138.

Selective reduction of COOH groups. The most convenient route to ethyl 4-hydroxycrotonate is reduction of ethyl fumarate with $BH_3 \cdot THF$ (equation I).[1]

$$\text{(I)}\quad HOOC\text{–}CH{=}CH\text{–}COOC_2H_5 \xrightarrow[50\text{–}60\%]{BH_3\cdot THF,\ -78\to25^\circ} HO\text{–}CH_2\text{–}CH{=}CH\text{–}COOC_2H_5$$

Hydroboration of hindered alkenes. Several sterically hindered alkenes can be hydroborated at high pressures (5000–6000 atm) to give the corresponding trialkylborane. For example, trithexylborane can be obtained by hydroboration of tetramethylethylene, but on releasing the pressure to 1 atm the boron isomerizes partially to a terminal carbon atom. Oxidation of the product with alkaline hydrogen peroxide gives a mixture of 2,3-dimethyl-2-butanol and 2,3-dimethyl-1-butanol in the ratio 72:1.

Hydroboration of 5β, 9(11)-cholene (**1**) is possible. Oxidation of the product with alkaline hydrogen peroxide leads to an isomeric cholene, probably the Δ^{11}-cholene (**2**).[2]

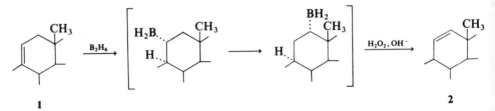

$$\mathbf{1} \xrightarrow{B_2H_6} \left[\cdots \right] \xrightarrow{H_2O_2,\ OH^-} \mathbf{2}$$

[1] A. S. Kende and P. Fludzinski, *Org. Syn.* submitted (1982).
[2] J. E. Rice and Y. Okamoto, *J. Org.*, **47**, 4189 (1982).

Borane–Triphenyl phosphite, $BH_3 \cdot P(OC_6H_5)_3$. Mol. wt. 324.12, m.p. 48–49°, stable to O_2 and H_2O. Preparation.[1]

Hydroboration.[2] The usual hydroboration reagents, $BH_3 \cdot THF$ and $BH_3 \cdot S(CH_3)_2$, are sensitive to oxygen and moisture and require special handling. The complexes of BH_3 and phosphorus compounds are generally stable, but much less reactive. The complex of BH_3 and triphenylphosphine, m.p. 189°, can be used for hydroboration if activated by addition of methyl iodide (to form a phosphonium iodide) or sulfur (to form a triphenylphosphine sulfoxide). The complex of borane and triphenyl phosphite does not require activation and hydroborates alkenes in a reasonable time in refluxing DME or THF. Trialkyl phosphite complexes are not useful.

[1] P. A. Chopard and R. F. Hudson, *J. Inorg. Nucl. Chem.*, **25**, 801 (1963).
[2] A. Pelter, R. Rosser, and S. Mills, *J.C.S. Chem. Comm.*, 1014 (1981).

Boric acid, 1, 63–66; **2,** 32; **3,** 29–30; **5,** 48–49; **9,** 59.

N-Acylation of indole.[1] Indole can be N-acylated by a carboxylic acid in the presence of boric acid (0.3 equiv.) in 45–80% yield.

[1] M. Terashima and M. Fujioka, *Heterocycles*, **19**, 91 (1982).

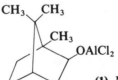

Bornyloxyaluminum dichloride, (**1**). Mol. wt. 251.12. The reagent is prepared from bornyl alcohol, aluminum chloride, and lithium aluminum hydride in the ratio 4:3.6:1 (**1,** 522).[1]

Diels-Alder reactions with aldehydes.[2] This catalyst is superior to zinc chloride[3] for promoting [4 + 2]cycloaddition of aldehydes with 1-alkoxy-3-trimethylsilyloxy-1,3-butadienes to form 2,3-dihydro-4*H*-pyrane-4-ones. The catalyst prepared from bornyl alcohol is somewhat more effective than similar catalysts from simple alcohols, and may be of value for asymmetric induction.

Example:

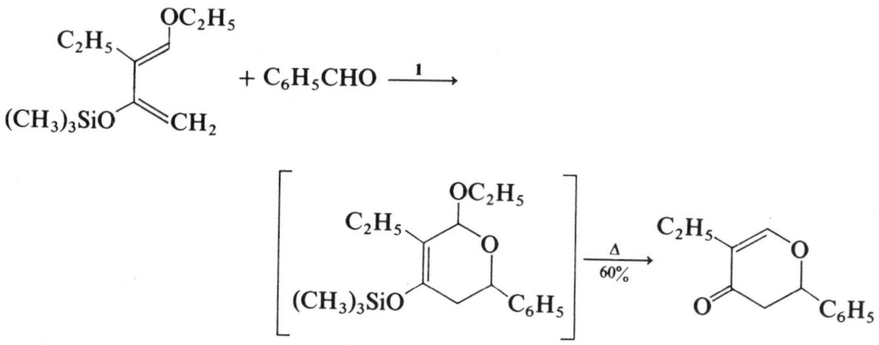

$$\text{C}_2\text{H}_5\text{C(=CHOC}_2\text{H}_5\text{)C(OSi(CH}_3)_3\text{)=CH}_2 + C_6H_5CHO \xrightarrow{1}$$

$$\left[\text{dihydropyran: } C_2H_5,\ OC_2H_5,\ (CH_3)_3SiO,\ C_6H_5 \right] \xrightarrow[60\%]{\Delta} \text{pyranone: } C_2H_5,\ O,\ C_6H_5$$

[1] E. L. Eliel and D. Nasipuri, *J. Org.*, **30**, 3812 (1965).
[2] R. W. Aben and H. W. Scheeren, *Synthesis*, 779 (1982).
[3] H. C. J. G. van Balen, A. A. Broekhuis, J. W. Scheeren, R. J. F. Nivard, *Rec. trav.*, **98**, 36 (1979).

Boron tribromide–Sodium iodide–15-Crown-5.

Cleavage of aliphatic methyl ethers.[1] Methyl ethers can be cleaved in moderate to high yield by addition of BBr_3 to a solution of the ether and the NaI complex with the crown ether in CH_2Cl_2. All three components are necessary for satisfactory results. The cleavage of primary and secondary methyl ethers proceeds satisfactorily, but yields are low with tertiary ethers.

[1] H. Niwa, T. Hida, and K. Yamada, *Tetrahedron Letters*, **22**, 4239 (1981).

Boron trifluoride, 1, 68–69; **3**, 32–33; **5**, 51–52; **7**, 31; **10**, 50–51.

Oxidation of glycal esters to ene-δ-lactones. Reaction of glycal esters with *m*-chloroperbenzoic acid involves the expected epoxidation followed by cleavage of the oxide by the *m*-chlorobenzoic acid formed. However, in the presence of BF_3, the reaction produces α,β-unsaturated δ-lactones, usually in excellent yield. This oxidation is also possible with pyridinium chlorochromate and BF_3 as catalyst, but in lower yield.[1]

Examples:

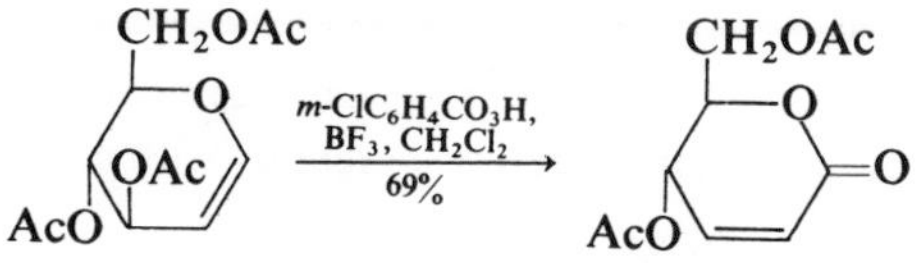

$$\text{glycal (CH}_2\text{OAc, OAc, AcO)} \xrightarrow[69\%]{m\text{-ClC}_6\text{H}_4\text{CO}_3\text{H},\ BF_3,\ CH_2Cl_2} \text{lactone (CH}_2\text{OAc, AcO, =O)}$$

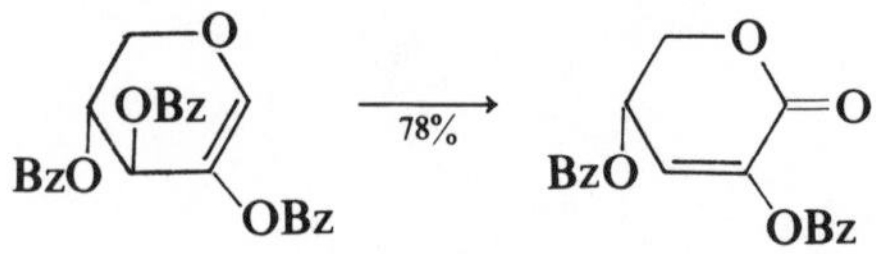

[Scheme: BzO, OBz, OBz substituted dihydropyran → (78%) BzO, OBz substituted pyranone]

[1] P. Jarglis and F. W. Lichtenthaler, *Tetrahedron Letters*, **23**, 3781 (1982).

Boron trifluoride etherate, 1, 70–72; **2**, 35–36; **3**, 33; **4**, 44–45; **5**, 54–55; **6**, 65–67; **7**, 31–32; **8**, 51–52; **9**, 64–65; **10**, 52–56.

Boron heterocycles.[1] A novel synthesis of the natural chalcone aurentiacin (**3**) involves the reaction of 2,4,6-trimethoxytoluene (**1**) with cinnamic acid and BF_3 etherate at 80° to form the novel heterocycle **2**, which is hydrolyzed to **3**. The formation of **2** involves a selective demethylation, since the reaction of **4** with 2-phenylpropionic acid and BF_3 followed by hydrolysis gives **5**, isomeric with dihydro-**3**.

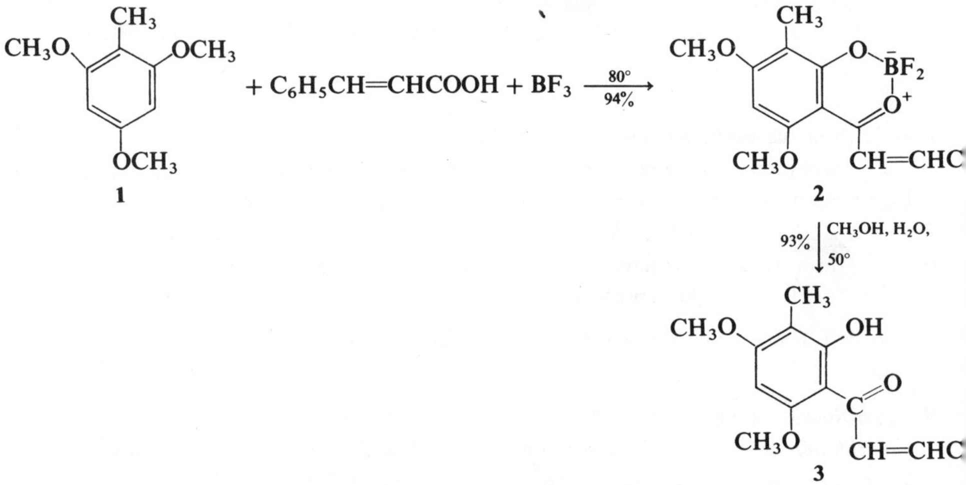

$$\mathbf{1} + C_6H_5CH{=}CHCOOH + BF_3 \xrightarrow[94\%]{80^\circ} \mathbf{2}$$

$$\mathbf{2} \xrightarrow[93\%]{CH_3OH,\ H_2O,\ 50^\circ} \mathbf{3}$$

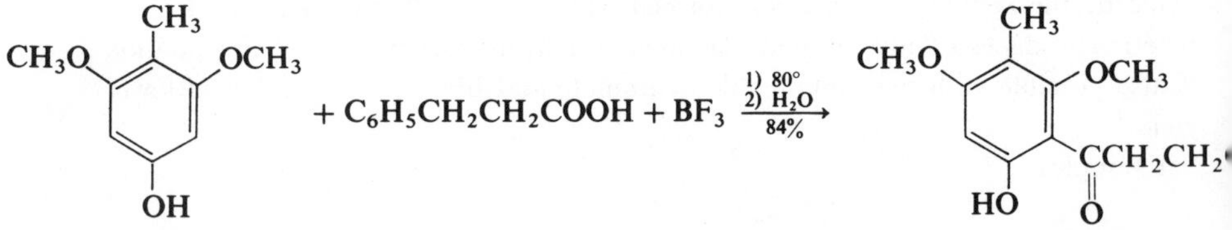

$$\mathbf{4} + C_6H_5CH_2CH_2COOH + BF_3 \xrightarrow[84\%]{1)\ 80^\circ\ \ 2)\ H_2O} \mathbf{5}$$

Stabilized carbocations. Acetals or ortho esters are converted by BF_3 etherate into methoxy-stabilized carbocations that react readily with methyl 5-lithiotetronates (butenolides).[2]

Examples:

OCH3 · C(OCH3)4, BF3·(C2H5)2O, 81% · HC(OCH3)3, BF3·(C2H5)2O, 85% · (CH3O)2CH · CH3OOC · Li · 74% · C6H5CH=CHCH(OCH3)2, BF3·(C2H5)2O · (E) · C6H5CH=CHCH

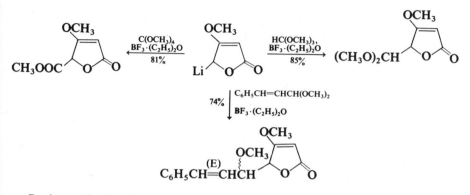

Regiospecific dimethoxymethylation of preformed lithium enolates is also possible by a related reaction.[3]

Examples:

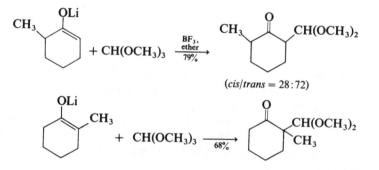

OLi · CH3 · + CH(OCH3)3 · BF3, ether, 79% · CH(OCH3)2

(*cis*/*trans* = 28:72)

OLi · CH3 · + CH(OCH3)3 · 68% · CH(OCH3)2 · CH3

Condensation of aldehydes with α-mercurio ketones.[4] In the presence of BF_3 etherate, α-mercurio ketones condense with aldehydes with moderate to high *erythro*-selection.

Examples:

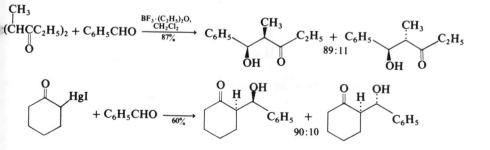

$Hg(CHCC_2H_5)_2$ + C_6H_5CHO · $BF_3\cdot(C_2H_5)_2O$, CH_2Cl_2, 87% · C_6H_5 · CH_3 · C_2H_5 · OH · O · 89:11

HgI · + C_6H_5CHO · 60% · H · OH · C_6H_5 · 90:10

This work raises the possibility that some *erythro*-selective aldol condensations with metal enolates may actually involve the α-metallo ketone.

Polyprenylquinones (*cf.* **9**, 499–500).[5] Coenzyme Q_n (**3**) can be obtained by rearrangement of the polyprenyl aryl ethers **1** with boron trifluoride etherate to the hydroquinones (**2**) followed by oxidation.

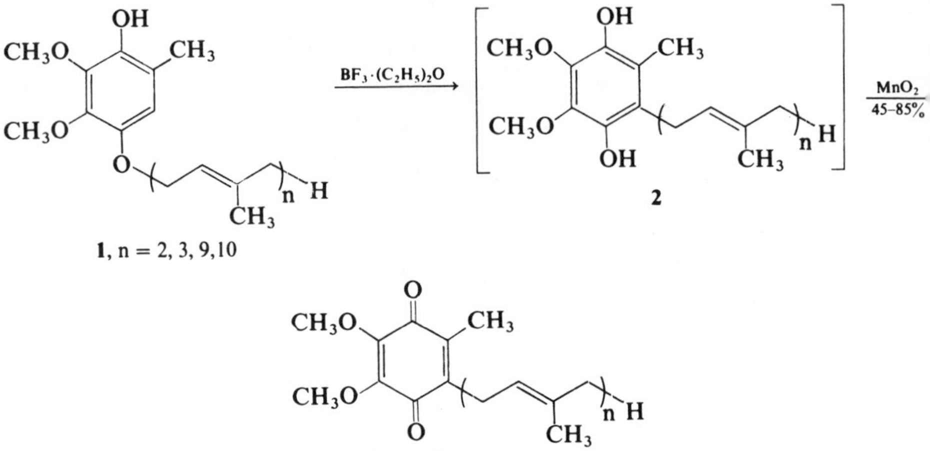

$BF_3 \cdot (C_2H_5)_2O$; MnO_2, 45–85%

1, n = 2, 3, 9, 10 **2** **3**

cis-***Hydroazulenes.***[6] The first step in a new route to this ring system involves the usual 1,8-addition of a Grignard reagent such as **1** to tropone to give the 2-substituted dihydrotropone **2**, easily converted to the aldehyde **3**. Cyclization of **3** with BF_3 etherate results in the tricyclic ether, in which the five- and seven-membered rings have the desired *cis*-relationship. The hydroazulenediol **5** is obtained by reduction of **4** with lithium–methylamine at 0°. The benzyl ether group of **4** can be cleaved selectively under milder conditions.

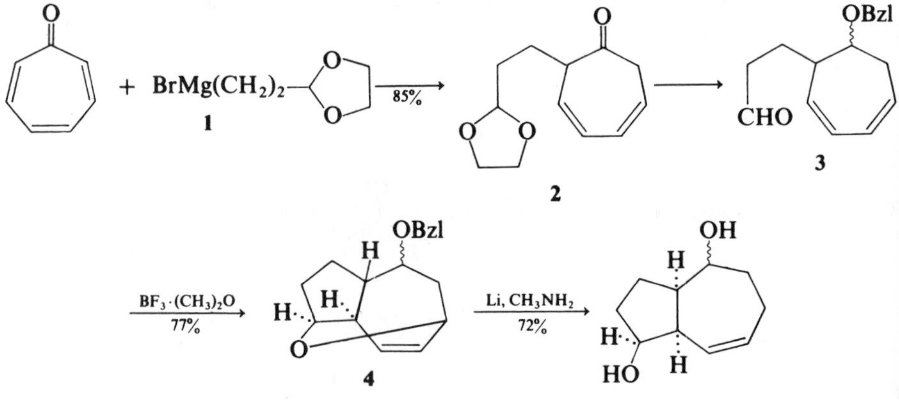

+ $BrMg(CH_2)_2$– **1** (85%) → **2** → **3**

$BF_3 \cdot (CH_3)_2O$, 77% → **4** ; Li, CH_3NH_2, 72% → **5**

Hydrolysis of a methyl ether.[7] Treatment of either **1** or **2** with BF_3 etherate yields exclusively **3** (aklavinone). One explanation for the result is formation of a cyclic borate ester intermediate involving the C_7- and C_9-hydroxyl groups. Under similar conditions, trifluoroacetic acid converts either **1** or **2** into an 8:1 mixture of **3** and 7-epi-**3**. BF_3 etherate was also used to convert the C_{10}-epimer of **2** into **3**.

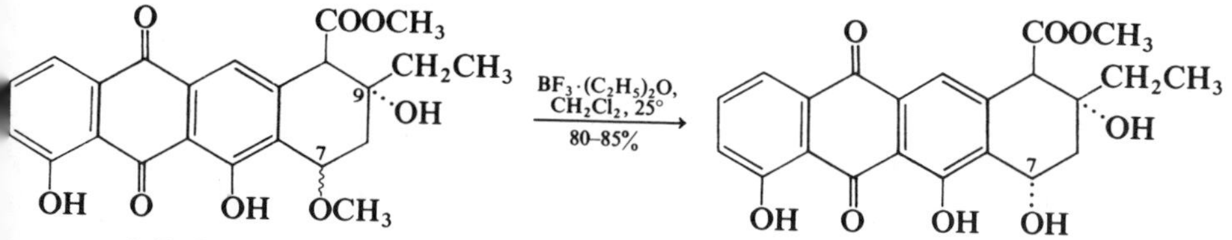

$BF_3 \cdot (C_2H_5)_2O$, CH_2Cl_2, 25°; 80–85%

1, 7α-OCH_3
2, 7β-OCH_3
→ **3**

Hydrolysis of dimethylhydrazones.[8] Addition of BF_3 etherate to an ethereal solution of a dimethylhydrazone precipitates a complex that is readily hydrolyzed by water to the parent ketone (65–75% yield). Enol acetates are stable to the reagent.

[1] G. P. Schiemenz and U. Schmidt, *Ann.*, 1509 (1982).
[2] A. Pelter and R. Al-Bayati, *Tetrahedron Letters*, **23**, 5229 (1982).
[3] M. Suzuki, A. Yanagisawa, and R. Noyori, *ibid.*, **23**, 3595 (1982).
[4] Y. Yamamoto and K. Maruyama, *Am. Soc.*, **104**, 2323 (1982).
[5] T. Yoshizawa, H. Toyofuku, K. Tachibana, and T. Kuroda, *Chem. Letters*, 1131 (1982).
[6] J. R. Rigby, *Tetrahedron Letters*, **23**, 1863 (1982).
[7] B. A. Pearlman, J. M. McNamara, I. Hasan, S. Hatakeyama, H. Sekizaki, and Y. Kishi, *Am. Soc.*, **103**, 4248 (1981).
[8] R. E. Gawley and E. J. Termine, *Syn. Comm.*, **12**, 15 (1982).

Bromine, 3, 34; **4**, 46–47; **5**, 55–57; **6**, 70–73; **7**, 33–35; **8**, 52–53; **9**, 65–66; **10**, 56.

Bromination-desilylbromination of β-silyl ketones. The preliminary work on protection of α,β-enones by this sequence (**8**, 196–197) has been perfected in several respects[1] and shown to be generally applicable to synthesis of α,β-enones.[2] Either bromine or phenyltrimethylammonium tribromide (PTAB, **1**, 855; **2**, 328; **4**, 386) can be used. In either case, bromine is introduced mainly at the α-position adjacent to the β-silyl group. Surprisingly, the undesired α′-bromo-β-silyl ketone is converted to the α-bromo-β-silyl ketone on exposure to hydrogen bromide at 20°. No other acids can effect this conversion. Desilylbromination is generally effected with DBU or F^-. The sequence is applicable to synthesis of α-methylene ketones (equation I) and lactones.

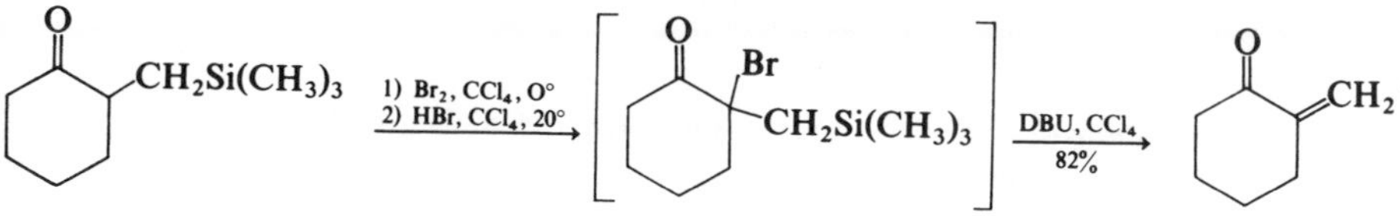

$CH_2Si(CH_3)_3$ —1) Br_2, CCl_4, 0°; 2) HBr, CCl_4, 20°→ [Br, $CH_2Si(CH_3)_3$] —DBU, CCl_4, 82%→ CH_2

This sequence can be used for conversion of a simple β-silyl-α,β-enone such as **1** into a more substituted α,β-unsaturated enone, as shown in equation (II).

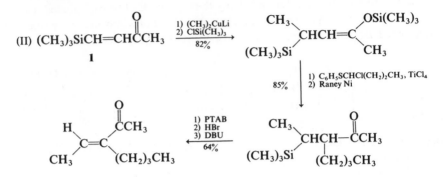

(II) $(CH_3)_3SiCH{=}CHC(O)CH_3$ (**1**) —[1) $(CH_3)_2CuLi$; 2) $ClSi(CH_3)_3$; 82%]→ $(CH_3)((CH_3)_3Si)CHCH{=}C(OSi(CH_3)_3)CH_3$ —[1) $C_6H_5SCHCl(CH_2)_2CH_3$, $TiCl_4$; 2) Raney Ni; 85%]→ $(CH_3)((CH_3)_3Si)CHCH((CH_2)_3CH_3)C(O)CH_3$ —[1) PTAB; 2) HBr; 3) DBU; 64%]→ $(H)(CH_3)C{=}C(C(O)CH_3)(CH_2)_3CH_3$

Halolactamization.[3] Unsaturated amides generally form lactones when cyclized by Br_2 or I_2. However the unsaturated N-tosyl amide **1**, when treated with Br_2 in the presence of $NaHCO_3$, forms the bromo N-tosyl β-lactam **2** (67% yield). Dehalogenation of **2** is best effected with Bu_3SnH.

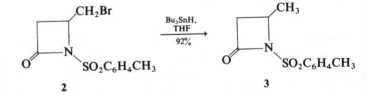

$CH_2{=}CHCH_2C(O)NHSO_2C_6H_4CH_3$-$p$ (**1**) —[Br_2, CH_2Cl_2, $NaHCO_3$; 67%]→ 4-(CH_2Br)-1-($SO_2C_6H_4CH_3$)azetidin-2-one (**2**) —[Bu_3SnH, THF; 92%]→ 4-(CH_3)-1-($SO_2C_6H_4CH_3$)azetidin-2-one (**3**)

This reaction is applicable to various N-sulfonylated β,γ-unsaturated amides. The products are analogs of natural β-lactams recently isolated from bacteria and known as monobactams.

[1] I. Fleming and J. Goldhill, *J.C.S. Perkin I*, 1493 (1980).
[2] I. Fleming and D. A. Perry, *Tetrahedron*, **37**, 4027 (1981).
[3] A. J. Biloski, R. D. Wood, and B. Ganem, *Am. Soc.*, **104**, 3233 (1982).

3-Bromo-4,5-dihydro-5-hydroperoxy-4,4-dimethyl-3,5-diphenyl-3H-pyrazole (1). Mol. wt. 361.25, m.p. 77–80°.

Preparation[1]:

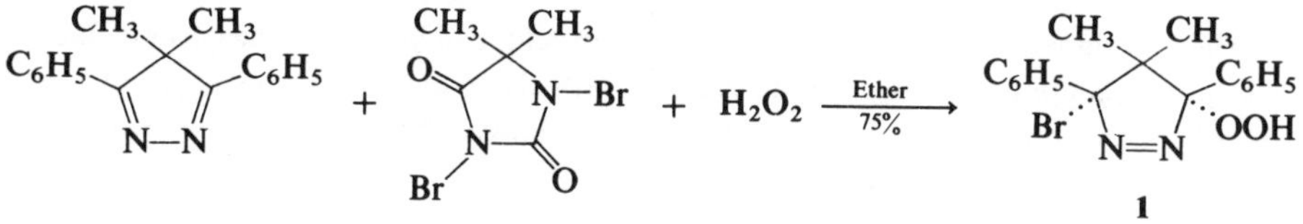

$$C_6H_5C(=N-N=)C(C_6H_5)C(CH_3)_2 + \text{1,3-dibromo-5,5-dimethylhydantoin} + H_2O_2 \xrightarrow[75\%]{\text{Ether}} \mathbf{1}$$

Oxidation of organic substrates. This hydroperoxide converts 2,3-dimethyl-2-butene into tetramethylethylene oxide with simultaneous formation of 3-bromo-4,5-dihydro-5-hydroxy-4,4-dimethyl-3,5-diphenyl-3H-pyrazole (**2**). Dialkyl olefins, however, are not epoxidized by **1**. Enol ethers are converted to a variety of epoxide rearrangement products.[2]

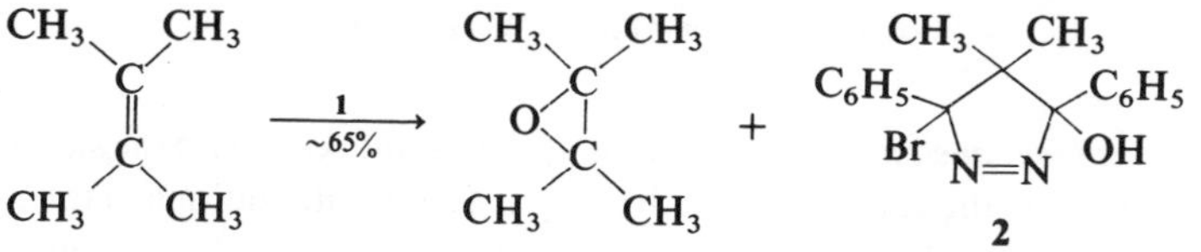

$$(CH_3)_2C=C(CH_3)_2 \xrightarrow[\sim 65\%]{\mathbf{1}} (CH_3)_2C\overset{O}{-}C(CH_3)_2 + \mathbf{2}$$

The reagent converts sulfides and tertiary amines into the corresponding sulfoxides and amine oxides in high yield.[3]

The hydroperoxide thus is similar to 2-hydroperoxyhexafluoro-2-propanol (**9**, 244–245) and to 4a-hydroperoxyflavins (**3**), known to be involved in some enzymic oxidations.

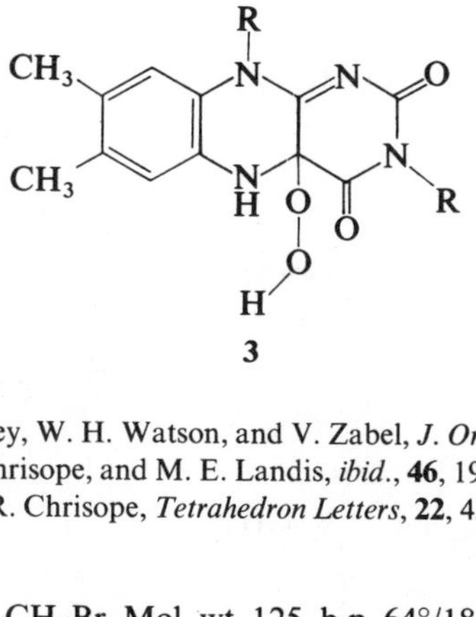

3

[1] M. E. Landis, R. L. Lindsey, W. H. Watson, and V. Zabel, *J. Org.*, **45**, 525 (1980).
[2] A. L. Baumstark, D. R. Chrisope, and M. E. Landis, *ibid.*, **46**, 1964 (1981).
[3] A. L. Baumstark, and D. R. Chrisope, *Tetrahedron Letters*, **22**, 4591 (1981).

2-Bromoethanol, $HOCH_2CH_2Br$. Mol. wt. 125, b.p. 64°/18 mm. Preparation.[1]

2-Bromoethyl esters of amino acids. N-Protected amino acids are esterified by 2-bromoethanol in THF or CH_2Cl_2 with DCC. These esters are stable in moderately acidic or basic media. They are cleaved by conversion to the 2-iodoethyl ester by Finkelstein exchange with NaI followed by zinc reduction in DMF. On treatment with trimethylamine, the esters are converted into hydrophilic choline esters, $RCOOCH_2CH_2\overset{+}{N}(CH_3)_3Br^-$, which are very stable to acids, but cleaved by OH^-.[2]

The 2-bromoethyl esters of N-protected amino acids can be used in the silver triflate method for coupling with a 1-bromo sugar to form glycopeptides. The ester group can be selectively removed by reaction with NaI followed by zinc reduction without cleavage of the sensitive glycosidic bond.[3]

[1] F. K. Thayer, C. S. Marvel, and G. S. Hiers, *Org. Syn. Coll. Vol. I*, 117 (1932).
[2] H. Kunz and M. Buchholz, *Ber.*, **112**, 2145 (1979).
[3] *Idem, Angew. Chem. Int. Ed.*, **20**, 894 (1981).

2-(2-Bromoethyl)-1,3-dioxane,

CH_2CH_2Br (**1**), **7**, 37.

Cyclopentane annelation.[1] A recent synthesis of the sesquiterpene silphinene (**7**) used two cyclopentane annelations, both involving conjugate addition of the Grignard reagent derived from **1** in the presence of $CuBr \cdot S(CH_3)_2$ to an α,β-enone. Thus **3**, obtained as shown, undergoes dehydration to **4**. This product was converted to **5**, which undergoes a second cyclopentane annelation to give eventually **6**, which has the carbon skeleton of **7**.

2 → 1) **1**, Mg; 2) $CuBr \cdot S(CH_3)_2$; 2) H_3O^+; 78% → **3** → 76% → **4**

→ 1) CH_3Li; 2) PCC → **5** → 1) **1**, $CuBr \cdot S(CH_3)_2$; 2) H_3O^+ → **6** → Several steps, 10% overall → **7**

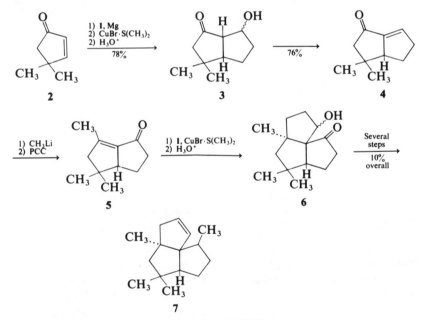

[1] A. Leone-Bay and L. A. Paquette, *J. Org.* **47**, 4173 (1982).

5-Bromo-7-nitroindoline, (1), N-acetate, m.p. 193–198°.[1]

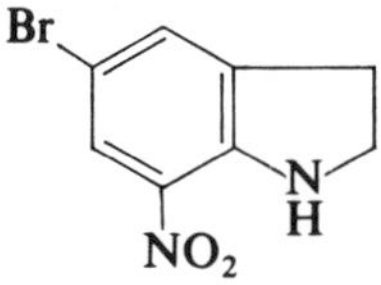

Protection of carboxylic acids.[2] N-Acyl derivatives of **1** are hydrolyzed in high yield (85–100%) when irradiated in CH_2Cl_2-dioxane-water at 350 nm (5–10 hours), with recovery of **1** in about 90% yield. The 7-nitro group is essential for this photoreaction.

Peptide synthesis.[3] The 5-bromo-7-nitroindoline (Bni) group can also be used to activate the terminal carboxyl group of a peptide to nucleophilic attack by the terminal amino group of another carboxyl-protected peptide by irradiation at 420 nm. Yields (70–80%) compare favorably with those obtained by the usual methods, and racemization is only slight when the photocoupling is conducted at −25 to −15°. Thus, the Bni group is unique in that it can be used for both masking and activation of a terminal carboxyl group.

[1] W. G. Gall, B. D. Astill, and V. Boekelheide, *J. Org.*, **20**, 1538 (1955).
[2] B. Amib, D. A. Ben-Efraim, and A. Patchornik, *Am. Soc.*, **98**, 843 (1976).
[3] Sh. Pass, B. Amib, and A. Patchornik, *ibid.*, **103**, 7674 (1981).

N-Bromosuccinimide, 1, 78–80; **2**, 40–42; **3**, 34–36; **4**, 49–53; **5**, 65–66; **6**, 74–76; **7**, 37–40; **8**, 54–56; **9**, 70–72; **10**, 57–59.

1-*Hydroxyanthraquinones.* Treatment of tetrahydroanthracenones **1**, prepared as formulated, with NBS in aqueous acetone, followed by quenching with triethylamine, results in 1-hydroxyanthraquinones (**2**). In the absence of the base, the quinone is formed in lower yield. The mechanism of this rearrangement is not known.[1]

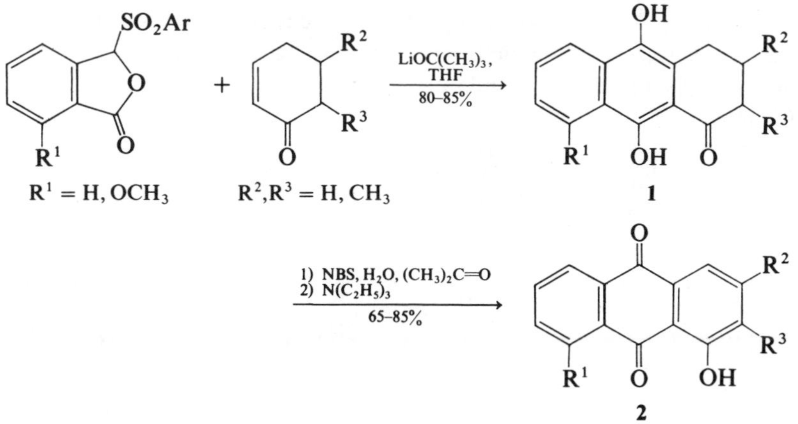

$R^1 = H, OCH_3$ $R^2,R^3 = H, CH_3$ $\xrightarrow[80–85\%]{LiOC(CH_3)_3,\ THF}$ **1**

$\xrightarrow[65–85\%]{1)\ NBS,\ H_2O,\ (CH_3)_2C{=}O\quad 2)\ N(C_2H_5)_3}$ **2**

[1] F. M. Hauser and S. Prasanna, *J. Org.*, **47**, 383 (1982).

1-Bromo-1-trimethylsilyl-1(Z),4-pentadiene (1). Mol. wt. 219.20.

Preparation:

$$HC{\equiv}CSi(CH_3)_3 + CH_2{=}CHCH_2Br \xrightarrow[94\%]{Br_2Pd(C_6H_5CN)_2,\ 20^\circ} \underset{\mathbf{1}}{(CH_2{=}CHCH_2)(H)C{=}C(Si(CH_3)_3)(Br)}$$

γ,δ-Enones. The Grignard reagent (**2**) derived from **1** couples with primary halides or tosylates in the presence of CuI to give a mixture of (E)- and (Z)-**3** (~ 4:1). The coupled products can be converted into γ,δ-unsaturated ketones (**5**).[1]

$$\underset{\mathbf{2}}{CH_2{=}CHCH_2CH{=}C(Si(CH_3)_3)(MgBr)} \xrightarrow[70\text{–}95\%]{RCH_2X,\ CuI} \underset{\mathbf{3}\ (E/Z = 4:1)}{CH_2{=}CHCH_2CH{=}C(Si(CH_3)_3)(CH_2R)}$$

$$\xrightarrow[80\text{–}95\%]{ClC_6H_4CO_3H,\ CH_2Cl_2} \underset{\mathbf{4}}{CH_2{=}CHCH_2\overset{O}{CH{-}C}(Si(CH_3)_3)(CH_2R)} \xrightarrow[80\text{–}90\%]{TFA} \underset{\mathbf{5}}{CH_2{=}CHCH_2CH_2\overset{O}{\overset{\|}{C}}CH_2R}$$

[1] R. Yamaguchi, H. Kawasaki, and M. Kawanisi, *Syn. Comm.*, **12**, 1027 (1982).

2-Bromo-3-(trimethylsilyl)propene, $H_2C{=}C(Br)CH_2Si(CH_3)_3$ **(1).**

The reagent is prepared by reaction of lithium (trimethylsilyl)cyanocuprate with 2,3-dibromopropene in THF/HMPT (3:1) in 63% yield.

α-Methylene-γ-butyrolactones; methylenecyclopentanes. In the presence of 1 equiv. of $TiCl_4$, **1** adds to both aliphatic aldehydes and ketones to form adducts, which are converted into 2-methylene-γ-butyrolactones on carbonylation with 1.5 equiv. of $Ni(CO)_2[P(C_6H_5)_3]_2$ and 2 equiv. of $N(C_2H_5)_3$.

Examples:

$$C_5H_{11}CHO \xrightarrow[88\%]{1,\ TiCl_4,\ CH_2Cl_2} C_5H_{11}\overset{OH}{CH}CH_2\overset{Br}{C}{=}CH_2 \xrightarrow[94\%]{Ni(CO)_2L_2,\ N(C_2H_5)_3} \text{5-}C_5H_{11}\text{-}\alpha\text{-methylene-}\gamma\text{-butyrolactone}$$

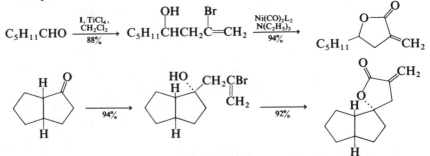

$$\text{cis-bicyclo[3.3.0]octan-2-one} \xrightarrow{94\%} \text{2-hydroxy-2-}(CH_2C(Br){=}CH_2)\text{ adduct} \xrightarrow{92\%} \text{spiro-}\alpha\text{-methylene-}\gamma\text{-butyrolactone}$$

α,β-Enones can be annelated by the Grignard reagent corresponding to **1** to give methylenecyclopentanes.[1]

Example:

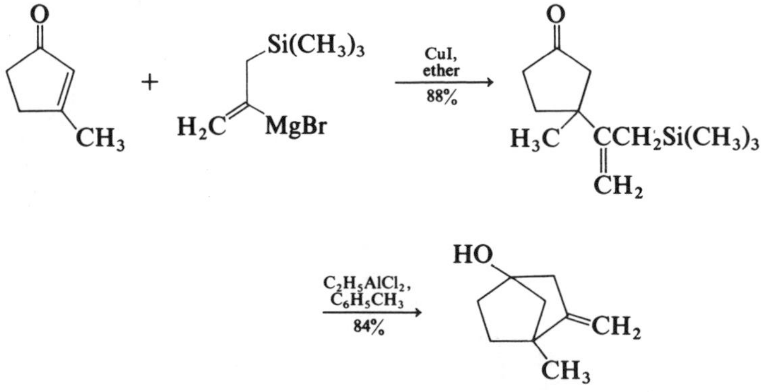

CuI, ether, 88%; $C_2H_5AlCl_2$, $C_6H_5CH_3$, 84%

Bicyclic methylenecyclopentanes. Trost and Chan[2] have extended their synthesis of methylenecyclopentanes by cycloaddition of trimethylenemethanepalladium complexes to alkenes (**9**, 454–455)[3] to an intramolecular [3 + 2] cycloaddition to give bicyclic methylenecyclopentanes. The substrates (**2**) can be prepared by reaction of the Grignard reagent prepared from 2-bromo-3-(trimethylsilyl)propene (**1**) with a suitable bifunctional aldehyde (equation I).

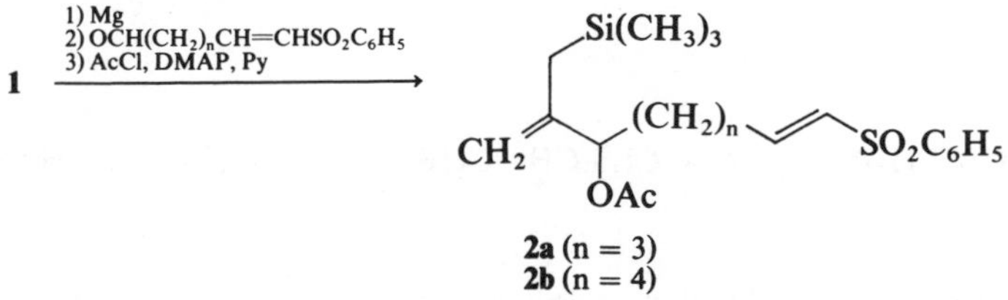

1 → 1) Mg; 2) $OCH(CH_2)_nCH{=}CHSO_2C_6H_5$; 3) AcCl, DMAP, Py

2a (n = 3)
2b (n = 4)

Treatment of **2a** with catalytic amounts of $Pd[P(C_6H_5)_3]_4$ and 1,2-bis(diphenylphosphine)ethane (dppe) results in the *cis*-fused bicyclic methylenecyclopentane (**3**) in yields as high as 65%. [3 + 2]Cycloadditions of these complexes involve two steps; conjugate addition followed by an S_N2-like displacement (equation II). Cyclization of **2b** in the presence of Pd(0) results in both the *cis*- and *trans*-bicyclo[4.3.0]nonanes **4** and **5**, with the *cis*-isomer predominating (equation III). Cyclization of **2** in which n = 8 fails.

(II) **2a** $\xrightarrow{PdL_4}$ [CH_2 $\overset{+}{P}dL_2$ CH_2 $SO_2C_6H_5$] $\longrightarrow$ [H CH_2 $\overset{+}{P}dL_2$ CH_2 H $SO_2C_6H_5$] $\xrightarrow{65\%}$

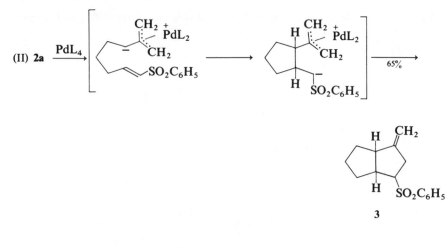

H CH_2 H $SO_2C_6H_5$

3

(III) **2b** $\xrightarrow[70\%]{}$ H CH_2 H $SO_2C_6H_5$ **4** + 2:1 H CH_2 H $SO_2C_6H_5$ **5**

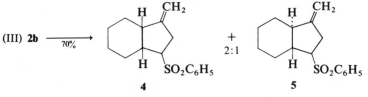

[1] B. M. Trost and B. P. Coppola, *Am. Soc.*, **104**, 6879 (1982).
[2] B. M. Trost and D. M. T. Chan, *Am. Soc.*, **104**, 3733 (1982).
[3] *Idem*, *ibid.*, **102**, 6359 (1980); **103**, 5972 (1981).

(2-Bromovinyl)trimethylsilane (1). Mol. wt. 179.15, b.p. 55°/42 mm.

Preparation[1]:

$$Cl_3SiCHBrCH_2Br \xrightarrow[28.7\%]{AlCl_3} Cl_3SiCH{=}CHBr \xrightarrow[42\%]{CH_3MgCl} (CH_3)_3SiCH{=}CHBr$$

1

Cyclopentenones. Trimethylvinylsilane has been used to prepare annelated cyclopentenones by cyclization of intermediate divinyl ketones (**9**, 498–499). The major limitation is that the double bond in the product is located at the most stable position (ring fusion). A modification using the Grignard reagent derived from **1** results in 4,5-annelated-2-cyclopentenones, as outlined in equation (I) for a typical case. The overall yields are in the range 39–65%. The *cis*-isomers are formed predominately or exclusively. The same sequence can be applied to acyclic α,β-unsaturated aldehydes to furnish 4- and 5-substituted 2-cyclopentenones, a cyclization that is not possible in the absence of the β-trimethylsilyl group.[2]

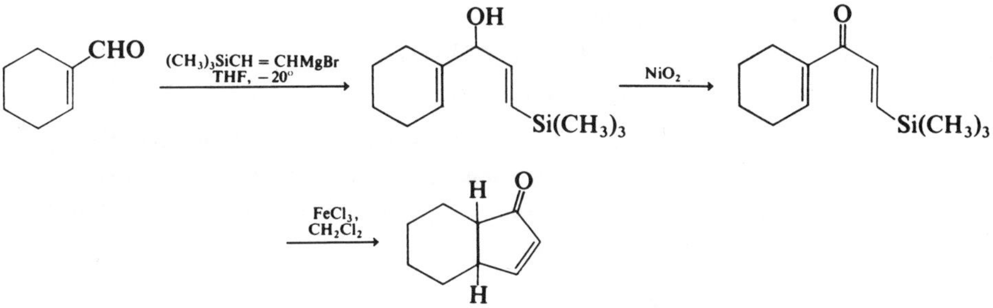

A similar effect of a silyl substituent on Nazarov cyclizations has been reported in an annelation with 1-phenylthio-1-trimethylsilylethylene (equation II).[3]

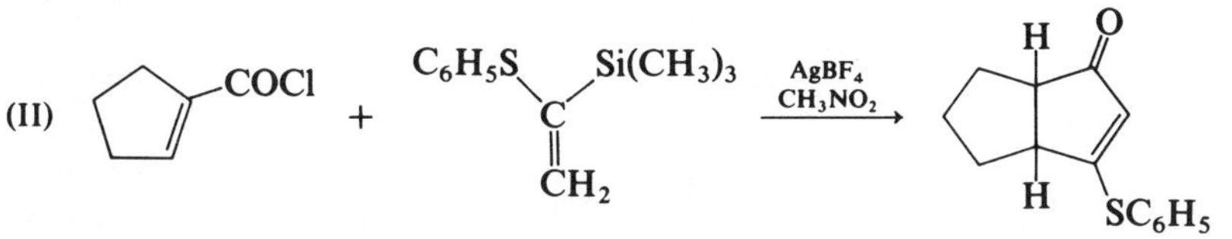

(II)

[1] V. F. Mirinov, A. D. Petrov, and N. G. Maskimova, *Bull. Acad. Sci. USSR, Div. Chem. Sci.* (English), 1864 (1959).
[2] S. E. Denmark and T. K. Jones, *Am. Soc.*, **104**, 2642 (1982).
[3] F. Cooke, R. Moerck, J. Schwindeman, and P. Magnus, *J. Org.*, **45**, 1046 (1980).

Brucine (1). Mol. wt. 394.45, $\alpha_D - 85°(C_2H_5OH)$.

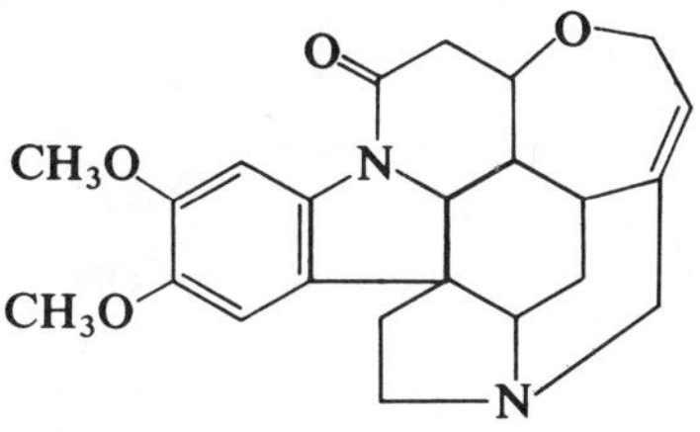

1

Resolution of **tert**-***acetylenic alcohols.*** Brucine forms stable 1:1 molecular complexes with only one enantiomer of several *tert*-acetylenic alcohols. In some favorable cases, complete resolution can be achieved by only one complexation; in other cases, repetition of complexation is necessary for complete resolution. The complexes are decomposed by dilute HCl. Complexation involves a hydrogen bond between the OH group and the N atom of brucine; in addition, the linearity of the acetylene group may be involved.[1]

[1] F. Toda and K. Tanaka, *Tetrahedron Letters*, **22**, 4669 (1981).

D-(−)- and L-(+)-2,3-Butanediol, $CH_3CH(OH)—CH(OH)CH_3$ **(1).**

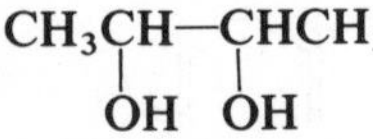

Asymmetric aldol reactions.[1] The acetal (**2**) obtained by reaction of benzaldehyde with D-(−)-**1** (available from Aldrich) reacts with the ketone **3** to give the two possible crossed aldols in the ratio 16:1. The products can be converted into optically active β-hydroxy ketones in three steps.

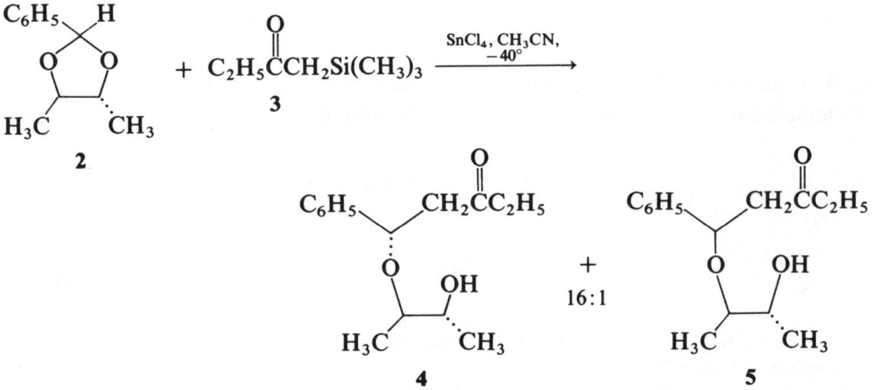

$$\mathbf{2} + C_2H_5COCH_2Si(CH_3)_3\ (\mathbf{3}) \xrightarrow[-40^\circ]{SnCl_4,\ CH_3CN} \mathbf{4} + \mathbf{5}\quad (16:1)$$

This reaction was used for an asymmetric synthesis of aklavinone (**8**), an aglycone of the anthracycline antibiotics. The key step was the reaction of the acetal **6** with **2**,

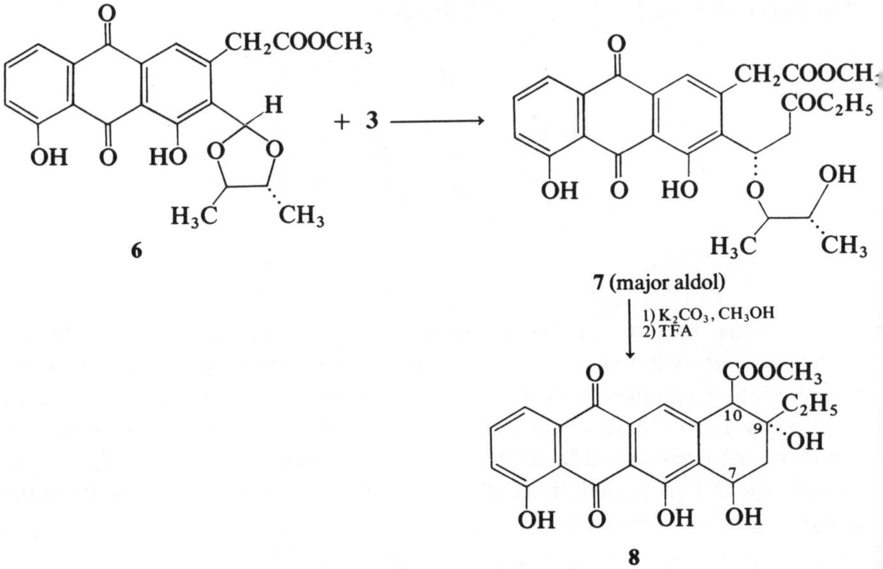

$$\mathbf{6} + \mathbf{3} \longrightarrow \mathbf{7}\ \text{(major aldol)} \xrightarrow{1)\ K_2CO_3,\ CH_3OH;\ 2)\ TFA} \mathbf{8}$$

which furnished the two possible crossed aldols as a 10:1 mixture in 83% combined yield. The major product **7** was converted into **8** in two steps.

A similar asymmetric aldol reaction was used for a synthesis of 11-deoxydaunomycinone (**11**) and related compounds. The key step was the reaction of the acetal **9** from L-(+)-2,3-butanediol with the ketone **10** in the presence of BF_3 at −78°. Two aldols were formed in the ratio 13:1 in 80–85% combined yield. The major aldol **11** was used for synthesis of **12**.

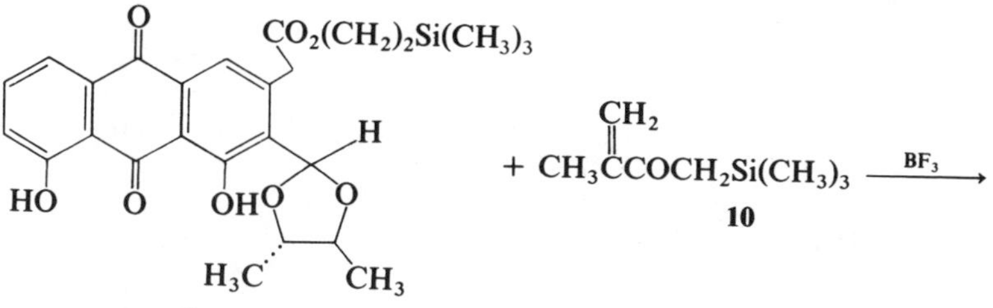

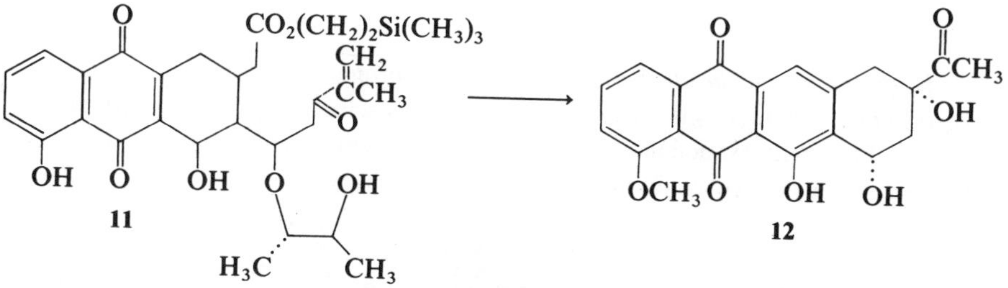

Asymmetric induction in cyclization of a dienic acetal. Johnson *et al.*[3] reported a very high degree of asymmetric induction in the $SnCl_4$-catalyzed cyclization of the acetal **1**, derived from (−)-2,3-butanediol. Cyclization results in two axial ethers and two equatorial ethers; cleavage and Jones oxidation converts these products into an octalone mixture consisting of **2a** and **2b** in the ratio 8:92.

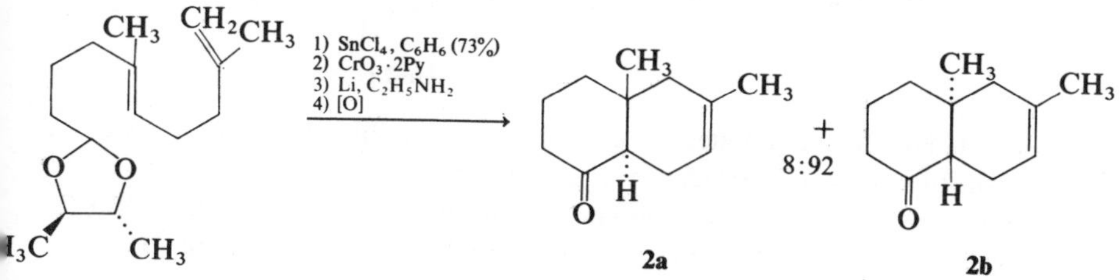

[1] J. M. McNamara and Y. Kishi, *Am. Soc.*, **104**, 7371 (1982).
[2] H. Sekizaki, M. Jung, J. M. McNamara, and Y. Kishi, *ibid.*, **104**, 7372 (1982).
[3] W. S. Johnson, C. A. Harbert, B. E. Ratcliffe, and R. D. Stipanovoic, *ibid.*, **98**, 6188 (1976).

(E)- and (Z)-2-(2-Butenyl)-4,4,5,5-tetramethyl-2-bora-1,3-dioxacyclopentane (1a and 1b).

Preparation[1]:

$$CH_2Cl_2 \xrightarrow[\text{3) HCl, }H_2O]{\text{1) }n\text{-BuLi; 2) B(OCH}_3)_3} Cl_2CHB(OH)_2 \xrightarrow[80\text{–}89\%]{(CH_3)_2C(OH)\text{—}C(OH)(CH_3)_2} Cl_2CHB(\text{pinacolato})$$

$$\xrightarrow[\text{AIBN}]{Bu_3SnH} ClCH_2B(\text{pinacolato}) \xrightarrow[56\%]{(E)\text{-}CH_3CH{=}CHLi} \textbf{1a} \qquad \xrightarrow[97\%]{(Z)\text{-}CH_3CH{=}CHLi} \textbf{1b}$$

1a

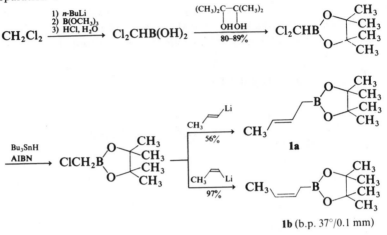

1b (b.p. 37°/0.1 mm)

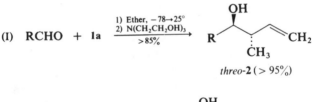

β-Methylhomoallylic alcohols.[2] These crotylboronates (**1**) react with aldehydes to give β-methylhomoallylic alcohols with a diastereoselectivity of > 95% (equations I and II). This condensation has certain advantages over the aldol condensation, since the products **2** can be converted to β-hydroxy aldehydes or carboxylic acids by

$$\text{(I)}\quad RCHO + \textbf{1a} \xrightarrow[>85\%]{\text{1) Ether, }-78\to25°;\ \text{2) }N(CH_2CH_2OH)_3} RCH(OH)CH(CH_3)CH{=}CH_2$$

threo-**2** (> 95%)

$$\text{(II)}\quad RCHO + \textbf{1b} \xrightarrow{>85\%} RCH(OH)CH(CH_3)CH{=}CH_2$$

erythro-**2** (> 95%)

oxidation. Highly enantioselective syntheses are possible with crotylboronates in which the pinacol unit is replaced by an optically active 1,2-diol (**9**, 363).[3]

Similar diastereoselectivity is observed in reactions of an aldehyde with [(Z)-γ-alkoxyallyl] boronates to form monoprotected *erythro*-1,2-diols (equation III).[4]

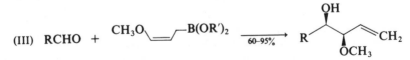

$$\text{(III)}\quad RCHO + CH_3OCH{=}CHCH_2B(OR')_2 \xrightarrow{60\text{–}95\%} RCH(OH)CH(OCH_3)CH{=}CH_2$$

[1] P. A. Thompson and P. G. M. Wuts, *Org. Syn.*, submitted (1982); P. G. M. Wuts and P. A. Thompson, *J. Organometal. Chem.*, **234**, 137 (1982).
[2] R. W. Hoffmann and H.-J- Seiss, *J. Org.*, **46**, 1309 (1981).
[3] R. W. Hoffmann, W. Ladner, K. Steinbach, W. Massa, R. Schmidt, and G. Snatzke, *Ber.*, **114**, 2786 (1981); R. W. Hoffmann and W. Helbig, *ibid.*, **114**, 2802 (1981).
[4] R. W. Hoffmann and B. Kemper, *Tetrahedron Letters*, **23**, 4883 (1980); P. G. M. Wuts and S. S. Bigelow, *J. Org.*, **47**, 2498 (1982).

(2S, 4S)-N-(*t*-Butoxycarbonyl)-4-(diphenylphosphino)-2-[(diphenylphosphino)methyl]-pyrrolidine (BPPM) (**1**). M.p. 102–103°; $\alpha_D - 40.44°$.

$(C_6H_5)_2P$ … $CH_2P(C_6H_5)_2$ … N … $COC(CH_3)_3$

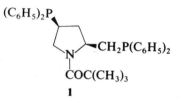

1

The pyrrolidine is prepared from 4-hydroxy-L-proline ethyl ester. Supplier: Chemical Dynamics Corp., South Plainfields, N.J.

Asymmetric hydrogenation. Prochiral α,β-unsaturated acids and their derivatives can be hydrogenated with high stereoselectivity by rhodium complexes with **1**, such as (BPPM)Rh(COD)Cl and $(BPPM)Rh(COD)^+ClO_4^-$, in which COD = 1,5-cyclooctadiene. The stereoselectivity is dependent in part on the hydrogen pressure, and the effect can be attenuated by addition of triethylamine, which also increases the optical yield. The stereoselectivity is markedly controlled by the stereochemistry of the double bond.[1]

The same catalyst effects asymmetric reduction of α-keto lactones (equation I)[2] and α-keto carboxylates.[3]

(I) $\xrightarrow[90\text{–}93\%]{H_2,\ (BPPM)Rh(COD)Cl,\ 50\ atm,\ 30\text{–}50°}$ (70–75%ee)

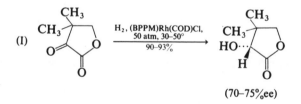

[1] I. Ojima, T. Kogure, and Y. Yoda, *J. Org.*, **45**, 4728 (1980).
[2] *Idem*, *Org. Syn.*, submitted (1983).
[3] I. Ojima, T. Kogure, and K. Achiwa, *J.C.S. Chem. Comm.*, 428 (1977).

***t*-Butylaminoborane,** $(CH_3)_3CNH_2 \cdot BH_3$ (**1**). Mol. wt. 86.97, m.p. 90–100° (dec.). Supplier: Aldrich.

Stereoselective reduction.[1] 12-Ketocholanic acids are reduced by this complex to 12β- and 12α-hydroxycholanic acids in ratios of 5.0–1.5. 3-Keto- and 7-ketocholanic acids are reduced with some preference to the α-hydroxy acids.

[1] F. C. Chang, *Syn. Comm.*, **11**, 875 (1981).

B-*n*-Butyl-9-borabicyclo[3.3.1]nonane, 9, 76–77.

Selective reduction of t-alkyl halides.[1] The ate complex of B-butyl-9-BBN with *n*-BuLi selectively reduces tertiary alkyl halides to hydrocarbons in high yield without effect on primary or secondary halides. It does not reduce aryl or vinyl halides, but does reduce benzyl and allyl halides. The reduction involves a carbonium ion, and thus can proceed with Wagner–Meerwein rearrangements in certain systems.

[1] H. Toi, Y. Yamamoto, A. Sonoda, and S.-I. Murahashi, *Tetrahedron*, **37**, 2261 (1981).

***t*-Butyl carbazate, 1**, 85; **2**, 46.

Monoalkylhydrazines. A new synthesis of these hydrazines is shown in equation (I).[1]

$$\text{(I)}\quad \begin{matrix}R^1\\ \quad\diagdown\\ \quad C{=}O\\ \quad\diagup\\ R^2\end{matrix} + H_2NNH\overset{O}{\overset{\|}{C}}OC(CH_3)_3 \xrightarrow[95\text{–}97\%]{C_6H_{14},\ \Delta} \begin{matrix}R^1\\ \quad\diagdown\\ \quad C{=}\end{matrix}NNH\overset{O}{\overset{\|}{C}}OC(CH_3)_3 \ (R^2)$$

$$\xrightarrow[90\text{–}95\%]{1)\ BH_3\cdot THF\quad 2)\ HCl,\ H_2O} \begin{matrix}R^1\\ \quad\diagdown\\ \quad CHNHNH_2\cdot HCl\\ \quad\diagup\\ R^2\end{matrix}$$

[1] N. I. Ghali, D. L. Venton, S. C. Hung, and G. C. Le Breton. *J. Org.*, **46**, 5413 (1981).

***t*-Butyldimethylchlorosilane, 4**, 57–58, 176–177; **5**, 74–75; **6**, 78–79; **7**, 59; **8**, 77; **9**, 77; **10**, 62.

α-Silyl aldehydes.[1] α-Trimethylsilyl aldehydes are difficult to prepare because of instability. However, α-*t*-butyldimethylsilyl aldehydes can be prepared by hydrolysis of the corresponding α-silyl imines by two-phase hydrolysis with aqueous HOAc and CH_2Cl_2 (equation I). C-Silylation of a long-chain aldehyde is not possible, but alkylation of the silyl imine provides a general route to α-silyl aldehydes.

$$\text{(I)}\quad CH_3CH{=}N{-}C_6H_{11}\text{-}c \xrightarrow[\text{2) } Cl(CH_3)_2SiC(CH_3)_3]{\text{1) LDA, THF, 0°}} (CH_3)_3CSi(CH_3)_2CH_2CH{=}N{-}C_6H_{11}\text{-}c \xrightarrow[\text{67\% overall}]{\text{HOAc, }H_2O,\ CH_2Cl_2} (CH_3)_3CSi(CH_3)_2CH_2CHO\ \ \mathbf{1}$$

$$(CH_3)_3CSi(CH_3)_2CH_2CH{=}N{-}C_6H_{11}\text{-}c \xrightarrow{\text{1) LDA; 2) RX}} (CH_3)_3CSi(CH_3)_2CH(R)CH{=}N{-}C_6H_{11}\text{-}c \xrightarrow{\text{HOAc, }H_2O,\ CH_2Cl_2} (CH_3)_3CSi(CH_3)_2{-}CH(R)CHO\ \ \mathbf{2}$$

The α-silyl aldehydes can serve as vinyl cation equivalents in reactions with Grignard or alkyllithium reagents (equation II).

$$\text{(II)}\quad (CH_3)_3CSi(CH_3)_2{-}CH(C_6H_{13}\text{-}n)CHO \xrightarrow[\text{79–85\%}]{C_2H_5MgBr} (CH_3)_3CSi(CH_3)_2{-}CH(n\text{-}C_6H_{13}){-}CH(OH)C_2H_5$$

$$(CH_3)_3CSi(CH_3)_2{-}CH(n\text{-}C_6H_{13}){-}CH(OH)C_2H_5 \xrightarrow[\text{84–89\%}]{BF_3\cdot(C_2H_5)_2O,\ CH_2Cl_2,\ 25°} n\text{-}C_6H_{13}CH{=}CHC_2H_5\ \ (98\%\ E)$$

$$(CH_3)_3CSi(CH_3)_2{-}CH(n\text{-}C_6H_{13}){-}CH(OH)C_2H_5 \xrightarrow[\text{72–79\%}]{\text{KH, THF, }25°} n\text{-}C_6H_{13}CH{=}CHC_2H_5\ \ (97\%\ Z)$$

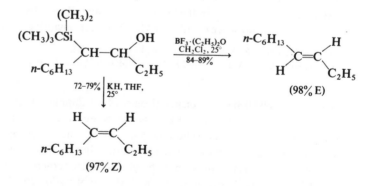

The aldehydes can also be used for α-vinylation of ketones and esters (equation III) and of amino acids (equation IV) via the stabase derivative (*see* 1,1,4,4-Tetramethyl-1,4-dichlorodisilethylene, this volume).

$$\text{(III)}\ CH_3\overset{O}{\overset{\|}{C}}OC(CH_3)_3 \xrightarrow[\ ]{1)\ LDA\ \ 2)\ \mathbf{1}} (CH_3)_3C\underset{CH_3}{\overset{CH_3}{Si}}CH_2\overset{OH}{CH}CH_2\overset{O}{\overset{\|}{C}}OC(CH_3)_3 \xrightarrow[64\%\ \text{overall}]{BF_3\cdot(C_2H_5)_2O} CH_2{=}CHCH_2\overset{O}{\overset{\|}{C}}OC(CH_3)_3$$

$$\text{(IV)}\ \text{[}(CH_3)_2Si\text{–}CH_2CH_2\text{–}Si(CH_3)_2\text{]}N{-}CH_2COOC_2H_5 \xrightarrow[93\%]{1)\ LDA\ \ 2)\ \mathbf{1}} \text{[}(CH_3)_2Si\text{–}CH_2CH_2\text{–}Si(CH_3)_2\text{]}N{-}CH(COOC_2H_5)CHOHCH_2Si(CH_3)_2C(CH_3)_3$$

$$\xrightarrow[48\%\ \text{overall}]{1)\ BF_3\cdot(C_2H_5)_2O\ \ 2)\ HCl,\ H_2O,\ \Delta} H_2N{-}\overset{COOH}{CH}{-}CH{=}CH_2$$

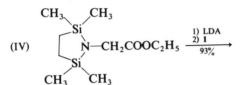

Selective silylation of ribonucleosides.[2] Only the 5′-hydroxyl group of ribonucleosides is silylated by reaction with the reagent in THF in the presence of silver nitrate. On addition of pyridine to the reaction, 2′,5′-disilyl derivatives are formed in 80–90% yield. The actual reagent may be *t*-butyldimethylsilyl nitrate. Highly selective 3′,5′-disilylation can be achieved in the presence of several silver salts ($AgNO_3$, $AgClO_4$, and AgOAc) in the presence of either DABCO or 4-nitropyridine N-oxide.

[1] P. F. Hudrlik and A. K. Kulkarni, *Am. Soc.*, **103**, 6251 (1981).
[2] G. H. Hakimelahi, Z. A. Proba, and K. K. Ogilvie, *Can. J. Chem.*, **60**, 1108 (1982).

_t_-Butyldimethylsilyl trifluoromethanesulfonate, 10, 63. Supplier: Fluka.

This triflate has been prepared by reaction of *t*-butyldimethylsilyl chloride with silver triflate (expensive)[1] or triflic acid (80% yield).[2] Since this triflate is sensitive to moisture, it should be used soon after preparation. In combination with 2,6-lutidine, it converts even tertiary alcohols into *t*-butyldimethylsilyl ethers in 70–90% yield.

4-Alkylpyridines.[3] Highly selective alkylation of pyridine at C_4 is possible by quaternization with this triflate followed by reaction with a Grignard reagent. Substitution occurs with almost complete regiospecificity (> 99%) to give 4-alkyl-1,4-dihydropyridines, which are oxidized by oxygen to 4-substituted pyridines (equation I).

(I) Pyridine $\xrightarrow[\text{Py, CH}_2\text{Cl}_2]{t\text{-BuMe}_2\text{SiOTf,}}$ N-(SiMe$_2$-*t*-Bu)pyridinium $^-$OTf $\xrightarrow{\text{RMgBr}}$ 4-H,4-R-1,4-dihydro-N-SiMe$_2$-*t*-Bu pyridine $\xrightarrow[\text{~60–70\% overall}]{O_2}$ 4-R-pyridine

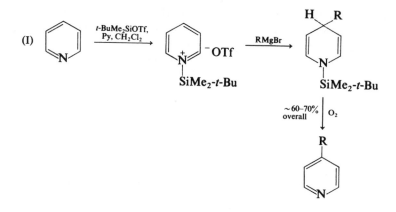

[1] M. Riediker and W. Graf, *Helv. Chim. Acta*, **62**, 205 (1979).
[2] E. J. Corey, H. Cho, C. Rucker, and D. Hua, *Tetrahedron Letters*, **22**, 3455 (1981).
[3] K. Akiba, Y. Iseki, and M. Wada, *Tetrahedron Letters*, **23**, 3935 (1982).

t-Butyl hydroperoxide–Bisoxobis(2,4-pentanedionate)molybdenum [$MoO_2(acac)_2$].

Cleavage of silyl enol ethers. The double bond of silyl enol ethers is cleaved to carbonyl compounds by reaction with *t*-butyl hydroperoxide in the presence of catalytic amounts of $MoO_2(acac)_2$.[1]

1-OSi(CH$_3$)$_3$-6-CH$_3$-cyclohexene $\xrightarrow[\text{90\%}]{\text{1) } t\text{-BuOOH, MoO}_2\text{(acac)}_2\text{, C}_6\text{H}_6\text{, 60°; 2) CH}_3\text{COOH, THF, H}_2\text{O, 60°}}$ HOOC(CH$_2$)$_3$CH(CH$_3$)COOH

(CH$_3$)(C$_6$H$_5$)C=CH–OSi(CH$_3$)$_3$ $\xrightarrow{98\%}$ $C_6H_5COCH_3$

CH$_3$ CH$_3$ / CH$_3$ cyclohexenyl–CH=CH–C(OSi(CH$_3$)$_3$)=CH$_2$ $\xrightarrow{85\%}$ CH$_3$ CH$_3$ / CH$_3$ cyclohexenyl–CH=CH–COOH

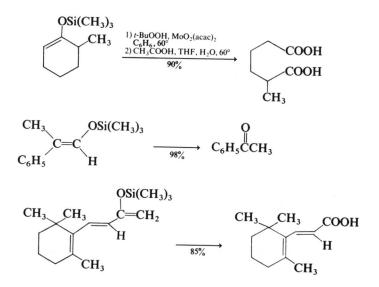

[1] K. Kaneda, N. Kii, K. Jitsukawa, and S. Teranishi, *Tetrahedron Letters*, **22**, 2595 (1981).

t-Butyl hydroperoxide–Dialkyl tartrate–Titanium (IV) isopropoxide, 10, 64–65.

Asymmetric epoxidation. Sharpless *et al.*[1] have reviewed the numerous applications of titanium-catalyzed asymmetric epoxidations developed in their own and other laboratories. All the reactions conform to the enantiomeric selectivity first observed and formulated as in Scheme (I).

D-(−)-diethyl tartrate (unnatural)

"$:\ddot{O}:$"

R^2 R^1 R^3 OH

"$:\ddot{O}:$"

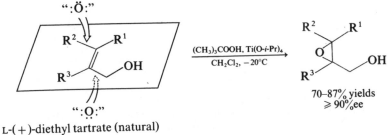

L-(+)-diethyl tartrate (natural)

$(CH_3)_3COOH$, Ti(O-*i*-Pr)$_4$, CH_2Cl_2, −20°C

R^2 R^1 O OH R^3

70–87% yields
≥ 90%ee

Scheme (I)

The value of this reaction is enhanced by the fact that the numerous selective transformations of the 2,3-epoxy alcohols at C_1, C_2 and C_3 with carbon, nitrogen, and oxygen nucleophiles are possible because of electronic and steric factors.

Synthesis of chiral monosaccharides. Masamune, Sharpless, and coworkers[2] have developed a general, iterative sequence for addition of two chiral hydroxymethylene units to an aldose. The key step involves regio- and stereoselective ring opening

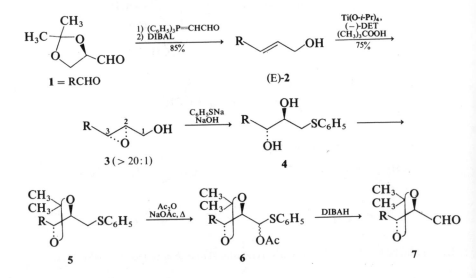

CH_3 H_3C O O CHO

1 = RCHO

1) $(C_6H_5)_3P{=}CHCHO$ 2) DIBAL 85%

R OH

(E)-**2**

Ti(O-*i*-Pr)$_4$, (−)-DET, $(CH_3)_3COOH$ 75%

R 3 2 1 OH O

3 (> 20:1)

C_6H_5SNa NaOH

R OH SC_6H_5 OH

4

CH_3 CH_3 O R SC_6H_5 O

5

Ac_2O NaOAc, Δ

CH_3 CH_3 O R SC_6H_5 O OAc

6

DIBAH

CH_3 CH_3 O R CHO O

7

of the chiral epoxides obtained by asymmetric epoxidation of allylic alcohols derived from the aldose by Wittig or related reactions. The route developed for the overall sequence is shown in Scheme (I) for conversion of D-glyceraldehyde acetonide (**1**) into the diacetonide of the 2,3-*erythro*-pentose **7**. The chiral epoxide **3** obtained by asymmetric epoxidation of (E)-**2** with (−)-diethyl tartrate (DET) is converted with inversion by C_6H_5SNa in an alkaline medium selectively into the *erythro*-diol **4**, which, after acetonation is converted into **5** by a Pummerer rearrangement. Hydrolysis with DIBAH gives the 2,3-*erythro*-D-pentose (**7**). The diacetonide of the other possible 2,3-*erythro*-D-pentose **9** is obtained by the same sequence from the *β*-oxide of (E)-**2**, prepared with (+)-DET.

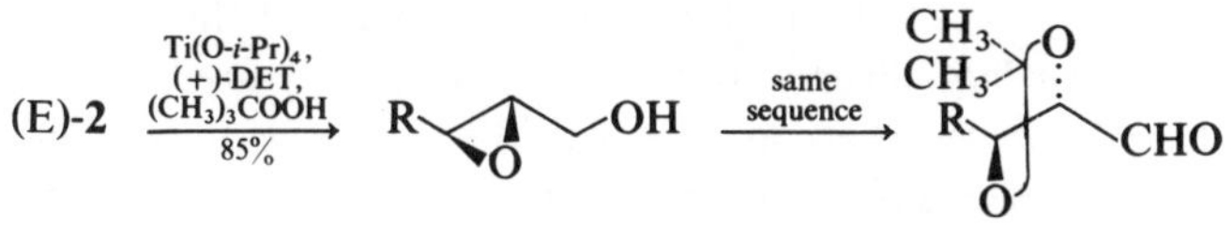

(E)-**2** —[$Ti(O\text{-}i\text{-}Pr)_4$, (+)-DET, $(CH_3)_3COOH$; 85%]→ **8** (> 20:1) —[same sequence]→ **9**

Scheme I

7 —[K_2CO_3, CH_3OH]→ **10**

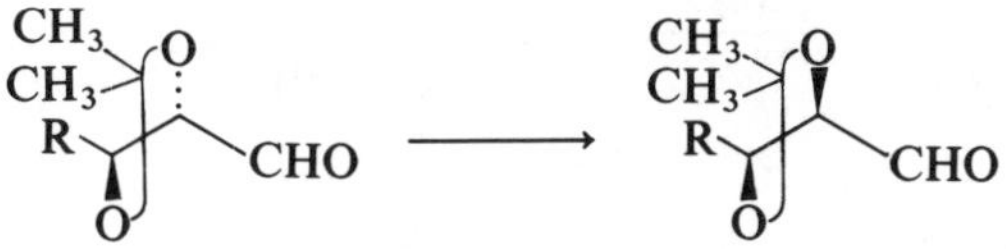

9 → **11**

Application of the same sequence to (Z)-**2**, however, is unsatisfactory because asymmetric epoxidation of (Z)-**2** proceeds very slowly and with low stereoselectivity. The difficulty has since been resolved by the observation that **7** and **9** are epimerized at C_2 to the corresponding 2,3-*threo*-aldoses **10** and **11**, respectively, by K_2CO_3 in methanol. As a result, the four possible D-pentoses can be prepared satisfactorily from (E)-**2**. The epimerization is equally effective in the tetrose, pentose, and hexose series.[3]

Other useful, regiospecific ring openings of these chiral *α*,*β*-epoxy alcohols have been reported. Thus, epoxides such as **3** and **8** are reduced by sodium bis(methyoxyethoxy)aluminum hydride (SMEAH) with high regioselectivity to chiral 1,3-diols, derivatives of 2-deoxy-D-pentitols,[4,5] This remarkable regiospecificity is unique to *α*,*β*-epoxy alcohols of this type and is not a general reaction. Other regiospecific ring openings at C_2 have been conducted by way of the benzyl carbonate

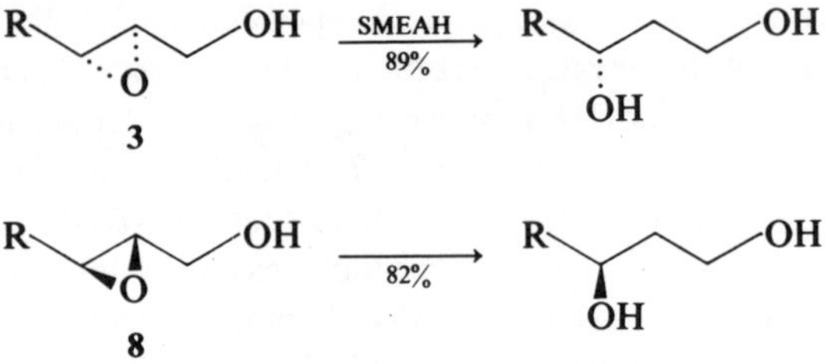

SMEAH 89%

3

82%

8

derivatives,[3] and the phenyl urethanes.[4,6] This last cleavage provides a general route to 2-amino-2-deoxy-D-pentitols.[4]

Kinetic resolution of chiral allylic alcohols.[7] Partial (at least 60% conversion) asymmetric epoxidation can be used for kinetic resolution of chiral allylic alcohols, particularly of secondary allylic alcohols in which chirality resides at the carbinol carbon such as **1**, drawn in accordance with the usual enantioface selection rule (Scheme I). (S)-**1** undergoes asymmetric epoxidation with L-diisopropyl tartrate (DIPT) 104 times faster than (R)-**1**. The optical purity of the recovered allylic alcohol after kinetic resolution carried to 60% conversion is often > 90%. In theory, any degree of enantiomeric purity is attainable by use of higher conversions. Secondary allylic alcohols generally conform to the reactivity pattern of **1**; the (Z)-allylic alcohols are less satisfactory substrates, particularly those substituted at the β-vinyl position by a bulky substituent.

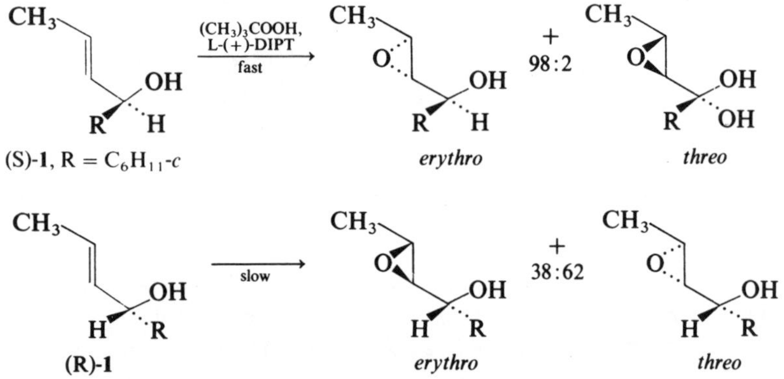

(S)-**1**, R = C_6H_{11}-*c* — $(CH_3)_3COOH$, L-(+)-DIPT, fast → *erythro* + *threo* (98:2)

(R)-**1** — slow → *erythro* + *threo* (38:62)

Scheme I

More recently, this kinetic resolution has been applied successfully to some primary allylic alcohols and an allenic alcohol.[7]

[1] K. B. Sharpless, C. H. Behrens, T. Katsuki, A. W. M. Lee, V. S. Martin, M. Takatani, S. M. Viti, F. J. Walker, and S. S. Woodard, *Pure and Applied Chem.*, **55**, 589 (1983).

[2] T. Katsuki, A. W. M. Lee, P. Ma, V. S. Martin, S. Masamune, K. B. Sharpless, D. Tuddenham, and F. J. Walker, *J. Org.*, **47**, 1373 (1982).

[3] A. W. M. Lee, V. S. Martin, S. Masamune, K. B. Sharpless, and F. J. Walker, *Am. Soc.*, **104**, 3515 (1982).
[4] N. Minami, S. S. Ko, and Y. Kishi, *Am. Soc.*, **104**, 1109 (1982).
[5] P. Ma, V. S. Martin, S. Masamune, K. B. Sharpless, and S. M. Viti, *J. Org.*, **47**, 1378 (1982).
[6] W. R. Roush and R. J. Brown, *J. Org.*, **47**, 1371 (1982).
[7] V. S. Martin, S. S. Woodard, T. Katsuki, Y. Yamada, M. Ikeda, and K. B. Sharpless, *Am. Soc.*, **103**, 6237 (1981).

t-Butyl hydroperoxide–Molybdenum carbonyl, 8, 62–63.

Selective epoxidation. One step in a total synthesis of DL-14α-methylestrone 3-methyl ester (**4**) is the epoxidation of the hindered C_{13}–C_{17} double bond of **1**. Epoxidation with *t*-butyl hydroperoxide catalyzed by molybdenum carbonyl gives a mixture of the desired α-epoxide (**2**) and the β-isomer (**3**) in a 5:1 ratio in almost quantitative yield. The same epoxides are obtained with *m*-chloroperbenzoic acid, but with opposite stereoselectivity. The α-epoxide (**2**) is rearranged to **4** by BF_3 etherate.[1]

CH_3O … **1** $\xrightarrow[\sim 100\%]{(CH_3)_3COOH,\ Mo(CO)_6}$ **2** + β-epoxide **3** (5:1)

2 $\xrightarrow[C_6H_5CH_3,\ 5°]{BF_3\cdot(C_2H_5)_2O}$ **4** (42% from **1**)

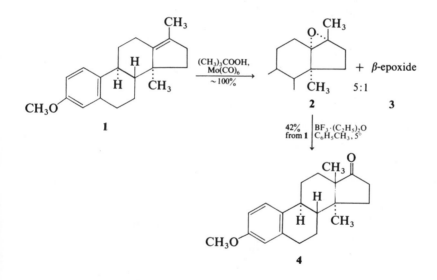

The final steps in a recent synthesis of the sesquiterpenoid mycotoxin verrucarol (**6**) required selective epoxidation of a 2,2-disubstituted double bond in the presence of a trisubstituted one. This reaction was effected, in 85% yield, by treatment of **4** with *t*-butyl hydroperoxide and $Mo(CO)_6$. Use of $VO(acac)_2$ as catalyst resulted in some decomposition in addition to the desired epoxidation. The final step was deprotection of **5** with fluoride ion to give **6**.[2]

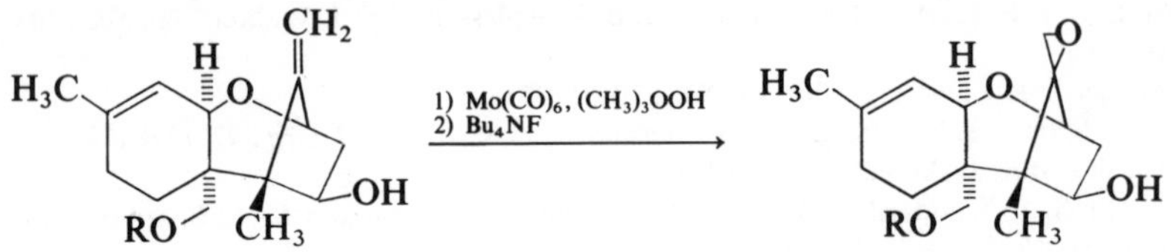

H_3C ... CH_2 ... H ... O ... OH ... RO ... CH_3

4, R = $Si(CH_3)_2C(CH_3)_3$

1) $Mo(CO)_6$, $(CH_3)_3OOH$
2) Bu_4NF

5, R = $Si(CH_3)_2C(CH_3)_3$
6, R = H

[1] M. B. Groen and F. J. Zeelen, *Tetrahedron Letters*, **23**, 3611 (1982).
[2] B. M. Trost and P. G. McDougal, *Am. Soc.*, **104**, 6110 (1982).

***t*-Butyl hydroperoxide–Selenium dioxide, 8**, 64–65; **9**, 79–80; **10**, 65–66.

Oxidation of allylic CH_3 groups.[1] This reaction is possible with minimal side reactions with SeO_2 supported on silica gel and *t*-butyl hydroperoxide (70%) in hexane or CH_2Cl_2. The main products are primary alcohols, sometimes accompanied by the corresponding α,β-enals. The allylic alcohol formed has the more stable configuration of the double bond (second example).

Examples:

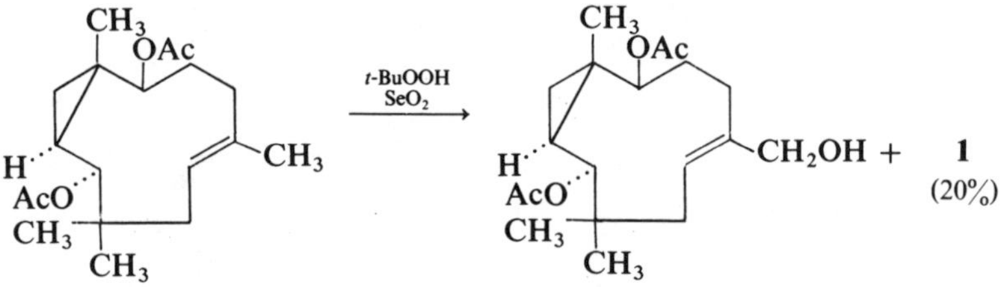

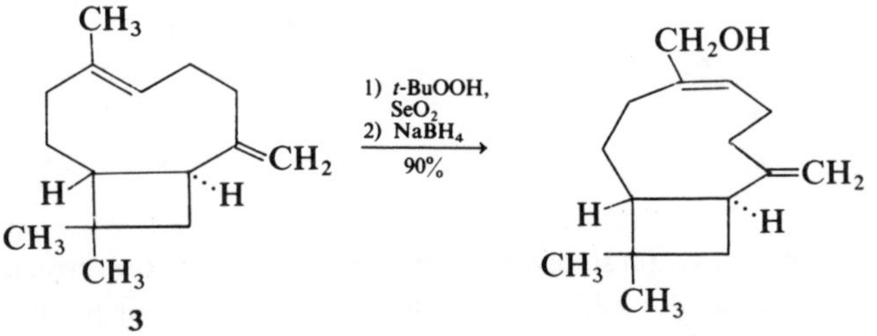

1 → (*t*-BuOOH, SeO_2) → **2** (80%) (CH_2OH) + **1** (20%)

3 → (1) *t*-BuOOH, SeO_2; 2) $NaBH_4$; 90%) → **4**

[1] B. R. Chhabra, K. Hayano, T. Ohtsuka, H. Shirahama, and T. Matsumoto, *Chem. Letters*, 1703 (1981).

_t_-Butyl hydroperoxide–Vanadyl acetoacetate, 2, 287; **4,** 346; **5,** 75–76; **7,** 43–44; **9,** 81–82; **10,** 66.

Epoxidation of β- or γ-trimethylsilyloxy allylic alcohols. The stereoselectivity of epoxidation of these allylic alcohols (followed by desilylation) can be controlled by substitution with a trimethylsilyl group. A β-silyl substituted allylic alcohol is converted mainly into an *erythro*-epoxy alcohol, whereas a γ-silyl substituent favors formation of a *threo*-epoxy alcohol. The stereoselectivity is usually the opposite to that obtained with *m*-chloroperbenzoic acid.[1]

Example:

$$CH_2{=}C(Si(CH_3)_3)CH(OH)C_4H_9\text{-}n \xrightarrow[2)\ CsF]{1)\ Epoxidation} \underset{erythro}{CH_2\text{—}CHCH(OH)C_4H_9\text{-}n \ (\text{epoxide})} + \underset{threo}{CH_2\text{—}CHCH(OH)C_4H_9\text{-}n \ (\text{epoxide})}$$

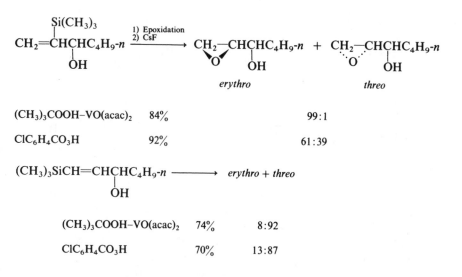

$(CH_3)_3COOH$–$VO(acac)_2$	84%	99:1
$ClC_6H_4CO_3H$	92%	61:39

$$(CH_3)_3SiCH{=}CHCH(OH)C_4H_9\text{-}n \longrightarrow erythro + threo$$

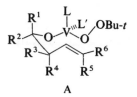

$(CH_3)_3COOH$–$VO(acac)_2$	74%	8:92
$ClC_6H_4CO_3H$	70%	13:87

Stereoselective epoxidation. A detailed study of epoxidation of homoallylic alcohols with this system indicates that the direction and degree of stereoselectivity can be predicted from a vanadate ester transition state with the chair comformation A. For example, the selectivity is > 100:1 when R^1 and R^4 = H and R^3 and R^5 = alkyl, since 1,3-interactions are minimal. R^1 can also be a methyl group, but the reaction is slowed. When R^1 = isopropyl and R^3 = methyl, severe 1,3-interactions in both chair forms result in low asymmetric induction (2:1 selectivity).[2]

[Structure A: chair vanadate ester transition state with substituents R^1–R^6, L, L′, OBu-*t*]

A

$(CH_3)_3COOH$, $VO(acac)_2$ 97% >400:1

91% 2.1:1

95% 211:1

81% 15.9:1

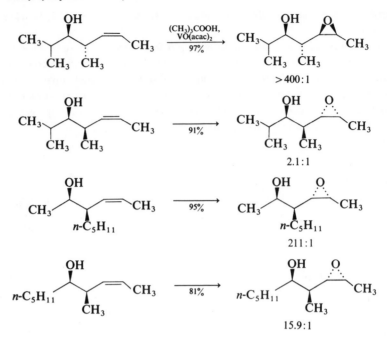

The final step in Danishefsky's synthesis of the sesquiterpene coriolin (2)[3] required bis epoxidation of **1**. All attempts to effect this reaction in one step resulted in a mixture of the desired diepoxide (**2**) and epicoriolin (**3**) in about equal parts under best conditions. However, a two-step epoxidation procedure (**1**→**4** and **5**→**6**) resulted in a stereoselective synthesis of **2**.

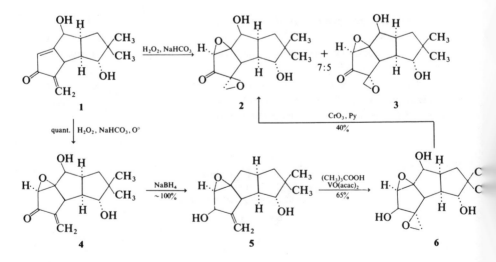

H_2O_2, $NaHCO_3$ → **2** + **3** (7:5)

1 → **4**: quant. H_2O_2, $NaHCO_3$, 0°

4 → **5**: $NaBH_4$, ~100%

5 → **6**: $(CH_3)_3COOH$, $VO(acac)_2$, 65%

6 → **2**: CrO_3, Py, 40%

[1] H. Tomioka, T. Suzuki, K. Oshima, and H. Nozaki, *Tetrahedron Letters*, **23**, 3387 (1982).
[2] E. D. Mihelich, K. Daniels, and D. J. Eickhoff, *Am. Soc.*, **103**, 7690 (1981).
[3] S. Danishefsky, R. Zamboni, M. Kahn, and S. J. Etheredge, *Am. Soc.*, **103**, 3460 (1981).

_t_-Butyl hydroperoxide–Vanadyl acetylacetonate–Titanium (IV) isopropoxide.

Diastereoselective epoxidation of an allylic alcohol. Epoxidation of either a *cis*- or *trans*-allylic alcohol substituted in the γ-position by an alkoxy function by either *m*-chloroperbenzoic acid or *t*-butyl hydroperoxide/VO(acac)$_2$ results mainly in the *anti*-epoxide (**9**, 109). Epoxidation of the allylic alcohol **1** with *m*-chloroperbenzoic acid conforms to this pattern, but epoxidation with *t*-butylhydroperoxide and VO(acac)$_2$ mediated by titanium (IV) isopropoxide favors formation of the *syn*-epoxide by a factor of 10:1. The methyl group attached to the double bond is necessary for this unusual *syn*-selectivity; when it is lacking, epoxidation with *t*-butyl hydroperoxide/Ti(IV) is *anti*-selective, but less so than epoxidation with the peracid.[1]

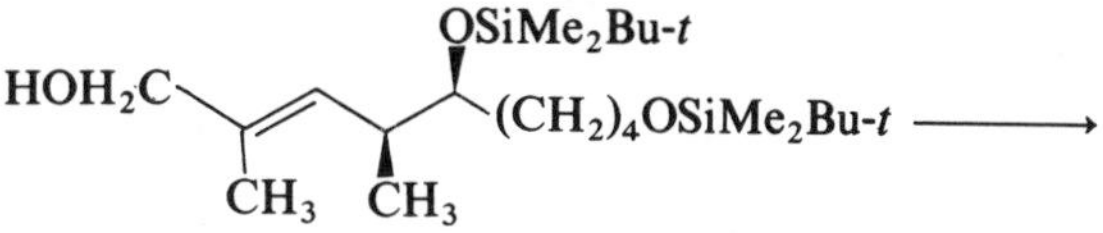

1

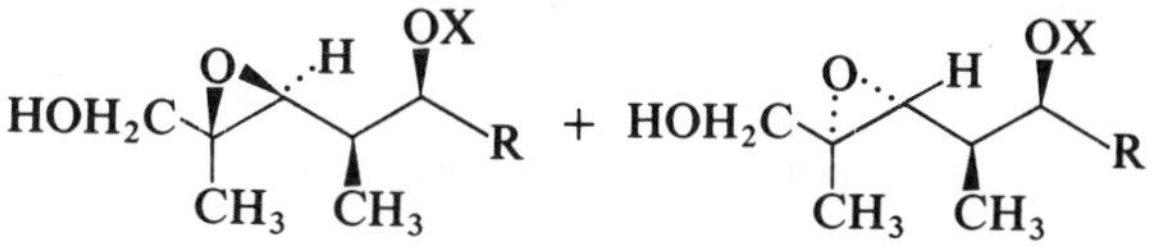

2 **3**

$(CH_3)_3COOH$-Ti(O-*i*-Pr)$_4$-VO(acac)$_2$	10:1
n-Cl-$C_6H_4CO_3H$	1:>25

[1] M. Isobe, M. Kitamura, S. Mio, and T. Goto, *Tetrahedron Letters*, **23**, 221 (1982).

_t_-Butyl isocyanide, 2, 50–51; **9,** 82; **10,** 67.

γ-Butyrolactones. In the presence of diethylaluminum chloride, *t*-butyl isocyanide undergoes cycloaddition to α,β-enones to give iminolactones, which can be converted into γ-butyrolactones (equation I).[1]

(I)

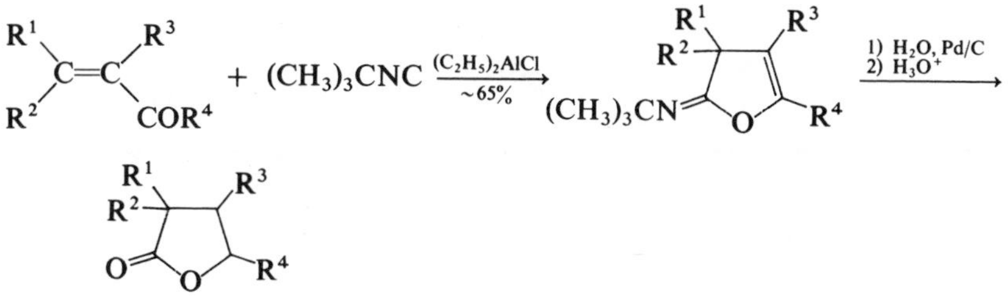

Conjugate hydrocyanation.[2] In the presence of $TiCl_4$, the isocyanide undergoes conjugate hydrocyanation to α,β-enones with elimination of isobutene. In this reaction, the reagent functions as a masked hydrogen cyanide. $TiCl_4$ does not promote the same reaction with α,β-unsaturated aldehydes or esters, but conjugate addition to these substrates is possible in moderate yield with $C_2H_5AlCl_2$ (enals) or with $AlCl_3$ (esters).

Examples:

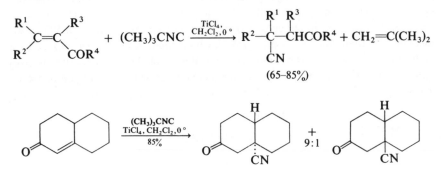

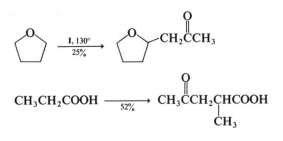

$$R^1R^2C{=}CR^3COR^4 + (CH_3)_3CNC \xrightarrow{TiCl_4,\ CH_2Cl_2,\ 0^\circ} R^2\text{—}C(R^1)(CN)\text{—}CH(R^3)COR^4 + CH_2{=}C(CH_3)_2$$

(65–85%)

(CH₃)₃CNC, TiCl₄, CH₂Cl₂, 0°, 85%; products (H, CN) 9:1

¹ Y. Ito, H. Kato, and T. Saegusa, *J. Org.*, **47**, 741 (1982).
² Y. Ito, H. Kato, H. Imai, and T. Saegusa, *Am. Soc.*, **104**, 6449 (1982).

O,O-*t*-Butyl O-isopropenyl peroxycarbonate,

$$CH_2{=}C(CH_3)OC({=}O)OOC(CH_3)_3 \quad \textbf{(1)}.$$

This peroxide is prepared by reaction of *t*-butyl hydroperoxide with isopropenyl carbonochloridate.

Acetonylation.[1] When heated, **1** is converted into a radical derived from acetone, which can acetonylate a variety of substrates (ketones, ethers, carboxylic acids, etc.) by a free radical mechanism.

Examples:

Tetrahydrofuran $\xrightarrow[25\%]{\mathbf{1},\ 130^\circ}$ 2-(CH_2COCH_3)tetrahydrofuran

$$CH_3CH_2COOH \xrightarrow[52\%]{} CH_3COCH_2CH(CH_3)COOH$$

$$CH_3CH_2C{\equiv}N \xrightarrow[32\%]{} CH_3\overset{O}{\overset{\|}{C}}CH_2\underset{CH_3}{\underset{|}{C}}HC{\equiv}N$$

[1] R. Jaouhari, B. Maillard, C. Filliatre, J.-J. Villenave, *Synthesis*, 760 (1982).

***n*-Butyllithium, 1**, 95–96; **2**, 51–53; **4**, 60–63; **5**, 78; **6**, 85–91; **7**, 45–47; **8**, 65–66; **9**, 83–87; **10**, 68–71.

[2,3]-Wittig rearrangement. The [2,3]-Wittig rearrangement of bis-allylic ethers with an alkyllithium is a useful route to 1,5-diene-3-ols.[1] The Wittig rearrangement of unsymmetrical bis-allylic ethers has now been examined.[2] A β-alkyl substituent has little effect on the site of lithiation; both α- and γ-substituents depress lithiation, but a γ-substituent has a greater effect. Thus, lithiation occurs on the less substituted allylic group. In cases where a new double bond is formed, the (E)-isomer is obtained with high selectivity. When a new chiral center is generated, a (Z)-substrate is converted mainly to the *erythro*-isomer and an (E)-substrate shows moderate *threo*-preference. Internal asymmetric induction has not been reported in other [2,3]-sigmatropic rearrangements.

Examples:

$$CH_2{=}CHCH_2O\underset{CH_3}{\underset{|}{C}}HCH{=}CH_2 \xrightarrow[79\%]{n\text{-BuLi}} CH_2{=}CH\underset{OH}{\underset{|}{C}}HCH_2CH{=}CHCH_3$$

(>95% E)

$$CH_2{=}CHCH_2OCH_2CH{=}CHCH_3 \longrightarrow CH_2{=}CH\underset{OH}{\underset{|}{C}}H{-}\overset{CH_3}{\overset{|}{C}}HCH{=}CH_2$$

E/Z = 93:7	81%	*threo/erythro* = 79:21
E/Z = 5:95	88%	*threo/erythro* = 12:88

The process has been used for a synthesis of (E)-6-nonene-2-one (**1**), a precursor of *exo*-brevicomin (**2**, equation I).[3]

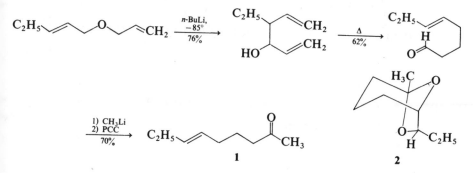

(I) C_2H_5–CH=CH–CH$_2$–O–CH$_2$–CH=CH_2 $\xrightarrow[76\%]{n\text{-BuLi},\ -85^\circ}$ [C_2H_5, HO, CH_2, CH_2 diene alcohol] $\xrightarrow[62\%]{\Delta}$ [C_2H_5, H, O aldehyde] $\xrightarrow[70\%]{1)\ CH_3Li\quad 2)\ PCC}$ C_2H_5–CH=CH–CH$_2$CH$_2$CH$_2$–CO–CH_3 (**1**)

2 (H_3C, O, O, H, C_2H_5)

Dimethylcarbene.[4] The usual methods for generation of carbenoids lead to poor results in the case of dimethylcarbene. The most satisfactory method involves generation in the presence of an alkene by α-elimination with *n*-butyllithium from 2,2-dibromopropane (which must be completely free from acetone or 2-bromoacetone), used after a detailed study of cyclopropanation of tetramethylethylene (equation I). *n*-Butyllithium is superior to methyl- or isopropyllithium. Addition of TMEDA or $KOC(CH_3)_3$ completely supresses cyclopropanation. Yields are improved by a lower reaction temperature. In all experiments, 35–45% of the alkene is recovered. Yields decrease with a decrease of alkyl substituents on the double bond; thus the yield is only 9% in the case of 1-dodecene. No cyclopropane was obtained from 2-norbornene.

$$\text{(I)}\ (CH_3)_2CBr_2 \xrightarrow[\text{ether, }-70^\circ]{n\text{-BuLi,}} \left[(CH_3)_2CBrLi \xrightarrow{-\text{LiBr}} (CH_3)_2C:\right] \xrightarrow[46\%]{(CH_3)_2C{=}C(CH_3)_2}$$

(hexamethylcyclopropane: cyclopropane ring bearing two CH_3 groups on each carbon)

Metalation of unsymmetrical imines. Pioneering studies on the metalation and subsequent alkylation of unsymmetrical imines indicated that the reaction occurs predominantly at the less substituted α-position.[5] This pattern has since been observed generally with lithium diethylamide, LDA, and ethylmagnesium bromide. Recent studies[6] indicate that the site of alkylation is independent of the alkylating group but is dependent on the substituent on the imine and particularly on the basicity of the base. Butyllithium (*n*-, *sec*-, and *t*-) can abstract a proton from the more substituted α-carbon of the acyclic imine **1** to some extent. In the case of the cyclic imine **2**, alkylation at the more substituted position is actually the main reaction. However, only substitution at the less substituted position of the dimethylhydrazone of 2-methylcyclohexanone is observed with either LDA or *sec*-butyllithium (**7**, 126–128).

$$\underset{\mathbf{1}}{CH_3(CH_2)_4C(=N\text{-}C_6H_{11}\text{-}c)CH_3} \xrightarrow[49\%]{\text{1) }sec\text{-BuLi; 2) }C_6H_5CH_2Cl} CH_3(CH_2)_4C(=O)(CH_2)_2C_6H_5 + CH_3(CH_2)_3CH(CH_2C_6H_5)C(=O)CH_3 \quad 74:26$$

$$\underset{\mathbf{2}}{\text{2-methyl-1-}(N\text{-}C_6H_{11}\text{-}c)\text{iminocyclohexane}} \xrightarrow[53\%]{\text{1) }n\text{-BuLi; 2) }C_6H_5CH_2Cl} \text{2-}CH_3\text{-6-}CH_2C_6H_5\text{-cyclohexanone} + \text{2-}CH_3\text{-2-}CH_2C_6H_5\text{-cyclohexanone} \quad 13:87$$

***Dimerization of pentene-3-yne-1* (1)** (*cf.*, **10**, 72).[7] Treatment of **1** with 2 equiv. of *n*-butyllithium in THF/hexane at 0° results in the dimer (**2**) in ~ 70% yield. The reaction is regiospecific and stereospecific. Either (E)- or (Z)-**1** is converted into (Z)-**2**. The addition of **1** to vinylacetylene in the presence of *n*-butyllithium gives (Z)-$HC{\equiv}CCH{=}CH(CH_2)_3C{\equiv}CH$ in 65% yield.

$$\underset{\mathbf{1}}{HC{\equiv}CCH{=}CHCH_3} \xrightarrow{2\,n\text{-BuLi}} \underset{\mathbf{a}}{\left[LiC{\equiv}CCH{=}CHCH_2Li\right]} \xrightarrow[\sim 70\%]{\mathbf{1}}$$

$$\underset{\mathbf{2}}{HC{\equiv}CCH\overset{(Z)}{=}CHCH_2\underset{\displaystyle CH_3}{\underset{|}{C}}HCH_2C{\equiv}CH}$$

[1] G. Wittig, *Experientia*, **14**, 389 (1958); R. W. Hoffmann, *Angew. Chem. Int. Ed.*, **18**, 563 (1979).
[2] T. Nakai, K. Mikami, S. Taya, and Y. Fujita, *Am. Soc.*, **103**, 6492 (1981).
[3] K. Mikami and T. Nakai, *Chem. Letters*, 1349 (1982).
[4] P. Fischer and G. Schaefer, *Angew. Chem. Int. Ed.*, **20**, 863 (1981).
[5] G. Stork and S. Dowd, *Am. Soc.*, **85**, 2178 (1963); G. Wittig, H. D. Frommeld, and P. Suchanek, *Angew. Chem. Int. Ed.*, **2**, 1683 (1963).
[6] A. Hosomi, Y. Araki, and H. Sakurai, *Am. Soc.*, **104**, 2081 (1982).
[7] L. Brandsma, H. D. Verkruÿsee, and H. Hommes, *J.C.S. Chem. Comm.*, 1214 (1982).

_n_-Butyllithium–Potassium _t_-butoxide, 5, 552; **8**, 67; **9**, 87; **10**, 72–73.

Butatrienylamines. Base-catalyzed rearrangement of enynamines (**1**) results in the isomers **3**. The intermediate butatrienylamines (**2**) in the isomerization can be isolated by deprotonation with *n*-BuLi and $KOC(CH_3)_3$ (1 : 1) followed by protonation with *t*-butyl alcohol (equation I).[1]

$$\text{(I)}\quad \underset{\mathbf{1}}{CH_2{=}CH{-}C{\equiv}CNR_2} \xrightarrow[66\text{–}90\%]{\substack{1)\ n\text{-BuLi, KOC(CH}_3)_3,\ -100^\circ \\ 2)\ \text{HOC(CH}_3)_3,\ -90 \to -25^\circ}}$$

$$\underset{\mathbf{2}}{CH_2{=}C{=}C{=}CHNR_2} \xrightarrow{\substack{KOC(CH_3)_3, \\ DMSO}} \underset{\mathbf{3}}{CH{\equiv}C{-}CH{=}CHNR_2}$$

[1] L. Brandsma, P. E. van Rijn, H. D. Verkruÿsee, and P. von R. Schleyer, *Angew. Chem. Int. Ed.*, **21**, 862 (1982).

_t_-Butyllithium, 1, 96–97; **5**, 79–80; **7**, 47; **8**, 70–72; **9**, 89; **10**, 76–77.

α-Keto dianions (**10**, 77). These dianions, prepared by reaction of bromo enol acetates with methyllithium and then with *t*-butyllithium, undergo an aldol-type reaction with aldehydes and ketones, even with very hindered ones. This addition is often possible when the corresponding monoanion fails to react. However, these α-keto dianions convert simple methyl ketones into the enolate.[1]

Examples:

$$C_6H_5C(OLi){=}CHLi + \text{2,2,6,6-tetramethylcyclohexanone} \xrightarrow{62\%} \text{1-}(C_6H_5COCH_2)\text{-2,2,6,6-tetramethylcyclohexanol}$$

$$\text{1-lithio-2-(lithiooxy)cyclohexene} + \text{2,4,6-}(CH_3)_3C_6H_2CHO \xrightarrow{58\%} \text{2-(2,4,6-trimethylbenzylidene)cyclohexanone}$$

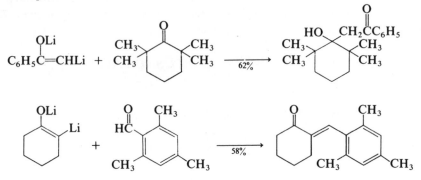

Alkynolate anions.[2] Treatment of the enolate anion **1** of α-chloroacetophenone with *t*-butyllithium (1.5–2.1 equiv.)[3] in THF involves deprotonation to the dianion **a** followed by loss of Cl^- and migration of the group on carbon to give the alkynolate anion **b**. The rearrangement was established by use of ^{13}C. The anion **b** reacts with benzaldehyde to form the *β*-lactone **c**, which can be isolated as such or as the product **2** of hydrolysis. Quenching **b** with methanol affords the ester $C_6H_5CH_2COOCH_3$ (57% yield).

$$\underset{\mathbf{1}}{C_6H_5C(OLi){=}CHCl} \xrightarrow[\text{THF, }-35^\circ]{t\text{-BuLi}} \left[\underset{\mathbf{a}}{C_6H_5C(OLi){=}CClLi} \xrightarrow{-Cl^-} \underset{\mathbf{b}}{LiOC{\equiv}CC_6H_5}\right] \xrightarrow{C_6H_5CHO,\ -78^\circ}$$

$$\left[\underset{\mathbf{c}}{\text{3,4-diphenyl-2-oxetanone}}\right] \xrightarrow[85\%]{H_3O^+,\ 25^\circ} \underset{\mathbf{2}}{(H)(C_6H_5)C{=}C(COOH)(C_6H_5)}$$

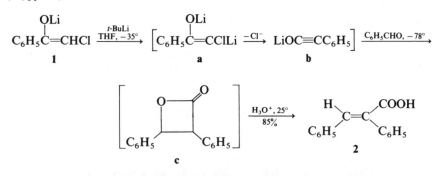

This rearrangement can be used to effect the equivalent of the Arndt-Eistert synthesis[4] by reaction of an ester with dibromomethyllithium (**5**, 403; **6**, 162–163) in the presence of a lithium dialkylamide and then with *t*-butyllithium (equation I).

$$\text{(I)}\quad C_6H_{11}COOC_2H_5 \xrightarrow[\text{LiN}(C_6H_{11})_2]{\text{LiCHBr}_2,} \left[RC(OLi){=}CBr_2 \xrightarrow{t\text{-BuLi}} RC{\equiv}C{-}OLi\right] \xrightarrow[63\%]{C_2H_5OH} C_6H_{11}CH_2COOC_2H_5$$

This new rearrangement is applicable in general to α-bromo-, α-chloro-, and α,α-dibromoketone enolates. It is formally isoelectronic with the Hofmann rearrangement of α-haloamides.[5]

Cycloalkanes.[6] Treatment of α,ω-diiodides with *t*-BuLi in pentane–ether at −23° provides an attractive route to three-, four-, and five-membered cycloalkanes. The corresponding dibromides do not undergo the same clean cyclization. The reaction is not useful for preparation of cycloalkanes containing more than five carbon atoms.

Examples:

$$\text{1,2-bis}(CH_2I)\text{cyclohexane} \xrightarrow[97\%]{t\text{-BuLi}} \text{bicyclo[4.2.0]octane (H, H)}$$

$$I(CH_2)_2C(CH_3)_2(CH_2)_2I \xrightarrow{93\%} \text{1,1-dimethylcyclopentane}$$

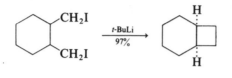

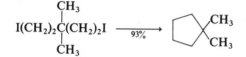

[1] C. J. Kowalski and K. W. Fields, *Am. Soc.*, **104**, 1777 (1982).
[2] C. J. Kowalski and K. W. Fields, *ibid.*, **104**, 321 (1982).
[3] *n*-BuLi can be used, but results are not easily reproducible.
[4] W. E. Bachmann and W. S. Strewe, *Org. Reactions*, **1**, 38 (1942).
[5] E. S. Wallis and J. F. Lane, *ibid.*, **3**, 267 (1949).
[6] W. F. Bailey and R. P. Gagnier, *Tetrahedron Letters*, **23**, 5123 (1982).

N^1-*t*-Butyl-N^2,N^2,N^3,N^3-tetramethylguanidine, (1).

Preparation[1,2]:

$$(CH_3)_2NC(=O)N(CH_3)_2 \xrightarrow[2)\ (CH_3)_3CNH_2]{1)\ COCl_2} (CH_3)_2NC(=NC(CH_3)_3)N(CH_3)_2 \quad \mathbf{1}$$

Synthetic uses. Barton *et al.*[2] have prepared a series of hindered pentaalkylguanidines such as **1**. The most hindered ones containing isopropyl groups are prepared from tetraalkylthiourea rather than the corresponding urea precursor. These strong bases are useful for several purposes. Hindered, tertiary carboxylic acids are converted into esters by alkyl iodides in the presence of these bases. The tosylate of 3β-cholestanol is converted into a mixture of Δ^2- and Δ^3-cholestenes (80–85% yield) when heated at 120° with these bases. Use of collidine as the base and a temperature of 170° in this elimination gives the cholestenes in 63% yield. The yield with DBN is only

31%. These bases are useful for C-alkylation of ethyl acetoacetate in high yield, because they are alkylated very slowly.

[1] H. Kessler, D. Leibfritz, and C. Burk, *Tetrahedron*, **26**, 1805 (1970).
[2] D. H. R. Barton, J. D. Elliott, and S. D. Géro, *J.C.S. Chem. Comm.*, 1136 (1981).

(R)-(+)-*t*-Butyl (*p*-tolylsulfinyl)acetate, 10, 405–407.

Chiral lactones. The aldol type condensation of this reagent with an aldehyde has been used to synthesize two chiral five- and six-membered lactones in >80% ee. The synthesis of the six-membered lactone (R)-(+)-δ-*n*-hexadecanolactone (**5**), a pheromone of the Oriental hornet, from the *t*-butyl ester (**1**) of the sulfoxide is formulated in equation (I).[1]

(I) $p\text{-Tolyl}-S(\rightarrow O)-CH_2COOC(CH_3)_3$ (**1**) $+ n\text{-}C_{11}H_{23}CHO \xrightarrow[\text{quant.}]{t\text{-BuMgBr, THF, }-78^\circ}$ $n\text{-}C_{11}H_{23}CH(OH)CH(S(\rightarrow O)\text{-}p\text{-Tolyl})COOC(CH_3)_3$ (**2**) $\xrightarrow[80\%]{\text{Al/Hg}}$ $n\text{-}C_{11}H_{23}CH(OH)CH_2COOC(CH_3)_3$ (**3**) $\xrightarrow{\text{Several steps}}$

$n\text{-}C_{11}H_{23}CH(OH)(CH_2)_3COOC(CH_3)_3$ (**4**) $\xrightarrow[\sim 40\%\text{ from } \mathbf{3}]{\text{TsOH, }C_6H_6}$ **5** (>80%ee)

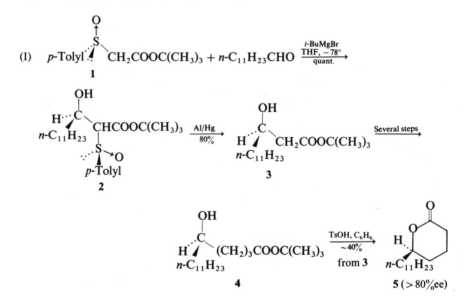

[1] G. Solladié and F. Matloubi-Moghadam, *J. Org.*, **47**, 91 (1982).

C

Calcium hypochlorite, $Ca(OCl)_2$. Mol. wt. 142.98. Suppliers: J. T. Baker, Fischer.

Oxidations. Calcium hypochlorite is comparable in efficiency to sodium hypochlorite (**10**, 365) for oxidation of secondary alcohols to ketones and has the advantage that it is commercially available as an inexpensive, stable solid, rather than in solution.[1] It is also useful for oxidation of aliphatic aldehydes to the corresponding acids; yields are generally about 75%. Benzaldehydes substituted by electron-withdrawing groups are oxidized by the reagent in high yield, but nuclear chlorination is the preferred reaction with benzaldehydes substituted by electron-donating groups.[2]

The reagent is effective for oxidative cleavage of α-glycols to aldehydes, of α-hydroxy ketones to an aldehyde and a carboxylic acid, and of α-diketones and α-keto acids to acids.[3]

Aqueous acetonitrile is generally used as solvent, with added acetic acid or methylene chloride for solubility purposes.

[1] S. O. Nwaukwa and P. M. Keehn, *Tetrahedron Letters*, **23**, 35 (1982).
[2] *Idem, ibid.*, **23**, 3131 (1982).
[3] *Idem, ibid.*, **23**, 3135 (1982).

***d*-10-Camphorsulfonic acid, 1,** 108–109; **2,** 58–59; **4,** 68; **6,** 611.

Cyclization of an α,β-epoxy ketone.[1] Treatment of the epoxy ketone **1** with *d*-10-camphorsulfonic acid in refluxing xylene produces the phenanthrene **2** as the only

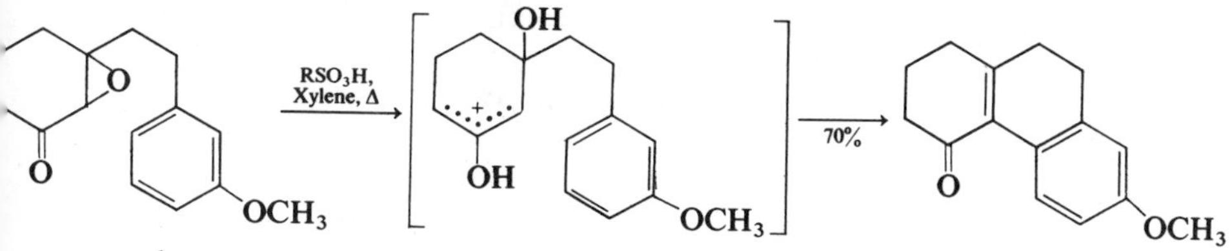

O OCH₃ **1** RSO_3H, Xylene, Δ OH OH OCH₃ **a** 70% O OCH₃ **2**

H_3C O O OCH₃ **3** $BF_3\cdot(C_2H_5)_2O$ H_3C OCH₃ O OH **4** (45%) + CH_3O H_3C O OH **5** (45%)

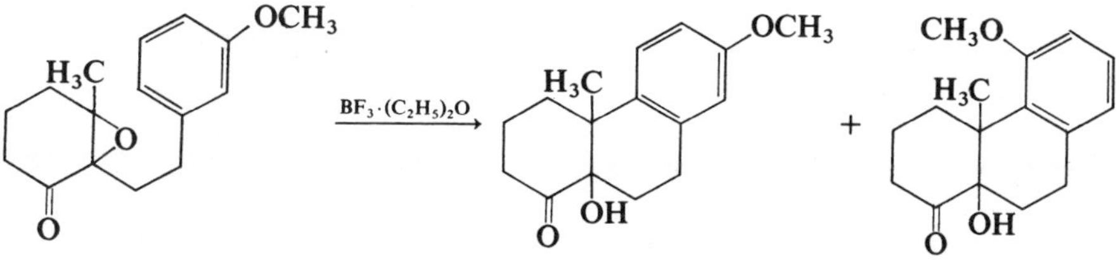

product of cyclization. The reaction proceeds through the hydroxyallyl cation **a**, which functions as an α-carbonyl cation. Cyclization of α,β-epoxy ketones with Lewis acids proceeds through a β-carbonyl cation, as in the cyclization of **3** to a mixture of **4** and **5**.[2] Cyclization of **1** with BF_3 etherate is not satisfactory.

[1] B. M. Trost and E. Jurayama, *Tetrahedron Letters*, **23**, 1047 (1982).
[2] J. K. Sutherland, *Chem. Soc. Rev.*, **9**, 265 (1980); J. Amupitan and J. K. Sutherland, *J.C.S. Chem. Comm.*, 398 (1980).

[(−)-Camphor-10-ylsulfonyl]-3-aryloxaziridines (1). Davis *et al.*[1,2] have prepared a series of these reagents by the general equation (I).

(I) CH_3 CH_3 CH_2—SO_2NH_2 + $ArCH(OC_2H_5)_2$ $\xrightarrow[60\text{–}70\%]{130\text{–}170°}$ R^*SO_2N=C(Ar)H

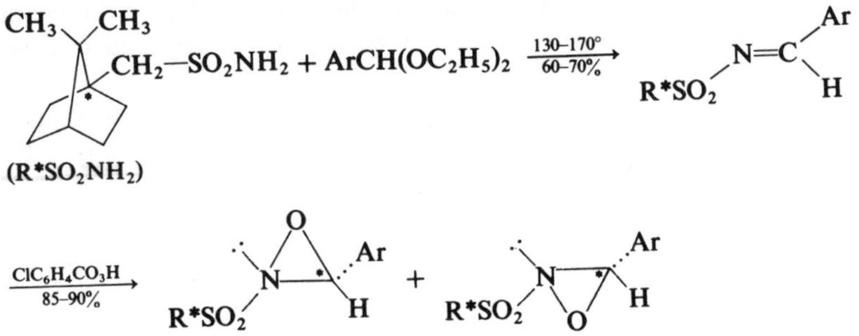

($R^*SO_2NH_2$)

$\xrightarrow[85\text{–}90\%]{ClC_6H_4CO_3H}$ (R,R)-**1** + (S,S)-**1**

The pure isomers have been obtained by crystallization when Ar = 2-chloro-5-nitrophenyl. The configuration has been determined by X-ray analysis of the corresponding reagent prepared from *d*-α-bromo-π-camphorsulfonamide.

Asymmetric oxidations. These reagents selectively oxidize sulfides and disulfides to sulfoxides and thiosulfinates [RS(O)SR], respectively. They also epoxidize olefins in a *syn*-stereospecific manner.[3]

The enolate of **2** can be oxidized by pure (S,S)-**1** to kjellmanianone (**3**), a cyclopentenoid antibiotic, in 44% chemical yield and 36.5% optical yield.[4]

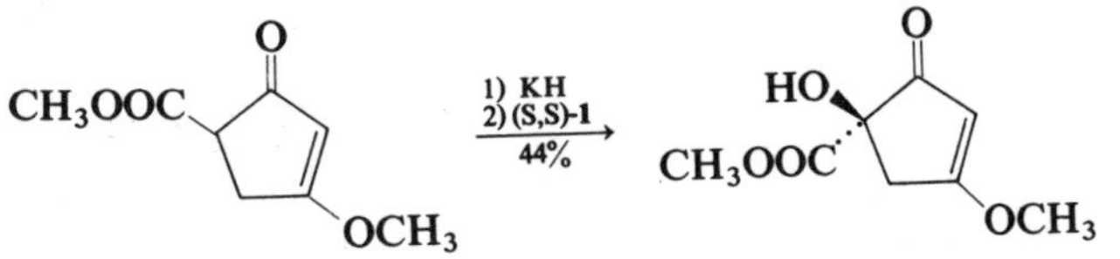

2 $\xrightarrow[44\%]{1)\ KH;\ 2)\ (S,S)\text{-}1}$ **3** (36.5%ee)

[1] F. A. Davis, R. H. Jenkins, Jr., S. Q. A. Rizvi, and T. W. Panunto, *J.C.S. Chem. Comm.*, 600, (1979).
[2] F. A. Davis, R. H. Jenkins, Jr., S. B. Awad, O. D. Stringer, W. H. Hanson, and J. Galloy, *Am. Soc.*, **104**, 5412 (1982).

[3] F. A. Davis, N. F. Abdul-Mali, S. B. Awad, and M. E. J. Harakal, *J.C.S. Chem. Comm.*, 917 (1981).
[4] D. Boschelli, A. B. Smith, III, O. D. Stringer, R. H. Jenkins, Jr., and F. A. Davis, *Tetrahedron Letters*, **22**, 4385 (1981).

***d*- and *l*-10-Camphorsulfonyl chloride, 1**, 109. Both *d*- and *l*-camphorsulfonic acid are commercially available.

2-Arylalkanoic acids. A new route to these acids is formulated in equation (I). The reaction involves a 1,2-shift of the aryl group. Overall yields are usually 65% and are higher when the aryl group is substituted by electron-donating groups.[1]

$$\text{(I)}\quad \mathrm{ArCCH_2CH_3}\ (\mathrm{C{=}O}) \xrightarrow[\text{dioxane}]{\mathrm{PyH\cdot Br_3}} \mathrm{Ar\overset{O\ Br}{C{=}C}HCH_3}\ \xrightarrow[\text{2) }\mathrm{RSO_2Cl,\ Py}]{\text{1) }\mathrm{CH_3ONa,\ CH_3OH}}$$

$$\mathrm{ArC(OCH_3)_2{-}CH(OSO_2R)CH_3} \xrightarrow{\mathrm{CaCO_3,\ DMF,\ H_2O,\ \Delta}} \mathrm{ArCH(CH_3)COOCH_3}$$

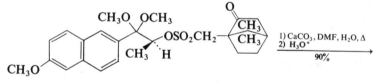

The reaction can be used to prepare optically active 2-arylalkanoic acids by use of *d*- or *l*-camphorsulfonyl chloride. The 1,2-rearrangement proceeds with complete inversion of the carbon at the 2-position. Thus, noproxen (**2**) was obtained in 90% yield by rearrangement of **1** followed by acid hydrolysis.[2]

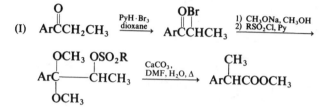

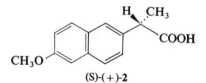

CH₃O-naphthyl–C(OCH₃)₂–CH(CH₃)(H)–OSO₂CH₂–(camphor, CH₃, CH₃, O) $\xrightarrow[90\%]{\text{1) }\mathrm{CaCO_3,\ DMF,\ H_2O,\ \Delta;}\ \text{2) }\mathrm{H_3O^+}}$

S-(−)-**1**

CH₃O-naphthyl–CH(CH₃)COOH (H, CH₃, COOH)

(S)-(+)-**2**

[1] G. Tsuchihashi, K. Kitajima, and S. Mitamura, *Tetrahedron Letters*, **22**, 4305 (1981).
[2] G. Tsuchihashi, S. Mitamura, K. Kitajima, and K. Kobayashi, *ibid.*, **23**, 5427 (1982).

Carboethoxyformimidate, $\mathrm{C_2H_5O{-}C({=}NH){-}C({=}O){-}OC_2H_5}$ (**1**). The reagent is prepared by reaction of ethyl cyanoformate in ethanol with hydrogen chloride. The resulting salt is decomposed with K_2CO_3.[1]

Heterocycles.[2] The reagent can be used for synthesis of a variety of mono- and bicyclic heterocycles. The imidate carbon is usually the preferred center for nucleophilic attack.

Examples:

$$CH_3CONHNH_2 \xrightarrow[70\%]{1} CH_3CONHN{=}C(NH_2)COOC_2H_5 \xrightarrow[52\%]{\Delta} \text{5-methyl-1,2,4-triazole-3-carboxylic acid ethyl ester (}CH_3\text{, }COOC_2H_5\text{)}$$

$$\text{o-phenylenediamine } (NH_2, NH_2) \xrightarrow[99\%]{1} \text{3-amino-2(1H)-quinoxalinone (H, N, O, N, } NH_2\text{)}$$

$$\text{o-aminothiophenol } (SH, NH_2) \xrightarrow{1} \text{benzothiazole-2-}COOC_2H_5\ (87\%) + \text{3-amino-2H-1,4-benzothiazin-2-one (S, O, N, } NH_2\text{)}\ (8\%)$$

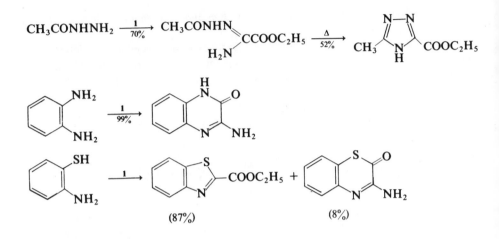

[1] J. W. Cornforth and R. H. Cornforth, *J. Chem. Soc.*, 96 (1947).

[2] A. McKillop, A. Henderson, P. S. Ray, C. Avendano, and E. G. Molinero, *Tetrahedron Letters*, **23**, 3357 (1982).

(Carbomethoxy)chloroketene, $(Cl)(CH_3OOC)C{=}C{=}O$ **(1).**

Preparation *in situ*:

$$CH_3OOCCH_2COOH \xrightarrow{SOCl_2} CH_3OOCCH_2COCl \xrightarrow{PCl_5} (Cl)(CH_3OOC)CHCOCl \xrightarrow{N(C_2H_5)_3} \mathbf{1}$$

vic-*Alkylation of cyclopentadiene.*[1] Unlike carbomethoxyketene, which does not add to cyclopentadiene, **1** adds to the diene with high stereoselectivity to form **2** (70% yield), in which the ester group has the less stable *exo*-configuration. This product is converted as shown into the *trans*-aldehyde **4** and further into the iodo lactone **5**, a common intermediate to various prostaglandins.

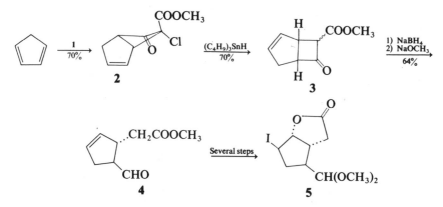

$$\text{cyclopentadiene} \xrightarrow[70\%]{1} \mathbf{2} \xrightarrow[70\%]{(C_4H_9)_3SnH} \mathbf{3} \xrightarrow[64\%]{1)\ NaBH_4;\ 2)\ NaOCH_3} \mathbf{4}\ (CH_2COOCH_3,\ CHO) \xrightarrow{\text{Several steps}} \mathbf{5}\ (I,\ CH(OCH_3)_2)$$

[1] S. Goldstein, P. Vannes, C. Houge, A. M. Frisque-Hesbain, C. Wiaux-Zamar, L. Ghosez, G. Germain, J. P. Declercq, M. Van Meerssche, and J. M. Arrieta, *Am. Soc.*, **103**, 4616 (1981).

Carbon monoxide, 2, 60, 204; **3,** 41–43; **5,** 96; **7,** 53; **8,** 76–77; **9,** 95.

Acyltrimethylsilanes.[1] The first intermediate in the reaction of CO with RLi is RCOLi, but attempts to trap this intermediate with chlorosilanes have not been successful.[2] Acyltrimethylsilanes can be obtained in high yield when the acyllithium is generated in the presence of $ClSiR_3$ (equation I).

$$\text{(I)}\quad R^1Li + CO \xrightarrow{-110^\circ} R^1\overset{O}{\overset{\|}{C}}Li \xrightarrow[\substack{45\text{–}70\% \\ \text{(isolated)}}]{ClSiR_3} R^1\overset{O}{\overset{\|}{C}}SiR_3$$

Limitations are that the method is not useful with aryllithiums or with hindered chlorosilanes such as $ClSi(CH_3)_2C_6H_5$.

[1] D. Seyferth and R. M. Weinstein, *Am. Soc.*, **104**, 5534 (1982).
[2] P. Jutzi and F.-W. Schroder, *J. Organometal. Chem.*, **24**, 1 (1970).

Carbon tetrabromide–Tin(II) fluoride.

2,2,2-Tribromoethanols. CBr_4 and SnF_2 react with aldehydes in DMSO at 25° to form one-carbon homologated 2,2,2-tribromoethanols in moderate to good yield. Presumably the reagents form an unstable adduct, such as Br_3CSnF_2Br, which adds to the carbonyl group. The acetate of the product is hydrolyzed to an α-acetoxy carboxylic acid by $AgNO_3$.[1]

Example:

$$C_6H_5CHO \xrightarrow[78\%]{CBr_4,\ SnF_2,\ DMSO} C_6H_5\underset{OH}{\underset{|}{C}}HCBr_3 \xrightarrow[40\%]{1)\ Ac_2O,\ Py;\ 2)\ AgNO_3,\ H_2O} C_6H_5\underset{OAc}{\underset{|}{C}}HCOOH$$

These reactions convert 2,3-O-isopropylidene-D-glyceraldehyde (**1**) into 2,3-diacetyl-D-erythronolactone (**3**).

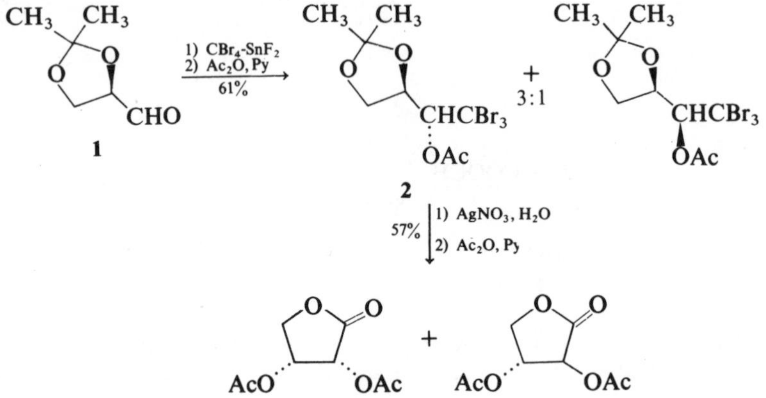

1) CBr_4-SnF_2 2) Ac_2O, Py 61%

1, **2** 3:1

57% 1) $AgNO_3$, H_2O 2) Ac_2O, Py

3 (major product)

[1] T. Mukaiyama, M. Yamaguchi, and J. Kato, *Chem. Letters*, 1505 (1981).

1,1′-Carbonylbis(3-imidazolium) bismethanesulfonate, $O{=}C(N\text{-imidazolium}\text{-}\overset{+}{N}H)_2\ 2CH_3SO_3^-$ **(1).** This salt is prepared by treatment of N,N′-carbonyldiimidazole with methanesulfonic acid (2 equiv.).

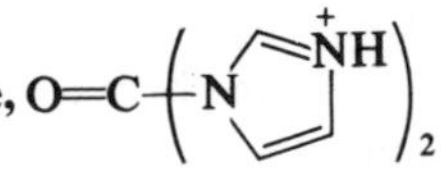

Dehydration of formamides.[1] This reagent converts formamides to isocyanides in 75–90% yield. It was developed particularly for use with chiral substrates that racemize readily in the presence of basic reagents. Thus it converts **2** into the isocyanide **3** with no significant racemization.

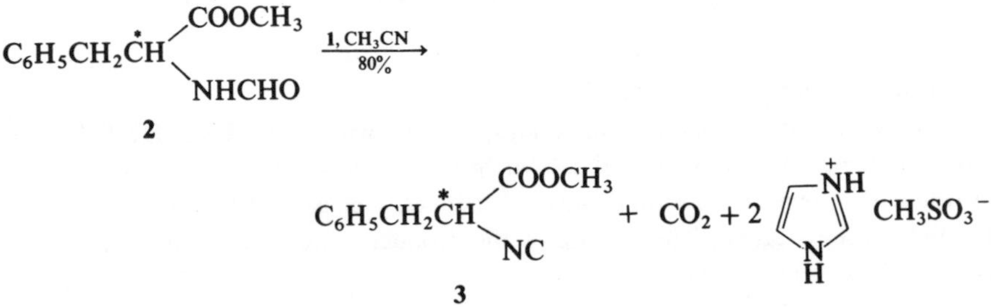

$$C_6H_5CH_2\overset{*}{C}H(COOCH_3)(NHCHO)\ (\mathbf{2}) \xrightarrow[80\%]{\mathbf{1},\ CH_3CN} C_6H_5CH_2\overset{*}{C}H(COOCH_3)(NC)\ (\mathbf{3}) + CO_2 + 2\ \text{imidazolium}\ CH_3SO_3^-$$

[1] G. Giesemann, E. von Hinrichs, and I. Ugi, *J. Chem. Research* (*S*), 79 (1982).

Carbonyl sulfide (Carbon oxysulfide), O=C=S. Mol. wt. 60.07, poisonous gas. Suppliers: K and K, ROC/RIC.

C-Carboxylation of enolates.[1] Carboxylation of potassium enolates generated from silyl enol ethers is not regioselective because of extensive enolate equilibration. Regiospecific C-carboxylation of lithium enolates is possible with carbonyl sulfide in place of carbon dioxide. The product is isolated as the thiol methyl ester. If simple esters are desired, transesterification can be effected with $Hg(OAc)_2$ (**8**, 444). Carboxylation of ketones in this way in the presence of NaH and DMSO is not satisfactory because of competing alkylation of the enolate.[2]

Example:

CH_3 / $OSi(CH_3)_3$ — 1) CH_3Li, DME 2) COS, −78° 3) CH_3I, 79% → CH_3 / O / O ‖ $CSCH_3$

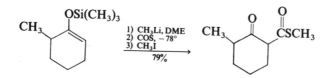

This procedure was used in a recent synthesis of the 11-deoxyanthracycline (**4**) to effect carboxylation of the enolate **1**, which was not possible by other means. The thiol ester (**2**) undergoes efficient cyclization to the tetracyclic diketone (**3**) with copper(I) triflate (**10**, 110).

CH_3O, $Si(CH_3)_3$, Li, CH_3O + O — THF, HMPT → [CH_3O, $Si(CH_3)_3$, CH_3O, O] **1**

1) COS 2) CH_3I, 66% → CH_3O, $Si(CH_3)_3$, O=C, CH_3O, CH_3S, O **2** — CuOTf, 77% → CH_3O, $Si(CH_3)_3$, CH_3O, O, O **3**

73% AgO ↓

O, O, OH, O **4**

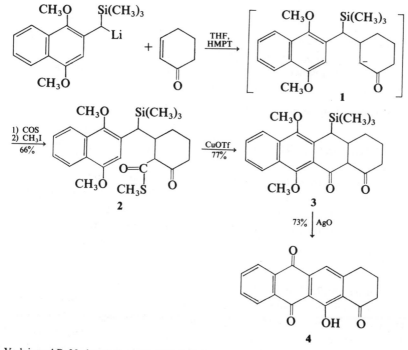

[1] E. Vedejs and B. Nader, *J. Org.*, **47**, 3193 (1982).

[2] C. Demuynck and A. Thuiller, *Bull. Soc.*, 2434 (1969).

Cerium amalgam, Ce/Hg. *Caution*: Ce/Hg burns immediately on contact with air. This amalgam is prepared by treatment of small pieces of the metal with a solution of $HgCl_2$ in C_2H_5OH (2 minutes) followed by several washes with C_2H_5OH under N_2. It is dried in vacuum.

Homoallylic alcohols.[1] Allyl iodide reacts with Ce/Hg to form allylcerium iodide, which reacts with ketones to form homoallylic alcohols in 50–95% yield. A similar reaction is possible with benzyl iodide. Use of unamalgamated Ce results in much lower yields.

Examples:

$$CH_2{=}CHCH_2I + C_6H_5CH{=}CHCOC_6H_5 \xrightarrow[70\%]{Ce/Hg,\ THF,\ 0^\circ} C_6H_5CH{=}CH\overset{\overset{\displaystyle OH}{|}}{\underset{\underset{\displaystyle C_6H_5}{|}}{C}}CH_2CH{=}CH_2$$

$$C_6H_5CH_2I + C_6H_5COCH_3 \xrightarrow[84\%]{} C_6H_5\overset{\overset{\displaystyle OH}{|}}{\underset{\underset{\displaystyle CH_3}{|}}{C}}CH_2C_6H_5$$

[1] T. Imamoto, Y. Hatanaka, Y. Tawarayama, and M. Yokoyama, *Tetrahedron Letters*, **22**, 4987 (1981).

Cerium(IV) ammonium nitrate–Sodium bromate.

Selective oxidation of sec-alcohols.[1] Secondary alcohols can be oxidized selectively in the presence of primary ones by $NaBrO_3$ (1 equiv.) in the presence of CAN or $Ce(SO_4)_2$ (0.1 equiv.) with 80–90% yields.

[1] H. Tomioka, K. Oshima, and H. Nozaki, *Tetrahedron Letters*, **23**, 539 (1982).

Cerium(III) iodide, CeI_3. The iodide is prepared in situ by reaction of Ce(0) with I_2 in THF at $0 \rightarrow 25^\circ$.

Organocerium reagents.[1] Organocerium reagents, obtained by transmetallation of alkyllithiums with CeI_3, undergo facile 1,2-addition to carbonyl compounds at -65°. Competitive enolization, observed with RLi reagents, is not observed, but reduction or reductive coupling can prevail at higher temperatures.

[1] T. Imamoto, T. Kusumoto, and M. Yokoyama, *J.C.S. Chem. Comm.*, 1042 (1982).

Cesium carbonate, 7, 57; **8**, 81; **9**, 99–100.

Imino Diels-Alder reactions.[1] The neurotoxic fungal toxin slaframine (**2**) has been synthesized by an intramolecular imino Diels-Alder reaction. The substrate is obtained from the amide **1** via a "methylol" derivative. These derivatives are best prepared by reaction of the amide with Cs_2CO_3 and paraformaldehyde in dry THF (equation I). When heated they lose acetic acid and undergo cyclization.

(I)

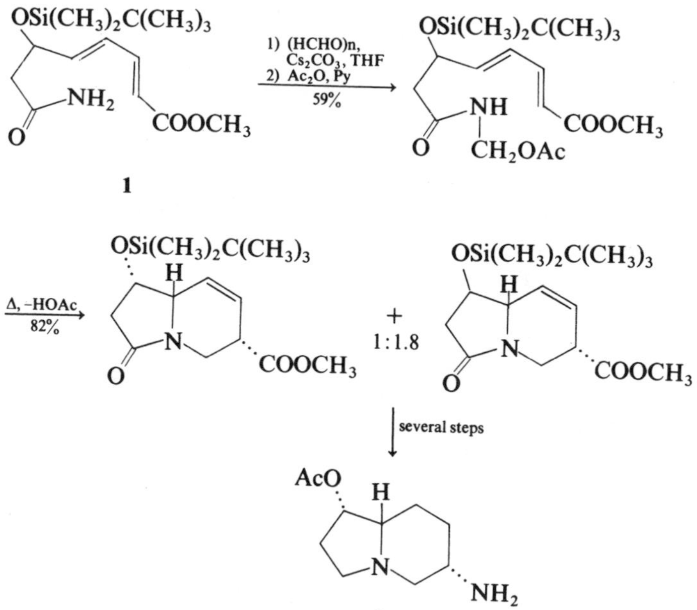
OSi(CH3)2C(CH3)3
NH2
COOCH3
O
1) (HCHO)n, Cs2CO3, THF
2) Ac2O, Py
59%
NH
CH2OAc
1
Δ, -HOAc
82%
H
N
+
1:1.8
several steps
AcO
NH2

2

Macrolides. Nine- and ten-membered lactones cannot be obtained from ω-halocarboxylic acids by treatment with Cs_2CO_3 in DMF because of preferred intermolecular cyclization (**9**, 100).[2] However, the mesylate of 9-hydroxydecanoic acid is cyclized by Cs_2CO_3 in DMF to phoracantholid I (**1**) in 45% yield together with the diolide (25% yield). Similar treatment of the mesylate of 10-hydroxydecanoic acid results in the lactone and the diolide in the ratio 7:3. Cyclization proceeds in very low yield when an α,β-double bond is introduced.[3]

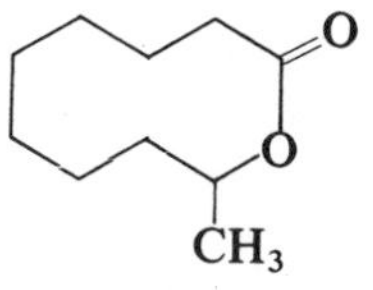
O
O
CH3

1

[1] R. A. Gobao, M. L. Bremmer, and S. M. Weinreb, *Am. Soc.*, **104**, 7065 (1982).
[2] W. H. Kruizinga and R. M. Kellogg, *ibid.*, **103**, 5183 (1981).
[3] M. Barbier, *J.C.S. Chem. Comm.*, 668 (1982).

Cesium fluoride, 7, 57–58; **8**, 81–82; **9**, 100; **10**, 81–84.

o-*Quinone methide imines.* CsF induces a 1,4-elimination from salts such as **1** to give a quinone methide imine (**a**), which dimerizes to **2**.[1]

$CH_2\overset{+}{N}(CH_3)_3I^-$ / $NSi(CH_3)_3$ / CH_3 (1) —CsF, 25°→ [a: CH_2, NCH_3] —77%→ 2 (NCH_3, N, CH_3)

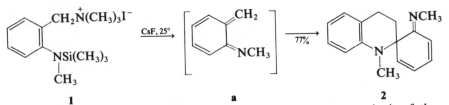

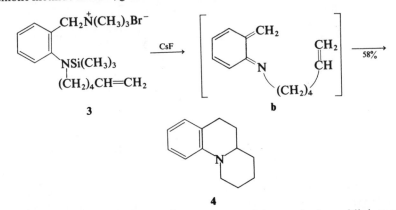

All attempts to trap **a** with dienophiles failed. However, a synthesis of the benzoquinolizidine **4** involves an intramolecular Diels-Alder reaction with an *o*-quinone methide imine **b**, generated from **3** by CsF.[1]

$CH_2\overset{+}{N}(CH_3)_3Br^-$ / $NSi(CH_3)_3$ / $(CH_2)_4CH{=}CH_2$ (3) —CsF→ [b: CH_2, N, CH_2, CH, $(CH_2)_4$] —58%→ 4

Silicon-induced fragmentations.[2] Treatment of the tetrahydropyridinium salt **1** with CsF in CH_3CN results in a stereoselective fragmentation to the (2Z,5E)-heptadiene **2**. In contrast, the N-oxide **3** fragments spontaneously to the (Z,Z)-heptadiene **4**. This fragmentation has been used to obtain (E,E)- and (E,Z)-dienes.[3]

1 (I^-, CH_3, CH_3, $\overset{+}{N}$, $Si(Me)_2$-*t*-Bu, $CHCH_3$) —CsF, CH_3CN, Δ, 77%→ $(CH_3)_2N$…CH_3 (2)

3 (CH_3, O^-, $\overset{+}{N}$, $Si(C_6H_5)_2$-*t*-Bu, $CHCH_3$) —20°→ CH_3, N, $(C_6H_5)_2SiO$, *t*-Bu, …CH_3 (4)

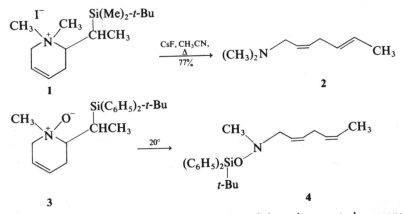

These dienes, when coupled with appropriate Grignard reagents by means of Li_2CuCl_4, provide access to pheromones of Lepidoptera.

Reaction of organosilanes with lactones and enones.[4] Certain carbanions, generated by desilylation, have been found to react not only with aldehydes and ketones, but also with some lactones and α,β-enones. Cesium fluoride or tetrabutylammonium fluoride supported on silica can be used as catalysts.

Examples:

$CH_2{=}CHCH_2Si(CH_3)_3$ + (cyclohexenone) $\xrightarrow[65\%]{CsF,\ THF}$ 3-($CH_2CH{=}CH_2$)cyclohexanone

(2-benzothiazolyl)$-Si(CH_3)_3 + C_6H_5CHO \xrightarrow[90\%]{}$ (2-benzothiazolyl)$-CH(OH)C_6H_5$

1

1 + (δ-valerolactone) $\xrightarrow[60\%]{}$ (2-benzothiazolyl)$-C(=O)(CH_2)_4OH$

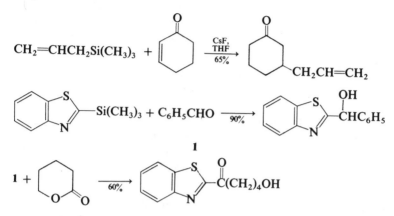

[1] Y. Ito, S. Miyata, M. Nakatsuka, and T. Saegusa, *Am. Soc.*, **103**, 5250 (1981).
[2] N. V. Bac and Y. Langlois, *ibid.*, **104**, 7666 (1982).
[3] G. Decodts, G. Dressaire, and Y. Langlois, *Synthesis*, 510 (1979); G. Dressaire and Y. Langlois, *Tetrahedron Letters*, **22**, 67 (1981).
[4] A. Ricci, M. Fiorenza, M. A. Grifagni, and G. Bartolini, *ibid.*, **23**, 5079 (1982).

Cesium fluoride–Silicon(IV) ethoxide, $CsF–Si(OC_2H_5)_4$, **10,** 82.

1,4-Addition to enamides.[1] In the presence of these reagents ketones, nitromethane and -ethane, and active methylene compounds undergo conjugate addition to tertiary enamides.

Examples:

2-CH_3-cyclohexanone + $C_6H_5CH{=}CHCON(C_2H_5)_2 \xrightarrow[86\%]{CsF,\ Si(OC_2H_5)_4,\ 80°}$ 2-CH_3-6-[$CH(C_6H_5)CH_2CON(C_2H_5)_2$]cyclohexanone

$H_2C(COOC_2H_5)(CN)$ + $CH_3CH{=}CHCON$(morpholine) $\xrightarrow[77\%]{}$ $NC(COOC_2H_5)CH–CH(CH_3)–CH_2CON$(morpholine)

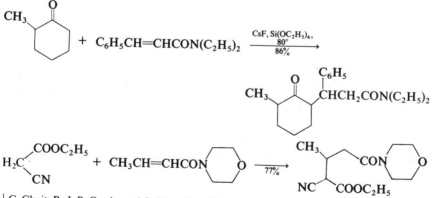

[1] C. Chuit, R. J. P. Corriu, and C. Reye, *Tetrahedron Letters*, **23**, 5531 (1982).

Cesium fluoroxysulfate, 10, 84–85.

Fluorination of alkenes.[1] Fluorination of some aliphatic alkenes with **1** has been reported.

Examples:

$$(C_6H_5)_2C{=}CH_2 \xrightarrow[70\%]{CsSO_4F,\ CH_2Cl_2,\ 25^\circ} (C_6H_5)_2C{=}CHF$$

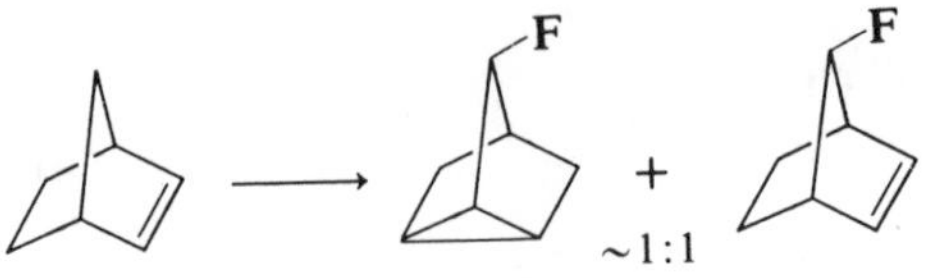

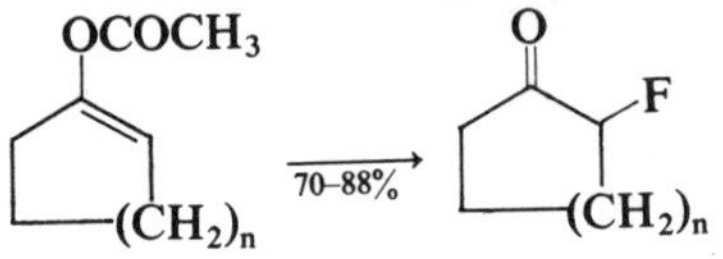

n = 1–4

[1] S. Stavber and M. Zupan, *J.C.S. Chem. Comm.*, 795 (1981).

Cesium propionate, $CsO_2CC_2H_5$ (**1**). The salt is prepared by addition of propionic acid in CH_3OH to Cs_2CO_3 dissolved in CH_3OH followed by removal of the CH_3OH; soluble in DMF.

S_N2-Substitutions. The configuration of secondary alcohols can be inverted by reaction of the mesylate with cesium propionate (65–100% yield). The reagent reacts with secondary alkyl halides to form the ester of the inverted alcohol (85–100% yield). Some elimination occurs in reaction of the mesylates of menthol and cholestanol.[1]

1,3-Dimethyl-2-imidazolidinone is superior to DMF as the solvent for this displacement.[2]

[1] W. H. Kruizinga, B. Strijtveen, and R. M. Kellogg, *J. Org.*, **46**, 4321 (1981).
[2] B. M. Trost and P. G. McDougal, *Am. Soc.*, **104**, 6110 (1982).

Chloramine-T, 4, 75, 445–446; **5,** 104; **7,** 58; **8,** 83; **9,** 101–102; **10,** 85.

Chlorolactonization.[1] Unsaturated carboxylic acids are converted into chloro lactones in moderate yield by reaction with chloramine-T and methanesulfonic acid.

Example:

$$CH_2{=}CH(CH_2)_2COOH + TsN^-ClNa^+ \xrightarrow[63\%]{CH_3SO_3H}$$

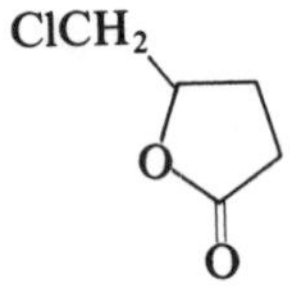

[1] B. Damin, A. Forestiere, J. Garapon, and B. Sillion, *J. Org.*, **46**, 3552 (1981).

Chlorine oxide, Cl_2O. Mol. wt. 86.91, yellow-red gas. *Caution*: Cl_2O is toxic and explosive, but solutions in CCl_4 can be handled safely and stored at −15 to 0°. It can be prepared by passing a mixture of chlorine and air over yellow HgO.

Chlorination.[1] Cl_2O is particularly useful for side-chain chlorination of deactivated aromatic substrates by a free-radical process involving ClO·. Although it can be used to convert methylarenes to mono- and dichloromethyl derivatives, it is particularly useful for conversion to trichloromethylarenes. Thus it converts *p*-nitrotoluene into *p*-nitrobenzotrichloride in 99.8% yield. It is also useful for selective replacement of benzylic hydrogens.

A powerful electrophilic species is obtained from Cl_2O and a strong protic acid such as trifluoroacetic acid. This reagent effects exclusive ring chlorination of deactivated arenes in high yield with the usual regioselectivity.

[1] F. D. Marsh, W. B. Farnham, D. J. Sam, and B. E. Smart, *Am. Soc.*, **104**, 4680 (1982).

Chlorobis(cyclopentadienyl)hydridozirconium (1), 6, 177–179; **7,** 101–102; **8,** 84–87; **9,** 104.

Hydrozirconation of imines.[1] Reaction of Schiff bases containing an α-hydrogen with **1** followed by acylation provides a route to enamides and/or imides and amides.

Examples:

$$(CH_3)_2CHCH{=}N\overset{CH_3}{\overset{|}{C}}HC_6H_5 \xrightarrow[2)\ CH_3COCl]{1)\ \mathbf{1}} (CH_3)_2C{=}CH\underset{COCH_3}{\underset{|}{N}}\overset{CH_3}{\overset{|}{C}}HC_6H_5 + C_6H_5\overset{CH_3}{\overset{|}{C}}HN(COCH_3)_2$$

2 (30%) **3** (20%)

With 1) CH_3COCl, 2) $N(C_2H_5)_3$: 82% **2**

(Scheme: N-(1-phenylethyl) imine of 2-methylcyclohexanone, 1) CH_3COCl, 2) **1** → N-acetyl-N-(1-phenylethyl) enamide of 2-methylcyclohexene (44%) + $CH_3CONH\overset{C_6H_5}{\overset{|}{C}}HCH_3$ (12%))

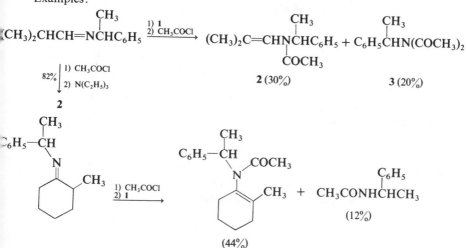

[1] K. S. Ng, D. E. Laycock, and H. Alper, *J. Org.*, **46**, 2899 (1981).

3-Chloro-4,5-dihydro-2-furylcopper, (structure: 2-Cu, 3-Cl dihydrofuran) The reagent is prepared *in situ* by treatment of 4-chloro-2,3-dihydrofurane in THF with *n*-BuLi at −75° and then with CuI at the same temperature.[1]

Vitamin A terpenes.[2] A novel entry to retinal (**5**) involves reaction of the chloride (**2**) with **1** to give an intermediate (**3**) in 55% yield. Oxidation of **4** with DMSO in the presence of N,N-dicyclohexylcarbodiimide gives rise to the four 11,13-stereoisomeric retinals (**5**). Addition of I_2 increases the proportion of the all-*trans*-retinal (E,E–**5**).

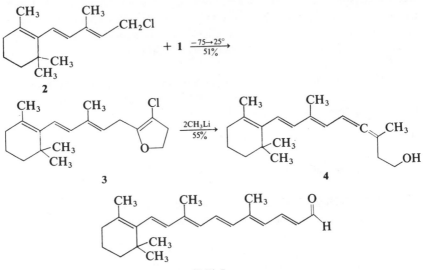

2 + **1** $\xrightarrow[51\%]{-75\rightarrow 25^\circ}$ **3** $\xrightarrow[55\%]{2CH_3Li}$ **4** → (E,E)-**5**

[1] M. Schlosser, B. Schaub, B. Spahić, and G. Sleiter, *Helv.*, **56**, 2166 (1973); B. Schaub and M. Schlosser, *ibid.*, **58**, 556 (1975).

[2] R. Ruzziconi and M. Schlosser, *Angew. Chem. Int. Ed.*, **21**, 855 (1982).

Chlorodiphenylphosphine, $(C_6H_5)_2PCl$ (**1**). Mol. wt. 220.64, b.p. 320°. Supplier: Aldrich.

$R^1S(O)CH_2R^2 \rightarrow R^1SC(O)R^2$.[1] The conversion of sulfoxides (**2**) into thiol esters (**5**) is possible via the sulfoxide phosphines (**3**), which rearrange slowly at 0–20° or, more efficiently, in the presence of iodine, to phosphine oxides **4**. These can be oxygenated via the α-lithio derivative to thiol esters **5**.

Example:

$$(CH_3)_3CS(O)CH_2CH(CH_3)_2 \; (\mathbf{2}) \xrightarrow[2)\ 1,\ THF]{1)\ n\text{-BuLi}} \left[(CH_3)_3CS(O)CH(P(C_6H_5)_2)CH(CH_3)_2\right] (\mathbf{3}) \xrightarrow[86\%]{I_2}$$

$$(CH_3)_3CSCH(P(O)(C_6H_5)_2)CH(CH_3)_2 \; (\mathbf{4}) \xrightarrow[80\%]{1)\ LDA;\ 2)\ O_2,\ -100^\circ} (CH_3)_3CSC(O)CH(CH_3)_2 \; (\mathbf{5})$$

Caution: Oxygenation of THF is inherently dangerous.

[1] E. Vedejs, H. Mastalerz, G. P. Meier, and D. W. Powell, *J. Org.*, **46**, 5253 (1981).

Chloro(methyl)aluminum amides,

$$CH_3\overset{\displaystyle Cl}{\overset{|}{A}}lNR^1R^2 \quad \textbf{(1)}.$$

Amides.[1] These reagents are prepared by reaction of $Al(CH_3)_3$ in toluene with the hydrochloride salts of NH_3, H_2NCH_3, and $HN(CH_3)_2$. They are easier to prepare than dimethylaluminum amides (**8**, 182), but they are equally suitable for conversion of esters to amides.

[1] J. I. Levin, E. Turose, and S. M. Weinreb, *Syn. Comm.*, **12**, 989 (1982).

Chloromethylcarbene, 10, 90.

α-Methyl-α,β-unsaturated esters. A new route to α-methyl-α,β-unsaturated esters is outlined in equation (I). A mixture of (E)- and (Z)-isomers is obtained, in which the (E)-isomer predominates.[1]

(I) $R^1R^2CHCOOR^3 \xrightarrow[80\text{–}90\%]{1)\ LDA,\ THF,\ -78°;\ 2)\ ClSi(CH_3)_3} R^1R^2C{=}C(OSi(CH_3)_3)(OR^3) \xrightarrow{:C(Cl)CH_3}$

$\left[\text{cyclopropane: } C(Cl)(CH_3),\ C(R^1)(R^2),\ C(OSi(CH_3)_3)(OR^3)\right] \xrightarrow[75\text{–}95\%]{N(C_2H_5)_3,\ CH_3OH,\ \Delta} R^1R^2C{=}C(CH_3)COOR^3$

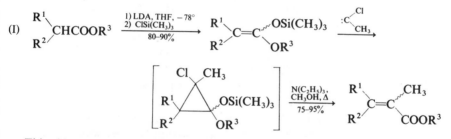

This chloromethylenation is also applicable to silyl enol ethers of aldehydes and ketones, as shown by a synthesis of manicone (**1**), the alarm pheromone of certain ants.[2]

$(C_2H_5)(CH_3)CHCH{=}C(OSi(CH_3)_3)C_2H_5 \xrightarrow{CHCl_2CH_3,\ n\text{-BuLi}} (C_2H_5)(CH_3)CHCH{-}C(OSi(CH_3)_3)(C_2H_5)$ [cyclopropane with $C(Cl)(CH_3)$] $\xrightarrow[75\%\ overall]{C_6H_5CH_3,\ \Delta}$

$(C_2H_5)(CH_3)CH(H)C{=}C(CH_3)COC_2H_5$

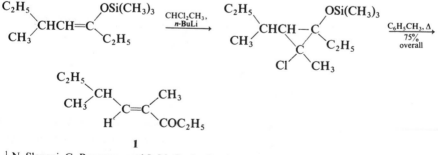

1

[1] N. Slougui, G. Rousseau, and J.-M. Conia, *Synthesis*, 58 (1982).
[2] L. Blanco, N. Slougui, G. Rousseau, and J.-M. Conia, *Tetrahedron Letters*, **22**, 645 (1981).

***m*-Chloroperbenzoic acid, 1**, 135–139; **2**, 68–69; **3**, 49–50; **6**, 110–114; **7**, 62–64; **8**, 97–102; **9**, 108–110; **10**, 92–93.

Heterocyclic nitroso compounds. Nitroso compounds can be prepared by reduction of nitro compounds, but many heterocycles can not be nitrated satisfactorily. A general route involves conversion of heterocyclic amines to sulfilimines, which need not be isolated, followed by oxidation with *m*-chloroperbenzoic acid (equation I).[1]

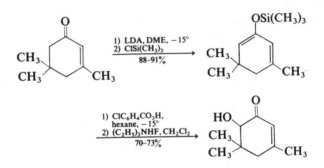

$$\text{Het—NH}_2 \xrightarrow[\text{3) NaOCH}_3]{\text{1) (CH}_3)_2\text{S, CH}_2\text{Cl}_2\text{; 2) NCS}} \text{Het—}\bar{\text{N}}\text{—}\overset{+}{\text{S}}(\text{CH}_3)_2 \xrightarrow[\text{45–60\% overall}]{\text{ClC}_6\text{H}_4\text{CO}_3\text{H}} \text{Het—NO}$$

α′-Hydroxy-α,β-enones. (**8**, 100–101; **10**, 92). Complete details for the α′-hydroxylation of α,β-enones via the kinetic enol silyl ether are available. DME is recommended as the solvent for preparation of the intermediate, which is best isolated by using a nonaqueous workup.[2]

3,5,5-Trimethylcyclohex-2-enone $\xrightarrow[\text{88–91\%}]{\text{1) LDA, DME, }-15^\circ\text{; 2) ClSi(CH}_3)_3}$ 1-$OSi(CH_3)_3$-3,5,5-trimethylcyclohexa-1,3-diene (with CH_3, CH_3, CH_3)

$\xrightarrow[\text{70–73\%}]{\text{1) ClC}_6\text{H}_4\text{CO}_3\text{H, hexane, }-15^\circ\text{; 2) (C}_2\text{H}_5)_3\text{NHF, CH}_2\text{Cl}_2}$ 6-HO-3,5,5-trimethylcyclohex-2-enone

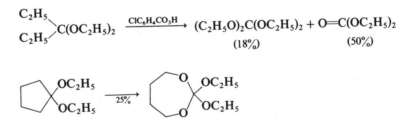

Double Baeyer-Villiger reactions.[3] Diethyl ketals undergo an exothermic reaction with excess *m*-chloroperbenzoic acid to give orthocarbonates, which are partly destroyed by the carboxylic acid formed in the process.

Examples:

$$(C_2H_5)_2C(OC_2H_5)_2 \xrightarrow{\text{ClC}_6\text{H}_4\text{CO}_3\text{H}} \underset{(18\%)}{(C_2H_5O)_2C(OC_2H_5)_2} + \underset{(50\%)}{O{=}C(OC_2H_5)_2}$$

Cyclopentanone diethyl ketal (OC_2H_5, OC_2H_5) $\xrightarrow{25\%}$ seven-membered cyclic orthocarbonate (ring O, O; OC_2H_5, OC_2H_5)

If the main objective is oxidative removal of the carbonyl group of the ketal, the mixture produced on peracid oxidation is hydrolyzed and then reduced in an ethereal solution with lithium aluminum hydride.

Example:

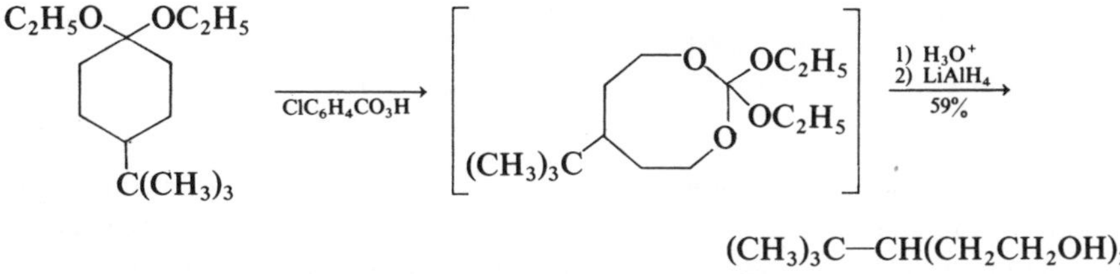

$$\text{4-}t\text{-butyl-1,1-diethoxycyclohexane} \xrightarrow{ClC_6H_4CO_3H} [\text{cyclic orthoester}] \xrightarrow[59\%]{1)\ H_3O^+\ \ 2)\ LiAlH_4} (CH_3)_3C{-}CH(CH_2CH_2OH)_2$$

Selenones.[4] γ-Hydroxy vinyl selenides (**1**) are labile and readily undergo dehydration on standing, but can be oxidized successfully to γ-hydroxy vinyl selenones (**2**) by *m*-chloroperbenzoic acid in the presence of K_2HPO_4 in CH_3OH at 25°.

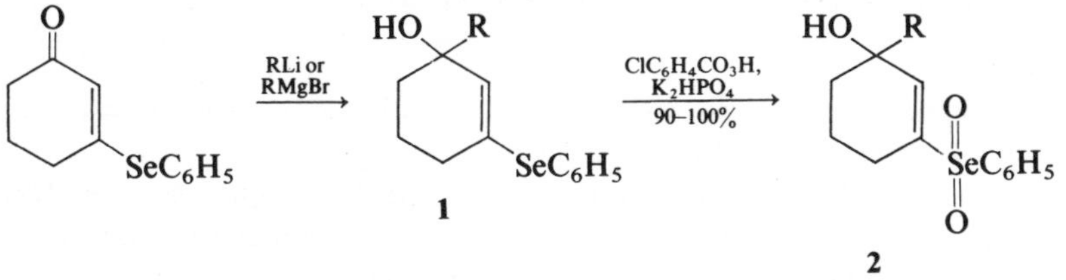

RLi or RMgBr; $ClC_6H_4CO_3H$, K_2HPO_4, 90–100%

These vinyl selenones are of interest because they undergo a 1,4-fragmentation at room temperature in the presence of bases to give an ethylenic ketone (equation I). If the double bond is tetrasubstituted, the fragmentation results in an acetylenic ketone (equation II).

(I) NaOCH₃; 86%; (Z/E = 80:20)

(II) NaH; 84%

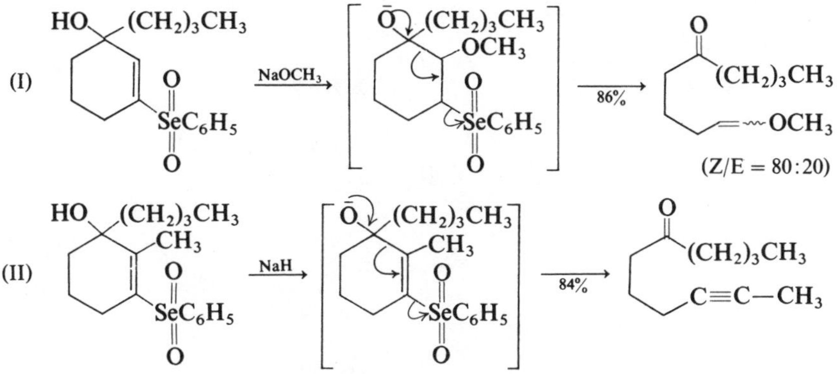

Oxidation of phosphites to phosphates.[5] This oxidation can be conducted with *m*-chloroperbenzoic acid in CH_2Cl_2. It is particularly useful in the phosphite triester synthesis of oligonucleotides. Previously, iodine in aqueous solution was used as oxidant; but use of aqueous solvents requires a drying step at a later stage.

[1] E. C. Taylor, C.-P. Tseng, and J. B. Rampal, *J. Org.*, **47**, 552 (1982).

[2] G. M. Rubottom, J. M. Gruber, H. D. Juve, Jr., and D. A. Charleson, *Org. Syn.* submitted (1982).
[3] W. F. Bailey and M.-J. Shih, *Am. Soc.*, **104**, 1769 (1982).
[4] M. Shimizu, R. Ando, and I. Kuwajima, *J. Org.*, **46**, 5246 (1981).
[5] K. K. Ogilivie and M. J. Nemer, *Tetrahedron Letters*, **22**, 2531 (1981).

***m*-Chloroperbenzoic acid–Potassium fluoride.** Potassium fluoride forms insoluble complexes with *m*-chloroperbenzoic acid and *m*-chlorobenzoic acid. It also increases the stability of hydroperoxides. Addition of KF improves the yields in the Baeyer-Villiger oxidation of benzaldehydes to aryl formates and in epoxidation of various olefins[1] including α-methylstyrene.[2]

[1] F. Camps, J. Colb, A. Messeguer, and M. A. Pericas, *Tetrahedron Letters*, **22**, 3895 (1981).
[2] F. Camps, J. Colb, A. Messeguer, and F. Pujol, *J. Org.*, **47**, 5402 (1982).

7-Chloro-5-phenyl-1-[(S)-α-phenylethyl]-1,3-dihydro-2H-1,4-benzodiazepine-2-one (1). Mol. wt. 390.86, m.p. 111–112°. Preparation.[1]

Chiral amino acids.[2] The stereochemistry of alkylation (quasi-equatorial) at C_3 of the anion of **1** results from the conformation of the seven-membered ring in a boat form. The products can be hydrolyzed to (S)-amino acids (**3**).

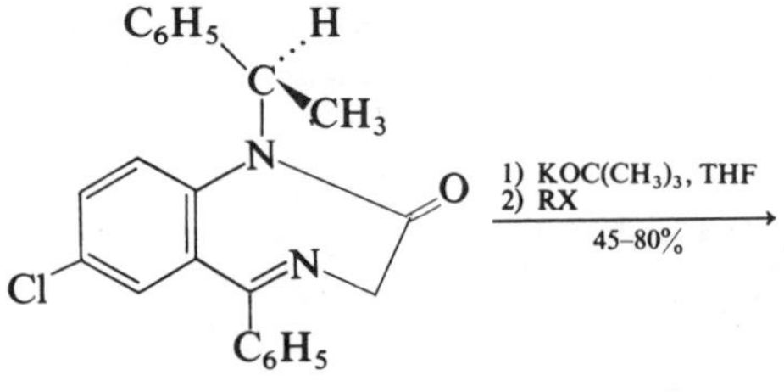

C_6H_5 H C CH_3 N O Cl N C_6H_5

1

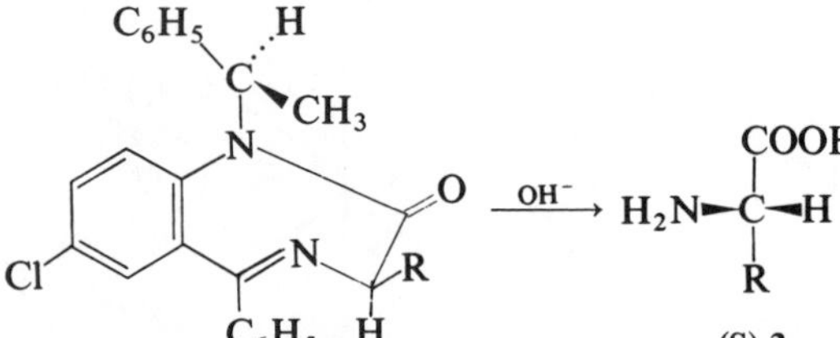

1) $KOC(CH_3)_3$, THF
2) RX
45–80%

C_6H_5 H C CH_3 N O Cl N R C_6H_5 H

2 (6–84%ee)

OH^-

COOH
H_2N—C—H
R

(S)-**3**

[1] V. Šunjić, M. Oklobžija, A. Lisini, A. Sega, F. Kajfez, D. Srzić, and L. Klasinc, *Tetrahedron*, **35**, 2531 (1979).
[2] E. Decorte, R. Toso, A. Sega, V. Šunjić, Ž. Ružić-Toroš, B. Kojić-Prodić, N. Bresciani-Pahor, G. Nardin, and L. Randaccio, *Helv.*, **64**, 1145 (1981).

Chlorosulfonyl isocyanate (CSI), 1, 117–118; **2**, 70; **3**, 51–53; **4**, 90–94; **5**, 132–136; **6**, 122; **7**, 65–66; **8**, 105–106; **10**, 94–95.

Azetidinones.[1] CSI and allyl iodide undergo [2 + 2]cycloaddition to form the azetidine-2-one (**1**), which can be converted into **3**, a potential precursor to carbapenams.

$$CH_2{=}CHCH_2I + O{=}C{=}NSO_2Cl \xrightarrow{37\%} \mathbf{1}\ (\text{4-}CH_2I\text{-azetidin-2-one}) \xrightarrow[98\%]{NaCN,\ DMSO} \mathbf{2}\ (\text{4-}CH_2CN\text{-azetidin-2-one}) \xrightarrow[13\%]{HSCH_2CH_2NH_2,\ C_2H_5OH} \mathbf{3}$$

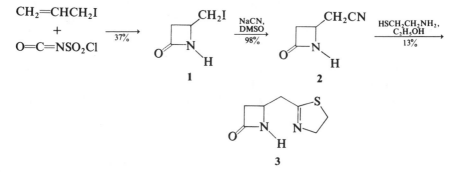

Chiral nitriles.[2] This reagent is recommended for the dehydration in ~60% yield of primary amides with an asymmetric center at the α-position to chiral nitriles with negligible racemization. The most satisfactory reagents for dehydration of optically active oximes to chiral nitriles is 1,1′-carbonyldiimidazole or N,N′-dicyclohexylcarbodiimide.

[1] T. Tanaka and T. Miyadera, *Synthesis*, 1497 (1982).
[2] C. Botteghi, G. Chelucci, and M. Marchetti, *Syn. Comm.*, **12**, 25 (1982).

Chlorotrimethylsilane, 1, 1232; **2**, 435–438; **3**, 310–312; **4**, 537–539; **5**, 709–713; **6**, 626–628; **7**, 66–67; **8**, 107–109; **9**, 112–113; **10**, 96.

Cleavage of epoxides.[1] Thermal cleavage of epoxides by silicon chlorides is usually not useful synthetically because of low regioselectivity. The cleavage is catalyzed by numerous nucleophilic reagents, particularly triphenylphosphine and tetra-*n*-butylammonium chloride. In the catalyzed reaction, the primary chloride is formed selectively at low temperatures. This cleavage is general for trialkylsilyl halides.

Examples:

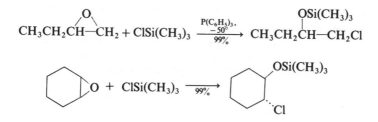

$$CH_3CH_2\overset{O}{CH{-}CH_2} + ClSi(CH_3)_3 \xrightarrow[99\%]{P(C_6H_5)_3,\ -50^\circ} CH_3CH_2\underset{}{\overset{OSi(CH_3)_3}{CH}}{-}CH_2Cl$$

$$\text{cyclohexene oxide} + ClSi(CH_3)_3 \xrightarrow{99\%} trans\text{-2-chloro-1-}OSi(CH_3)_3\text{-cyclohexane}$$

Vinylallenes.[2] Chlorotrimethylsilane couples with 5-chloro-3-en-1-ynes (**1**) to form the vinylallenes **2** in the presence of Li–ether–THF or Mg–HMPT. Yields are similar with both systems.

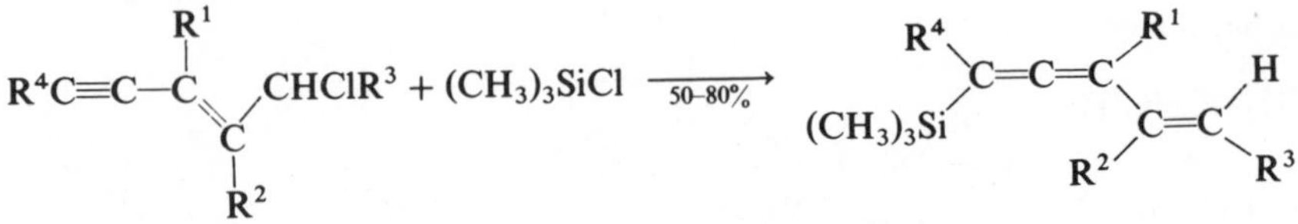

$$R^4C{\equiv}C{-}C(R^1){=}C(R^2){-}CHClR^3 + (CH_3)_3SiCl \xrightarrow{50\text{–}80\%} (CH_3)_3Si(R^4)C{=}C{=}C(R^1){-}C(R^2){=}C(H)R^3$$

1 **2**

Dienamides.[3] A key step in a synthesis of *cis*-dihydroxylycoricidine (**3**) involved *in situ* generation of a dienamide (**a**) for an intramolecular [4 + 2]cycloaddition to form **2**.

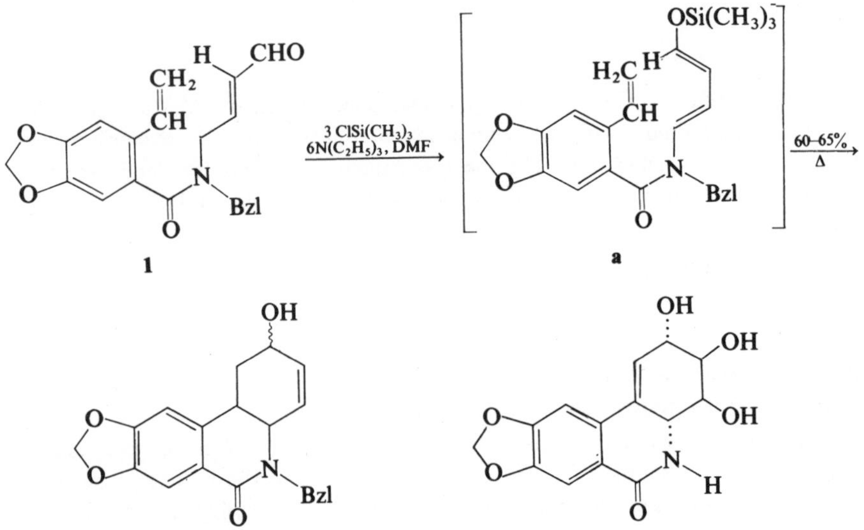

$$\mathbf{1} \xrightarrow[6N(C_2H_5)_3,\ DMF]{3\ ClSi(CH_3)_3} [\mathbf{a}] \xrightarrow[\Delta]{60\text{–}65\%} \mathbf{2} \longrightarrow \mathbf{3}$$

1 **a** **2** **3**

Trimethylsilyl ethers.[4] These protective derivatives of alcohols are conveniently prepared with K_2CO_3 or Na_2CO_3 as base and Aliquat 336 as the phase-transfer catalyst (65–95% yield). *t*-Butyldimethylsilyl ethers can be prepared in the same way.

Trimethylsilyl esters.[5] These esters are inert to diborane in THF and thus are useful for protection of carboxylic acids during hydroboration.

Esterification.[6] Chlorotrimethylsilane is an efficient reagent for esterification of carboxylic acids by primary and secondary alcohols. It is converted into hexamethyldisiloxane, $(CH_3)_3SiOSi(CH_3)_3$. Yields are typically 70–95%.

[1] G. C. Andrews, T. C. Crawford, and L. G. Contillo, Jr., *Tetrahedron Letters*, **22**, 3803 (1981).
[2] J.-P. Dulcere, J. Grimaldi, and M. Santelli, *ibid.*, **22**, 3179 (1981).

[3] G. E. Keck, E. Boden, and U. Sonnewald, *ibid.*, **22**, 2615 (1981).
[4] M. Lissel and J. Weiffen, *Syn. Comm.*, **11**, 545 (1981).
[5] G. L. Larson, M. Ortiz, and M. Rodriquez de Roca, *ibid.*, **11**, 583 (1981).
[6] R. Nakao, K. Oka, and T. Fukumoto, *Bull. Chem. Soc. Japan*, **54**, 1267 (1981).

Chlorotrimethylsilane–Sodium, 2, 435–436; **3,** 311–312; **4,** 537; **5,** 711–712.

Cyclopropanone ethyl hemiketal.[1] This useful precursor to 1-cyclopropanols is readily prepared by a procedure (equation I) adapted from Rühlmann's reaction (**4**, 537).

(I) $ClCH_2CH_2COOC_2H_5 \xrightarrow[60-85\%]{Na,\ ClSi(CH_3)_3}$ [cyclopropane with $OSi(CH_3)_3$, OC_2H_5] $\xrightarrow[78-95\%]{CH_3OH}$ [cyclopropane with OH, OC_2H_5]

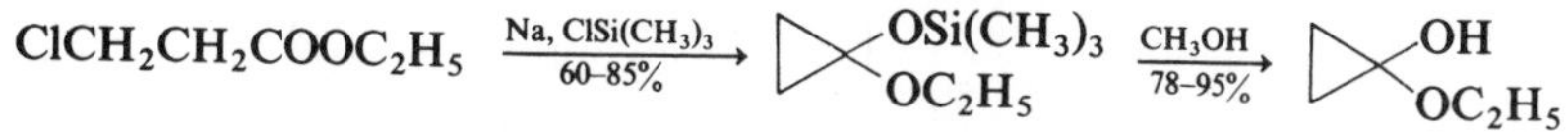

[1] J. Salaun and J. Marguerite, *Org. Syn.*, submitted (1981).

Chlorotrimethylsilane–Sodium iodide, 9, 251–252; **10,** 97–98.

Dehydroxylation.[1] Treatment of the photoecdysteroid muristerone (**1**) with $ClSi(CH_3)_3$ and NaI results in 14-desoxymuristerone (**2**). Unlike 14-desoxysteroids, which generally are less active than the 14α-hydroxy counterparts, **2** exhibits enhanced activity as an insect molting hormone.

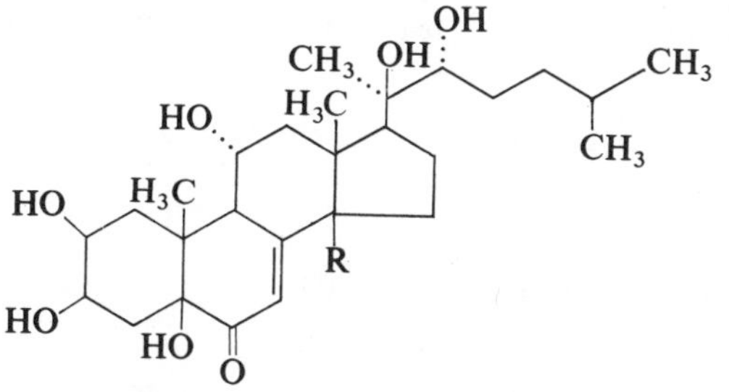

1), R = OH
2), R = H

Deoxygenation of epoxides.[2] Oxiranes are converted into alkenes with the same configuration by $ISi(CH_3)_3$ generated *in situ*.

Examples:

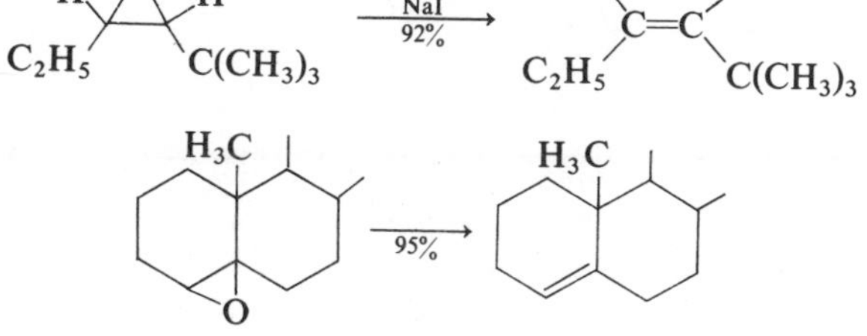

[epoxide: H, C_2H_5 / H, $C(CH_3)_3$] $\xrightarrow[92\%]{ClSi(CH_3)_3,\ NaI}$ [C_2H_5(H)C=C(H)$C(CH_3)_3$]

[decalin epoxide with H_3C] $\xrightarrow{95\%}$ [octalin with H_3C]

Deoxygenation of* vic*-diols.[3] Secondary-tertiary (both *cis* and *trans*) vicinal diols are converted into alkenes directly by treatment with chlorotrimethylsilane and sodium iodide in CH_3CN at room temperature (80–95% yield, seven examples).

Allylic bis-secondary *vic*-diols are converted into 1,3-dienes in 80–98% yield under the same conditions. The reagents also effect conversion of allylic alcohols into 1,3-dienes under these conditions (75–90% yield).[4]

Example:

$$\xrightarrow[90\%]{ClSi(CH_3)_3,\ NaI,\ CH_3CN}$$

[1] P. Cherbas, D. A. Trinor, R. J. Stonard, and K. Nakanishi, *J.C.S. Chem. Comm.*, 1307 (1982).
[2] R. Caputo, L. Mangoni, O. Neri, and Z. Palumbo, *Tetrahedron Letters*, **22**, 3551 (1981).
[3] N. C. Barua and R. P. Sharma, *Tetrahedron Letters*, **23**, 1365 (1982).
[4] N. C. Barua, J. C. Sarma, R. P. Sharma, and J. N. Barua, *Chem. Ind.*, 956 (1982).

Chlorotrimethylsilane–Sodium iodide–Zinc.

Deoxygenation of N-oxides.[1] Pyridine N-oxides are deoxygenated by this combination generally in 80–90% yield. The coproducts are $(CH_3)_3SiOSi(CH_3)_3$, NaCl, and $ZnCl_2$.

[1] T. Morita, K. Kuroda, Y. Okamoto, and H. Sakurai, *Chem. Letters*, 921 (1981).

1-Chloro-1-(trimethylsilyl)ethyllithium, $(CH_3)_3SiC(CH_3)(Li)Cl$ (1).

The reagent is obtained by deprotonation of 1-chloroethyltrimethylsilane (Petrarch Systems) with *sec*-butyllithium in THF at $-78 \rightarrow -55°$.

Methyl ketones.[1] This carbanion adds to aldehydes and ketones to give epoxytrimethylsilanes in moderate to high yield. The products are hydrolyzed to homologated methyl ketones in high yield (equation I).

$$\text{(I)}\quad R^1R^2C{=}O + \mathbf{1} \xrightarrow[40\text{–}95\%]{THF,\ -78\rightarrow20°} R^1R^2C\overset{O}{—}C(CH_3)Si(CH_3)_3 \xrightarrow[75\text{–}90\%]{H_2SO_4,\ CH_3OH} R^1R^2CHCOCH_3$$

The reagent was used to obtain (R)-(+)-frontalin (**5**) from the aldehyde **2**, available in two steps from (3R)-(−)-linalool. The adduct **3** from reaction of **2** with **1** was converted to the diol **4** by ozonation and reduction. Treatment of **4** with BF_3 etherate effects conversion to the methyl ketone and then cyclization to **5**.

$$(CH_3)_3SiO\cdots C(CH_3)(CH{=}CH_2)CH_2CH_2CHO \ (\mathbf{2}) \xrightarrow[95\%]{\mathbf{1}} \mathbf{3} \xrightarrow[65\%]{1)\ O_3;\ 2)\ NaBH_4} \mathbf{4} \xrightarrow[74\%]{BF_3\cdot(C_2H_5)_2O} \mathbf{5}$$

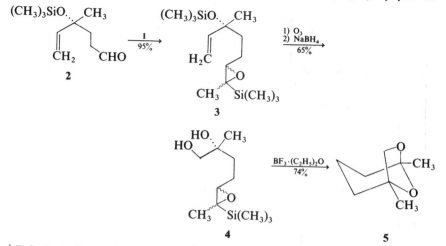

[1] F. Cooke, G. Roy, and P. Magnus, *Organometallics*, **1**, 893 (1982).

3-Chloro-2-(trimethylsilyloxy)-1-propene,

$$CH_2{=}C(OSi(CH_3)_3)CH_2Cl \quad \mathbf{(1).}$$

Mol. wt. 164.70, b.p. 61–63°. Preparation (equation I).[1]

$$\text{(I)}\ (CH_3)_3SiCH_2Cl \xrightarrow{1)\ Mg,\ (C_2H_5)_2O;\ 2)\ (ClCH_2CO)_2O} (CH_3)_3SiCH_2COCH_2Cl \xrightarrow[62\%]{HgI_2} \mathbf{1}$$

Acetonylation.[2] The reagent acetonylates α-metallated imines and hydrazones to give, after hydrolysis, 1,4-diketones.

Examples:

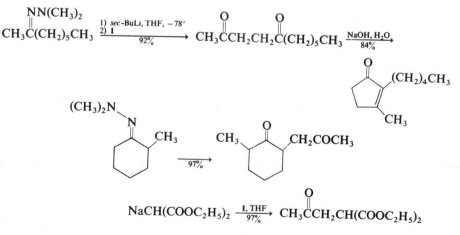

$$CH_3C(=NN(CH_3)_2)(CH_2)_5CH_3 \xrightarrow[92\%]{1)\ sec\text{-BuLi, THF, }-78^\circ;\ 2)\ \mathbf{1}} CH_3\overset{O}{\overset{\|}{C}}CH_2CH_2\overset{O}{\overset{\|}{C}}(CH_2)_5CH_3 \xrightarrow[84\%]{NaOH,\ H_2O} \text{2-(CH}_2)_4CH_3\text{-3-CH}_3\text{-cyclopent-2-enone}$$

$$\text{2-methylcyclohexanone }(CH_3)_2N{-}N{=}\text{ hydrazone} \xrightarrow{97\%} \text{2-CH}_3\text{-6-CH}_2COCH_3\text{-cyclohexanone}$$

$$NaCH(COOC_2H_5)_2 \xrightarrow[97\%]{\mathbf{1},\ THF} CH_3\overset{O}{\overset{\|}{C}}CH_2CH(COOC_2H_5)_2$$

[1] H. Sakurai, A. Shirahata, Y. Araki, and A. Hosomi, *Tetrahedron Letters*, **21**, 2325 (1980).
[2] A. Hosomi, A. Shirahata, Y. Araki, and H. Sakurai, *J. Org.*, **46**, 4631 (1981).

Chlorotris(triphenylphosphine)rhodium(I), 1, 1252; **2**, 248–253; **3**, 325–329; **4**, 559–562; **5**, 736–740; **6**, 562–563; **7**, 68; **8**, 109; **9**, 113–114; **10**, 98–99.

[2 + 2 + 2]Cycloadditions.[1] This rhodium complex effects trimerization of a 1,6-heptadiyne with a terminal alkyne to form a benzene derivative (equation I). The paper reports one example of an intramolecular [2 + 2 + 2]cycloaddition (equation II).

(I) O(CH$_2$C≡CH)$_2$ + n-C$_3$H$_7$C≡CH $\xrightarrow[58\%]{ClRh[P(C_6H_5)_3]_3}$ 5-(n-propyl)-1,3-dihydroisobenzofuran

(II) HC≡CCH$_2$OCH$_2$C≡CCH$_2$OCH$_2$C≡CH $\xrightarrow{75\%}$ bis(dihydrofuran)-fused benzene

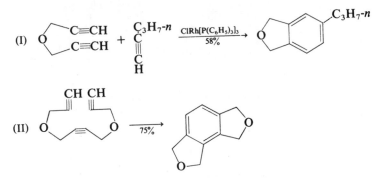

[1] R. Grigg, R. Scott, and P. Stevenson, *Tetrahedron Letters*, **23**, 2691 (1982).

Chlorotris(triphenylphosphine)rhodium–Hydrosilanes, 4, 562–568.

Hydrosilylation of carbonyl compounds. The definitive report on reduction of carbonyl compounds by hydrosilylation catalyzed by Wilkinson's catalyst is available.[1] Hydrosilylation can be used to effect regioselective 1,2- or 1,4-reduction of α,β-enals or -enones by proper choice of the hydrosilane. In general, monohydrosilanes favor 1,4-adducts, whereas dihydrosilanes favor 1,2-adducts. The regioselectivity is also influenced by the substituents on silicon and on the substrates. The presence of a phenyl group on the enone system can effect dramatic changes in the selectivity.

Examples:

$$(CH_3)_2C{=}CHCOCH_3 \xrightarrow{Rh(I)} (CH_3)_2CHCH_2COCH_3 + (CH_3)_2C{=}CHCH(OH)CH_3$$

$(C_2H_5)_3SiH$	99:1	95%
$C_6H_5SiH_3$	1:99	95%

$$C_6H_5CH{=}CHCHO \xrightarrow{Rh(I)} C_6H_5CH_2CH_2CHO + C_6H_5CH{=}CHCH_2OH$$

$(C_2H_5)_3SiH$	100%	—
$(C_6H_5)_2SiH_2$	—	100%

This reaction was used to synthesize faranal (**1**), the trail pheromone of the pharaoh ant.[2]

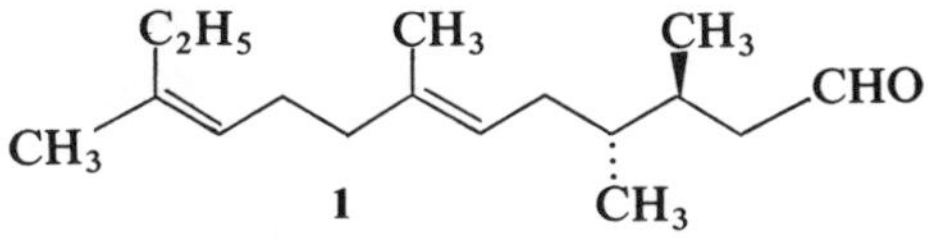

[1] I. Ojima and T. Kogure, *Organometallics*, **1**, 1390 (1982).
[2] M. Kobayashi, T. Koyama, K. Ogura, S. Sato, F. T. Ritter, and I. E. Bruggemann-Rotgans, *Am. Soc.*, **102**, 6602 (1980).

Chromic anhydride–Dimethylformamide, 1, 147.

Oxidation of π-allylpalladium chloride complexes. Snatzke's reagent is superior to other chromium(VI) reagents for oxidation of steroidal π-allylpalladium chloride complexes to α,β-enones.[1]

Examples:

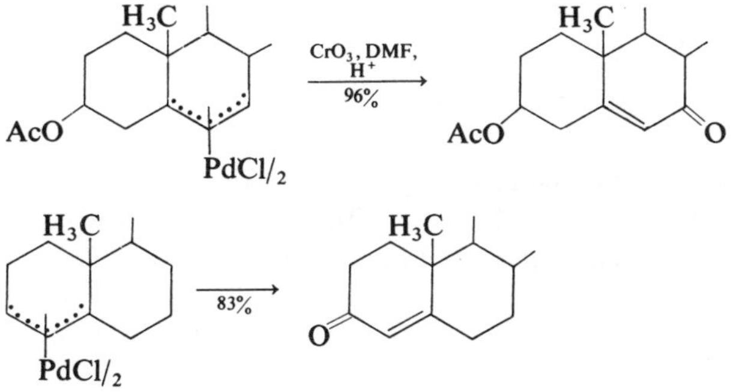

CrO_3, DMF, H^+; 96%

83%

[1] J. Y. Satoh and C. A. Horiuchi, *Bull. Chem. Soc. Japan*, **54**, 625 (1981).

Chromium carbonyl, 5, 142–143; **6**, 125–126; **7**, 71–72; **8**, 110; **9**, 117–119; **10**, 100.

Halogenation catalyst.[1] Chromium carbonyl catalyzes the monohalogenation of cyclohexane by CCl_4 (78% yield). Other cycloalkanes undergo the same reaction. Bromination can be effected in this way with $CBrCl_3$. Other metal carbonyl complexes are less active. $Cr(CO)_6$ is actually more efficient than di-*t*-butyl peroxide. A free radical mechanism is involved.

Arene(tricarbonyl)chromium complexes. The $Cr(CO)_3$ group can be used to enhance a benzylic position to deprotonation.[2] Thus, the complex **1** from toluene when treated with potassium *t*-butoxide in DMSO reacts with benzaldehyde to give **2** in 86% yield. The stereochemistry of this reaction was investigated with the complex **3** from indane. The product **5** of hydroxymethylation followed by reduction is the *anti*-isomer (^{1}H NMR data).

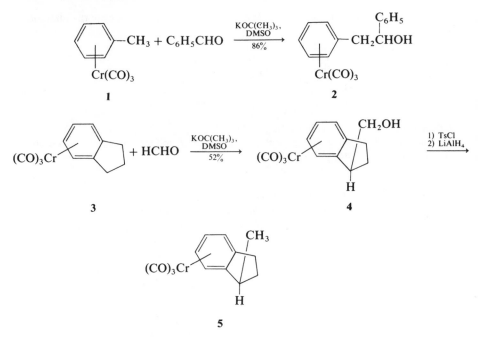

$$\text{C}_6\text{H}_5\text{CH}_3\text{Cr(CO)}_3\ (\mathbf{1}) + \text{C}_6\text{H}_5\text{CHO} \xrightarrow[86\%]{\text{KOC(CH}_3)_3,\ \text{DMSO}} \text{C}_6\text{H}_5\text{CH}_2\text{CH(C}_6\text{H}_5)\text{OH}\cdot\text{Cr(CO)}_3\ (\mathbf{2})$$

$$(\text{CO})_3\text{Cr-indane}\ (\mathbf{3}) + \text{HCHO} \xrightarrow[52\%]{\text{KOC(CH}_3)_3,\ \text{DMSO}} (\text{CO})_3\text{Cr-1-(CH}_2\text{OH)indane}\ (\mathbf{4}) \xrightarrow{1)\ \text{TsCl};\ 2)\ \text{LiAlH}_4} (\text{CO})_3\text{Cr-1-methylindane}\ (\mathbf{5})$$

Electrophilic substitution of indole occurs preferentially at C_3; metallation can activate C_2. Complexation with $Cr(CO)_3$ results in a marked preference for substitution at C_4; steric factors can favor C_7- over C_4-substitution.[3]

Example:

$$(\text{CO})_3\text{Cr-(1-methylindole)} \xrightarrow[92\%]{1)\ \text{LiC(CH}_3)_2\text{COOC(CH}_3)_3;\ 2)\ \text{I}_2} \text{4-[(CH}_3)_2\text{CCOOC(CH}_3)_3\text{]-1-methylindole} + \text{C}_7\text{-isomer}\quad 99:1$$

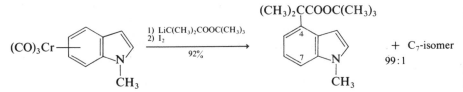

[1] R. Davis, I. F. Groves, and C. R. Rowland, *J. Organometal. Chem.*, **239**, C9 (1982).
[2] J. Brocard, J. Lebibi, and D. Couturier, *J.C.S. Chem. Comm.*, 1264 (1981).
[3] M. F. Semmelhack, W. Wulff, and J. L. Garcia, *J. Organometal. Chem.*, **240**, C5 (1982).

Chromium(II) chloride, 8, 110–112; **9,** 119.

Homoallylic alcohols (**8**, 111–112). The high *threo*-selectivity observed in the reaction of benzaldehyde with crotyl bromide (either *trans* or *cis*) is general for relatively unhindered aldehydes (equation I). High *threo* selectivity is still observed in the reaction with an α-methyl substituted aldehyde, but α-asymmetric induction (at C_3) is rather low (2.2:1) with simple aldehydes (equation II).[1]

(I) 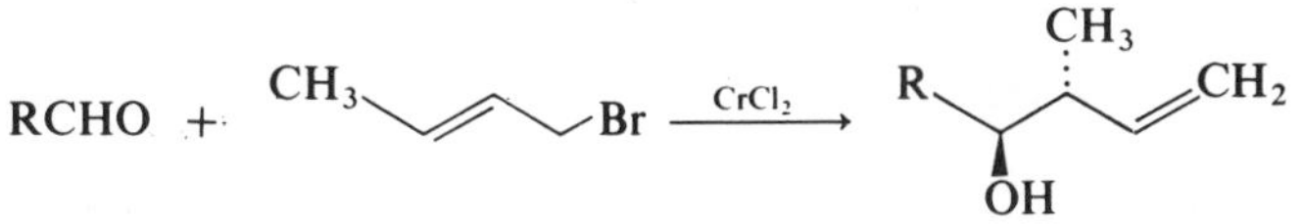 RCHO + $CH_3CH{=}CHCH_2Br$ $\xrightarrow{CrCl_2}$ [product: R–CH(OH)–CH(CH_3)–CH=CH_2]

(II) 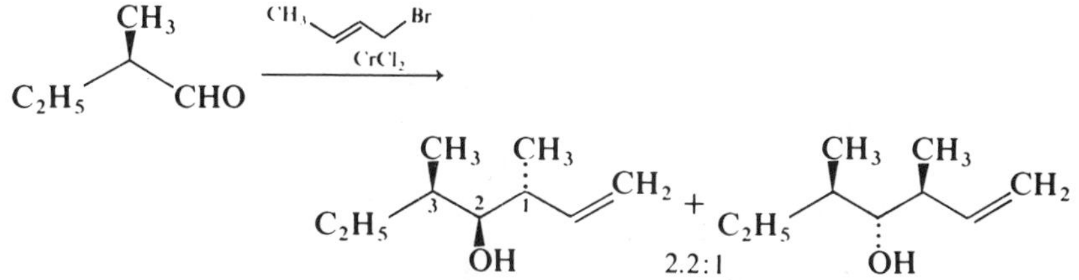 C_2H_5–CH(CH_3)–CHO $\xrightarrow[CrCl_2]{CH_3CH=CHCH_2Br}$

[products: C_2H_5–CH(CH_3)–CH(OH)–CH(CH_3)–CH=CH_2 (C-1, C-2, C-3 numbered) + C_2H_5–CH(CH_3)–CH(OH)–CH(CH_3)–CH=CH_2 (diastereomer), 2.2:1]

This reaction played a key role in a highly stereocontrolled synthesis of the aliphatic segment **1** of the antibiotic rifamycin S (**2**),[2] as formulated in equation (III). In each of the two Cr(II) mediated reactions, the desired aldol is essentially the only product isolated. Further studies[3] indicate that the 2,3-stereoselectivity is sensitive to the large substituent α to the aldehyde. A cyclic acetal group appears to play a specific role in contrast to an acyclic group, but the factors controlling stereoselectivity are not well understood.

(III) 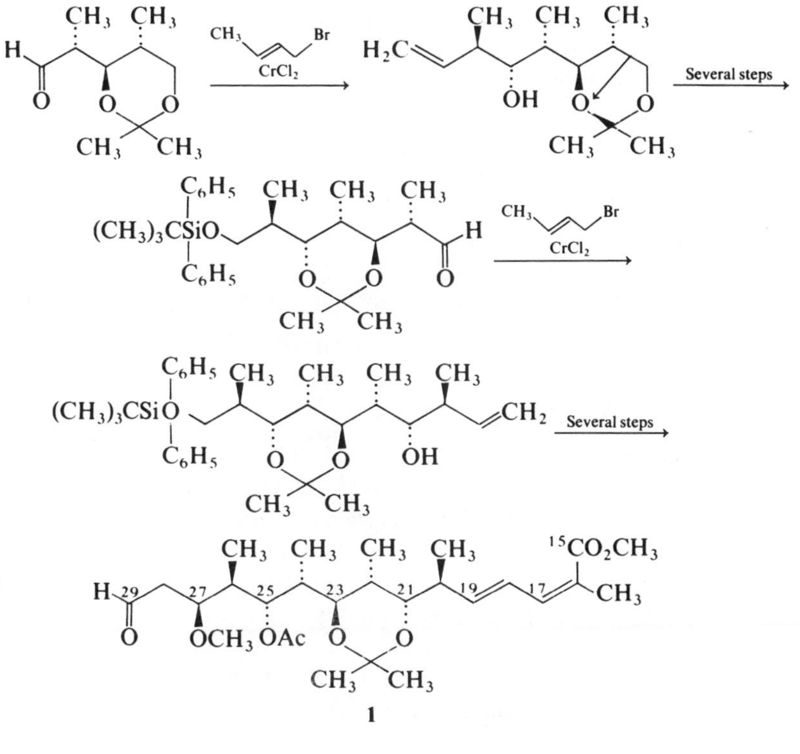 [aldehyde with CH_3, CH_3 substituents and acetonide (O, O, CH_3, CH_3)] $\xrightarrow[CrCl_2]{CH_3CH=CHCH_2Br}$ [H_2C=CH–CH(CH_3)–CH(OH)–CH(CH_3)–acetonide (CH_3, CH_3)] $\xrightarrow{\text{Several steps}}$

[$(CH_3)_3CSi(C_6H_5)_2O$–CH_2–CH(CH_3)–(acetonide, CH_3, CH_3)–CH(CH_3)–CHO] $\xrightarrow[CrCl_2]{CH_3CH=CHCH_2Br}$

[$(CH_3)_3CSi(C_6H_5)_2O$–CH_2–CH(CH_3)–(acetonide)–CH(CH_3)–CH(OH)–CH(CH_3)–CH=CH_2] $\xrightarrow{\text{Several steps}}$

[H–C29(=O)–C27(OCH_3)–CH(CH_3)–C25(OAc)–CH(CH_3)–C23/C21 (acetonide, CH_3, CH_3)–CH(CH_3)–C19=…C17=C(CH_3)–$^{15}CO_2CH_3$]

1

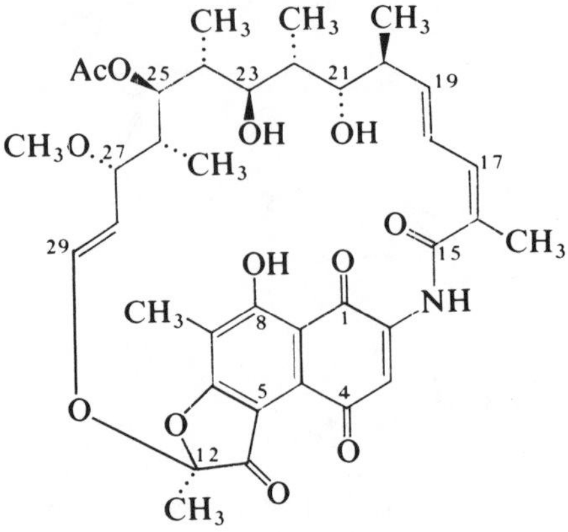

2

Coupling of alkyl halides (*cf.* **8**, 111).[4] *t*-Alkyl, benzylic, and allylic halides form symmetrical dimers in about 85–95% yield when treated with $CrCl_2$.[5] The dimerization involves monoalkylchromium(III) complexes followed by a radical reaction. Cross-coupling of these same halides is also possible, but yields are lower (40–90%).

[1] T. Hiyama, K. Kimura, and N. Nozaki, *Tetrahedron Letters*, **22**, 1037 (1981).
[2] H. Nagaoka and Y. Kishi, *Tetrahedron*, **37**, 3873 (1981).
[3] M. D. Lewis and Y. Kishi, *Tetrahedron Letters*, **23**, 2343 (1982).
[4] R. Sustmann and R. Altevogt, *Tetrahedron Letters*, **22**, 5165, 5167 (1981).
[5] Prepared by reduction of $CrCl_3$ with $Li(C_2H_5)_3BH$.

Chromyl chloride, 1, 151; **2**, 79; **3**, 62; **4**, 98–99; **5**, 144–145; **6**, 126–127; **8**, 112.

α-Hydroxy ketones.[1] Reaction of CrO_2Cl_2 with enol silyl ethers in CH_2Cl_2 at $-78°$ results in α-hydroxy ketones in 70–80% yield. α-Chloro ketones are not formed.

[1] T. V. Lee and J. Toczek, *Tetrahedron Letters*, **23**, 2917 (1982).

Cinchona alkaloids, 6, 501; **7**, 311; **8**, 430–431; **9**, 403; **10**, 338.

[2 + 2]*Cycloaddition of ketene to chloral.*[1] Quinidine (**1**) catalyzes the addition of ketene to chloral to give the β-propiolactone (**2**) in 89% yield and in 98% ee. As expected, quinine (C_9-epimer of **1**) also catalyzes this reaction to give (R)-(+)-**2** in 76% ee. Examination of 15 chiral tertiary amines has led to certain conclusions: the chirality of the product depends on the chirality of the carbon adjacent to the amine function, and the presence or absence of a hydroxyl group β to the amine is unimportant in this case. Even simple amines such as 1,2-dimethylpyrrolidine or N,N-dimethyl-α-phenylethylamine are fairly effective (60 and 77% ee, respectively).

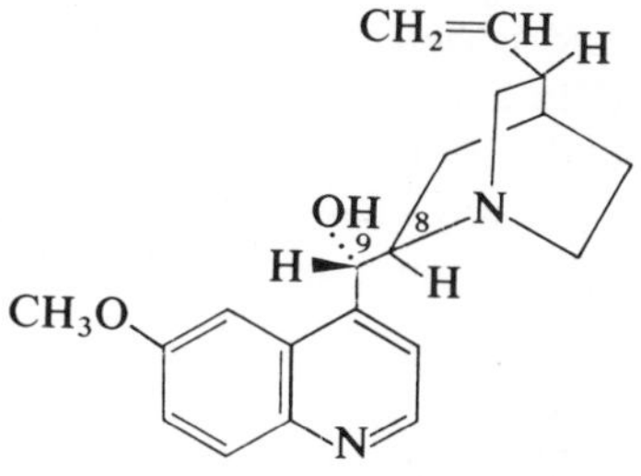

1 (9R)

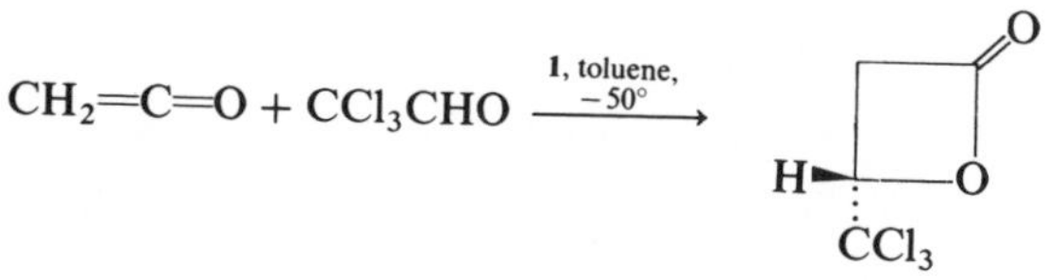

$$CH_2{=}C{=}O + CCl_3CHO \xrightarrow[-50^\circ]{\text{1, toluene,}}$$

(S)-(−)-2

Known reactions convert **2** into malic acid **3**. Thus pure natural (S)-(−)-malic acid can be obtained from (S)-(−)-**2** in 79% yield.

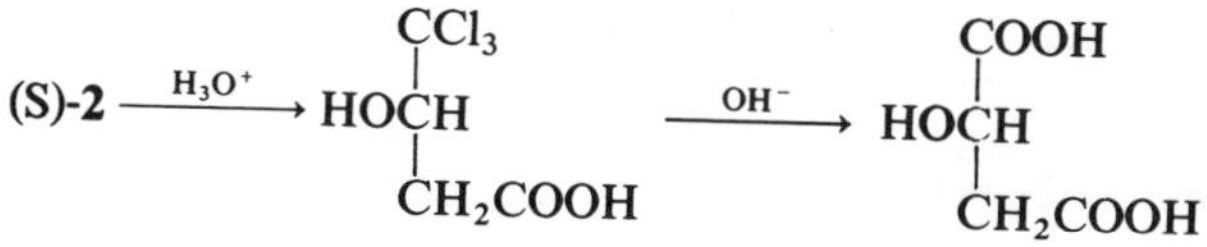

$$\textbf{(S)-2} \xrightarrow{H_3O^+} HOCH(CCl_3)CH_2COOH \xrightarrow{OH^-} HOCH(COOH)CH_2COOH$$

(S)-(−)-3

[1] H. Wynberg and E. G. J. Staring, *Am. Soc.*, **104**, 166 (1982).

Cobaloxime(I), (1).

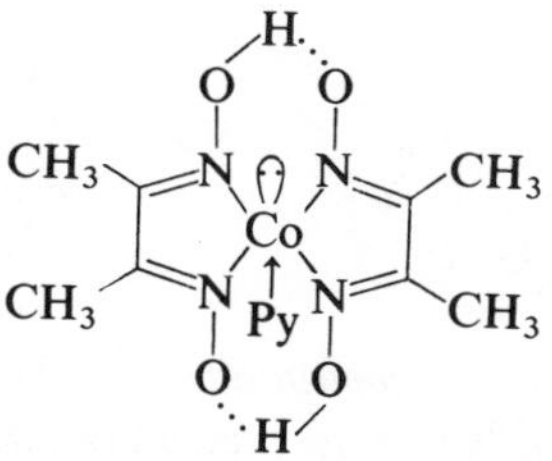

1

Cobaloxime refers to the bis(dimethylglyoximate)cobalt group. Several of these cobalt complexes have been prepared because they serve as models for vitamin B_{12}.

Cobaloxime(I) is unstable and is best prepared *in situ* by $NaBH_4$ reduction of chloro(pyridine)cobaloxime(III), a stable complex formed from dimethylglyoxime, $CoCl_2 \cdot 6H_2O$, and pyridine (70% yield).[1]

Reductive cyclization.[2] β-Methylene-γ-butyrolactones (**5**) can be prepared by reductive cyclization with **1** of 2-(2-propynyloxy)ethyl bromides (**3**) to 3-methyleneoxolanes (**4**), followed by oxidation.

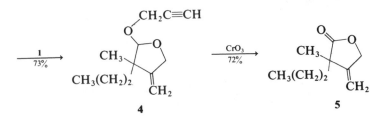

$$\underset{\mathbf{2}}{CH_3(CH_3(CH_2)_2)CH{-}CHO} \xrightarrow[88\%]{1)\ Br_2;\ 2)\ HOCH_2C{\equiv}CH,\ HC(OCH_2C{\equiv}CH)_3} \underset{\mathbf{3}}{CH_3(CH_3(CH_2)_2)C(Br){-}CH(OCH_2C{\equiv}CH)_2} \xrightarrow[73\%]{\mathbf{1}} \mathbf{4} \xrightarrow[72\%]{CrO_3} \mathbf{5}$$

[1] C. N. Schranzer, *Inorg. Syn.*, **11**, 62 (1968).
[2] M. Okabe and M. Tada, *J. Org.*, **47**, 5382 (1982).

Cobaltacyclopentane-2-ones,

L(Cp)Co(CO)(CH₂)₃ metallacycle (**1**).

Preparation[1]:

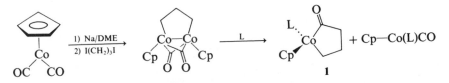

$$CpCo(CO)_2 \xrightarrow{1)\ Na/DME;\ 2)\ I(CH_2)_3I} [Cp_2Co_2(CO)_2(CH_2)_4] \xrightarrow{L} \mathbf{1} + Cp{-}Co(L)CO$$

Asymmetric alkylation and aldol condensations.[2] The enolate (**2**) of **1** reacts with primary iodides to give essentially a single product (**3**), in which the alkyl group is *syn* to the cyclopentadienyl ring. Aldol condensation with acetone leads to only one observable product (**4**). Only two isomeric products are obtained on aldol condensation with prochiral aldehydes and ketones; as expected for a *trans*-enolate, the *threo*-aldol predominates or is the exclusive product (**5**) as in the case of pivaldehyde.

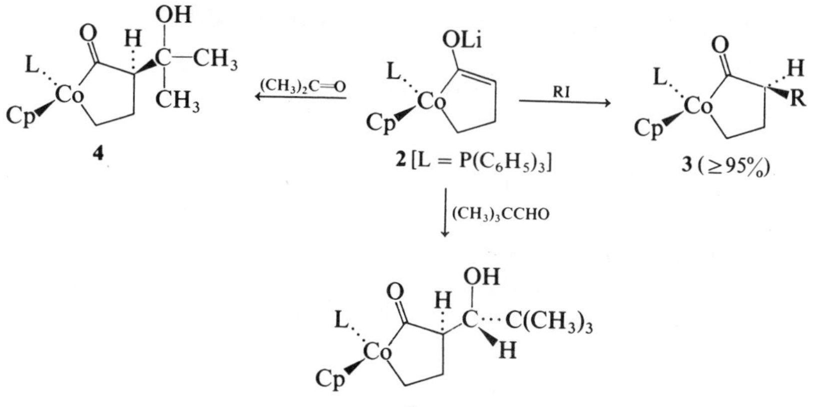

4

$(CH_3)_2C{=}O$

$2\ [L = P(C_6H_5)_3]$

RI

$3\ (\geq 95\%)$

$(CH_3)_3CCHO$

5

These reactions lead to optically active products when an optically active phosphine is used in place of triphenylphosphine. Thus treatment of the enol of optically active **6** with pivaldehyde gives the aldol **7** in 98% isolated yield. The absolute configuration was confirmed by X-ray diffraction. Only one method is known at the present time for removal of the metal moiety. Thus oxidation of **7** with ferric chloride affords the optically pure cyclobutanone **8** in 70% yield.

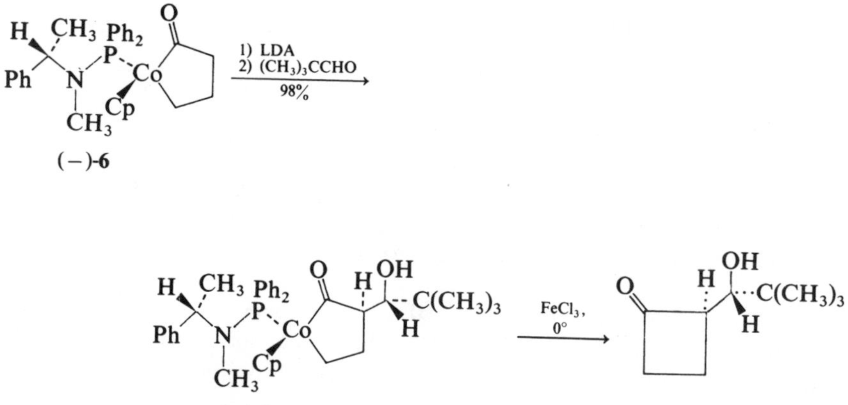

(−)-**6**

1) LDA
2) $(CH_3)_3CCHO$
98%

(−)-**7**

$FeCl_3$, 0°

8

[1] K. H. Theopold and R. G. Bergman, *Am. Soc.*, **102**, 5694 (1980); *Organometallic*, **1**, 1571 (1982).

[2] K. H. Theopold, P. N. Becker, and R. G. Bergman, *ibid.*, **104**, 5250 (1982).

Cobalt(II) bis(salicyidene-γ-iminopropyl)methylamine (CoSalMDPT), (1).

Preparation.[1]

$$\mathrm{Co}\left(\left(o\text{-}\mathrm{HOC_6H_4CH{=}N(CH_2)_3}\right)_2\mathrm{NCH_3}\right)$$

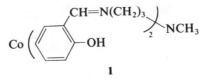

1

***Catalytic oxidation of* 1-*alkenes*.**[2] The Co(II) complex catalyzes the oxygenation of terminal alkenes to the methyl ketone and the corresponding alcohol. The reaction is not a radical-initiated autoxidation; it probably involves a hydroperoxide intermediate.

[1] L. Sacconi and I. Bertini, *Am. Soc.*, **88**, 5180 (1966).
[2] A. Zombeck, D. E. Hamilton, and R. S. Drago, *ibid.*, **104**, 6782 (1982).

Cobalt boride–*t*-Butylamine-borane, $Co_2B–(CH_3)_3CNH_2 \cdot BH_3$.

Reduction of nitriles. It has been recognized for some time that the combination of $NaBH_4$ with $CoCl_2$, $NiCl_2$, and $RbCl_3$ can reduce certain functional groups that are inert to $NaBH_4$ alone. In a study of the actual reducing species involved in the case of $NaBH_4–CoCl_2$, Heinzman and Ganem[1] investigated the properties of Co_2B, a black granular precipitate formed by the reaction of $NaBH_4$ with $CoCl_2$ in CH_3OH. *Caution*: Co_2B is pyrophoric when dry, but it can be stored moist with solvent. Co_2B even in combination with H_2 does not effect reductions possible with $NaBH_4–CoCl_2$; rather, its function is to coordinate with certain functional groups and, thereby, to catalyze reductions with $NaBH_4$.

$NaBH_4$–Co_2B reductions:

$$C_6H_5CH_2CN \xrightarrow[74\%]{NaBH_4,\ Co_2B,\ CH_3OH,\ 25^\circ} C_6H_5CH_2CH_2NH_2$$

$$C_6H_5CH{=}CHCOOCH_3 \xrightarrow[85\%]{} C_6H_5CH_2CH_2COOCH_3$$

$$CH_3(CH_2)_5C{\equiv}CH \xrightarrow[83\%]{} CH_3(CH_2)_6CH_3$$

Co_2B in combination with $t\text{-}BuNH_2 \cdot BH_3$ is a new, superior reagent for selective reduction of nitriles to amines in 75–90% yield. It does not reduce α,β-unsaturated esters to any extent and reduces alkenes and alkynes only slowly.

[1] S. W. Heinzman and B. Ganem, *Am. Soc.*, **104**, 6801 (1982).

Cobalt(II) phthalocyanine, 9, 119–121; **10**, 102–103.

Reduction catalyst.[1] This metal-phthalocyanine (CoPc) as well as several other metal-Pcs catalyze the reduction of various groups by $NaBH_4$ in ethanol at 20–25°.

The reaction rates decrease in the series C=C ~ RX ~ RNO_2 and $ArNO_2$ ~ RNO and ArNO > C=NOH ~ C=NR > RCN. Arenes, aryl halides, esters, and amides are not reduced. The usual catalyst poisons have no effect or only slight effect. The catalyst is recovered quantitatively and can be reused without purification.

Vitamin B_{12} also catalyzes the reduction of a nitro group by $NaBH_4$, but less efficiently.

[1] H. Eckert and Y. Kiesch, *Angew. Chem. Int. Ed.*, **20**, 473 (1981).

Collins reagent [Dipyridine chromium(VI) oxide], 2, 74–75; **3,** 55–56; **4,** 216–217; **9,** 121.

Oxidative cyclization of 5,6-dihydroxyalkenes. Oxidation of unsaturated diols **1a** and **1b** with either Collins reagent or pyridinium chlorochromate results in the corresponding *cis*-tetrahydrofuranediols (**2**) as the major product.[1] The selective stereochemistry is very similar to that observed in the oxidative cyclization of 1,5-hexadienes with potassium permanganate (**9**, 388–389) to *cis*-2,5-disubstituted tetrahydrofuranes.

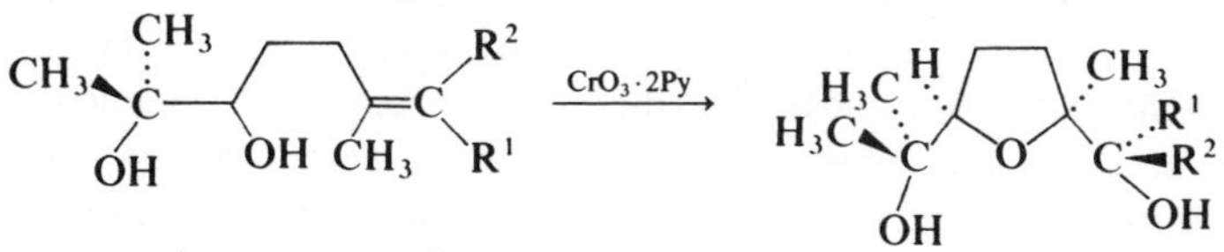
CH3
CH3
C
OH
OH CH3
R2
C
R1
CrO3·2Py
H3C
H3C
H
C
O
CH3
R1
C
R2
OH
OH

1a (R^1 = CH_2OAc, R^2 = H)
1b (R^1 = H, R^2 = CH_2OAc)

2

[1] D. M. Walba and G. S. Stoudt, *Tetrahedron Letters*, **23**, 727 (1982).

Copper(II) acetate–Copper(II) tetrafluoroborate.

Diels-Alder reaction of furanes.[1] A number of copper(I) and copper(II) salts can catalyze the cycloaddition of α-acetoxy- and α-chloroacrylonitrile to furane at 20–35°.

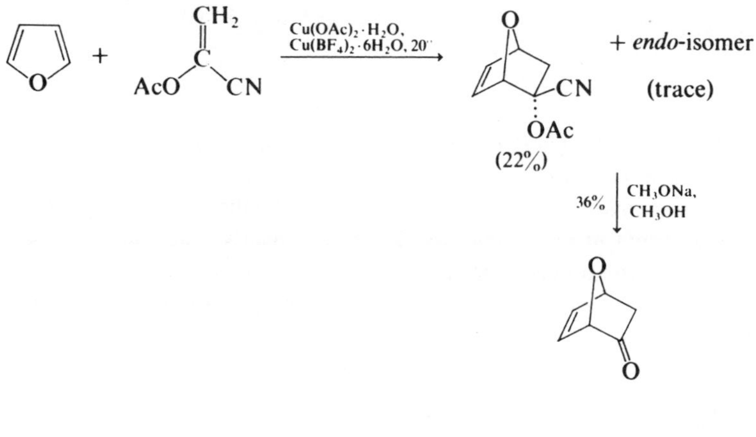
O
+
CH2
C
AcO
CN
Cu(OAc)2·H2O,
Cu(BF4)2·6H2O, 20°
O
CN
OAc
(22%)
+ endo-isomer
(trace)
36%
CH3ONa,
CH3OH
O
O

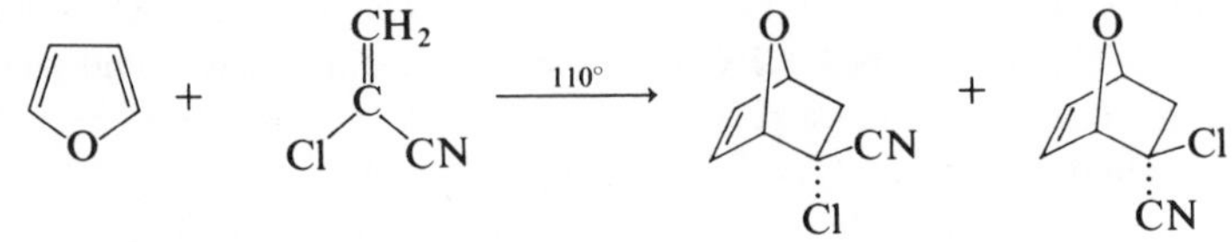

none	5%	50:50
$Cu(OAc)_2 \cdot H_2O$	62%	50:50
Cu(tartrate)$\cdot 3H_2O$	28%	70:30

[1] E. Vieira and P. Vogel, *Helv.*, **65**, 1700 (1982).

Copper(I) bromide, 1, 165–166; **2**, 90–91; **3**, 67; **4**, 108; **5**, 163–164; **6**, 143–144; **7**, 79–80; **8**, 116–117.

Coupling of aryl bromides with diethyl sodiomalonate. CuBr is the most effective Cu(I) catalyst for effecting coupling of aryl iodides or bromides with the sodium salts of active methylene compounds such as diethyl sodiomalonate. Coupling is facilitated by *ortho*-substituents in the halide, particularly nitro, carbomethoxy, and methoxy groups.[1] This reaction has been adapted to synthesis of benzofurane-2-ones.[2]

Example:

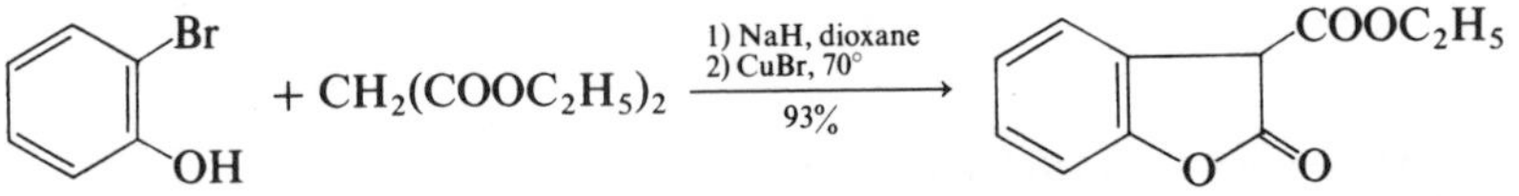

$+ CH_2(COOC_2H_5)_2$ $\xrightarrow[93\%]{1)\ NaH,\ dioxane;\ 2)\ CuBr,\ 70°}$

[1] J. Setsune, K. Matsukawa, H. Wakemoto, and T. Kitao, *Chem. Letters*, 367 (1981).
[2] J. Setsune, K. Matsukawa, and T. Kitao, *Tetrahedron Letters*, 663 (1982).

Copper(I) chloride, 1, 166–169; **2**, 91–92; **3**, 67–69; **4**, 109–110; **5**, 164–165; **6**, 145–146; **7**, 80–81; **8**, 118–119; **9**, 123.

2,2′-Biphenyldicarboxylic acids.[1] Phenanthrenequinone (**1**) is oxidized by oxygen in the presence of CuCl in pyridine to 2,2′-biphenyldicarboxylic acid (**2**) in high yield. Under the same conditions the monoimine **3**, prepared *in situ*, is oxidized to **4**. Amino and hydroxyl substituents interfere with this reaction, but alkyl, nitro, and halo groups are well-tolerated.

[1] E. Balogh-Hergovich, G. Speier, and Z. Tyeklár, *Synthesis*, 731 (1982).

Copper(I) iodide, 1, 169; **2,** 92; **3,** 69–71; **5,** 167–168; **6,** 147; **8,** 81–83; **9,** 124–125; **10,** 107.

Exaltolide. The key step in a new synthesis of exaltolide (**4**) is the three-carbon homologation of a Grignard reagent (**1**) with *β*-propiolactone (**2**) catalyzed by CuI. The product (**3**) is converted into **4** in two steps.[1]

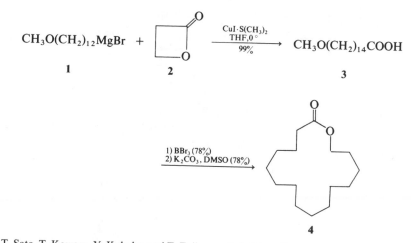

[1] T. Sato, T. Kawara, Y. Kokubu, and T. Fujisawa, *Bull. Chem. Soc. Japan*, **54**, 945 (1981).

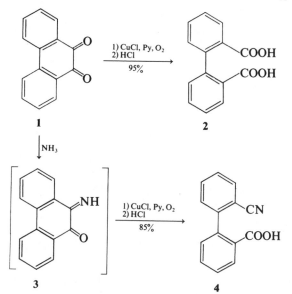

Copper permanganate, $Cu(MnO_4)_2 \cdot 8H_2O$. Supplier: Carus Chemical Company.

Oxidation.[1] Hydrated copper permanganate oxidizes alcohols in CH_2Cl_2 rapidly and in high yields. Ketones are obtained from secondary alcohols, and carboxylic acids from primary and benzylic alcohols. Primary or secondary allylic alcohols are also oxidized efficiently. Anhydrous reagent is much less active.

[1] N. A. Noureldin and D. G. Lee, *J. Org.*, **47**, 2790 (1982).

Copper(II) sulfate–Sodium borohydride.

Hydration of nitriles. The hydration of nitriles to amides is commonly carried out under acid or basic catalysis. The reaction can be carried under neutral conditions by catalysis with a black powder consisting essentially of Cu(0), obtained by reduction of $CuSO_4$ with $NaBH_4$ in aqueous NaOH. The method is applicable even to sparingly water-soluble nitriles, and yields range from 50–95% (crude).[1]

[1] M. Ravindranathan, N. Kalyanam, and S. Sivaram, *J. Org.*, **47**, 4812 (1982).

Copper(I) trifluoromethanesulfonate, 5, 151–152; **6,** 130–133; **7,** 75–76; **8,** 125–126; **10,** 108–110.

Photobicyclization of diunsaturated ethers. The first examples of this reaction were reported in 1978 (equations I and II).[1]

$$\text{(I)}\ CH_2{=}CHCH_2OH \xrightarrow[254\ \text{nm.}]{CuOTf,} \begin{array}{l} CH_2{=}CHCH_2 \diagdown \\ \qquad\qquad\qquad O \\ CH_2{=}CHCH_2 \diagup \end{array} \xrightarrow[254\ \text{nm.}]{CuOTf,} \begin{array}{l} CH_2{-}CHCH_2 \diagdown \\ \ |\qquad\ | \qquad\quad O \\ CH_2{-}CHCH_2 \diagup \end{array}$$

$$\text{(II)}\ \begin{array}{l} CH_2{=}CHCH_2 \diagdown \\ \qquad\qquad\qquad CHOH \\ CH_2{=}CHCH_2 \diagup \end{array} \xrightarrow[254\ \text{nm.}]{CuOTf,} \begin{array}{l} CH_2{-}CHCH_2 \diagdown \\ \ |\qquad\ | \qquad\quad CHOH \\ CH_2{-}CHCH_2 \diagup \end{array}$$

(2 isomers)

This reaction is applicable to both homoallyl and diallyl ethers and results in multicyclic furanes. These products can be oxidized to novel butyrolactones.[2]

Examples:

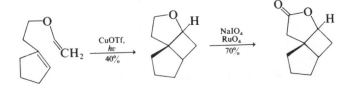

$$\text{O, } CH_2 \xrightarrow[40\%]{CuOTf,\ h\nu} \text{O, H} \xrightarrow[70\%]{NaIO_4,\ RuO_4} \text{O, O, H}$$

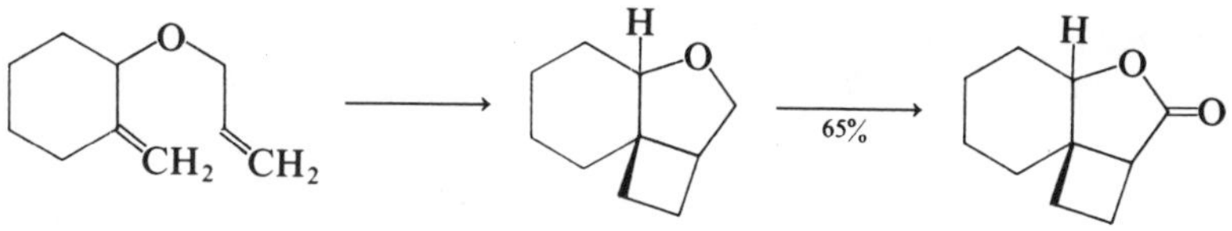

CH_2 CH_2 → H, O → (65%) H, O, =O

[1] J. Th. Evers and A. Mackor, *Tetrahedron Letters*, 821 (1978).
[2] S. R. Raychandhuri, S. Ghosh, and R. G. Salomon, *Am. Soc.*, **104**, 6841 (1982).

Crotyltri-*n*-butyltin, $CH_3CH{=}CHCH_2Sn(C_4H_9\text{-}n)_3$ **(1)**. $(n\text{-}Bu)_3SnLi$ reacts with (E)- or (Z)-crotyl chloride in THF at −20° to form (E)- or (Z)-crotyltri-*n*-butyltin with complete retention of the stereochemistry of the double bond.

erythro-*Selective addition to aldehydes.*[2] Irrespective of the geometry, (E)- or (Z)-crotyltributyltin (**1**) reacts with aldehydes in the presence of BF_3 to form almost exclusively *erythro-β*-methylhomoallyl alcohols (**2**). The products are useful precursors to the *erythro*-carboxylic acids (**3**).

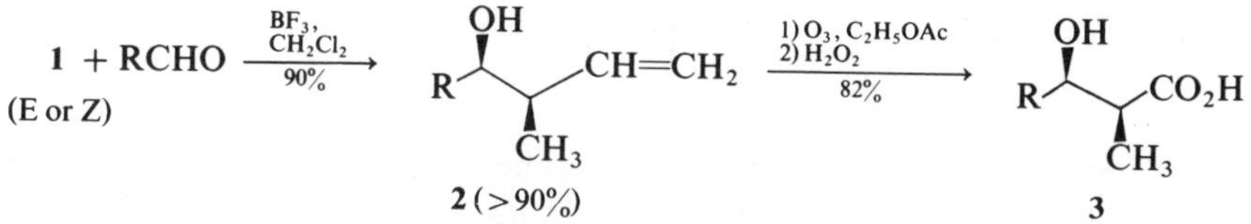

1 + RCHO (E or Z) —(BF_3, CH_2Cl_2; 90%)→ R–CH(OH)–CH(CH_3)–CH=CH_2 **2** (>90%) —(1) O_3, C_2H_5OAc; 2) H_2O_2; 82%)→ R–CH(OH)–CH(CH_3)–CO_2H **3**

This reaction provides a stereoselective, short synthesis of the Prelog-Djerassi lactonic acid **4**.[3]

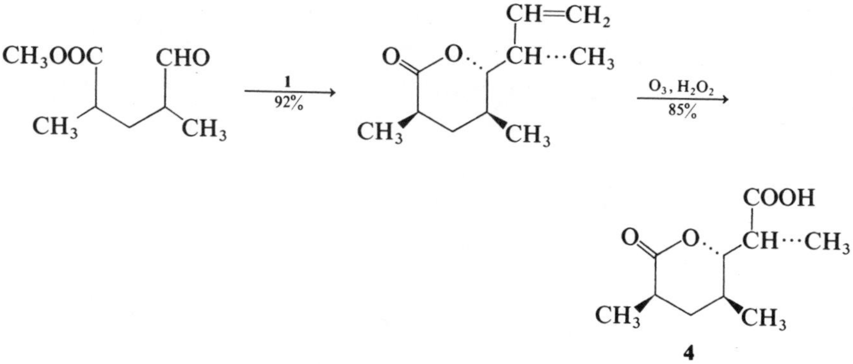

CH_3OOC, CHO, CH_3, CH_3 —(**1**; 92%)→ O, O, CH=CH_2, CH···CH_3, CH_3, CH_3 —(O_3, H_2O_2; 85%)→ O, O, COOH, CH···CH_3, CH_3, CH_3 **4**

[1] E. Matarasso-Tchiroukhine and P. Cadiot, *J. Organometal. Chem.*, **121**, 155, 169 (1976).
[2] Y. Yamamoto, H. Yatagai, Y. Naruta, and K. Maruyama, *Am. Soc.*, **102**, 7109 (1980).
[3] K. Maruyama, Y. Ishihara, and Y. Yamamoto, *Tetrahedron Letters*, **22**, 4235 (1981).

Crown ethers, 4, 142–145; **5**, 152–155; **6**, 133–137; **7**, 76–79; **8**, 128–130; **9**, 126–127; **10**, 110–112.

Xanthates (**3**, 123; **4**, 195). Xanthates can be prepared by a one-pot procedure using potassium *t*-butoxide as base and 18-crown-6 as catalyst[1] (equation I).

$$\text{(I)}\quad ROH + CS_2 \xrightarrow[CH_2Cl_2]{KOC(CH_3)_3,\ \text{Crown ether},} RO\overset{\overset{S}{\|}}{C}SK \xrightarrow[\substack{72\text{–}89\%\\ \text{overall}}]{CH_3I} RO\overset{\overset{S}{\|}}{C}SCH_3$$

Dibromocarbene.[2] 1-Bromobenzocyclobutene can be prepared conveniently by reaction of cycloheptatriene with dibromocarbene generated from bromoform with base in the presence of 18-crown-6. No reaction occurs in the absence of the crown ether.

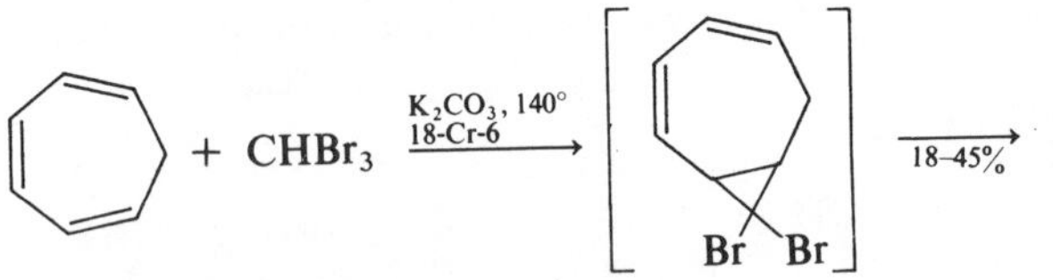

$$\text{cycloheptatriene} + CHBr_3 \xrightarrow[\text{18-Cr-6}]{K_2CO_3,\ 140°} [\text{dibromocarbene adduct}] \xrightarrow{18\text{–}45\%} \text{1-bromobenzocyclobutene}$$

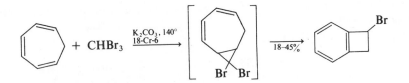

Wittig-Horner reactions.[3] The reaction of the phosphonate **2** with the aldehyde **1** under literature conditions gives the *trans*-stilbene **3** in only about 10% yield. Addition of 15-crown-5 raises the yield to 45%. This step was used in a synthesis of pallescensin E (**4**), a furanosesquiterpene in a marine sponge.

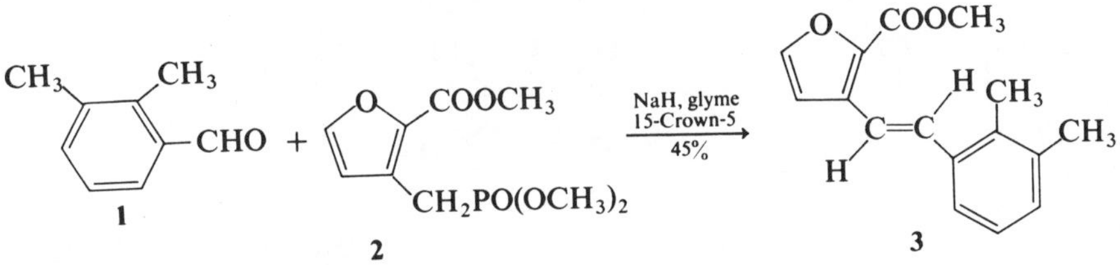

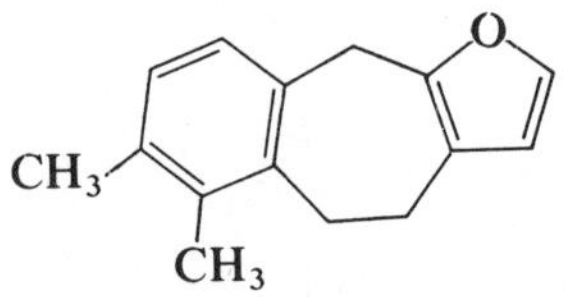

$$\mathbf{1} + \mathbf{2}\ (CH_2PO(OCH_3)_2,\ COOCH_3) \xrightarrow[45\%]{\text{NaH, glyme}\ \text{15-Crown-5}} \mathbf{3}$$

4

Chiral crown ethers. Cram and Sogah[4] have observed that potassium bases [$KOC(CH_3)_3$ or KNH_2] complexed by the chiral crown ethers **1** or **2** catalyze asymmetric Michael additions to methyl vinyl ketone and to methyl acrylate to give adducts in 60–99% optical purity.

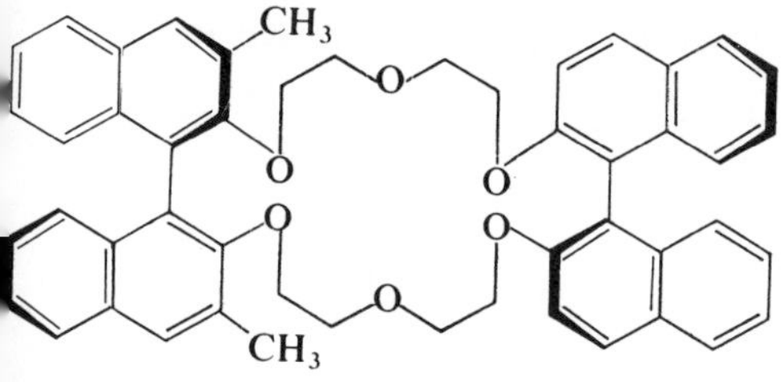

CH$_3$ O O O O O O CH$_3$

(R,R)-**1**

CH$_3$ O O O O O O CH$_3$

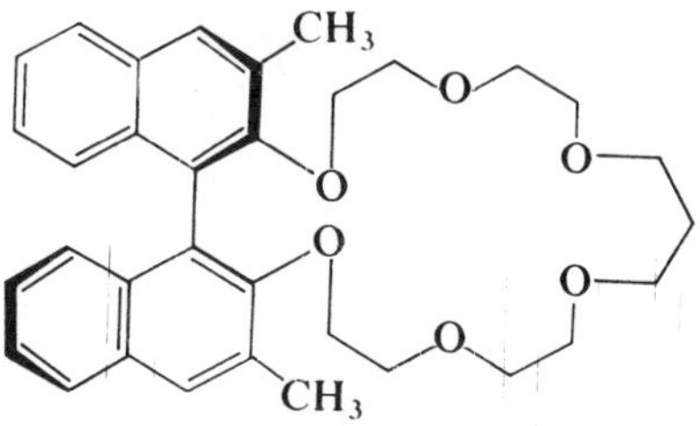

(R)-**2**

[1] R. Chênevert, R. Paquin, and A. Rodrigue, *Syn. Comm.*, **11**, 817 (1981).
[2] M. R. DeCamp and L. A. Viscogliosi, *J. Org.*, **46**, 3918 (1982).
[3] R. Baker and R. J. Sims, *J.C.S. Perkin I*, 3087 (1981).
[4] D. J. Cram and G. D. Y. Sogah, *J.C.S. Chem. Comm.*, 625 (1981).

Cyanoacetic acid, N≡CCH$_2$COOH. Mol. wt. 85.08, m.p. 65–67°.

cis-*Dihydroxylation.* *trans*-1,2-Bromohydrins can be converted into *cis*-1,2-diols by the sequence formulated for *trans*-2-bromocyclohexanol. The cyanoacetate ester **1**, prepared in 85% yield by reaction with cyanoacetic acid (tosyl chloride in pyridine-methylene chloride, **1**, 1183–1184), on treatment with sodium hydride is converted into the ketene acetal **2**, which is hydrolyzed by acid to the monocyanoacetate of *cis*-cyclohexane-1,2-diol (**3**). The diol itself is obtained by hydrolysis with potassium carbonate in methanol (95% yield). This methodology is particularly attractive because the starting bromohydrins are available from cyclohexenes by two stereochemically complementary routes. The reaction with hypobromous acid adds oxygen to the more hindered face of the bond, whereas epoxidation followed by cleavage with HBr adds oxygen to the less hindered face. In contrast, dihydroxylation with OsO_4 occurs preferentially at the less hindered face of the alkene. An example is the dihydroxylation of **4**. Reaction of **4** with OsO_4–pyridine results in a 1:1 mixture of the 2α,3α-diol and the 2β,3β-diol. Epoxidation ($ClC_6H_4CO_3H$) results in the 2α,3α-epoxide (95% yield), which is converted into the *trans*-bromohydrin **5** (90% yield). This in turn is converted through the cyanoester into the 2α,3α-diol (**6**) in 72.5% overall yield.

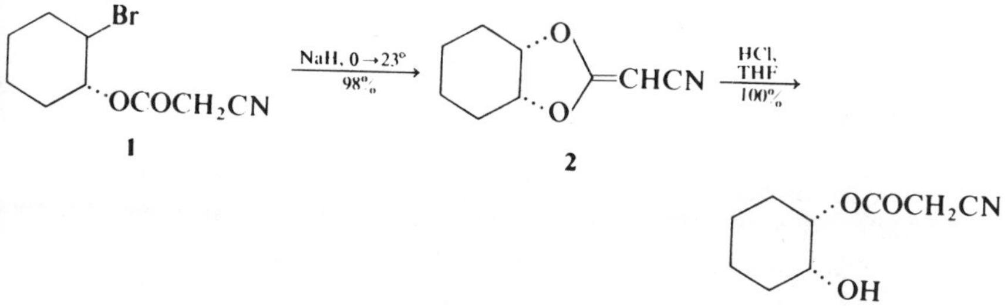

Br OCOCH$_2$CN **1** $\xrightarrow[98\%]{\text{NaH, } 0\rightarrow 23^\circ}$ O O =CHCN **2** $\xrightarrow[100\%]{\text{HCl, THF}}$

OCOCH$_2$CN OH **3**

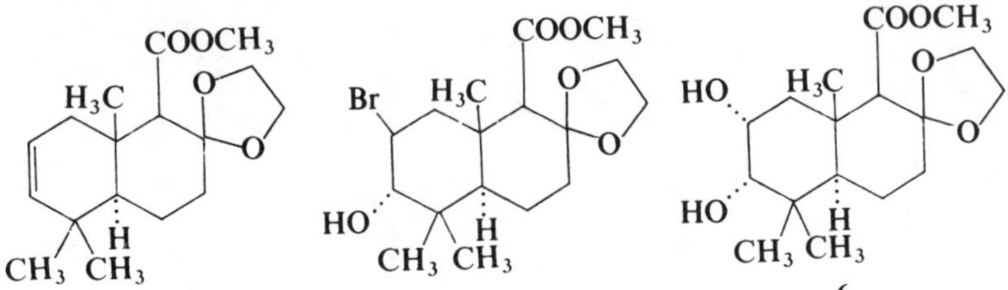

[1] E. J. Corey and J. Das, *Tetrahedron Letters*, **23**, 4217 (1982).

Cyanogen chloride N-oxide, $ClC{\equiv}N \rightarrow O$ **(1)**. The reagent can be generated *in situ* by treatment of dichloroformaldoxime with silver nitrate.

3-Chloroisoxazolines.[1] The reagent forms 1,3-cycloadducts with mono- and disubstituted alkenes (equation I).

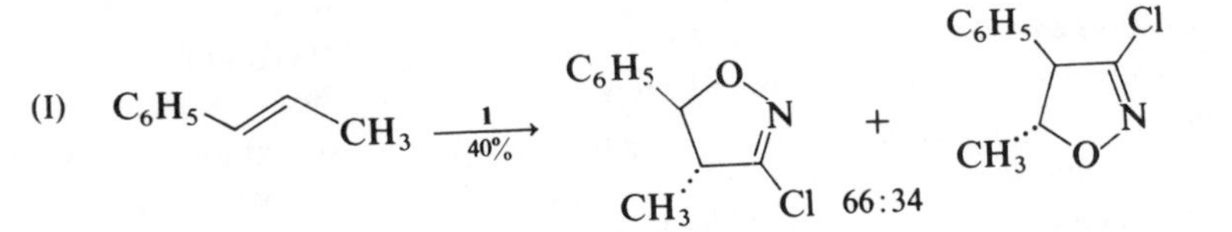

This reaction has been used for preparation of the antimetabolite AT-125 (**2**, acivin) from (S)-vinylglycine (equation II).

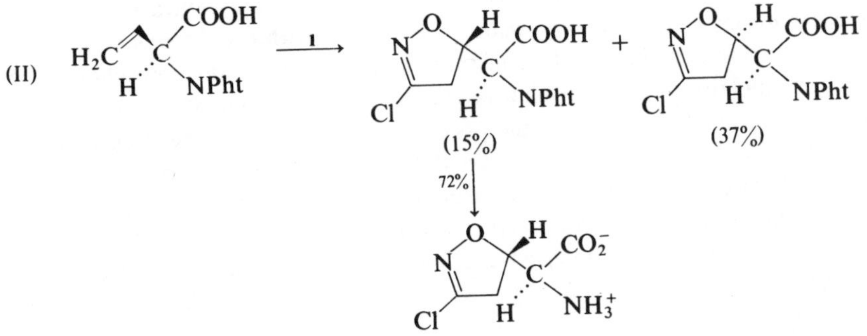

[1] P. A. Wade, M. K. Pillay, and S. M. Singh, *Tetrahedron Letters*, **23**, 4563 (1982).

Cyanomethylenetriphenylphosphorane, $(C_6H_5)_3P{=}CHCN$ **(1)**. Preparation.[1]

Selective reactions. Wittig reactions of **1** with an aldehyde are possible in the presence of keto, ester, and amino groups.[2] Wittig reagents react under normal conditions with acid chlorides. The stabilized ylid **1** reacts preferentially with the acid chloride group of 4-formylbenzoyl chloride (**2**) to give **3** as the major product.[3]

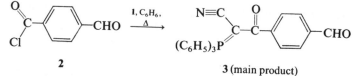

2 → ($1, C_6H_6, \Delta$) → 3 (main product)

In contrast, the nonstabilized ylids $(C_6H_5)_3P{=}CH_2$ and $(C_6H_5)_3P{=}C(CH_3)_2$ and even the moderately stabilized ylid $(C_6H_5)_3P{=}CHC_6H_5$ react preferentially with the formyl group of **2**, even under salt-free conditions.[4] Inverse addition is desirable to avoid presence of excess ylid.

[1] G. P. Schiemenz and H. Engelhard, *Ber.*, **94**, 578 (1961).
[2] A. Maercker, *Org. React.*, **14**, 270 (1965).
[3] F. Wätjen, O. Dahl, and O. Buchardt, *Tetrahedron Letters*, **23**, 4741 (1982).
[4] H. J. Bestmann, *Angew. Chem.*, **77**, 651 (1965).

Cyanotrimethylsilane, 4, 542–543; **5,** 720–722; **6,** 632–633; **7,** 397–399; **8,** 133; **9,** 127–129; **10,** 112–114. The reagent can be prepared in high yield by reaction of chlorotrimethylsilane and potassium cyanide in acetonitrile in the presence of catalytic amounts of sodium iodide and pyridine. Iodotrimethylsilane is generated *in situ* as an intermediate.[1,2]

β-Hydroxy isocyanides; β-amino alcohols.[3] The reagent reacts with epoxides in the presence of zinc iodide to give β-hydroxy isocyanides in contrast to the usual β-hydroxy nitriles (**5**, 720). The products are useful intermediates to oxazolines and β-amino alcohols. The reaction is highly regio- and stereoselective (equation I).

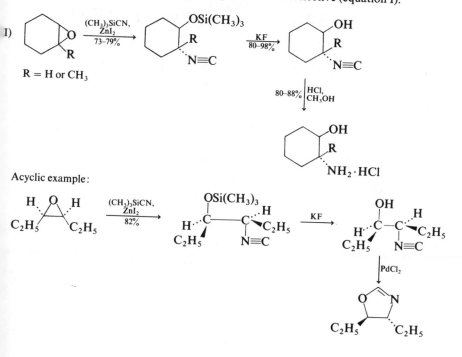

I) epoxide → ($(CH_3)_3SiCN$, ZnI_2, 73–79%) → $OSi(CH_3)_3$, R, $N{\equiv}C$ → (KF, 80–98%) → OH, R, $N{\equiv}C$ → (HCl, CH_3OH, 80–88%) → OH, R, $NH_2 \cdot HCl$

R = H or CH_3

Acyclic example:

H, O, H, C_2H_5, C_2H_5 → ($(CH_3)_3SiCN$, ZnI_2, 82%) → $OSi(CH_3)_3$, H, H, C_2H_5, C_2H_5, $N{\equiv}C$ → (KF) → OH, H, H, C_2H_5, C_2H_5, $N{\equiv}C$ → ($PdCl_2$) → O, N, C_2H_5, C_2H_5

Arylacetic acids.[4] A new synthesis of phenylacetic acids from aromatic ketones is illustrated for conversion of 6-methoxy-1-tetralone (**1**) into 6-methoxy-1,2,3,4-tetrahydronaphthalene-1-carboxylic acid (**3**).

$$\mathbf{1} \xrightarrow[83.2\%]{NCSi(CH_3)_3,\ ZnI_2} \mathbf{2}\ (NC,\ OSi(CH_3)_3) \xrightarrow[68\%]{1)\ SnCl_2;\ 2)\ HOAc;\ 3)\ HCl,\ \Delta} \mathbf{3}\ (COOH)$$

CH_3O **1** CH_3O **2** CH_3O **3**

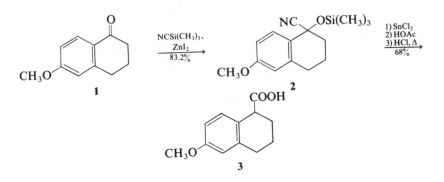

Cleavage of epoxides and oxetanes.[5] In the presence of diethylaluminum chloride as catalyst, cyanotrimethylsilane reacts with epoxides and oxetanes regioselectively to give open-chain products in which the CN group is attached to the less substituted α-carbon of the cyclic ether.

Examples:

$$\text{(2-methyloxirane)} + (CH_3)_3SiCN \xrightarrow[83\%]{(C_2H_5)_2AlCl,\ 20^\circ} N{\equiv}CCH_2CH(CH_3)OSi(CH_3)_3$$

$$\text{(2-methyloxetane)} + (CH_3)_3SiCN \xrightarrow{47\%} N{\equiv}C(CH_2)_2CH(CH_3)OSi(CH_3)_3$$

Enals → enones.[6] The adducts of enals with $(CH_3)_3SiCN$ can be alkylated regioselectively at the carbonyl carbon to afford, after hydrolysis, enones (equation I).

$$\text{(I)}\quad R^1R^2C{=}CR^3CHO \xrightarrow{(CH_3)_3SiCN,\ ZnI_2} R^1R^2C{=}CR^3CH(CN)OSi(CH_3)_3 \xrightarrow{1)\ LDA;\ 2)\ R^4Br}$$

$$R^1R^2C{=}CR^3C(R^4)(CN)OSi(CH_3)_3 \xrightarrow[60\text{–}85\%\ \text{overall}]{H_3O^+\ \text{or}\ F^-} R^1R^2C{=}CR^3C(=O)R^4$$

Cyanomethyl ethers, $ROCH_2CN$.[7] Conventional methods for preparation of these ethers with $ClCH_2CN$ fail. A recent method involves reaction of $(CH_3)_3SiCN$ and ZnI_2 (**9**, 127–128) with methylsulfoxymethyl ethers in CH_2Cl_2.

Example:

CH_3SCH_2O–(steroid-like bicycle, H_3C) $\xrightarrow[81\%]{m\text{-}ClC_6H_4CO_3H,\ CH_2Cl_2,\ 0^\circ}$ $CH_3S(=O)CH_2O$–(bicycle, H_3C) $\xrightarrow[80\%]{(CH_3)_3SiCN,\ ZnI_2}$ $NCCH_2O$–(bicycle, H_3C)

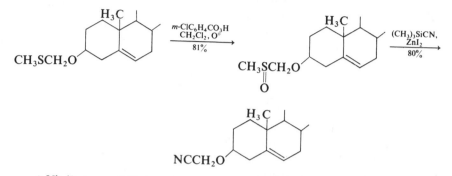

t-Nitriles; gem-dialkylation of ketones.[8] *t*-Alkyl halides undergo dehydrohalogenation when treated with cyanide ion. However, they react with cyanotrimethylsilane in the presence of $SnCl_4$ as catalyst to form *t*-nitriles in 75–90% yield. The selective reaction of the primary, tertiary dichloride (**1**) is noteworthy.

$$\underset{\mathbf{1}}{(CH_3)_2C(Cl)(CH_2)_3Cl} \xrightarrow[72\%]{(CH_3)_3SiCN,\ SnCl_4,\ CH_2Cl_2} \underset{\mathbf{2}}{(CH_3)_2C(CN)(CH_2)_3Cl}$$

Since *t*-alkyl chlorides can be obtained from ketones, this cyanation represents a method for *gem*-dialkylation of ketones.

Aryl nitriles.[9] The most satisfactory route to the naphthalenecarbonitrile **3** is addition of $(CH_3)_3SiCN$ to the 1-tetralone **1** followed by conversion to the α,β-unsaturated nitrile **2**. Dehydrogenation to the fully aromatic nitrile can be effected with DDQ, but is effected preferably with palladium-on-charcoal in combination with sulfur.

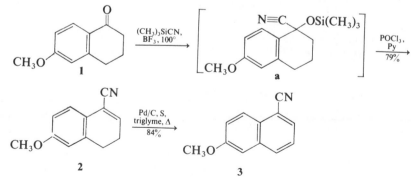

CH_3O-tetralone **1** $\xrightarrow{(CH_3)_3SiCN,\ BF_3,\ 100^\circ}$ [CH_3O-tetralin with $N{\equiv}C$ and $OSi(CH_3)_3$ at C-1, **a**] $\xrightarrow[79\%]{POCl_3,\ Py}$

CH_3O-dihydronaphthalene-CN **2** $\xrightarrow[84\%]{Pd/C,\ S,\ triglyme,\ \Delta}$ CH_3O-naphthalene-CN **3**

2-Alkoxy or 2,2-dialkoxy nitriles.[10] Ketals or orthoesters are converted into 2-alkoxy or 2,2-dialkoxy nitriles, respectively, by reaction with cyanotrimethylsilane in the presence of $SnCl_2$ or $BF_3 \cdot O(C_2H_5)_2$ as catalyst. A typical reaction is the preparation of dimethoxyacetonitrile (**1**), which is a useful reagent for conversion of alkyl halides to the homologous methyl carboxylate (equation I).

$$HC(OCH_3)_3 + NCSi(CH_3)_3 \xrightarrow[80\%]{BF_3 \cdot O(C_2H_5)_2,\ 25^\circ} \underset{\mathbf{1}}{(CH_3O)_2CHCN}$$

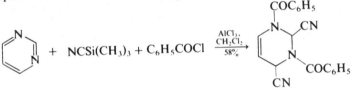

$$\text{(I)}\quad \mathbf{1} \xrightarrow[55\%]{1)\ LDA,\ THF,\ HMPT;\ 2)\ n\text{-}C_8H_{17}Br} n\text{-}C_8H_{17}C(OCH_3)_2CN \xrightarrow[96\%]{p\text{-}TsOH,\ (CH_3)_2C{=}O,\ H_2O,\ \Delta} n\text{-}C_8H_{17}\overset{O}{\overset{\|}{C}}OCH_3$$

Reissert compounds.[11] Pyridine does not form a Reissert compound, but pyridazine and pyrimidine react with cyanotrimethylsilane and benzoyl chloride to form a monocyclic Reissert compound.[12]

Example:

$$\text{pyrimidine} + NCSi(CH_3)_3 + C_6H_5COCl \xrightarrow[58\%]{AlCl_3,\ CH_2Cl_2} \text{1,3-bis}(COC_6H_5)\text{-2,4-bis}(CN)\text{-1,2,3,4-tetrahydropyrimidine}$$

[1] P. Cazeau. F. Moulines, O. Laporte, and F. Duboudin, *J. Organometal. Chem.*, **201**, C9 (1980); F. Duboudin, P. Cazeau, F. Moulines, and O. Laporte, *Synthesis*, 213 (1982).
[2] M. T. Reetz and I. Chatziiosifidis, *Synthesis*, 330 (1982).
[3] P. G. Gassman and T. L. Guggenheim, *Am. Soc.*, **104**, 5849 (1982).
[4] J. L. Belletire, H. Howard, and K. Donahue, *Syn. Comm.*, **12**, 763 (1982).
[5] J. C. Mullis and W. P. Weber, *J. Org.*, **47**, 2873 (1982).
[6] U. Hertenstein, S. Hünig, and M. Öller, *Ber.*, **113**, 3783 (1980).
[7] J. A. Schwindeman and P. D. Magnus, *Tetrahedron Letters*, **22**, 4925 (1981).
[8] M. T. Reetz and I. Chatziiosifidis, *Angew. Chem. Int. Ed.*, **20**, 1017 (1981).
[9] S. A. Jacobs and R. G. Harvey, *Org. Syn.*, submitted (1982).
[10] K. Utimoto, Y. Wakabayashi, Y. Shishiyama, M. Inoue, and H. Nozaki, *Tetrahedron Letters*, **22**, 4279 (1981).
[11] W. E. McEwen and R. L. Cobb, *Chem. Rev.*, **55**, 511 (1955).
[12] S. Veeraraghavan, D. Bhattacharju, and F. D. Popp, *J. Heterocyclic Chem.*, **18**, 443 (1981).

β-Cyclodextrin, 6, 151–152; 8, 133–135; 9, 129.

Diels-Alder catalysis. Water can accelerate the Diels-Alder reaction of cyclopentadiene with methyl vinyl ketone or acrylonitrile by a hydrophobic interaction. β-Cyclodextrin can increase this effect, possibly because the components can fit into the hydrophobic cavity.[1]

β-Cyclodextrin also catalyzes the intramolecular Diels-Alder cycloaddition of the furane **1** in water at 89° to give **2** in 91% yield. In the absence of the catalyst, the yield of **2** is only 20%. No yield enhancement is observed with α-cyclodextrin or with Brij-35.[2]

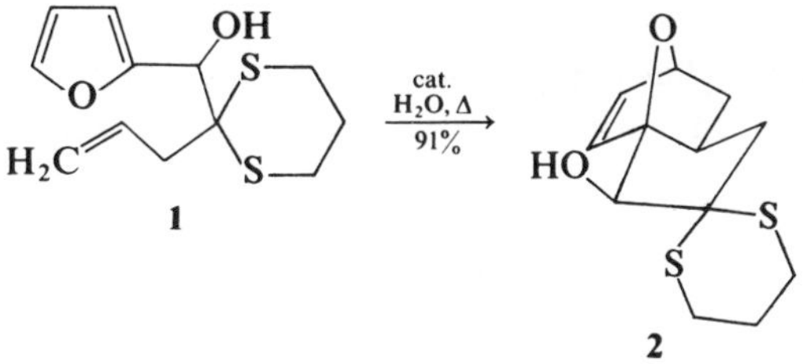

[1] R. Breslow and D. C. Rideout, *Am. Soc.*, **102**, 7816 (1980).
[2] D. D. Sternbach and D. M. Rossana, *ibid.*, **104**, 5853 (1982).

1-Cyclohexyl-3-(2-morpholinoethyl)carbodiimide metho-*p*-toluenesulfonate, $C_6H_{11}N{=}C{=}NCH_2CH_2\overset{+}{N}CH_3(CH_2CH_2)_2O\ CH_3C_6H_4SO_3^-$ **(1)**, **1**, 181. Supplier: Aldrich.

Cyclodehydration of N-acyl-α-amino acids.[1] 2-Oxazoline-5-ones are prepared conveniently by dehydration of N-acyl-α-amino acids with **1**. This reagent is superior to acetic acid, phosphorus halides, or DCC.

Example:

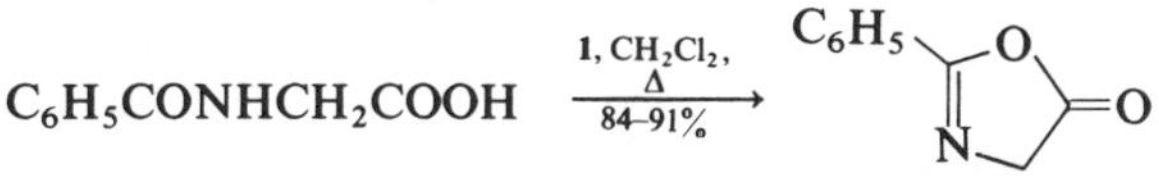

$C_6H_5CONHCH_2COOH \xrightarrow[84\text{–}91\%]{1,\ CH_2Cl_2,\ \Delta}$

Macrocylic lactonization.[2] Lactonization of **1** was effected most efficiently with this carbodiimide. Subsequent cleavage of the methoxymethyl ether (NaI in acetone) results in the antibiotic milbemycin β_3 **(2)** in 85% overall yield.

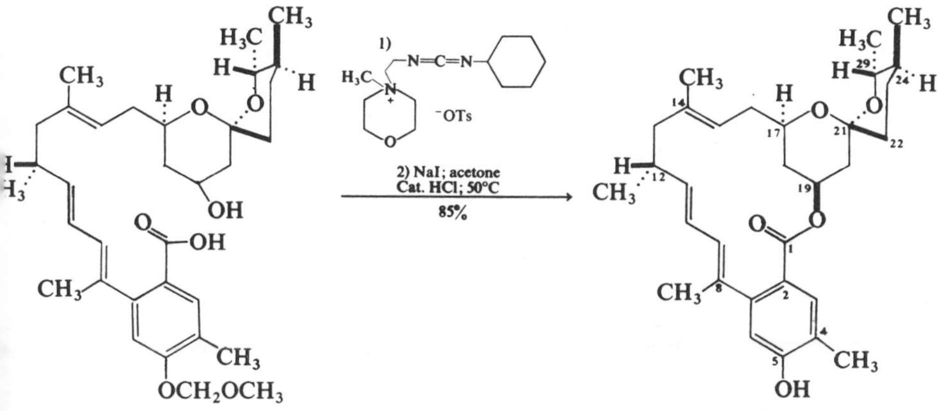

[1] C. F. Hoyng, T. Tiner-Harding, and M. G. McKenna, *Org. Syn.*, submitted (1981).
[2] D. R. Williams, B. A. Barner, K. Nishitani, and J. G. Phillips, *Am. Soc.*, **104**, 4708 (1982).

Cyclopentadienylmagnesium hydride (CpMgH). The hydride is prepared by reaction of cyclopentadiene in THF with magnesium hydride; it exists as the dimer (**1**) in this solvent.

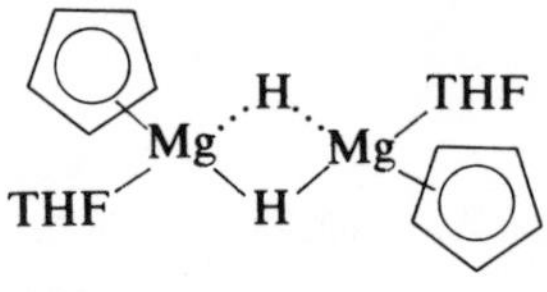

1

The hydride converts cyclopentadiene into Cp_2Mg. It converts aromatic ketones into deep blue radical anion-radical cation pairs. It also converts polynuclear arenes into radical pairs $[ArH^{\dot{-}}—CpMgH^{\dot{+}}]$.[1]

[1] A. B. Goel and E. C. Ashby, *J. Organometal. Chem.*, **214**, Cl (1981).

D

Darvon alcohol [(+)-(2S,3R)-4-(Dimethylamino)-3-methyl-1,2-diphenyl-2-butanol], **5**, 231; **8**, 184–185.

Chiral* 1,2-*aminodiols. Morrison *et al.*[1] have prepared five chiral 1,2-aminodiols related to Darvon alcohol. The only useful one for asymmetric reduction of ketones in conjunction with $LiAlH_4$ is **1**, prepared from (S)-propylene oxide and (S)-α-methylbenzylamine. Acetophenone and propiophenone are reduced by $LiAlH_4$-**1** to the corresponding (R)-alcohols in 77–82% ee. In this case, the three (S)-centers reinforce one other.

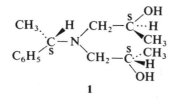

1

[1] J. D. Morrison, E. R. Grandbois, S. I. Howard, and G. R. Weisman, *Tetrahedron Letters*, **22**, 2619 (1981).

1,4-Diazabicyclo[2.2.2]octane (DABCO), 2, 99–101; **4**, 119; **5**, 176–177; **7**, 86–87.

Michael addition. DABCO is a particularly effective catalyst for Michael addition of β-keto thiolesters to α,β-unsaturated esters and ketones. The addition to enones provides an attractive route to 1,5-diketones that can undergo annelation to fused or

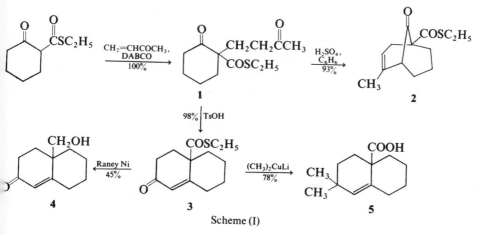

$CH_2{=}CHCOCH_3$, DABCO, 100%; H_2SO_4, C_6H_6, 93%; 98% TsOH; Raney Ni, 45%; $(CH_3)_2CuLi$, 78%

Scheme (I)

bridged ring systems with substituents at the angular position (Scheme I).[1] The thiolester substituent in the products can be reduced selectively to the primary alcohol by Raney nickel. The conversion of the keto group of **3** to a *gem*-dimethyl group by reaction with lithium dimethyl cuprate is without precedent. The related O-ester merely undergoes 1,2-addition to give the expected epimeric carbinols.

In the presence of a tertiary amine, particularly DABCO, S,S′-diethyl dithiomalonate undergoes 1,4-addition to α,β-enones and α,β-unsaturated esters that are not disubstituted in the β-position. The products are reduced by acid-washed Raney nickel to 1,5-ketols and δ-hydroxy esters.[2]

Examples:

$COOC_2H_5$, CH_3, CH_3 (unsaturated lactone) $\xrightarrow[100\%]{CH_2(COSC_2H_5)_2,\ DABCO}$ lactone with $COOC_2H_5$ and $CH(COSC_2H_5)_2$ $\xrightarrow[75\%]{\text{Raney Ni}}$ lactone with $COOC_2H_5$ and CH_2CH_2OH

$$(C_2H_5OOC)_2C{=}CHCH(CH_3)_2 \xrightarrow{95\%} (C_2H_5OOC)_2CHCH(CH(CH_3)_2)CH(COSC_2H_5)_2 \xrightarrow[2)\ OH^-]{1)\ \text{Raney Ni}}$$ lactone with $COOC_2H_5$ and $CH(CH_3)_2$

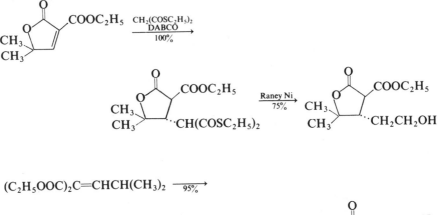

Lithium trialkylborohydrides.[3] Trialkylboranes react rapidly with lithium aluminum hydride in ether in the presence of DABCO to form trialkylborohydrides and aluminum hydride, which is precipitated as a complex with DABCO. The method is suitable for preparation of even highly hindered complex hydrides. Yields are quantitative.

$$R_3B + LiAlH_4 \xrightarrow{\text{quant.}} LiR_3BH + AlH_3$$

$$AlH_3 \xrightarrow{\text{DABCO}} \text{Complex}$$

[1] H.-J. Liu, L.-K. Ho, and H. K. Lai, *Can. J. Chem.*, **59**, 1685 (1981).
[2] H.-J. Liu and I. V. Oppong, *ibid.*, **60**, 94 (1982).
[3] H. C. Brown, J. L. Hubbard, and B. Singaram, *Tetrahedron*, **37**, 2359 (1981).

1,8-Diazabicyclo[5.4.0]-7-undecene (DBU), 2, 101; **4,** 16–18; **5,** 177–178; **6,** 158; **7,** 87–88; **8,** 141; **9,** 132–133.

Michael additions. Michael addition of nitroalkanes to vinyl sulfoxides (equation I) and to ketene diethyl dithioacetal S-monooxide (equation II) proceeds in high yield when DBU is used as base. The adducts can be converted into α,β-enals.[1]

$$\text{(I)}\quad R^1R^2CHNO_2 + R^3CH{=}CH\overset{O}{\overset{\uparrow}{S}}C_6H_5 \xrightarrow[CH_3CN]{DBU,} R^1R^2C(NO_2)-CH(R^3)CH_2\overset{O}{\overset{\uparrow}{S}}C_6H_5$$

$$\xrightarrow[\substack{50-70\% \\ \text{overall}}]{\substack{1)\,(CH_3CO)_2O \\ 2)\,DBU,\,(C_2H_5)_2O}} R^1R^2C{=}C(R^3)CHO$$

$$\text{(II)}\quad R^1R^2CHNO_2 + CH_2{=}C(SC_2H_5)(\overset{\downarrow}{S}C_2H_5 \to O) \xrightarrow{DBU} R^1R^2C(NO_2)CH_2CH(SC_2H_5)-\underset{\downarrow O}{S}C_2H_5$$

$$\xrightarrow[\substack{50-80\% \\ \text{overall}}]{HClO_4} R^1R^2C{=}CHCHO$$

t-Butyl esters.[2] The reaction of N-acylimidazoles with *t*-butyl alcohol is accelerated by DBU. This modified esterification is useful for acid-sensitive substrates. It fails, however, with pivalic acid.

$$RCOOH + (C_3H_3N_2)-\overset{O}{\overset{\|}{C}}-(C_3H_3N_2) \longrightarrow \left[R\overset{O}{\overset{\|}{C}}-N(C_3H_3N)\right] \xrightarrow[60-90\%]{(CH_3)_3COH,\ DBU} RCOOC(CH_3)_3$$

[1] N. Ono, H. Miyake, R. Tanikaga, and A. Koji, *J. Org.*, **47**, 5017 (1982).
[2] S. Ohta, A. Shimabayashi, M. Aono, and M. Okamoto, *Synthesis*, 833 (1982).

1-Diazolithioacetone, (1). The reagent is prepared *in situ* by reaction of LDA with diazoacetone in THF at −78°.

$$N_2{=}\overset{Li}{\overset{|}{C}}-\overset{O}{\overset{\|}{C}}CH_3$$

β-Diketones.[1] The reagent reacts with aldehydes to form an α-diazo-β-hydroxy ketone (**2**), which is transformed into a β-diketone (**3**) by a catalytic amount of $Rh_2(OAc)_4$.

Example:

$$C_6H_5CH_2CHO + \mathbf{1} \xrightarrow{61\%} C_6H_5CH_2CH(OH)C(=N_2)COCH_3 \ (\mathbf{2}) \xrightarrow[77\%]{Rh_2(OAc)_4} C_6H_5CH_2COCH_2COCH_3 \ (\mathbf{3})$$

[1] R. Pellicciari, R. Fringuelli, E. Sisani, and M. Curini, *J.C.S. Perkin I*, 2566 (1981).

Dibenzo-18-crown-6–Bromine (DBC·Br$_2$). The yellow, crystalline complex separates on addition of bromine to the polyether dissolved in CCl_4. The molar ratio of polyether/Br_2 is about 1.85.[1]

Bromination.[2] This bromine–crown ether complex, like dioxane–bromine (**5**, 58), can brominate alkenes, but the stereoselectivity is greater than that with free bromine and is less sensitive to solvent effects. Thus, bromination of *trans*-(β-methylstyrene with DBC·Br_2 occurs exclusively by *anti*-addition and bromination of *cis*-β-methylstyrene occurs by *anti*-addition to the extent of 95–100%. The bromine complex of polydibenzo-18-crown-6[3] is a particularly useful reagent because it can be packed as a slurry in a chromatography column. The alkene is then placed on the column and eluted with CCl_4.

[1] E. Schori and J. Jagur-Grodzinski, *Israel J. Chem.*, **10**, 935 (1972).
[2] K. H. Pannell and A. J. Mayr, *J.C.S. Perkin I*, 2153 (1982).
[3] *Parish Chem.*, Provo, Utah.

Diborane, 1, 199–207; **2**, 106–108; **3**, 76–77; **4**, 124–126; **5**, 184–186; **6**, 161–162; **7**, 89; **8**, 141–143; **9**, 136–138.

Reduction of β-lactams.[1] Reduction of β-lactams and penicillins results in 1,3-amino alcohols rather than azetidines.

Examples:

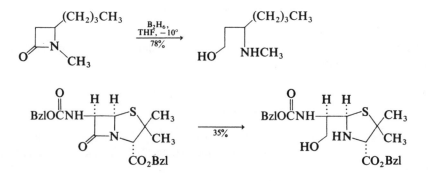

1-Methyl-4-butylazetidin-2-one $\xrightarrow[78\%]{B_2H_6,\ THF,\ -10^\circ}$ $HOCH_2CH_2CH((CH_2)_3CH_3)NHCH_3$

BzlOCNH-penicillin ester (CO_2Bzl) $\xrightarrow{35\%}$ ring-opened amino alcohol (HO, HN, CO_2Bzl)

Catechol estrogens.[2] Hydroboration-oxidation of the 2-chloromercuric derivative (**2**) of an estrogen (**1**) results in the catechol estrogen **3** in 45% overall yield. An alternate route to **3** is also shown in equation (I).

(I) **1** $\xrightarrow[80\%]{HgCl_2}$ **2** $\xrightarrow[\text{45\% overall}]{\text{1) } B_2H_6\text{; 2) } H_2O_2\text{; 3) } Ac_2O\text{, Py}}$ **3**

1 $\xrightarrow[80\%]{BF_3,\ Ac_2O,\ 0°}$ **4** $\xrightarrow{\text{1) } H_2O_2\text{; 2) } Ac_2O\text{, Py}}$ **3** $\xrightarrow[75\%]{\text{1) Py·HCl; 2) } Ac_2O\text{, Py}}$ **5**

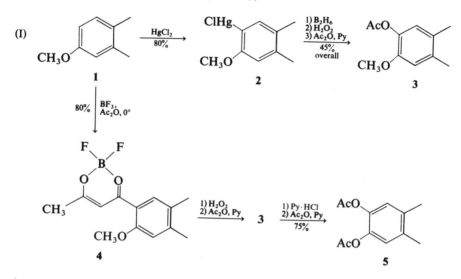

[1] P. G. Sammes and S. Smith, *J.C.S. Chem. Comm.*, 1143 (1982).
[2] E. Santaniello, A. Fiecchi, and P. Ferraboschi, *ibid.*, 1157 (1982).

Dibromobis(triphenylphosphine)nickel(II)–Zinc.

α,β-Unsaturated nitriles.[1] A Ni(0) catalyst generated *in situ* from $Br_2Ni[P(C_6H_5)_3]_2$, Zn, and $P(C_6H_5)_3$ (1:3:2) effects cyanation of vinyl halides (equation I). The stereochemistry is mainly retained; the (E)-isomers are formed in 95% selectivity and the (Z)-isomers in 69–88% selectivity.

(I) $R^1R^2C{=}CR^3X + KCN \xrightarrow{\text{Ni(0), DMF}} R^1R^2C{=}CR^3CN$

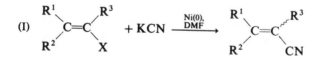

[1] Y. Sakakihara, N. Yadani, I. Ibuki, M. Sakai, and N. Uchino, *Chem. Letters*, 1565 (1982).

Dibromomethane, CH_2Br_2. Mol. wt. 173.85, b.p. 96–98°. Suppliers: Aldrich, Fluka.

Methylene acetals.[1] Carbohydrate methylene acetals are prepared conveniently from 1,3-diols and *cis*- or *trans*-1,2-diols by reaction with dibromomethane in dry DMSO in the presence of finely powdered potassium hydroxide. Yields are 35–90%.

[1] A. Lipták, V. A. Oláh, and J. Kerékgyártó, *Synthesis*, 420 (1982).

Dibromomethyllithium, Br_2CHLi (**1**), **5**, 398; **7**, 90.

1-*Bromo-1-alkenes.*[1] These products can be prepared from aldehydes or ketones by a two-step process: addition of dibromomethyllithium (prepared *in situ*) to form dibromo alcohols, followed by reduction with zinc dust and acetic acid in refluxing methylene chloride. The method is particularly useful for preparation of 1-bromo-1,3-dienes. The only drawback is the lack of stereoselectivity.

Examples:

$$\text{C}_6\text{H}_{11}\text{CHO} \xrightarrow[83\%]{CH_2Br_2,\ LiN(C_6H_{11})_2} \text{C}_6\text{H}_{11}\text{CH(OH)CHBr}_2 \xrightarrow[75\%]{Zn,\ HOAc,\ CH_2Cl_2,\ \Delta} \text{C}_6\text{H}_{11}\text{CH=CHBr}$$

E/Z = 55:45

$$\text{C}_6\text{H}_5\text{CHO} \xrightarrow{98\%} \text{C}_6\text{H}_5\text{CH(OH)CHBr}_2 \xrightarrow{95\%} \text{C}_6\text{H}_5\text{CH=CHBr}$$

E/Z = 55:45

The products are useful precursors to 1-alkynes. The 1-bromo-1,3-dienes can be converted to alkenyllithiums by *tert*-butyllithium (2 equiv.) at $-120°$ in THF/pentane. These products react with aldehydes with retention of olefin geometry.

Example:

$$\text{CH}_3\text{CH=CHCH=CHCH=CHBr} \xrightarrow[71\%]{1)\ tert\text{-BuLi};\ 2)\ (CH_3)_2CHCHO} \text{CH}_3\text{CH=CHCH=CHCH=CHCH(OH)CH(CH}_3)_2$$

E/Z = 55:45 E/Z = 57:43

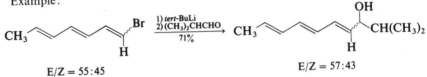

Perhydroazulenones.[2] The key step in a synthesis of carpesiolin (**3**), a pseudoguaianolide lactone, is a regioselective ring expansion of a perhydroindanone (**1**) to **2** by dibromomethyllithium.

$$\mathbf{1} \xrightarrow[40\%]{1)\ Br_2CHLi;\ 2)\ n\text{-BuLi}} \mathbf{2}$$

1 (substituents: CH_3, O, $(CH_3)_3CO$, CH_3)

2 (substituents: CH_3, H, O, $(CH_3)_3CO$, CH_3)

3 (substituents: CH_3, H, O, O, O, CH_3, OH, CH_2)

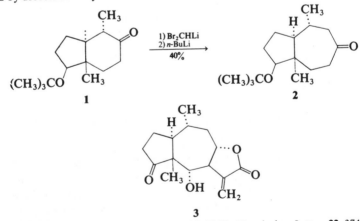

[1] D. R. Williams, K. Nishitani, W. Bennett, and S. Y. Sit, *Tetrahedron Letters*, **22**, 3745 (1981).

[2] K. Nagao, M. Chiba, I. Yoshimura, and S.-W. Kim, *Chem. Pharm. Bull. Japan*, **29**, 2733 (1981).

Di-*n*-butylboryl trifluoromethanesulfonate (1), **7**, 91–92; **9**, 140.

Aldol-type reactions.[1] In the presence of **1** or of 9-BBN triflate and a tertiary amine, acetonitriles and benzaldehydes react to form aldol products.

Examples:

$$CH_3CN + C_6H_5CHO \xrightarrow[90\%]{\text{1, base}\; CH_2Cl_2} C_6H_5CH(OH)CH_2CN$$

$$CH_3CH{=}CHCH_2CN + C_6H_5CHO \xrightarrow[82\%]{} C_6H_5CH(OH)CH(CH{=}CHCH_3)CN$$

[1] H. Hamana and T. Sugasawa, *Chem. Letters*, 1401 (1982).

Di-*t*-butyldichlorosilane, $[(CH_3)_3C]_2SiCl_2$ (**1**). Mol. wt. 213.2, b.p. 212°. Supplier: Petrarch. The reagent can be prepared by chlorination of di-*t*-butylchlorosilane (Cl_2,CCl_4, 8–10°).

Protection of* 1,2- *and* 1,3-*diols.[1] 1,2- and 1,3-Diols are converted into di-*t*-butylsilylene derivatives by reaction with **1** with triethylamine as base and 1-hydroxybenzotriazole as catalyst in CH_3CN. The protective group is removed most conveniently by desilylation with pyridinium fluoride.

[1] B. M. Trost and C. G. Caldwell, *Tetrahedron Letters*, **22**, 4999 (1981).

(2,6-Di-*t*-butyl-4-methylphenoxy)methylaluminum trifluoromethanesulfonate (1).

$$4\text{-}CH_3\text{-}2,6\text{-}[C(CH_3)_3]_2C_6H_2{-}O{-}Al(CH_3)(OSO_2CF_3) \quad \textbf{(1)}$$

The reagent is prepared by addition of trimethylaluminum to 2,6-di-*t*-butyl-4-methylphenol in CH_2Cl_2 at −78°. After a reaction period of 30 minutes at 0°, triflic acid is added at −78°. The mixture is then stirred for 30 minutes at 0°.

Rearrangement of allyl ethers.[1] Treatment of isopentenyl prenyl ether **2** with **1** results in formation of the two C_{10}-alcohols **3** and **4** in a 1:3 ratio. Use of other Lewis acids results in various side reactions. This rearrangement to **4** mimics the biosynthesis of acyclic terpenes by head to tail condensation of isoprene units.

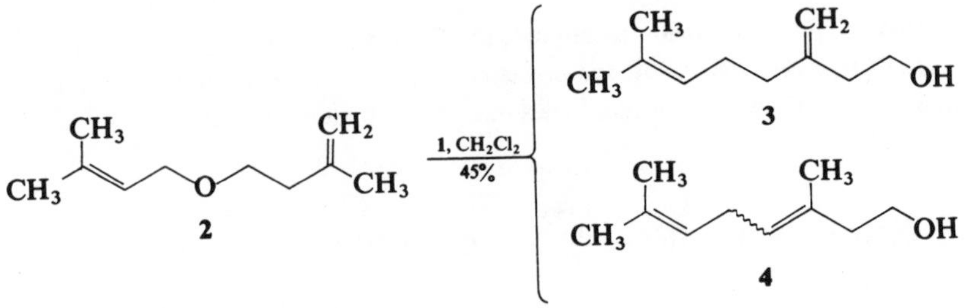

CH_3 CH_2 CH_3 O CH_3 **2** $\xrightarrow[45\%]{1,\ CH_2Cl_2}$ CH_3 CH_2 CH_3 OH **3**; CH_3 CH_3 CH_3 OH **4**

The organoaluminum reagent can also effect a biomimetic head to head condensation of C_5 units. Thus, diprenyl ether (**5**) is converted by treatment with **1** into lavandulol (**6**) in 33% yield.

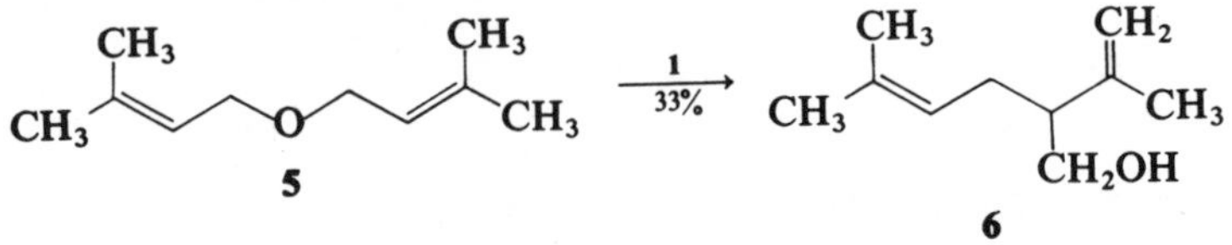

CH_3 CH_3 CH_3 O CH_3 **5** $\xrightarrow[33\%]{1}$ CH_3 CH_2 CH_3 CH_3 CH_2OH **6**

[1] Y. Yamamura, K. Umeyama, K. Maruoka, and H. Yamamoto, *Tetrahedron Letters*, **23**, 1933 (1982).

2,6-Di-*t*-butyl-4-methylpyridine, 8, 145–146; **9,** 141; **10,** 123.

Deoxygenation of ketones. Enol triflates, readily available by reaction of ketones and triflic anhydride catalyzed by this sterically hindered base, undergo rapid hydrogenolysis to the saturated hydrocarbon. Overall yields are 65–90%.[1]

Examples:

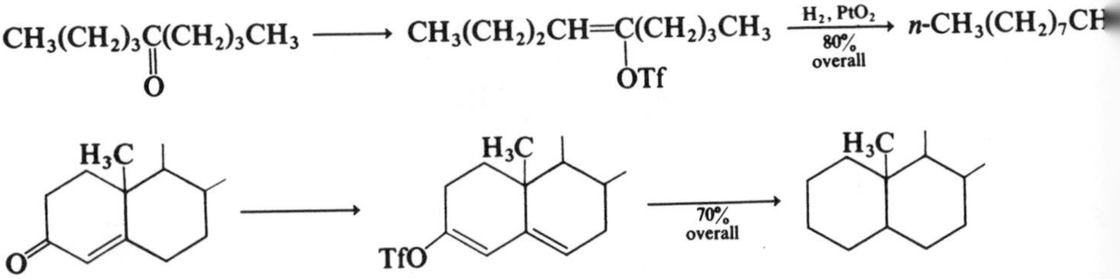

$CH_3(CH_2)_3C(=O)(CH_2)_3CH_3 \longrightarrow CH_3(CH_2)_2CH{=}C(OTf)(CH_2)_3CH_3 \xrightarrow[\text{80\% overall}]{H_2,\ PtO_2} n\text{-}CH_3(CH_2)_7CH$

H_3C (enone) $\longrightarrow$ H_3C, TfO (dienyl triflate) $\xrightarrow{\text{70\% overall}}$ H_3C (decalin)

[1] V. B. Jigajinni and R. H. Wightman, *Tetrahedron Letters*, **23**, 117 (1982).

Di-*t*-butyl nitroxide, 1, 211; **2,** 81; **4,** 129; **2,2,6,6-Tetramethylpiperidine N-oxyl, 6,** 110–111.

Reactions with arenediazonium fluoroborates. Treatment of arenediazonium tetrafluoroborates with these nitroxides, (**1**) or (**2**), generates aryl radicals (Ar·), which can then undergo useful synthetic reactions.[1]

Examples:

BF_4^- $\overset{+}{N_2}$ $CH{=}CH_2$ + $[(CH_3)_3C]_2N{-}O\cdot$ (**1**) $\xrightarrow{76\%}$ $CH_2O{-}N[C(CH_3)_3]_2$

BF_4^- $\overset{+}{N_2}$ $C{\equiv}CH$, N–Ac + CH_3, CH_3, N–O·, CH_3, CH_3 (**2**) $\xrightarrow{48\%}$ CHO, N–Ac + CH_3, CH_3, N–H, CH_3, CH_3

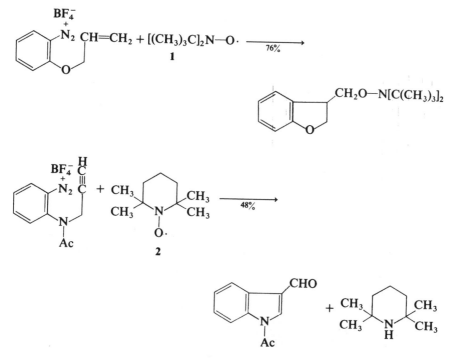

[1] A. L. J. Beckwith and G. F. Meijs, *J.C.S. Chem. Comm.*, 595 (1981).

Di-*n*-butyltin dichloride, $(C_4H_9)_2SnCl_2$. Mol. wt. 303.83, m.p. 39–41°. Supplier: Alfa.

Ketimines.[1] Di-*n*-butyltin dichloride is an efficient catalyst for condensation of cyclic ketones with a primary amine to form an imine. However, ketones with an optically active center at the α-position undergo partial racemization. Extensive loss of optical activity is observed when titanium(IV) chloride is used as catalyst.

[1] C. Stetin, B. de Jeso, and J. C. Pommier, *Syn. Comm.*, **12**, 495 (1982).

Dicarbonyl(cyclopentadienyl)cobalt, 5, 172–173; **6,** 153–154; **7,** 94–95; **8,** 146–147; **9,** 142–143; **10,** 126–127.

Biphenylenes.[1] *o*-Diethynylbenzene (**1**)[2] cocyclizes with monoalkynes in the presence of this cobalt catalyst to furnish biphenylenes (**2**). The yield of **2** is high when

R^1 and $R^2 = Si(CH_3)_3$, but is only moderate when R^1 and/or $R^2 = H, C_5H_{11}, C_6H_5$, or $COOCH_3$. Even so, this reaction provides access to novel biphenylene derivatives.

$$\mathbf{1} + R^1C{\equiv}CR^2 \xrightarrow[25\text{–}95\%]{CpCo(CO)_2,\ h\nu} \mathbf{2}$$

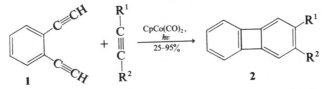

[1] B. C. Berris, Y.-H. Lai, and K. P. C. Vollhardt, *J.C.S. Chem. Comm.*, 953 (1982).
[2] S. Takahashi, Y. Kuroyama, K. Sonogashira, and N. Hagihara, *Synthesis*, 627 (1980).

Di-μ-carbonylhexacarbonyldicobalt, 1, 224–225; **3,** 89; **4,** 139; **5,** 204–205; **6,** 172; **7,** 99–100; **8,** 148–150; **9,** 144–145; **10,** 129–130.

$Co_2(CO)_6$-complexes of enynes.[1] The complex of isopropenylacetylene **1** reacts consecutively with an electrophile and a nucleophile via a stabilized intermediate carbenium ion.

Example:

$$HC{\equiv}C{-}C(CH_3){=}CH_2\,[Co_2(CO)_6] + CH_3\overset{+}{C}OBF_4^- \longrightarrow \left[HC{\equiv}C{-}\overset{+}{C}(CH_3)CH_2COCH_3\,[Co_2(CO)_6]\right] \xrightarrow{CH_3OH}$$

$$HC{\equiv}CC(CH_3)(OCH_3)CH_2COCH_3\,[(CO)_6Co_2] \xrightarrow[\text{65\% overall}]{Fe(III)} HC{\equiv}CC(CH_3)(OCH_3)CH_2COCH_3$$

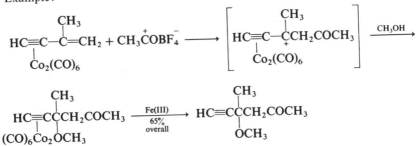

Cyclopentenone synthesis.[2] The key step in a regiospecific synthesis of methylenemycin B (**4**) involves the reaction of (2-butyne)hexacarbonyldicobalt (**1**) with tetrahydro-2-(2-propenyloxy)pyrane (**2**). Although the reaction of complexes

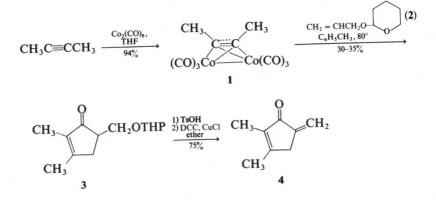

$$CH_3C{\equiv}CCH_3 \xrightarrow[94\%]{Co_2(CO)_8,\ THF} \mathbf{1}\ [CH_3C{\equiv}CCH_3\cdot(CO)_3Co{-}Co(CO)_3] \xrightarrow[30\text{–}35\%]{CH_2{=}CHCH_2O\text{-}THP\ (\mathbf{2}),\ C_6H_5CH_3,\ 80^\circ}$$

$$\mathbf{3}\ (CH_2OTHP) \xrightarrow[75\%]{\text{1) TsOH; 2) DCC, CuCl, ether}} \mathbf{4}\ (=CH_2)$$

such as **1** with simple alkenes shows little regiospecificity, only one cyclopentenone was formed in this instance.

Reductive carbonylation and alkylation of imines.[3] This cobalt carbonyl catalyzes a reaction of imines with organoboranes and carbon monoxide that results in amides. β-Keto amides are obtained from α-keto imines by this reaction.

Example:

$$C_6H_5COC(C_6H_5)=NC_6H_5 + CO + (n\text{-}C_4H_9)_3B \xrightarrow[\text{THF}]{Co_2(CO)_8,} C_6H_5COCH(C_6H_5)N(C_6H_5)COC_4H_9\text{-}n\ (82\%) + C_6H_5COCH(C_6H_5)NHC_6H_5\ (18\%)$$

[1] A. A. Schegolev, W. A. Smit, Y. B. Kalyan, M. Z. Krimer, and R. Caple, *Tetrahedron Letters*, **23**, 4419 (1982).

[2] D. C. Billington and P. L. Pauson, *Organometallics*, **1**, 1560 (1982).

[3] H. Alper and S. Amaratunga, *J. Org.*, **47**, 3593 (1982).

Dichlorobis(cyclopentadienyl)titanium, 10, 130–131.

Hydromagnesiation of 2-propynylic alcohols. The hydromagnesiation of alkynes (**10**, 130) has been extended to propynylic alcohols (equation I), which provides a new route to terpenoids.[1]

$$\text{(I)}\quad RC{\equiv}CCH_2OH \xrightarrow[Cp_2TiCl_2]{2\ i\text{-BuMgCl},} \underset{ClMg}{\overset{R}{}}C{=}C\underset{H}{\overset{CH_2OMgCl}{}} \xrightarrow{I_2} \underset{I}{\overset{R}{}}C{=}C\underset{H}{\overset{CH_2OH}{}}$$

$$\xrightarrow{CH_3I} \underset{CH_3}{\overset{R}{}}C{=}C\underset{H}{\overset{CH_2OH}{}}$$

The reaction when applied to **1** involves a facile rearrangement of the initial adduct (**a**) leading to (E)-3-trimethylsilyl-2-alkene-1-ols (**3**).[2]

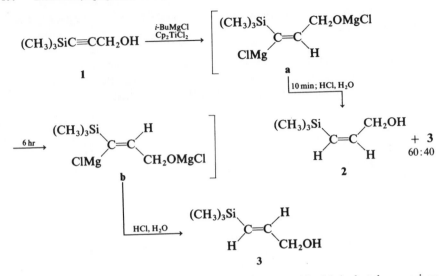

$$(CH_3)_3SiC{\equiv}CCH_2OH \xrightarrow[Cp_2TiCl_2]{i\text{-}BuMgCl} \left[(CH_3)_3Si(ClMg)C{=}C(H)CH_2OMgCl \ (\mathbf{a}) \xrightarrow{6\,hr} (CH_3)_3Si(ClMg)C{=}C(H)CH_2OMgCl \ (\mathbf{b}) \right]$$

1 → a; a $\xrightarrow{10\ min;\ HCl,\ H_2O}$ $(CH_3)_3Si(H)C{=}C(H)CH_2OH$ (2) + 3, 60:40

b $\xrightarrow{HCl,\ H_2O}$ $(CH_3)_3Si(H)C{=}C(H)CH_2OH$ (3)

—COOH → —CHO.[3] Reaction of a carboxylic acid with isobutylmagnesium bromide and catalytic amounts of titanocene dichloride affords aldehydes in fair to moderate yield (equation I). The reaction fails with α,β-unsaturated acids.

$$\text{(I)}\quad RCOOH + 2(CH_3)_2CHCH_2MgCl \xrightarrow[48\text{–}73\%]{Cp_2TiCl_2,\ ether} RCHO$$

[1] F. Sato, H. Ishikawa, H. Watanabe, T. Miyake, and M. Sato, *J.C.S. Chem. Comm.*, 718 (1981).
[2] F. Sato, H. Watanabe, Y. Tanaka, and M. Sato, *ibid.*, 1126 (1982).
[3] F. Sato, T. Jinbo, and M. Sato, *Synthesis*, 871, (1981).

Dichlorobis(isopropoxy)titanium(IV), $TiCl_4$–$Ti(O\text{-}i\text{-}Pr)_4$ (1:1). This is a milder Lewis acid than $TiCl_4$.[1]

Diels-Alder catalyst.[2] The acrylates (**4**) of three chiral alcohols (**1–3**) have been found to undergo asymmetric Diels-Alder reactions with cyclopentadiene (equation I) in the presence of a Lewis acid catalyst. For this purpose, catalysts of the type $TiCl_2(X_2)$ are superior to $TiCl_4$ because they do not promote polymerization of the acrylate. The final products (**6**) all have the *endo*-orientation; the configuration (R) or (S) depends upon the chiral alcohols. Those derived from **1** all have (S)-configuration;

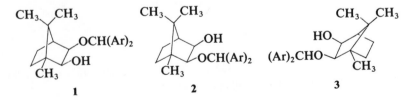

CH₃ CH₃, OCH(Ar)₂, OH, CH₃ — **1**; CH₃ CH₃, OH, OCH(Ar)₂, CH₃ — **2**; CH₃ CH₃, HO, (Ar)₂CHO, CH₃ — **3**

those from **2** have (R)-configuration; those from **3** (only one example) have the (S)-configuration.

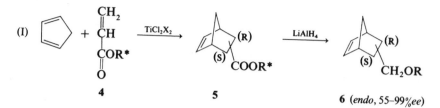

(I) cyclopentadiene + $CH_2{=}CH{-}COR^*$ (**4**, $C{=}O$) $\xrightarrow{TiCl_2X_2}$ (S),(R) bicyclic ester, COOR* (**5**) $\xrightarrow{LiAlH_4}$ (S),(R) bicyclic, CH_2OR

4 **5** **6** (*endo*, 55–99%*ee*)

[1] T. Mukaiyama, *Angew. Chem. Int. Ed.*, **16**, 817 (1977).
[2] W. Oppolzer, C. Chapuis, G. M. Dạo, D. Reichlin, and T. Godel, *Tetrahedron Letters*, **23**, 4781 (1982).

Dichlorobis(methyldiphenylphosphine)palladium(II), $PdCl_2[PCH_3(C_6H_5)_2]_2$.

α-Keto amides. This Pd(II) complex is a particularly effective catalyst for biscarbonylation of ArBr in the presence of secondary amines to form α-keto amides.[1]

Example:

$$C_6H_5Br + 2CO + HN(C_2H_5)_2 \xrightarrow[95\%]{100°} C_6H_5COCON(C_2H_5)_2 + C_6H_5CON(C_2H_5)_2 \quad (86:14)$$

[1] F. Ozawa, H. Soyama, T. Yamamoto, and A. Yamamoto, *Tetrahedron Letters*, **23**, 3383 (1982).

Dichlorobis(triphenylphosphine)nickel(II), 9, 147–148.

Alkyl aryl sulfides. In the presence of this nickel catalyst, Grignard reagents react selectively at the C–Cl bond of alkyl chlorophenyl sulfides. Disubstituted benzenes can be prepared by this sequence.[1]

Example:

$$Cl{-}C_6H_4{-}SCH(CH_3)_2 \xrightarrow[60\%]{CH_3MgBr,\ Ni(II)} CH_3{-}C_6H_4{-}SCH(CH_3)_2 \xrightarrow[51\%]{CH_2{=}CHCH_2MgBr,\ Ni(II)} CH_3{-}C_6H_4{-}CH_2CH{=}CH_2$$

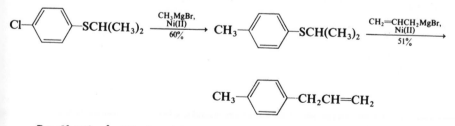

Desulfuration.[2] Vinyl sulfides are reduced to the corresponding alkene by isopropylmagnesium bromide in the presence of $[(C_6H_5)_3P]_2NiCl_2$. No overreduction is observed, as frequently noted in reductions with Raney nickel.

Examples:

C_6H_5S(H)C=C($CH_2C_6H_5$) $(CH_3O)_2CH(CH_2)_2$ —[$(CH_3)_2CHMgBr$, Ni(II), 76%]→ (H)C=C(H) $(CH_3O)_2CH(CH_2)_2$ $CH_2C_6H_5$

SC_6H_5, CH_3O —[70%]→ CH_3O

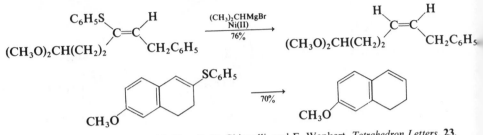

[1] M. Tiecco, L. Testaferri, M. Tingoli, D. Chianelli, and E. Wenkert, *Tetrahedron Letters*, **23**, 4629 (1982).

[2] B. M. Trost and P. L. Ornstein, *ibid.*, **22**, 3463 (1981).

2,3-Dichloro-5,6-dicyano-1,4-benzoquinone (DDQ), 1, 215–219; **2**, 112–117; **3**, 83–84; **4**, 120–134; **5**, 193–194; **6**, 168–170; **7**, 96–97; **8**, 153–156; **9**, 148–151; **10**, 135–136.

Aromatization to furanes.[1] The dihydrofurane group in **1** can be aromatized with the quinone to form the benzofurane **2** in 48% yield.

$NHCOC(CH_3)_3$, CH_3O, O —[DDQ, Dioxane, 48%]→ $NHCOC(CH_3)$, CH_3O, O

1 **2**

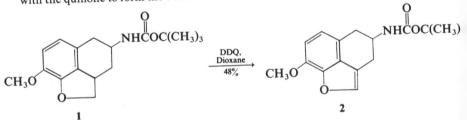

Protection of hydroxyl groups. Hydroxyl groups are often protected as the benzyl ethers. The benzyl group has one limitation however; it is removed by hydrogenolysis or by reduction with sodium in liquid ammonia, reactions that can also affect other groups. *p*-Methoxybenzyl ethers offer the advantage that they are removed by oxidation with DDQ, a reaction that does not affect other common protecting groups or a number of functional groups. The oxidation is particularly facile in CH_2Cl_2 containing a trace of water.[2]

Example:

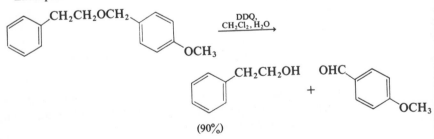

$C_6H_5CH_2CH_2OCH_2C_6H_4OCH_3$ —[DDQ, CH_2Cl_2, H_2O]→ $C_6H_5CH_2CH_2OH$ (90%) + $OHC\text{-}C_6H_4\text{-}OCH_3$

The *p*-methoxybenzyl group has also been recommended as an N-protecting group. In this use, it is cleaved by trifluoroacetic acid at 65°, conditions to which an N-benzyl group is stable.[3]

Oxidation of an allylic ether. Oxidation of aryltetralin lignanes such as **1** results in two products, **2** and **3**, which are believed to be formed as indicated in equation (I).[4]

(I)

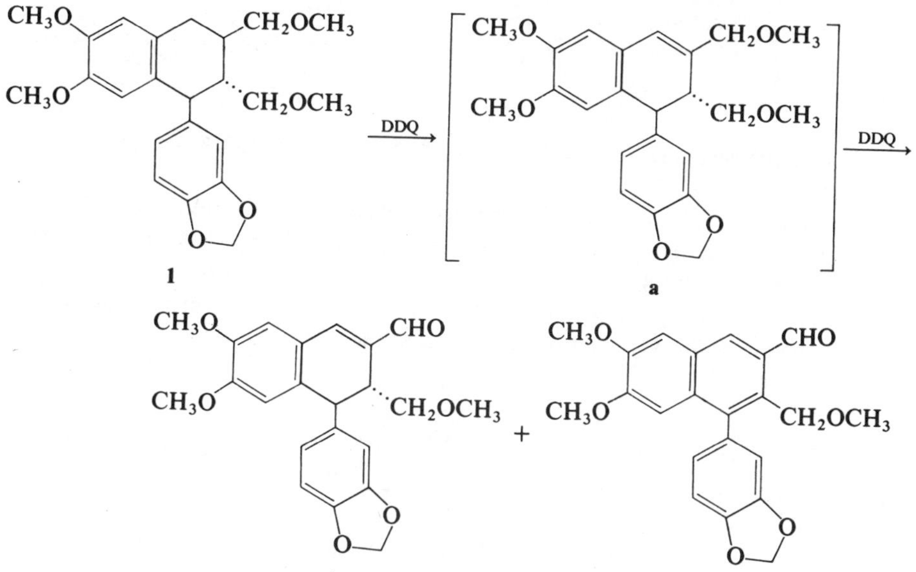

CH_3O, CH_3O, CH_2OCH_3, CH_2OCH_3 **1** $\xrightarrow{DDQ}$ [CH_3O, CH_3O, CH_2OCH_3, CH_2OCH_3 **a**] $\xrightarrow{DDQ}$

CH_3O, CH_3O, CHO, CH_2OCH_3 **2** + CH_3O, CH_3O, CHO, CH_2OCH_3 **3**

o-*Quinones.* *cis*-1,2-Tetrahydrodiols of type **1** are oxidized to *o*-quinones (**2**) by DDQ in dioxane.[5]

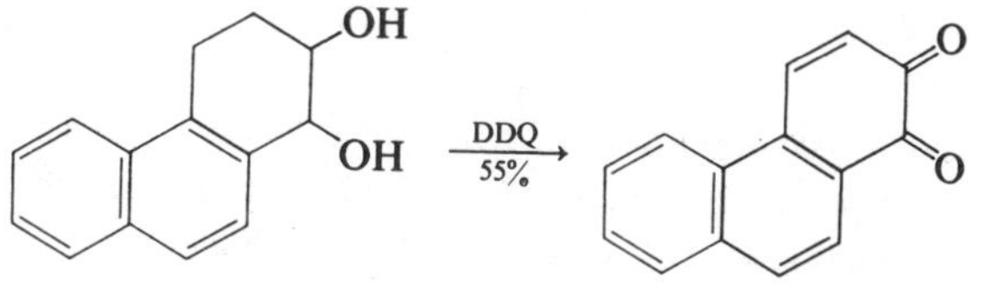

OH, OH **1** $\xrightarrow[55\%]{DDQ}$ O, O **2**

[1] W. Haefliger and E. Kloppner, *Helv*, **65**, 1837 (1982).
[2] Y. Oikawa, T. Yoshioka, and O. Yonemitsu, *Tetrahedron Letters*, **23**, 885, 889 (1982).
[3] D. R. Buckle and C. J. M. Rockell, *J.C.S. Perkin I*, 627 (1982).
[4] R. S. Ward, P. Satyanarayana, and B. V. G. Rao, *Tetrahedron Letters*, **22**, 3021 (1981).
[5] K. L. Platt and F. Oesch, *ibid.*, **23**, 163 (1982).

Dichloro[(1,3-diphenylphosphino)propane]nickel(II), $NiCl_2$(dppp), red crystals. Preparation.[1]

Cross-coupling of Grignard reagents and aromatic and heterocyclic halides. This nickel complex is one of the most effective catalysts for coupling of Grignard reagents with aryl[2], alkenyl[2] and heterocyclic[3] halides. Both alkyl and aryl Grignard reagents can be coupled. The reaction provides a useful route to isoquinoline alkaloids.

[1] G. R. Van Hecke and W. De W. Horrocks, Jr., *Inorg. Chem.*, **5**, 1968 (1966).
[2] K. Tamao, K. Sumitani, Y. Kiso, M. Zembayashi, A. Fujioka, S. Kodama, I. Nakajima, A. Minato, and M. Kumada, *Bull. Chem. Soc. Japan*, **49**, 1958 (1976).
[3] K. Tamao, S. Kodama, I. Nakajima, M. Kumada, A. Minato, and K. Suzuki, *Tetrahedron*, **38**, 3347 (1982).

Dichloroketene, 1, 221–222; **2,** 118; **3,** 87–88; **4,** 134–135; **8,** 156; **9,** 152–154; **10,** 139–140.

cis-*Bicyclo*[4.2.0]*octanones*; cis-*octahydroindoles*. Dichloroketene (generated from trichloroacetyl chloride) undergoes [2 + 2]cycloaddition to 1-substituted cycloalkenes (**4**, 135–136; **8**, 156) with high regio- and stereoselectivity. The products are reduced by zinc and acetic acid or ammonium chloride to *cis*-bicyclic butanones, which undergo Baeyer-Villiger oxidation to *cis*-bicyclic lactones.[1]

Example:

C_6H_5 ; $Cl_2C{=}C{=}O$; H_5C_6 Cl Cl O H ; Zn, HOAc

C_6H_5 O H ; CF_3CO_3H ; C_6H_5 O O H

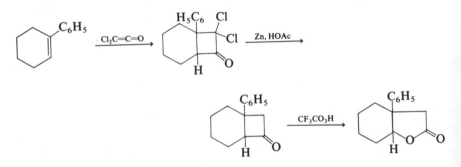

These bicyclo[4.2.0]octanones also undergo aza-ring expansion to lactams. Beckmann rearrangement of the oxime of a typical bicyclic butanone (**1**) results in an octahydroisoindolone (**2**), whereas rearrangement of the methylnitrone (**3**) with tosyl chloride (**4**, 510–511; **6**, 598) proceeds in the opposite direction to give an octahydroindolone (**4**) as the only product.[2]

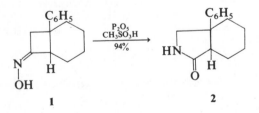

C_6H_5 N OH H ; P_2O_5, CH_3SO_3H, 94% ; C_6H_5 HN O H

1 **2**

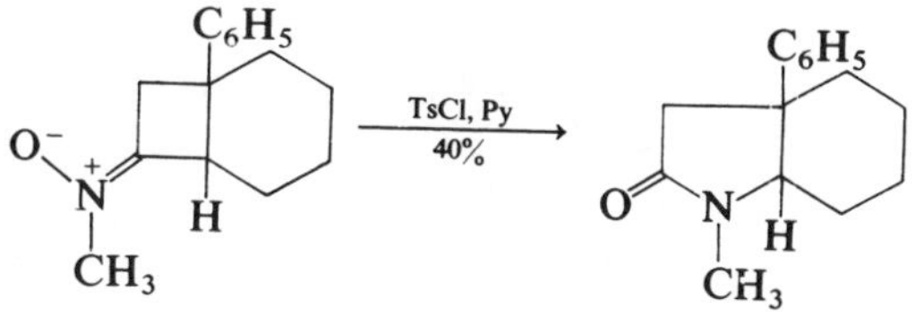

C_6H_5 … TsCl, Py 40% … 3 → 4

This sequence was used for a synthesis of the alkaloid mesembranol (5) in 7% overall yield from 4-hydroxycyclohexanone.[3]

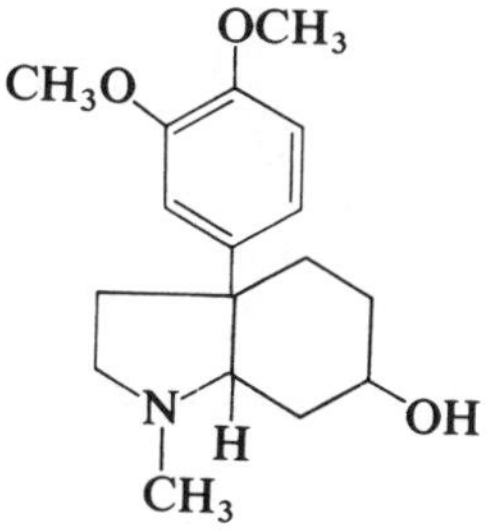

5

1,2-*Cyclobutanediones*.[4] An attractive route to the diketone **4** involves addition of dichloroketene to **1** in the presence of $POCl_3$ (**8**, 156). Remaining steps involve dechlorination followed by oxidation with selenium dioxide.

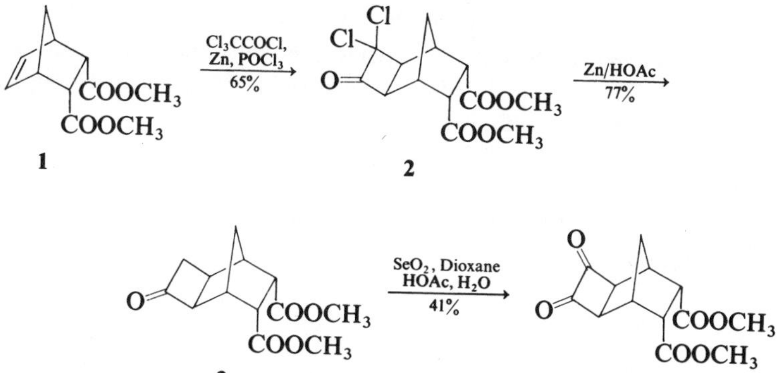

1 → (Cl_3CCOCl, Zn, $POCl_3$, 65%) → 2 → (Zn/HOAc, 77%) → 3 → (SeO_2, Dioxane, HOAc, H_2O, 41%) → 4

***Cycloaddition to vinyl sulfoxides; γ-butyrolactones*.**[5] Dichloroketene (best generated by zinc reduction of trichloroacetyl chloride, **8**, 156) reacts with vinyl sulfoxides to form α-dichloro-γ-butyrolactones in ~50–80% yield. A polar mechanism involving a Pummerer-type rearrangement has been suggested.

Examples:

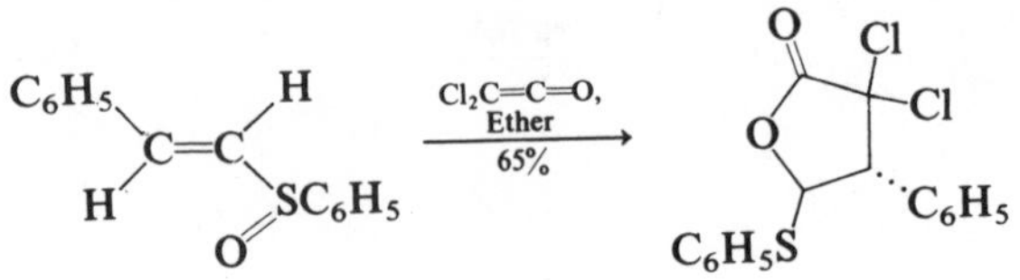

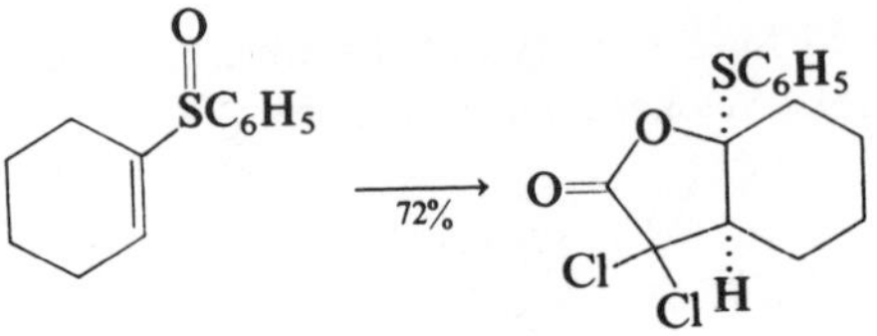

$C_6H_5CH{=}CHS(O)C_6H_5 \xrightarrow[65\%]{Cl_2C{=}C{=}O,\ Ether}$ (lactone with Cl, Cl, C_6H_5, C_6H_5S)

(1-cyclohexenyl phenyl sulfoxide, SC_6H_5) $\xrightarrow{72\%}$ (bicyclic dichlorolactone with SC_6H_5, Cl, Cl, H)

[1] P. W. Jeffs, G. Molina, M. W. Cass, and N. A. Cortese, *J. Org.*, **47**, 3871 (1982).
[2] P. W. Jeffs, G. Molina, N. A. Cortese, P. R. Hauck, and J. Wolfram, *ibid.*, **47**, 3876 (1982).
[3] P. W. Jeffs, N. A. Cortese, and J. Wolfram, *ibid.*, **47**, 3881 (1982).
[4] W. Reid and O. Bellinger, *Synthesis*, 729 (1982).
[5] J. P. Marino and M. Neisser, *Am. Soc.*, **103**, 7687 (1981).

Dichlorotris(dimethylamino)phosphine (1). Mol. wt. 308.0, m.p. 240–245°.

Preparation[1]:

$$P[N(CH_3)_2]_3 + CCl_3CCl_3 \longrightarrow \underset{\mathbf{1}\ (92.5\%)}{Cl_2P[N(CH_3)_2]_3} + Cl_2C{=}CCl_2$$

$-NH-\overset{O}{\overset{\|}{C}}- \rightarrow -N{=}C(Cl)-$.[2] A key step in a synthesis of the antimetabolite (αS,5S)-α-amino-3-chloro-4,5-dihydro-5-isoxazoleacetic acid (AT-125, **4**) from L-glutamic acid involved reaction of the protected derivative (**2**) of tricholomic acid with **1** in refluxing THF to give **3** in 54% isolated yield. The protective groups were then removed with TFA–thioanisole in 87% yield.

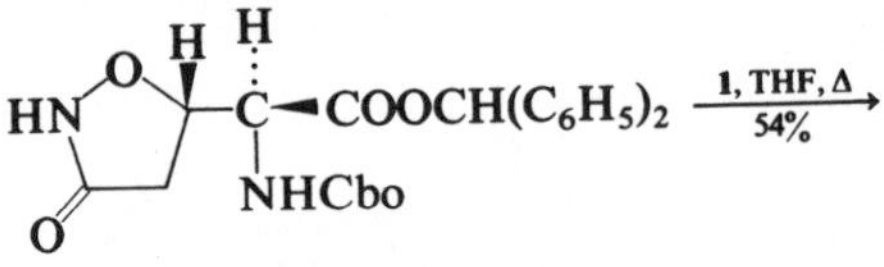

2 ($COOCH(C_6H_5)_2$, NHCbo) $\xrightarrow[54\%]{1,\ THF,\ \Delta}$ **3** ($COOCH(C_6H_5)_2$, NHCbo, Cl) $\xrightarrow[87\%]{TFA,\ C_6H_5SCH_3}$ **4** (COO^-, NH_3^+, Cl)

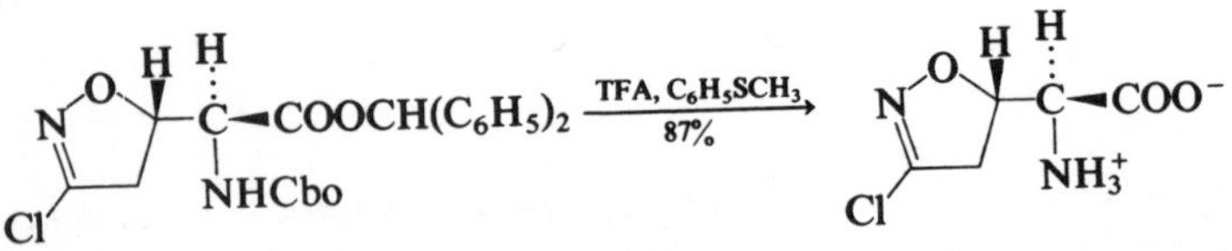

[1] R. Appel and H. Schöler, *Ber.*, **110**, 2382 (1977).
[2] R. B. Silverman and M. W. Holladay, *Am. Soc.*, **103**, 7357 (1981).

Dichlorotris(triphenylphosphine)ruthenium(II), 4, 564; **5**, 740–741; **6**, 654–655; **7**, 99; **8**, 159–161; **10**, 141–142.

N-Alkylation of aniline by alcohols. The Ru(II) complex catalyzes the alkylation of aniline by alcohols. If the reaction is prolonged, dialkylation is observed. Alkylation with allylic alcohols results in 2,3-dialkylquinolines (equation I).[1]

(I) 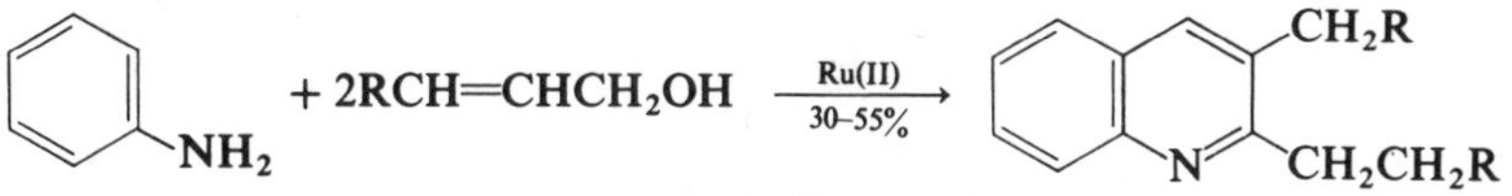 $C_6H_5NH_2$ + 2RCH=CHCH$_2$OH $\xrightarrow[30\text{–}55\%]{\text{Ru(II)}}$ 2-(CH_2CH_2R)-3-(CH_2R)quinoline

Oxygenation of allylic alcohols. This Ru(II) complex, as well as $RuBr_2[P(C_6H_5)_3]_3$ and $RuH(OAc)[P(C_6H_5)_3]_3$, catalyzes the oxidation of allylic alcohols to α,β-unsaturated ketones or aldehydes by molecular oxygen with retention of configuration. The oxidation of retinol to retinal (second example) requires addition of 2,6-dimethylpyridine to prevent side reactions.[2]

Examples:

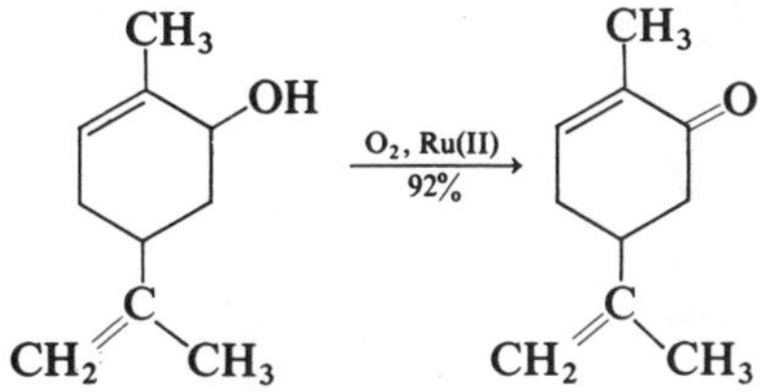

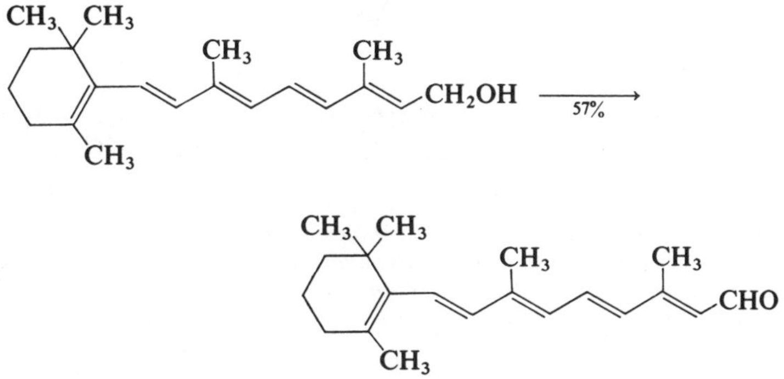

(carveol) $\xrightarrow[92\%]{O_2,\ Ru(II)}$ (carvone)

(retinol, ...CH$_2$OH) $\xrightarrow[57\%]{}$ (retinal, ...CHO)

Ru(II)-Catalyzed reactions of endoperoxides.[3] Although the Ru(II)-catalyzed reaction of endoperoxides usually results in complex mixtures, the reaction of the prostaglandin endoperoxide (**1**) is a remarkably clean fragmentation to the triene **2**, as the only isolable product. Endoperoxides containing a double bond undergo useful isomerizations to diepoxides (equations I and II).

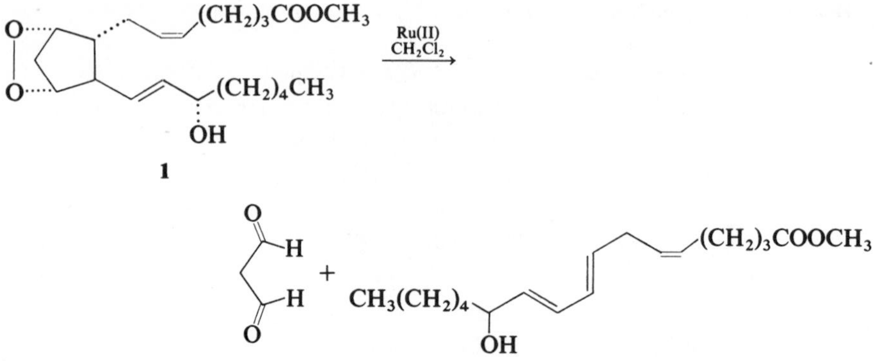

$$\xrightarrow[CH_2Cl_2]{Ru(II)}$$

1

$(CH_2)_3COOCH_3$, $(CH_2)_4CH_3$, OH

2

$CH_3(CH_2)_4$, OH, $(CH_2)_3COOCH_3$

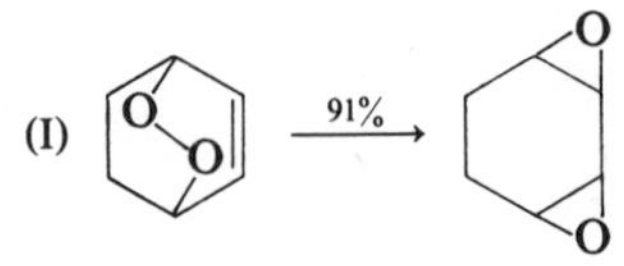

(I) 91%

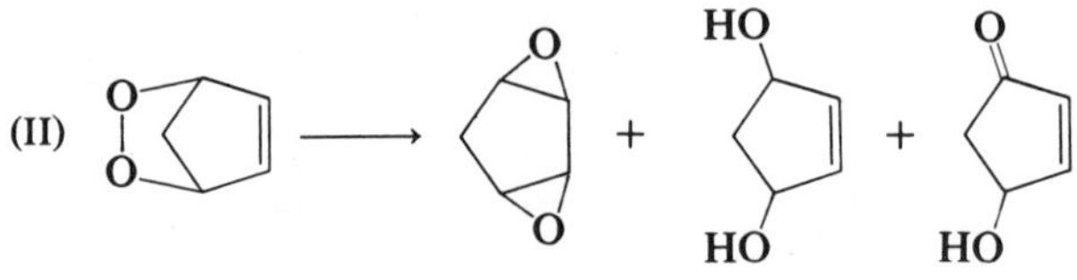

(II) (78%) (3%) (5%)

Transfer hydrogenation of carbonyl compounds.[4] Carbonyl compounds are reduced to alcohols by formic acid with this ruthenium catalyst in high yield without a solvent.

[1] Y. Watanabe, Y. Tsuji, and Y. Ohsugi, *Tetrahedron Letters*, **22**, 2667 (1981).
[2] M. Matsumoto and S. Ito, *J.C.S. Chem. Comm.*, 907 (1981).
[3] M. Suzuki, R. Noyori, and N. Hamanaka, *Am. Soc.*, **104**, 2024 (1982).
[4] Y. Watanabe, T. Ohta, and Y. Tsuji, *Bull. Chem. Soc. Japan*, **55**, 2441 (1982).

Dicyclohexylborane, 3, 90–91; **4**, 141; **8**, 162.

β-Hydroxy carboxylic acids. Optically active β-hydroxy carboxylic acids can be obtained from optically active propargyl alcohols (**10**, 320) such as **1** by the sequence shown in equation (I). The alcohols (**1**) can also be converted into optically active α-hydroxy carboxylic acids (equation II).[1]

$$\text{(I)}\quad \underset{\mathbf{1}}{CH_3(CH_2)_4\overset{OH}{C}{}^*HC{\equiv}CH} \xrightarrow[3)\ HCl]{1)\ n\text{-BuLi};\ 2)\ (CH_3)_3SiCl} \underset{\mathbf{2}}{CH_3(CH_2)_4\overset{OH}{C}{}^*HC{\equiv}CSi(CH_3)_3} \xrightarrow[82\%]{1)\ (C_6H_{11})_2BH;\ 2)\ H_2O_2,\ OH^-} \underset{\mathbf{3}\ (86\%\ ee)}{CH_3(CH_2)_4\overset{OH}{C}{}^*HCH_2COOH}$$

$$\text{(II)}\quad \mathbf{1} \xrightarrow[92\%]{1)\ Ac_2O;\ 2)\ KMnO_4} \underset{\mathbf{4}}{CH_3(CH_2)_4\overset{OAc}{C}{}^*HCOOH} \xrightarrow{1)\ NaOH;\ 2)\ CH_2N_2} \underset{\mathbf{5}\ (92\%\ ee)}{CH_3(CH_2)_4\overset{OH}{C}{}^*HCOOCH_3}$$

[1] M. M. Midland and P. E. Lee, *J. Org.*, **46**, 3933 (1981).

Dicyclohexylcarbodiimide (DCC), 1, 231–236; **2,** 126; **3,** 91; **4,** 141; **5,** 206–207; **6,** 174; **7,** 100–101; **8,** 162–163; **9,** 156–157; **10,** 142.

Nitroalkenes. Nitroaldols are dehydrated to nitroalkenes by reaction with DCC and a catalytic amount of CuCl in ether at 20° (equation I).[1]

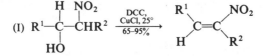

$$\text{(I)}\quad R^1\text{—}\overset{H}{\underset{HO}{C}}\text{—}\overset{NO_2}{C}HR^2 \xrightarrow[65\text{–}95\%]{DCC,\ CuCl,\ 25^\circ} (R^1)(H)C{=}C(NO_2)(R^2)$$

Imines. Imines are usually prepared by acid-catalyzed condensation of aldehydes or ketones with primary amines with azeotropic removal of water. This method gives poor yields with low-boiling imines.[2] A new method involves dehydrocyanation of α-cyanoamines with DCC and Dabco as catalyst (equation I).[3]

$$R^1R^2C(NHR^3)(CN) + C_6H_{11}N{=}C{=}NC_6H_{11} \xrightarrow[75\text{–}80\%]{Dabco} R^1R^2C{=}NR^3 + C_6H_{11}NH\underset{CN}{C}{=}NC_6H_5$$

The method can even be used to prepare acetonimine (**1**) from 2-amino-2-methylpropanenitrile in 64% yield. The product is stable at 0° for several days, but gradually loses NH_3 to form **2**.

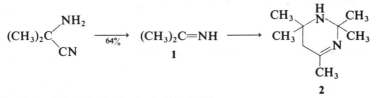

$$(CH_3)_2C(NH_2)(CN) \xrightarrow{64\%} \underset{\mathbf{1}}{(CH_3)_2C{=}NH} \longrightarrow \mathbf{2}$$

(**2**: ring with H–N flanked by $C(CH_3)_2$ groups, ring N=C bearing CH_3)

[1] P. Knochel and D. Seebach, *Synthesis*, 1017 (1982).

[2] R. W. Layer, *Chem. Rev.*, **63**, 489 (1963).
[3] K. Findeisen, H. Heitzler, and K. Dehnicke, *Synthesis*, 702 (1981).

Dicyclopentadienyl(isoprene)zirconium, $(C_5H_5)_2Zr(CH_2{=}CH\underset{\underset{CH_3}{|}}{C}{=}CH_2)$ **(1)**; orange, air-sensitive. Preparation.[1]

Coupling with alkenes and alkynes.[2] The reagent reacts with these substrates to form metallacyclic compounds, which are hydrolyzed to acyclic compounds. The adducts are formed with high regioselectivity.

Examples:

$$CH_2{=}CHC_2H_5 + \mathbf{1} \longrightarrow \left[Cp_2Zr\text{(metallacycle with } CH_3 \text{ and } C_2H_5) \right] \xrightarrow[70\%]{H_3O^+} (CH_3)_2C{=}CH(CH_2)_3C_2H_5$$

$$CH_3C{\equiv}CCH_3 + 1 \xrightarrow[90\%]{} Cp_2Zr\text{(metallacycle with } CH_3, CH_3, CH_3) \xrightarrow{H_3O^+} (CH_3)_2C{=}CHCH_2\underset{\underset{CH_3}{|}}{C}{=}CHCH_3$$

$$C_6H_5C{\equiv}CC_6H_5 + 1 \xrightarrow[70\%]{-\text{Isoprene}} Cp_2Zr\text{(zirconacyclopentadiene with four } C_6H_5) \xrightarrow{H_3O^+} C_6H_5CH{=}\overset{\overset{C_6H_5}{|}}{C}\underset{\underset{C_6H_5}{|}}{C}{=}CHC_6H_5$$

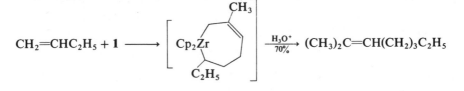

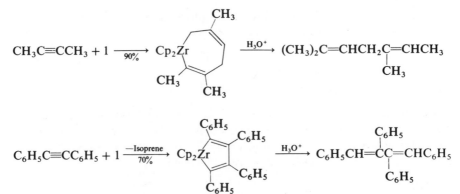

The Zr-complex **1** catalyzes the tail-to-tail linear dimerization of isoprene.

[1] H. Yasuda, Y. Kajihara, K. Mashima, K. Lee, and A. Nakamura, *Chem. Letters*, 519 (1981).
[2] H. Yasuda, Y. Kajihara, K. Nagasuna, K. Mashima, and A. Nakamura, *ibid.*, 719 (1981).

Dicyclopentadienyl(η^3-1-trimethylsilylallyl)titanium,

$$CH_2{\cdots}\underset{\underset{TiCp_2}{|}}{CH}{\cdots}CHSi(CH_3)_3$$

The reagent is prepared *in situ* by reaction of trimethylsilylallyllithium with Cp_2TiCl_2.

1,3-Dienes.[1] The reagent reacts with various aldehydes to give 3-trimethylsilyl-4-hydroxy-1-alkenes in 80–98% yield. The products can be converted into either (E)- or (Z)-terminal 1,3-dienes.

Example:

$$C_6H_5CHO \xrightarrow[95\%]{1)\ \mathbf{1};\ 2)\ H_3O^+;\ 3)\ O_2} C_6H_5CH(OH)CH(Si(CH_3)_3)CH{=}CH_2 \xrightarrow[87\%]{KH,\ THF} CH_2{=}CH\text{–}CH{=}CH\text{–}C_6H_5\ (Z)$$

$$\xrightarrow[\ ]{86\%,\ H_2SO_4,\ THF} CH_2{=}CH\text{–}CH{=}CH\text{–}C_6H_5\ (E)$$

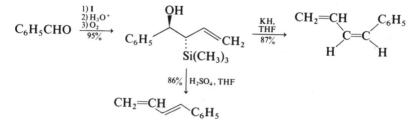

The reagent reacts to only a minor extent with ketones, esters, or halides.

[1] F. Sato, Y. Suzuki, and M. Sato, *Tetrahedron Letters*, **23**, 4589 (1982).

Diethoxycarbenium tetrafluoroborate, $(C_2H_5O)_2CH^+BF_4^-$ **(1)**. The reagent is generated as a slurry by reaction of triethyl orthoformate at $-30°$ with a solution of BF_3 etherate in CH_2Cl_2.

α-Diethoxymethylation of ketones.[1] Ketones react with **1** in CH_2Cl_2 at $-78°$ in the presence of ethyldiisopropylamine (3 equiv.) to form β-keto acetals. The reagent is comparable to the Vilsmeier reagent for formylation. Yields are typically 45–80%. With unsymmetrical ketones, reaction at the less substituted α-position is favored, but α,α′-bis(diethoxymethyl) ketones can be prepared. A hindered nonnucleophilic base is essential for the transformation.

[1] W. L. Mock and H.-R. Tsou, *J. Org.*, **46**, 2557 (1981).

Diethoxyvinylidentriphenylphosphorane (1). Mol. wt. 376.42, m.p. 80–81°.

Preparation[1]:

$$(C_6H_5)_3P{=}CHC(O)OC_2H_5 \xrightarrow[68\%]{(C_2H_5)_3O^+BF_4^-} (C_6H_5)_3\overset{+}{P}\text{—}CH{=}C(OC_2H_5)_2BF_4^-$$

$$\xrightarrow[48\%]{NaNH_2} \underset{\mathbf{1}}{(C_6H_5)_3P{=}C{=}C(OC_2H_5)_2}$$

Ethyl esters. This phosphorane reacts with substrates with a strongly acidic group by transfer of an ethyl group.[2] This same reaction is possible with carboxylic acids, which are converted into ethyl esters (THF, Et_2O, or C_6H_6, 25°).[3]

$$\mathbf{1} + CH_2(COOCH_3)_2 \longrightarrow \underset{(70\%)}{C_2H_5CH(COOCH_3)_2} + (C_6H_5)_3P{=}CHCOOC_2H_5$$

Example:

$$C_6H_5COOH + \mathbf{1} \longrightarrow \underset{(77\%)}{C_6H_5COOC_2H_5} + \underset{(70\%)}{(C_6H_5)_3P{=}CHCOOC_2H_5}$$

[1] H. J. Bestmann, R. W. Saalfrank, and J. P. Snyder, *Ber.*, **106**, 2601 (1973).
[2] H. J. Bestmann and R. W. Saalfrank, *ibid.*, **109**, 403 (1976).
[3] H. J. Bestmann and K. Roth, *Synthesis*, 998 (1981).

2-Diethylamino-4-phenylthio-2-butenenitrile,

$$\begin{matrix} C_6H_5SCH_2 & & CN \\ & C{=}C & \\ H & & N(C_2H_5)_2 \end{matrix} \quad (\mathbf{1}).$$

Preparation:

$$C_6H_5SCH_2CH_2CON(C_2H_5)_2 \xrightarrow[60\%]{1)\ P_4S_{10},\ DME;\ 2)\ CH_3I,\ CH_3CN} C_6H_5SCH_2CH_2\overset{SCH_3}{\overset{|}{C}}{=}\overset{+}{N}(C_2H_5)_2\ I^- \xrightarrow[86\%]{1)\ KCN;\ 2)\ \Delta(-CH_3SH)} \mathbf{1}$$

Homologated (E)-α,β-unsaturated carboxylic acids.[1] The anion (**2**) of **1**, formed with LDA in THF at $-50°$, is alkylated exclusively at the γ-position. The products (**3**) are converted into α,β-unsaturated acids (**4**) on oxidation and thermolysis. This behavior contrasts with that of the anion of the corresponding silyl enol ether, $CH_3SCH_2CH{=}C(CN)OSi(CH_3)_3$, which undergoes exclusively α-alkylation.[2]

$$\underset{\mathbf{2}}{\begin{matrix} C_6H_5\overset{Li}{\overset{|}{S}CH} & & CN \\ & C{=}C & \\ H & & N(C_2H_5)_2 \end{matrix}} \xrightarrow{RX,\ THF,\ 50°} \underset{\mathbf{a}}{\left[R\overset{SC_6H_5}{\overset{|}{C}}HCH{=}C\begin{matrix} CN \\ N(C_2H_5)_2 \end{matrix} \right]} \xrightarrow[55\text{–}78\%]{H_3O^+}$$

$$\underset{\mathbf{3}}{R\overset{SC_6H_5}{\overset{|}{C}}HCH_2COOH} \xrightarrow[70\text{–}90\%]{1)\ ClC_6H_4CO_3H;\ 2)\ 60°,\ CCl_4} \underset{\mathbf{4}}{\begin{matrix} R & & H \\ & C{=}C & \\ H & & COOR \end{matrix}}$$

A similar reaction with aldehydes results in γ-hydroxy-α,β-unsaturated acids (66% yield, 1 example). The anion **2** undergoes exclusive 1,4-addition to cyclopentenone and cyclohexenone (yields of products, 67% and 85%, respectively). The anion thus is the equivalent of a β-carboxyl vinyl anion.

[1] S. De Lombaert, B. Lesur, and L. Ghosez, *Tetrahedron Letters*, **23**, 4251 (1982).
[2] S. Hünig, *Chimia*, **36**, 1 (1982).

(Diethylamino)sulfur trifluoride (DAST), **6**, 183–184; **8**, 166–167; **10**, 142–143.

α-Fluoro-β-amino acids.[1] Reaction of N,N-dibenzyl-L-serine benzyl ester (**1**) with DAST provides **2**, which is converted into optically active α-fluoro-β-alanine (**3**) on hydrogenolysis.

$$\underset{\mathbf{1}}{\underset{|\atop NBzl_2}{HOCH_2CHCOOBzl}} \xrightarrow[90\%]{\text{DAST, THF}} \underset{\mathbf{2}}{\underset{|\atop F}{Bzl_2NCH_2CHCOOBzl}}$$

$$\xrightarrow{H_2,\ Pd/C} \underset{\mathbf{3},\ \alpha_D + 29°}{\underset{|\atop F}{H_2NCH_2CHCOOH}}$$

[1] L. Somekh and A. Shanzer, *Am Soc.*, **104**, 5836 (1982).

Diethyl azodicarboxylate, 1, 245–247; **2**, 128–129; **4**, 148–149; **5**, 212–213; **6**, 185; **9**, 160.

1,3-*Diene* → 1,4-*diene*.[1] An interesting way to convert 3β-acetoxy-5,7-cholestadiene (**1**) into the 5,8-cholestadiene (**4**) proceeds via the adduct (**2**) of the 5,7-diene with diethyl azodicarboxylate (**1**, 246). The adduct is converted to the 5,8-diene by reduction with lithium in ethylamine (**1**, 574–581).

C_8H_{17} H_3C H_3C AcO **1** (1g) $\xrightarrow{C_2H_5O_2CN{=}NCO_2C_2H_5}$

H_3C AcO $NCO_2C_2H_5$ / $HNCO_2C_2H_5$ **2**(0.77g) + HO $NCO_2C_2H_5$ / $HNCO_2C_2H_5$ **3** (0.095g)

$\xrightarrow[\sim 50\%]{Li,\ H_2NC_2H_5}$

H_3C HO **4**

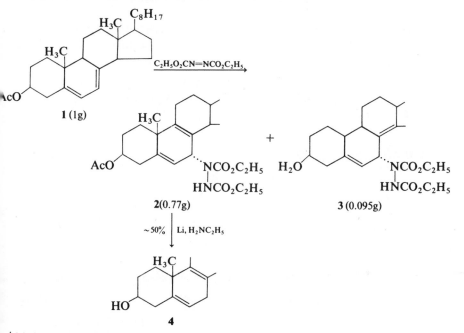

[1] M. Anastasia, A. Fiecchi, and G. Galli, *J. Org.*, **46**, 3421 (1981).

N,N-Diethylbenzeneselenenamide, $C_6H_5SeN(C_2H_5)_2$ (**1**). Mol. wt. 228.19, liquid, stable at low temperatures and to mildly basic conditions. The selenenamide is prepared in 59% yield by reaction of C_6H_5SeCl with diethylamine in hexane at 50°.[1]

Selenenylation of aldehydes.[2] The reagent reacts readily even with nonactivated aldehydes to give α-phenylseleno aldehydes in 60–85% yield. Ketones are unreactive under comparable conditions.

Example:

$$CH_3\overset{O}{\overset{\|}{C}}(CH_2)_4CHO \xrightarrow[72\%]{\mathbf{1},\ CH_2Cl_2,\ 25°,\ 5\ min.} CH_3\overset{O}{\overset{\|}{C}}(CH_2)_3\overset{SeC_6H_5}{\overset{|}{C}}HCHO$$

[1] H. J. Reich and J. M. Renga, *J. Org.*, **40**, 3313 (1975).
[2] M. Jefson and J. Meinwald, *Tetrahedron Letters*, **22**, 3561 (1981).

Diethyl benzoylphosphonate, $C_6H_5\overset{O}{\overset{\|}{C}}-\overset{O}{\overset{\|}{P}}(OC_2H_5)_2$ (**1**). Mol. wt. 242.20, b.p. 141–147°/3 mm. The reagent is prepared by reaction of benzoyl chloride with triethyl phosphite (72% yield).

Benzoylation of amines.[1] The reagent (1.3 equiv.) converts primary amines into amides in good yield. 4-(Dimethylamino)pyridine does not catalyze this reaction noticeably. Diethyl acetylphosphonate can be used in the same way for acetylation of primary amines.

Benzoylation of alcohols.[2] This reaction with **1** is markedly catalyzed by DBU, but not DMAP. Tertiary alcohols react slowly and it is possible to effect selective benzoylation of primary alcohols in the presence of secondary alcohols. The selectivity can be enhanced by use of diisopropyl benzoylphosphonate.

[1] M. Sekine, M. Satoh, H. Yamagata, and T. Hata, *J. Org.*, **45**, 4162 (1980).
[2] M. Sekine, A. Kume, and T. Hata, *Tetrahedron Letters*, **22**, 3617 (1981).

Diethyl N-benzylideneaminomethylphosphonate (1), 9, 161–162.

Preparation[1]:

$$\text{(phthalimido)}N-CH_2Br \xrightarrow[80-90\%]{P(OC_2H_5)_3} \text{(phthalimido)}N-CH_2\overset{O}{\overset{\|}{P}}(OC_2H_5)_2 \xrightarrow{N_2H_4}$$

$$\left[(C_2H_5O)_2\overset{O}{\overset{\|}{P}}CH_2NH_2\right] \xrightarrow[84-90\%]{C_6H_5CHO} (C_2H_5O)_2\overset{O}{\overset{\|}{P}}CH_2N{=}CHC_6H_5$$

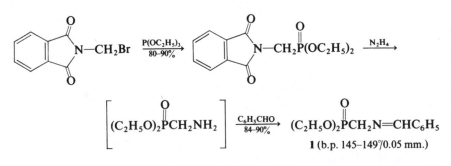

1 (b.p. 145–149°/0.05 mm.)

gem-Acylation–alkylation of carbonyl compounds.[2] The anion of the reagent reacts with carbonyl compounds to form a 2-azadiene (**a**) which usually is not isolated but converted by reaction with an alkyllithium into a lithioenamine (**b**). Lithioenamines undergo regioselective reaction at carbon with a variety of electrophiles. The net result is replacement of the carbonyl group by an acyl group and an alkyl group, which can be suitably functionalized for further transformations. This methodology has been used in the synthesis of complex alkaloids.[3]

(I) $$1 + R^1R^2C{=}O \xrightarrow{n\text{-BuLi}} \left[R^1R^2C{=}CHN{=}CHC_6H_5 \; (\mathbf{a}) \xrightarrow{n\text{-BuLi}} R^1R^2C{\cdots}\bar{C}{\cdots}NCH(Bu\text{-}n)C_6H_5 \; Li^+ \; (\mathbf{b}) \right] \xrightarrow[35\text{–}80\%]{1)\,R^3X;\ 2)\,H_3O^+} R^1R^2C(CHO)R^3$$

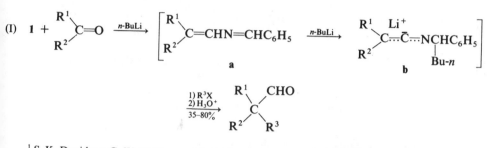

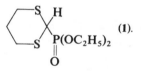

[1] S. K. Davidsen, G. W. Phillips, and S. F. Martin, *Org. Syn.* submitted (1982).
[2] S. F. Martin, G. W. Phillips, T. A. Puckette, and J. A. Colapret, *Am. Soc.*, **102**, 5866 (1980).
[3] S. F. Martin, T. A. Puckette, and J. A. Colapret, *J. Org.*, **44**, 3391 (1979); S. F. Martin and P. J. Garrison, *ibid.*, **47**, 1513 (1982).

Diethyl (1,3-dithian-2-yl)phosphonate, (1).

(2-H-1,3-dithian-2-yl with $P(OC_2H_5)_2$ and P=O)

Preparation (**8**, 88–89).[1]

Homologation of lactones.[2] An efficient method for homologation of five- and six-membered lactones is formulated in equation (I).

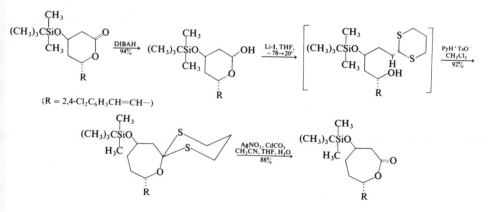

$(CH_3)_3CSi(CH_3)_2O$-substituted δ-lactone $\xrightarrow[94\%]{DIBAH}$ lactol (OH) $\xrightarrow[-78\to 20°]{Li\text{-}1,\ THF}$ [ketene dithioacetal hydroxy intermediate (H, OH)] $\xrightarrow[92\%]{PyH^+TsO^-,\ CH_2Cl_2}$ spiro dithiane oxepane $\xrightarrow[88\%]{AgNO_3,\ CdCO_3,\ CH_3CN,\ THF,\ H_2O}$ seven-membered lactone

(R = $2,4\text{-}Cl_2C_6H_3CH{=}CH{-}$)

[1] C. G. Kruse, N. L. J. M. Broekhof, A. Wijsman, and A. van der Gen, *Tetrahedron Letters*, 885 (1977).
[2] T.-J. Lee, W. J. Holtz, and R. L. Smith, *J. Org.*, **47**, 4750 (1982).

Diethyl oxomalonate, 6, 188; **8,** 170; **10,** 143–144.

Improved preparation[1]:

$$CH_2(COOC_2H_5)_2 \xrightarrow[60\to 90°]{Br_2,} [Br_2C(COOC_2H_5)_2] \xrightarrow[C_2H_5OH]{KOAc,} (AcO)(Br)C(COOC_2H_5)_2$$

$$\xrightarrow[\text{41–47\% overall}]{Bu_4NBr,\ \Delta} O{=}C(COOC_2H_5)_2 + AcBr$$

[1] S. N. Pardo and R. G. Salomon, *J. Org.*, **46**, 2598 (1981).

(Diethylphosphinyl)difluoromethyllithium (1).

Preparation:

$$HCF_2Cl + NaOP(OC_2H_5)_2 \longrightarrow HCF_2\overset{O}{\overset{\|}{P}}(OC_2H_5)_2 \xrightarrow[-78°]{LDA,\ THF,} \underset{\mathbf{1}}{LiCF_2\overset{O}{\overset{\|}{P}}(OC_2H_5)_2}$$

1,1-*Difluoroalkenes.* The reagent reacts with aldehydes and ketones to form adducts, which are converted into 1,1-difluoroalkenes when heated.[1]

Examples;

$$p\text{-}(CH_3)_2NC_6H_4CHO \xrightarrow[71\%]{\mathbf{1}} p\text{-}(CH_3)_2NC_6H_4CH(OH)CF_2\overset{O}{\overset{\|}{P}}(OC_2H_5)_2 \xrightarrow[67\%]{THF,\ \Delta} p\text{-}(CH_3)_2NC_6H_4CH{=}CF_2$$

$$C_6H_5COC_3H_7\text{-}n \xrightarrow[62\%]{1)\ \mathbf{1}\quad 2)\ THF,\ \Delta} (C_6H_5)(n\text{-}C_3H_7)C{=}CF_2$$

The same transformation can be effected under neutral conditions with $(CH_3)_3SiCF_2\overset{O}{\overset{\|}{P}}(OC_2H_5)_2$, which forms similar adducts with aldehydes and ketones in the presence of cesium fluoride.[2]

[1] M. Obayashi, E. Ito, K. Matsui, and K. Kondo, *Tetrahedron Letters*, **23**, 2323 (1982).
[2] M. Obayashi and K. Kondo, *ibid.*, **23**, 2327 (1982).

Diethyl phosphorocyanidate, 5, 217; **6**, 192–193; **7**, 107; **9**, 163–167; **10**, 145.

Thiocyanates.[1] Sodium arenesulfinates, $ArSO_2Na$, are converted into aryl thiocyanates, ArSCN in about 60–75% yield by reaction with diethyl phosphorocyanidate in THF. This reaction is also possible with benzyl and adamantyl sulfinates but yields are only about 40%. Alkyl thiocyanates cannot be obtained in this way.

[1] S. Harusawa and T. Shioiri, *Tetrahedron Letters*, **23**, 447 (1982).

L-(+)-and D-(−)-Diethyl tartrate (DET).

Chiral allenylboronic esters.[1] These reagents can react with aldehydes to provide β-acetylenic alcohols enantioselectivity (60–95% ee). Thus, allenylboronic acid **1**, prepared as shown, reacts with aldehydes in the presence of (+)-DET or (+)-DIPT to give (S)-alcohols (**2**), whereas (R)-alcohols (**2**) are obtained in the presence of (−)-DET or (−) DIPT, both in optical yields of 85–95%. The selectivities are lower in reaction with aryl and α,β-unsaturated aldehydes.

$$HC{\equiv}CCH_2MgBr \xrightarrow[30\text{–}35\%]{1)\,B(OCH_3)_3,\,-78\to 0^\circ;\ 2)\,H_3O^+} CH_2{=}C{=}C(H)B(OH)_2 \quad \mathbf{1}$$

$$\mathbf{1} \xrightarrow[85\text{–}95\%\,ee]{1)\,(+)\text{-DET};\ 2)\,RCHO} \text{(S)-}\mathbf{2}\ \ RCH(OH)CH_2C{\equiv}CH$$

$$\mathbf{1} \xrightarrow[85\text{–}95\%\,ee]{1)\,(-)\text{-DET};\ 2)\,RCHO} \text{(R)-}\mathbf{2}\ \ RCH(OH)CH_2C{\equiv}CH$$

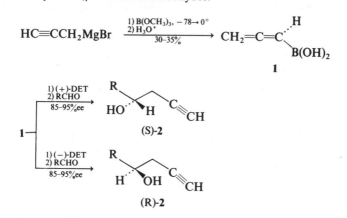

[1] R. Haruta, M. Ishiguro, N. Ikeda, and H. Yamamoto, *Am. Soc.*, **104**, 7667 (1982).

Diethyl (2-tetrahydropyranyloxy)methylphosphonate (1).

Preparation[1]:

$$(C_2H_5O)_2\overset{O}{\overset{\|}{P}}H + (CH_2O)_n \xrightarrow[50\text{–}65\%]{N(C_2H_5)_3,\ 120\text{–}130^\circ} (C_2H_5O)_2\overset{O}{\overset{\|}{P}}CH_2OH \xrightarrow[84\text{–}93\%]{\text{dihydropyran},\ POCl_3} (C_2H_5O)_2\overset{O}{\overset{\|}{P}}CH_2O\text{-THP}$$

1 (b.p. 110°/0.05 mm.)

Homologated aldehydes.[2] The anion of the reagent converts carbonyl compounds into the THP enol ether of one-carbon homologated aldehydes (equation I).

(I) $R^1R^2C{=}O + \mathbf{1} \xrightarrow{\text{LDA}} R^1R^2C(OH)CH(OTHP)P(O)(OC_2H_5)_2 \xrightarrow[\text{overall}]{\text{KO-}t\text{-Bu, 67–85\%}} R^1R^2C{=}CH(OTHP)$ (ag.) $\xrightarrow{\text{HCl, H}_2\text{O, THF}} R^1R^2CHCHO$

[1] A. F. Kluge, *Org. Syn.*, submitted (1982).
[2] A. F. Kluge and I. S. Cloudsdale, *J. Org.*, **44**, 4847 (1979).

Diethylzinc–Methylene iodide, 1, 253; **2**, 134; **4**, 153. Diethylzinc is available from Aldrich and Alfa.

Methylidenation of allylic thioethers.[1] Methylidenation of an allylic phenylthioether with methylene iodide–diethylzinc is accompanied by a 2,3-sigmatropic rearrangement to a homologous allylic phenylthioether. The rearrangement is also initiated by ethylidene iodide. Cyclopropanation is not observed. The Simmons–Smith reaction with allylic sulfides results only in formation of an insoluble polymer.

Examples:

$C_6H_5SCH_2CH{=}C(CH_3)_2 \xrightarrow{CH_2I_2,\ (C_2H_5)_2Zn} [H_2C{=}S^+(C_6H_5)CH_2CH{=}C(CH_3)_2] \xrightarrow{80\%} C_6H_5SCH_2C(CH_3)_2CH{=}CH_2$

3-Methyl-1-(phenylthio)cyclohex-2-ene $\xrightarrow{72\%}$ 1-methyl-1-($CH_2SC_6H_5$)cyclohex-2-ene

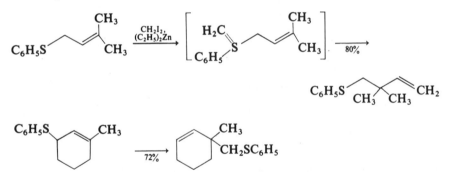

[1] Z. Kosarych and T. Cohen, *Tetrahedron Letters*, **23**, 3019 (1982).

Dihydrotetrakis(triphenylphosphine)ruthenium, $RuH_2[P(C_6H_5)_3]_4$ **(1)**. The complex is prepared by addition of a solution of $RuCl_3$ and $NaBH_4$ in ethanol to triphenylphosphine dissolved in ethanol (94% yield).[1]

Isomerization of 2-butyne-1,4-diol. In the presence of this catalyst, 2-butyne-1,4-diol isomerizes to butyrolactone. $RuH_2[P(C_6H_5)_3]_3$ is somewhat less active.[2]

$$HOCH_2C{\equiv}CCH_2OH \xrightarrow[49\%]{\text{Ru(II), 145°, Diglyme}} \gamma\text{-butyrolactone}$$

Unsymmetrical secondary and tertiary amines. Unsymmetrical amines are obtained in 50–95% yield by reaction of an alcohol and amine in the presence of this ruthenium catalyst. The intramolecular version of this reaction provides an efficient synthesis of cyclic amines either from α,ω-amino alcohols and an alcohol or from α,ω-diols and an amine. The cyclization is useful for preparation of tetrahydroisoquinolines.[3]

Example:

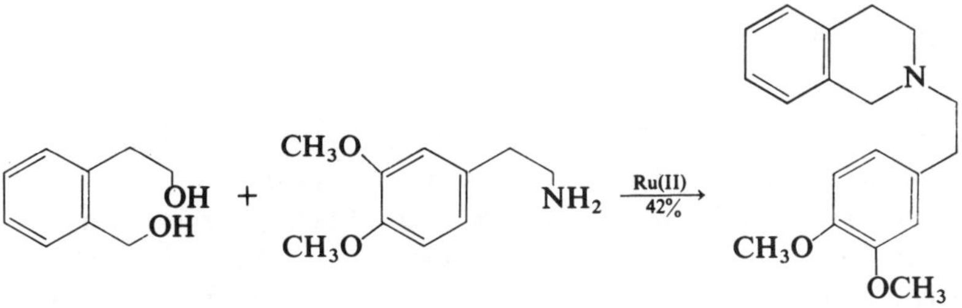
OH
OH
+
CH3O
CH3O
NH2
Ru(II)
42%
N
CH3O
OCH3

Esters; lactones.[4] In the presence of this catalyst, primary alcohols condense oxidatively to form esters.

Examples:

$$2\ n\text{-}C_4H_9OH \xrightarrow[98\%]{Ru(II),\ 180^\circ} n\text{-}C_3H_7CO_2C_4H_9\text{-}n + H_2$$

$$2\ C_6H_5CH_2OH \xrightarrow[97\%]{} C_6H_5CO_2CH_2C_6H_5$$

Under these conditions, primary diols are converted into lactones. The use of acetone as a hydrogen acceptor is beneficial in this variation.

Examples;

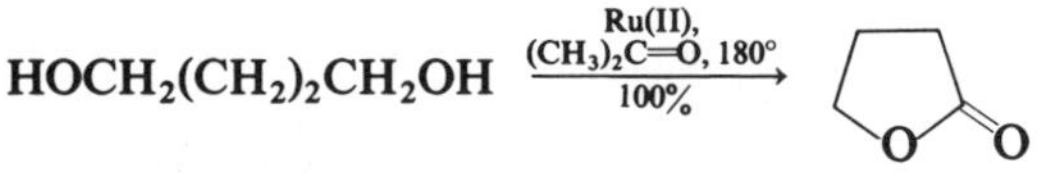
HOCH2(CH2)2CH2OH
Ru(II),
(CH3)2C=O, 180°
100%
O
O

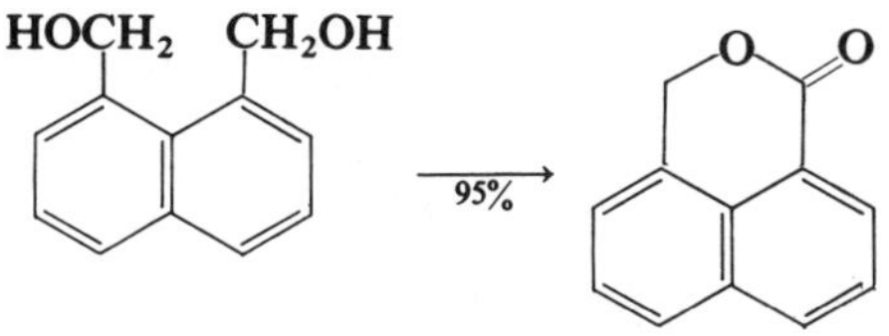
HOCH2
CH2OH
95%
O
O

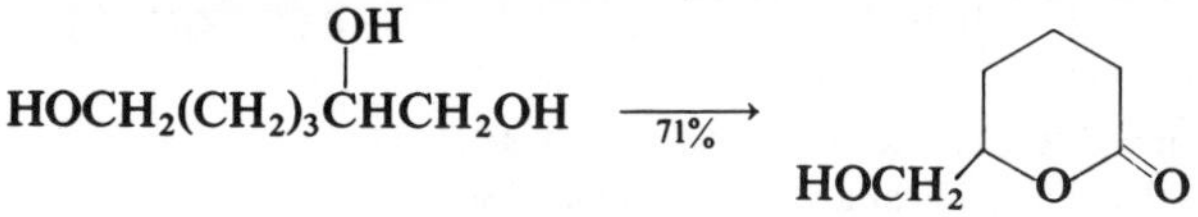
OH
HOCH2(CH2)3CHCH2OH
71%
HOCH2
O
O

Presumably, the condensation proceeds by an initial oxidation to an aldehyde.

[1] J. J. Levison and S. D. Robinson, *J. Chem. Soc., A*, 2947 (1970).
[2] Y. Shvo, Y. Blum, and D. Reshef, *J. Organometal. Chem.*, **238**, C79 (1982).
[3] S.-I. Murahashi, K. Kondo, and T. Hakata, *ibid.*, **23**, 229 (1982).
[4] S.-I. Murahashi, K. Ito, T. Naota, and Y. Maeda, *Tetrahedron Letters*, **22**, 5327 (1981).

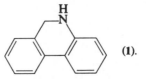

5,6-Dihydrophenanthridine, **(1).**

Mol. wt. 181·233, m.p. 124°. The reagent is prepared by reduction of phenanthridine with lithium aluminum hydride in refluxing ether (75% yield).[1]

Protection[2] and activation[3] of carboxylic acids. Carboxylic acids react with **1** in the presence of a 2-chloropyridinium salt, proton sponge, and DMAP to form amides (**2**). These amides are stable to acids and bases but deprotection is possible with oxidative hydrolysis with ceric ammonium nitrate (CAN). If the oxidation is carried out in the presence of an amine, an amide is obtained in 70–95% yield. For this purpose, the combination of copper(II) oxide and ceric pyridinium chloride is far superior to CAN.[4] No racemization was observed in the benzoylation of an α-amino ester.

RCOOH + 1 —(2-chloro-6-methyl-3-phenyl-1-phenylpyridinium BF_4^-, DMAP; ~95%)→ RC(=O)—N (**2**) —CAN→ RCOOH (~98%) + **3**

2 + R′NH_2 —($(C_5H_5NH)_2CeCl_6$, CuO; CH_3CN; 70–95%)→ RC(=O)NHR′ + **3**

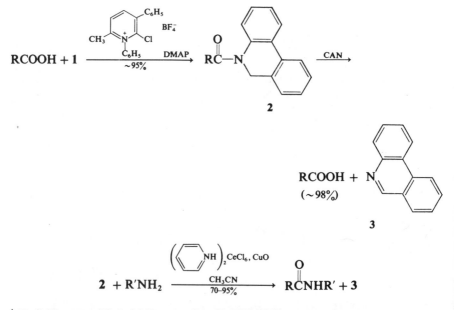

[1] W. C. Wooten and R. L. McKee, *Am. Soc.*, **71**, 2946 (1949).
[2] T. Uchimaru, K. Narasaka, and T. Mukaiyama, *Chem. Letters*, 1551 (1981).

[3] K. Narasaka, T. Hirose, T. Uchimaru, and T. Mukaiyama, *ibid.*, 991 (1982).
[4] D. C. Bradley, A. K. Chatterjee, and W. Wardlaw, *J. Chem. Soc.*, 2260 (1956).

Diisobutylaluminum hydride (DIBAH), **1**, 260–262; **2**, 140–142; **3**, 101–102; **4**, 158–161; **5**, 224–225; **6**, 198–201; **7**, 111–113; **8**, 173–174; **9**, 171–172; **10**, 149.

1,4-Reduction of α-oxoketene dithioacetals.[1] DIBAH in combination with triethylamine reduces these β-heteroatom enones in a conjugate manner.

Examples:

CH_3O–C_6H_4–$COCH{=}C(SCH_3)_2$ $\xrightarrow[65\text{–}83\%]{DIBAH,\ N(C_2H_5)_3}$ CH_3O–C_6H_4–$COCH_2CH(SCH_3)_2$

$C_6H_5COCH{=}$(1,3-dithiane) $\xrightarrow{69\%}$ $C_6H_5COCH_2$–(1,3-dithian-2-yl)

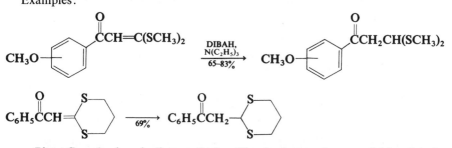

Pictet-Spengler isoquinoline synthesis. The final steps in a synthesis of (−)-antirhine (**2**), a Corynanthe-type indole alkaloid, involved partial reduction of the lactam **1** to a hemiaminal, which cyclizes to **2** in the presence of dilute HC1.[2]

1 $\xrightarrow{1)\ DIBAH;\ 2)\ HCl,\ H_2O,\ 25°}$ **2**

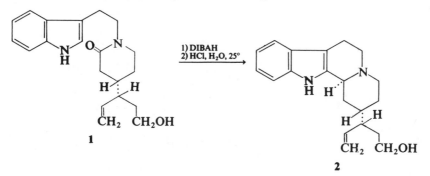

The same strategy was used successfully to cyclize **3** to **4**, the diastereoisomers of (±)-decarbomethoxynauclechine.[3].

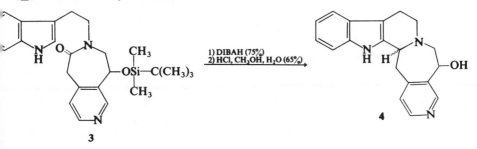

3 $\xrightarrow{1)\ DIBAH\ (75\%);\ 2)\ HCl,\ CH_3OH,\ H_2O\ (65\%)}$ **4**

α-Alkyl-substituted aldehydes.[4] Primary and secondary nitriles can be converted into α-alkylated aldehydes by a one-pot process involving conversion to an aluminum imide, deprotonation, and alkylation. If the α-hydrogen is benzylic, *n*-butyllithium can be used as the base; in general, LDA in combination with a slight excess of HMPT is most satisfactory.

Examples:

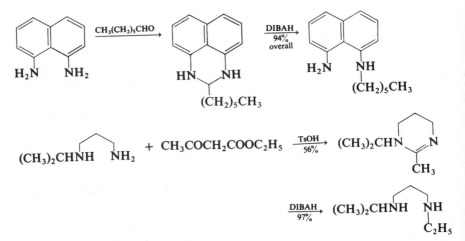

$$(CH_3)_2CHCN \xrightarrow{\text{DIBAH}} \left[(CH_3)_2CHCH{=}NAl(i\text{-}Bu)_2\right] \xrightarrow[65\%]{\text{1) LDA, HMPT; 2) } C_6H_5CH_2Br} (CH_3)_2C(CH_2C_6H_5){-}CHO$$

$$\text{cyclopentyl-CN} \xrightarrow[72\%]{\text{1) DIBAH; 2) LDA; 3) } C_5H_{11}X} \text{1-}(C_5H_{11}\text{-}n)\text{-cyclopentane-1-CHO}$$

Monoalkylation of polyamines. Reductive cleavage of aminals and amidines by DIBAH can be used to effect monoalkylation of polyamines.[5]

Examples:

$$\text{1,8-}(H_2N)_2C_{10}H_6 \xrightarrow{CH_3(CH_2)_5CHO} \text{perimidine aminal (HN–CH((CH}_2)_5CH_3)\text{–NH)} \xrightarrow[\text{94\% overall}]{\text{DIBAH}} \text{8-}H_2N\text{-}C_{10}H_6\text{-1-NH}(CH_2)_5CH_3$$

$$(CH_3)_2CHNH(CH_2)_3NH_2 + CH_3COCH_2COOC_2H_5 \xrightarrow[56\%]{\text{TsOH}} \text{1-}(CH_3)_2CH\text{-2-}CH_3\text{-tetrahydropyrimidine} \xrightarrow[97\%]{\text{DIBAH}} (CH_3)_2CHNH(CH_2)_3NHC_2H_5$$

1-Halo-1-alkenes.[6] Some years ago, Zweifel's group reported the preparation of (E)-1-alkenyl bromides and iodides by *cis*-hydroalumination of 1-alkynes followed by reaction with Br_2 or I_2 (**2**, 141). The method is not applicable to preparation of the (Z)-isomers. The general approach becomes much more useful when applied to 1-alkynylsilanes (**1**), obtained by sequential treatment of 1-alkynes in THF with *n*-butyllithium and chlorotrimethylsilane.[7] Hydroalumination of **1** followed by reaction of the resultant 1-alkenylalanes (**2**) with NCS, Br_2, or I_2 produces (E)-1-halo-1-alkenylsilanes (**3**) in high isomeric purity and yield. These products can be isomerized to the (Z)-isomers by bromine under irradiation when X = Cl or Br. (E)-Iodo-1-alkenylsilanes are isomerized in the presence of catalytic amounts of *t*-butyllithium.[8]

$RC{\equiv}CSi(CH_3)_3$ (**1**) $\xrightarrow[\text{ether}]{(i\text{-}C_4H_9)_2AlH,}$ R(H)C=C(Si(CH_3)_3)(Al(i-C_4H_9)_2) (**2**) $\xrightarrow[\text{85–95\% overall}]{\text{NCS, }Br_2\text{, or }I_2}$ R(H)C=C(Si(CH_3)_3)X, (E)-**3** (X = Cl, Br, I) $\xrightarrow{NaOCH_3,\ CH_3OH}$ R(H)C=C(H)X, (E)-**4**

(E)-**3** (X = Cl, Br) $\xrightarrow{Br_2,\ h\nu}$ R(H)C=C(Br)Si(CH_3)_3, (Z)-**3** $\xrightarrow{NaOCH_3,\ CH_3OH}$ R(H)C=C(Br)H, (Z)-**4**

(E)-**3** (X = I) $\xrightarrow[-78\to25^\circ]{t\text{-}C_4H_9Li}$ R(H)C=C(I)Si(CH_3)_3, (Z)-**3**

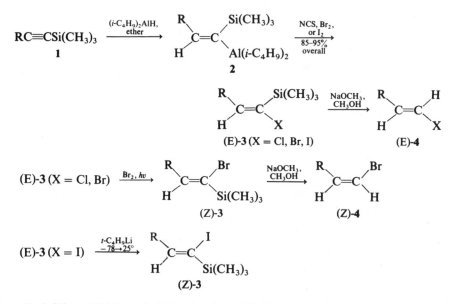

Both (E)- and (Z)-**3** are desilylated regiospecifically to (E)- and (Z)-1-halo-1-alkenes (**4**) by methanolic sodium methoxide at 25–65° in yields of 70–95%.[9]

Reduction of isoflavenones. Isoflavenones (**1**) can be reduced to 4-isoflavanones (**2**) by DIBAH. Use of lithium aluminum hydride results in ring cleavage by a retro-Michael reaction. This reduction, however, is general for only α,β-enones of this type (coplanar, six-membered, endocyclic).[10]

1 (R^1, CH_3O, $C_6H_4R^3$, R^2) $\xrightarrow[\text{50–90\%}]{\text{DIBAH, THF, }-65^\circ}$ **2** (R^1, CH_3O, $C_6H_4R^3$, R^2)

Selective reduction of α,α-dihalo ketones.[11] Reduction of α,α-dihalo ketones can be effected without hydrogenolysis of the halo groups with either DIBAH or borane–dimethyl sulfide. Reactions with the former reagent are generally faster but work-up can be complicated by gelatinous aluminum salts. In general, the yields are roughly comparable.

[1] R. B. Gammill, D. M. Sobieray, and P. M. Gold, *J. Org.*, **46**, 3555 (1981).

[2] S. Takano, N. Tamura, and K. Ogasawara, *J. C. S. Chem. Comm.*, 1155 (1981).
[3] A. Shariff and S. McLean, *Tetrahedron Letters*, **23**, 4895 (1982).
[4] H. L. Goerning and C. C. Tseng, *J. Org.*, **46**, 5250 (1981).
[5] H. Yamamoto and K. Maruoka, *Am. Soc.*, **103**, 4186 (1981).
[6] H. P. On, W. Lewis, and G. Zweifel, *Synthesis*, 999 (1981).
[7] G. Zweifel and W. Lewis, *J. Org.*, **43**, 2739 (1978).
[8] G. Zweifel, R. E. Murray, and H. P. On, *ibid.*, **46**, 1292 (1981).
[9] G. Zweifel and N. R. Pearson, *ibid.*, **46**, 829 (1981).
[10] S. Antus, A. Gottsegen, and M. Nógrádi, *Synthesis*, 574 (1981).
[11] B. L. Jensen, J. Jewett-Bronson, S. B. Hadley, and L. G. French, *ibid.* 732 (1982).

Diisopinocampheylborane (1), **1**, 262–263; **4**, 161–162; **6**, 202; **8**, 174. This borane can be prepared in high optical purity (99% ee) from commercially available α-pinene of 92% ee by hydroboration with $BH_3 \cdot S(CH_3)_2$ (2:1 ratio) at 0°. After 1 hour, $S(CH_3)_2$ is removed under vacuum and 15% excess α-pinene is introduced. After an equilibration of 72 hours at 0°, the diisopinocampheylborane has an optical purity of 99.1% ee.[1]

Asymmetric hydroboration.[2] In a review of asymmetric hydroboration, Brown *et al.* conclude that this is the preferred reagent for asymmetric hydroboration of unhindered *cis*-alkenes. Thus, (R)-(−)-2-butanol can be prepared from *cis*-2-butene with (−)-**1** in 98% ee and (S)-(+)-2-butanol is obtained using (+)-**1** in 95% ee. The alcohols obtained in this way have the same absolute configuration.

[1] H. C. Brown, M. C. Desai, and P. K. Jadhav, *J. Org.*, **47**, 5065 (1982).
[2] H. C. Brown, P. K. Jadhav and A. K. Mandal, *Tetrahedron*, **37**, 3547 (1981).

Diisopropylamine-borane, $[(CH_3)_2CH]_2NH:BH_3$. Supplier: Callery Chem. Co.

Stereoselective reduction of a ketone.[1] The first step in a conversion of the

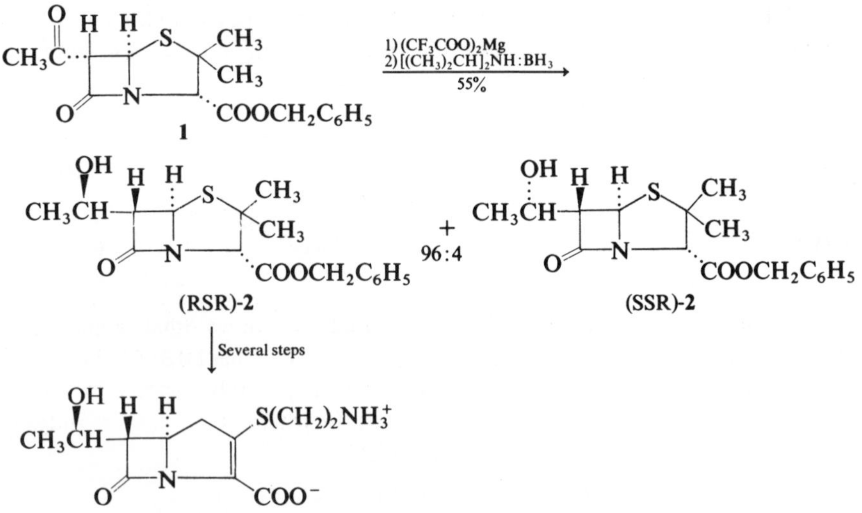

1 —— 1) $(CF_3COO)_2Mg$; 2) $[(CH_3)_2CH]_2NH:BH_3$; 55% ——> (RSR)-**2** + (SSR)-**2** (96:4)

(RSR)-**2** —— Several steps ——> **3**

penicillin derivative **1** to thienamycin (**3**) involves reduction of the carbonyl group to the alcohol **2**. This reaction was conducted with notably high selectivity on the magnesium chelate with diisopropylamine-borane.

[1] S. Karady, J. S. Amato, R. A. Reamer, and L. M. Weinstock, *Am. Soc.*, **103**, 6765 (1981).

Diisopropylsilyl ditriflate; Di-*t*-butylsilyl ditriflate, $R_2Si(OSO_2CF_3)_2$. The reagents are prepared by reaction of diisopropylchlorosilane or di-*t*-butylchlorosilane with 2 equiv. of triflic acid at 22°→reflux. Yields are 71–77%.

Protection of* 1,2-, 1,3-, *and* 1,4-*diols.[1] The reagents in the presence of 2,6-lutidine convert diols into dialkysilylene derivatives in good yields. Deprotection of these derivatives is effected with aqueous HF in CH_3CN. Some typical products and yields are formulated.

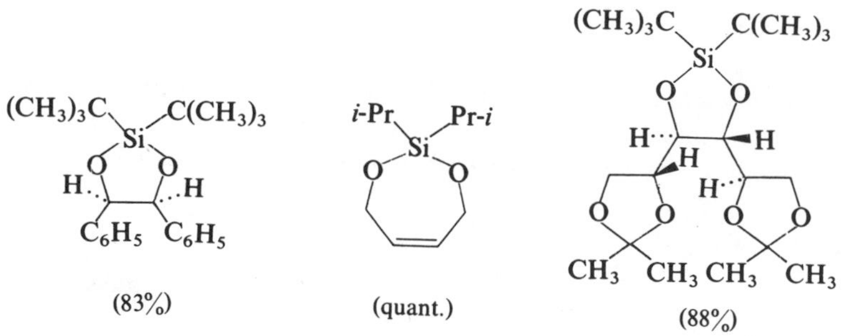

(83%) (quant.) (88%)

[1] E. J. Corey and P. B. Hopkins, *Tetrahedron Letters*, **23**, 4871 (1982).

Diketene, 1, 264–266; **5**, 225–226; **6**, 202.

Anthraquinones. A new regioselective route to highly substituted anthraquinones (**4**) involves the reaction of diketene in the presence of sodium hydride with ethyl 4-aryl-3-oxobutanoates (**1**) prepared as shown from arylacetic acids. The products, after methylation, are cyclized to anthrones (**3**), which are oxidized to anthraquinones.[1]

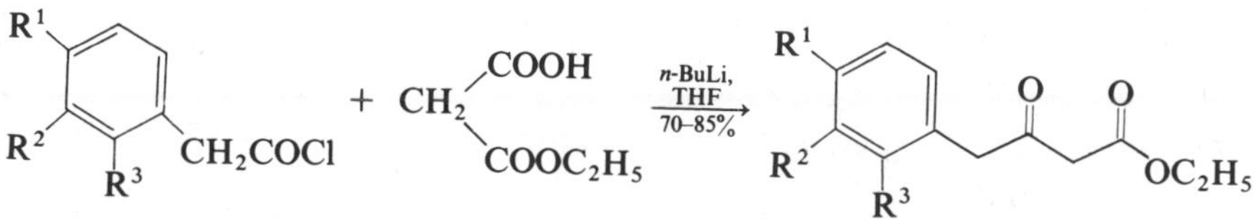

n-BuLi, THF, 70–85%

1

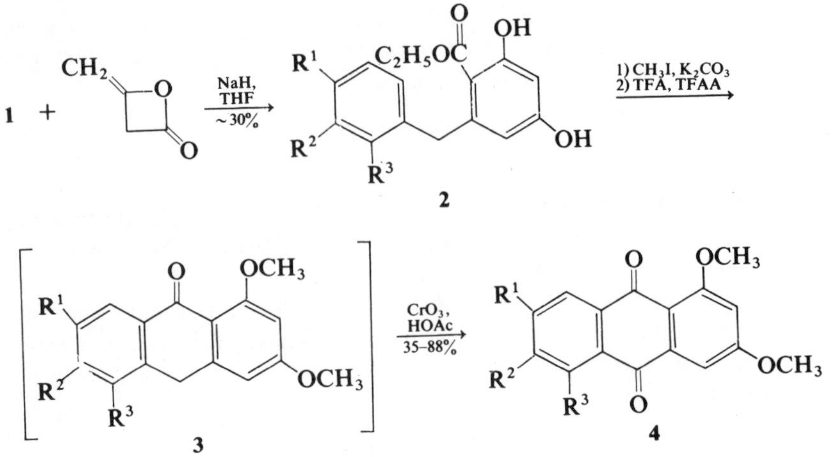

Fluorenes can be prepared by a similar reaction of diketene with ethyl 3-aryl-3-oxopropionates.[2]

[2+2]*Spirocycloaddition.*[3] The photoreaction of diketene with uracil results in an adduct, which loses CO_2 when heated to give a mixture of two products (**1** and **2**). The latter product is converted to the pyrimidine **3** by dehydrogenation.

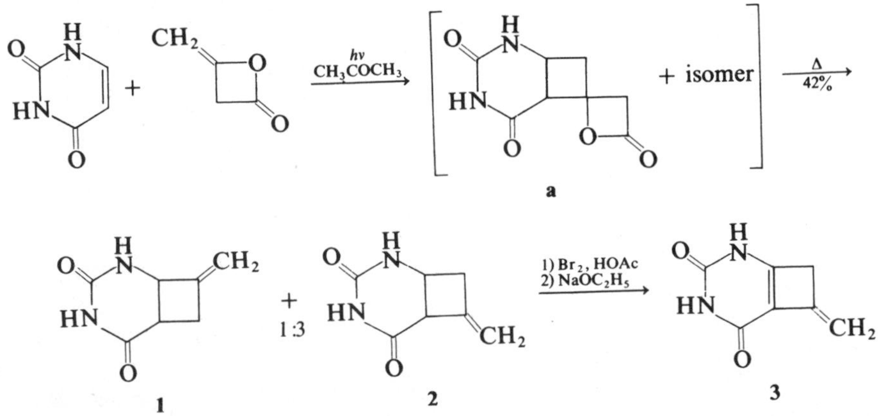

[1] N. Katagiri, T. Kato, and J. Nakano, *Chem. Pharm. Bull.*, **30**, 2440 (1982).
[2] J. Nakano, N. Katagiri, and T. Kato, *ibid.*, **30**, 2590 (1981).
[3] T. Chiba, H. Takahashi, and T. Kato, *Heterocycles*, **19**, 703 (1982).

Dilithium tetrachlorocuprate, 4, 163–164; **5,** 226; **7,** 114; **8,** 176. Aldrich supplies the reagent as a 0.1 *M* solution in THF. *Caution*: flammable, toxic.

Grignard coupling.[1] The Grignard reagent **1** couples with the allylic chloride **2**

in the presence of Li_2CuCl_4 to give the THP ether (**3**) of lactarol as the sole product. Deprotection of **3** followed by oxidation provides lactaral (**4**) in high yield.

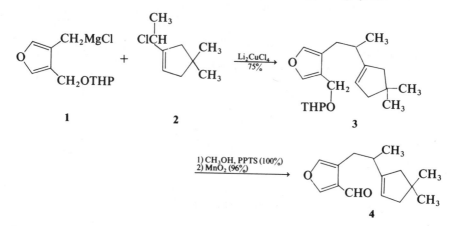

CH3, CH2MgCl, ClCH, CH3, CH3, CH2OTHP, O, +, Li_2CuCl_4 75%, O, CH3, CH2, THPO, CH3, CH3

1 2 3

1) CH_3OH, PPTS (100%)
2) MnO_2 (96%)

O, CHO, CH3, CH3, CH3

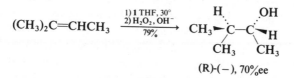

4

[1] S. P. Tanis and D. B. Head, *Tetrahedron Letters*, **23**, 5509 (1982).

Dilongifolylborane (Lgf_2BH), **1**, m. p. 160–161°.

H3C CH3, HB_2, CH3

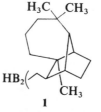

1

The borane is prepared by hydroboration of (+)-longifolene with $BH_3 \cdot S(CH_3)_2$ in ethyl ether. The material is dimeric; it is only sparingly soluble in common organic solvents.

Asymmetric hydroboration.[1] This borane effects enantioselective hydroboration of *cis*-trisubstituted acyclic and cyclic olefins to provide, after oxidation, (R)-alcohols with optical purities of 60–78% ee. The steric requirements of **1** are less than those of diisopinocampheylborane (**1**, 262–263; **4**, 161), but greater than those of isopinocampheylborane (**8**, 267).

Example;

$$(CH_3)_2C{=}CHCH_3 \xrightarrow[79\%]{\text{1) 1 THF, 30°; 2) } H_2O_2,\ OH^-} CH_3\text{—}CH(CH_3)\text{—}CH(OH)CH_3$$

(R)-(−), 70%ee

[1] P. K. Jadhav and H. C. Brown, *J. Org.*, **46**, 2988 (1981).

1(E),3-Dimethoxybutadiene, (1). Mol. wt. 114.14, b.p. 53–57°/21–23 mm, slightly impure.

Preparation[1]:

$$CH_3CH{=}CHCHO \xrightarrow[81\%]{Br_2} CH_3CH{=}C(Br)CHO \xrightarrow[\text{quant.}]{HC(OCH_3)_3,\ HCl} CH_3CH{=}C(Br)CH(OCH_3)_2$$

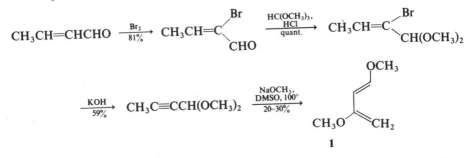

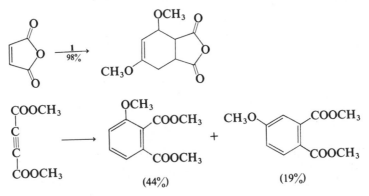

$$\xrightarrow[59\%]{KOH} CH_3C{\equiv}CCH(OCH_3)_2 \xrightarrow[20\text{–}30\%]{NaOCH_3,\ DMSO,\ 100^\circ} CH_3OCH{=}CH{-}C(OCH_3){=}CH_2 \ (\mathbf{1})$$

Diels-Alder reactions.[2] This diene provides an alternative to 1-methoxy-3-trimethysilyloxy-1,3-butadiene (Danishefsky diene) in Diels-Alder reactions.

Examples:

Maleic anhydride $\xrightarrow[98\%]{\mathbf{1}}$ adduct bearing OCH_3 and CH_3O groups on the cyclohexene ring fused to the anhydride

$CH_3OOC{-}C{\equiv}C{-}COOCH_3 \longrightarrow$ 3-methoxy-1,2-benzenedicarboxylic acid dimethyl ester (OCH_3, $COOCH_3$, $COOCH_3$) (44%) + 4-methoxy isomer (CH_3O, $COOCH_3$, $COOCH_3$) (19%)

[1] P. Dowd and W. Weber, *J. Org.*, **47**, 4774 (1982).
[2] *Idem, ibid.*, **47**, 4777 (1982).

(3S,6S)-(+)-2,5-Dimethoxy-3,6-dimethyl-3,6-dihydropyrazine, 10, 151–152.

(R)-α-Methyl-α-amino acids.[1] The asymmetric synthesis of (R)-α-methylamino acids starting with this pyrazine derived from cyclo(L-Ala-L-Ala) has the disadvantage that only half of the L-Ala is recovered, the other half being incorporated into the amino acid. The drawback can be avoided by use of the bis-lactam ether (**2**) of cyclo(L-Val-DL-Ala). The anion of **2** is alkylated almost exclusively *trans* to the isopropyl group. When D-Val is used as the chiral auxiliary, the opposite configuration at C_3 is induced. Hydrolysis gives the (R)-α-methylamino acid (**4**) with recovery of L-ValOCH$_3$.

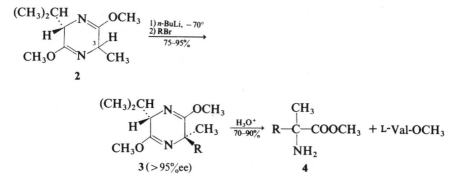

(CH$_3$)$_2$CH N OCH$_3$ H H 3 CH$_3$O N CH$_3$ **2** —1) n-BuLi, −70°; 2) RBr; 75–95%→

(CH$_3$)$_2$CH N OCH$_3$ H CH$_3$ CH$_3$O N R **3** (>95%ee) —H$_3$O$^+$, 70–90%→ R—C(CH$_3$)(NH$_2$)—COOCH$_3$ **4** + L-Val-OCH$_3$

[1] U. Schöllkopf, U. Groth, K.-O. Westphalen, and C. Deng, *Synthesis*, 969 (1981).

Dimethoxymethane (Methylal), **1**, 671–672; **6**, 374; **7**, 115–116.

6-Methylene-3-keto-Δ^4-steroids.[1] 3-Keto-Δ^4-steroids are methylenated at C_6 by reaction with dimethoxymethane and phosphoryl chloride at 65° in chloroform containing sodium acetate. Yields are usually 50–75%.

[1] K. Annen, H. Hofmeister, H. Laurent, and R. Weichert, *Synthesis*, 34 (1982).

Dimethyl 1,3-acetonedicarboxylate (Dimethyl 3-ketoglutarate), CH$_3$OOCCH$_2$COCH$_2$COOCH$_3$ (**1**). Mol. wt. 174.15, b.p. 150°/25 mm. Supplier: Aldrich.

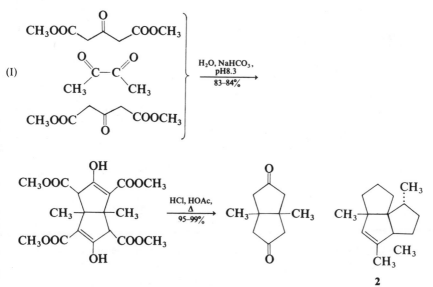

(I) CH$_3$OOC–CH$_2$–CO–CH$_2$–COOCH$_3$ + CH$_3$CO–COCH$_3$ + CH$_3$OOC–CH$_2$–CO–CH$_2$–COOCH$_3$ —H$_2$O, NaHCO$_3$, pH8.3; 83–84%→

[bicyclic diol tetraester: OH, CH$_3$OOC, COOCH$_3$, CH$_3$, CH$_3$, CH$_3$OOC, COOCH$_3$, OH] —HCl, HOAc, Δ; 95–99%→ [dione: O, CH$_3$, CH$_3$, O]   CH$_3$ CH$_3$ CH$_3$ CH$_3$ **2**

Bicyclo*[3.3.0]*octane-3,7-dione derivatives. The condensation of this reagent with 1,2-dicarbonyl compounds in an alkaline medium to give this ring system was reported some time ago[1] and has since been explored in detail, mainly by Cook and Weiss.[2] Detailed directions are now available for preparation of bicyclo[3.3.0]octane-3, 7-dione from glyoxal and that of the 1,5-dimethyl derivative from biacetyl (equation I).[3] The condensation is applicable to a variety of diketones and yields are generally >60%. It has been used for preparation of several natural products such as isocomene (**2**).[4]

[1] U. Weiss and J. M. Edwards, *Tetrahedron Letters*, 4885 (1968).
[2] S. Yang-Lan, M. Mueller-Johnson, J. Oehldrich, D. Wichman, J. M. Cook, and U. Weiss, *J. Org.*, **41**, 4053 (1976).
[3] S. H. Bertz, J. M. Cook, A. Gawish, and U. Weiss, *Org. Syn.*, submitted (1981).
[4] W. G. Dauben and D. M. Walker, *J. Org.*, **46**, 1103 (1981).

Dimethylaluminum benzenethiolate, $(CH_3)_2AlSC_6H_5$ (**1**). Mol. wt. 166.22, white powder. The reagent is prepared from $Al(CH_3)_3$ and C_6H_5SH.[1]

1,3-*Cycloalkanediones.*[2] Treatment of the δ-lactone **2** in THF with **1** [or $(CH_3)_2AlSeCH_3$, **8**, 182–183)] effects cyclization to the 1, 3-dione **3**, probably via an aluminum enolate (**a**) (*cf.*, **9**, 172–173).

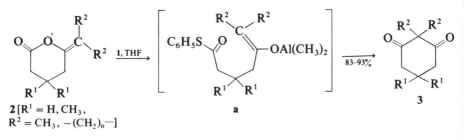

2 [R^1 = H, CH_3, R^2 = CH_3, $-(CH_2)_n-$] —(1, THF)→ [**a**: C_6H_5S, O, R^2, R^2, $OAl(CH_3)_2$, R^1, R^1] —(83–93%)→ **3** (R^2, R^2, O, O, R^1, R^1)

Another example:

(CH₃ cyclohexanone, =C(CH₃)₂) —($ClC_6H_4CO_3H$, HOAc, H_2SO_4, 25%)→ (CH₃ lactone, =C(CH₃)₂) —(1, 98%)→ (CH₃ cycloheptanedione, CH₃, CH₃, O)

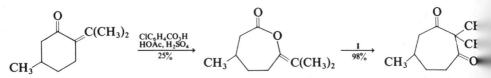

[1] N. Davidson and H. C. Brown, *Am. Soc.*, **64**, 316 (1942).
[2] H. Tomioka, K. Oshima, and H. Nozaki, *Tetrahedron Letters*, **23**, 99 (1982).

3-(Dimethylamino)-2-azaprop-2-en-1-ylidene dimethylammonium chloride (1). Mol. wt. 149.65, m.p. 101–103°, hygroscopic. The reagent is prepared by reaction of cyanuric chloride with DMF in dioxane (35°, 2–3 hours), equation (I).[1]

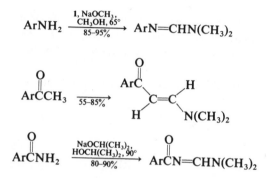

$$\text{(I)}\quad \text{2,4,6-trichloro-1,3,5-triazine} + 6\,(CH_3)_2NCHO \xrightarrow[\sim 95\%\ \text{crude}]{-3CO_2} 3\,(CH_3)_2\overset{+}{N}{=}CH{-}N{=}CHN(CH_3)_2\ Cl^- \quad \mathbf{1}$$

β-Dimethylaminomethylenation. In the presence of base, Gold's reagent converts amines into amidines, ketones into enaminones, and amides into acyl amidines.[2].

Examples;

$$ArNH_2 \xrightarrow[85\text{–}95\%]{\mathbf{1},\ NaOCH_3,\ CH_3OH,\ 65°} ArN{=}CHN(CH_3)_2$$

$$Ar\overset{O}{\overset{\|}{C}}CH_3 \xrightarrow{55\text{–}85\%} Ar\overset{O}{\overset{\|}{C}}(H)C{=}C(H)N(CH_3)_2$$

$$Ar\overset{O}{\overset{\|}{C}}NH_2 \xrightarrow[80\text{–}90\%]{NaOCH(CH_3)_2,\ HOCH(CH_3)_2,\ 90°} Ar\overset{O}{\overset{\|}{C}}N{=}CHN(CH_3)_2$$

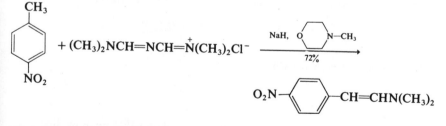

The reagent reacts with unsymmetrical ketones fairly selectively to form the less substituted enaminone.[3]

The reaction is also applicable to esters and lactones in the presence of NaOR in ROH (55–85% yield).

o- and *p*-Nitroarenes react with Gold's reagent in the presence of NaH and N-methylmorpholine to form enamines.[4]

Example:

$$p\text{-}CH_3C_6H_4NO_2 + (CH_3)_2NCH{=}NCH{=}\overset{+}{N}(CH_3)_2Cl^- \xrightarrow[72\%]{NaH,\ \text{N-methylmorpholine}} O_2N{-}C_6H_4{-}CH{=}CHN(CH_3)_2$$

[1] H. Gold, *Agnew. Chem.*, **72**, 956 (1960).
[2] J. T. Gupton, C. Colon, C. R. Harrison, M. J. Lizzi, and D. Polk, *J. Org.*, **45**, 4522 (1980), J. T. Gupton and S. A. Andrews, *Org. Syn.*, submitted (1981).
[3] J. T. Gupton, S. A. Andrews, and C. Colon, *Syn. Comm.*, **12**, 35 (1982).
[4] J. T. Gupton, M. J. Lizzi, and D. Polk, *Syn. Comm.*, **12**, 939 (1982).

2-Dimethylamino-N,N'-diphenyl-1,3,2-diazaphospholidine. Mol wt. 285, 22, m.p. 140°. The phospholidine is prepared by reaction of N,N'-diphenyl-1,2-diaminoethane and tris(dimethylamino)phosphine (85% yield).

C_6H_5–N / $PN(CH_3)_2$ / N–C_6H_5 (cyclic) **(1)**.

Alkyl halides.[1] The reagent reacts with primary and secondary alcohols to form highly crystalline 2-alkoxy N, N'-diphenyl-1,3-2-diazaphospholanes, which are useful derivatives for characterization. The alcohol is regenerated by treatment with acetic acid or methanol.

In addition, these derivatives can be converted efficiently into alkyl halides. Typical procedures are formulated.

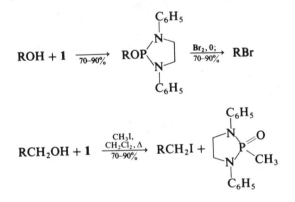

$$\text{ROH} + \mathbf{1} \xrightarrow[70\text{–}90\%]{} \text{ROP}(\text{N}(C_6H_5)CH_2CH_2\text{N}(C_6H_5)) \xrightarrow[70\text{–}90\%]{Br_2,\ 0;} \text{RBr}$$

$$\text{RCH}_2\text{OH} + \mathbf{1} \xrightarrow[70\text{–}90\%]{CH_3I,\ CH_2Cl_2,\ \Delta} \text{RCH}_2\text{I} + (\text{N}(C_6H_5)CH_2CH_2\text{N}(C_6H_5))\text{P}(=\text{O})CH_3$$

[1] S. Hanessian, Y. Leblanc, and P. Lavallée, *Tetrahedron Letters*, **23**, 4411 (1982).

4-(Dimethylamino)pyridinium chlorochromate, 4-$N(CH_3)_2$-pyridinium ($\overset{+}{N}$H) CrO_3Cl^- **(1)**.

DMAP·$HCrO_3Cl$ is obtained by reaction of 4-dimethylaminopyridine with chromium trioxide (1 equiv.) in aqueous hydrochloric acid. It is a yellow-orange, nonhygroscopic solid that is moderately sensitive to light.

Oxidation of allylic and benzylic alcohols.[1] The reagent oxidizes primary and secondary alcohols very slowly, but allylic and benzylic alcohols are oxidized to the corresponding aldehydes in moderate to high yield (3–20 hours). The initially (Z)-unsaturated aldehydes are isomerized by the acidic reagent to the (E)-isomers.

[1] F. S. Guziec, Jr., and F. A. Luzzio, *J. Org.*, **47**, 1787 (1982).

4-Dimethylamino-1,1,2-trimethoxybutadiene (1). Mol. wt. 187.2, yellow oil, b.p. 78°/0.15 mm.

Preparation:

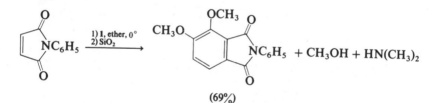

$$(CH_3O)_2CHC(=O)CH{=}CHN(CH_3)_2 \xrightarrow[95\%]{(CH_3)_3O^+BF_4^-} (CH_3O)_2CHCH\cdots\overset{+}{C}(OCH_3)\cdots CHN(CH_3)_2\ BF_4^-$$

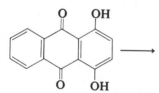

$$\xrightarrow[77\%]{KOC(CH_3)_3,\ THF} (CH_3O)_2C{=}C(OCH_3){-}CH{=}CHN(CH_3)_2$$

1

Diels-Alder aromatization.[1] The reagent forms aromatic cycloadducts from Diels-Alder reactions.

Examples:

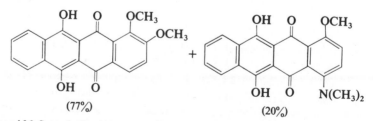

N-Phenylmaleimide $\xrightarrow{1)\ \mathbf{1},\ ether,\ 0°;\ 2)\ SiO_2}$ 3,4-dimethoxy-N-phenylphthalimide (69%) $+ CH_3OH + HN(CH_3)_2$

1,4-Dihydroxyanthraquinone $\longrightarrow$ 6,11-dihydroxy-1,2-dimethoxy-5,12-naphthacenedione (77%) + 6,11-dihydroxy-1-methoxy-4-dimethylamino-5,12-naphthacenedione (20%)

[1] R. Gompper and M. Sramek, *Synthesis*, 649 (1981).

Dimethylbromosulfonium bromide, $(CH_3)_2SBr^+Br^-$ (**1**). The reagent is prepared as a yellow solid in 91% yield by reaction of $(CH_3)_2S$ in CH_2Cl_2 with Br_2 at $-40°$; or, more conveniently, *in situ* in acetonitrile.

α-Bromo-α,β-enones.[1] The *in situ* reagent reacts with α,β-enones at -40 to 0° to form salts, which lose a proton and dimethyl sulfide on treatment with potassium carbonate to give α-bromo enones in high yield. The reaction fails with hindered enones and is sluggish with α,β-unsaturated esters.

Examples:

$$CH_2{=}CHCOCH_3 \xrightarrow[84\%]{1} \underset{{}^+S(CH_3)_2Br^-}{\underset{|}{C}}H_2CHBrCOCH_3 \xrightarrow[85\%]{K_2CO_3} CH_2{=}\underset{Br}{\underset{|}{C}}COCH_3$$

$$\text{2-cyclohexen-1-one} \xrightarrow[96\%]{} \xrightarrow[92\%]{} \text{2-bromo-2-cyclohexen-1-one}$$

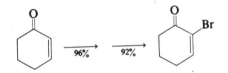

[1] Y. L. Chow and B. H. Bakker, *Can. J. Chem.*, **60**, 2268 (1982).

Dimethylformamide, 1, 278–281; **2**, 153–154; **3**, 115; **4**, 184; **5**, 247–249; **7**, 124; **8**, 189–190; **9**, 182.

Formylation of an alkyllithium (**1**, 280).[1] Formylation of an alkyllithium or a Grignard reagent with DMF (Bouveault reaction) is generally unsatisfactory because of side reactions. However, sonication of the mixture of an alkyl or aryl halide, lithium, and DMF substantially improves the rate and the yield. The method is applicable to primary, secondary, and tertiary bromides or chlorides. Typical yields are in the range 65–85%.

[1] C. Pétrier, A. L. Gemal, and J.-L. Luche, *Tetrahedron Letters*, **23**, 3361 (1982).

Dimethylformamide dimethyl acetal, 1, 281–282; **2**, 154; **3**, 115–116; **4**, 184–185; **5**, 254; **6**, 221–222; **8**, 191–192; **10**, 158–159.

(E)-1,3-Disubstituted-1,3-dienes. The reaction of α,β-enones (**1**) with the dianion of acetic acid (**2**) results almost exclusively in 1,2-addition to give 4,5-unsaturated-3-hydroxycarboxylic acids (**3**).[1] These products on dehydrative decarboxylation with DMF dimethyl acetal (**6**, 221–222) give pure (E)-1,3-dienes (**4**).[2]

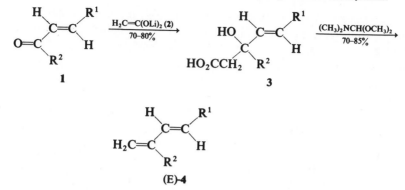

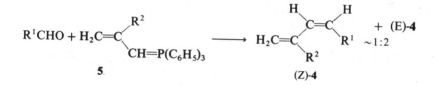

$$\underset{\mathbf{1}}{\text{O=C}(R^2)\text{–CH=CHR}^1} \xrightarrow[70\text{–}80\%]{H_2C{=}C(OLi)_2\ (\mathbf{2})} \underset{\mathbf{3}}{HO_2CCH_2C(OH)(R^2)\text{–CH=CHR}^1} \xrightarrow[70\text{–}85\%]{(CH_3)_2NCH(OCH_3)_2} \underset{(E)\text{-}\mathbf{4}}{H_2C{=}C(R^2)\text{–CH=CHR}^1}$$

The (Z)-isomers are prepared by a nonstereoselective Wittig reaction of the phosphorane **5** with aldehydes. On thermolysis of the resulting (E)/(Z) mixture, (E)-**4** is converted into a dimer, and (Z)-**4** is unchanged and easily separated from the dimer.

$$R^1CHO + \underset{\mathbf{5}}{H_2C{=}C(R^2)CH{=}P(C_6H_5)_3} \longrightarrow \underset{(Z)\text{-}\mathbf{4}}{H_2C{=}C(R^2)\text{–CH=CHR}^1} + (E)\text{-}\mathbf{4} \quad \sim 1:2$$

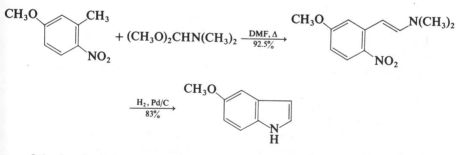

Indole synthesis.[3] A general indole synthesis involves reaction of an *o*-nitrotoluene derivative with dimethylformamide dimethyl acetal in refluxing DMF with removal of the methanol to form a nitro N,N-dimethyl enamine. Reduction of the nitro group is accompanied by spontaneous cyclization to an indole. Catalytic hydrogenation is generally preferred for this step.

Example:

$$CH_3O\text{-}C_6H_3(CH_3)(NO_2) + (CH_3O)_2CHN(CH_3)_2 \xrightarrow[92.5\%]{DMF,\ \Delta} CH_3O\text{-}C_6H_3(NO_2)\text{–CH=CH–}N(CH_3)_2 \xrightarrow[83\%]{H_2,\ Pd/C} \text{5-}CH_3O\text{-indole (N–H)}$$

One drawback in this synthesis is that electron-donating groups retard the aminomethylenation step. A simple expedient is the addition of pyrrolidine to the reaction mixture. A more reactive reagent is obtained, which forms mainly a pyrrolidine enamine, which is also converted to an indole on reduction.

Butenolides. A new method was used to convert the allylic alcohol **1** into the butenolide (**3**). When **1** is heated with dimethylformamide dimethyl acetal, the allylic amide **2** is formed (**5**, 253; **8**, 191–192). The epoxide of **2** on treatment with base is converted to an α,β-unsaturated amide (**a**), which on acid hydrolysis yields the butenolide (**3**).[4]

$CH(CH_3)_2$ H_3C OCH_3 H CH_2OH **1** —[$(CH_3)_2NCH(OCH_3)_2$, Xylene, Δ; 80%]→ CH_3 O $(CH_3)_2NC$ H CH_2 **2** —[1) m-$ClC_6H_4CO_3H$ (100%) 2) $LiN[Si(CH_3)_3]_2$]→

[CH_3 O $(CH_3)_2NC$ H CH_2OH] **a** —[HCl, H_2O; 80%]→ $CH(CH_3)_2$ H_3C OCH_3 O O H **3**

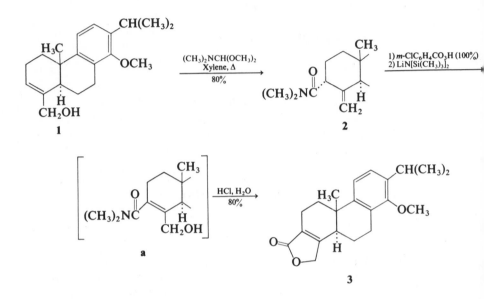

[1] J. Mulzer, U. Kuhl, and G. Bruntrup, *Tetrahedron Letters*, 2949, 2953 (1978).
[2] J. Mulzer, G. Bruntrup, U. Kuhl, and G. Hartz, *Ber.*, **115**, 3453 (1982).
[3] A. D. Batcho and W. Leimgruber, *Org. Syn.* submitted (1982).
[4] L. C. Garver and E. E. van Tamelen, *Am. Soc.*, **104**, 867 (1982).

N,N-Dimethylhydrazine, 1, 289–290; **2,** 154–155; **3,** 117; **5,** 254; **6,** 223; **7,** 126–130; **8,** 192–193; **9,** 184–185.

Diels-Alder reactions of α,β-unsaturated N,N-dimethylhydrazones.[1] These readily available hydrazones can function as 1-amino-1-aza-1,3-dienes in Diels-Alder reactions. Thus, **1** undergoes regioselective cycloaddition with various electrophilic dienophiles to give tetrahydropyridines such as **2** and **3**. Unfortunately, removal of the dimethylamino group with zinc and acetic acid (or other reagents) also effects reduction of the double bond. The initial adduct from cycloaddition of **1** with naphthoquinone is unstable and undergoes spontaneous elimination of the elements of dimethylamine to give the aromatic adduct **4**.

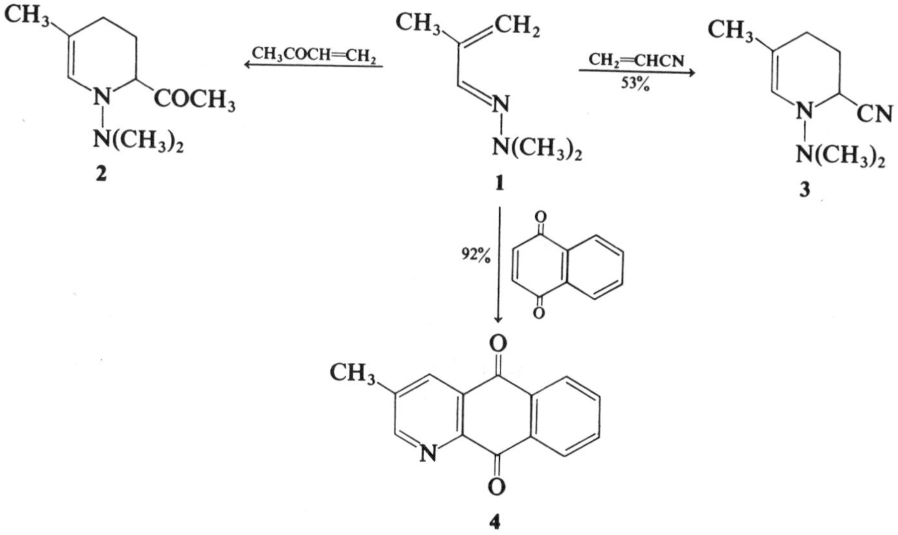

[1] B. Serckx-Poncin, A.-M. Hesbain-Frisque, and L. Ghosez, *Tetrahedron Letters*, **23**, 3261 (1982).

N,O-Dimethylhydroxylamine hydrochloride, $CH_3ONHCH_3 \cdot HCl$ (**1**). Mol. wt. 97.55, m.p. 112–115°. Suppliers: Aldrich, Pfaltz, and Bauer.

$R^1COOH \rightarrow R^1CHO$ or R^1COR^2.[1] Acid chlorides react with **1** to form amides (**2**), which are reduced by $LiAlH_4$ or DIBAH almost exclusively to the corresponding aldehyde (equation I).

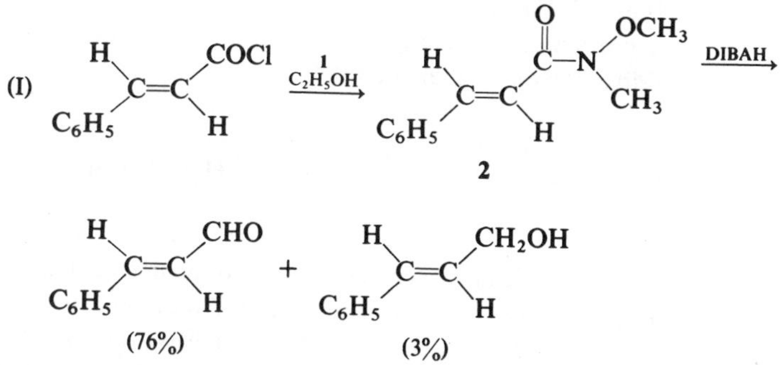

The amides (**2**) also react with Grignard or alkyllithium reagents to form ketones in generally high yield.

Examples:

$$C_6H_5C(=O)-N(CH_3)-OCH_3 \longrightarrow C_6H_5C(=O)-C\equiv CC_6H_5$$

$C_6H_5C\equiv CMgBr$	65°	92%
$C_6H_5C\equiv CLi$	20°	90%

$$CH_3OC(=O)(CH_2)_3C(=O)-N(CH_3)-OCH_3 + 6C_6H_5MgBr \xrightarrow[98\%]{65^\circ} (C_6H_5)_2C(OH)(CH_2)_3C(=O)C_6H_5$$

[1] S. Nahm and S. M. Weinreb, *Tetrahedron Letters*, **22**, 3815 (1981).

1,3-Dimethyl-2-imidazolidinone (DMI),

$$\text{cyclic } CH_3N-C(=O)-NCH_3 \text{ (CH}_2CH_2\text{ ring)}$$

Mol. wt. 114.15, b.p. 221–223°. Suppliers: Aldrich, Fluka.

Dipolar aprotic solvent. DMI is a useful solvent for organometallic reactions involving strong bases or alkali metals, to which it is stable.[1] Dehydrations and dehydrohalogenations with methyltriphenoxyphosphonium iodide can be conducted equally well in DMI as in HMPT.[2]

DMI is the solvent of choice for the conversion of allylic iodides to homoallylic alcohols via an allyltin dihaloiodide (equation I).[3]

$$\text{(I)}\quad CH_2{=}CHCH_2I + R^1COR^2 \xrightarrow{SnF_2} \xrightarrow[60\text{–}95\%]{H_2O} CH_2{=}CHCH_2C(R^1)(R^2)-OH$$

It can be used in nucleophilic displacement reactions of unactivated aryl halides[4] that originally were only possible in solutions in HMPT. An example is the reaction of C_6Cl_6 or C_6F_6 with the sodium salts of thiols to give hexakis(alkylthio)benzenes in >90% yield[4, 5]. Indeed, hexakis(β-naphthylthio)benzene (**2**) was prepared by reaction of hexachlorobenzene and the sodium salt of β-mercaptonaphthalene in **1** in 72% yield. Hexasubstituted benzenes such as **2** are of interest because they can function as inclusion hosts related to clathrates of phenol.[6]

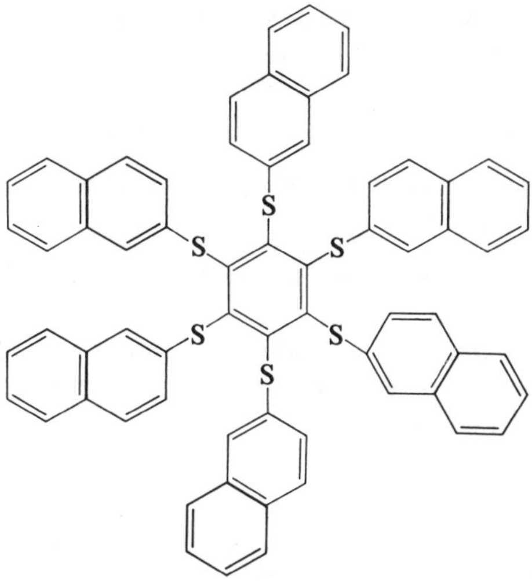

2

[1] H. Sakurai and F. Kondo, *J. Organometal. Chem.*, **92**, C46 (1975); **117**, 149 (1976).
[2] C. W. Spangler, D. P. Kjell, L. L. Wellander, and M. A. Kinsella, *J. C. S. Perkin I*, 2287 (1981).
[3] T. Mukaiyama, T. Harada, and S. Shoda, *Chem. Letters*, 1507 (1980).
[4] D. D. MacNicol, P. R. Mallinson, A. Murphy, and G. J. Sym, *Tetrahedron Letters*, **23**, 4131 (1982).
[5] P. Cogolli, F. Maiolo, L. Testaferri, M. Tingoli, and M. Tiecco, *J. Org.*, **44**, 2642 (1979); L. Testaferri, M. Tingoli, and M. Tiecco, *ibid.*, **45**, 4376 (1980).
[6] D. D. MacNicol, J. J. McKendrick, and D. R. Wilson, *Chem. Soc. Rev.*, **7**, 65 (1978).

Dimethyl methylphosphonate, 3, 117.

Methylenation. This phosphonate ester is generally regarded as unsuitable for methylenation of ketones, even though the anion forms the intermediate adduct in satisfactory yield. However, under suitable experimental conditions, this reagent was used for methylenation of **1** to give **2** in 90% yield, higher than that obtained with methylphosphonic acid bis(dimethylamide) (78%).[1] The optimum yield is obtained if exactly 2 equiv. of base and 1 equiv. of water are used to quench the adduct before cycloelimination.

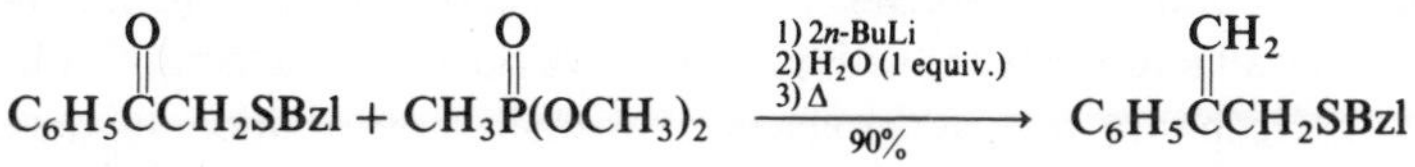

$$\underset{\mathbf{1}}{C_6H_5\overset{O}{\overset{\|}{C}}CH_2SBzl} + CH_3\overset{O}{\overset{\|}{P}}(OCH_3)_2 \xrightarrow[90\%]{1)\ 2n\text{-BuLi};\ 2)\ H_2O\ (1\ \text{equiv.});\ 3)\ \Delta} \underset{\mathbf{2}}{C_6H_5\overset{CH_2}{\overset{\|}{C}}CH_2SBzl}$$

[1] M. A. Fox, C. A. Triebel, and R. Rogers, *Syn. Comm.*, **12**, 1055 (1982).

Dimethyl(methylthio)sulfonium tetrafluoroborate, $CH_3S\overset{+}{S}(CH_3)_2BF_4^-$ (**1**). Mol. wt. 196.1, m.p. 82–84°. The salt is prepared by reaction of dimethyl disulfide with trimethyloxonium tetrafluoroborate.[1]

Cyclization of thioketals. The reagent (**1**) converts thioacetals and thioketals into thiosulfonium ions which dissociate rapidly into alkylthio carbocations, $RS\overset{+}{C}\langle$, and $RSSCH_3$. The addition of water to the carbocations leads to the parent ketone, but this method of deprotection results in only moderate yields because of other intervening reactions.[2]

This reaction can initiate intramolecular cyclization of unsaturated thioketals and of enol silyl ethers containing a thioketal group to provide an equivalent of a directed intramolecular aldol reaction.[3]

Examples;

Si(CH3)3, SCH3, C—CH3, SCH3 —(1, CH2Cl2, 25°)→ [SCH3, CH3] —(68%)→ CH3S, CH3

OSi(CH3)3, CH2=C—(CH2)4C(SCH3)2, CH3 —(62%)→ CH3, CH3S, O —(Oxid., 90%)→ CH3, O

OSi(CH3)3, CH2=C(CH2)3C(CH2)3CH=CHSi(CH3)3, CH3S SCH3 —(86%)→ SCH3, O —(94%)→ O

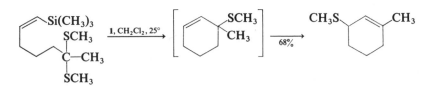

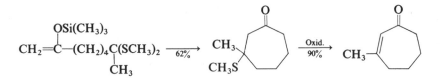

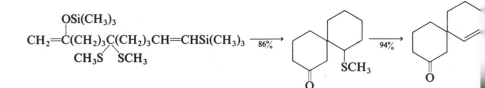

Azasulfenylation of alkenes.[4, 5] The adducts of dimethyl(methylthio)-sulfonium tetrafluoroborate (**1**) with alkenes react slowly but smoothly with various nitrogen nucleophiles to give products of overall *trans*-addition to the alkene. Regioselectivity depends on the substitution pattern of the alkene and on the nucleophilicity of the attacking reagent, and is subject to some control. The mechanism of this azasulfenylation is not certain; it may involve an episulfonium ion.

Examples:

Cyclohexene $\xrightarrow[92\%]{1)\ \mathbf{1};\ 2)\ NH_4OH,\ CH_3CN}$ *trans*-2-(methylthio)cyclohexylamine (NH_2, SCH_3)

1-Methylcyclohexene $\xrightarrow{1)\ \mathbf{1};\ 2)\ KN_3,\ CH_3NO_2}$ 1-azido-1-methyl-2-(methylthio)cyclohexane (N_3, CH_3, SCH_3) (71%) + 1-methyl-1-(methylthio)-2-azidocyclohexane (SCH_3, CH_3, N_3) (29%)

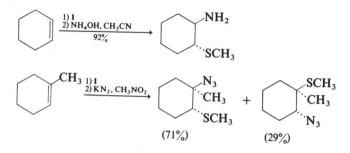

Desulfurization or sulfoxide elimination of the adducts effects addition of the nitrogen nucleophile to an olefin or a substitution reaction with double-bond migration.

Oxy- and oxosulfenylation of alkenes.[6] Similar reactions can be effected with oxygen nucleophiles such as ^-OH or ^-OAc, resulting in oxysulfenylation. DMSO also reacts to give an adduct that is converted to the β-keto sulfide on addition of diisopropylethylamine.

Examples:

1-Methylcyclohexene $\xrightarrow[93\%]{1)\ \mathbf{1};\ 2)\ CaCO_3,\ H_2O,\ CH_2Cl_2}$ 1-methyl-2-(methylthio)cyclohexanol (OH, CH_3, SCH_3)

$$(CH_3CH_2CH_2)_2C{=}CH_2 \xrightarrow[97\%]{1)\ \mathbf{1};\ 2)\ KOAc,\ CH_3CN}$$

$$(CH_3CH_2CH_2)_2\underset{\displaystyle SCH_3}{\underset{|}{C}}{-}CH_2OAc + (CH_3CH_2CH_2)_2\underset{\displaystyle OAc}{\underset{|}{C}}{-}CH_2SCH_3 \qquad 84:16$$

Cyclohexene $\xrightarrow{1)\ \mathbf{1};\ 2)\ (CH_3)_2SO}$ [2-(methylthio)cyclohexyl-$O\overset{+}{S}(CH_3)_2$ (SCH_3)] $\xrightarrow[72\%]{C_2H_5N(i\text{-}C_3H_7)_2}$ 2-(methylthio)cyclohexanone (O, SCH_3)

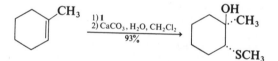

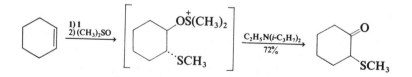

These reactions, followed by desulfurization of the adducts, provide a net hydration of an alkene or conversion of an alkene to a ketone. Control of the regioselectivity of the nucleophilic attack provides useful flexibility.

Cyanosulfenylation is generally conducted by mixing the alkene with **1** in CH_3CN followed by addition of finely powdered NaCN at 25°. Addition of dimethyl sulfide sometimes improves the yield. The adducts are all assumed to be *trans*, but regioselectivity is a more critical question. With monosubstituted alkenes, only *anti*-Markovnikov addition is observed. Markovnikov adducts are favored by use of the less nucleophilic $(CH_3)_3SiCN$.

Examples:

$$CH_2{=}CH(CH_2)_3COOCH_3 \xrightarrow[81\%]{1)\,\mathbf{1};\ 2)\,NaCN} NCCH_2{-}\underset{}{\overset{SCH_3}{\overset{|}{C}}}H(CH_2)_3COOCH_3$$

$$(CH_3CH_2CH_2)_2C{=}CH_2 \longrightarrow (CH_3CH_2CH_2)_2\underset{SCH_3}{\underset{|}{C}}{-}CH_2CN + (CH_3CH_2CH_2)_2\underset{CN}{\underset{|}{C}}{-}CH_2SCH_3$$

NaCN	98%	2%
$(CH_3)_3SiCN$	35%	52%

[1] H. Meerwein, K. F. Zenner, and R. Gipp, *Ann.*, **688**, 67 (1965).
[2] J. K. Kim, J. K. Pau, and M. C. Caserio, *J. Org.*, **44**, 1544 (1979).
[3] B. M. Trost and E. Murayama, *Am. Soc.*, **103**, 6529 (1981).
[4] B. M. Trost and T. Shibata, *ibid.*, **104**, 3225 (1982).
[5] M. C. Caserio and J. K. Kim, *ibid.*, **104**, 3231 (1982).
[6] B. M. Trost, T. Shibata, and S. J. Martin, *ibid.*, **104**, 3228 (1982).

Dimethyl 2-oxoglutaconate (1). Mol. wt. 174.13, liquid.

Preparation[1]:

$$CH_3OOC(CH_2)_3COOCH_3 \xrightarrow[97\%]{1)\,Br_2,\,CH_2Cl_2,\,\Delta;\ 2)\,N(C_2H_5)_3} CH_3OOC\overset{O}{\overset{\|}{C}}CH{=}CHCOOCH_3$$

1

Doebner-Miller[2] annelation.[1] In a synthesis of methoxatin (**6**), the coenzyme of the alcohol dehydrogenase of methylotrophic bacteria, the third ring was added by annelation of **2** by a remarkably facile Doebner-Miller reaction with **1** to give **3**, which undergoes aromatization to **4** in the presence of dry HCl. Oxidative demethylation converted **4** into the quinone **5**. Hydrolysis of **5** to the free triacid presented a problem because of the ester group on the pyrrole ring, but was solved by first conversion to a monoketal followed by treatment with K_2CO_3.

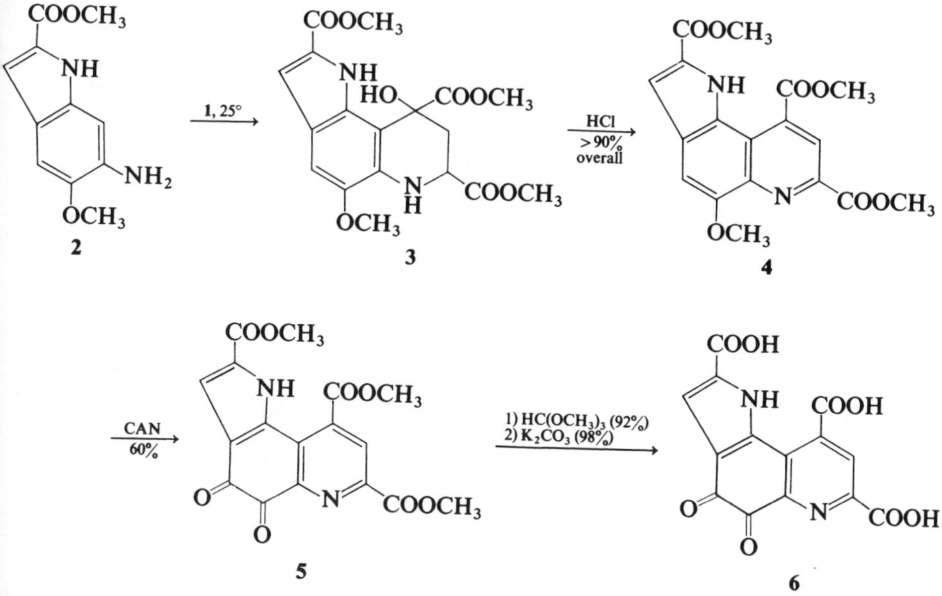

$\xrightarrow{1,\ 25°}$ 3 $\xrightarrow[\text{overall}]{\text{HCl}\ >90\%}$ 4 $\xrightarrow[60\%]{\text{CAN}}$ 5 $\xrightarrow{1)\ HC(OCH_3)_3\ (92\%);\ 2)\ K_2CO_3\ (98\%)}$ 6

[1] E. J. Corey and A. Tramontano, *Am. Soc.*, **103**, 5599 (1981).
[2] F. W. Bergström, *Chem. Rev.*, **35**, 153 (1944).

1,3-Dimethyl-2-oxohexahydropyrimidine (DMPU), b.p. 54°/0.05 mm. Supplier: Fluka. (**1**),

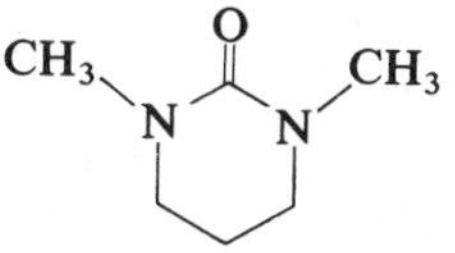

Dipolar aprotic solvent. This cyclic urea can serve as a substitute for the carcinogenic hexamethylphosphoric triamide (HMPT) in reactions of highly nucleophilic and basic reagents. It mimics the effect of HMPT in Wittig olefination and in selective generation of various enolates. It forms homogeneous solutions with THF even at −78°.[1]

Caution: explosive decomposition occurs on attempts to prepare solutions of CrO_3 in DMPU (also possible with HMPT).

[1] T. Mukhopadhyay and D. Seebach, *Helv.*, **65**, 385 (1982).

Dimethyl oxomalonate, 6, 188; **10**, 143–144.

Enol pyruvates. In connection with a synthesis of chorismic acid (**1**) McGowan and Berchtold[1] developed a new synthesis of enol pyruvates, as outlined in chart (I).

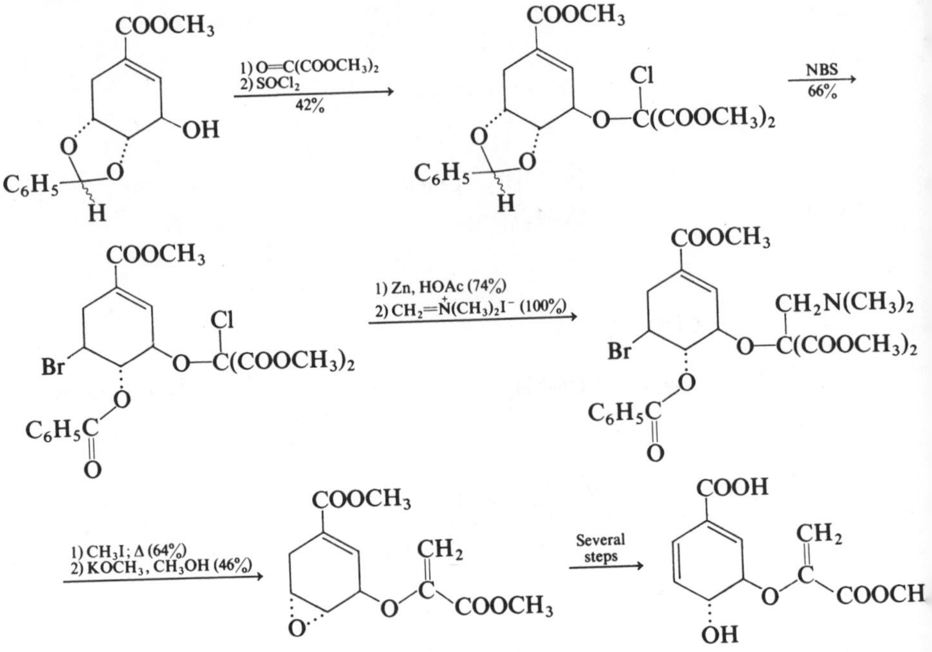

1) $O{=}C(COOCH_3)_2$
2) $SOCl_2$
42%

NBS
66%

1) Zn, HOAc (74%)
2) $CH_2{=}\overset{+}{N}(CH_3)_2I^-$ (100%)

1) CH_3I; Δ (64%)
2) $KOCH_3$, CH_3OH (46%)

Several steps

1

Chart (I)

A recent synthesis of **1** by Ganem *et al.*[2] uses dimethyl diazomalonate for introduction of the enol pyruvate side chain.

[1] D. A. McGowan and G. A. Berchtold, *Am. Soc.*, **104**, 7036 (1982).
[2] B. Ganem, N. Ikota, V. B. Muralidharan, W. S. Wade, S. D. Young, and Y. Yukimoto, *ibid.*, **104**, 6787 (1982).

1,3-Dimethyl-2-phenyl-1,3,2-diazaphospholidine, $CH_3{-}N\langle\ \rangle N{-}CH_3$, P–$C_6H_5$ (**1**).

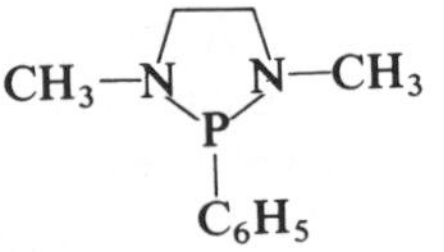

Mol. wt. 194. 22, b.p. 95°/0.7 mm. The reagent is prepared from *sym*-dimethylethylenediamine and dichlorophenylphosphine (70% yield). It should be stored under N_2 or argon at −20°.[1]

Alkenes from 1,2-diols. Corey and Hopkins[2] have reported two improvements in the preparation of alkenes from 1,2-diols via the thionocarbonate (**1**, 1233–1234; **3**, 315–316). The thionocarbonate is prepared with thiophosgene and 4-dimethylaminopyridine, a method generally successful except for ditertiary diols. The thionocarbonate is converted into an alkene when treated with **1** at 25–40° for 2–24 hours.

Example:

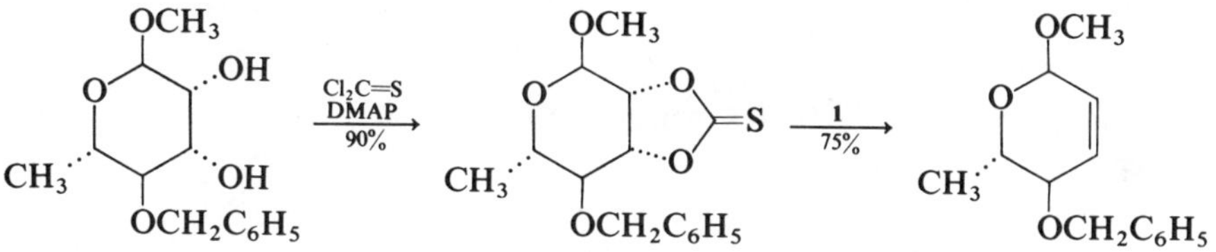

[1] M. K. Das and J. J. Zuckerman, *Inorg. Chem.*, **10**, 1028 (1971).
[2] E. J. Corey and P. B. Hopkins, *Tetrahedron Letters*, **23**, 1979 (1982).

Dimethylphenylsilyllithium, $LiSi(CH_3)_2C_6H_5$ (**1**).

Preparation[1]:

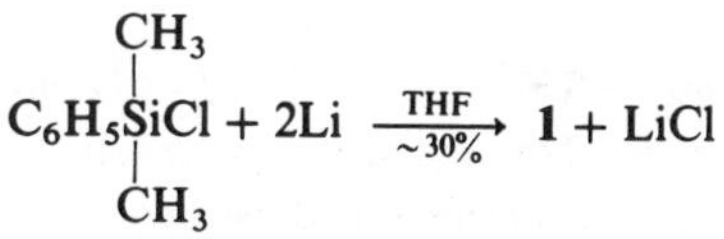

$$C_6H_5Si(CH_3)_2Cl + 2Li \xrightarrow[\sim 30\%]{THF} \mathbf{1} + LiCl$$

C–S Transposition.[2] A synthesis of the cycloundecanone **5** from **2** involves reaction with **1** to give the carbinol **3**. Treatment of **3** with CsF gives a product tentatively formulated as **4**. The synthesis of **5** is completed by desulfuration with Raney nickel.

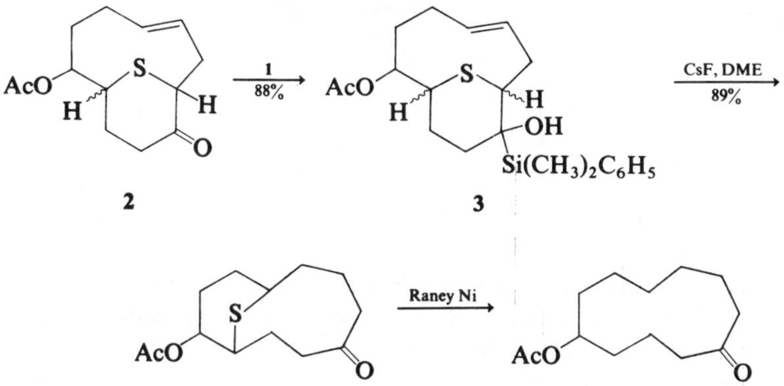

[1] M. V. George, D. J. Peterson, and H. Gilman, *Am. Soc.*, **82**, 403 (1960).
[2] E. Vedejs, M. J. Arnost, J. M. Eustache, and G. A. Krafft, *J. Org.*, **47**, 4384 (1982).

(S)-(+)-N,S-Dimethyl-S-phenylsulfoximine, $C_6H_5S^*(=O)(=NCH_3)CH_3$ **(1)**.

Mol. wt. 169. 24, b.p. 115–118°/0.4 mm, α_D +183°.

Preparation[1]:

$$C_6H_5S(=O)CH_3 \xrightarrow[92\%]{HN_3} C_6H_5S(=O)(=NH)CH_3 \xrightarrow[(80\%)]{\text{1) (+)-10-camphorsulfonic acid; 2) }(CH_3)_3O^+BF_4^-}$$

$$C_6H_5\overset{+}{S}(=O)(NHCH_3)CH_3BF_4^- \xrightarrow[80\%]{\text{aq. NaOH}} (S)\text{-}(+)\text{-}\mathbf{1}$$

Optically active alcohols.[2] The anion of **1** adds to a prochiral ketone to form a mixture of two diastereomers (**2a** and **2b**) which can usually be separated by chromatography on silica gel to give the pure diastereomers. Optically active tertiary methylcarbinols (**3**) are obtained by Raney nickel desulfuration of the pure adducts (equation I).

$$\mathbf{1} + R^1COR^2 \xrightarrow[83\text{–}100\%]{n\text{-BuLi, THF}} C_6H_5S^*(=O)(=NCH_3)\text{—}CH_2\text{—}C(R^1)(OH)\text{—}R^2 \;(\mathbf{2a} + \mathbf{2b}) \xrightarrow{\text{1) separation (67–94\%); 2) Raney Ni (43–83\%)}} CH_3\text{—}C(R^1)(OH)\text{—}R^2 \;(\mathbf{3})$$

Application of this sequence to alkyl aryl ketones generally gives the tertiary alcohols in 85–99% optical purity. However, separation of the adducts of prochiral dialkyl ketones is generally difficult. The resolution of adducts of aldehydes by chromatography is impractical; desulfuration of the unresolved adducts results in optically active secondary methylcarbinols in only 25–45% optical purity.

Asymmetric cyclopropanation.[3] The anion of (+)-**1** reacts with 2-cyclohexene-1-ones to form two diastereomeric adducts (**2**), which are separated by chromatography on silica gel. Cyclopropanation of optically pure **2a** or **2b** with the Simmons-Smith reagent occurs exclusively *cis* to the hydroxyl group to give optically pure **3a** or **3b**. The optically pure cyclopropyl ketone (**4**) is released on mild thermolysis of **3** with regeneration of (+)-**1**.

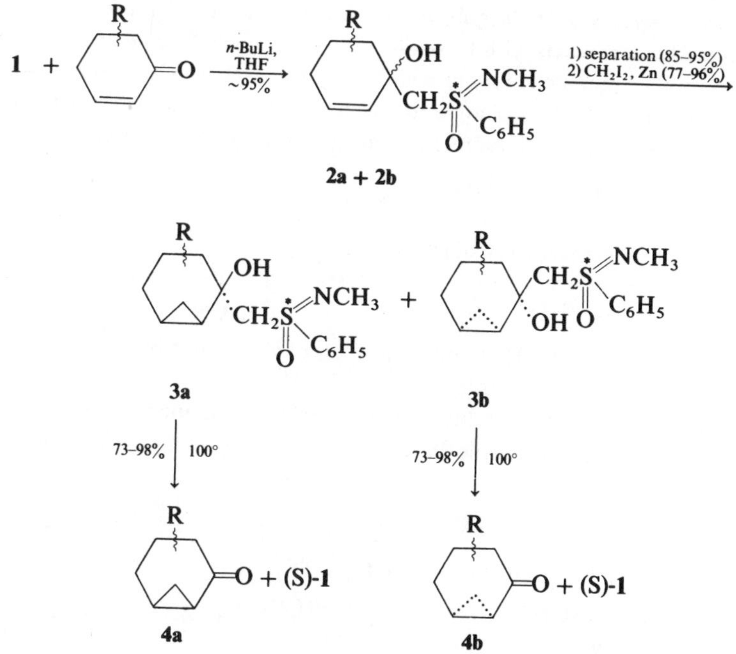

This sequence was used to convert (±)-**5** into (+)-**6** (94% ee), a precursor to (−)-thujopsene (**7**), a tricyclic sesquiterpene.

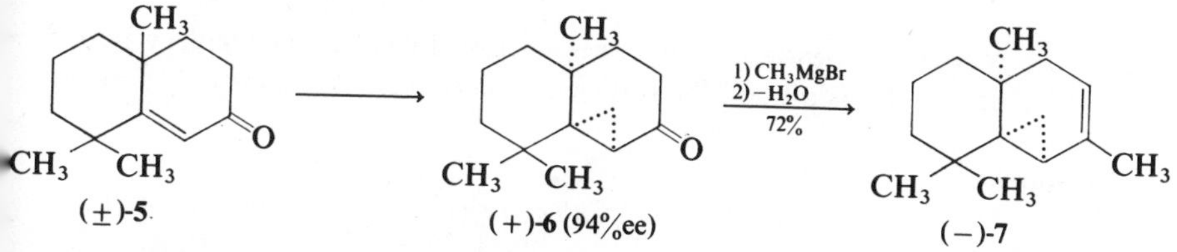

The sequence can be applied to acyclic enones, but in this case each diastereomeric adduct undergoes cyclopropanation to give two cyclopropyl diastereomers.

[1] C. R. Johnson, M. Haake, and C. W. Schroeck, *Am. Soc.*, **92**, 6594 (1970).
[2] C. R. Johnson and C. J. Stark, Jr., *J. Org.*, **47**, 1193 (1982).
[3] C. R. Johnson and M. R. Barbachyn, *Am. Soc.*, **104**, 4290 (1982).

Dimethyl phosphite–Chloroacetic acid, $(CH_3O)_2P(O)H–ClCH_2COOH$. In the presence of a base, this combination of reagents results in the phosphonate reagent $(CH_3O)_2P(O)CH_2CO_2^-M^+$ (**1**).

α,β-Unsaturated acids (acrylic acids).[1] The reagent **1**, prepared *in situ* with sodium methoxide, reacts with benzaldehydes to produce (E)-cinnamic acids in 70–90% yield. When prepared in glyme with NaH, the reagent can be used generally with both aldehydes and ketones to form the unsaturated acids. Moreover, many 2-haloalkanoic acids and their esters and amides can replace the haloacetic acid.

[1] D. R. Britelli, *J. Org.*, **46**, 2514 (1981).

2-(2,6-Dimethylpiperidino)acetonitrile, (**1**).

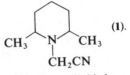

CH$_3$ N CH$_3$
CH$_2$CN

Mol. wt. 452.24, b.p. 99–102°/15 mm. The reagent is available by condensation of the piperidine and chloroacetonitrile.[1]

Aldehyde synthesis.[2] The lithio derivative (**2**) of **1** can function as an alternative to N,N-diethylaminoacetonitrile (**9**, 159) for synthesis of aldehydes (equation I).

(I) CH$_3$ N CH$_3$ / LiCHCN (**2**) $\xrightarrow{\text{RX, THF}}$ CH$_3$ N CH$_3$ / RCHCN $\xrightarrow[\text{35–65\% overall}]{CuSO_4\cdot 5H_2O,\ C_2H_5OH}$ RCHO

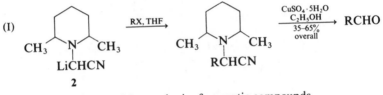

The reagent has been used for synthesis of aromatic compounds.

Example:

2 + CH$_3$O, CH$_3$O, (CH$_2$)$_3$Br $\xrightarrow[\text{70\%}]{\text{1) THF, }-78°\text{; 2) (HOOC)}_2}$ CH$_3$O, CH$_3$O

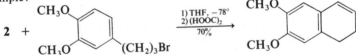

Formylation of α, β-enones.[3] In the presence of HMPT, **2** reacts with cyclic α, β-enones mainly by 1,4-addition. The adduct can be converted to a hydroxy dimethyl acetal.

Example:

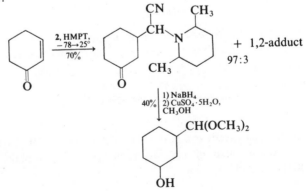

O $\xrightarrow[\text{70\%}]{\text{2, HMPT, }-78\rightarrow 25°}$ CN, CH, N, CH$_3$, CH$_3$, O + 1,2-adduct 97:3

$\downarrow$ 40% 1) NaBH$_4$ 2) CuSO$_4\cdot$5H$_2$O, CH$_3$OH

CH(OCH$_3$)$_2$, OH

[1] F. Hoffman-La Roche, *Neth. Appl.* **6**, 409, 619 [C.A., **63**, 2960b (1965)].
[2] T. Wakamatsu, J. Kondo, S. Hobara, and Y. Ban, *Heterocycles*, **19**, 481 (1982).
[3] T. Wakamatsu, S. Hobara, and Y. Ban, *ibid.*, **19**, 1395 (1982).

Dimethylsulfonium methylide, 1, 296–310, **2**, 157–158; **3**, 119–123; **4**, 192–199; **5**, 263–266; **6**, 225–229; **7**, 133–135; **8**, 198–199; **9**, 189.

Butenolides. A synthesis of the isodehydroabietenolide (**5**) used a new method for preparation of a butenolide (**2**) which was then transformed into the benzenoid ring of **5**. Thus, reaction of the ketene dithioketal **1** with dimethylsulfonium methylide followed by acid hydrolysis produced the butenolide **2** in high yield. This product was then converted via the furane **3** into the salicyclic ester **4** via a Diels-Alder reaction.[1]

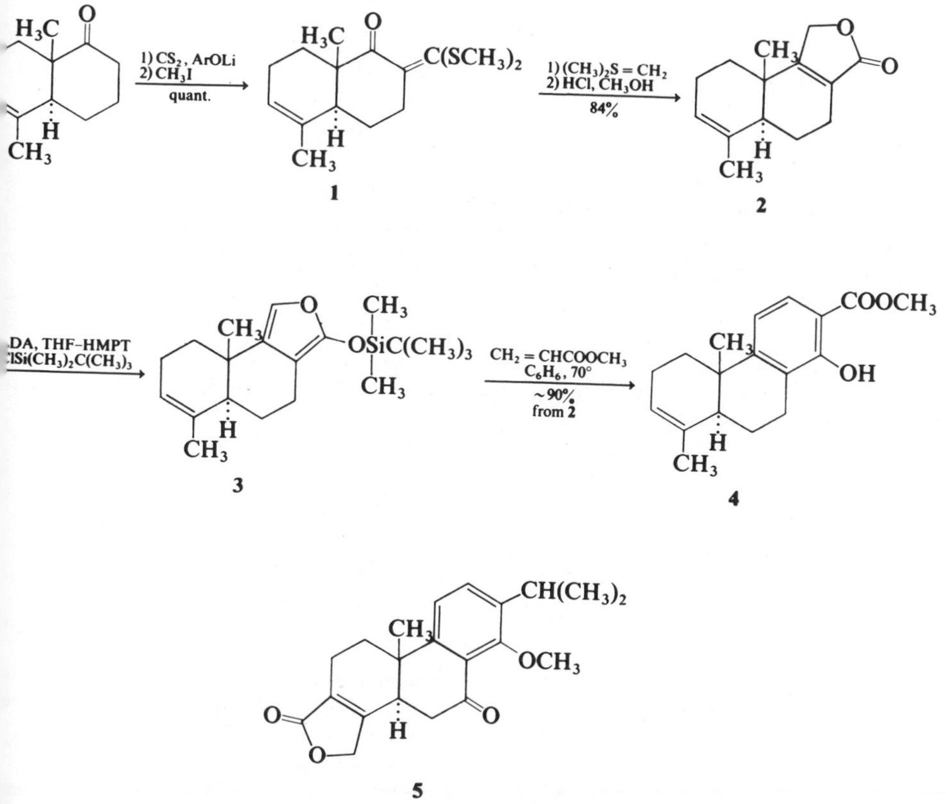

1) CS_2, ArOLi 2) CH_3I quant. → **1**; 1) $(CH_3)_2S=CH_2$ 2) HCl, CH_3OH 84% → **2**; LDA, THF–HMPT, $ClSi(CH_3)_2C(CH_3)_3$ → **3**; $CH_2=CHCOOCH_3$, C_6H_6, 70° ~90% from **2** → **4**; **5**

Dihydrofuranes. The reaction of dimethylsulfonium methylide (**1**) with α-keto ketene dithioacetals (**2**) provides a simple synthesis of dihydrofuranes (**3**), which can be converted into various furanes and butenolides as shown in Scheme (I).[2]

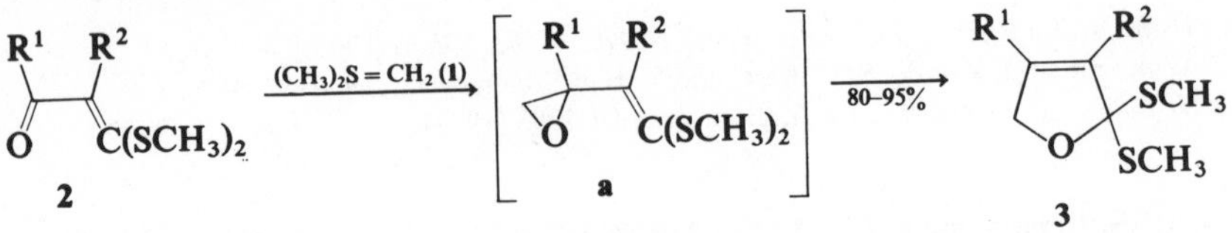

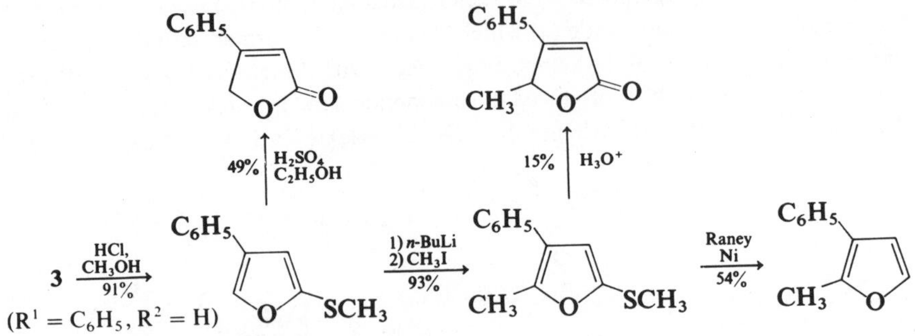

Scheme I

Epoxidation.[3] Trimethylsulfonium iodide reacts with aldehydes in the presence of solid KOH in CH_3CN to form epoxides in high yield. The reaction with ketones proceeds in lower yields.

Examples:

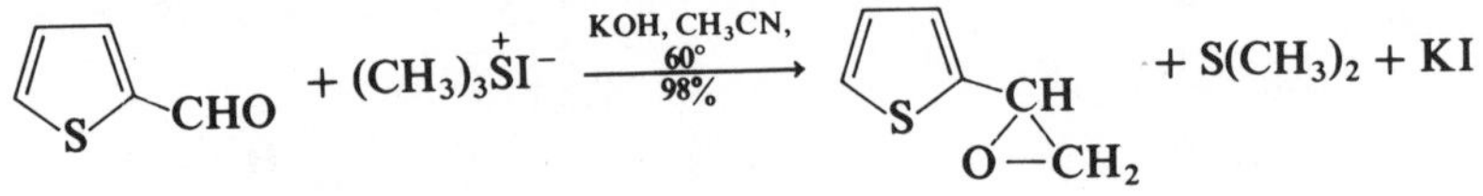

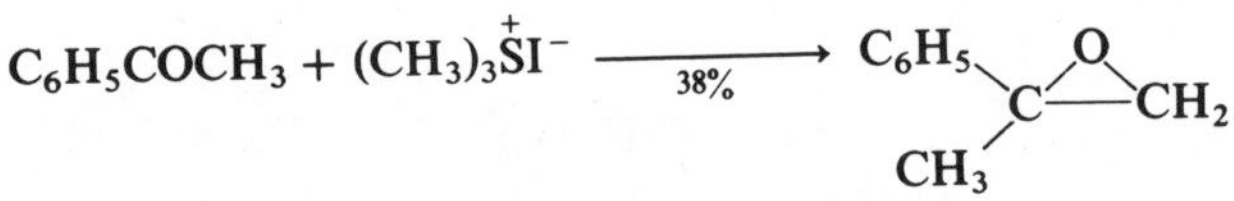

$$C_6H_5COCH_3 + (CH_3)_3\overset{+}{S}I^- \xrightarrow{38\%}$$

[1] L. C. Garver and E. E. van Tamelen, *Am. Soc.*, **104**, 867 (1982).
[2] R. Okazaki, Y. Negishi, and N. Inamoto, *J.C.S. Chem. Comm.*, 1055 (1982).
[3] E. Borredon, M. Delmas, and A. Gaset, *Tetrahedron Letters*, **23**, 5283 (1982).

Dimethyl sulfoxide, 1, 296–310; **2**, 157–158; **3**, 119–123; **4**, 192–194; **5**, 263–266; **7**, 133–135; **8**, 198–199; **9**, 189; **10**, 166–167.

α-Amino aldehydes. An important step in a morphine synthesis involves introduction of an α-formyl group on the intermediate enamine **1**. Thus, reaction of diazomethane with the immonium salt **2** gives the aziridinium salt **3**, which is oxidized by anhydrous DMSO (Kornblum oxidation) in very high yield to the α-amino aldehyde **4**. The cyclopropane ring is also easily cleaved by hydroxide and chloride ions.[1]

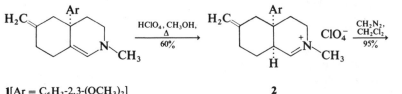

$$\mathbf{1} \xrightarrow[60\%]{HClO_4,\ CH_3OH,\ \Delta} \mathbf{2}\ (ClO_4^-) \xrightarrow[95\%]{CH_2N_2,\ CH_2Cl_2}$$

1[Ar = C_6H_3-2,3-$(OCH_3)_2$] **2**

$$\mathbf{3}\ (ClO_4^-) \xrightarrow[95\%]{(CH_3)_2S=O,\ 25^\circ} \mathbf{4} + (CH_3)_2S$$

3 **4**

[1] D. A. Evans and C. H. Mitch, *Tetrahedron Letters*, **23**, 285 (1982).

Dimethyl sulfoxide–Oxalyl chloride, 8, 200; **9**, 192; **10**, 167–168.

1,3-Diketones; β-keto esters. Smith and Levenberg[1] have examined the oxidation of β-hydroxy ketones (aldols) with a number of oxidizing reagents and find that only DMSO–$(COCl)_2$ and Collins reagent give satisfactory results. In general, the former reagent affords higher yields, usually in the range 55–85%.

The Swern reagent is also generally superior for oxidation of *β*-hydroxy esters to *β*-keto esters (50–90% yield).

Examples:

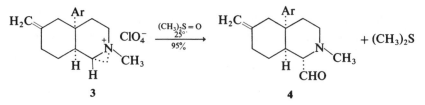

$$C_2H_5C(=O)-CH(CH_3)-CH(OH)C_2H_5 \longrightarrow C_2H_5C(=O)-CH(CH_3)-C(=O)C_2H_5$$

DMSO—$(ClOC)_2$	85%
CrO_3/Py	74%

$$C_2H_5OC(=O)-CH(CH_3)-CH(OH)C_2H_5 \longrightarrow C_2H_5OC(=O)-CH(CH_3)-C(=O)C_2H_5$$

DMSO—$(ClOC)_2$	91%
CrO_3/Py	72%

Oxidation of 1,4-diols. The 1,4-diol **1** is transformed directly into the 1,4-dialdehyde polygodial (**2**) by oxidation with Swern's reagent.[2]

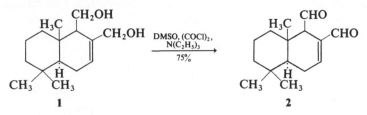

$$\mathbf{1} \xrightarrow[75\%]{DMSO,\ (COCl)_2,\ N(C_2H_5)_3} \mathbf{2}$$

[1] A. B. Smith, III, and P. A. Levenberg, *Synthesis*, 567 (1981).
[2] M. Jallali-Naini, G. Boussac, P. Lamaitre, M. Larcheveque, D. Guillerm, and J.-Y. Lallemand, *Tetrahedron Letters*, **22**, 2995 (1981).

Dimethyl sulfoxide–Sulfur trioxide, 2, 165; **4,** 200–201; **6,** 229–230.

α-Amino aldehydes.[1] Optically active N-protected α-amino aldehydes can be prepared with little or no racemization by reduction of N-protected α-amino esters to the corresponding α-amino alcohols with $NaBH_4$-LiCl (1 : 1)[2] followed by oxidation with the Parikh-Doering reagent (equation I).

$$\text{(I)}\quad \underset{\text{NHBoc}}{\text{RCHCOOCH}_3} \xrightarrow{NaBH_4,\ LiCl,\ C_2H_5OH,\ THF} \underset{\text{NHBoc}}{\text{RCHCH}_2\text{OH}} \xrightarrow[65\text{–}90\%]{DMSO\text{—}SO_3\cdot Py,\ N(C_2H_5)_3} \underset{\text{NHBoc}}{\text{RCHCHO}}$$

In an alternative route,[3] Boc-protected amino acids are reduced to the protected amino alcohol with borane–THF (45–95% yield) and pyridinium dichromate is used for oxidation to the aldehyde (75–90% yield). The optical rotations of the aldehydes obtained by these two procedures differ considerably, presumably owing to racemization encountered in the PDC oxidation.

Oxidation of a chiral alcohol. The (S)-alcohol **1** is oxidized in high yield by DMSO–$C_5H_5N \cdot SO_3$ to the aldehyde **2** with no more than 0.1% racemization. The same oxidation when effected with Collins reagent or with H_2CrO_4-SiO_2 (**8**, 110) proceeds with 5% and 22% racemization, respectively.[4]

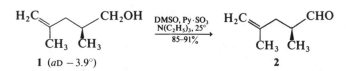

$$\mathbf{1}\ (\alpha_D\ -3.9^\circ) \xrightarrow[85\text{–}91\%]{DMSO,\ Py\cdot SO_3,\ N(C_2H_5)_3,\ 25^\circ} \mathbf{2}$$

[1] Y. Hamada and T. Shioiri, *Chem. Pharm. Bull.*, **30**, 1921 (1982).
[2] *Idem*, *Tetrahedron Letters*, **23**, 1193 (1982).
[3] C. F. Stanfield, J. E. Parker, and P. Kanillis, *J. Org.*, **46**, 4797, 4799 (1981).
[4] D. A. Evans and J. Bartroli, *Tetrahedron Letters*, **23**, 807 (1982).

Dimethyl(2,4,6-tri-*t*-butylphenoxy)chlorosilane (1). This silane is suggested as an alternative to the expensive *t*-butyldimethylchlorosilane. It is more easily prepared (equation I) and is comparable as a silylation reagent. The resulting silyl enol ethers

(I) $(CH_3)_3C$–[2,6-$(C(CH_3)_3)_2$-phenyl]–OH + $(CH_3)_2SiCl_2$ $\xrightarrow[93\%]{N(C_2H_5)_3,\ CH_3CN,\ \Delta}$ $(CH_3)_3C$–[2,6-$(C(CH_3)_3)_2$-phenyl]–$OSi(CH_3)_2Cl$

1, m.p. 80–81°

are more resistant to acid hydrolysis than *t*-butyldimethylsilyl enol ethers, and they are particularly useful substrates for the oxidation with trityl tetrafluoroborate to *α,β-enones* (**8**, 524–525).[1]

Example:

$OSi(CH_3)_2OC_6H_2[C(CH_3)_3]_3$ (cyclohexenyl ether) $\xrightarrow{(C_6H_5)_3C^+BF_4^-}$ cyclohexenone (91%) + cyclohexanone (6%)

[1] P. A. Manis and M. W. Rathke, *J. Org.*, **46**, 5348 (1981).

3,5-Dinitroperbenzoic acid, 9, 196.

Wagner-Meerwein rearrangement.[1] Epoxidation of the cycloalkene (**1**) with buffered *m*-chloroperbenzoic acid or with the more reactive 3,5-dinitroperbenzoic

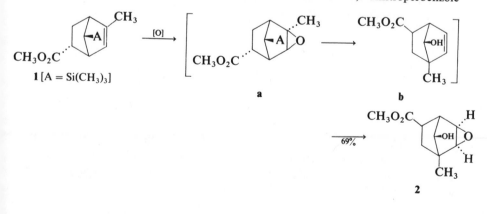

CH_3, CH_3O_2C, A; **1** [A = $Si(CH_3)_3$] $\xrightarrow{[O]}$ [CH_3, CH_3O_2C, A, O (**a**) ⟶ CH_3O_2C, OH, CH_3 (**b**)] $\xrightarrow{69\%}$ CH_3O_2C, H, OH, O, H, CH_3 (**2**)

acid is slow. With the latter reagent, the major product is not the expected epoxide (**a**) but the epoxide (**2**) of an alcohol (**b**), formed by a Wagner-Meerwein rearrangement of **a**. The rearrangement of **a** to **b** under such mild conditions is evidently triggered by the trimethylsilyl group. The product was used to prepare the lactone **3**, a promising precursor to verrucarol (**4**), a terpene antibiotic.

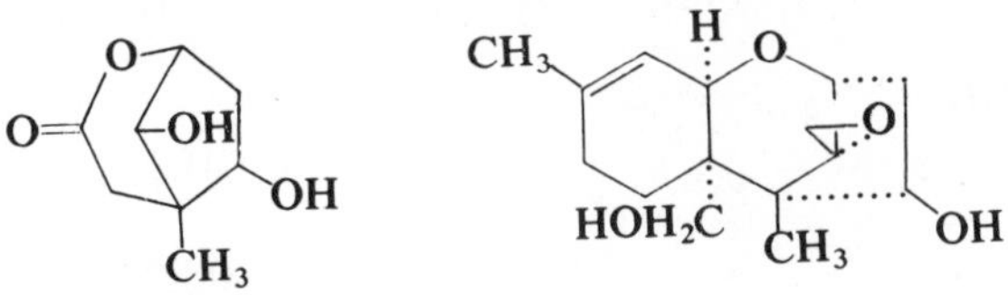

[1] W. R. Roush and T. E. D'Ambra, *J. Org.*, **46**, 5045 (1981).

Diperoxo-oxohexamethylphosphoramidomolybdenum(VI), 4, 203–204; **5**, 269–270; **7**, 136; **8**, 206–208; **9**, 197–198; **10**, 169–170.

Indoles; benzofuranes.[1] The sulfene cycloadducts (**1**) from 8-azaheptafulvenes rearrange on oxidation of the anion with MoO_5·HMPT–LDA to 1,2-disubstituted indoles (**2**).

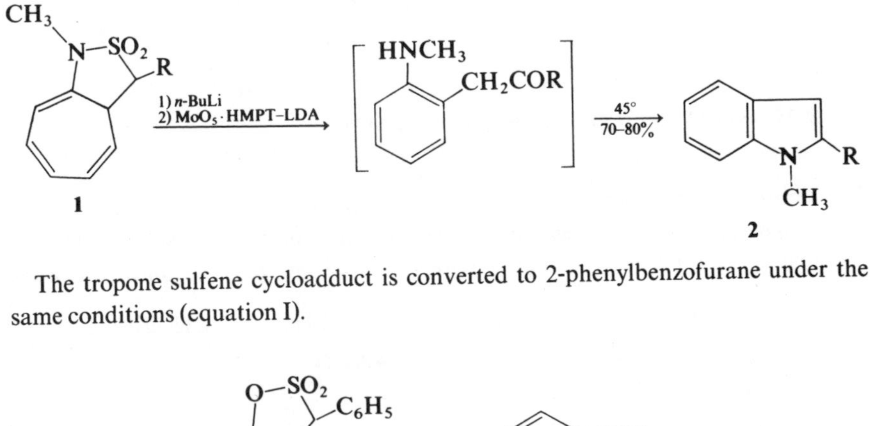

The tropone sulfene cycloadduct is converted to 2-phenylbenzofurane under the same conditions (equation I).

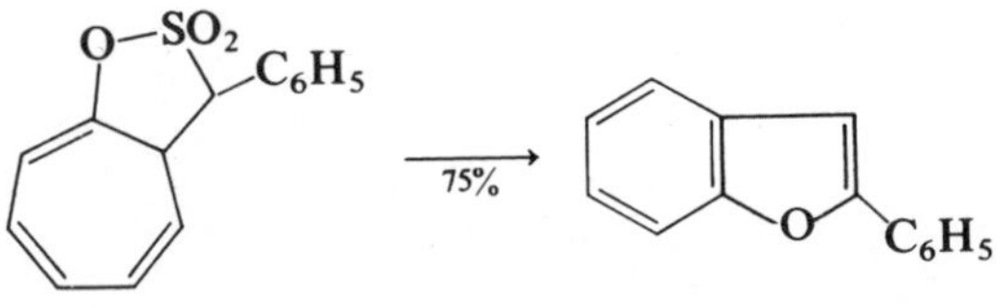

>CHNO₂→ >C=O. This transformation can be effected by treatment with base to generate the nitronate anion followed by oxidation with MoO_5·Py·HMPT. Primary nitro compounds under these conditions are converted to carboxylic acids via an aldehyde. This method is an alternative to the Nef reaction.[2]

Examples:

$$\text{2-methylnitrocyclohexane} \xrightarrow[85\%]{1)\ LDA\quad 2)\ MoO_5\cdot Py\cdot HMPT} \text{2-methylcyclohexanone}$$

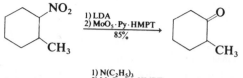

$$C_6H_5CH_2NO_2 \xrightarrow[79\%]{1)\ N(C_2H_5)_3\quad 2)\ MoO_5\cdot Py\cdot HMPT} C_6H_5COOH$$

[1] B. D. Dean and W. E. Truce, *J. Org.*, **46**, 3575 (1981).
[2] M. R. Gabobardes and H. W. Pinnick, *Tetrahedron Letters*, **22**, 5235 (1981).

Diphenylamine·Borane, $(C_6H_5)_2NH\cdot BH_3$(**1**), m.p. 86°.

Preparation:

$$4HN(C_6H_5)_2 + 4BF_3\cdot O(C_2H_5)_2 + 3NaBH_4 \xrightarrow[99\%]{THF,\ 0°} \mathbf{1} + 3NaBF_4 + 4(C_2H_5)_2O$$

Hydroboration; reduction of $\mathbf{>C{=}O}$**.**[1] This complex is as efficient as $BH_3\cdot THF$ for hydroboration; it forms diborane when dissolved in benzene. It reduces carbonyl compounds to alcohols at 0° in hexane or CH_2Cl_2. The stereoselectivity resembles that of $BH_3\cdot THF$. It reduces carboxylic acids to primary alcohols in 60–75% yield.

[1] C. Camacho, G. Uribe, and R. Contreras, *Synthesis*, 1027 (1982).

Diphenyl diselenide–Selenium dioxide.

α-Phenylselenenylation.[1] Dialkyl ketones, aliphatic aldehydes, and β-keto esters undergo α-phenylselenenylation on reaction with diphenyl diselenide and selenium dioxide and a trace of H_2SO_4 at 10° for 10–15 hours. α-Phenylselenenylation of alkyl aryl ketones requires reflux in ethanol (20 hours). The actual reagent is believed to be $C_6H_5Se^+$, generated by oxidation of $(C_6H_5Se)_2$.

Examples:

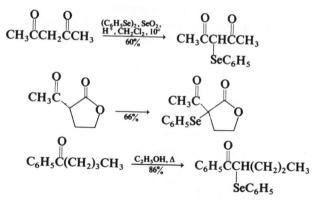

$$CH_3\overset{O}{\overset{\|}{C}}CH_2\overset{O}{\overset{\|}{C}}CH_3 \xrightarrow[60\%]{(C_6H_5Se)_2,\ SeO_2,\ H^+,\ CH_2Cl_2,\ 10°} CH_3\overset{O}{\overset{\|}{C}}\underset{SeC_6H_5}{\underset{|}{C}H}\overset{O}{\overset{\|}{C}}CH_3$$

$$\text{2-acetyl-}\gamma\text{-butyrolactone} \xrightarrow{66\%} \text{2-acetyl-2-(}C_6H_5Se\text{)-}\gamma\text{-butyrolactone}$$

$$C_6H_5\overset{O}{\overset{\|}{C}}(CH_2)_3CH_3 \xrightarrow[86\%]{C_2H_5OH,\ \Delta} C_6H_5\overset{O}{\overset{\|}{C}}\underset{SeC_6H_5}{\underset{|}{C}H}(CH_2)_2CH_3$$

[1] N. Miyoshi, T. Yamamoto, N. Kambe, S. Murai, and N. Sonada, *Tetrahedron Letters*, **23**, 4813 (1982).

Diphenyl disulfide, 5, 276–277; **6,** 235–238; **7,** 137; **8,** 210; **10,** 171.

4-Substituted 2-alkylfuranes.[1] Electrophilic substitution of 2-alkylfuranes occurs at C_5. Introduction of a phenylthio group at C_5 blocks the position and also activates C_4 to electrophilic attack.

Example:

2-methylfuran $\xrightarrow[85\%]{1)\ n\text{-BuLi};\ 2)\ (C_6H_5)_2S_2}$ 2-methyl-5-(phenylthio)furan (CH_3, SC_6H_5) $\xrightarrow[86\%]{Br_2}$ 4-bromo derivative (CH_3, SC_6H_5, Br) $\xrightarrow[91\%]{1)\ t\text{-BuLi};\ 2)\ ClSi(CH_3)_3}$ 4-$Si(CH_3)_3$ derivative (CH_3, SC_6H_5, $Si(CH_3)_3$) $\xrightarrow[50\%]{\text{Raney Ni}}$ 2-methyl-4-(trimethylsilyl)furan (CH_3, $Si(CH_3)_3$)

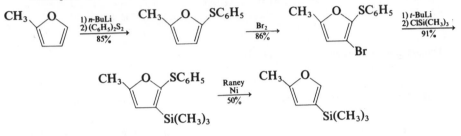

[1] S. M. Nolan and T. Cohen, *J. Org.*, **46**, 2473 (1981).

Diphenyl 2-keto-3-oxazolinylphosphonate (1).

Synthesis of amides and peptides.[1] The reagent, prepared as shown, converts carboxylic acids into 3-acyl-2-oxazolones (**2**), which acylate amines under mild conditions in high yield. A one-step synthesis of dipeptides in high yields is possible

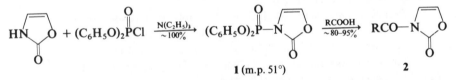

2-oxazolone (HN) + $(C_6H_5O)_2P(O)Cl$ $\xrightarrow[\sim 100\%]{N(C_2H_5)_3}$ $(C_6H_5O)_2P(O)$–N(oxazolone) **1** (m.p. 51°) $\xrightarrow[\sim 80\text{–}95\%]{RCOOH}$ RCO–N(oxazolone) **2**

by reaction of an N-protected amino acid with an amino acid ester in acetonitrile at room temperature in the presence of **1** and triethylamine.

Thiol esters.[2] Thiol esters are formed in good to high yield by treatment of a mixture of an acid and a thiol with **1** in the presence of triethylamine (equation I).

$$\text{(I)}\quad R^1COOH + R^2SH \xrightarrow[70\text{–}95\%]{\mathbf{1},\ N(C_2H_5)_3,\ CH_3CN} R^1COSR^2$$

[1] T. Kunieda, Y. Abe, T. Higuchi, and M. Hirobe, *Tetrahedron Letters*, **22**, 1257 (1981).
[2] T. Kunieda, Y. Abe, and M. Hirobe, *Chem. Letters*, 1427 (1981).

N,N-Diphenyl-p-methoxyphenylchloromethyleniminium chloride,

$$p\text{-}CH_3OC_6H_4(Cl)C{=}\overset{+}{N}(C_6H_5)_2Cl^- \quad (\mathbf{1}).$$

The salt is prepared by reaction of $p\text{-}CH_3OC_6H_4\overset{\overset{\displaystyle O}{\|}}{C}N(C_6H_5)_2$ with oxalyl chloride at 60° overnight.

Ketone synthesis from R¹COOH and R²MgX.[1] This reaction is generally not satisfactory because the Grignard reagent reacts with the ketone as formed to produce a tertiary alcohol. This reaction can be useful, however, when **1** is used to effect the condensation by activation of the carboxyl group. When the free acids are used, 2 equiv. of the Grignard reagent are required because of the HCl liberated. Yields are higher when metal carboxylates are employed. Steric effects in the acid are not too important, but are in the case of the Grignard reagent. In addition, catalytic amounts of added CuI improve the yield.

Examples:

$$CH_3(CH_2)_4COOLi + C_6H_5(CH_2)_2MgBr \xrightarrow[94\%]{\mathbf{1},\ CuI,\ THF,\ CH_2Cl_2} CH_3(CH_2)_4\overset{\overset{O}{\|}}{C}(CH_2)_2C_6H_5$$

$$(CH_3)_2CHCOOLi + C_6H_5(CH_2)_2MgBr \xrightarrow[88\%]{} (CH_3)_2CH\overset{\overset{O}{\|}}{C}(CH_2)_2C_6H_5$$

$$CH_3O\overset{\overset{O}{\|}}{C}(CH_2)_3COOLi + C_6H_5(CH_2)_2MgBr \xrightarrow[69\%]{} CH_3O\overset{\overset{O}{\|}}{C}(CH_2)_3\overset{\overset{O}{\|}}{C}(CH_2)_2C_6H_5$$

[1] T. Fujisawa, T. Mori, and T. Sato, *Tetrahedron Letters*, **23**, 5059 (1982).

O-(Diphenylphosphinyl)hydroxylamine, $(C_6H_5)_2\overset{\overset{O}{\|}}{P}ONH_2$ (**1**). Mol. wt. 233. 21, m.p. > 130° (dec.) Prepared in 92% yield by reaction of diphenylphosphinic acid with hydroxylamine.[1]

Amination. The carbanions of organometallic reagents react with **1** to give primary amines in moderate yield. Yields are higher in the case of stabilized carbanions.[2] The reagent also aminates N-, P-, and S-nucleophiles.[1,3]

Examples:

$$C_6H_5CH_2MgCl \xrightarrow[70\%]{1)\ n\text{-BuLi};\ 2)\ \mathbf{1}} C_6H_5CH_2NH_2$$

$$C_6H_5CH{=}C\begin{matrix}\diagup OLi \\ \diagdown OC_2H_5\end{matrix} \xrightarrow[45\%]{\mathbf{1}} C_6H_5CH\begin{matrix}\diagup NH_2 \\ \diagdown COOC_2H_5\end{matrix}$$

A chiral reagent **2** of the same type has been prepared from (−)-ephedrine and used to obtain chiral amines in ~ 20–45% ee.[4]

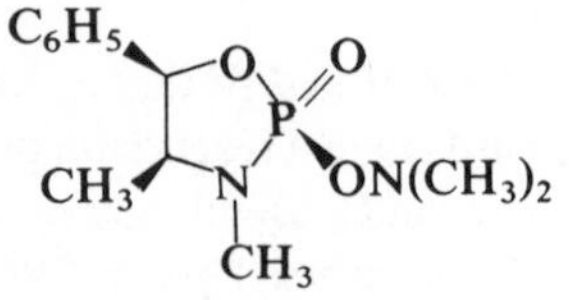

C_6H_5, O, P, O, CH_3, N, $ON(CH_3)_2$, CH_3

2

[1] M. J. P. Harges, *J. C. S. Perkin I*, 3284 (1981).
[2] G. Boche, M. Bernheim, and W. Schrott, *Tetrahedron Letters*, **23**, 5399 (1982).
[3] W. Klötzer, H. Baldinger, E. M. Karpitschka, and J. Knoflach, *Synthesis*, 592 (1982).
[4] G. Boche and W. Schrott, *Tetrahedron Letters*, **23**, 5403 (1982).

Diphenyl phosphorazidate (DPPA), **4**, 210–211; **5**, 200; **6**, 193; **7**, 138; **8**, 211–212; **10**, 173.

4-Methoxycarbonyloxazoles. Condensation of carboxylic acids with methyl isocyanoacetate in the presence of K_2CO_3 and DPPA (1 equiv.) leads to 5-substituted 4-methoxycarbonyloxazoles. Racemization is only slight when this C-acylation is applied to Boc-amino acids.[1]

Example:

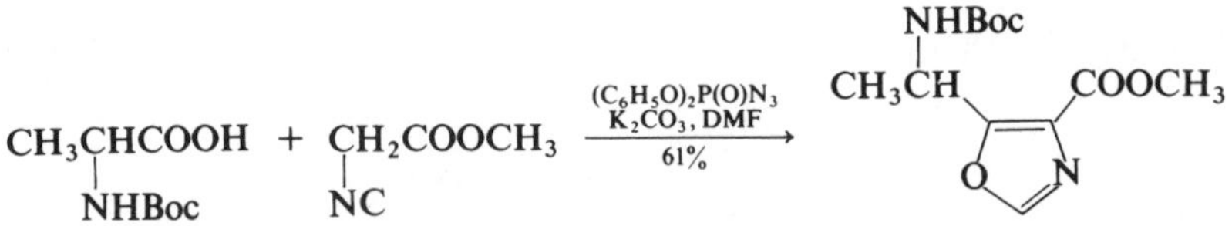

$$CH_3CH(NHBoc)COOH + CH_2(NC)COOCH_3 \xrightarrow[61\%]{(C_6H_5O)_2P(O)N_3,\ K_2CO_3,\ DMF} \text{5-}(CH_3CH(NHBoc))\text{-4-}(COOCH_3)\text{oxazole}$$

αD + 19.3°

Thiol esters. DPPA in combination with triethylamine promotes coupling of carboxylic acids and thiols to form thiol esters, R^1COSR^2 (50–95% yield). Only slight racemization occurs during reaction with amino acids.[2]

Curtius degradation.[3] The reaction of the carboxylic acid **1** with this azide in *t*-butyl alcohol gives the Boc-protected amine **2** directly.

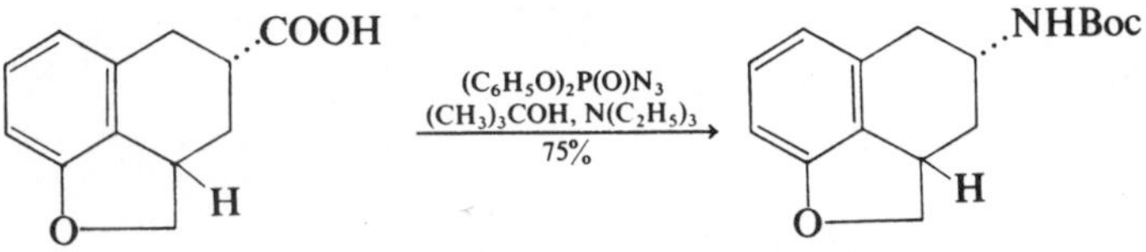

$$\mathbf{1}\ (\text{COOH, H}) \xrightarrow[75\%]{(C_6H_5O)_2P(O)N_3,\ (CH_3)_3COH,\ N(C_2H_5)_3} \mathbf{2}\ (\text{NHBoc, H})$$

[1] Y. Hamada and T. Shioiri, *Tetrahedron Letters*, **23**, 235 (1982).
[2] T. Shioiri and S. Yamada, *Chem. Pharm. Bull.*, **25**, 2423 (1972).
[3] W. Haefliger and E. Klöppner, *Helv.*, **65**, 1837 (1982).

Diphenyl phosphorochloridate, 1, 345–346; **2,** 180; **3,** 133.

Macrolactonization.[1] The hydroxy acid **1** does not undergo cyclization with the usual methods for carboxyl activation, but cyclization is possible in modest yield by conversion to the phosphoric acid mixed anhydride with $(C_6H_5O)_2POCl$, followed by treatment with 4-dimethylaminopyridine in benzene at 80°. The desired 14-membered lactone and a dimer are obtained in 32% and 25% yield, respectively. The diol **2** is then generated in quantitative yield with TFA in aqueous acetonitrile (100% yield). The last step in the synthesis of the natural macrolide narbonolide is the selective oxidation of the C_3-hydroxyl group. Most of the known oxidants favor oxidation of the C_5-hydroxyl group. Only two reagents are useful, $RuCl_2[P(C_6H_5)_3]_3$ (**10,** 141) and $DMSO/DCC/H_3PO_4$, both of which give about equal amounts of the C_3- and C_5-ketone. By reduction and recycling, the desired oxidation to the C_3-ketone is possible in 65% yield.

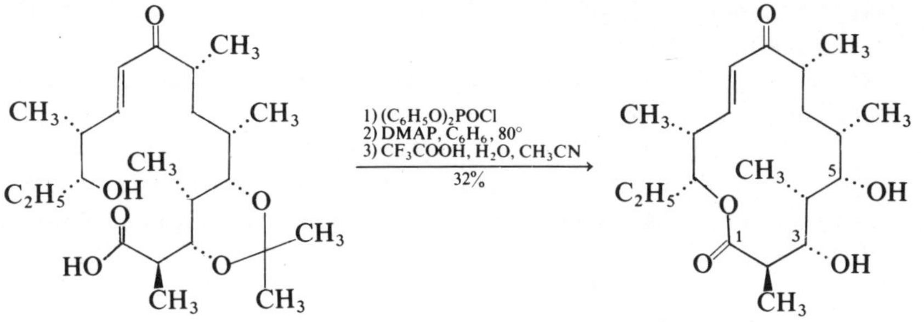

1) $(C_6H_5O)_2POCl$
2) DMAP, C_6H_6, 80°
3) CF_3COOH, H_2O, CH_3CN
32%

1 **2**

[1] T. Kaiho, S. Masamune, and T. Toyoda, *J. Org.*, **47**, 1612 (1982).

Diphenylselenium bis(trifluoro)acetate (1). Mol. wt. 459.22, m.p. 175–176°.

Preparation:

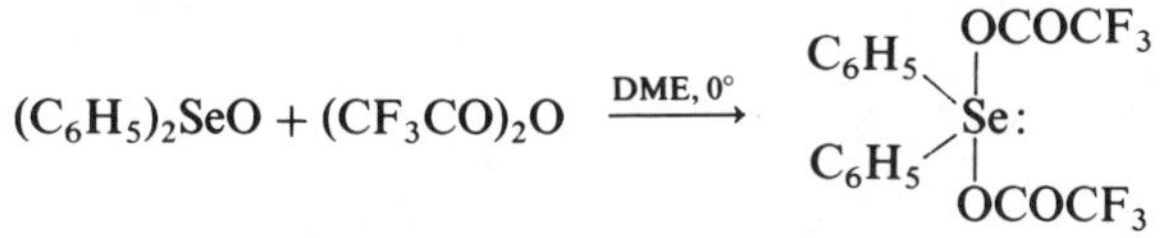

$$(C_6H_5)_2SeO + (CF_3CO)_2O \xrightarrow{\text{DME, 0°}} (C_6H_5)_2Se(OCOCF_3)_2$$

1

Oxidation of amines.[1] This selenurane mimics some two-electron biological oxidations of heterocyclic amines.

Examples:

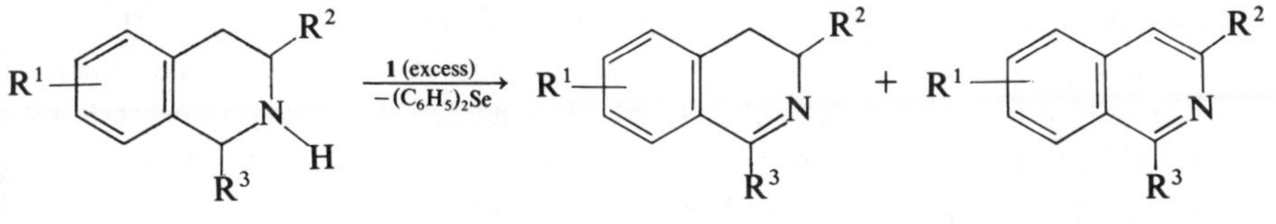

1 (excess), $-(C_6H_5)_2Se$

(80–100%) (0–10%)

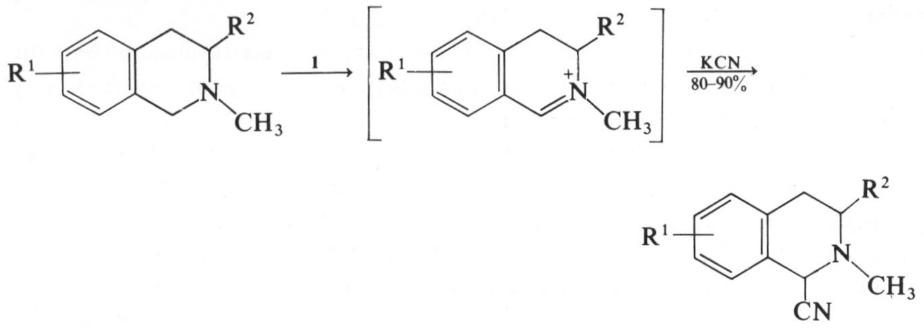

[1] J. P. Marino and R. D. Larsen, Jr., *Am. Soc.*, **103**, 4642 (1981).

Diphenyltin dihydride, $(C_6H_5)_2SnH_2$, **1**, 349. The reagent is prepared *in situ* by reaction of $(C_6H_5)_2SnCl_2$ with $LiAlH_4$ (1 equiv.) in ether.

Reduction of aldehydes and ketones.[1] The reactivity of tin hydrides for this reduction follows the order:

$$(C_6H_5)_2SnH_2 > (n\text{-}C_4H_9)_2SnH_2 > (C_6H_5)_3SnH > (n\text{-}C_4H_9)_3SnH$$

The final steps in a synthesis of the marine antibiotic (−)-malyngolide (**2**) from D-glucose involves oxidative cleavage of the triol (**1**) with $Pb(OAc)_4$ to an aldehyde, which is reduced directly to **2** with diphenyltin dihydride.[2]

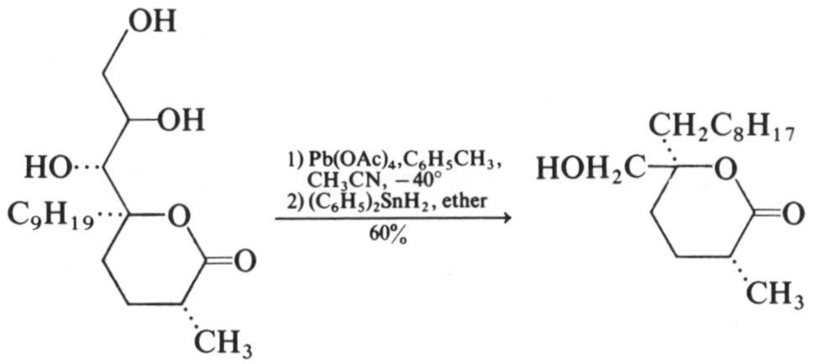

[1] H. G. Kuivila, *Synthesis*, 504 (1970).
[2] J.-R. Pougny, P. Rollin, and P. Sinaÿ, *Tetrahedron Letters*, **23**, 4929 (1982).

Diphosphorus tetraiodide, 1, 349–350; **9**, 203–209; **10**, 174.

Vinyl sulfides; ketene thiocetals.[1] These products, as well as the seleno anologs, are readily prepared from β-hydroxy orthothioesters or β-hydroxy thioacetals by reaction with P_2I_4 or PI_3 and $N(C_2H_5)_3$ in CH_2Cl_2. Rearrangement products are encountered as by-products when the hydroxyl group is secondary and become the primary products when $SOCl_2$ is used.

Examples:

$$n\text{-}C_5H_{11}\underset{\displaystyle OH}{\underset{|}{C}H}\text{—}C(SCH_3)_3 \xrightarrow{PI_3} \underset{63\%}{n\text{-}C_5H_{11}CH{=}C(SCH_3)_2} + \underset{13\%}{n\text{-}C_5H_{11}\underset{\displaystyle SCH_3}{\underset{|}{C}}{=}C(SCH_3)_2}$$

$$\triangleright\text{—}\underset{\displaystyle OH}{\underset{|}{C}H}\text{—}CH(SeCH_3)_2 \xrightarrow[65\%]{} \underset{(E + Z)}{\triangleright\text{—}CH{=}CHSeCH_3}$$

Vinyl selenides, ketene selenoacetals. Methyl selenoacetals (**1**) derived from ketones are converted into vinyl selenides on reaction with P_2I_4 or PI_3 and triethylamine in CH_2Cl_2 at 25° (equation I). The sulfur analogs undergo the same reaction.

$$\text{(I)}\quad R^1CH_2\underset{\displaystyle R^2}{\underset{|}{C}}(SeCH_3)_2 \xrightarrow[\sim 70\text{–}80\%]{N(C_2H_5)_3} \underset{\mathbf{1}}{R^1CH{=}C\begin{matrix} SeCH_3 \\ R^2 \end{matrix}}$$

Ketene dithioacetals and diselenoacetals are obtained under the same conditions from orthothioesters and orthoselenothioesters (equation II).[2] These products undergo selective C–Se bond cleavage on treatment with *n*-butyllithium to give a lithio

$$\text{(II)}\quad RCH_2C(SeCH_3)_3 \xrightarrow[65\text{–}90\%]{} RCH{=}C(SeCH_3)_2$$

derivative that can be trapped by various electrophiles to give vinyl selenides as a mixture of (Z)- and (E)-isomers (equation III). These products are convertible to selenium-free alkenes (Bu_3SnH) and vinyl bromides (Br_2).[3]

$$\text{(III)}\quad RCH{=}C(SeCH_3)_2 \xrightarrow{n\text{-BuLi}} RCH{=}C\begin{matrix} SeCH_3 \\ Li \end{matrix} \xrightarrow{E^+} RCH{=}C\begin{matrix} SeCH_3 \\ E \end{matrix}$$

[1] J. N. Denis, S. Desauvage, L. Hevesi, and A. Krief, *Tetrahedron Letters*, **22**, 4009 (1981).
[2] J. N. Denis and A. Krief, *ibid.*, **23**, 3407 (1982).
[3] *Idem, ibid.*, **23**, 3411 (1982).

(+)-Di-3-pinanylborane, 1, 262–263; **4,** 161–162; **6,** 202.

Asymmetric hydroboration. Details are available for the preparation of the reagent and for the asymmetric hydroboration of methyl 2,4-cyclopentadiene-1-acetate (**1**) to give (2R, 5R)-methyl 5-hydroxy-2-cyclopentene-1-acetate (**2**).[1]

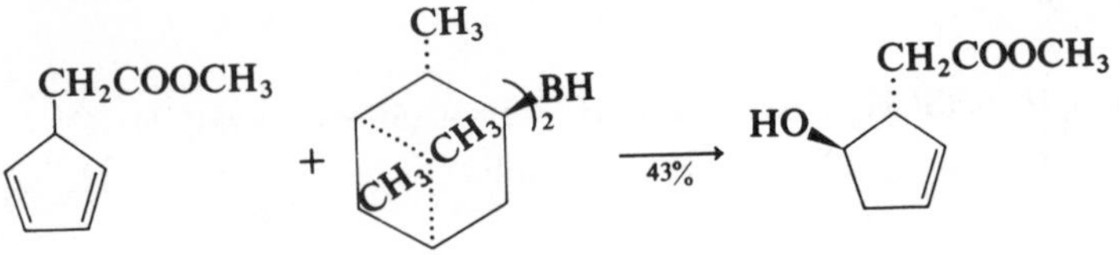

$\text{1} + \text{((+)-Ipc)}_2\text{BH} \xrightarrow{43\%}$ **2** (94–96%ee)

This reaction is useful for asymmetric hydroboration of *cis*-disubstituted alkenes to give highly pure optically active alcohols, but is less useful in the case of 1,1-disubstituted, *trans*-disubstituted, or trisubstituted alkenes.

[1] J. J. Partridge, N. K. Chadha, and M. R. Uskoković, *Org. Syn.* submitted (1982).

Disiamylborane (Sia_2BH, Bis-3-methyl-2-butylborane), **1**, 57–59; **3**, 22; **4**, 37; **5**, 39–41; **6**, 62; **8**, 41; **10**, 40.

Selective hydroboration. The final step in a new synthesis of homogeraniol (**2**) involves selective hydroboration of the triene **1** with a slight excess of Sia_2BH at 0°.[1]

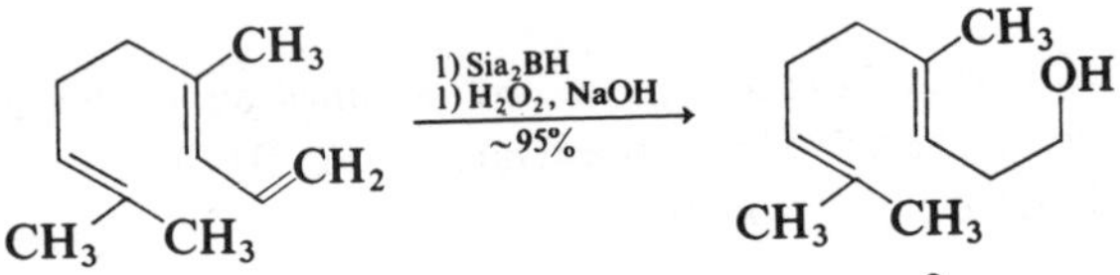

$\mathbf{1} \xrightarrow[\sim 95\%]{1)\ Sia_2BH;\ 1)\ H_2O_2,\ NaOH} \mathbf{2}$

[1] E. J. Leopold, *Org. Syn.* submitted (1981).

N,N′-Disuccinimidyl carbonate (1), 9, 208.

α,β-Unsaturated amino acids.[1] The reaction of **1** with N-protected β-hydroxy α-amino acids with **1** and triethylamine at 25° causes β-elimination of the hydroxyl group.

Examples:

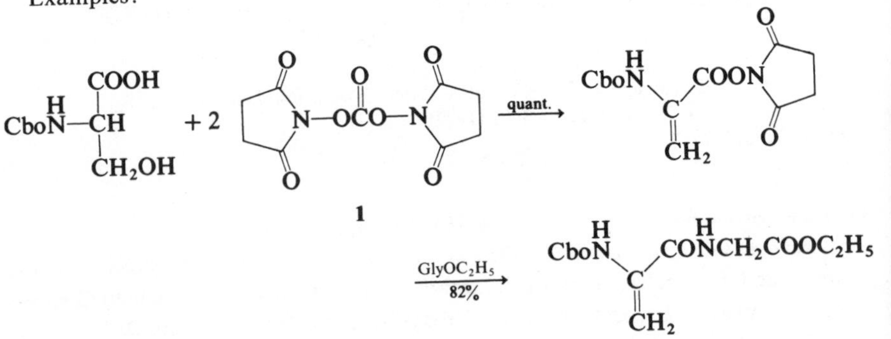

$\text{CboNH–CH(COOH)CH}_2\text{OH} + 2\ \mathbf{1} \xrightarrow{\text{quant.}} \text{CboNH–C(=CH}_2\text{)–COON(succinimide)}$

$\xrightarrow[82\%]{GlyOC_2H_5} \text{CboNH–C(=CH}_2\text{)–CONHCH}_2\text{COOC}_2\text{H}_5$

CboNH–CH(COOH)–CH(OH)–CH$_3$ $\xrightarrow[\text{quant.}]{1}$ CboNH–C(COON-succinimide)=C(CH$_3$)H $\xrightarrow[86\%]{\text{AlaOCH}_3}$ CboNH–C(CONHCH(CH$_3$)COOCH$_3$)=C(CH$_3$)H

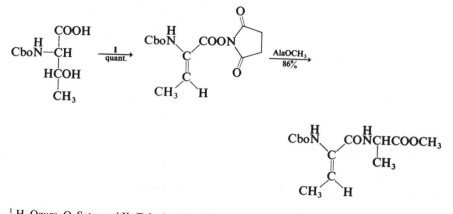

[1] H. Ogura, O. Sato, and K. Takeda, *Tetrahedron Letters*, **22**, 4817 (1981).

1,3-Dithienium tetrafluoroborate (1), 4, 218.

β-Dicarbonyl compounds.[1] Enol silyl ethers of carbonyl compounds are alkylated by this reagent (**1**) to give a protected β-dicarbonyl compound. The reaction is regiospecific and also kinetically stereoselective.

Examples:

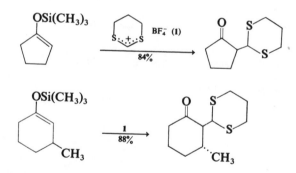

1-(trimethylsilyloxy)cyclopentene + 1,3-dithienium BF_4^- (**1**) $\xrightarrow{84\%}$ 2-(1,3-dithian-2-yl)cyclopentanone

1-(trimethylsilyloxy)-3-methylcyclohexene $\xrightarrow[88\%]{1}$ 2-(1,3-dithian-2-yl)-3-methylcyclohexanone (trans)

1,5-Dicarbonyl compounds.[2] The alkylation of O-silyl dienolates is γ-regioselective and thus provide a useful route to protected 1,5-dicarbonyl compounds.

Examples:

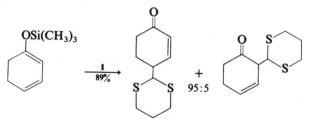

1-(trimethylsilyloxy)-1,3-cyclohexadiene $\xrightarrow[89\%]{1}$ 4-(1,3-dithian-2-yl)cyclohex-2-enone + 6-(1,3-dithian-2-yl)cyclohex-2-enone, 95:5

$$(CH_3)_3SiO\text{-[bicyclic diene, }CH_3\text{]} \xrightarrow[84\%]{1} \text{[bicyclic enone, }CH_3\text{, 1,3-dithian-2-yl]}$$

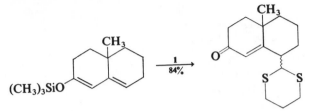

β,γ-Unsaturated-1,3-dithianes.[3] The reagent reacts with allyltrimethylsilane rapidly with evolution of BF_3 and formation of **2** in 78% yield.

$$CH_2{=}CHCH_2Si(CH_3)_3 \xrightarrow{1,\ CH_2Cl_2,\ 20^\circ} \text{(1,3-dithian-2-yl)}{-}CH_2CH{=}CH_2 + FSi(CH_3)_3 + BF_3$$

2 (78%)

Substitution in the double bond of the allylsilane decreases reactivity somewhat, but the reaction is still regioselective (equation I).

$$\text{(I)}\ C_6H_5CH{=}CHCH_2Si(CH_3)_3 \xrightarrow[60\%]{1,\ CH_2Cl_2,\ \Delta} \text{(1,3-dithian-2-yl)}{-}CH_2CH{=}CHC_6H_5$$

The 1,3-dithiane substituent can be hydrolyzed to a formyl group or converted into a methyl group.

[1] I. Paterson and L. G. Price, *Tetrahedron Letters*, **22**, 2829 (1981).

[2] *Idem, ibid.*, **22**, 2833 (1981).

[3] C. Westerlund, *ibid.*, **23**, 4835 (1982).

E

NB-Enantrane (1). This is the trademark of Aldrich for the 9-BBN adduct of the benzyl ether of 6,6-dimethylbicyclo [3.1.1] hept-2-ene-2-ethanol (nopol).

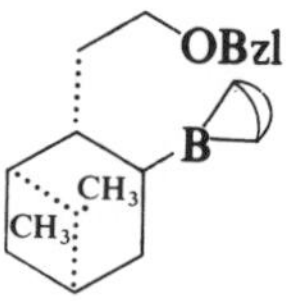

1

Reduction of α,β-ynones.[1] α,β-Acetylenic ketones are reduced by **1** to (S)-propargyl alcohols in 86–96% enantioselective purity. It is thus comparable to B-3-pinanyl-9-BBN in enantiomeric yields, although the reductions require 1–2 days and an excess of neat reagent for complete reaction.

[1] M. M. Midland and A. Kazubski, *J. Org.*, **47**, 2814 (1982).

NB-Enantride (2). This name is the trademark of Aldrich for the borohydride prepared by treatment of the 9-BBN adduct of the benzyl ether of 6,6-dimethylbicyclo [3.1.1] hept-2-ene-2-ethanol (nopol, supplier: Aldrich) with *t*-BuLi.

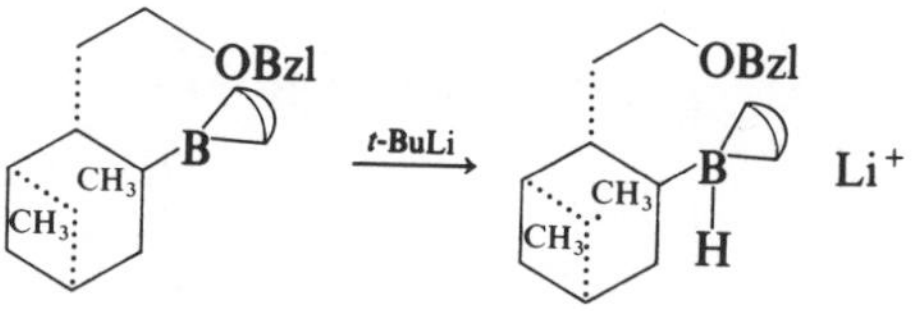

t-BuLi → Li^+

1 (NB-Enantrane.) **2**

Asymmetric reduction of prochiral ketones.[1] Chiral metal hydrides previously investigated have been effective only for asymmetric reduction of aromatic or α,β-acetylenic ketones. This new reagent unexpectedly reduces straight-chain aliphatic ketones such as 2-butanone and 2-octanone to the corresponding (S)-alcohols in 76%

and 79% ee, respectively. However, branching in one of the alkyl groups decreases the enantioselectivity. It apparently is somewhat more effective for enantioselective reduction of aliphatic prochiral ketones than B-3-pinanyl-9-borabicyclo[3.3.1]nonane (this volume).

[1] M. M. Midland and A. Kazubski, *J. Org.*, **47**, 2495 (1982).

Ephedrine, 5, 289–290; **9**, 209.

(R)-α-Phenylalkylamines. (R)-α-Phenylethylamine (**4**) can be prepared in about 97% optical purity by addition of methylmagnesium bromide to the chiral hydrazone obtained from benzaldehyde and the N-amino derivative (**1**) of D-(−)-ephedrine followed by hydrogenolysis (equation I). Acetophenone is not a suitable starting material because it reacts with **1** to form two isomeric hydrazones.[1]

(I) C_6H_5CHO + **1** [CH_3–CH(N(CH_3)NH_2)–CH(OH)–C_6H_5] $\xrightarrow{\sim 100\%}$ **2** [CH_3–CH(N(CH_3)N=CHC_6H_5)–CH(OH)–C_6H_5] $\xrightarrow[79\%]{CH_3MgBr}$

3 [CH_3–CH(N(CH_3)NH–CH(CH_3)C_6H_5)–CH(OH)–C_6H_5] $\xrightarrow{H_2, Pd/C}$ CH_3···CH(NH_2)–C_6H_5 (**(R)-4** (97%ee)) + CH_3–CH($NHCH_3$)–CH(OH)–C_6H_5

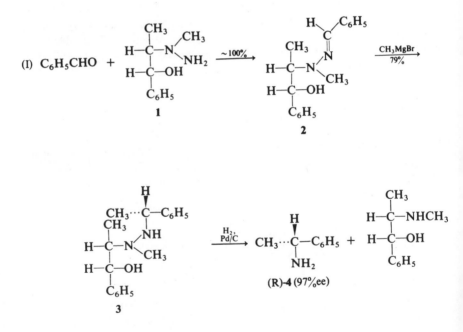

Chiral formylcyclopropanes.[2] The oxazolidine (**1**), obtained by reaction of cinnamaldehyde with D-(−)-ephedrine, reacts with diazomethane [$Pd(OAc)_2$ catalysis] to give **2** in at least 90% ee. Hydrolysis of **2** to **3** is effected by moist SiO_2. It is also possible to prepare cyclopropanes with three chiral centers, such as **5**.

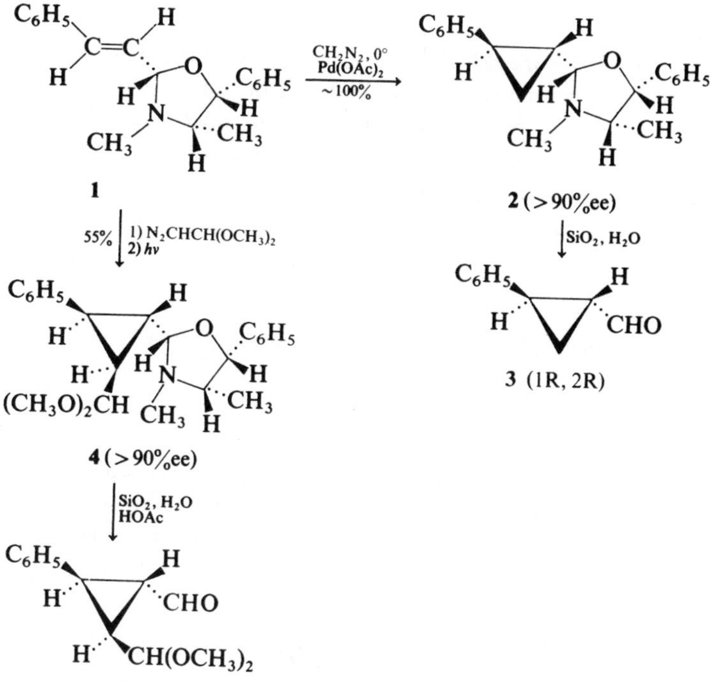

CH_2N_2, 0°
$Pd(OAc)_2$
~100%

1

2 (>90%ee)

55% 1) $N_2CHCH(OCH_3)_2$ 2) $h\nu$

SiO_2, H_2O

3 (1R, 2R)

4 (>90%ee)

SiO_2, H_2O HOAc

5

Chiral β-alkylalkanoic acids.[3] Grignard reagents add stereoselectively in a conjugate fashion to α,β-unsaturated amides derived from L-(+)-ephedrine (equation I). Diethyl ether is the solvent of choice. Alkyllithium reagents are less effective than RMgBr. The high stereoselectively results from formation of rigid chelated intermediates.

(I)

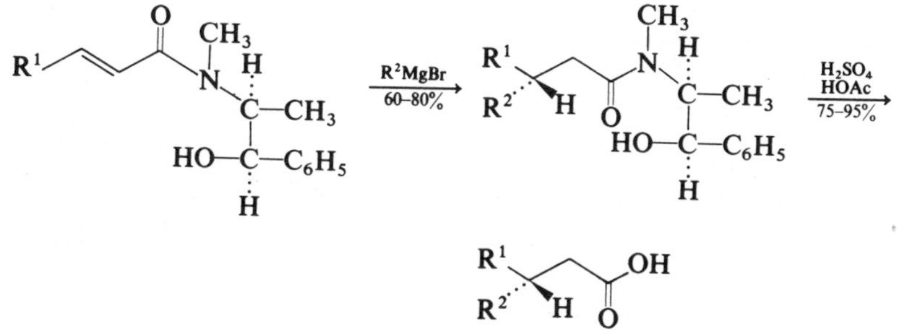

R^2MgBr 60–80%

H_2SO_4 HOAc 75–95%

(80–99%ee)

[1] H. Takahashi, K. Tomita, and H. Otomasu, *J.C.S. Chem. Comm.*, 668 (1979); H. Takahashi, K. Tomita, and H. Noguchi, *Chem. Pharm. Bull. Japan*, **29**, 3387 (1981).

[2] H. Abdallah, R. Grée, and R. Carrié, *Tetrahedron Letters*, **23**, 503 (1982).

[3] T. Mukaiyama and N. Iwasawa, *Chem. Letters*, 913 (1981).

Erbium chloride, 9, 210–211.

Selective reduction of aldehydes.[1] Conjugated aldehydes are selectively reduced in the presence of nonconjugated ones by $ErCl_3$ in aqueous ethanol. The effect is probably a result of preferential hydration of nonconjugated aldehydes.

Selective reduction of ketones in the presence of aldehydes is also possible with $ErCl_3$. $CrCl_3$ is almost as satisfactory for this selective reduction.

[1] A. L. Gemal and J. L. Luche, *Tetrahedron Letters*, **22**, 4077 (1981).

α-Ethoxyvinyllithium, $CH_2{=}C(OC_2H_5)Li$ The reagent is prepared by reaction of ethyl vinyl ether with *t*-butyllithium–TMEDA.[1]

γ-Ethoxy dienones.[2] The γ-ethoxy dienone **3** can be obtained by 1,2-addition of 2-ethoxyvinyllithium to 3-ethoxy-2-cyclohexenone (**1**) followed by selective hydrolysis of the adduct (**2**) by silica gel chromatography. This sequence is generally applicable to cyclic enones, but is not useful for acyclic enones.

1 (OC_2H_5) + $CH_2{=}C(OC_2H_5)Li$ $\xrightarrow[100\%]{\text{ether, } -78\rightarrow 0^\circ}$ **2** (OH; $C{=}CH_2$; OC_2H_5) $\xrightarrow{SiO_2}$ **3** (O; CH_2; OC_2H_5)

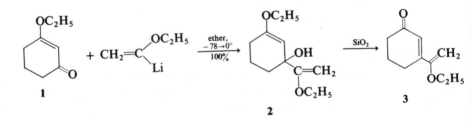

[1] U. Schöllkopf and P. Hansck, *Ann.*, **763**, 208 (1972).
[2] G. A. Kraus and M. E. Krolski, *Syn. Comm.*, **12**, 521 (1982).

(4S, 5S)-2-Ethyl-4-(methoxymethyl)-5-phenyloxazoline, C_2H_5– (oxazoline ring; O, N; C_6H_5; CH_2OCH_3) (1

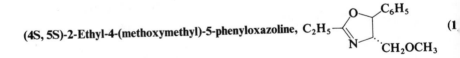

Mol. wt. 203.28, b.p. 91–93°/0.25 mm, $\alpha_D - 84.2°$. Preparation.[1]

erythro-*Selective aldol reactions.*[2] This oxazoline has been used as the chiral auxiliary in aldol condensation. Thus, the boron azaenolate **2**, prepared from **1** and 9-BBN triflate (**9**, 131–132), reacts with aldehydes to give, after acid hydrolysis and esterification, predominately *erythro*-*β*-hydroxy esters (equation I). Enantioselectivity, however, is only moderate.

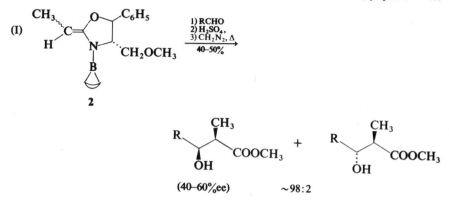

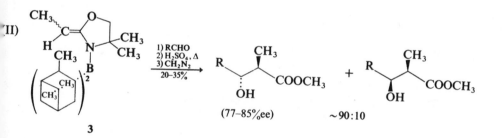

(I) [Oxazoline 2, B-substituted] $\xrightarrow[\text{40–50\%}]{\text{1) RCHO; 2) } H_2SO_4\text{; 3) } CH_2N_2, \Delta}$ R–CH(OH)–CH(CH_3)–$COOCH_3$ (40–60%ee) + R–CH(OH)–CH(CH_3)–$COOCH_3$ ~98:2

2

Use of the boron azaenolate **3**, prepared from achiral 2-ethyl-4,4-dimethyloxazoline and the chiral boryl triflate, undergoes aldol condensation to give mainly *threo-β*-hydroxy esters with enantioselectivity of about 80% (equation II).

(II) **3** $\xrightarrow[\text{20–35\%}]{\text{1) RCHO; 2) } H_2SO_4, \Delta\text{; 3) } CH_2N_2}$ R–CH(OH)–CH(CH_3)–$COOCH_3$ (77–85%ee) + R–CH(OH)–CH(CH_3)–$COOCH_3$ ~90:10

3

[1] A. I. Meyers, G. Knaus, K. Kamata, and M. E. Ford, *Am. Soc.*, **98**, 567 (1976).
[2] A. I. Meyers and Y. Yamamoto, *ibid.*, **103**, 4278 (1981).

Ethyl β-phenylsulfonylpropiolate (1). Mol. wt. 238.26, oil, unstable at 25°.
Preparation:

$$HC{\equiv}CCOOC_2H_5 \xrightarrow[86\%]{\text{1) LDA, THF, }-78°\text{; 2) } C_6H_5SSO_2C_6H_5} C_6H_5SC{\equiv}CCOOC_2H_5 \xrightarrow{ClC_6H_4CO_3H} C_6H_5SO_2C{\equiv}CCOOC_2H_5$$

1

Arenes.[1] This reagent is a reactive dienophile; moreover, the regioselectivity on addition to 1-substituted butadienes is extremely high. The adducts are easily converted to arenes by loss of $C_6H_5SO_2H$ in the presence of base.

Example:

$$\text{CH}_2\text{=CH–CH=CHCH}_3 \xrightarrow[74\%]{\mathbf{1},\ 25°} \text{(cyclohexadiene: }CH_3\text{, }COOC_2H_5\text{, }SO_2C_6H_5) \xrightarrow[74\%]{KOC(CH_3)_3,\ DMF,\ 80°} \text{(benzene: }CH_3\text{, }COOC_2H_5)$$

$$\text{(cyclohexadiene adduct)} \xrightarrow[58\%]{NiO_2,\ C_6H_6} \text{(benzene: }CH_3\text{, }COOC_2H_5\text{, }SO_2C_6H_5)$$

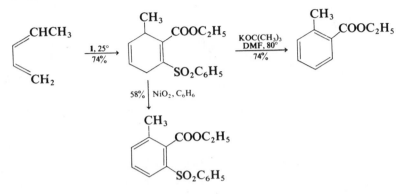

[1] M. Shen and A. G. Schultz, *Tetrahedron Letters*, **22**, 3347 (1981).

Ethyl trimethylsilylacetate, 7, 150–152; **8**, 226–227.

$\mathbf{RCH_2COOR'} \longrightarrow \mathbf{RC(=O)CH{=}CHCOOR''}$. This conversion can be effected as outlined in equation (I). In three cases examined,

$$\text{(I)}\quad RCH_2COOR' \xrightarrow{LDA,\ CH_3SO_2SCH_3} \xrightarrow[95\%]{LDA,\ CH_3SO_2SCH_3} RC(SCH_3)_2\text{—}COOR' \xrightarrow[82\%]{DIBAH,\ C_6H_5CH_3}$$

$$RC(SCH_3)_2\text{—}CHO \xrightarrow[86\%]{1)\ (CH_3)_3SiCH(Li)COOC_2H_5;\ 2)\ \text{Several steps}} RC(=O)CH{=}CHCOOR''\ (E)$$

only the (E)-isomer of the γ-ketocrotonate was obtained.[1]

Silylation of acidic compounds.[2] The combination of ethyl trimethylsilylacetate and 2–5 mole % of tetra-*n*-butylammonium fluoride (TBAF) is a markedly efficient reagent for transfer of trimethylsilyl groups to relatively acidic substrates such as ketones, alcohols, thiols, phenols, and carboxylic acids, and even 1-alkynes. $KOCH_3$/(18-crown-6) and Triton B are also effective catalysts.

Examples:

$$C_6H_5C{\equiv}CH + (CH_3)_3SiCH_2COOC_2H_5 \xrightarrow[88\%]{F^-,\ THF} C_6H_5C{\equiv}CSi(CH_3)_3$$
1

$$C_6H_5CH_2SH + \mathbf{1} \xrightarrow[87\%]{} C_6H_5CH_2SSi(CH_3)_3$$

(Menthol-type cyclohexanol: ring bearing CH_3, OH and $CH(CH_3)_2$) $+ \mathbf{1} \xrightarrow[92\%]{}$ (corresponding $OSi(CH_3)_3$ ether bearing CH_3 and $CH(CH_3)_2$)

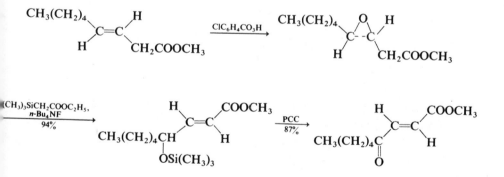

Indirect allylic oxidation of α,β-unsaturated esters.[3] No satisfactory method is known for effecting this oxidation directly, but it can be accomplished indirectly in three steps from a β,γ-unsaturated ester.

Example:

$$CH_3(CH_2)_4CH{=}CHCH_2COOCH_3 \xrightarrow{ClC_6H_4CO_3H} CH_3(CH_2)_4\overset{O}{\overbrace{CH{-}CH}}CH_2COOCH_3$$

$$\xrightarrow[94\%]{(CH_3)_3SiCH_2COOC_2H_5,\ n\text{-}Bu_4NF} CH_3(CH_2)_4CH(OSi(CH_3)_3)CH{=}CHCOOCH_3 \xrightarrow[87\%]{PCC} CH_3(CH_2)_4C(=O)CH{=}CHCOOCH_3$$

[1] C. Le Drian and A. E. Greene, *Am. Soc.*, **104**, 5473 (1982).
[2] E. Nakamura, K. Hashimoto, and I. Kuwajima, *Bull. Chem. Soc. Japan*, **54**, 805 (1981).
[3] T. Fujisawa, M. Takeuchi, and T. Sato, *Chem. Letters*, 1795 (1982).

Ethyl vinyl ether, 1, 386–388; **2**, 198; **4**, 234–235.

Lactone annelation.[1] A new lactone synthesis from an allylic alcohol such as **1** involves a Claisen rearrangement (**4**, 234–235) followed by an intramolecular Michael addition, as formulated in equation (I).

(I)

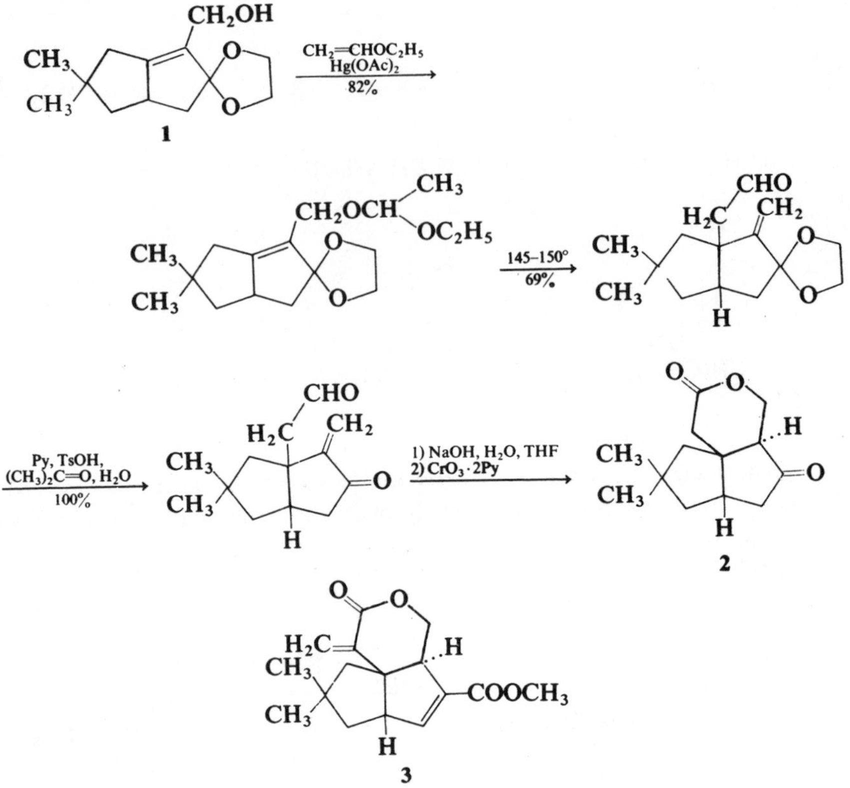

The product (**2**) is an intermediate in a synthesis of pentalenolactone E methyl ester (**3**), a sesquiterpene antibiotic.[2]

[1] L. A. Paquette, G. D. Annis, H. Schostarez, and J. F. Blount, *J. Org.*, **46**, 3768 (1981).

[2] L. A. Paquette, H. Schostarez, and G. D. Annis, *Am. Soc.*, **103**, 6526 (1981).

F

Ferric chloride–Silica, 10, 185.

Quinones from disilyloxyarenes.[1] 1,4-Disilyloxybenzenes are converted by $FeCl_3/SiO_2$ into 1,4-benzoquinones in moderate to good yield. This reaction can be conducted with catalytic amounts of the oxidant if air is present. However, PCC is superior for this reaction.

Cleavage of aromatic benzyl ethers.[1] In this reaction, the reagent acts as a Lewis acid. Yields are greatly improved if the solvent is completely removed from the slurry. A high surface area is important to this cleavage, and the Chromosorbs and Celite Filter Aid can replace silica.

[1] T. C. Jempty, K. A. Z. Gogins, Y. Mazur, and L. L. Miller, *J. Org.*, **46**, 4545 (1981).

Ferric nitrate/K10 bentonite clay (Clayfen). The reagent is prepared by addition of the clay to $Fe(NO_3)_3$ in acetone followed by evaporation of the solvent at temperatures below 50°. *Warning*: An unstable reagent is formed at higher temperatures.[1]

Oxidation of alcohols.[2] Secondary alcohols are oxidized to carbonyl compounds by Clayfen in pentane or hexane under vigorous stirring. Primary benzyl alcohols are oxidized satisfactorily, but primary aliphatic alcohols are oxidized to complex mixtures. Isolated yields are generally $>80\%$. Nitrite esters (RONO) have been identified as intermediates.

[1] A. Cornelius, and P. Laszlo, *Synthesis*, 849 (1980).
[2] A. Cornelius, P.-Y. Herze, and P. Laszlo, *Tetrahedron Letters*, **23**, 5035 (1982).

Ferrocenylphosphines, chiral. Hayashi et al.[1] have prepared a large number of chiral ferrocenylophosphines, which have planar chirality owing to 1,2-unsymmetrical substitution on the ferrocene group. In addition, one of the substituents contains a chiral carbon atom. The most interesting phosphines have amino or hydroxyl groups in the side chain. These phosphines, in combination with transition metals, can promote highly efficient enantioselective reactions.

(S)-α-(R)-2-diphenylphosphinoferrocencyl ethyldimethylamine,

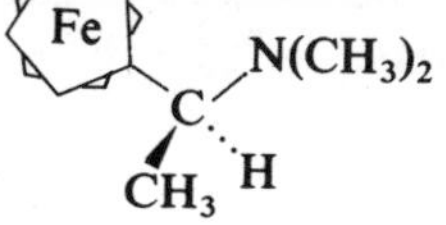

Fe; $P(C_6H_5)_2$; $N(CH_3)_2$; C; CH_3; H (**1**).

m.p. 139°, $\alpha_D -361°$. Use of this chiral ligand in the nickel-catalyzed coupling of 1-phenylethylmagnesium chloride with vinyl bromide gives optically active 3-phenyl-1-butene in 52–63% optical purity (equation I). In this case both the ferrocene planar

$$\text{(I)}\quad C_6H_5\overset{\displaystyle CH_3}{\overset{|}{C}}HMgCl + CH_2{=}CHBr \xrightarrow[83\%]{Ni/1^*} C_6H_5\overset{\displaystyle CH_3}{\overset{|}{C}}HCH{=}CH_2 \quad (52\text{–}63\%ee)$$

chirality and the amino group play an important part in the induction of asymmetry.[2]

This coupling reaction was used in a synthesis of (R)-(−)-curcumene (**2**) in 66% optical yield. (equation II).[3]

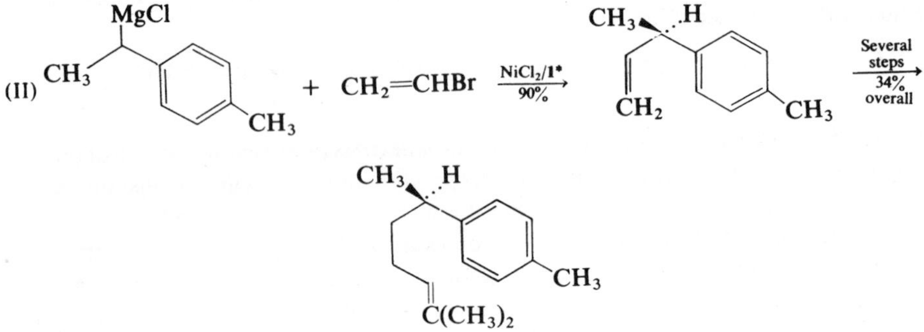

(II) MgCl, CH_3, CH_3 + $CH_2{=}CHBr$ $\xrightarrow[90\%]{NiCl_2/1^*}$ CH_3, H, CH_2, CH_3 $\xrightarrow[34\%\ \text{overall}]{\text{Several steps}}$

CH_3, H, CH_3, $C(CH_3)_2$

2 (66%ee)

(R)-1-(S)-1′, 2-Bis(diphenylphosphino)ferrocenyl ethanol

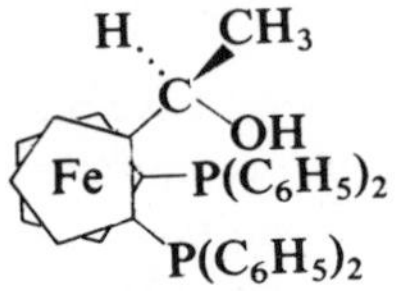

H, CH_3, C, OH, Fe, $P(C_6H_5)_2$, $P(C_6H_5)_2$ (**3**).

m.p. 155°, $\alpha_D -285°$. Complexes of Rh with this phosphine catalyze the enantioselective hydrogenation of aminomethyl aryl ketones to 2-amino-1-arylethanols in optical yields of 50–85%ee (equation III).[4]

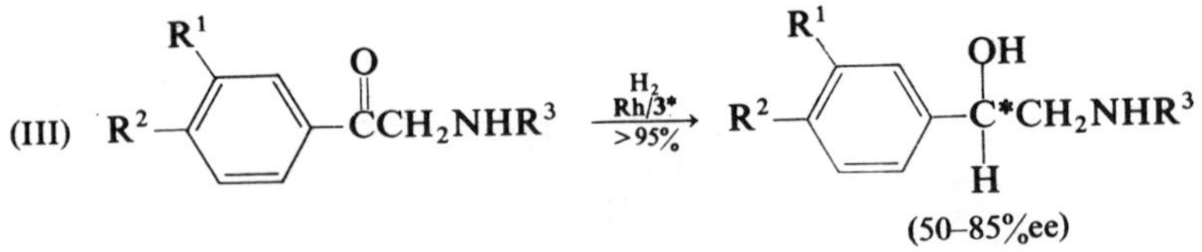

(III) R^1, R^2–C_6H_3–$\overset{O}{\overset{\|}{C}}CH_2NHR^3$ $\xrightarrow[>95\%]{H_2,\ Rh/3^*}$ R^1, R^2–C_6H_3–$\overset{OH}{C^*}(H)CH_2NHR^3$

(50–85%ee)

This ligand, in combination with a rhodium complex prepared *in situ* from $Rh(NBD)_2{}^+ClO_4{}^-$ (NBD = norbornadiene), can be used for an enantioselective hydrogenation of prochiral ketones to secondary alcohols (equation IV).[5]

$$\text{(IV)}\quad R^2CH_2COR^1 \xrightarrow[\text{2) ClP(O)(C}_6\text{H}_5)_2]{\text{1) LDA}} \begin{matrix} R^2 & & OP(O)(C_6H_5)_2 \\ & C{=}C & \\ H & & R^1 \end{matrix} \xrightarrow[\text{Rh/3*}]{H_2}$$

$$R^2CH_2\overset{*}{C}HR^1(OP(O)(C_6H_5)_2) \xrightarrow{CH_3Li} R^2CH_2\overset{*}{C}HR^1(OH)$$

(40–78%ee)

Asymmetric Grignard cross-coupling.[6] Palladium and nickel complexes have been used to effect cross-coupling of RMgX with aryl or vinyl halides. Chiral ferrocenylphosphines complexed with a metal are useful for asymmetric coupling of Grignard reagents as a consequence of kinetic resolution. The highest optical yields reported to date are obtained from the coupling shown in equation (I), effected with $NiCl_2$ complexed with **1a** or **1b** (2 equiv.). Use of **1a** gives (S)-**2** in 68% ee, use of **1b** gives (R)-**2** in 66% ee, and use of **1c** gives (R)-**2** in 54% ee. The ligands are about as

$$\text{(I)}\quad CH_3CH(C_6H_5)MgCl + CH_2{=}CHBr \xrightarrow[95\%]{NiCl_2L_2{}^*} CH_3\overset{*}{C}H(C_6H_5)CH{=}CH_2$$

2

effective when complexed with $PdCl_2$. The data obtained from these and other ferrocenylphosphines permit some generalizations of the requisites for high stereoselectivity. Substituents on the $P(C_6H_5)_2$ group have little effect, but the substituents on the amino group exert a large effect and can even alter the configuration of the final product. The paper ascribes the stereocontrol to coordination of the amino group of **1** to the magnesium atom of the racemic Grignard reagent.

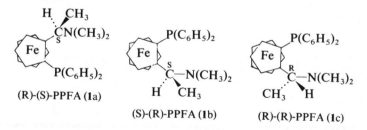

(R)-(S)-PPFA (**1a**)

(S)-(R)-PPFA (**1b**)

(R)-(R)-PPFA (**1c**)

[1] T. Hayashi, T. Mise, M. Fukushima, M. Kagotani, N. Nagashima, Y. Hamada, A. Matsumoto, S. Kawakami, M. Konishi, K. Yamamoto, and M. Kumada, *Bull. Chem. Soc. Japan*, **53**, 1138 (1980).

[2] T. Hayashi, M. Tajika, K. Tamao, and M. Kumada, *Am. Soc.*, **98**, 3718 (1976).
[3] K. Tamao, T. Hayashi, H. Matsumoto, H. Yamato, and M. Kumada, *Tetrahedron Letters*, **23**, 2155 (1979).
[4] T. Hayashi, A. Katsumura, M. Konoishi, and M. Kumada, *ibid.*, 425 (1979).
[5] T. Hayashi, K. Kanehira, and M. Kumada, *ibid.*, **22**, 4417 (1981).
[6] T. Hayashi, M. Konishi, M. Fukushima, T. Mise, M. Kagotani, M. Tajika, and M. Kumada, *Am. Soc.*, **104**, 180 (1982).

Fluorodimethoxyborane, 6, 261–262; **10**, 186.

Homoallylic alcohols.[1] Alkenyldimethoxyboranes add to aldehydes regioselectively to form homoallylic alcohols. The reaction is also diastereoselective: (Z)-alkenylboranes react to form *erythro*-adducts, and (E)-alkenylboranes produce *threo*-adducts. The boranes are prepared by treatment of alkenylpotassium compounds with fluorodimethoxyborane (**6**, 261–262).

Examples:

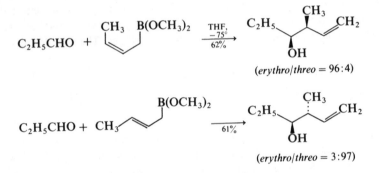

C_2H_5CHO + CH_3 $B(OCH_3)_2$ (Z-alkenylborane) $\xrightarrow[62\%]{\text{THF, } -75^\circ}$ C_2H_5–CH(OH)–CH(CH_3)–CH=CH_2

(*erythro*/*threo* = 96:4)

C_2H_5CHO + CH_3–CH=CH–CH_2–$B(OCH_3)_2$ (E) $\xrightarrow{61\%}$ C_2H_5–CH(OH)–CH(CH_3)–CH=CH_2

(*erythro*/*threo* = 3:97)

Variations of the aldehydes and alkenylboranes affect only slightly the stereoselectivity, but not the regioselectivity.

[1] M. Schlosser and K. Fujita, *Angew. Chem. Int. Ed.*, **21**, 309 (1982); K. Fujita and M. Schlosser, *Helv.*, **65**, 1258 (1982).

Formaldehyde, 1, 397–402; **2**, 200–201; **4**, 238–239; **6**, 264–267; **7**, 158–160; **8**, 231–232; **9**, 224–225; **10**, 186.

Iminium ion–vinylsilane cyclization. This reaction can be used to control the configuration of exocyclic double bonds β to the nitrogen function of alkaloids. Thus, reaction of the (Z)- and (E)-vinylsilanes **1** and **2** with paraformaldehyde (excess) and *d*-camphor-10-sulfonic acid proceeds with >98% retention of configuration to give **3** and **4**, respectively. The latter product is identical with the natural indoloquinolizidine alkaloid deplancheine.[1]

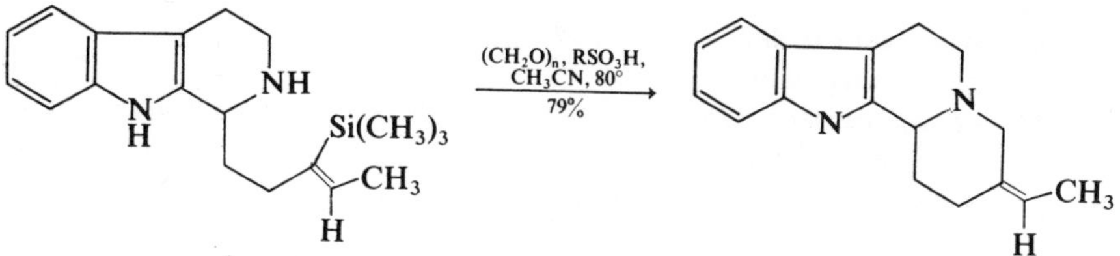

(CH2O)n, RSO3H, CH3CN, 80°, 79%: 1 → 3

83%: 2 → 4

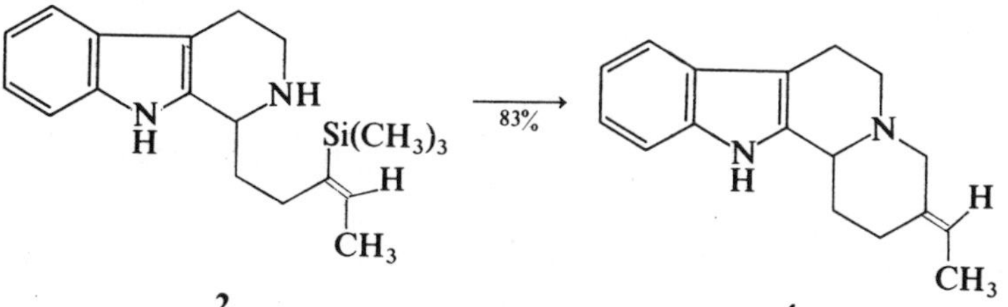

This cyclization has been used in a stereospecific synthesis of a pumiliotoxin alkaloid (**6**).[2]

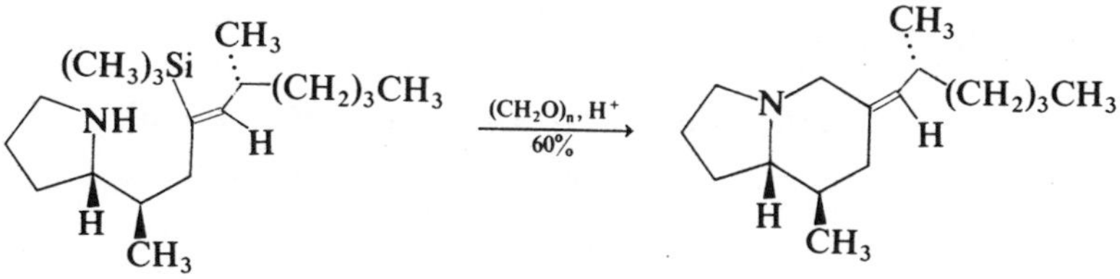

$(CH_2O)_n$, H^+, 60%: 5 → 6

[1] L. E. Overman and T. C. Malone, *J. Org.*, **47**, 5297 (1982).
[2] L. E. Overman and K. L. Ball, *Am. Soc.*, **102**, 1851 (1981).

Formaldehyde diethyl acetal, $H_2C(OC_2H_5)_2$. Mol. wt. 104,15, b.p. 87.5°.

Ethoxymethyl ethers, $ROCH_2OC_2H_5$.[1] These ethers are readily prepared by reaction of an alcohol or phenol with excess ethylal and an acid catalyst (H_2SO_4, HCl, TsOH, ion exchange resin), equation (I). The excess reagent and its azeotrope with

$$\text{(I)}\quad ROH + H_2C(OC_2H_5)_2 \xrightarrow{H^+} ROCH_2OC_2H_5 + C_2H_5OH$$

C_2H_5OH (b.p. 74°) are easily removed by distillation. The ethers are stable to base, but cleaved by acid.

[1] U.-A Schaper, *Synthesis*, 794 (1981).

Formaldehyde dimethyl dithioacetal S,S-dioxide, $CH_3SCH_2SO_2CH_3$ (**1**). Mol. wt. 140.22, m.p. 51°. Preparation.[1]

Methyl carboxylates.[2] The reagent (**1**) is alkylated by alkyl halides under phase-transfer conditions with NaOH as base. The product is converted to a methyl ester by oxidation followed by a Pummerer-type rearrangement (equation I).

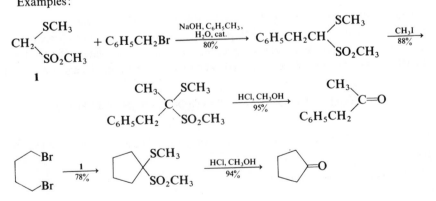

$$\text{(I)}\quad \mathbf{1} + RBr \xrightarrow[60\text{–}70\%]{NaOH,\ H_2O,\ C_6H_5CH_3,\ cat.} RCH(SCH_3)(SO_2CH_3) \xrightarrow[85\text{–}95\%]{H_2O_2,\ HOAc} RCH(SOCH_3)(SO_2CH_3) \xrightarrow[\sim 85\%]{HCl,\ CH_3OH,\ \Delta} RCOOCH_3$$

The reaction of **1** with an allylic bromide is interesting because a β,γ-carboxylic acid ester or an α-substituted vinylacetic acid ester can be obtained (equation II).

$$\text{(II)}\quad RCH{=}CHCH_2Br + \mathbf{1} \xrightarrow[75\%]{NaOH,\ H_2O,\ C_6H_5CH_3,\ cat.} RCH{=}CHCH_2CH(SCH_3)(SO_2CH_3) \xrightarrow{\sim 80\%} RCH{=}CHCH_2COOCH_3$$

$$RCH{=}CHCH_2Br + \mathbf{1} \xrightarrow[45\text{–}60\%]{K_2CO_3,\ KI,\ HMPT} RCH(CH(SCH_3)SO_2CH_3){-}CH{=}CH_2 \xrightarrow{70\text{–}80\%} RCH(COOCH_3)CH{=}CH_2$$

Ketone synthesis. Dialkylation of **1** under phase-transfer conditions results in the dimethyl dithioacetal S,S-dioxide derivative of a ketone, which is hydrolyzed to the ketone when heated with hydrochloric acid in methanol. The sequence provides a route to both acyclic and cyclic ketones.[3]

Examples:

$$\underset{\mathbf{1}}{CH_2(SCH_3)(SO_2CH_3)} + C_6H_5CH_2Br \xrightarrow[80\%]{NaOH,\ C_6H_5CH_3,\ H_2O,\ cat.} C_6H_5CH_2CH(SCH_3)(SO_2CH_3) \xrightarrow[88\%]{CH_3I}$$

$$(CH_3)(C_6H_5CH_2)C(SCH_3)(SO_2CH_3) \xrightarrow[95\%]{HCl,\ CH_3OH} (CH_3)(C_6H_5CH_2)C{=}O$$

$$Br(CH_2)_4Br \xrightarrow[78\%]{\mathbf{1}} \text{1-(methylthio)-1-(methylsulfonyl)cyclopentane}\ (SCH_3,\ SO_2CH_3) \xrightarrow[94\%]{HCl,\ CH_3OH} \text{cyclopentanone}\ ({=}O)$$

[1] K. Ogura, M. Suzuki, and G. Tsuchihashi, *Bull. Chem. Soc. Japan*, **53**, 1414 (1980).
[2] K. Ogura, J. Watanabe, and H. Iida, *Tetrahedron Letters*, **22**, 4499 (1981).
[3] K. Ogura, M. Suzuki, J. Watanabe, M. Yamashita, H. Iida, and G. Tsuchihashi, *Chem. Letters*, 813 (1982).

Formic acid, 1, 404–407; **2**, 202–203; **3**, 147; **4**, 239–240; **5**, 316–319; **7**, 160; **8**, 232; **9**, 226–227.

2,2-Dialkyl-δ-valerolactones.[1] Reaction of primary, tertiary-1,4-diols (**1**) with formic acid and 97–100% sulfuric acid (Koch-Haaf carboxylation[2]) provides 2,2-dialkyl-δ-valerolactones (**2**) in high yield.

R^1, R^2, OH, OH, **1**; HCOOH, H_2SO_4, 85–95%; R^1, R^2, O, O, **2**

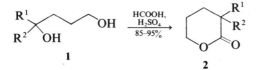

[1] Y. Takahashi, N. Yoneda, and H. Nagai, *Chem. Letters*, 1187 (1982).
[2] W. Haaf, *Org. Syn., Coll. Vol.*, **5**, 739 (1973).

(S)-N-Formyl-2-methoxymethylpyrrolidine, (H, N, CH_2OCH_3, CHO) (**1**). Preparation from (S)-prolinol, **8**, 16.

(R)- and (S)-α-Hydroxy ketones; vic-diols. Reaction of **1** with an aldehyde or ketone (LiTMP) at −100° results in a mixture of (S,S)- and (S,R)-hydroxy amides (**2**) with only slight diastereomeric excess (5–20%). Reaction of the separated amides with methyllithium gives both the (S)- and (R)-hydroxy ketones (**3**) in optically pure form. On further reaction with methyllithium, optically pure *vic*-diols (**4**) can be obtained.[1]

1 + $R^1R^2C{=}O$ —LiTMP, −100°, 70–80%→ (S,S)-**2** + (R,S)-**2**; each —CH_3Li→ (S)-**3** / (R)-**3** ($R^1R^2C(OH)COCH_3$); —CH_3Li→ (S)-**4** / (R)-**4** ($R^1R^2C(OH)C(OH)(CH_3)_2$)

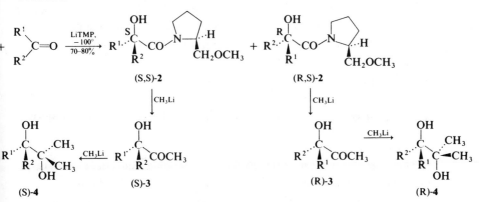

[1] D. Enders and H. Lotter, *Angew. Chem. Int. Ed.*, **20**, 795 (1981).

4-Formyl-1-methylpyridinium benzenesulfonate,

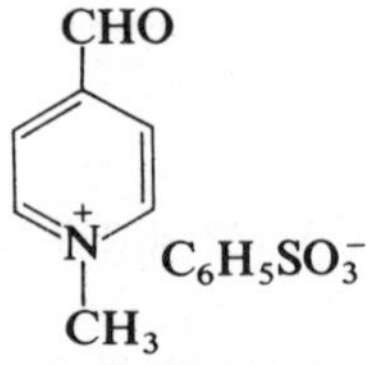

Mol. wt. 279.30, m.p. 100–101°. The reagent is obtained by reaction of pyridine-4-carboxaldehyde with methyl benzenesulfonate in refluxing benzene (83% yield).

Conversion of amines to carbonyl compounds.[1] This derivative of vitamin B_6 converts primary amines to aldehydes or ketones by a process that mimics the biological process of this conversion mediated by vitamin B_6 (equation I). The process

(I) 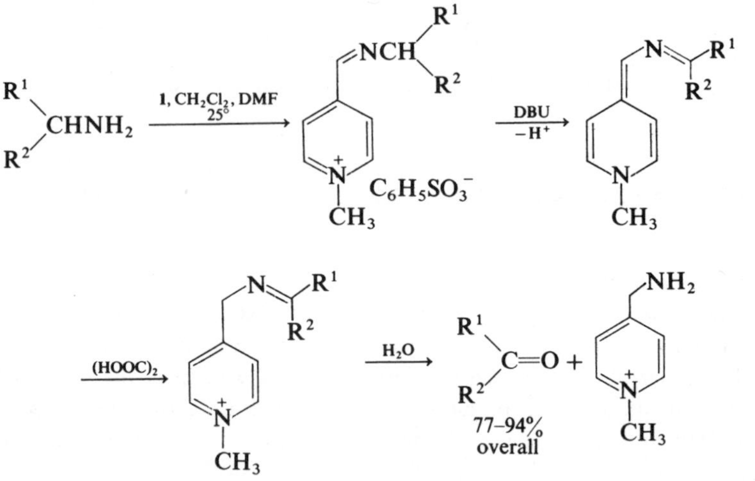 $R^1R^2CHNH_2 \xrightarrow[25°]{1,\ CH_2Cl_2,\ DMF}$ … $\xrightarrow[-H^+]{DBU}$ … $\xrightarrow{(HOOC)_2}$ … $\xrightarrow{H_2O}$ $R^1R^2C{=}O$ (77–94% overall) +

involves imine formation, prototropic rearrangement, and finally hydrolysis, but is conducted in one flask without isolation of intermediates. The mild conditions permit the presence of even sensitive functional groups. Oxidative deamination of the ε-amino group of lysine was effected with no loss of chirality.

[1] T. F. Buckley and H. Rapoport, *Am. Soc.*, **104**, 4446 (1982).

N-Formylpiperidine,  NCHO (**1**). Mol. wt. 113.16, b.p. 222°. Suppliers: Aldrich, Reilly Tar and Chemicals.

RMgX→RCHO (*cf.*, **8**, 341). N-Formylpiperidine converts a variety of organolithium or Grignard reagents into aldehydes in about 75–95% yield.[1]

[1] G. A. Olah and M. Arvanaghi, *Angew. Chem. Int. Ed.*, **20**, 878 (1981); *idem*, *Org. Syn.*, submitted (1982).

G

Grignard reagents, 1, 415–424; **2**, 205; **5**, 321; **6**, 269–270; **7**, 163–164; **8**, 235–238; **9**, 229–233; **10**, 189–194.

Ketene dithioacetals.[1] These useful synthetic intermediates can be prepared from Grignard reagents in a one-pot sequence (equation I). The method fails with allylmagnesium bromide, and only low yields of ketene dithioacetals are obtained from cyclohexylmagnesium chloride and benzylmagnesium chloride (5–10%).

$$\text{(I)}\quad R^1R^2CHMgX \xrightarrow{CS_2} \left[R^1R^2CH{-}C(=S)SMgX \xrightarrow{LDA} R^1R^2C{=}C(SLi)(SMgX) \right] \xrightarrow[60\text{–}80\%]{(R^3O)_2SO_2} R^1R^2C{=}C(SR^3)_2$$

Allyl ethers. In the presence of $TiCl_4$, Grignard reagents couple with dialkyl acetals of α,β-enals to give allyl ethers in high yield. $TiCl_4$ also promotes coupling of Grignard reagents with alkyl aryl acetals to give unsymmetrical ethers.[2]

Examples:

$$C_6H_5CH{=}CHCH(OCH_3)_2 + C_6H_5CH_2CH_2MgBr \xrightarrow[81\%]{TiCl_4,\ THF,\ -78^\circ} C_6H_5CH{=}CHCH(OCH_3)CH_2CH_2C_6H_5$$

$$CH_3CH(OC_2H_5)(OC_6H_5) + C_6H_5CH_2MgBr \xrightarrow[51\%]{TiCl_4} CH_3CH(OC_2H_5)(CH_2C_6H_5)$$

Beckmann rearrangement.[3] Reaction of oxime sulfonates with simple Grignard reagents in dry toluene at -78° results in imines, which can be reduced to α-alkylamines or alkylated with allylic or propargylic Grignard reagents to furnish α,α-dialkylamines.

Example:

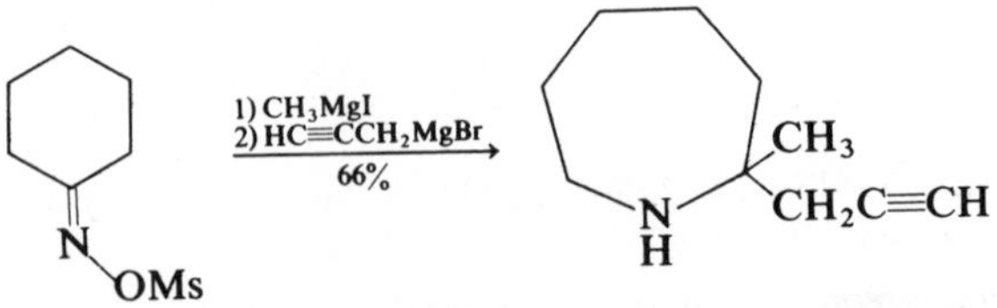

$$\text{cyclohexanone } O\text{-mesyloxime} \xrightarrow[66\%]{1)\ CH_3MgI;\ 2)\ HC{\equiv}CCH_2MgBr} \text{2-methyl-2-(}CH_2C{\equiv}CH\text{)azepane (NH)}$$

Organoaluminum reagents have also been employed for alkylative rearrangement, but use of Grignard reagents permits greater flexibility. Reagents of the type RLi or RZnX are either ineffective or less satisfactory.

Mg-ene cyclization.[4] A recent synthesis of the sesquiterpene $\Delta^{9(12)}$-capnellene (**5**) involves two intramolecular Mg-ene reactions, **1**→**2** and **3**→**4**.[5] Surprisingly, the cyclization of **3** to **4** occurs at 20°.

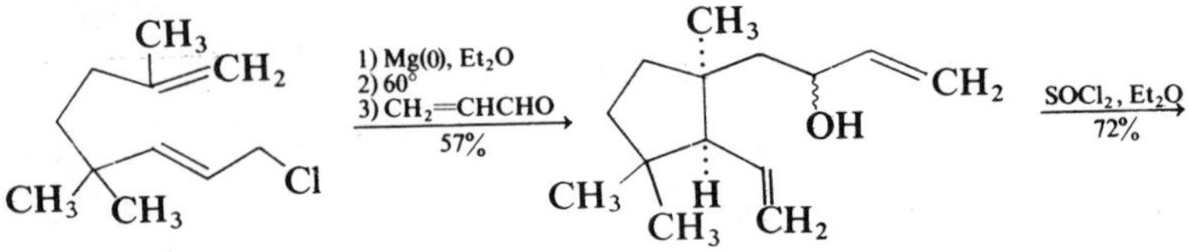

$$\mathbf{1} \xrightarrow[57\%]{1)\ Mg(0),\ Et_2O;\ 2)\ 60^\circ;\ 3)\ CH_2{=}CHCHO} \mathbf{2} \xrightarrow[72\%]{SOCl_2,\ Et_2O}$$

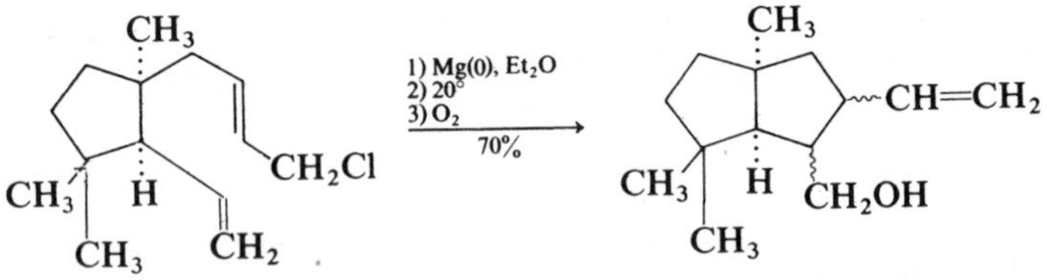

$$\mathbf{3} \xrightarrow[70\%]{1)\ Mg(0),\ Et_2O;\ 2)\ 20^\circ;\ 3)\ O_2} \mathbf{4}\ (cis/trans = 3:2)$$

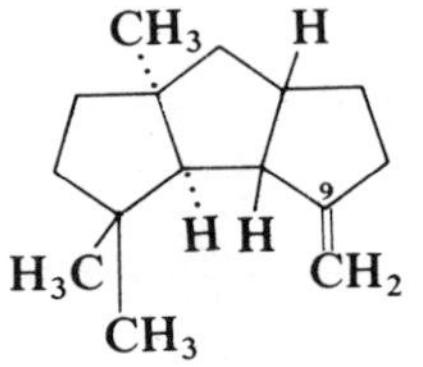

5

A stereoselective synthesis of sinularene (**8**) and the 5-epimer (**9**) from a norbornene also involves as the key step the Mg-ene reaction **6**→**7**. The product can be converted by known reactions into either **8** or **9**.[5]

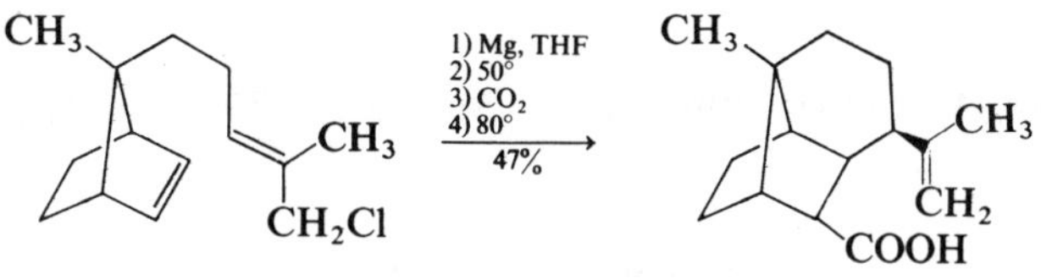

$\xrightarrow[\text{47\%}]{\text{1) Mg, THF; 2) 50°; 3) CO}_2\text{; 4) 80°}}$

6 (with CH_3, CH_3, CH_2Cl) → **7** (with CH_3, CH_3, CH_2, COOH)

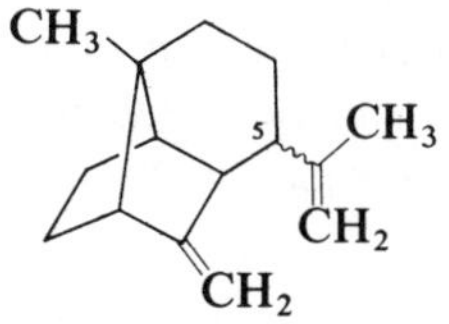

8 (5β)
9 (5α)

(with CH_3, CH_3, CH_2, CH_2; position 5)

4-*Keto carboxylic acids.* The first step in a new synthesis of these acids is silylation of the lithium enolate of butyrolactone with methyldiphenylchlorosilane to give the α-silyl lactone **1** in high yield.[6] This product reacts with various Grignard reagents to form an adduct that loses methyldiphenylsiloxide to give a dihydrofurane.[7]

Example:

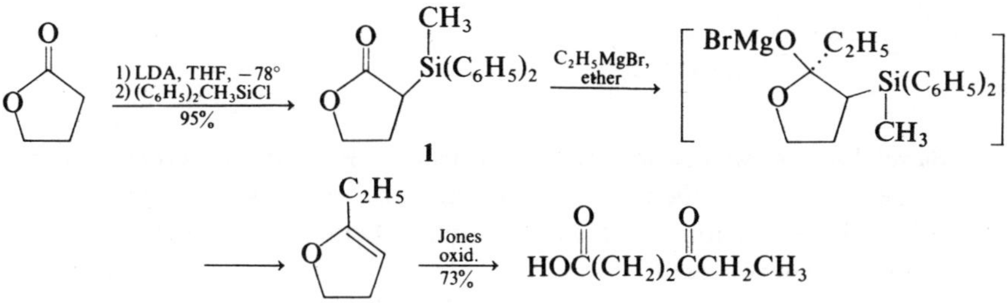

$\xrightarrow[\text{95\%}]{\text{1) LDA, THF, }-78°\text{; 2) }(C_6H_5)_2CH_3SiCl}$ **1** (CH_3, $Si(C_6H_5)_2$) $\xrightarrow{C_2H_5MgBr,\ \text{ether}}$ [BrMgO, C_2H_5, $Si(C_6H_5)_2$, CH_3]

$\longrightarrow$ (C_2H_5) $\xrightarrow[\text{73\%}]{\text{Jones oxid.}}$ $HO\overset{O}{\overset{\|}{C}}(CH_2)_2\overset{O}{\overset{\|}{C}}CH_2CH_3$

α-Keto esters.[8] A convenient route to a α-keto esters involves addition of primary or aryl Grignard reagents to diethyl oxalate (equation I). Only mediocre yields are obtained from reactions in ether; use of THF as solvent gives yields of 40–70%.

$$\text{(I)}\quad RCH_2MgBr + C_2H_5O\overset{O}{\overset{\|}{C}}-\overset{O}{\overset{\|}{C}}OC_2H_5 \xrightarrow[\text{40–70\%}]{\text{THF, }-10°} RCH_2\overset{O}{\overset{\|}{C}}-\overset{O}{\overset{\|}{C}}OC_2H_5$$

1,1-*Dialkylethylenes.* The reaction of Grignard reagents with ethyl trimethylsilylacetate followed by acidic work-up results in symmetrically substituted 1-alkenes (equation I). The reaction is not useful with hindered Grignard reagents such as isobutylmagnesium bromide.[9]

$$\text{(I)}\quad (CH_3)_3SiCH_2\overset{O}{\overset{\|}{C}}OC_2H_5 \xrightarrow[\text{ether}]{RMgX} \left[(CH_3)_3SiCH_2\overset{O}{\overset{\|}{C}}R \xrightarrow{RMgX} (CH_3)_3SiCH_2\underset{R}{\underset{|}{\overset{OH}{\overset{|}{C}}}}{-}R\right]$$

$$\xrightarrow[45\text{–}95\%]{H_2SO_4} CH_2{=}CR_2$$

Homologation of α,β-unsaturated carboxylic acids. Ethyl crotonate can be alkylated at C_3 by CuCl-catalyzed addition of Grignard reagents followed by treatment with methanesulfinyl chloride. Subsequent desulfinylation yields the homologated α,β-unsaturated ester (equation I). The method is also applicable to α,β-enones (equation II).[10]

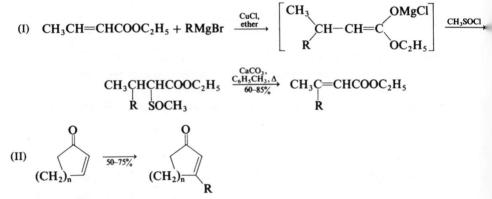

$$\text{(I)}\quad CH_3CH{=}CHCOOC_2H_5 + RMgBr \xrightarrow[\text{ether}]{CuCl,} \left[\begin{matrix}CH_3 & & OMgCl\\ & CH{-}CH{=}C & \\ R & & OC_2H_5\end{matrix}\right] \xrightarrow{CH_3SOCl}$$

$$CH_3\underset{R}{\underset{|}{C}}H\underset{SOCH_3}{\underset{|}{C}}HCOOC_2H_5 \xrightarrow[60\text{–}85\%]{CaCO_3,\ C_6H_5CH_3,\ \Delta} CH_3\underset{R}{\underset{|}{C}}{=}CHCOOC_2H_5$$

$$\text{(II)}\quad \text{cycloalkenone } ((CH_2)_n) \xrightarrow{50\text{–}75\%} \text{3-R-cycloalkenone } ((CH_2)_n,\ R)$$

Bicyclobutanolides; 4-substituted-2-butenolides.[11] Primary and secondary Grignard reagents react with the bicyclic anhydride **1** to form bicyclobutanolides (**2**) in 80–100% yield. Distillation of **2** effects a retro Diels-Alder reaction to give 4-disubstituted 2-butenolides (**3**).

Example:

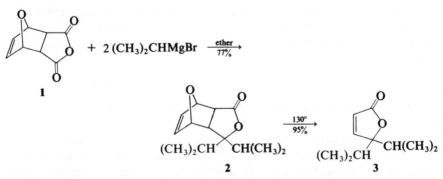

$$\mathbf{1} + 2\,(CH_3)_2CHMgBr \xrightarrow[77\%]{\text{ether}} \mathbf{2}\ ((CH_3)_2CH,\ CH(CH_3)_2) \xrightarrow[95\%]{130^\circ} \mathbf{3}\ ((CH_3)_2CH,\ CH(CH_3)_2)$$

This synthesis can be extended to di-Grignard reagents (equation I).

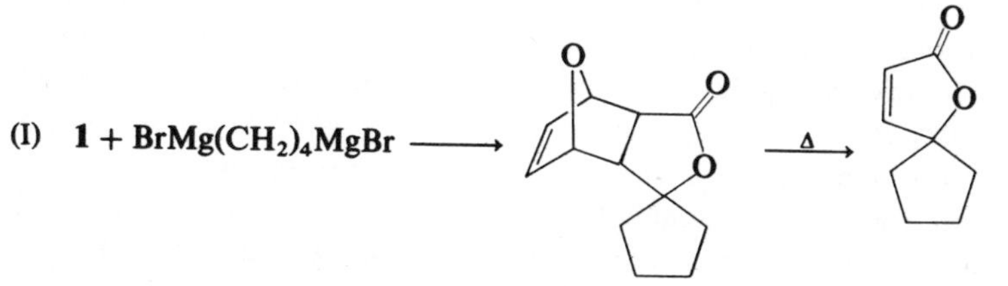

Phenylytterbium iodide, C_6H_5YbI. This reagent can be prepared by reaction of C_6H_5I with Yb powder in THF. In reactions with carbonyl compounds, it behaves in the same way as Grignard reagents, except that it reacts more readily with esters than with ketones.[12]

1 R. Kaya and R. N. Beller, *Synthesis*, 814 (1981).
2 H. Ishikawa, T. Mukaiyama, and S. Ikeda, *Bull. Chem. Soc. Japan*, **54**, 776 (1981).
3 K. Hattori, K. Maruoka, and H. Yamamoto, *Tetrahedron Letters*, **23**, 3395 (1982).
4 Review: H. Lehmkuhl, *Bull. Soc.*, **II**, 87 (1981).
5 W. Oppolzer and K. Battig, *Tetrahedron Letters*, **23**, 4669 (1982); W. Oppolzer, H. F. Strauss, and D. P. Simmons, *ibid.*, **23**, 4673 (1982).
6 G. L. Larson and L. M. Fuentes, *Am. Soc.*, **103**, 2418 (1981).
7 L. M. Fuentes and G. L. Larson, *Tetrahedron Letters*, **23**, 271 (1982).
8 L. M. Weinstock, R. B. Currie, and A. V. Lovell, *Syn. Comm.*, **11**, 943 (1981).
9 G. L. Larson and D. Hernandez, *Tetrahedron Letters*, **23**, 1035 (1982).
10 T. Fujisawa, A. Noda, T. Kawara, and T. Sato, *Chem. Letters*, 1159 (1981).
11 P. Canonne, M. Akssira, and G. Lemay, *Tetrahedron Letters*, **22**, 2611 (1981).
12 T. Fukagawa, Y. Fujiwara, K. Yokoo, and H. Taniguchi, *Chem. Letters*, 1771 (1981).

Guanidines, $R_2NC(=NR)NR_2$. Barton and coworkers[1] have synthesized a series of hindered pentaalkylguanidines for use as proton acceptors in organic reactions. A typical synthesis of a particularly useful one, 2-*t*-butyl-1,1,3,3-tetramethylguanidine (**1**), is shown in equation (I). The more hindered pentaisopropylguanidine (**2**) is prepared from tetraisopropylthiourea (equation II).

$$\text{(I)}\quad (CH_3)_2N\overset{O}{\overset{\|}{C}}N(CH_3)_2 \xrightarrow{COCl_2} \left[(CH_3)_2N\overset{Cl}{\overset{|}{C}}{}^{+}N(CH_3)_2 \quad Cl^- \right] \xrightarrow[85\%]{(CH_3)_3CNH_2} (CH_3)_2N\overset{NC(CH_3)_3}{\overset{\|}{C}}N(CH_3)_2$$

1, b.p. 80–82°, $pK_a \sim 14$

$$\text{(II)}\quad (i\text{-}Pr)_2N\overset{S}{\overset{\|}{C}}N(i\text{-}Pr)_2 \xrightarrow[69\%]{1)\,COCl_2;\ 2)\,i\text{-}PrNH_2} (i\text{-}Pr)_2N\overset{N\text{-}i\text{-}Pr}{\overset{\|}{C}}N(i\text{-}Pr)_2$$

2, b.p. 60°/0.01 mm

The guanidine **1** can be alkylated more readily than **2**, but nevertheless is a very effective proton acceptor. It is preferred to DBU as the base for esterification of carboxylic acids by an alkyl halide. Thus severely hindered tertiary carboxylic acids can be alkylated by isopropyl iodide in about 90% yield in the presence of **1** (1.5 equiv.). Selective C-monoalkylation of a typical β-keto ester was effected in 80% yield in the presence of **1** (1.0 equiv.). Preliminary experiments suggest that **1** is not particularly useful for base-promoted elimination reactions, but that the more hindered **2** is superior to collidine or DBN for this purpose.

[1] D. H. R. Barton, J. D. Elliott, and S. D. Géro, *J.C.S. Perkin I*, 2085 (1982).

H

2,3,4,5,6,6-Hexachloro-2,4-cyclohexadiene-1-one (1); 2,3,4,4,5,6-hexachloro-2,5-cyclohexadiene-1-one (2).

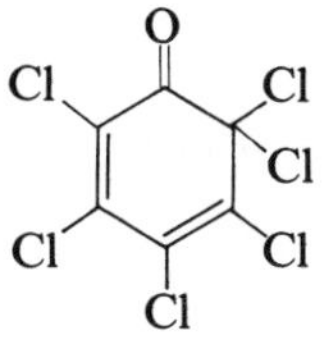

1 m.p. 51°

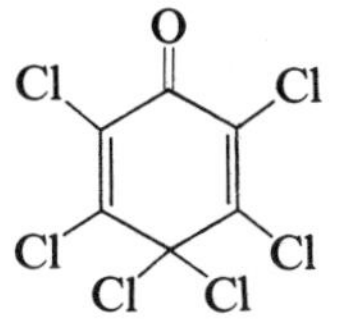

2, m.p. 106°

Reagent **1** is prepared by chlorination of sodium pentachlorophenolate in CCl_4 at 0^0. Reagent **2** is prepared by chlorination of pentachlorophenol in the presence of $AlCl_3$.[1]

Regioselective chlorination of aromatics. Phenol undergoes regioselective chlorination at the *ortho*-position with **1**, but selective *para*-chlorination with **2**. Chlorination of anisole by either reagent results exclusively in *para*-chlorination, because of steric effects.[2] Chlorination of naphthol with **1** and **2** is more regioselective than that of phenol.[3]

α-Chlorination of alkyl aryl ketones[4] Reagent **1** is not satisfactory for chlorination of dialkyl ketones, but it is useful for α-chlorination of alkyl aryl ketones (equation I). α,α-Dichlorination is possible by use of 2 equiv. of **1**.

(I) 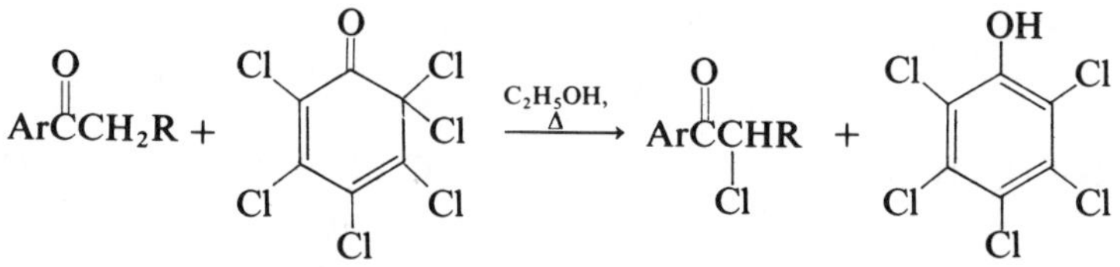 $ArC(O)CH_2R$ + **1** $\xrightarrow{C_2H_5OH,\ \Delta}$ $ArC(O)CHClR$ (50–100%) + pentachlorophenol

[1] R. Fort and L. Denivelle, *Bull. Soc.*, 1834 (1956).
[2] A. Guy, M. Lemaire, and J.-P. Guette, *Tetrahedron*, **38**, 2339 (1982).
[3] *Idem, Ibid.*, **28**, 2347 (1982).
[4] A. Guy, M. Lemaire, and J.-P. Guette, *Synthesis*, 1018 (1982).

Hexafluoroantimonic acid, 5, 309–310; **6**, 272–273; **7**, 166–167; **8**, 239–240; **10**, 195.

Carboxylation of γ-butyrolactones.[1] γ-Butyrolactones react with carbon monoxide in HF–SbF_5, followed by an aqueous workup, to form dicarboxylic acids in quantitative yield.

Examples:

$$\text{5-methyl-}\gamma\text{-butyrolactone} + CO \xrightarrow[\text{quant.}]{1)\ HF\text{-}SbF_5;\ 2)\ H_2O} HOOCCH(CH_3)CH_2CH_2COOH$$

$$\text{5,5-dimethyl-}\gamma\text{-butyrolactone} + CO \xrightarrow{\text{quant.}} (CH_3)_2C(COOH)CH_2CH_2COOH$$

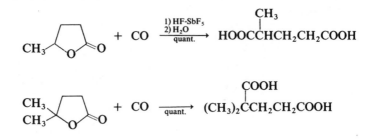

[1] N. Yoneda, A. Suzuki, and Y. Takahasi, *Chem. Letters*, 767 (1981).

Hexakis(acetato)trihydrato-μ$_3$-oxotrisrhodium acetate, $[Rh_3O(OCOCH_3)_6(H_2O)_3]$-$OCOCH_3$ (**1**). The complex is obtained in high yield by ozonization of $Rh(OAc)_2$ in acetic acid.[1]

Allylic oxidation.[2] *t*-Butyl hydroperoxide (0.5 equiv.) in acetic acid in the presence of catalytic amounts of this "Rh_3O" complex (**1**) oxidizes cycloalkenes to α,β-enones and the corresponding allylic acetates in the ratio of 6–7:1. Other rhodium complexes are less effective. Allylic alcohols (but not the acetates) are oxidized by this reaction to α,β-enones.

Examples:

$$\text{cyclohexene} \xrightarrow[\mathbf{1}]{(CH_3)_3COOH,\ HOAc} \text{2-cyclohexenone} + \text{3-acetoxycyclohexene} \quad (7.7:1)$$

$$C_6H_5CH_2CH{=}CH_2 \longrightarrow C_6H_5CH{=}CHCHO + C_6H_5CH{=}CHCH_2OH \quad (5.6:1)$$

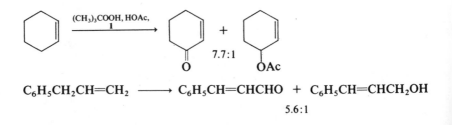

Prins reaction.[3] Conjugated dienes and activated alkenes react with aldehydes and carboxylic acids in the presence of ruthenium catalysts to give 1,3-diols. Variable amounts of acetoxybutene and higher molecular weight compounds can be formed. $RuCl_3 \cdot 3H_2O$ can be used, but the best results are obtained with **1**.

Example:

$$CH_2{=}CHCH{=}CH_2 + (CH_2O)_n + CH_3COOH \xrightarrow{1,\ 80^\circ}$$

$$\underset{(59\%)}{CH_2{=}CH\underset{OAc}{\underset{|}{C}}HCH_2CH_2OH} + \underset{(14\%)}{CH_2{=}CH\underset{OAc}{\underset{|}{C}}HCH_2CH_2CH{=}CHCH_2CH_2OH}$$

[1] S. Uemera, A. Spencer, and G. Wilkinson, *J.C.S. Dalton*, 2665 (1973).
[2] S. Uemura, S. R. Patil, *Tetrahedron Letters*, **23**, 4353 (1982).
[3] J. Thivolle-Cazat and I. Tkatchenko, *J.C.S. Chem. Comm.*, 1128 (1982).

Hexamethyldisilane, $(CH_3)_3SiSi(CH_3)_3$. Mol. wt. 146. 38, b. p. 112–114^9.

Reductive silylation of* p-*quinones. This reagent in combination with a catalytic amount of iodine converts *p*-quinones into 1,4-bis(trimethylsilyloxy)arenes in almost quantitative yield.[1]

Iodotrimethylsilane may be the actual reagent. The products are useful for protection of quinones since they are reconverted to the parent quinones in 60–90% yield by oxidation with PCC in CH_2C1_2 at 25°.[2]

[1] H. Matsumoto, S. Koike, I. Matsubara, T. Nakano, Y. Nagai, *Chem. letters*, 533 (1982).
[2] J. P. Willis, K. A. Z. Cogins, and L. L. Miller, *J. Org.*, **46**, 3215 (1981).

Hexamethylphosphoric triamide (HMPT), **1**, 425; **2**, 207; **3**, 148–149; **6**, 279–280; **9**, 235–236, **10**, 199.

γ-Functionalization of tiglaldehyde.[1] Anions of α,β-unsaturated aldimines are known to react with electrophiles at both the α- and γ-position. Thus, the lithium salt (**1**) of tiglaldehyde cyclohexylimine reacts with aldehydes to give, after acid hydrolysis of the imine, mainly the product of α-substitution. Addition of HMPT (1 equiv.) to the reaction followed by equilibration at 0° (2 hours) results mainly in formation of the product of γ-subsitution.

$$(CH_3)(H)C{=}C(CH_3)CH{=}NC_6H_{11} \xrightarrow{LDA} \underset{\mathbf{1}}{CH_2{\cdots}\underset{H}{\bar{C}}{\cdots}\underset{CH_3}{C}{\cdots}CH{\cdots}NC_6H_{11}}\ Li^+$$

$$\xrightarrow[2)\ H_3O^+]{1)\ C_6H_5(CH_2)_2CHO} \underset{(\alpha)}{CH_2{=}CHC(CHO)(H_3C){-}CH(OH)(CH_2)_2C_6H_5} + \underset{(\gamma,\ >95\%\ E)}{C_6H_5(CH)_2CH(OH)CH_2CH{=}C(CH_3)CHO}$$

	(α)	(γ, >95% E)
THF/Hexane = 1:5, −78°	54%	3%
+ HMPT (1 eq.) − 78→0°	15%	66%

HMPT exerts a similar effect in the reactions of the dianion of tiglic acid with aldehydes, but addition of HMPT to reactions of the lithium salt of crotonaldehyde cyclohexylimine with aldehydes results in a complex mixture of products.

[1] E. Vedejs and D. M. Gapinski, *Tetrahedron Letters*, **22**, 4913 (1981).

Homophthalic anhydride, 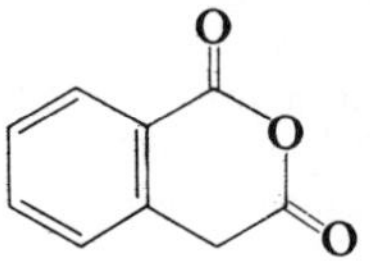 (**1**). Mol. wt. 162.14, m.p. 142°.

Preparation.[1] Supplier: Aldrich.

Diels-Alder reactions. Homophthalic anhydride (**1**) undergoes Diels-Alder reactions with some alkynes and benzoquinones at 150–200° to give linear phenols, probably via the tautomer **a**.[2]

Examples:

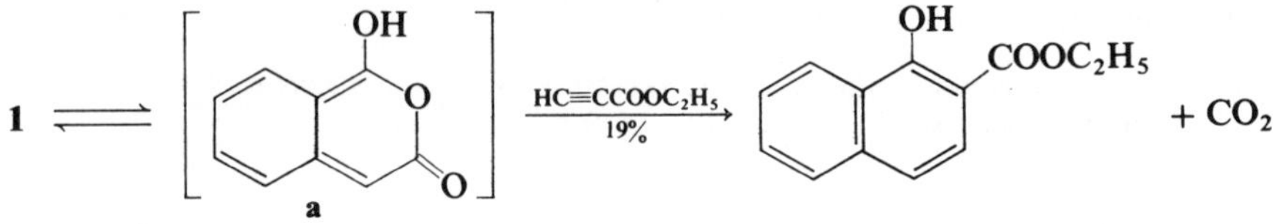

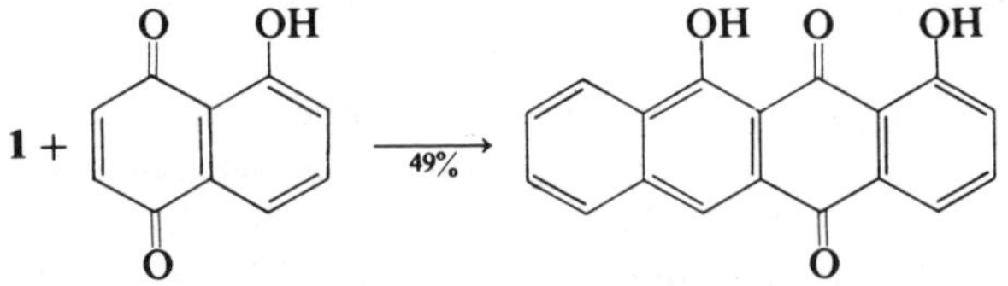

1 ⇌ [**a**] $\xrightarrow[19\%]{HC{\equiv}CCOOC_2H_5}$ (OH, $COOC_2H_5$ naphthalene) + CO_2

1 + (O, OH, O naphthoquinone) $\xrightarrow{49\%}$ (OH, O, OH, O tetracyclic product)

This methodology provides a regiocontrolled synthesis of anthracyclinones such as daunomycinone (**2**) as outlined in Scheme (I).[3]

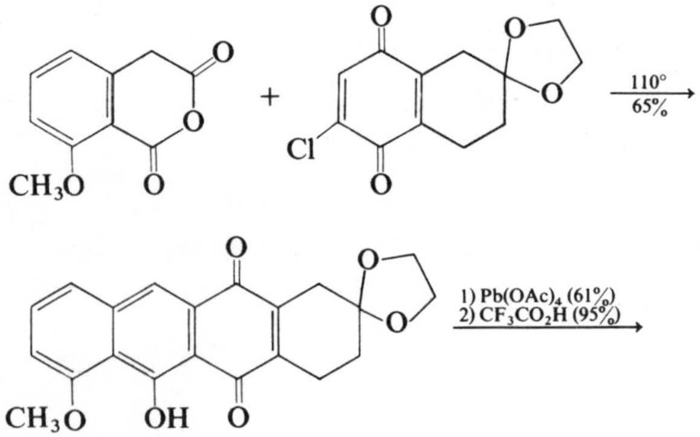

(CH_3O, O, O anhydride) + (O, O, O, Cl, O quinone ketal) $\xrightarrow[65\%]{110°}$

(CH_3O, OH, O, O, O, O) $\xrightarrow{\text{1) Pb(OAc)}_4\text{ (61\%) 2) CF}_3\text{CO}_2\text{H (95\%)}}$

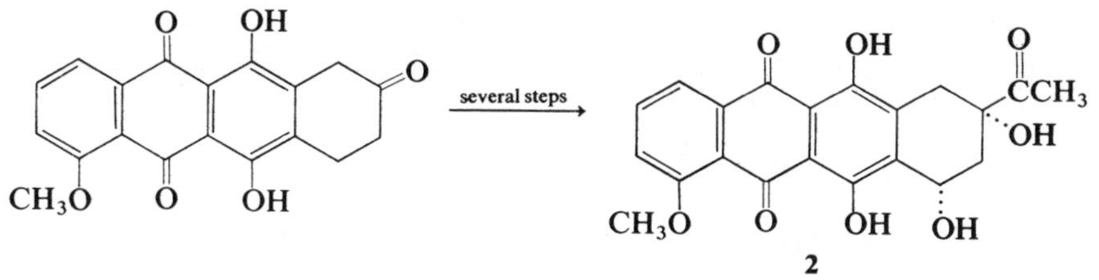

Scheme (I)

[1] O. Grummitt, R. Egan, and A. Buck, *Org. Syn. Coll. Vol.*, **3**, 449 (1955).
[2] Y. Tamura, A. Wada, M. Sasho, and Y. Kita, *Tetrahedron Letters*, **22**, 4283 (1981).
[3] Y. Tamura, A. Wada, M. Sasho, K. Fukunaga, H. Maeda, and Y. Kita, *J. Org.*, **47**, 4376 (1982).

Hydrazine, 1, 434–445; **2**, 211; **3**, 153; **4**, 248; **5**, 327–329; **6**, 280–281; **7**, 170–171; **8**, 245; **9**, 236–237.

Reduction of nitroarenes (**6**, 281; **9**, 237). Ethanol–dichloroethane (1:1) is recommended as the solvent for reduction of nitroarenes to aminoarenes by hydrazine hydrate in combination with Raney nickel. Yields are nearly quantitative.[1].

[1] N. R. Ayyangar, A. C. Lugada, P. V. Nikrad, and V. K. Sharma, *Synthesis*, 640 (1981).

Hydridotetrakis(triphenylphosphine)rhodium(I), $HRh[P(C_6H_5)_3]_4$(**1**).

Transfer hydrogenation.[1] This complex catalyzes hydrogen transfer between an alcohol and an α,β-enone (equation I). The H originally attached to the carbon bearing the OH group is transferred regioselectively to the β-carbon of the enone. The rate-limiting step is the cleavage of the O—H bond.

$$\text{(I)}\quad C_6H_5CHOHCH_3 + C_6H_5CH{=}CHCOCH_3 \xrightarrow{\mathbf{1}} C_6H_5COCH_3 + C_6H_5CH_2CH_2COCH_3.$$

[1] D. Beaupere, P. Bauer, L. Nadjo, and R. Uzan, *J. Organometal. Chem.*, **238**, C12 (1982).

Hydriodic acid, 1, 449–450; **2**, 213–214; **8**, 246; **9**, 238; **10**, 199–200.

Polycyclic arenes (*cf.* **9**, 238). Harvey's reductive cyclization reaction is also applicable to *o*-naphthoylbenzoic and -naphthoic acids.[1]

Examples:

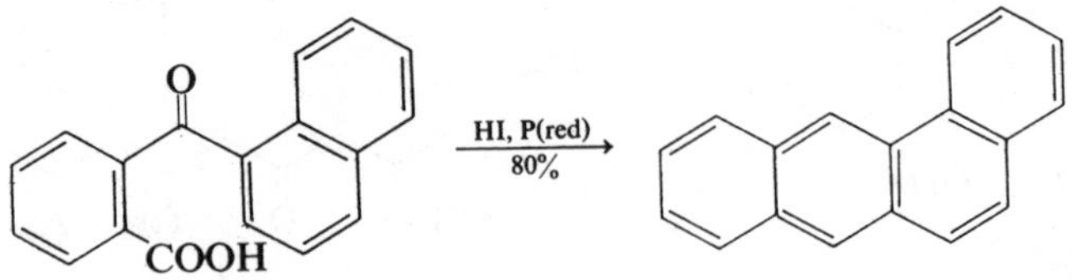

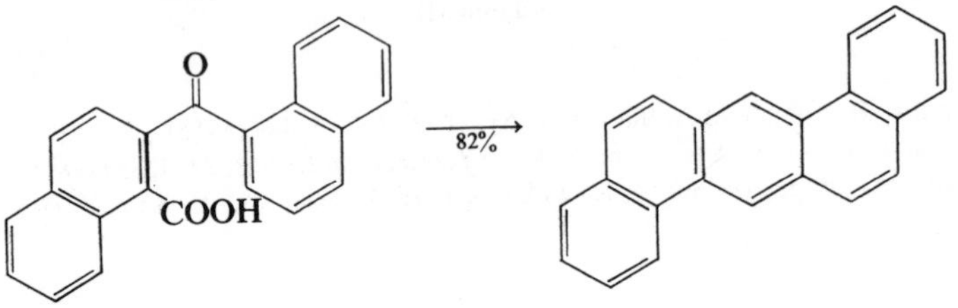

[1] K. L. Platt and F. Oesch, *J. Org.*, **46**, 2601 (1981).

Hydrogen fluoride–Dimethyl sulfide.

Deprotection of synthetic peptides.[1] The protecting groups of synthetic peptides are generally removed by treatment with anhydrous HF and anisole. Most of the protecting groups can be removed by a low concentration of HF in dimethyl sulfide (1:3, v/v), and this method also decreases various side reactions caused by strong acids. Rate studies indicate that low-concentration HF cleavage is S_N2, whereas an S_N1 mechanism is involved in absolute HF. The mixture HF–$S(CH_3)_2$–*p*-cresol (25:65:10) is somewhat more potent.

[1] J. P. Tam, W. F. Heath, and R. B. Merrifield, *Tetrahedron Letters*, **23**, 4435 (1982).

Hydrogen peroxide–Hydrogen fluoride/Boron trifluoride etherate.

Hydroxylation of arenes.[1] Benzene and alkyl derivatives are oxidized to monophenols in 35–60% yield by 30% H_2O_2 in combination with HF/BF_3 etherate at −78–60°. The reactive reagent is probably $H_3O_2{}^+BF_4{}^-$.

[1] G. A. Olah, A. P. Fung, and T. Keumi, *J. Org.*, **46**, 4305 (1981).

Hydrogen peroxide–Vilsmeier reagent.

Epoxidation.[1] The combination of hydrogen peroxide (30%) and the Vilsmeier reagent presumably affords the salt $[(CH_3)_2N{=}CHOOH]^+\ \bar{P}O_2Cl_2$ (**1**). When **1** is generated in methanol in the presence of an olefin at −20°, *trans*-1,2-dichloro compounds are formed in high yield. If the reaction is conducted at −80° and in the presence of sodium carbonate, epoxides are formed in high yield.

[1] J.-P. Dulcere and J. Rodriguez, *Tetrahedron Letters*, **23**, 1887 (1982).

Hydroxylamine, 1, 478–481; **7,** 176–177; **9,** 245; **10,** 206–207.

β-Diketones.[1] β-Diketones react readily with hydroxylamine to form isoxazoles and can be reformed by acid hydrolysis and thus can function as β-diketone synthons. For example, 3,5-dimethylisoxazole, readily obtained from acetylacetone, can be metallated by a variety of bases, and the anion so obtained can be alkylated at the C_5-methyl group. A second alkylation occurs only at the C_3-methyl group. Thus it is possible to prepare selectively 3,5-dialkylisoxazoles, from which the corresponding β-diketones are obtained in 70–95% yield by acid hydrolysis in ethanol.

Example:

$CH_3COCH_2COCH_3$ —(H_2NOH, ~85%)→ 3,5-dimethylisoxazole —(1) n-BuLi, THF; 2) n-BuI; 90%)→ 5-$CH_3(CH_2)_4$-3-CH_3-isoxazole —(1) sec-BuLi, $(C_2H_5)_2O$; 2) CH_3I; 92%)→ 5-$CH_3(CH_2)_4$-3-CH_2CH_3-isoxazole —(H^+)→ $CH_3(CH_2)_4COCH_2COCH_2CH_3$

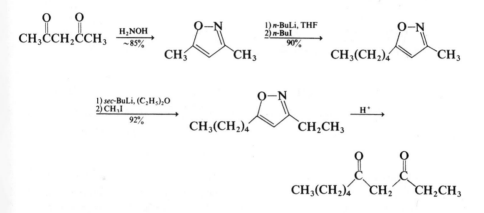

N⁷-Hydroxycephalosporins.[2] Reaction of the 7-oxocephem **1** with hydroxylamine gives a single oxime (**2**). Reduction of **2** with $BH_3 \cdot$pyridine leads stereoselectively to the N^7-hydroxycephalosporin **3**.

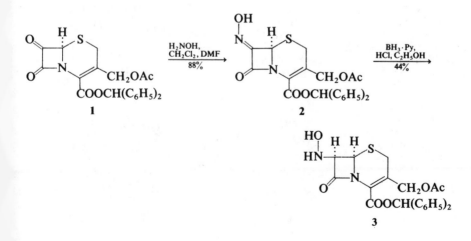

1 (7-oxocephem, CH_2OAc, $COOCH(C_6H_5)_2$) —(H_2NOH, CH_2Cl_2, DMF; 88%)→ **2** (oxime, N–OH) —($BH_3 \cdot Py$, HCl, C_2H_5OH; 44%)→ **3** (HO–HN, CH_2OAc, $COOCH(C_6H_5)_2$)

N^7-Methoxycephalosporins can be prepared in this way by use of methoxylamine (**1**, 670).

[1] D. J. Brunelle, *Tetrahedron Letters*, **22**, 3699 (1981).
[2] D. Hagiwara, K. Sawada, T. Ohnami, and M. Hashimoto, *Chem. Pharm. Bull.*, **30**, 3061 (1982).

2-Hydroxymethyl-3-trimethylsilylpropene (1), 9, 454–455. The reagent is prepared by conversion of α-methallyl alcohol into the allyl dianion with *n*-BuLi and TMEDA in a mixture of ether and THF followed by trimethylsilylation and acid hydrolysis of the silyl ether group (equation I).[1]

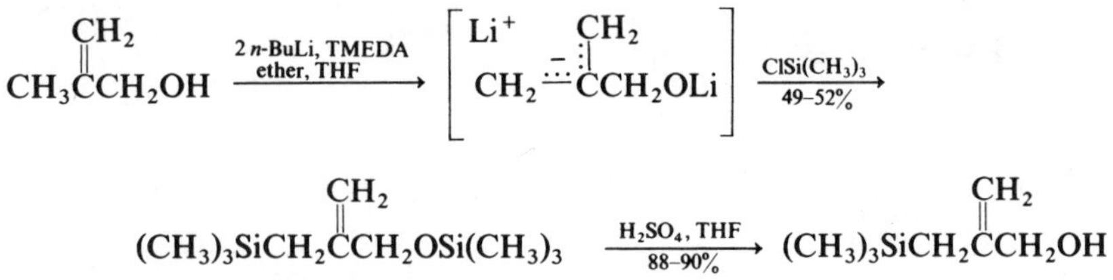

$$CH_3C(=CH_2)CH_2OH \xrightarrow[\text{ether, THF}]{2\ n\text{-BuLi, TMEDA}} \left[Li^+ \ CH_2 \overset{-}{\cdots} C(=CH_2)CH_2OLi \right] \xrightarrow[49\text{–}52\%]{ClSi(CH_3)_3}$$

$$(CH_3)_3SiCH_2C(=CH_2)CH_2OSi(CH_3)_3 \xrightarrow[88\text{–}90\%]{H_2SO_4,\ THF} (CH_3)_3SiCH_2C(=CH_2)CH_2OH$$

1 (b.p. 53–54°/1.6 mm)

Methylenecyclopentane annelation (**9**, 454–455). The mesylate (**2**) of **1** has been used as an electrophilic equivalent of trimethylenemethane for methylenecyclopentane annelation of a cyclopentanone in a synthesis of coriolin (**7**).[2] Thus, reaction of the enolate of **4**, a protected equivalent of **3**, reacts with **2** to give **5**, which after oxidation to the disulfone is cyclized by fluoride ion to the tricyclic methylenecyclopentane derivative **6**. The product is converted into **7** by several known transformations.

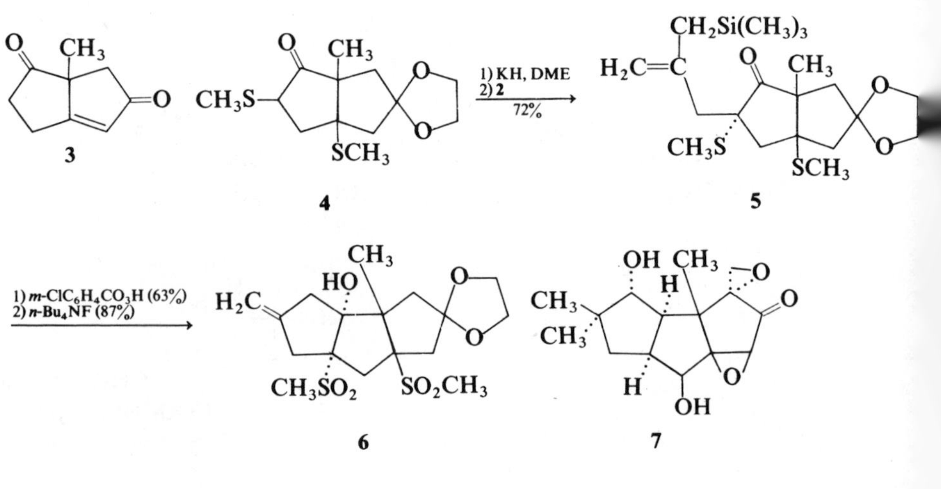

3 → **4** $\xrightarrow[72\%]{1)\ KH,\ DME;\ 2)\ \mathbf{2}}$ **5** $\xrightarrow{1)\ m\text{-}ClC_6H_4CO_3H\ (63\%);\ 2)\ n\text{-}Bu_4NF\ (87\%)}$ **6** → **7**

Fused methylenecyclopentanes of type **9** in which n = 3 or >6 when treated with KH and 18-crown-6 undergo a three-carbon ring expansion to give products of type **10** in about 90% yield. This ring expansion was used to convert the *β*-keto sulfone (**11**)

$(CH_2)_n$ O $SO_2C_6H_5$ **8** —1) **2**, NaH 2) F^-→ $(CH_2)_n$ OH $=CH_2$ $SO_2C_6H_5$ **9**, n = 3 or >6 —KH, 18-Crown-6, ~90%→ $(CH_2)_n$ O $-CH_3$ $SO_2C_6H_5$ **10**

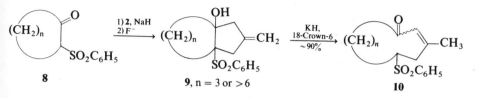

into muscone (**14**). In this case, ring expansion accompanies cyclization of the intermediate **12**.[3]

O $SO_2C_6H_5$ **11** —1) NaH 2) **2** 83%→ O $-Si(CH_3)_3$ CH_2 $SO_2C_6H_5$ **12** —F^- 92%→

O $-CH_3$ $SO_2C_6H_5$ **13** —1) H_2, Pd 2) Na/Hg 95%→ O $-CH_3$ **14**

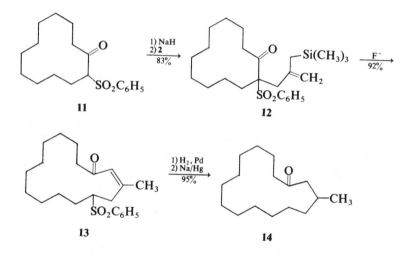

2-Choromethyl-3-trimethylsilylpropene,

$$(CH_3)_3SiCH_2C(=CH_2)CH_2Cl \quad \textbf{(2)}.$$

This related conjunctive reagent has been used for methylenecyclopentane annelation of a cyclic *α,β*-enone by conjugate addition of the nucleophilic allylsilane unit of **2** to the enone followed by cyclization (equation I).[4]

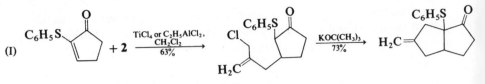

(I) C_6H_5S-cyclopentenone + 2 $\xrightarrow[63\%]{TiCl_4\ or\ C_2H_5AlCl_2,\ CH_2Cl_2}$ Cl, C_6H_5S, O, H_2C $\xrightarrow[73\%]{KOC(CH_3)_3}$ C_6H_5S, O, H_2C

[1] B. M. Trost, D. M. T. Chan, and T. N. Nanninga, *Org. Syn.*, submitted (1981).
[2] B. M. Trost and D. P. Curran, *Am. Soc.*, **103**, 7380 (1981).
[3] B. M. Trost and J. E. Vincent, *ibid.*, **102**, 5680 (1980).
[4] S. Knapp, U. O'Connor, and D. Mobilio, *Tetrahedron Letters*, **21**, 4557 (1980).

Hypochlorous acid, 1, 245, 487–488; **10**, 208–209.

α-Chloro-β,γ-unsaturated ketones. α,β-Enones that can exist in the *s-cis* conformation react with HOCl to form α-chloro-β,γ-enones.[1]

Examples;

$$(CH_3)_2C{=}CHCOCH_3 \xrightarrow[71\%]{HOCl} CH_2{=}C(CH_3){-}CHClCOCH_3$$

$$C_6H_5(CH_3)C{=}CHCOCH_3 \longrightarrow C_6H_5C({=}CH_2)CHClCOCH_3 + C_6H_5(CH_3)C{=}C(Cl)COCH_3 \quad (1:1)$$

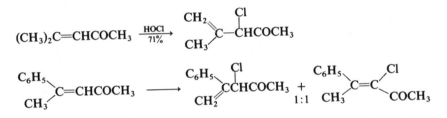

Isoprenylation of isopropenyl groups.[2] Cyclic terpenes substituted by an isopropenyl group can be converted into sesquiterpenes with the bisabolane skeleton by addition of HOCl under biphasic conditions to give an allylic chloride (**10**, 208–209) followed by a zinc-induced reaction with isovaleraldehyde to form an isomeric mixture of homoallylic alcohols.

Example:

CH_3, O, CH_3, CH_2 $\xrightarrow[72\%]{HOCl}$ CH_3, O, CH_2, CH_2Cl $\xrightarrow[84\%]{Zn,\ (CH_3)_2CHCH_2CHO}$ CH_3, O, OH, $CH_2{=}C{-}CH_2CHCH_2CH(CH_3)_2$

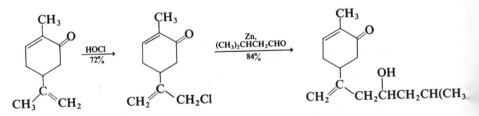

[1] S. G. Hegde and J. Wolinsky, *Tetrahedron Letters*, **22**, 5019 (1981).
[2] *Idem*, *J. Org.*, **47**, 3148 (1982).

I

Iodine, 1, 495–500; **2**, 220–222; **3**, 159–160; **4**, 258–260; **5**, 346–347; **6**, 293–295; **7**, 179–181; **8**, 256–260; **9**, 248–249; **10**, 210–211.

Di- and trisubstituted alkenes. A few years ago Zweifel and coworkers[1] reported a stereospecific synthesis of *cis*-alkenes by treatment of the adduct of a 1-alkyne and a dialkylborane with I_2 and NaOH (equation I). The reaction is believed to involve an iodonium ion, transfer of one of the R groups, and *trans*-elimination of I^- and $\overset{+}{B}ROH$.

$$\text{(I)}\quad R^1C{\equiv}CH + R_2BH \xrightarrow{\text{THF}} \left[\begin{matrix} R^1 & & H \\ & C{=}C & \\ H & & BR_2 \end{matrix}\right] \xrightarrow[65\text{–}85\%]{I_2,\ NaOH} \begin{matrix} R^1 & & R \\ & C{=}C & \\ H & & H \end{matrix}$$

The usefulness of this synthesis has been limited by the limited availability of dialkylboranes. Recently these boranes have been obtained *in situ* by hydridation of dialkylhaloboranes (equation II).[2]

$$\text{(II)}\quad R_2BX + HC{\equiv}CR^1 \xrightarrow[(\text{-LiX, -AlX}_3)]{1/4\ LiAlH_4,\ THF} \begin{matrix} R^1 & & H \\ & C{=}C & \\ H & & BR_2 \end{matrix} \xrightarrow[60\text{–}75\%]{I_2,\ NaOCH_3,\ -78^\circ} \begin{matrix} R^1 & & R \\ & C{=}C & \\ H & & H \end{matrix}$$

Brown *et al.*[3] have now extended the Zweifel synthesis to a stereospecific synthesis of trisubstituted alkenes from an internal alkyne (equation III). Evidently a similar mechanism is involved: *trans*-addition of I^+ and $^-OCH_3$ followed by *trans*-elimination of I^- and $\overset{+}{B}ROCH_3$ resulting in *trans*-orientation of the two alkyl groups of the alkyne.

$$\text{(III)}\quad R^1C{\equiv}CR^1 + R_2BCl(Br) \xrightarrow{1/4\ LiAlH_4,\ THF} \left[\begin{matrix} R_2B & & H \\ & C{=}C & \\ R^1 & & R^1 \end{matrix}\right] \xrightarrow[70\text{–}75\%]{I_2,\ NaOCH_3} \begin{matrix} R^1 & & H \\ & C{=}C & \\ R & & R^1 \end{matrix}$$

Another stereospecific route to trisubstituted alkenes is outlined in equation (IV). The synthesis involves migration of an alkyl group with inversion at the migration terminus and *trans*-deiodoboronation.[4]

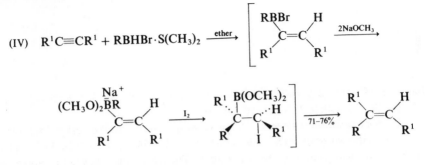

(IV) $R^1C{\equiv}CR^1 + RBHBr \cdot S(CH_3)_2 \xrightarrow{\text{ether}} \left[(RBBr)(R^1)C{=}C(H)(R^1) \right] \xrightarrow{2NaOCH_3}$

$Na^+ \; (CH_3O)_2\bar{B}R(R^1)C{=}C(H)(R^1) \xrightarrow{I_2} \left[R^1(B(OCH_3)_2)(R)C{-}C(H)(R^1)(I) \right] \xrightarrow{71\text{–}76\%} (R^1)(R)C{=}C(H)(R^1)$

cis- and trans-Alkenes. Zweifel's original procedure utilizing a dialkylborane for hydroboration suffers from two disadvantages: one alkyl group is not utilized, and dihydroboration competes with monohydroboration. Both difficulties are surmounted by use of an alkylbromoborane, $RBHBr \cdot S(CH_3)_2$ (**1**), prepared as shown in equation (I).[5] Reaction of a 1-alkyne with **1** provides a vinylborane, which rearranges to a *cis*-alkene on treatment with a base and iodine (equation II). Vinyl iodides are formed to a minor extent (3–16% yield). This procedure was used to synthesize muscalure (**2**), a sex pheromone of the housefly, from 1-tridecene and 1-decyne in 59% yield.[6]

(I) $HBBr_2 \cdot S(CH_3)_2 \xrightarrow{\text{alkene}} R^1BBr_2 \cdot S(CH_3)_2 \xrightarrow{0.25\ LiAlH_4} \underset{\mathbf{1}}{R^1BHBr \cdot S(CH_3)_2}$

(II) $R^2C{\equiv}CH \xrightarrow{\mathbf{1}} (R^1BBr)(H)C{=}C(H)(R^2) \xrightarrow{CH_3ONa,\ CH_3OH} (R^1BOCH_3)(H)C{=}C(H)(R^2) \xrightarrow[60\text{–}75\%\ \text{overall}]{CH_3ONa,\ I_2,\ CH_3OH} (H)(R^1)C{=}C(H)(R^2)$

$(H)(CH_3(CH_2)_{12})C{=}C(H)((CH_2)_7CH_3)$

2

Alkylbromoboranes (**1**) can also be used to provide *trans*-alkenes by hydroboration of a 1-bromo-1-alkyne. The adducts rearrange with sodium methoxide to (*trans*-1-alkyl-1-alkenyl)boronate esters (**3**), which can be converted into either *trans*-alkenes (**4**) or ketones (**5**), as shown in equation III.[7]

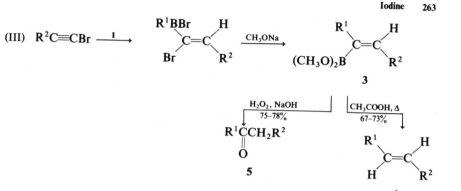

(III) $R^2C\equiv CBr \xrightarrow{1}$ $R^1BBr(Br)C{=}C(H)R^2$ $\xrightarrow{CH_3ONa}$ $(CH_3O)_2B(R^1)C{=}C(H)R^2$ **3**

3 $\xrightarrow[75\text{–}78\%]{H_2O_2,\ NaOH}$ $R^1C(=O)CH_2R^2$ **5**

3 $\xrightarrow[67\text{–}73\%]{CH_3COOH,\ \Delta}$ $R^1(H)C{=}C(H)R^2$ **4**

Cyclic iodocarbonates (**10**, 210–211). Cyclic iodocarbonates are useful precursors to epoxy alcohols, obtained by hydrolysis with Amberlyst A26, OH^- form, and to triols, formed on treatment with Amberlyst A26 in the CO_3^{-2} form.[8]

Example:

$CH_3C(=CH_2)CH_2OH \xrightarrow[85\%]{1)\ n\text{-BuLi};\ 2)\ CO_2;\ 3)\ I_2}$ (cyclic carbonate with ICH_2, CH_3) $\xrightarrow[85\%]{OH^-}$ (epoxide) CH_3, CH_2OH $\xrightarrow[80\%]{CO_3^{-2}}$ $HOCH_2C(OH)(CH_3)CH_2OH$

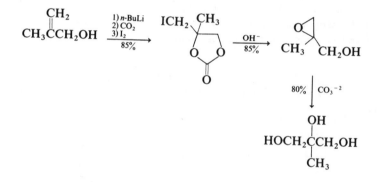

Iodolactonization (*cf.* **9**, 248). Iodolactonization of 3-hydroxy-4-alkenoic acids (**1**) under kinetic conditions proceeds with high asymmetric induction to give the unstable 3,4-*cis*-iodo-γ-lactones (**2**). Methanolysis converts **2** into the alcohols **3** and **4**. These same products result from Sharpless epoxidation (**5**, 75–76; **9**, 81–82) of the corresponding methyl esters,

$CH_3OC(=O)CH_2CH(OH)(R^1)C{=}C(H)R^2$ but with reversed selectivity.[9]

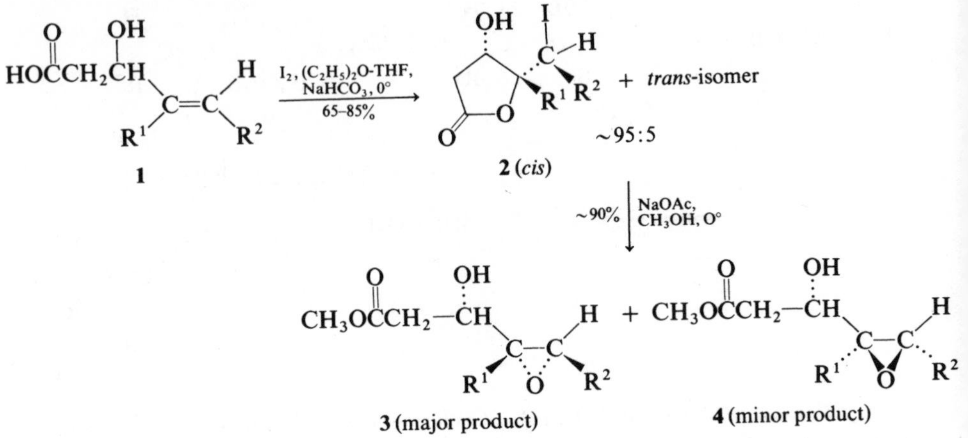

$HOCCH_2CH(OH)C(R^1)=CHR^2$ (1) $\xrightarrow[65\text{–}85\%]{I_2, (C_2H_5)_2O\text{-THF}, NaHCO_3, 0^\circ}$ 2 (*cis*) + *trans*-isomer, ~95:5

2 $\xrightarrow[\sim 90\%]{NaOAc, CH_3OH, 0^\circ}$ 3 (major product) + 4 (minor product)

Iodolactonization of γ, δ-unsaturated alcohols results in preferential formation of *trans*-2, 5-disubstituted tetrahydrofuranes. However, the corresponding benzyl ethers cyclize preferentially to the *cis*-isomers. The alkyl group must be bulky enough to exert a steric effect, but not to prevent cyclization. Substituted benzyl ethers are particularly useful. Examples of this steric control are illustrated for the preparation of *trans*- and *cis*-linalyl oxide (equations I and II).[10]

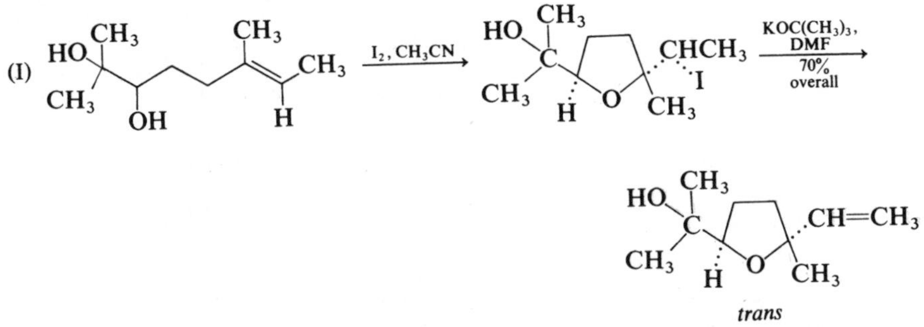

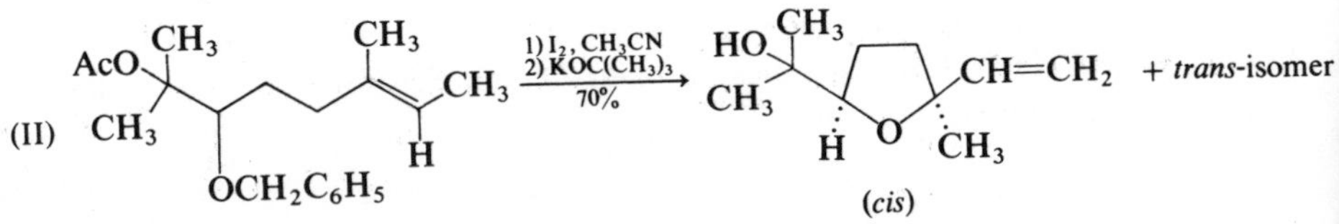

(I) $\xrightarrow{I_2, CH_3CN}$ $\xrightarrow[70\%\ \text{overall}]{KOC(CH_3)_3, DMF}$ *trans*

(II) $\xrightarrow[70\%]{1)\ I_2, CH_3CN;\ 2)\ KOC(CH_3)_3}$ (*cis*) + *trans*-isomer

Iodocarbamation.[11] In connection with a synthesis of the chiral β-amino acid negomycin (**4**, an antibiotic with high activity against gram-negative bacteria but with low toxicity), Japanese chemists have developed a novel 1,3-asymmetric induction by iodocarbamation of a chiral acyclic homoallylamine (**1**) prepared in several steps from (S)-β-aminoglutarate. Treatment of **1a** with iodine affords the cyclic carbamate **2** in excellent yield, but the *trans/cis* ratio is 3:7. However, this ratio is reversed by use of a bulky group on nitrogen. The product was converted to **4** by known reactions.

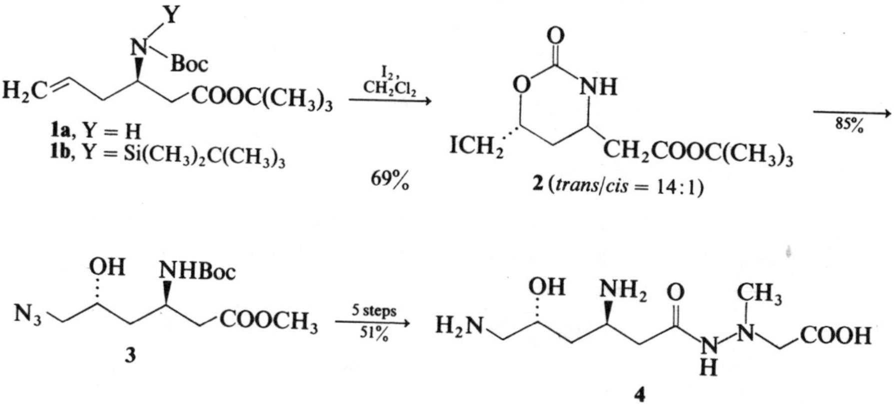

1a, Y = H
1b, Y = $Si(CH_3)_2C(CH_3)_3$

I_2, CH_2Cl_2, 69%

2 (*trans/cis* = 14:1)

85%

5 steps, 51%

3 **4**

Iodoamination. 1,3-Oxazolines (**2**) are obtained on treatment of allylic trichloroacetimidates (**1**) with I_2–pyridine. These products are hydrolyzed to iodoamino alcohols (**3**) in quantitative yield by HCl in CH_3OH or, more slowly, by aqueous CH_3OH.

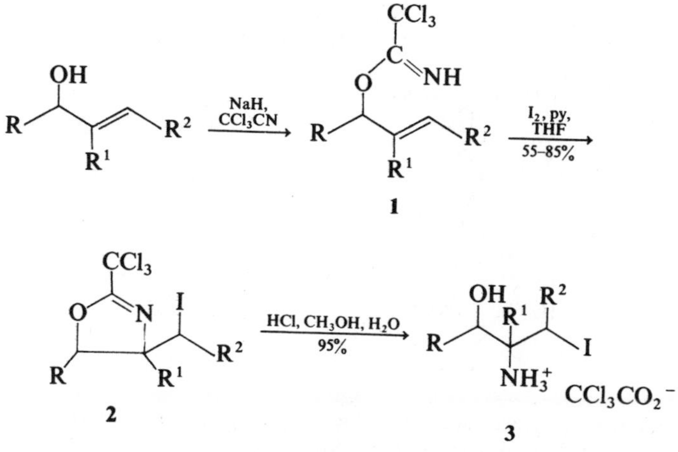

NaH, CCl_3CN

I_2, py, THF, 55–85%

HCl, CH_3OH, H_2O, 95%

1 **2** **3**

The same sequence can be applied to homoallylic alcohols (equation I).[12]

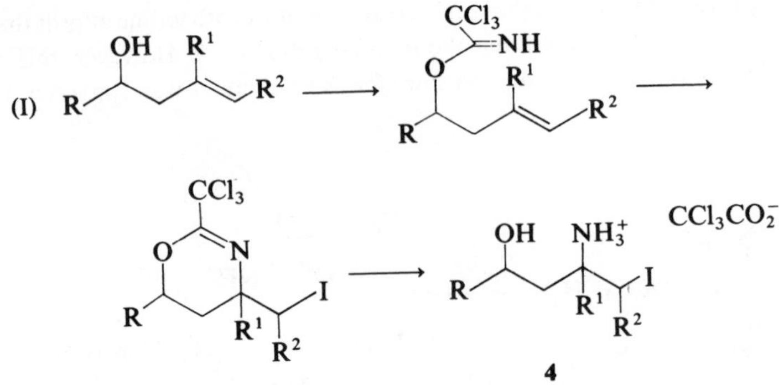

(I)

4

The resulting salts, **3** and **4**, can be hydrolyzed to 2-amino-1,3-diols and 2-amino-1,4-diols, respectively, by treatment with acetate ions supported on Amberlyst A 26 (equation II).[13]

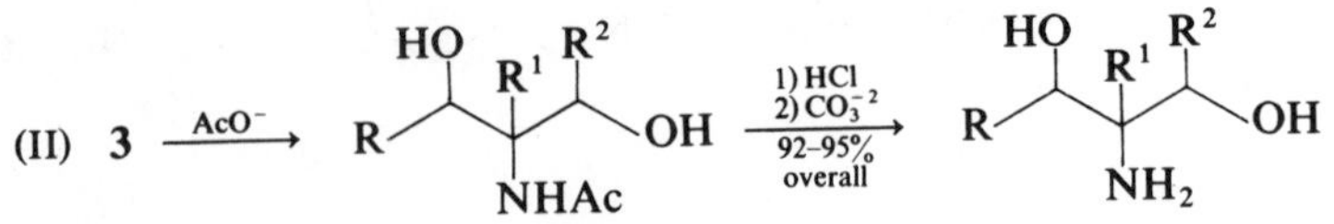

(II) **3** $\xrightarrow{AcO^-}$ … $\xrightarrow[\text{92–95\% overall}]{\text{1) HCl, 2) } CO_3^{-2}}$ …

Cyclization of unsaturated benzyl carbamates. Iodine induces cyclization of unsaturated benzyl carbamates. Benzeneselenenyl chloride is less effective for this reaction.[14]

Examples:

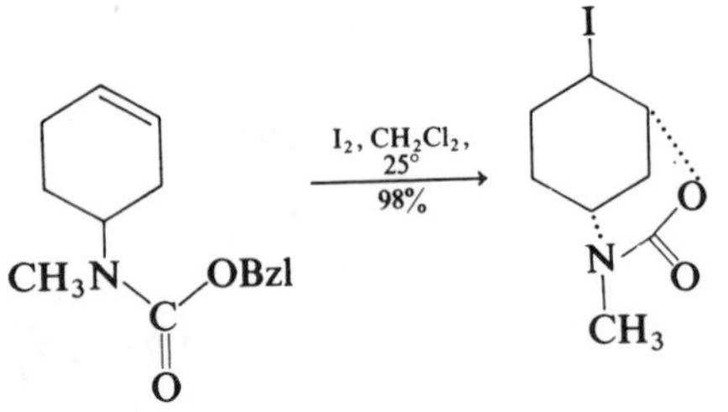

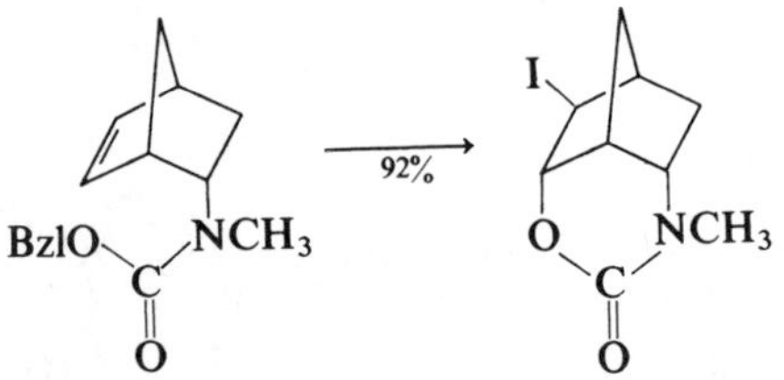

$\xrightarrow[98\%]{I_2, CH_2Cl_2, 25°}$

$\xrightarrow{92\%}$

[1] G. Zweifel, H. Arzoumanian, and C. C. Whitney, *Am. Soc.*, **89**, 3652 (1967).
[2] S. U. Kulkarni, D. Basavaiah, and H. C. Brown, *J. Organometal. Chem.*, **225**, C1 (1982).
[3] H. C. Brown, D. Basavaiah, S. U. Kulkarni, *J. Org.*, **47**, 171 (1982).
[4] H. C. Brown and D. Basavaiah, *J. Org.*, **47**, 5407 (1982).

[5] S. U. Kulkarni, D. Basavaiah, M. Zaidlewicz, and H. C. Brown, *Organometallics*, **1**, 212 (1982).
[6] H. C. Brown and D. Basavaiah, *J. Org.*, **47**, 3806 (1982).
[7] H. C. Brown, D. Basavaiah, and S. U. Kulkarni, *ibid.*, **47**, 3808 (1982).
[8] A. Bongini, G. Cardillo, M. Orena, G. Porzi, and S. Sandri, *J. Org.*, **47**, 4626 (1982).
[9] A. R. Chamberlin, M. Dezube, and P. Dussault, *Tetrahedron Letters*, **22**, 4611 (1981).
[10] S. D. Rychnovsky and P. A. Bartlett, *Am. Soc.*, **103**, 3963 (1981).
[11] Y.-F. Wang, T. Izawa, S. Kobayashi, and M. Ohno, *Am. Soc.*, **104**, 6465 (1982).
[12] G. Cardillo, M. Orena, G. Porzi, and S. Sandri, *J.C.S. Chem. Comm.*, 1308 (1982).
[13] *Idem, ibid.*, 1309 (1982).
[14] S. Takano and S. Hatakeyama, *Heterocycles*, **19**, 1243 (1982).

Iodine–Aluminum chloride–Copper(II) chloride.

Aryl iodides.[1] Aryl iodides can be prepared by reaction of arenes with I_2 and a mixture of $AlCl_3$ and $CuCl_2$. Yields are moderate to good, but the reaction fails with ethyl benzoate, acetophenone, and *o*-nitroanisole. The reagents are used in the ratio $I_2:AlCl_3:CuCl_2 = 0.5:1:1$.

[1] T. Sugita, M. Idei, Y. Ishibashi, and Y. Takegami, *Chem. Letters*, 1418 (1982).

Iodine–Copper(II) acetate (10, 211).

α-Iodination of phenols.[1] Estrogens are selectively iodinated at C_2 by I_2–$Cu(OAc)_2$ in HOAc. This reaction converts estrone into the enol acetate of 2-iodoestrone (90% yield), which is reduced to 2-iodoestradiol by $NaBH_4$ (95% yield).[1]

[1] C. A. Horiuchi and J. Y. Satoh, *J.C.S. Chem. Comm.*, 671 (1982).

Iodine–Mercury(II) oxide.

Oxasteroids.[1] Hydroxy steroids are converted by this system into hypoiodites, which on irradiation are cleaved to formyl esters. These products are cyclized to oxasteroids with sodium borohydride.

Examples:

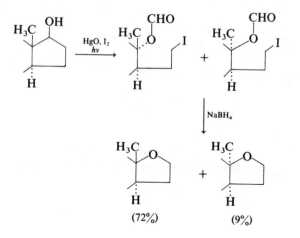

H_3C OH, H $\xrightarrow{HgO,\ I_2,\ h\nu}$ H_3C O–CHO, I, H + H_3C O–CHO, I, H $\xrightarrow{NaBH_4}$ H_3C O, H (72%) + H_3C O, H (9%)

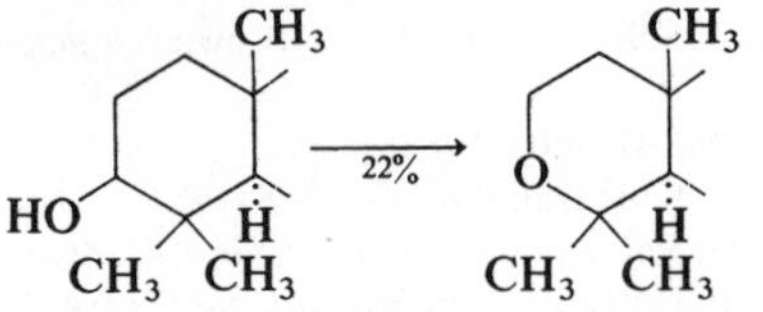

$$\text{(HO-substituted trimethylcyclohexane)} \xrightarrow{22\%} \text{(trimethyltetrahydropyran)}$$

[1] H. Suzinome and S. Yamada, *Chem. Letters*, 1233 (1982).

Iodine–Silver carboxylates, 9, 249.

α-Acyloxy ketones.[1] Reaction of enol silyl ethers with I_2 and AgOCOR followed by desilylation [$(C_2H_5)_3NHF$] provides a convenient synthesis of α-acyloxy ketones in 70–92% yield.

Example:

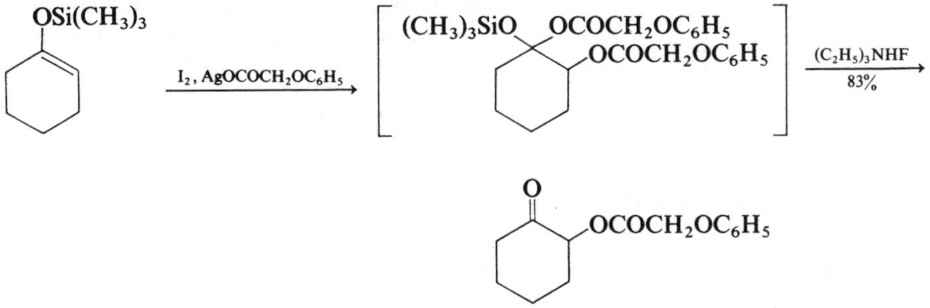

$$\text{1-}OSi(CH_3)_3\text{-cyclohexene} \xrightarrow{I_2,\ AgOCOCH_2OC_6H_5} \left[(CH_3)_3SiO,\ OCOCH_2OC_6H_5,\ OCOCH_2OC_6H_5\text{-cyclohexane} \right] \xrightarrow[83\%]{(C_2H_5)_3NHF} \text{2-}OCOCH_2OC_6H_5\text{-cyclohexanone}$$

[1] G. M. Rubottom, R. C. Mott, and H. D. Juve, Jr., *J. Org.*, **46**, 2717 (1981).

Iodine-Silver nitrate.

Methyl α-arylacetates. These esters have been obtained by oxidative rearrangement of alkyl aryl ketones with thallium(III) nitrate in acidic methanol or trimethyl orthoformate (**4**, 496; **5**, 656; **7**, 362). A new method, which avoids the toxic TTN, is based on the Woodward version[1] of the Prévost reaction. Thus, treatment of the ketone with iodine (or bromine) and silver nitrate (2 equiv.) in refluxing methanol containing trimethyl orthoformate results in methyl α-arylacetates in ~90% yield from simple substrates. Yields are lowered by electron-withdrawing substituents on the aromatic group and by α-branching in the alkyl group.[2]

[1] R. B. Woodward and F. V. Brutcher, *Am. Soc.*, **80**, 209 (1958).
[2] S. D. Higgins and C. B. Thomas, *J.C.S. Perkin I*, 235 (1982).

Iodine Monochloride, 1, 502; **10**, 212.

Alkyl iodides (**10**, 212). The original procedure for preparation of alkyl iodides from triallyboranes (equation I) is limited to a yield of 66% based on the alkene. The

$$\text{(I)}\quad R_3B + 2ICl \longrightarrow 2RI$$

yield is improved by use of alkyldicyclohexylboranes (equation II).[1]

(II) 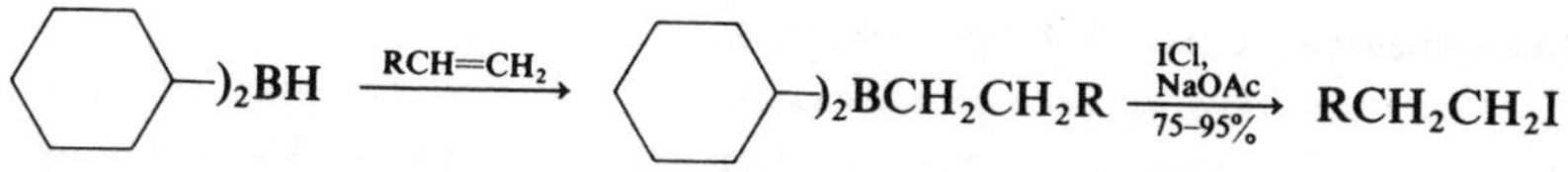 $(C_6H_{11})_2BH \xrightarrow{RCH{=}CH_2} (C_6H_{11})_2BCH_2CH_2R \xrightarrow[75\text{–}95\%]{ICl,\ NaOAc} RCH_2CH_2I$

[1] E. E. Gooch and G. W. Kabalka, *Syn. Comm.*, **11**, 521 (1981).

Iodonium di-*sym*-collidine perchlorate (1), 10, 212–213.

cis-*Oxyamination* (**10**, 213). In the first use of **1** for *cis*-oxyamination of an allylic amine, the double bond was an exocyclic methylene group. The same methodology can also convert an internal allylic amine into a *cis*-β-hydroxy amine, as illustrated in a synthesis of holacosamine (**2**), a component of some glycosteroids.[1]

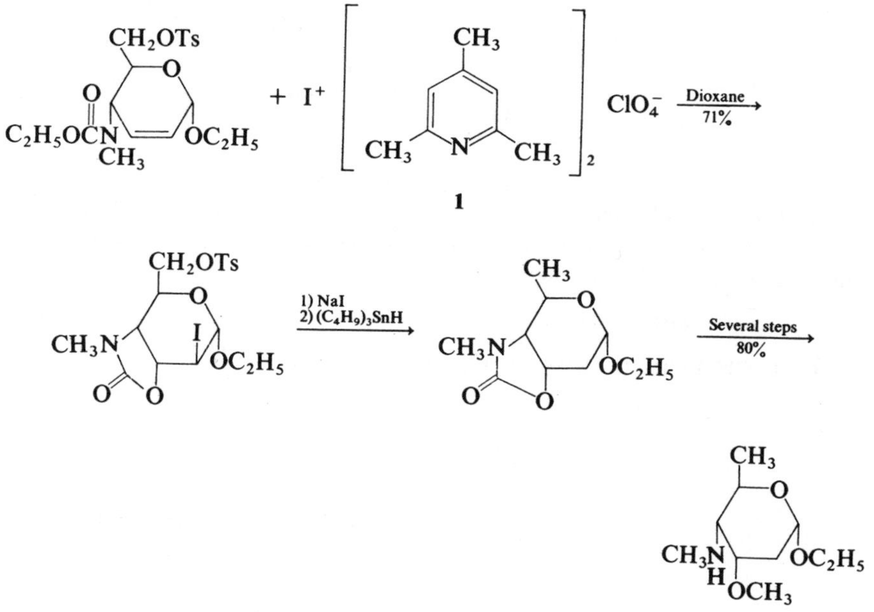

CH_2OTs, $C_2H_5OC(O)N(CH_3)$, OC_2H_5 + I^+ [collidine]$_2$ ClO_4^- $\xrightarrow[71\%]{Dioxane}$

1

$\xrightarrow{1)\ NaI;\ 2)\ (C_4H_9)_3SnH}$ $\xrightarrow[80\%]{Several\ steps}$ CH_3, CH_3NH, OCH_3, OC_2H_5

2

[1] M. Georges and B. Fraser-Reid, *Tetrahedron Letters*, **22**, 4635 (1981).

Iodophenylbis(triphenylphosphine)palladium, $C_6H_5PdI[P(C_6H_5)_3]_2$ (**1**). Mol. wt. 834.99. Preparation.[1]

Cyanocarbonylation.[2] Aromatic iodides are converted into aroyl nitriles by carbonylation catalyzed by **1** and in the presence of KCN. Isolated yields are about 50–85%.

$$\text{(I) ArI + CO + KCN} \xrightarrow[50\text{–}85\%]{\text{1, THF}} \text{ArCOCN + KI}$$

[1] P. Fitton and E. A. Rick, *J. Organometal. Chem.*, **28**, 287 (1971).
[2] M. Tanaka, *Bull. Chem. Soc. Japan*, **54**, 637 (1981).

Iodosylbenzene, C_6H_5IO, **1**, 507–508; **10**, 213–214.

Steroidal dihydroxyacetone side chains.[1] The C_{17}-acetyl group can be converted into the dihydroxyacetone group as formulated in equation (I). The advantage of this approach is that the Δ^5-3β-OH group is not affected.

(I) [scheme: C_{17}-acetyl (CH₃–CO, H₃C, ···H) →($C_6H_5I{=}O$, KOH, CH_3OH; 67%) CH₂OH–C(OCH₃)₂ (H₃C, ···H) →(1) Ac_2O, Py; 2) 138°; 80%) CH₃O, CH₂OAc enol ether (H₃C) →(1) $ClC_6H_4CO_3H$; 2) OH^-; 60%) CH₂OH–C=O (H₃C, ···OH)]

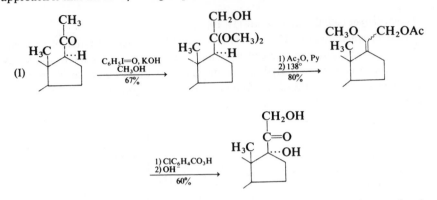

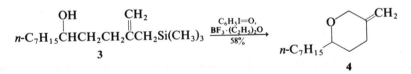

β-Methylene cyclic ethers.[2] Allylsilanes such as **1** and **3** undergo intramolecular cyclization to 5- or 6-membered β-methylene cyclic ethers in the presence of iodosylbenzene and BF_3 etherate.

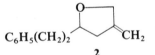

$$\underset{\mathbf{1}}{C_6H_5(CH_2)_2CH(OH)CH_2C(=CH_2)CH_2Si(CH_3)_3} + C_6H_5I{=}O \xrightarrow[68\%]{BF_3\cdot(C_2H_5)_2O} \mathbf{2}$$

2: 2-$C_6H_5(CH_2)_2$-4-methylenetetrahydrofuran ($C_6H_5(CH_2)_2$, O, CH_2)

$$\underset{\mathbf{3}}{n\text{-}C_7H_{15}CH(OH)CH_2CH_2C(=CH_2)CH_2Si(CH_3)_3} \xrightarrow[58\%]{C_6H_5I{=}O,\ BF_3\cdot(C_2H_5)_2O} \mathbf{4}$$

4: 2-n-C_7H_{15}-4-methylenetetrahydropyran (n-C_7H_{15}, O, CH_2)

[1] R. M. Moriarty, L. S. John, and P. C. Du, *J.C.S. Chem. Comm.*, 641 (1981).
[2] M. Ochiai, E. Fujita, M. Arimoto, and H. Yamaguchi, *ibid.*, 1108 (1982).

Iodosylbenzene–Ruthenium catalysts.

Oxidation. In the presence of $Cl_2Ru[P(C_6H_5)_3]_3$, iodosylbenzene oxidizes alcohols to carbonyl compounds.[1] The oxidation is similar to oxidation with N-methylmorpholine, which is also catalyzed by Ru(II) (**7**, 244). Thus oxidation of primary alcohols can proceed to carboxylic acids, but stops at the aldehyde if

$C_6H_5I(OAc)_2$ is used. Oxidation of unsaturated alcohols can be complex. With the exception of RuO_2, almost any ruthenium complex is effective.[1]

In conjunction with several ruthenium catalysts, C_6H_5IO oxidizes internal alkynes to α-diketones in 65–85% yield and terminal alkynes to carboxylic acids (70–80% yield).[2]

C_6H_5IO-Ru(II) also oxidizes 1-alkynyl ethers and amines to the corresponding α-keto esters and amides in fair to high yield.[3]

Examples:

$$C_6H_5C{\equiv}COC_2H_5 \xrightarrow[70\%]{C_6H_5IO,\ RuCl_2[P(C_6H_5)_3]_3} C_6H_5\overset{O}{\overset{\|}{C}}COOC_2H_5$$

$$C_6H_5C{\equiv}C{-}N(CH_3)_2 \xrightarrow[84\%]{} C_6H_5\overset{O}{\overset{\|}{C}}CON(CH_3)_2$$

The same oxidations can also be conducted in comparable yield with the catalytic RuO_4 procedure of Sharpless (this volume).

[1] P. Müller and J. Godoy, *Tetrahedron Letters*, **22**, 2361 (1981).
[2] P. Müller and J. Godoy, *Helv.*, **64**, 2531 (1981).
[3] P. Müller and J. Godoy, *Tetrahedron Letters*, **23**, 3661 (1982).

Iodotrimethylsilane, 8, 261–263; **9**, 251–252; **10**, 216–219.

Nazarov-type cyclization of α,α′-dienones.[1] The cyclization of α,α′-dienones of type **1** to annelated cyclopentenones (**2**) is effected in yields of 10–20% with the usual Lewis acid catalysts: $SnCl_4$, CF_3COOH, or BF_3 etherate. However, reaction of **1** with 2 equiv. of $ISi(CH_3)_3$ effects this cyclization in 48% yield. However $SnCl_4$ is more satisfactory for cyclization to ring-fused cyclopentenoid systems such as **4**.

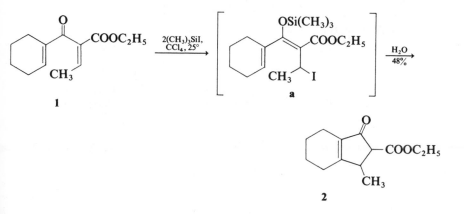

O; $COOC_2H_5$; CH_3; **1**; $2(CH_3)_3SiI$, CCl_4, 25°; $OSi(CH_3)_3$; $COOC_2H_5$; CH_3; I; **a**; H_2O, 48%; O; $COOC_2H_5$; CH_3; **2**

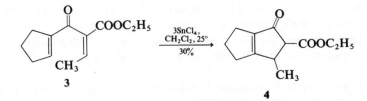

$\xrightarrow[30\%]{3SnCl_4,\ CH_2Cl_2,\ 25°}$

3 COOC₂H₅ CH₃ → 4

Spirocyclization of amines. Two new syntheses of desamylhistrionicotoxin (**1**) involve as the key reactions a Michael addition catalyzed by $(CH_3)_3SiI$ (equation I) and a spirocyclization of an allylic alcohol amine catalyzed by $(CH_3)_3SiI$ (equation II).[2]

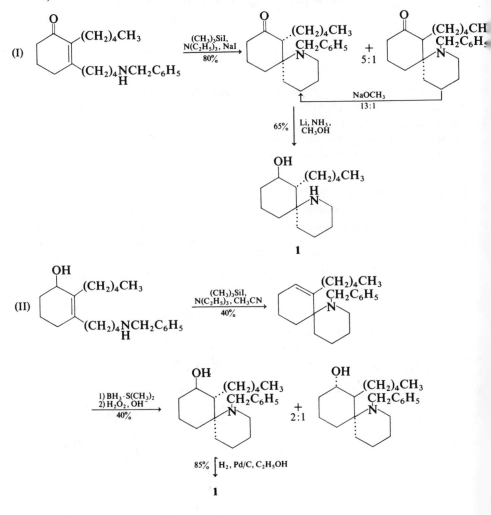

(I) $\xrightarrow[80\%]{(CH_3)_3SiI,\ N(C_2H_5)_3,\ NaI}$ 5:1; NaOCH₃ 13:1; 65% Li, NH₃, CH₃OH → **1**

(II) $\xrightarrow[40\%]{(CH_3)_3SiI,\ N(C_2H_5)_3,\ CH_3CN}$

$\xrightarrow[40\%]{1)\ BH_3\cdot S(CH_3)_2\ \ 2)\ H_2O_2,\ OH^-}$ 2:1

85% H_2, Pd/C, C_2H_5OH → **1**

Reductive rearrangement of bicyclo [4.2.0] octane-2,5-diones.[3] The photochemical cycloadduct (**1**) of 2-cyclohexene-1,4-dione with ethylene on treatment with $ISi(CH_3)_3$ (2 equiv.) rearranges to bicyclo [3.3.0] oct-1(5)-ene-2-one (**2**) (equation I). Mixtures are obtained from derivatives of **1** substituted in the four-membered ring.

(I) **1** $\xrightarrow[96\%]{2ISi(CH_3)_3}$ **2** $+ I_2 + [(CH_3)_3Si]_2O$

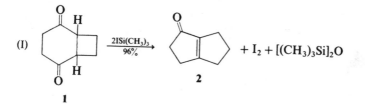

γ-Butyrolactones.[4] γ-Butyrolactones can be prepared from alkenes by addition of ethyl diazoacetate followed by cleavage of the cyclopropane ring and then ring closure. Iodotrimethylsilane is used to effect the ring cleavage (**9**, 252).

Example:

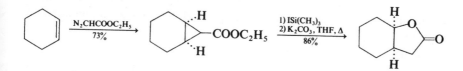

$\xrightarrow[73\%]{N_2CHCOOC_2H_5}$ $COOC_2H_5$ $\xrightarrow[86\%]{1)\ ISi(CH_3)_3\quad 2)\ K_2CO_3,\ THF,\ \Delta}$

C_3-Iodomethylcephalosporins. C_3-Acetoxymethyl- or C_3-carbamoylmethylcephalosporins (**1**) are converted to C_3-iodomethylcephalosporins (**2**) on reaction with $ISi(CH_3)_3$ in CH_2Cl_2 at 20°. The ester group in **1** is not cleaved if it is benzyl or *t*-butyl, but *p*-methoxybenzyl or benzhydryl esters are cleaved faster than the allylic acetate group. This reaction is observed with both Δ^2- and Δ^3-cephems.[5]

$\xrightarrow[\sim 95\%]{ISi(CH_3)_3}$

1, $R^1 = CH_2C_6H_5, C(CH_3)_3$
$R^2 = CH_3, NH_2$

2

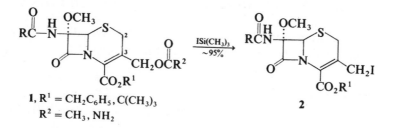

Methyl enol ethers. Reaction of $ISi(CH_3)_3$ and $HN[Si(CH_3)_3]_2$ with dimethyl acetals or ketals in CH_2Cl_2 or $CHCl_3$ results in methyl enol ethers.[6] Cyclopropylcarbinyl methyl ketals under these conditions undergo ring opening to give ω-iodo methyl enol ethers.

Examples:

$$C_6H_5C(OCH_3)_2CH_3 \xrightarrow{ISi(CH_3)_3} \left[C_6H_5(CH_3)C(OCH_3)\overset{+}{O}(CH_3)Si(CH_3)_3 \right] I^- \xrightarrow[81\%]{HN[Si(CH_3)_3]_2} C_6H_5(CH_3O)C{=}CH_2$$

$$\text{1-methyl-2,2-dimethoxycyclohexane} \xrightarrow{83\%} \text{1-methoxy-6-methylcyclohexene} + \text{1-methoxy-2-methylcyclohexene} \quad (67:33)$$

$$\text{cyclopropyl-}C(C_6H_5)(OCH_2)_2 \xrightarrow{65\%} H(I(CH_3)_2)C{=}C(OCH_3)(C_6H_5) + H(I(CH_2)_2)C{=}C(C_6H_5)(OCH_3) \quad (43:57)$$

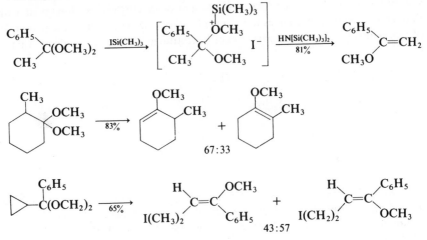

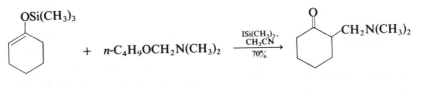

α-Aminomethyl ketones. Iodotrimethylsilane (or trimethylsilyl triflate) promotes reaction of enol silyl ethers with aminomethyl ethers to give α-aminomethyl ketones. Presumably, the effective reagent is an oxonium salt formed from the aminomethyl ether and the catalyst.[7]

Examples:

$$\text{1-}OSi(CH_3)_3\text{-cyclohexene} + n\text{-}C_4H_9OCH_2N(CH_3)_2 \xrightarrow[70\%]{ISi(CH_3)_3,\ CH_3CN} \text{2-}CH_2N(CH_3)_2\text{-cyclohexanone}$$

$$C_6H_5C(OSi(CH_3)_3){=}CH_2 + CH_3OCH_2N(C_2H_5)_2 \xrightarrow{36\%} C_6H_5\overset{O}{\overset{\|}{C}}{-}CH_2CH_2N(C_2H_5)_2$$

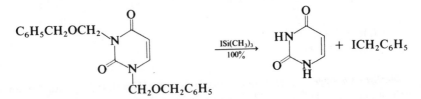

NH-Protection of uracils (**10**, 248). Benzyloxymethyl derivatives of uracils are removed quantitatively by reaction with $ISi(CH_3)_3$ in refluxing $CHCl_3$.[8]

$$\text{1,3-bis}(C_6H_5CH_2OCH_2)\text{uracil} \xrightarrow[100\%]{ISi(CH_3)_3} \text{uracil} + ICH_2C_6H_5$$

Reduction of sulfonyl halides to disulfides. Two laboratories[9, 10] have reported the reduction of sulfonyl chlorides to disulfides by $ISi(CH_3)_3$ or $ClSi(CH_3)_3$-NaI in yields of 80–100%. The actual mechanism of this reduction is evidently more complex than that originally assumed since thiolsulfonates, RSO_2SR, have been identified as intermediates.

Review (156 references).[11]

[1] J. P. Marino and R. J. Linderman, *J. Org.*, **46**, 3696 (1981).
[2] S. A. Godleski and D. J. Heacock, *ibid.*, **47**, 4820 (1982).
[3] K. Sasaki, T. Kushida, M. Iyoda, and M. Oda, *Tetrahedron Letters*, **23**, 2117 (1982).
[4] S. P. Brown, B. S. Bal, and H. W. Pinnick, *ibid.*, **22**, 4891 (1981).
[5] R. Bonjouklian and M. L. Phillips, *ibid.*, **22**, 3915 (1981).
[6] R. D. Miller and D. R. McKean, *ibid.*, **23**, 323 (1982).
[7] A. Hosomi, S. Iijima, and H. Sakurai, *ibid.*, **23**, 547 (1982).
[8] N. G. Kundu, *Syn. Comm.*, **11**, 787 (1981).
[9] G. A. Olah, S. C. Narang, L. D. Field, and G. F. Salem, *J. Org.*, **45**, 4792 (1980).
[10] P. Kiełbasinski, J. Drabowicz, and M. M. Rołajczyk, *ibid.*, **47**, 4806 (1982).
[11] G. A. Olah and S. C. Narang, *Tetrahedron Rpt.*, **38**, 2225 (1982).

Iodylbenzene (Iodoxybenzene), $C_6H_5IO_2$, **1**, 511. The reagent is prepared conveniently by oxidation of iodobenzene with NaOCl (m.p. 250–253°, 75% yield). It deflagrates at 230°.

Dehydrogenation of* 3-*keto steroids. The dehydrogenation of 3-keto steroids to 1,4-diene-3-ones with benzeneseleninic anhydride (**8**, 31) can be carried out in comparable yield by use of a process in which the benzeneseleninic anhydride is used in catalytic amounts and is continuously regenerated from diphenyl diselenide by oxidation with iodylbenzene. In practice, *m*-iodylbenzoic acid is a more convenient reagent, since *m*-iodobenzoic acid is easily recovered. 12-Keto and 12-hydroxy steroids are oxidized by the catalytic system to $\Delta^{9(11)}$-12-keto steroids in high yield. In fact, methyl desoxycholate (**1**) can be oxidized in this way directly to the trienedione **2** in 64% yield.[1]

OH $C_4H_8COOCH_3$ CH_3 CH_3 HO H **1** —64%→ O $C_4H_8COOCH_3$ CH_3 CH_3 O **2**

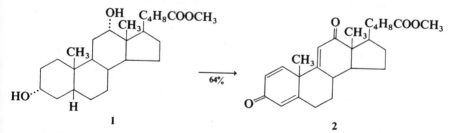

Lewis acid catalyzed oxidations.[2] Oxidations with $C_6H_5IO_2$ alone generally proceed rather slowly at room temperature, but are strongly catalyzed by various Lewis acids. Thus, in the presence of acetic acid, benzylic alcohols are oxidized to aldehydes; α-naphthol can be oxidized to the *o*-quinone under these conditions. Glycol cleavage occurs efficiently with trichloroacetic acid catalysis.

Thioacetals are oxidized at room temperature to the corresponding carbonyl compound by the reagent and a catalytic amount of *p*-toluenesulfonic acid. Oxidation of sulfides to sulfoxides can be effected with several catalysts (acetic acid, trichloroacetic anhydride).

[1] D. H. R. Barton, C. R. A. Godfrey, J. W. Morzycki, W. B. Motherwell, and S.V. Ley, *J.C.S. Perkin I*, 1947 (1982).
[2] D. H. R. Barton, C. R. A. Godfrey, J. W. Morzycki, W. B. Motherwell, and A. A Stobie, *Tetrahedron Letters*, **23**, 957 (1982).

Ion-exchange resins, 1, 511–517; **2**, 227–228; **4**, 266–267; **5**, 355–356; **6**, 302–304; **7**, 182; **8**, 263–264, **9**, 256–257; **10**, 220–221.

Hydrolysis of amides and hydrazides.[1] Unsubstituted amides ($RCONH_2$) are hydrolyzed to acids in refluxing water when mixed with a 15-fold excess of Amberlyst 15 or Amberlite-IR-120 resins generally in ~90% yield. Esters are formed when an alcohol is used instead of water. Hydrazides are hydrolyzed more rapidly than amides under these conditions.

[1] W. J. Greenlee and E. D. Thorsett, *J. Org.*, **46**, 5351 (1981).

Iron–Graphite, $C_{24}Fe$. This black powder is prepared by addition of $FeCl_3$ to potassium–graphite in THF. It is prepared and used under argon, because of ready oxidation by air.[1]

Debromination. $C_{24}Fe$ effects stereospecific *anti*-debromination of *vic*-dibromides (THF, 70°).[1] It is an effective substitute for $Fe_2(CO)_9$[2] for generation of 2-oxyallyl cations from α,α′-dibromo ketones (**5**, 222–223; **6**, 195–196; **9**, 477–478).

[1] D. Savoia, E. Tagliavini, C. Trombini, and A. Umani-Ronchi, *J. Org.*, **47**, 876 (1982).
[2] R. Noyori, *Acc. Chem. Res.*, **12**, 61 (1979).

Iron(II) sulfate, $FeSO_4$.

Cleavage of oxaziridines.[1] Bicyclic oxaziridines such as **2** are cleaved to keto amides (**3**) by reaction with $FeSO_4 \cdot 7H_2O$ in aqueous ethanol in moderate to high yield (equation I).

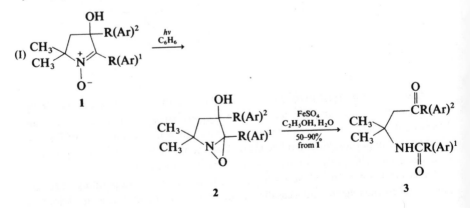

(I) **1** $\xrightarrow{h\nu,\ C_6H_6}$ **2** $\xrightarrow[\text{50–90\% from 1}]{FeSO_4,\ C_2H_5OH,\ H_2O}$ **3**

OH, $R(Ar)^2$, CH_3, CH_3, $\overset{+}{N}$, $R(Ar)^1$, O^- — **1**

OH, $R(Ar)^2$, $R(Ar)^1$, CH_3, CH_3, N, O — **2**

O, $CR(Ar)^2$, CH_3, CH_3, $NHCR(Ar)^1$, O — **3**

This reaction has been applied to tricyclic oxaziridines to obtain monocyclic lactams (**6**) by a three-membered ring expansion of cyclic ketones **4** (equation II).

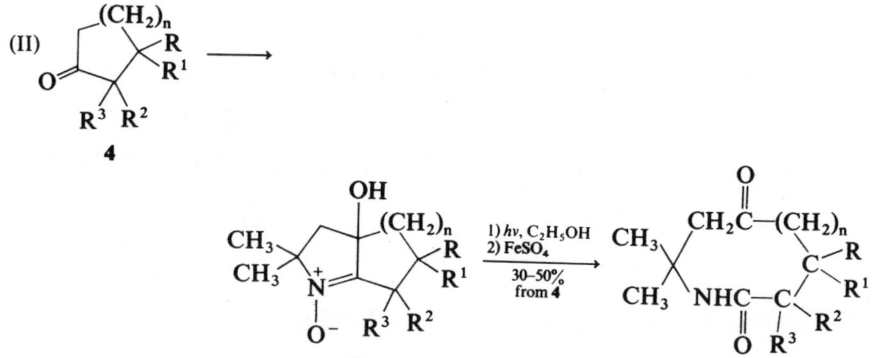

(II) **4** → **5** (1) $h\nu$, C_2H_5OH; 2) $FeSO_4$; 30–50% from **4**) → **6**

[1] D. S. C. Black and L. M. Johnstone, *Angew. Chem. Int. Ed.*, **20**, 669, 670 (1981).

2,3-O-Isopropylidene-2,3-dihydroxy-1,4-bis(diphenylphosphino)butane (DIOP), **4**, 273; **5**, 360–361; **9**, 259–260.

γ-and δ-Lactones.[1] Optically active γ- and δ-lactones can be prepared by hydrogenation of prochiral cyclic anhydrides with $Ru_2Cl_4[(-)\text{-DIOP}]_3$ as catalyst.

Example:

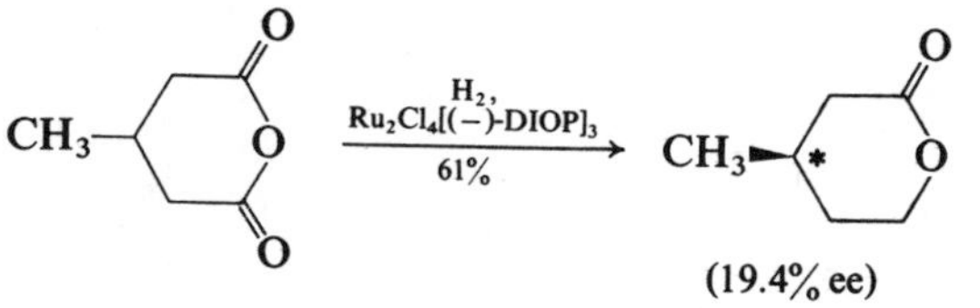

H_2, $Ru_2Cl_4[(-)\text{-DIOP}]_3$, 61% (19.4% ee)

[1] K. Osakada, M. Obana, T. Ikariya, M. Saburi, and S. Yoshikawa, and S. Yoshikawa, *Tetrahedron Letters*, **22**, 4297 (1981).

Isopropyl isocyanate, $(CH_3)_2CHN{=}C{=}O$ (1). Mol. wt. 85.11, b.p. 74–75°. Supplier: Aldrich.

Chromatographic separation of enantiomers.[1] The carbamate, ureido, and amide derivatives obtained without racemization from enantiomeric amines, alcohols, and carboxylic acids, respectively (equations I–III), with this isocyanate are stable for months and are suitable for gas chromatographic separation using a polymeric chiral stationary phase (derived, for example, from L-valine-(S)-α-phenylethylamide). This methodology permits separation of chiral α- and β-hydroxy acids and also N-methylamino acids.

$$\text{(I)}\quad R^1\text{—}\underset{}{\overset{R^2}{\overset{|}{C}}}HNH_2 \xrightarrow[CH_2Cl_2]{\mathbf{1},\ 20°} R^1\text{—}\overset{R^2}{\overset{|}{C}}H\text{—}NH\overset{O}{\overset{\|}{C}}NHCH(CH_3)_2$$

$$\text{(II)}\quad R^1\text{—}\overset{R^2}{\overset{|}{C}}HOH \xrightarrow{\mathbf{1},\ 100°} R^1\text{—}\overset{R^2}{\overset{|}{C}}HO\overset{O}{\overset{\|}{C}}\text{—}NHCH(CH_3)_2$$

$$\text{(III)}\quad CH_3NH\overset{R^1}{\overset{|}{C}}HCOOH \xrightarrow{\mathbf{1},\ 100°} (CH_3)_2CHNH\overset{O}{\overset{\|}{C}}\text{—}\underset{CH_3}{\underset{|}{N}}\text{—}\overset{R^1}{\overset{|}{C}}H\overset{O}{\overset{\|}{C}}\text{—}NHCH(CH_3)_2$$

[1] I. Benecke and W. A. König, *Angew. Chem. Int. Ed.*, **21**, 609 (1982).

K

Ketene *t*-butyldimethylsilyl methyl acetal (1).

Preparation:

$$CH_3COOCH_3 \xrightarrow[72\%]{1)\,LDA,\,HMPT;\ 2)\,ClSiMe_2\text{-}t\text{-}Bu} CH_2{=}C(OCH_3)(OSiMe_2\text{-}t\text{-}Bu)$$

1 (b.p. 76°/24 mm)

Michael addition.[1] This ketene silyl acetal undergoes Michael addition to α,β-enones in acetonitrile in the absence of a Lewis acid to afford the corresponding O-silylated Michael adduct in high yield. These O-silyl enolates undergo site-specific electrophilic substitution. This sequence was used for vicinal dialkylation of cyclohexanone (equation I) and of cyclopentanone. It is particularly useful for synthesis of methyl jasmonate and related compounds from cyclopentenone.

(I) Cyclohexenone $\xrightarrow[90\%]{1,\ CH_3CN}$ **2** (1-OSi(CH$_3$)$_2$C(CH$_3$)$_3$-cyclohexene, 3-CH_2COOCH_3) $\xrightarrow[51\%]{C_6H_5SCH_2Cl,\ TiCl_4,\ CH_2Cl_2}$ **3** (2-$CH_2SC_6H_5$-3-CH_2COOCH_3-cyclohexanone) $\xrightarrow[34\%\ \text{from}\ 2]{\text{Raney Ni}}$ 2-CH_3-3-CH_2COOCH_3-cyclohexanone

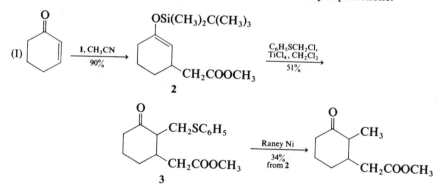

Silylation.[2] In the presence of TsOH, **1** silylates a variety of substrates: alcohols, acids, mercaptans, phenols, and imides. Isolated yields are 70–85%.

[1] K. Yasuyuki, J. Segawa, J. Haruta, H. Yasuda, and Y. Tamura, *J.C.S. Perkin I*, 1099 (1982).
[2] Y. Kita, J. Haruta, T. Fujii, J. Segawa, and Y. Tamura, *Synthesis*, 451 (1981).

Ketene dimethyl acetal (1,1-Dimethoxyethylene), 9, 262–264; **10,** 226.

Salicylate annelation.[1] A new method involves conversion of a cyclohexanone into the annelated 3-carbomethoxy-2-pyrone by reaction of the enolate with dimethyl methoxymethylenemalonate[2] to form a 3-carbomethoxy-2-pyrone. This product undergoes a facile Diels-Alder reaction with 1,1-dimethoxyethylene with loss of CO_2 and CH_3OH to give a salicylate.

Example:

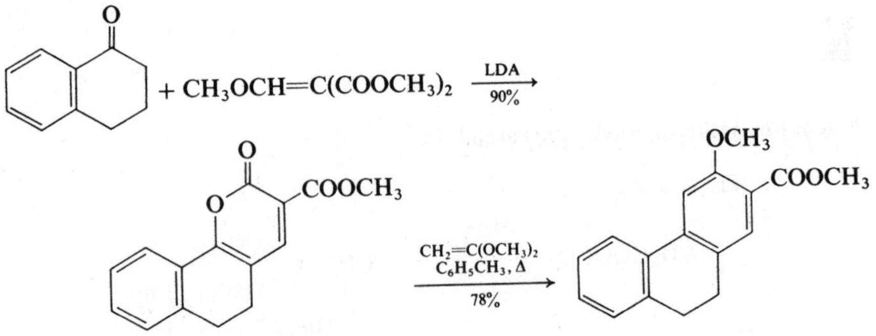

$$\text{1-tetralone} + CH_3OCH{=}C(COOCH_3)_2 \xrightarrow[90\%]{LDA}$$

$$\xrightarrow[78\%]{CH_2{=}C(OCH_3)_2,\ C_6H_5CH_3,\ \Delta}$$

COOCH$_3$ OCH$_3$ COOCH$_3$

This reaction has been used in a regiospecific synthesis of juncusol (**1**) (a natural dihydrophenanthrene), as outlined in Scheme I.[3]

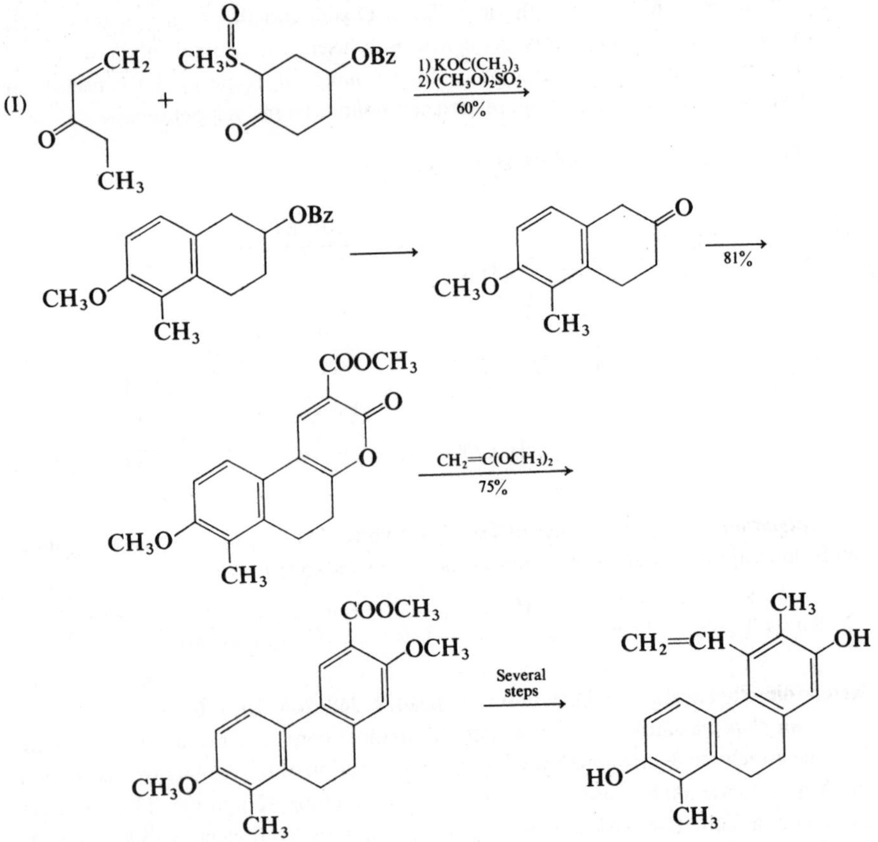

(I) CH$_2$ CH$_3$S OBz + O O CH$_3$

$$\xrightarrow[60\%]{1)\ KOC(CH_3)_3\quad 2)\ (CH_3O)_2SO_2}$$

OBz CH$_3$O CH$_3$ ⟶ O CH$_3$O CH$_3$ $\xrightarrow[81\%]{}$

COOCH$_3$ O O CH$_3$O CH$_3$

$$\xrightarrow[75\%]{CH_2{=}C(OCH_3)_2}$$

COOCH$_3$ OCH$_3$ CH$_3$O CH$_3$ $\xrightarrow{\text{Several steps}}$ CH$_3$ CH$_2$=CH OH HO CH$_3$

1

Scheme I

[1] D. L. Boger and M. D. Mullican, *Tetrahedron Letters*, **23**, 4551 (1982).
[2] R. C. Fuson, W. E. Parham, and L. J. Reed, *J. Org.*, **11**, 194 (1946).
[3] D. L. Boger and M. D. Mullican, *Tetrahedron Letters*, **23**, 4555 (1982).

Ketenylidenetriphenylphosphorane (1).

Preparation[1]:

$$(C_6H_5)_3P{=}CHCOOCH_3 \xrightarrow{NaN[Si(CH_3)_3]_2} (C_6H_5)_3P{=}C{=}C{=}O + NaOCH_3 + HN[Si(CH_3)_3]_2$$

1 (80%)

Butenolides. 21-Hydroxy-20-keto steroids react with **1** to form cardenolides by an intramolecular Wittig reaction (equation I).[2]

(I) CH_2OH, $C{=}O$, CH_3 $\xrightarrow{\mathbf{1},\ C_6H_6,\ \Delta}$ [O, CH_2, $C{=}O$, $C{=}O$, $HC{=}P(C_6H_5)_3$, CH_3] $\xrightarrow{70\text{–}80\%}$ O, O, CH_3

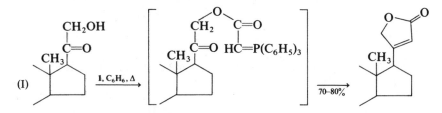

McMurry *et al.*[3] have used the mixed anhydride of trifluoroacetic acid and diethylphosphonoacetic acid, $CF_3COOCOCH_2PO(OC_2H_5)_2$(**2**) for the same reaction (70–80% yield).

In a synthesis of the natural insecticide ajugarin-IV (**4**),[4] the butenolide portion was prepared from the acid chloride **3** in two steps. Reaction with tris(trimethylsilyloxy)ethylene (**8**, 523; **9**, 512)[5] led to a hydroxymethyl ketone, which on reaction with **1** was converted into the butenolide **4**. In this series, **1** was vastly superior to the alternative Wittig-Horner reagent **2**.

COCl, H, CH_3, CH_3, H_3OOC, CH_3, OAc

3

$\xrightarrow{\text{1) } (CH_3)_3SiOCH{=}C[OSi(CH_3)_3]_2 \text{ (56\%)} \quad \text{2) } \mathbf{1} \text{ (86\%)}}$

O, H, CH_3, O, CH_3, CH_3OOC, CH_3, OAc

4

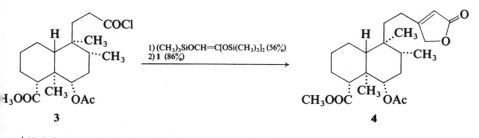

[1] H. J. Bestmann, *Angew. Chem. Int. Ed.*, **16**, 349 (1977).
[2] K. Nickisch, W. Klose, and F. Bohlmann, *Ber.*, **113**, 2038 (1980).
[3] S. F. Donovan, M. A. Avery and J. E. McMurry, *Tetrahedron Letters*, **20**, 3287 (1979).
[4] A. S. Kende, B. Roth, and I. Kubo, *ibid.*, **23**, 1751 (1982).
[5] A. Wissner, *J. Org.*, **44**, 4617 (1979).

α-Ketoketene dithioacetals (1).

Preparation:

$$\mathrm{RCOCH_3} \xrightarrow[]{\substack{1)\ \mathrm{NaH,\ DMSO} \\ 2)\ \mathrm{CS_2} \\ 3)\ \mathrm{CH_3I}}} \mathrm{RCOCH{=}C(SCH_3)_2}$$

1

1,5-*Enediones.* These compounds can be prepared in about 50–100% yield by conjugate addition of the anion of a methyl ketone [$KOC(CH_3)_3$ in THF] to **1** at 25° (equation I).[1]

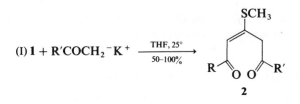

$$\text{(I)}\ \mathbf{1} + \mathrm{R'COCH_2^-K^+} \xrightarrow[50\text{–}100\%]{\mathrm{THF,\ 25^\circ}} \mathrm{RCOCH{=}C(SCH_3)CH_2COR'}$$

2

The products can be converted into pyridines and pyrylium salts, as formulated in equations (II) and (III).

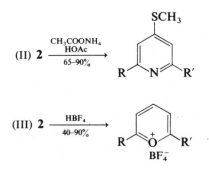

$$\text{(II)}\ \mathbf{2} \xrightarrow[65\text{–}90\%]{\mathrm{CH_3COONH_4},\ \mathrm{HOAc}} \text{2-R-6-R'-4-(SCH}_3\text{)pyridine}$$

$$\text{(III)}\ \mathbf{2} \xrightarrow[40\text{–}90\%]{\mathrm{HBF_4}} \text{2-R-6-R'-pyrylium } \mathrm{BF_4^-}$$

Both of these reactions are advantageously carried out as a one-pot procedure without isolation of **2**.[2]

[1] K. T. Potts, M. J. Cipullo, P. Ralli, and G. Theodoridis, *Am. Soc.*, **103**, 3584 (1981).

[2] *Idem, Ibid.*, **103**, 3585 (1981).

L

Lanthanum nitrate, $La(NO_3)_3 \cdot 6H_2O$. Mol. wt. 433.02, m.p. 40°.

Nitration of phenols.[1] Phenols can be nitrated in a two-phase system (ether–water) by $NaNO_3$(1 equiv.) and HC1 (excess) in the presence of a catalytic amount of several rare earth nitrates in yields generally >80%. The *o/p* ratio can be controlled to some extent by change in the acidity, but *ortho*-nitration generally predominates. Aromatic hydrocarbons are not nitrated under these conditions.

[1] M. Overtani, P. Girard, and H. B. Kagan, *Tetrahedron Letters*, **23**, 4315 (1982).

Lead(IV) acetate–Metal halides.

α-Halo ketones.[1] Enol ethers or esters are converted into α-halo ketones by reaction with lead(IV) acetate (1.5 equiv.) and a metal halide (1 equiv.) in CH_3OH at 0–25°; yields are 50–95%.

[1] S. Motohashi, M. Satomi, Y. Fujimoto, and T. Tatsuno, *Synthesis*, 1021 (1982).

Lead dioxide–Boron trifluoride etherate.

Oxidative hydrolysis of* 1,3-*dithiane derivatives. The oxidative cleavage of 1,3-dithiane derivatives to the parent carbonyl compounds can be effected by reaction with PbO_2 in aqueous THF in the presence of BF_3 etherate or $HC1O_4$ (equation I).[1]

(I) 2-R^1,2-R^2-1,3-dithiane + PbO_2 + H_2O $\xrightarrow[\text{THF}]{BF_3 \cdot (C_2H_5)_2O,}$ R^1COR^2 (80–95%)

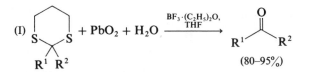

In one case, this reagent was superior to mercuric oxide/boron trifluoride (**3**, 136), ceric ammonium nitrate (**4**, 74), or sulfuryl chloride (**7**, 350).

[1] D. Ghiringhelli, *Synthesis*, 580 (1982).

Levulinic anhydride, 6, 318–319; **10**, 230–231.

N-Levulinyl nucleosides. N-Levulinylcytidine (**1**) can be prepared by reaction of the free nucleoside with levulinic anhydride in refluxing methanol (70% yield), but higher yields (86%) are obtained by reaction of the trisilyl derivative with levulinic acid, DCC, and DMAP, followed by desilylation (F^-). N-Levulination of trisilyladenosine is best effected with levulinic acid and N-ethoxycarbonyl-2-ethoxy-1,2-

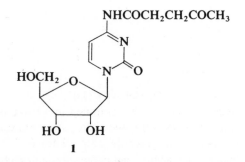

$NHCOCH_2CH_2COCH_3$ ... HOCH₂ ... HO OH

1

dihydroquinoline (EEDQ, **2**, 191). N-Levulinylation of trisilylguanosine is possible, but in low yield, by deprotonation with *n*-butyllithium followed by reaction with the pentafluorophenyl ester of levulinic acid.[1]

The N-levulinyl nucleosides are cleaved rapidly by hydrazine hydrate. They have been used successfully in nucleotide synthesis.

[1] K. K. Ogilvie, M. J. Nemer, G. H. Hakimelahi, Z. A. Proba, and M. Lucas, *Tetrahedron Letters*, **23**, 2615 (1982).

1-Lithiocyclopropyl phenyl sulfide, 5, 372–373; **6**, 319–320; **7**, 190; **9**, 271–272; **10**, 231.

Alkylation. Reaction of this reagent (**1**) with primary alkyl iodides or bromides results in derivatives which are cleaved by either HBr/HOAc or $HgCl_2$ to ketones (equation I). The products can also be desulfurized to substituted cyclopropanes.

(I) 1-(C_6H_5S)-1-Li-cyclopropane (**1**) $\xrightarrow[60\text{–}85\%]{RCH_2X}$ 1-(C_6H_5S)-1-(RCH_2)-cyclopropane (**2**)

2 $\xrightarrow[40\text{–}60\%]{HBr,\ HOAc}$ $C_2H_5\overset{O}{\overset{\|}{C}}CH_2R$

2 $\xrightarrow[50\text{–}77\%]{HgCl_2,\ CH_3CN,\ H_2O}$ $RCH_2\overset{O}{\overset{\|}{C}}CH_2CH_2SC_6H_5$

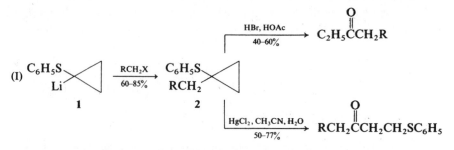

The reagent cleaves epoxides regioselectively to give 1-(β-hydroxy)cyclopropyl phenyl sulfides, which are cleaved by $HgCl_2$ to α,β-enones (equation II).[1]

(II) epoxide with $CH_2CH(CH_3)_2$ side chain $\xrightarrow[87\%]{1}$ 1-(SC_6H_5)-1-[$CH_2CH(OH)CH_2CH(CH_3)_2$]cyclopropane $\xrightarrow[42\%]{HgCl_2}$ $(CH_3)_2CHCH_2CH{=}CH\overset{O}{\overset{\|}{C}}CH_2CH_2SC_6H_5$

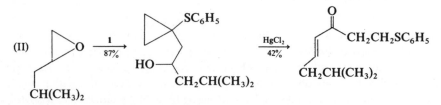

[1] C. L. Bumgardner, J. R. Lever, and S. T. Purrington, *Tetrahedron letters*, **23**, 2379 (1982).

2-Lithio-1,3-dithiane (1), 6, 248; **10**, 231.

Reaction with sugar lactones. The reaction of sugar lactones with **1** provides a preparative route to higher sugars.[1]

Example:

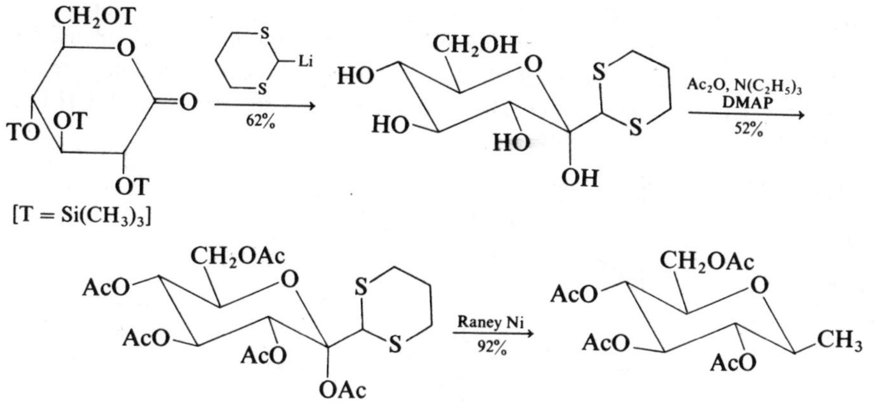

[T = $Si(CH_3)_3$]

[1] D. Horton and W. Priebe, *Carbohydrate Res.*, **94**, 27 (1981).

Lithiomethyl isocyanide, 4, 272; **10**, 231–232.

Furanosesquiterpenes.[1] The furanosesquiterpene petasalbine (**1**) has been synthesized by application of "bis heteroannelation" (equation I).

(I)

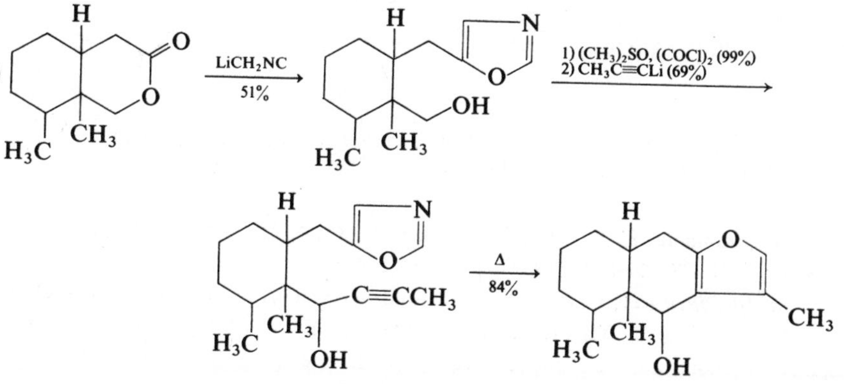

[1] P. A. Jacobi and D. G. Walker, *Am. Soc.*, **103**, 4611 (1981).

α-Lithiovinyltrimethylsilane, $CH_2{=}C(Li)Si(CH_3)_3$ (**1**). The reagent is prepared by reaction of α-bromovinyltrimethylsilane (**8**, 56–57) with *t*-butyllithium at −78° in ether.

Cyclopentene annelation. The known rearrangement of vinyl cyclopropanes to cyclopentenes (*cf.* **7**, 190; **9**, 83, 265) can be used to obtain silyl-substituted cyclopentenes. The precursors, (1-trimethylsilylcyclopropyl)ethylenes (**2**), are usually prepared by addition of **1** to an aldehyde or a ketone followed by cyclopropanation (C_2H_5ZnI, CH_2I_2) of the adduct. The products are then dehydrated by TsOH to **2** (equation I).[1]

(I) R¹, R² ketone —1→ adduct with OH, $Si(CH_3)_3$, CH_2 —C_2H_5ZnI, CH_2I_2→ cyclopropyl carbinol with OH, $Si(CH_3)_3$ —TsOH, 50–75% overall→ **2** (R¹, R², $Si(CH_3)_3$)

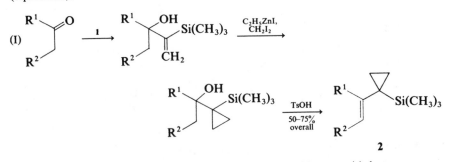

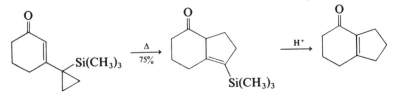

These silyl-substituted vinyl cyclopropanes are stable to acid but rearrange thermally to silyl-substituted cyclopentenes in which the double bond occupies the more substituted position.[2]

Example:

$Si(CH_3)_3$ —Δ, 75%→ $Si(CH_3)_3$ —H^+→

[1] L. A. Paquette, K. A. Horn, and G. J. Wells, *Tetrahedron Letters*, **23**, 259 (1982).
[2] L. A. Paquette, G. J. Wells, K. A. Horn, and T.-H. Yan, *ibid.*, **23**, 263 (1982).

Lithium–Ammonia, 1, 601–603; **2**, 205; **3**, 179–182; **4**, 288–290; **5**, 379–381; **6**, 322–323; **7**, 195; **8**, 282–284; **9**, 273–274; **10**, 234–236.

1α, 3β-Dihydroxy-Δ⁵-steroids. The final step in a synthesis of 1α-hydroxycholesterol (**3**) conceived by Barton *et al.* (**5**, 380–381) involves Li-NH_3 reduction of 1α, 2α-epoxy-4,6-cholestadiene-3-one (**1**) directly to **3** in one operation (~60% yield). Other laboratories have since reported yields of 25–75%. A careful study by chemists at Hoffmann-LaRoche[1] has led to a reproducible procedure giving yields of about 80% in the cholestane, pregnane, and androstane series. The epoxydienone (**1**) is reduced first with a stoichimetric amount of lithium in ammonia to give the 1α-hydroxy-Δ^5-3-one (**2**). This intermediate need not be isolated, but is reduced by alternating treatment with NH_4Cl and 1 equiv. of lithium.

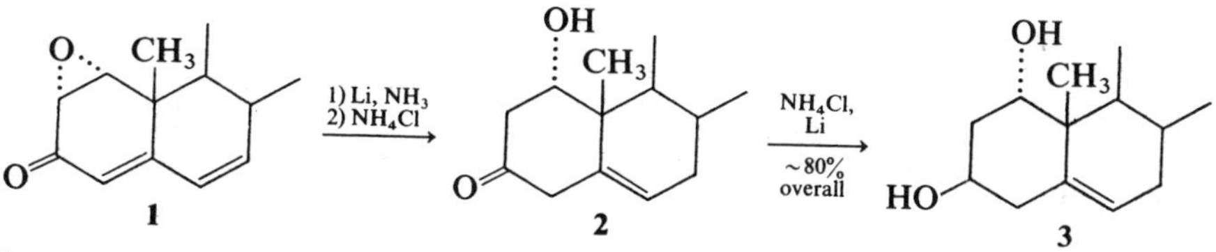

1) Li, NH_3 2) NH_4Cl → ; NH_4Cl, Li, ~80% overall →

1 **2** **3**

[1] A. Fürst, L. Labler, and W. Meier, *Helv.*, **64**, 1870 (1981).

Lithium–Ethylamine, 1, 574-581; **2,** 241–242; **3,** 175; **4,** 287–288; **5,** 377–379; **6,** 322; **7,** 194–195; **8,** 284–285; **10,** 136.

Reductive desulfonylation.[1] A stereocontrolled method for addition of the steroid side chain to a 17-keto steroid is outlined in scheme (I). The various steps proceed selectively to the sulfone **5**. Reductive desulfonylation of **5** with Na/Hg, Na_2HPO_4 in CH_3OH gives the desired **6** (57% yield) and the undesired alkene in a 2:1 ratio. The desired stereoselectivity was obtained with lithium in ethylamine. The final step was hydrogenation of the 17(20)-double bond to give a protected cholesterol (**7**).

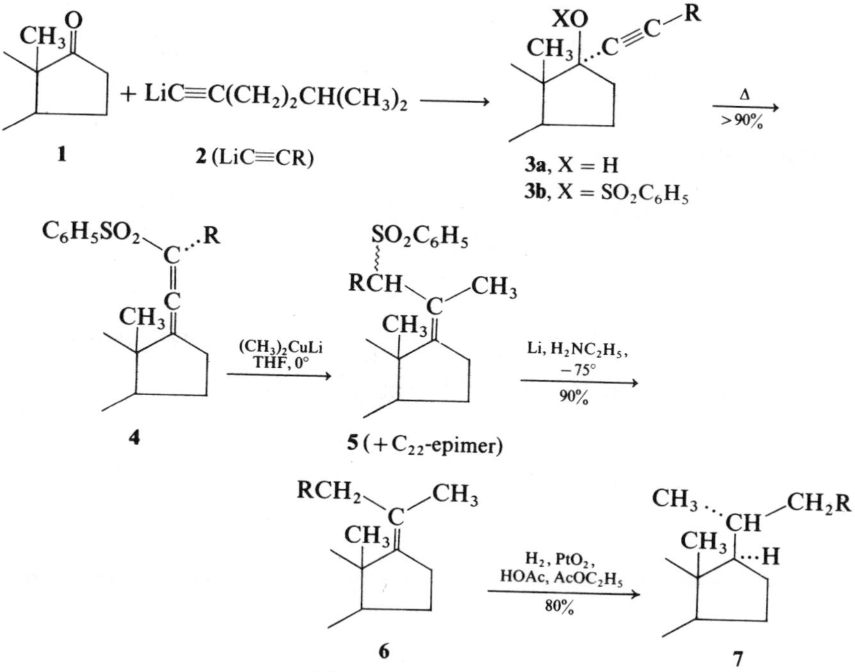

1 + $LiC{\equiv}C(CH_2)_2CH(CH_3)_2$ → **3a**, X = H; **3b**, X = $SO_2C_6H_5$ —Δ, >90%→

2 ($LiC{\equiv}CR$)

4 —$(CH_3)_2CuLi$, THF, 0°→ **5** (+C_{22}-epimer) —Li, $H_2NC_2H_5$, −75°, 90%→

6 —H_2, PtO_2, HOAc, $AcOC_2H_5$, 80%→ **7**

Scheme I

Δ⁵- and Δ⁷-Steroids. The adducts (**2**) of ergosteryl acetate (**1**) and 4-phenyl-1,2,4-triazoline-3,5-dione (PTAD) are reduced to the Δ^5- and the Δ^7-sterol in a 3:2 ratio.[2]

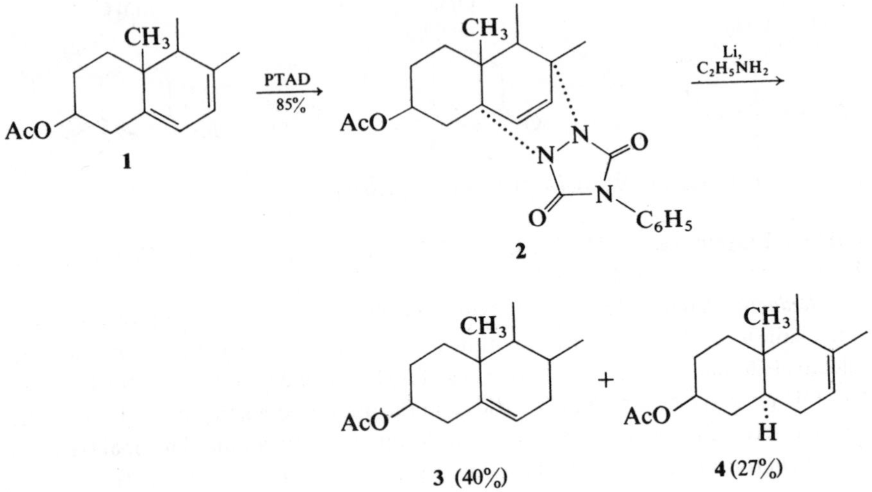

1 → (PTAD, 85%) → **2** → (Li, $C_2H_5NH_2$) →

3 (40%) + **4** (27%)

[1] M. Ohmori, S. Yamada, and H. Takayama, *Tetrahedron Letters*, **23**, 4709 (1982).
[2] M. Anastasia, P. Ciuffreda, and A. Fiecchi, *J.C.S. Chem. Comm.*, 1169 (1982).

Lithium acetylides, LiC≡CR.

Conjugate addition. Lithium acetylides undergo conjugate addition to the sterically hindered α,β-unsaturated trityl ketone **1**. The products (**2**) are cleaved by $LiBH(C_2H_5)_3$ (**4**, 313–314) to triphenylmethane and primary alcohols (**3**) in satisfactory yield.[1]

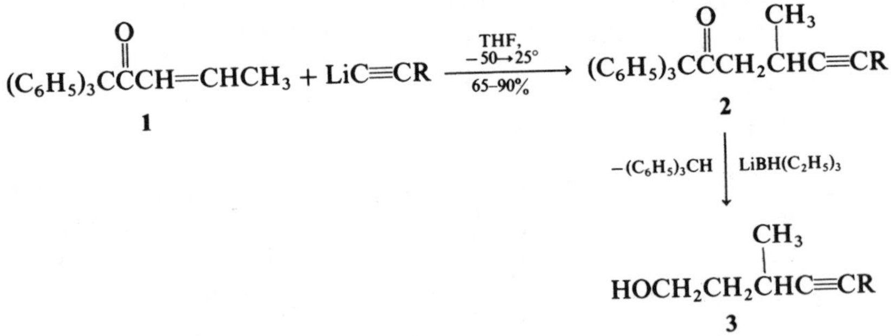

$$\underset{\mathbf{1}}{(C_6H_5)_3C\overset{O}{\overset{\|}{C}}CH{=}CHCH_3} + LiC{\equiv}CR \xrightarrow[65\text{–}90\%]{THF,\ -50\rightarrow25^\circ} \underset{\mathbf{2}}{(C_6H_5)_3C\overset{O}{\overset{\|}{C}}CH_2\overset{CH_3}{\overset{|}{C}}HC{\equiv}CR}$$

$$\downarrow -(C_6H_5)_3CH,\ LiBH(C_2H_5)_3$$

$$\underset{\mathbf{3}}{HOCH_2CH_2\overset{CH_3}{\overset{|}{C}}HC{\equiv}CR}$$

[1] R. Locher and D. Seebach, *Angew. Chem. Int. Ed.*, **20**, 569 (1981).

Lithium aluminum hydride, 1, 581-595; **2,** 292; **3,** 176–177; **4,** 291–293; **5** 382–389; **6,** 325–326; **7,** 196; **8,** 286–289, **9,** 274–277; **10,** 236–237.

Reduction of isoxazolines*; *3-amino alcohols. Lithium aluminum hydride reduction of the ready available isoxazolines **1** is effected with unusually high 1,3-asymmetric induction (equation I). The stereoselectivity is not affected drastically by the presence of a hydroxyl group in the side chains at C_3 or C_5, but isoxazolines bearing a 4α-hydroxyl group are reduced almost entirely to the *erythro* (*β*)-diastereomer.[1]

(I) $C_{14}H_{29}CH{=}CH_2 + O{\leftarrow}N{\equiv}C{-}CH_3 \longrightarrow$ **1** $\xrightarrow[95\%]{LiAlH_4,\ ether}$ *β*-**2** (*erythro*) + *α*-**2** (88:12)

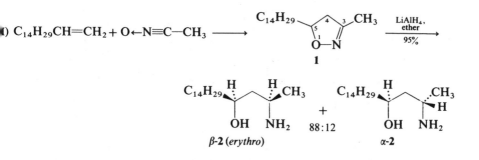

This effect was used to prepare the tetrahydropyranyl ether (**3**) of racemic phytosphingosine,[2] a base characteristic of plant sphingolipids (equation II).

(II) $C_{14}H_{29}CH{=}CH_2 + O_2NCH_2CH_2OTHP \xrightarrow[74\%]{C_6H_5NCO}$ 3-(CH_2OTHP)-5-($C_{14}H_{29}$)-isoxazoline $\xrightarrow[55.5\%]{1)\ LDA;\ 2)\ B(OCH_3)_3;\ 3)\ H_2O_2,\ N(C_2H_5)_3}$ 4-OH isoxazoline $\xrightarrow[\sim 95\%]{LiAlH_4}$ **3**

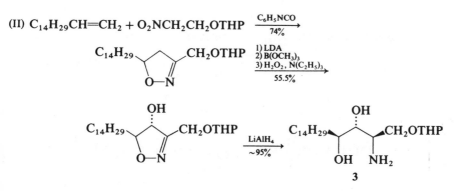

Allenic alcohols.[3] Treatment of the propargylic alcohol **1** with lithium aluminum hydride does not lead to the expected vinyl allenic alcohol **2** (**3**, 99). However, addition of solid iodine to the intermediate aluminate does effect the desired elimination to give **2**. This result was unexpected because addition of a solution of iodine in THF leads to a vinyl iodide (**3**). This sequence provides a general synthesis of (hydroxymethyl)allenes (second example).

Examples:

$$HOCH_2C{\equiv}C{-}CH(OCH_3){-}CH{=}CH{-}CH_3 \ (\mathbf{1}) \xrightarrow[60\%]{1)\ LiAlH_4;\ 2)\ I_2\ (solid)} HOCH_2CH{=}C{=}CH{-}CH{=}CH{-}CH_3 \ (\mathbf{2})$$

$$\mathbf{1} \xrightarrow{1)\ LiAlH_4;\ 2)\ I_2\ in\ THF} HOCH_2CH{=}CI{-}CH(OCH_3){-}CH{=}CH{-}CH_3 \ (\mathbf{3})$$

$$HOCH_2C{\equiv}CC(OCH_3)(CH_3)CH_2CH_3 \xrightarrow{98\%} HOCH_2CH{=}C{=}C(CH_3)CH_2CH_3$$

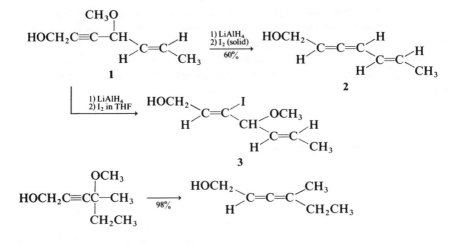

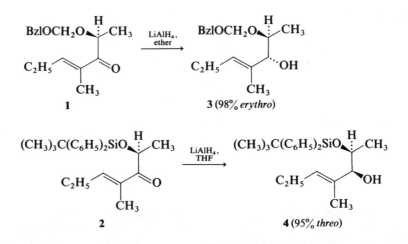

***Allylic* 1,2-*diols*.** Stereoselective reduction of α′-alkoxyenones depends in part on the reducing agent but mainly on the nature of the alkoxy group. Thus the benzyloxymethyl ether **1** is reduced by $LiAlH_4$ almost exclusively to the *erythro*-diol **3**. Chelation may be a factor. In contrast, the highly hindered *t*-butyldiphenylsilyl ether **2** is reduced by the same reagent with high *threo*-selectivity.[4]

$$BzlOCH_2O{-}CH(CH_3){-}CO{-}C(CH_3){=}CH{-}C_2H_5 \ (\mathbf{1}) \xrightarrow{LiAlH_4,\ ether} BzlOCH_2O{-}CH(CH_3){-}CH(OH){-}C(CH_3){=}CH{-}C_2H_5$$

3 (98% *erythro*)

$$(CH_3)_3C(C_6H_5)_2SiO{-}CH(CH_3){-}CO{-}C(CH_3){=}CH{-}C_2H_5 \ (\mathbf{2}) \xrightarrow{LiAlH_4,\ THF} (CH_3)_3C(C_6H_5)_2SiO{-}CH(CH_3){-}CH(OH){-}C(CH_3){=}CH{-}C_2H_5$$

4 (95% *threo*)

This work and parallel studies by another group[5] have been used to assign the *threo*-configuration to the allylic diol group in the side chain of the frog poison pumiliotoxin B (**5**).

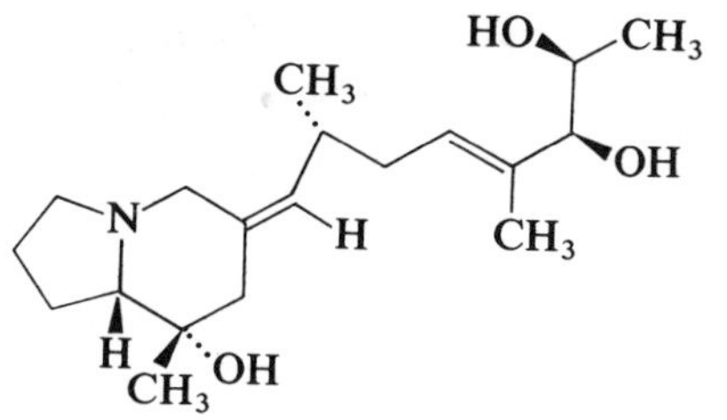

5

γ-Lactone annelation.[6] Although an α,β-unsaturated nitrile normally undergoes 1,2-reduction with this hydride, intramolecular conjugate reduction is possible by an adjacent alkoxy hydride generated from an acetoxy or hydroxy group. This strategy is used for conversion of the acetoxy ketone **1** into the *trans*-fused γ-butyrolactone **4**, whose relative stereochemistry is fixed by the configuration of the acetoxy group.

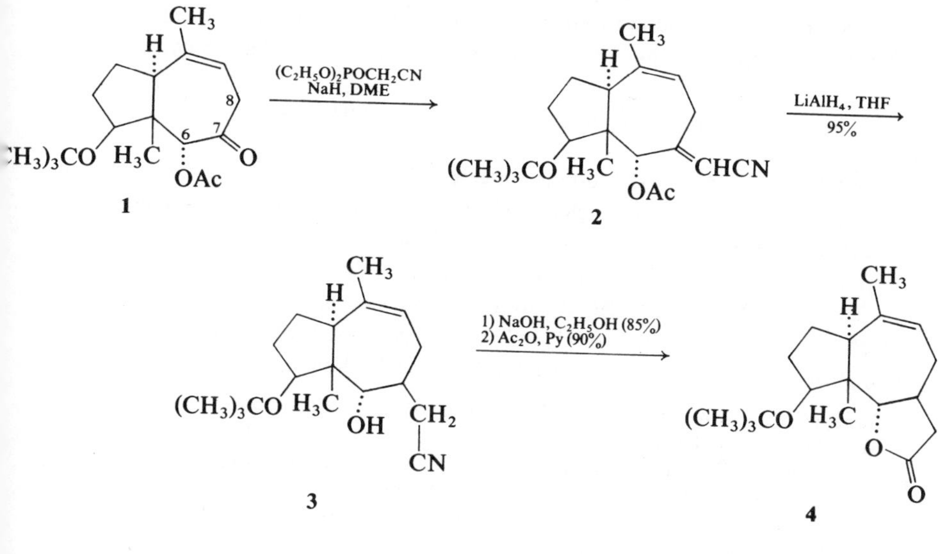

$(C_2H_5O)_2POCH_2CN$, NaH, DME → LiAlH₄, THF 95%

1) NaOH, C_2H_5OH (85%) 2) Ac_2O, Py (90%)

1 **2** **3** **4**

The annelation can be directed to either side of an unsymmetrical ketone by controlling the site of hydroxylation.

Diastereoselective reduction of chiral β-keto sulfoxides.[7] The chiral β-keto sulfoxides (**1**) are reduced by $LiAH_4$ and several borohydrides mainly to **2**. The stereoselectivity is increased as the temperature is lowered. In contrast, DIBAH and diborane reduction results mainly in the alcohol **3**.

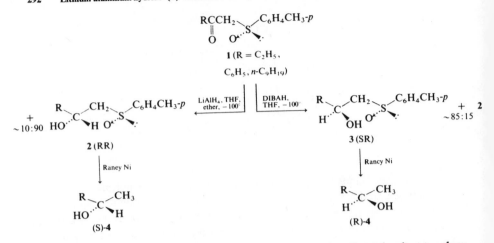

$RCOCH_2S(O)C_6H_4CH_3$-*p*

1 (R = C_2H_5, C_6H_5, *n*-C_9H_{19})

LiAlH$_4$, THF, ether, −100° → **2** (RR) + **3** ~10:90

DIBAH, THF, −100° → **3** (SR) + **2** ~85:15

2 (RR) —Raney Ni→ (S)-**4**

3 (SR) —Rancy Ni→ (R)-**4**

Hydrodehalogenation. Unlike slurries of $LiAlH_4$ in ethereal solvents, clear solutions of the hydride in THF reduce alkyl halides to the corresponding alkanes rapidly and often in practically quantitative yield.[8] However at least 1 molar equiv. of $LiAlH_4$ per RX is required (equation I). The rate of reduction for the halides is I > Br > Cl. The effect of the structure on the rate is in the order *n*-butyl bromide >

$$\text{(I)}\ LiAlH_4 + RX \longrightarrow RH + AlH_3 + LiX$$

isobutyl bromide > *t*-butyl bromide ≫ neopentyl bromide. The ring size also has a pronounced effect on rate: cyclopentyl > cycloheptyl > cyclooctyl > cyclohexyl. This effect is the order for S_N2 exchange of cyclic bromides with I^-. $LiAlH_4$ also reduces aryl, tertiary bridgehead, cyclopropyl, and vinyl halides.

Ultrasound markedly accelerates the rate of heterogeneous reduction of aryl halides, even deactivated ones, by $LiAlH_4$ in DME. Yields of the corresponding arenes of > 90% can often be obtained after reaction times of 4–7 hours.[9]

[1] V. Jäger, W. Schwab, and V. Buss, *Angew. Chem. Int. Ed.*, **20**, 601 (1981).
[2] W. Schwab and V. Jäger, *ibid.*, **20**, 603 (1981).
[3] G. E. Keck and R. R. Webb, II, *Tetrahedron Letters*, **23**, 3051 (1982).
[4] L. E. Overman and R. S. McCready, *ibid.*, **23**, 2355 (1982).
[5] T. Tokuyama, K. Shimada, M. Uemura, and J. W. Daley, *ibid.*, **23**, 2121 (1982).
[6] P. T. Lansbury and J. P. Vacca, *ibid.*, **23**, 2623 (1982).
[7] G. Solladié, C. Greck, G. Demailly, and A. Solladié-Cavallo, *ibid.*, **23**, 5047 (1982).
[8] S. Krishnamurthy and H. C. Brown, *J. Org.*, **47**, 276 (1982).
[9] B. H. Han and P. Boudjouk, *Tetrahedron Letters*, **23**, 1643 (1982).

Lithium aluminum hydride–(S)-4-Anilino-3-methylamino-1-butanol.

Asymmetric reduction of alkyl aryl ketones. The complex (**2**) derived from lithium aluminum hydride and this diamino alcohol (**1**), prepared from (S)-(−)-

aspartic acid, reduces alkyl aryl ketones to the corresponding (S)-carbinols in yields of 51–88%ee. Dialkyl ketones are reduced, as expected, in much lower optical yield (33%ee, one example).[1]

$$\mathrm{LiAlH_4 + H{-}\underset{CH_2NHC_6H_5}{\overset{CH_2CH_2OH}{\underset{|}{\overset{|}{C}}}}{-}NHCH_3 \longrightarrow \underset{\mathbf{2}}{Complex} \xrightarrow{ArCOR} \underset{(51\text{–}88\%\ ee)}{(S)\text{-}ArCH(OH)R}}$$

1, $\alpha_D - 13.7°$

[1] T. Sato, Y. Goto, and T. Fujisawa, *Tetrahedron Letters*, **23**, 4111 (1982).

Lithium aluminum hydride–Silica gel.

Selective reduction of carbonyl groups in the presence of an ester group. Selective reduction of various keto esters to hydroxy esters is possible in fair yield with $LiAlH_4$ in the presence of a small amount of silica gel (equation I).[1]

$$\text{(I)}\quad \mathrm{CH_3\overset{\overset{O}{\|}}{C}(CH_2)_nCOOR} \xrightarrow[40\text{–}85\%]{\mathrm{LiAlH_4{-}SiO_2},\ \text{ether}} \mathrm{CH_3\underset{OH}{\underset{|}{C}}H(CH_2)_nCOOR}$$

$n = 0\text{–}3$

[1] Y. Kamitori, M. Hojo, R. Masuda, T. Inone, and T. Izumi, *Tetrahedron Letters*, **23**, 4585 (1982).

Lithium borohydride, 1, 603; **4**, 296. Solutions of the hydride in ether or THF can be prepared readily by treatment of $NaBH_4$ (suspension) with LiBr (mechanical stirring with glass beads present).[1]

Reduction of esters.[2] Lithium borohydride is more effective than $Ca(BH_4)_2$ or $NaBH_4$ for reduction of esters in ethyl ether. It is less active in THF than in ether. Alcohol solvents are less useful for this reduction because of competitive solvolysis. Selective reduction is possible in the presence of nitro, halo, cyano, and alkoxy groups.[3]

[1] H. C. Brown, Y. M. Choi, and S. Narasimhan, *Inorg. Chem.*, **20**, 4454 (1981).
[2] H. C. Brown and S. Narasimhan, *J. Org.*, **47**, 1604 (1982).
[3] H. C. Brown, S. Narasimhan, and Y. M. Choi, *ibid.*, **47**, 4702 (1982).

Lithium bronze, Li · 4NH_3. The reagent is prepared by condensation of a slight excess of NH_3 into a flask containing Li metal. It resembles mercury in appearance, and can be kept at 20° for 2–3 weeks if protected from air. *Follow instructions for handling.*[1]

Conjugate reduction of α,β-enones.[1] Lithium bronze can be used in place of lithium blue for conjugate reduction of cyclic α,β-enones to the saturated ketone in high yield. It requires an added proton source (*t*-butyl alcohol).

Example:

$$\text{3,5,5-trimethylcyclohex-2-enone} \xrightarrow[90\%]{\text{Li}\cdot 4\text{NH}_3,\ (\text{C}_2\text{H}_5)_2\text{O},\ (\text{CH}_3)_3\text{COH}} \text{3,3,5-trimethylcyclohexanone}$$

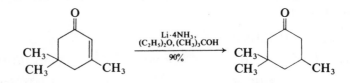

***Reduction of* $-C{\equiv}C(CH_2)_2OH$.**[2] Lithium bronze is the most effective reagent for conjugate reduction of homopropargylic alcohols to homoallylic alcohols.

Example:

$$C_{10}H_{21}C{\equiv}C(CH_2)_2OH \xrightarrow{\text{Li}\cdot 4\text{NH}_3,\ (C_2H_5)_2O,\ (CH_3)_3COH} \underset{(95\%)}{C_{10}H_{21}CH{=}CH(CH_2)_2OH\ (trans)} + \underset{(1.3\%)}{C_{14}H_{29}OH}$$

[1] R. H. Mueller and J. G. Gillick, *J. Org.*, **43**, 4647 (1978).
[2] R. E. Doolittle, *Org. Prep. Proc. Int.*, **13**, 179 (1981).

Lithium *n*-butylborohydride, Li-*n*-$BuBH_3$. Mol. wt. 77.89. The reagent is prepared by reaction of *n*-butyllithium with borane–dimethyl sulfide.

Selective reductions.[1] This complex borohydride is particularly useful for selective 1,2-reduction of acyclic α,β-enones and of conjugated cyclohexenones to allylic alcohols. However, the 1,2-selectivity is less marked with conjugated cyclopentenones. The reagent reduces unhindered cyclic ketones to the more stable (equatorial) alcohols with stereoselectivity greater than that of sodium borohydride.

[1] S. Kim, Y. C. Moon, and K. H. Ahn, *J. Org.*, **47**, 3311 (1982).

Lithium chloropropargylide, 5, 397; **9**, 279–280.

1-Ene-5-ynes.[1] 2-Alkynylboranes (**1**), prepared by reaction of lithium chloropropargylide with trialkylboranes, react with allylic bromides in the presence of CuI to form 1,5-enynes regioselectively (equation I).

$$\text{(I)}\quad LiC{\equiv}CCH_2Cl + R_3B \xrightarrow{-78\rightarrow 25^\circ} \underset{\mathbf{1}}{RC{\equiv}CCH_2BR_2} \xrightarrow[60\text{–}70\%]{BrCH_2C(R'){=}CH_2,\ CuI,\ LiOCH_3} \underset{\mathbf{2}\ (R' = H, CH_3)}{RC{\equiv}CCH_2CH_2C(R'){=}CH_2}$$

[1] S. Hara, Y. Satoh, and A. Suzuki, *Chem. Letters*, 1289 (1982).

Lithium 9,9-di-*n*-butyl-9-borabicyclo[3.3.1]nonanate (1).

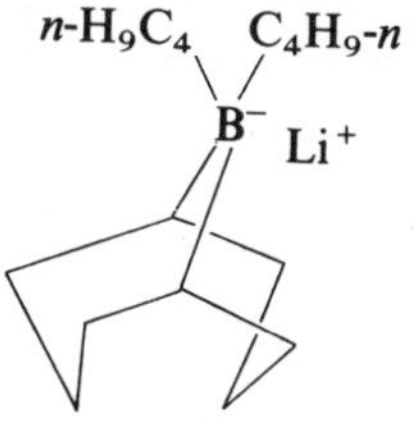

$n\text{-}H_9C_4$ $C_4H_9\text{-}n$ B^- Li^+

1

The ate complex is prepared by reaction of B-*n*-butyl-9-borabicyclo[3.3.1]nonane (**9**, 76–77) with *n*-butyllithium in pentane or hexane.

Selective reduction of halides.[1] The complex **1** reduces tertiary halides to alkanes without attack on primary or secondary halides. It is more selective than organotin hydrides. Benzylic and allyl halides are also reduced easily. The reagent reduces a mixture of the halides **2** mainly to *trans*-**3**; thus, a partial inversion of configuration is involved.

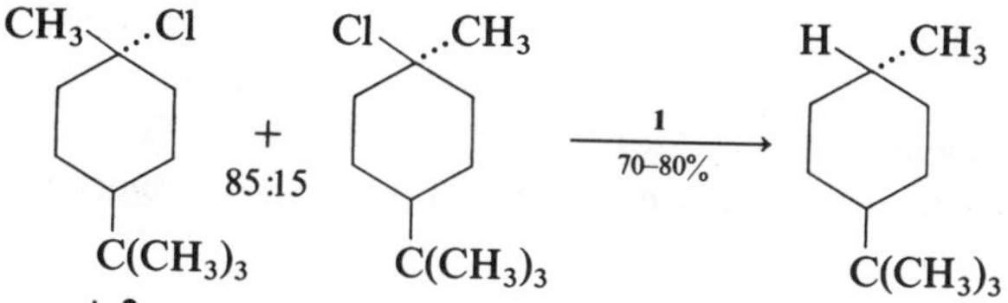

CH_3, Cl, $C(CH_3)_3$ (*cis*-**2**) + Cl, CH_3, $C(CH_3)_3$ (*trans*-**2**) (85:15) $\xrightarrow[70\text{–}80\%]{\mathbf{1}}$ H, CH_3, $C(CH_3)_3$ (**3**)

[1] H. Toi, Y. Yamamoto, A. Sonoda, and S.-I. Murahashi, *Tetrahedron*, **37**, 2261 (1981).

Lithium diethylamide, 1, 610–611; **2**, 247–248; **4**, 298; **5**, 398–399; **6**, 331.

Alkylation of amino acids.[1] This base is preferable to LDA for dideprotonation of di-*t*-butyl N-formylaspartate (**1**) to give **2**. The dianion is alkylated by CH_3I, $CH_2{=}CHCH_2Br$, and $C_6H_5CH_2Br$ to give a mixture of β- and α-alkylated aspartate derivatives (**3** and **4**) in a ratio of *ca.* 7:2. As expected, the β-alkylated α-amino acid esters **3** are optically pure; surprisingly, the products **4** are also optically active and, after crystallization, can be obtained in optical yields of ~60%.

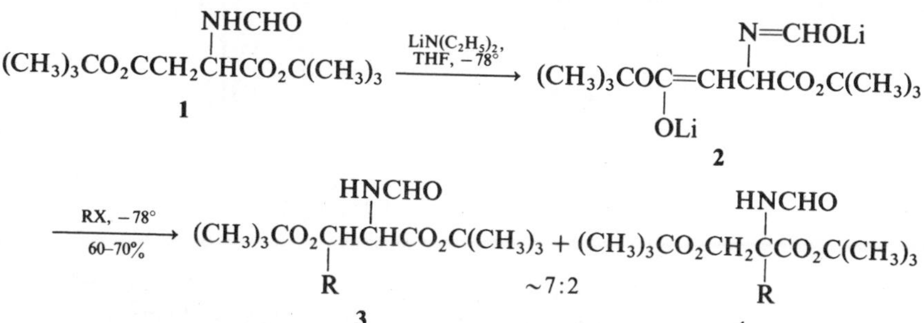

$$\underset{\mathbf{1}}{(CH_3)_3CO_2CCH_2\overset{NHCHO}{CH}CO_2C(CH_3)_3} \xrightarrow[THF,\ -78°]{LiN(C_2H_5)_2,} \underset{\mathbf{2}}{(CH_3)_3CO\underset{OLi}{C}{=}CH\overset{N=CHOLi}{C}HCO_2C(CH_3)_3}$$

$$\xrightarrow[60\text{–}70\%]{RX,\ -78°} \underset{\mathbf{3}}{(CH_3)_3CO_2\underset{R}{C}H\overset{HNCHO}{C}HCO_2C(CH_3)_3} + \underset{\mathbf{4}}{(CH_3)_3CO_2CH_2\overset{HNCHO}{\underset{R}{C}}CO_2C(CH_3)_3} \quad \sim 7:2$$

[1] D. Seebach and D. Wasmuth, *Agnew. Chem. Int. Ed.*, **20**, 971 (1981).

Lithium diisobutyl-t-butylaluminum hydride (1). The hydride is prepared by addition of *t*-butyllithium in pentane to diisobutylaluminum hydride in hexane.

Reduction of cyclic and bicyclic ketones.[1] This hydride reduces 4-*t*-butyl- and 3-methylcyclohexanone with some bias favoring formation of the more stable isomer. In contrast, 2-methylcyclohexanone is reduced preferentially to the less stable *cis*-isomer (72%). Essentially, only the *cis*-isomer is formed on reduction of 2-*t*-butylcyclohexanone (98% stereoselectivity). Norcamphor is reduced to the *endo*-alcohol in >99% stereoselectivity, whereas the more hindered camphor is reduced to the *exo*-alcohol with 98% stereoselectivity.

[1] S. Kim, K. H. Ahn, and Y. W. Chung, *J. Org.*, **47**, 4581 (1982).

Lithium diisopropylamide, 1, 611; **2**, 249; **3**, 184–185; **4**, 298–302; **5**, 400–406; **6**, 334–339; **7**, 204–207; **8**, 292; **9**, 280–283; **10**, 241–243.

Reaction with a hindered epoxide.[1] The *trans*-epoxide (**1**) of tetramethyllimonene is inert to KOH (130°), $LiAlH_4$ (THF, 90°), and even lithium triethylborohydride. It is opened by aluminum isopropoxide (110°) to give 70:30 mixture of **2** and **3**. Reaction with LDA is more selective and gives **2** in 95% yield. In contrast, reaction with N-lithioethylenediamine (**1**, 567–570) gives **3** in 90% yield. The 70:30 mixture of **2** and **3** is converted by N-lithioethylenediamine to the more stable isomer **3**.

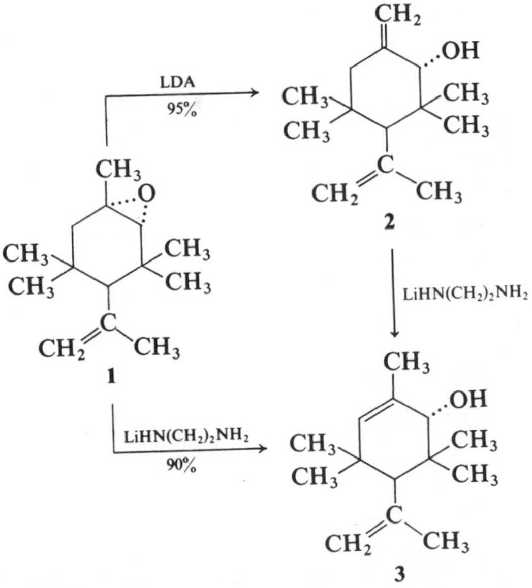

LDA 95%

CH_2 OH CH_3 CH_3 CH_3 CH_3 C CH_2 CH_3

2

CH_3 O CH_3 CH_3 CH_3 CH_3 C CH_2 CH_3

1

$LiHN(CH_2)_2NH_2$

$LiHN(CH_2)_2NH_2$ 90%

CH_3 OH CH_3 CH_3 CH_3 CH_3 C CH_2 CH_3

3

1-Alkynes from $RCOCH_3$ (**10**, 243). Complete details are available for conversion of methyl ketones to 1-alkynes via enol phosphates (equation I)[2]. The method is general (eight other examples). LDA is satisfactory when the R group of

$RCOCH_3$ is either bulky or does not contain an α-methylene or α-methine group. In such cases, lithium 2,2,6,6-tetramethylpiperidide is a satisfactory substitute. Isolated yields range from 61–85%.

(I) [structure] 1) LDA, THF 2) $ClPO(OC_2H_5)_2$ 3) LDA, 72% → [structure]

Claisen rearrangement of allyl α-hydroxy acetates; α-hydroxy-γ,δ-unsaturated acids. Allyl α-hydroxy acetates undergo Claisen rearrangement when treated with LDA (2 equiv.) to give, after aqueous workup, α-hydroxy-γ,δ-unsaturated acids.[3]

Example:

[structure] 1) LDA, THF, −78° 2) −78→20° 3) NH_4OH, H_2O, 71% → [structure]

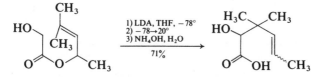

Deconjugation of α,β-unsaturated esters. It has been known for some time that deprotonation and reprotonation (or alkylation) of α,β-unsaturated esters is accompanied by migration of the double bond to the β,γ-position (thermodynamically the less stable position). The stereochemistry of the isomerization has now been elucidated by two groups.[4, 5] A (Z)-2-alkenoate isomerizes almost exclusively to the (E)-3-alkenoate. The (E)-2-alkenoate isomerizes to (Z)- and (E)-3-alkenoates in a ratio of about 85:15. Both the (E)- and (Z)-3-alkenoates are unchanged on re-exposure to this sequence. Both papers suggest possible reasons for this stereochemical outcome.

Apparently, the aldol condensation proceeds with the same stereochemical results. In any case, this approach permits a stereospecific synthesis of litsenolide B_2(**1**) and related Lauraceae lactones, as outlined in equation (I).[5]

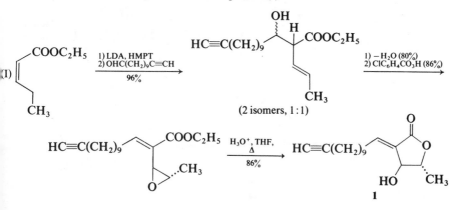

(I) [structure] 1) LDA, HMPT 2) $OHC(CH_2)_9C{\equiv}CH$, 96% → [structure] (2 isomers, 1:1) 1) $-H_2O$ (80%) 2) $ClC_6H_4CO_3H$ (86%) →

[structure] H_3O^+, THF, Δ, 86% → [structure] **1**

Alkylation of cyclopropanecarboxylic acid esters.[6] Deprotonation of methyl silyloxycyclopropanecarboxylates (**1**), prepared as shown, is possible with LDA in THF at $-78°$. The resulting anions react with primary alkyl iodides and benzylic or allylic bromides to give **2** in high yield. These products are cleaved by F^- to methyl 4-ketocarboxylates **3**.

$$(CH_3)_3SiO(R^1)C{=}C(R^2)(R^3) \xrightarrow[60\text{–}85\%]{N_2CHCOOCH_3} \underset{\mathbf{1}}{(CH_3)_3SiO(R^1)\text{-cyclopropane}(H)(COOCH_3)(R^2)(R^3)} \xrightarrow[70\text{–}90\%]{1)\,LDA;\ 2)\,RX}$$

$$\underset{\mathbf{2}}{(CH_3)_3SiO(R^1)\text{-cyclopropane}(R)(COOCH_3)(R^2)(R^3)} \xrightarrow[70\text{–}95\%]{(C_2H_5)_3\overset{+}{N}HF^-} \underset{\mathbf{3}}{R^1C(=O)\text{—}C(R^2)(R^3)\text{—}CH(R)COOCH_3}$$

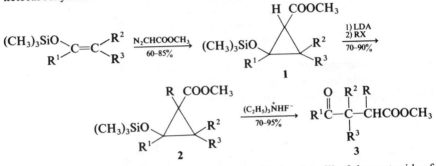

Diastereoselective alkylation of tartaric acid. The enolate (**2**) of the acetonide of dimethyl (R, R)-tartrate (**1**) can be generated with LDA in THF–HMPT at $-70°$ and is sufficiently stable for alkylation with allyl and benzyl halides, but not with other simple alkyl halides, and for addition to acetone (60% yield). The main products (**3**) of allylation and benzylation have the *trans*-configuration, and thus the substitution occurs with retention of configuration.[7]

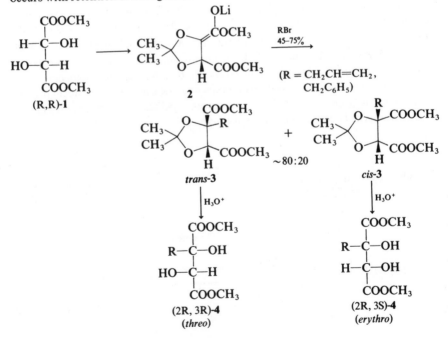

$$\underset{(R,R)\text{-}\mathbf{1}}{CH_3OOC\text{—}CH(OH)\text{—}CH(OH)\text{—}COOCH_3} \longrightarrow \underset{\mathbf{2}}{(CH_3)_2C\langle O\text{—}C(=C(OLi)OCH_3)\text{—}CH(COOCH_3)\text{—}O\rangle} \xrightarrow[45\text{–}75\%]{RBr}$$

$$(R = CH_2CH{=}CH_2,\ CH_2C_6H_5)$$

$$\textit{trans-}\mathbf{3} + \textit{cis-}\mathbf{3} \quad (\sim 80:20)$$

$$\textit{trans-}\mathbf{3} \xrightarrow{H_3O^+} (2R, 3R)\text{-}\mathbf{4}\ (\textit{threo}):\ COOCH_3\text{—}C(R)(OH)\text{—}C(H)(OH)\text{—}COOCH_3 \quad [R\text{—C—OH};\ HO\text{—C—H}]$$

$$\textit{cis-}\mathbf{3} \xrightarrow{H_3O^+} (2R, 3S)\text{-}\mathbf{4}\ (\textit{erythro}):\ COOCH_3\text{—}C(R)(OH)\text{—}C(H)(OH)\text{—}COOCH_3 \quad [R\text{—C—OH};\ H\text{—C—OH}]$$

α-Dialkylamino ketones.[8] The anions of α-dialkylamino nitriles (**1**) condense with aldehydes to form adducts (**2**) that lose HCN on distillation to give α-amino ketones (**3**). The method is particularly useful for regiospecific synthesis of unsymmetrical α-amino ketones.

$$R^1CHO + R_2NH \xrightarrow[72\text{–}92\%]{KCN,\ H_2O} \underset{\mathbf{1}}{(R^1)(R_2N)CHCN} \xrightarrow{1)\ LDA\quad 2)\ R^2CHO} \left[\underset{\mathbf{2}}{(R^1)(R_2N)C(CN){-}CH(OH)R^2}\right] \xrightarrow[70\text{–}95\%]{\Delta\ (-HCN)} \underset{\mathbf{3}}{(R^1)(R_2N)CH{-}\overset{O}{\overset{\|}{C}}R^2}$$

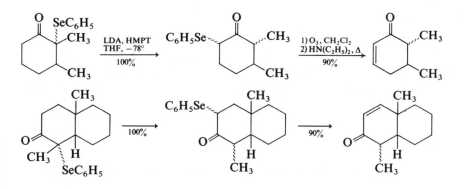

Rearrangement of α-phenylselenenyl ketones. Various α-phenylselenenyl ketones rearrange in the presence of 0.5 equiv. of LDA and HMPT to α′-phenylselenenyl ketones, apparently by an intermolecular process. This rearrangement is observed particularly with substrates substituted by alkyl groups in the α- and β-positions. The reaction provides a route to α′, β′-unsaturated ketones.[9]

Examples:

2-SeC_6H_5-2,3-dimethylcyclohexanone $\xrightarrow[100\%]{LDA,\ HMPT\ \ THF,\ -78^\circ}$ 6-C_6H_5Se-2,3-dimethylcyclohexanone $\xrightarrow[90\%]{1)\ O_3,\ CH_2Cl_2\quad 2)\ HN(C_2H_5)_2,\ \Delta}$ 2,3-dimethylcyclohex-5-enone

(decalone: CH_3, SeC_6H_5, CH_3, H) $\xrightarrow{100\%}$ (C_6H_5Se, CH_3, H, CH_3) $\xrightarrow{90\%}$ (enone: CH_3, H, CH_3)

[1] R. J. Giguere and H. M. R. Hoffmann, *Tetrahedron Letters*, **22**, 5039 (1981).
[2] E. Negishi, A. O. King, and J. J. M. Tour, *Org. Syn.*, submitted (1982).
[3] D. J. Ager and R. C. Cookson, *Tetrahedron Letters*, **23**, 3419 (1982).
[4] E.-P. Krebs, *Helv.*, **64**, 1023 (1981).
[5] A. S. Kende and B. H. Toder, *J. Org.*, **47**, 163 (1982).
[6] I. Böhm, E. Hirsch, and H.-U. Reissig, *Angew. Chem. Int. Ed.*, **20**, 574 (1981).
[7] R. Naef and D. Seebach, *ibid.*, **20**, 1030 (1981).
[8] D. Enders and H. Lotter, *Tetrahedron Letters*, **23**, 639 (1982).
[9] D. Liotta, M. Saindane, and D. Brothers, *J. Org.*, **47**, 1598 (1982).

Lithium hexamethyldisilazide, 4, 296; **5,** 393–394.

trans-*Selective Wittig reaction.*[1] The nonstabilized ylide **1a** or **1b**, generated from $(C_6H_5)_3\overset{+}{P}(CH_2)_4COOH\,Br^-$ with dimsylsodium or potassium *t*-butoxide, shows slight stereoselectivity in reactions with aromatic aldehydes. Use of lithium hexamethyldisilazide, $LiN[Si(CH_3)_3]_2$, as the base, however, results in a marked preference for the *trans*-isomer. This stereoselectivity is general for aromatic aldehydes but not for aliphatic aldehydes. Use of lithium hexamethyldisilazide for generation of $(C_6H_5)_3P{=}CHCH_3$ shows no *trans*-bias. Evidently the effect is associated with a terminal carboxy group.

$$(C_6H_5)_3P{=}CH(CH_2)_3COOM + C_6H_5CHO \xrightarrow[83\%]{} C_6H_5CH{=}CH(CH_2)_3COOH$$

		cis/trans
1a,	M = Na	50:50
1b,	M = K	40:60
1c,	M = Li	20:80

[1] B. E. Maryanoff and B. A. Duhl-Emswiler, *Tetrahedron letters*, **22,** 4185 (1981).

Lithium iodide, 5, 410–411; **7,** 208; **9,** 283; **10,** 245–246.

Expansion of cyclobutanones to cyclopentanones. 1-Oxaspiro[2.3]hexanes are isomerized to cyclopentanones in quantitative yield in the presence of LiI. The reaction is a particularly attractive route to 3,3,4,4-tetrasubstituted cyclopentanes (equation I).[1]

(I) [scheme: spiro epoxide (R, CH_3, H_3C, CH_3) $\xrightarrow{LiI,\ CH_2Cl_2,\ \Delta}$ [intermediate: OLi, $CH_2{-}I$, R, CH_3, H_3C, CH_3] $\xrightarrow{78\%}$ cyclopentanone (O, R, CH_3, H_3C, CH_3)]

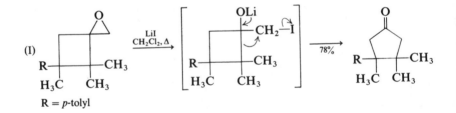

R = *p*-tolyl

Lithium bromide in HMPT also effects this transformation (equation II).[2]

(II) [scheme: $(CH_2)_2OC(C_6H_5)_3$-substituted spiro epoxide, O $\xrightarrow[78\%]{LiBr,\ HMPT}$ $(CH_2)_2OC(C_6H_5)_3$-substituted spiro ketone, O]

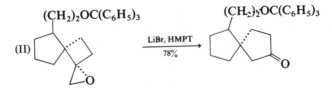

The reaction is also applicable when the epoxide is trisubstituted (equation III).[3]

(III) [cyclobutane with CH_3Se and C(OH)(CH_3)(C_9H_{19})] $\xrightarrow[70\%]{1)\ CH_3I;\ 2)\ KOH}$ [spiro-epoxide with CH_3, C_9H_{19}] $\xrightarrow[77\%]{LiI,\ CH_2Cl_2}$ [2-methyl-2-C_9H_{19}-cyclopentanone]

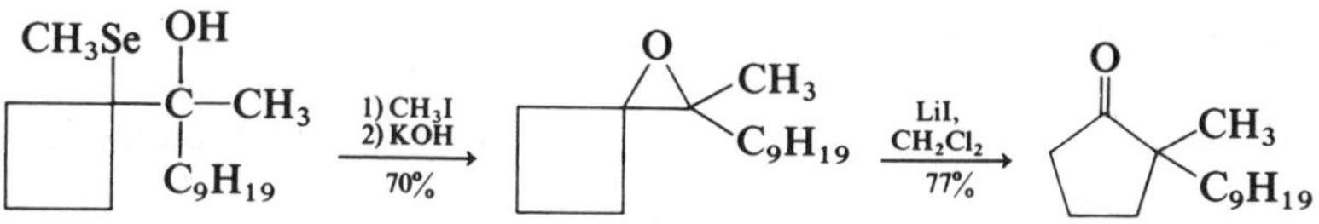

Cleavage of alkyl aryl ethers (**4**, 305). The most difficult step in a recent synthesis of aklavinone (**3**) is demethylation of **1** to **2**. Lewis acid reagents (BBr_3) were useless because of preferential attack of the allylic hydroxyl; nucleophilic demethylation ($LiSCH_3$) results in concomitant aromatization of ring A. Demethylation with LiI buffered with benzoic acid in pyridine–collidine at 145° was successful and proceeded in 92% yield.[4] Aklavinone is the aglycone of an antitumor anthracycline that is less toxic than adriamycin.

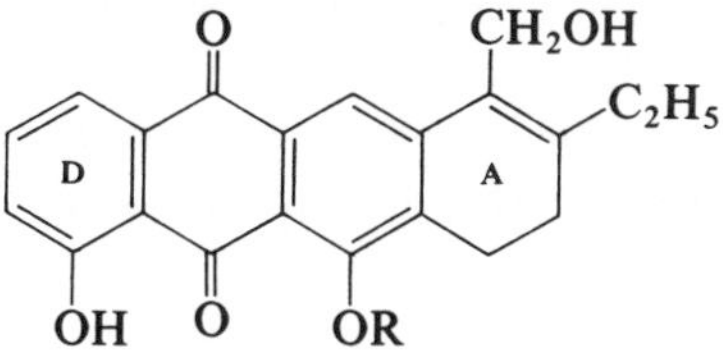

1, R = CH_3
2, R = H

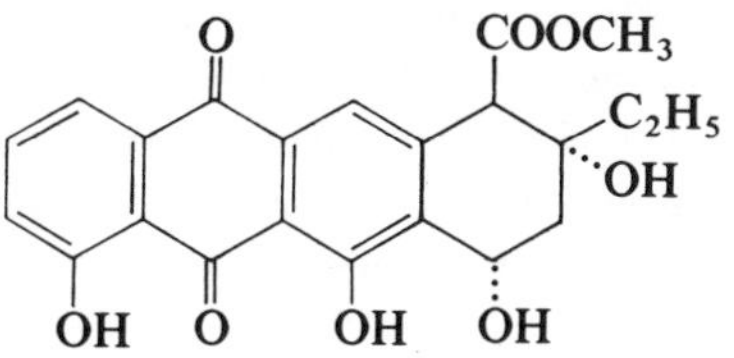

3

Decarboalkoxylation (**9**, 283).[5] The α-methoxycarbonyl group of **1** is removed cleanly by LiI in boiling collidine with concomitant hydrolysis of the ester group, possibly via an anhydride. The product (**2**) is a precursor to methyl jasmonate (**3**).

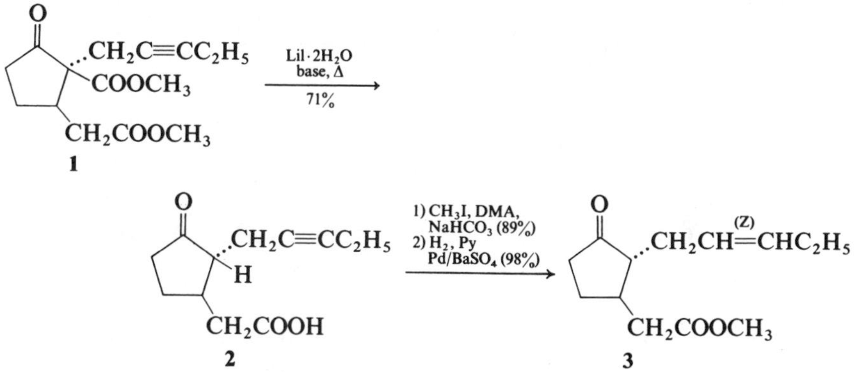

1 $\xrightarrow[71\%]{LiI\cdot 2H_2O,\ base,\ \Delta}$ **2** $\xrightarrow{1)\ CH_3I,\ DMA,\ NaHCO_3\ (89\%);\ 2)\ H_2,\ Py,\ Pd/BaSO_4\ (98\%)}$ **3**

[1] M. L. Leriverend and P. Leriverend, *Compt. Rend.* (*C*), **280**, 791 (1975).
[2] B. M. Trost and L. H. Latimer, *J. Org.*, **43**, 1031 (1978).
[3] S. Halazy and A. Krief, *J.C.S. Chem. Comm.*, 1200 (1982).
[5] A. S. Kende and J. T. Rizzi, *Am. Soc.*, **103**, 4247 (1981).
[5] F. Johnson, K. G. Paul, and D. Favara, *J. Org.*, **47**, 4254 (1982).

Lithium methylsulfinylmethylide–Tri-*n*-butylborane.

Methylthiomethyl esters.[1] Treatment of DMSO with *n*-butyllithium followed by tri-*n*-butylborane affords the ate complex $CH_3S(O)CH_2\bar{B}Bu_3Li^+$ (**1**), which converts acid chlorides into methylthiomethyl esters.

$$RCOCl + \mathbf{1} \xrightarrow[30\text{–}70\%]{THF} RC(=O)OCH_2SCH_3$$

[1] M. J. Kukla, *Tetrahedron Letters*, **23**, 4539 (1982).

Lithium methylthioformaldine (1), 8, 305.

Addition to aldehydes; α-hydroxy aldehydes. A key step in a stereocontrolled synthesis of α-multistriatin (**6**) from *meso*-2,4-dimethylglutaric acid (**2**) is stereoselective addition of **1** to the aldehyde **4** to give **5**. The thioacetal group of **5** is hydrolyzed with HgI_2 and HgO in aqueous THF; reduction of the resulting α-hydroxy aldehyde and hydrolysis of the ethylene ketal group gives **6**. Other formylation reagents were not useful in this synthesis.[1]

2 → 3: 1) CH_3OH (98%) 2) $(COCl)_2$ (95%) 3) CH_3CH_2Cu (97%)

3 → 4: 1) $(HOCH_2)_2$ 2) $LiAlH_4$ 3) $CrO_3\cdot 2Py$; 91%

4 → 5: **1**; 89%

5 → 6: 1) HgI_2, HgO (95%) 2) $NaCNBH_3$, pH 2–3 (48%)

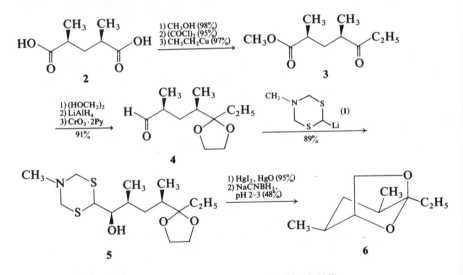

[1] D. M. Walba, and M. D. Wand, *Tetrahedron Letters*, **23**, 4995 (1982).

Lithium naphthalenide, 2, 288–289; **3,** 208; **4,** 348–349; **5,** 468; **6,** 415; **8,** 305–306; **9,** 285; **10,** 247.

Organolithium compounds. Reaction of the chlorohydrin (**2**) with *n*-butyllithium at −78° and then lithium naphthalenide (**1**) results in the organolithium compound **a**, which decomposes at temperatures above $-78°$ by β-elimination to an alkene. At lower temperatures, **a** reacts readily with various electrophiles to give products of substitution, such as **3**.[1]

Examples;

$$\underset{\mathbf{2}}{C_6H_5CH(OH)CH_2Cl} \xrightarrow[2)\ \mathbf{1},\ -78°]{1)\ n\text{-BuLi},\ -78°} \left[\underset{\mathbf{a}}{C_6H_5CH(OLi){-}CH_2Li}\right] \xrightarrow[79\%]{C_6H_5CHO} \underset{\mathbf{3}}{C_6H_5CH(OH)CH_2CH(OH)C_6H_5}$$

$$\text{2-chlorocyclohexanone} \longrightarrow \left[\text{1-}n\text{-Bu-1-OLi-2-Li-cyclohexane}\right] \xrightarrow{47\%} \text{1-}n\text{-butylcyclohexene}$$

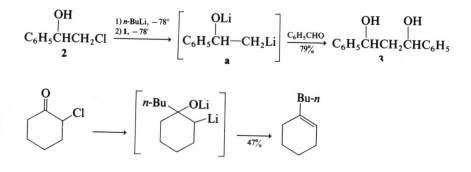

Peterson reaction (**5**, 724; **9**, 493). This reaction is limited by the fact that chloromethyl-trimethylsilane is the only commercially available suitable silane. Ager[2] has now found two general routes to a α-trimethylsilylcarbanions **3**, which undergo the Peterson reaction to give alkenes (**4**) in satisfactory yield.

$$C_6H_5SCH_2SC_6H_5 \xrightarrow{1)\ n\text{-BuLi, RX}} \underset{\mathbf{1}}{C_6H_5S{-}CH(R)SC_6H_5} \xleftarrow{H^+} RCHO + C_6H_5SH$$

$$\mathbf{1} \xrightarrow[70\text{–}90\%]{1)\ [C_{10}H_8]^{\dot{-}}\ Li^+\ \ 2)\ ClSi(CH_3)_3} \underset{\mathbf{2}}{(C_6H_5S)((CH_3)_3Si)CHR}$$

$$\xrightarrow{[C_{10}H_8]^{\dot{-}}\ Li^+} \left[\underset{\mathbf{3}}{(CH_3)_3SiCH(R)Li}\right] \xrightarrow[60\text{–}85\%]{R^1COR^2}$$

$$\left[\underset{\mathbf{a}}{R{-}CH(Si(CH_3)_3){-}C(OLi)(R^1)R^2}\right] \xrightarrow{60\text{–}85\%} \underset{\mathbf{4}}{R^1R^2C{=}CHR}$$

[1] J. Barluenga, J. Flórez, and M. Yus, *J.C.S. Chem. Comm.*, 1153 (1982).
[2] D. J. Ager, *Tetrahedron Letters*, **22**, 2923 (1981).

Lithium triethylborohydride, 4, 313–314; **6,** 348–349; **7,** 215–216; **8,** 309–310; **9,** 286; **10,** 249–250.

Reduction of allylic derivatives (**10,** 385–386). A variety of allylic functional groups can be reduced by the combination of $Pd[P(C_6H_5)]_4$ and this bulky hydride with good to excellent regio- and stereoselectivity. Groups that can be displaced include OR, SR, SO_2R, SeR, and $OSi(CH_3)_2C(CH_3)_3$.[1]

Examples:

$$\text{(E)-}CH_3(CH_2)_6CH{=}CHCH_2OCH_3 \xrightarrow[89\%]{Pd(0),\ Li(C_2H_5)_3BH,\ THF,\ 66^\circ} \text{(E)-}CH_3(CH_2)_6CH{=}CHCH_3 + \text{(Z)-isomer}\quad 94:3$$

$$\text{(E)-}CH_3(CH_2)_6CH{=}CHCH_2SO_2C_6H_5 \xrightarrow[65\%]{} \underset{(86\%)}{\text{(E)-}CH_3(CH_2)_6CH{=}CHCH_3} + \underset{(13\%)}{\text{(Z)-isomer}} + \underset{(1\%)}{\text{1-alkene}}$$

Thiiranes.[2] Lithium triethylborohydride converts *trans*-iodothiocyanates of cyclic and acyclic alkenes into thiiranes. The thiirane is formed with clean inversion at the quarternary carbon.

Examples:

trans-1-Iodo-2-thiocyanatocyclohexane (SCN, I) $\xrightarrow[93\%]{LiB(C_2H_5)_3H,\ ether,\ 25^\circ}$ cyclohexene sulfide (S)

$CH_3(CH_2)_5CH(I)CH_2SCN \xrightarrow[87\%]{} CH_3(CH_2)_5$-thiirane (S)

$(CH_3)_3C$-cyclohexane(I)(CH_2SCN) $\xrightarrow[100\%]{}$ $(CH_3)_3C$-cyclohexane spiro-thiirane (S)

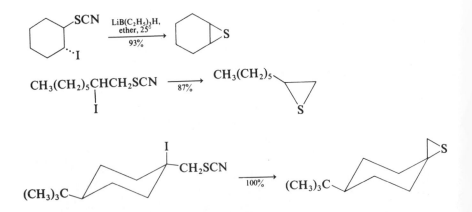

Thiiranes can also be prepared by treatment of oxiranes with the thiation reagent *cis*-3a, 4, 5, 6, 7, 7a-hexahydrobenzothiazolidine-2-thione (**1**) and TFA.[3]

Example:

$(CH_3)_3C$ + 1 $\xrightarrow[86\%]{CF_3COOH,\ CH_2Cl_2}$

CH_2S N S OH $(CH_3)_3C$ $\xrightarrow[quant.]{CF_3COOH}$

S $(CH_3)_3C$ + H N O S

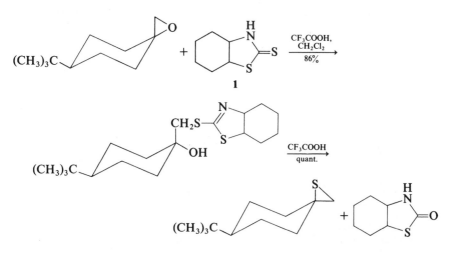

[1] R. O. Hutchins and K. Learn, *J. Org.*, **47**, 4380 (1982).

[2] R. C. Cambie, P. S. Rutledge, G. A. Strange, and P. D. Woodgate, *Heterocycles*, **19**, 1501 (1982).

[3] R. C. Cambie, G. D. Mayer, P. S. Rutledge, and P. D. Woodgate, *J.C.S. Perkin*, **1**, 52 (1981).

Lithium tris(3-ethyl-3-pentyloxy)aluminum hydride, $Li[(C_2H_5)_3CO]_3AlH$ (**1**). The hydride is prepared in quantitative yield by reaction of 3-ethyl-3-pentanol with $LiAlH_4$.

Chemoselective reduction of aldehydes.[1] Aldehydes can be reduced in the presence of ketones by **1** with 98–100% chemoselectivity. This chemoselectivity is the highest reported for this reduction.

[1] S. Krishnamurthy, *J. Org.*, **46**, 4628 (1981).

Lithium tris(phenylthio)methanide, $LiC(SC_6H_5)_3$ (**1**). This anion of tris(phenythio)methane[1] is prepared with *n*-butyllithium in THF at $-78°$.

Conversion of trialkylboranes into ketones and* t*-alcohols.[2] This anion reacts with trialkylboranes at room temperature to give **2**, formed by two spontaneous rearrangements of R groups from boron to carbon. Ketones are obtained by alkaline oxidation of **2**. A third migration is possible by treatment of **2** with $HgCl_2$ or FSO_3CH_3 (**3**, 202). The product, **3**, is oxidized to a tertiary alcohol. The reaction is shown using R_3B, but a mixed trialkylborane can be used; in this case two different ketones are possible.

$$R_3B + \mathbf{1} \longrightarrow [R_3BC(SC_6H_5)_3]^- \ Li^+ \longrightarrow \underset{\mathbf{2}}{RB(SC_6H_5)\overset{R}{\underset{R}{|}}{C}SC_6H_5}$$

$$\mathbf{2} \xrightarrow[70\text{–}80\%]{[O]} R_2C{=}O$$

$$\mathbf{2} \xrightarrow{HgCl_2 \text{ or } FSO_3CH_3} \underset{\mathbf{3}}{C_6H_5SBX \cdot CR_3} \xrightarrow[70\text{–}90\%]{[O]} R_3COH$$

The reaction is sensitive to the size of R; thus no ketone was obtained from bis(2-methylcyclohexyl)hexylborane.

[1] J. Hine, *Am. Soc.*, **72**, 2438 (1950).
[2] A. Pelter and J. M. Rao, *J.C.S. Chem. Comm.*, 1149 (1981).

M

Magnesium, 1, 627–629; **2,** 254; **3,** 189; **4,** 315; **5,** 419; **6,** 351–352; **7,** 218; **10,** 251.

Highly reactive powder (**5**, 419; **7**, 218). An improved procedure uses lithium together with an electron carrier (naphthalene or biphenyl, 10% based on Li) in THF, which need not be removed before preparation of Grignard reagents or other reagents.[1]

Phenanthrene synthesis. Treatment of (Z)-2-chlorostilbenes, available by Wittig reactions with 2-chlorobenzaldehydes, with Riecke activated magnesium in refluxing THF (12 hours) produces phenanthrenes (64–83% yield). (Z)-2-Bromostilbenes are reduced and isomerized to (E)-stilbenes by activated magnesium. The phenanthrene synthesis involves radical intermediates.[2]

This reaction has been used for synthesis of juncunone (**1**), an unusual natural acetyl-9,10-dihydrophenanthrene.[3]

CH_3O / CH_3 -substituted phenyl–CH=CH–(Cl, CH_3, $OCH(CH_3)_2$-substituted phenyl) $\xrightarrow[85\%]{\text{Mg, THF, }\Delta}$ phenanthrene (CH_3O, CH_3, $OCH(CH_3)_2$, CH_3) $\xrightarrow{\text{several steps}}$ **1** (Ac, OCH_3, CH_3, CH_3O, CH_3)

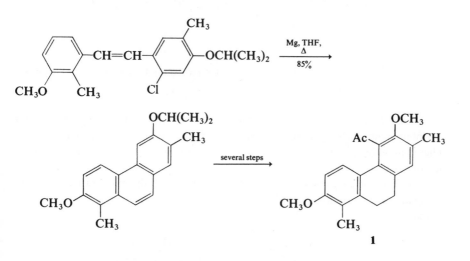

Barbier cyclization.[4] The Barbier reaction has been used for cyclopentenone annelation.[5] Thus α-allylcycloalkanones are converted into the iodo ketones **1**, which on treatment in THF with magnesium activated by $HgCl_2$ are converted into bicycloalkanols (**2**) in moderate to good yield. Formation of *cis*-fused products is greatly favored. The reaction gives poor yields when extended to annelation of six- and seven-membered rings.

$$\text{(CH}_2)_n\text{-fused 2-methylene cyclic ketone} \xrightarrow[\text{2) NaI, (CH}_3)_2\text{C=O}]{\text{1) HBr, } h\nu} \underset{\mathbf{1}}{\text{(CH}_2)_n\text{ 2-(3-iodopropyl) ketone}} \xrightarrow[\text{ether}]{\text{Mg/Hg}}$$

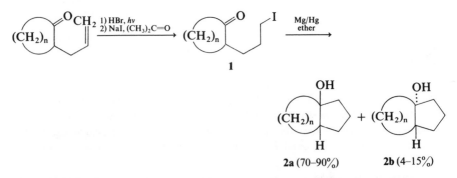

2a (70–90%) + **2b** (4–15%)

Mg–THF slurries. A clean, alkali-free slurry of Mg in THF can be obtained by cocondensation in a rotating solution reactor.[6] This form of magnesium is successful for preparation of hitherto inaccessible Grignard reagents. For example, cyclopropylmethylmagnesium bromide, as usually prepared in ether, rearranges to 4-butenylmagnesium bromide. This rearrangement occurs to only a slight extent when the Mg-THF slurry is used (equation I).

$$\text{(I)}\quad \text{c-C}_3\text{H}_5\text{–CH}_2\text{Br} \xrightarrow{\text{Mg, THF, }-75^\circ} \text{c-C}_3\text{H}_5\text{–CH}_2\text{MgBr} + \text{CH}_2\text{=CHCH}_2\text{CH}_2\text{MgBr}$$

$$\xrightarrow[78\%]{\text{CO}_2} \text{c-C}_3\text{H}_5\text{–CH}_2\text{COOH} + \text{CH}_2\text{=CHCH}_2\text{CH}_2\text{COOH} \quad 92:8$$

A magnesium slurry in ether can be used to prepare benzolcyclobutenylmethylmagnesium bromide at $-50°$ without the rearrangement to *o*-vinylphenylmethylmagnesium bromide, usually encountered with other preparations of the metal.[7]

This form of magnesium is also recommended for the preparation of allylic Grignard reagents in high yield.[8]

Examples:

$$\text{CH}_2\text{=C(CH}_3\text{)CH}_2\text{Br} \xrightarrow[92\%]{\text{1) Mg, THF; 2) CO}_2} \text{CH}_2\text{=C(CH}_3\text{)CH}_2\text{COOH}$$

$$\text{CH}_3\text{CH=CHCH}_2\text{Br} \xrightarrow[94\%]{} \text{(HOOC)(CH}_3\text{)CHCH}_2\text{=CH}_2$$

[1] R. D. Rieke, P. Z.-J. Li, T. P. Burns, and S. T. Uhm, *J. Org.*, **46**, 4323 (1981).
[2] C. Brown, B. J. Sikkel, C. F. Carvalho, and M. V. Sargent, *J.C.S. Perkin I*, 3007 (1982).
[3] C. F. Carvalho and M. V. Sargent, *J.C.S. Chem. Comm.*, 1198 (1982).

[4] Y. Leroux, *Bull.Soc.*, 359 (1968).
[5] J. K. Crandall and H. S. Magaha, *J. Org.*, **47**, 5368 (1982).
[6] K. J. Klabunde, H. F. Efner, L. Satek, and W. J. Donley, *J. Organometal. Chem.*, **71**, 309 (1974).
[7] E. P. Kündig and C. Perret, *Helv.* **64**, 2606 (1981).
[8] W. Oppolzer, E. P. Kündig, P. M. Bishop, and C. Perret, *Tetrahedron Letters*, **23**, 3901 (1982).

Magnesium methoxide, $Mg(OCH_3)_2$.

γ-Hydroxy-α,β-unsaturated esters.[1] Condensation of methyl phenylsulfinylacetate (**1**) with aldehydes [LDA, $ZnCl_2$, $(CH_3)_3CMgBr$] generally results only in the adducts **2**, which are difficult to dehydrate without further undesired reactions.[2]

$$\underset{\mathbf{1}}{C_6H_5\overset{O}{\overset{\|}{S}}CH_2COOCH_3} + RCHO \longrightarrow \underset{\mathbf{2}}{C_6H_5\overset{O}{\overset{\|}{S}}\underset{RCHOH}{\underset{|}{C}}HCOOCH_3}$$

If the condensation is effected with magnesium methoxide in methanol, γ-hydroxy-α,β-unsaturated esters (**3**) are obtained directly via isomerization of the double bond, a [2,3] sigmatropic rearrangement of the sulfoxide group, and cleavage of the sulfenate. The same transformation of **1** to **3** can be conducted in two steps, as formulated in equation (I). The overall yield is about the same as in the direct method.

$$\mathbf{1} \xrightarrow[CH_3OH]{Mg(OCH_3)_2} \left[\text{Mg chelate: } (CH_3O)_2Mg(O{=}S(C_6H_5)CH{=}C(OCH_3)O) \xrightarrow{RCH_2CHO} C_6H_5\overset{O}{\overset{\|}{S}}\underset{CHCH_2R}{\underset{\|}{C}}COOCH_3 \right] \xrightarrow{25\text{–}65\%} \underset{\mathbf{3}}{RCH(OH)CH{=}CHCOOCH_3}$$

$$\text{(I)}\quad \mathbf{1} \xrightarrow[20\text{–}50\%]{1)\ NaH,\ ZnCl_2;\ 2)\ RCH_2CHO} C_6H_5\overset{O}{\overset{\|}{S}}\underset{CHCH_2R}{\underset{\|}{C}}COOCH_3 \xrightarrow[35\text{–}50\%]{Py\ or\ N(C_2H_5)_3} \mathbf{3}$$

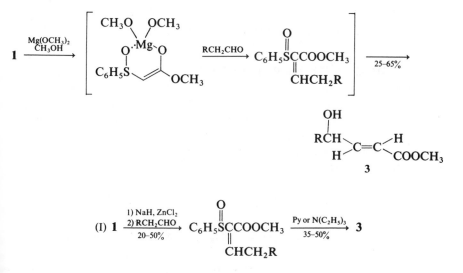

[1] Q. B. Cass, A. A. Jaxa-Chamiec, and P. G. Sammes, *J.C.S. Chem. Comm.*, 1248 (1981).
[2] T. R. Hoye and M. J. Kurth, *J. Org.*, **45**, 3549 (1980).

Magnesium methyl carbonate, 1, 631–633; **2,** 256; **3,** 190–191; **5,** 420–421.

α-Methylene-γ-butyrolactones (**3**, 190; **5**, 420). The lactone **1** has been converted into the α-methylene lactone **3** by carboxylation to give **2** followed by reaction with Eschenmoser's salt (**7**, 130–131). The product (**3**), obtained in 75% overall yield, is converted by known reactions into the pseudoguaianolide aromatin (**4**).[1]

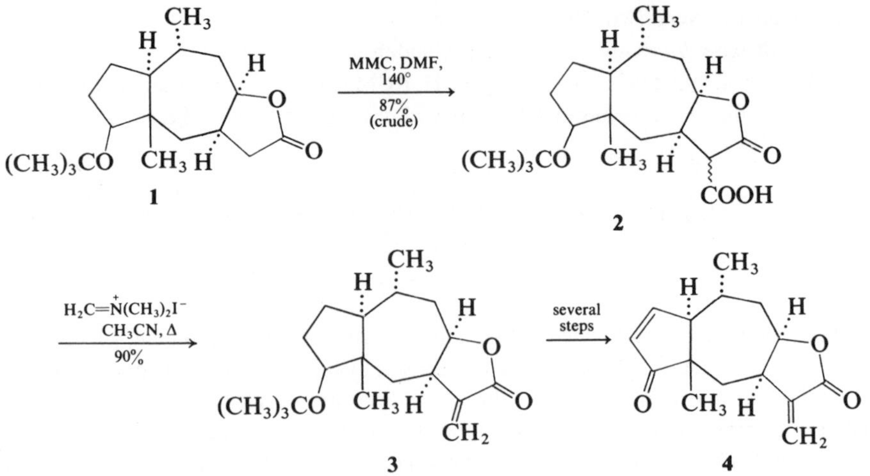

MMC, DMF, 140°; 87% (crude)

1 → **2**

$H_2C{=}\overset{+}{N}(CH_3)_2I^-$, CH_3CN, Δ; 90%

2 → **3**

several steps

3 → **4**

[1] P. T. Lansbury and J. P. Vacca, *Tetrahedron*, **38**, 2797 (1982).

Manganese(II) chloride–Lithium aluminum hydride.

Homoallylic alcohols.[1] A lower valent manganese species, generated from $MnCl_2$ and $LiAlH_4$ (1 : 1), effects addition of allylic bromides to aldehydes and ketones to give homoallylic alcohols in yields of about 50–90%. Addition of crotyl bromide occurs exclusively at the γ-position as in reactions of crotylchromium and crotyltin reagents, but shows only slight stereoselectivity. Equatorial addition is favored in the addition of the crotyl reagent to cyclohexanones.

Example:

$$C_6H_5CHO + CH_3CH{=}CHCH_2Br \xrightarrow[78\%]{MnCl_2\text{-}LiAlH_4,\ THF} C_6H_5\underset{OH}{CH}\text{—}\underset{CH_3}{CH}CH{=}CH_2$$

(*threo*/*erythro* = 35:65)

[1] T. Hiyama, M. Obayashi, and A. Nakamura, *Organometallics*, **1**, 1249 (1982).

Manganese dioxide, 1, 637–643; **2**, 257–263; **3**, 191–194; **4**, 317–318; **5**, 422–424; **6**, 357; **8**, 312.

Dehydrogenation of hydroaromatics.[1] Various hydroaromatics can be aromatized with active MnO_2.

Examples:

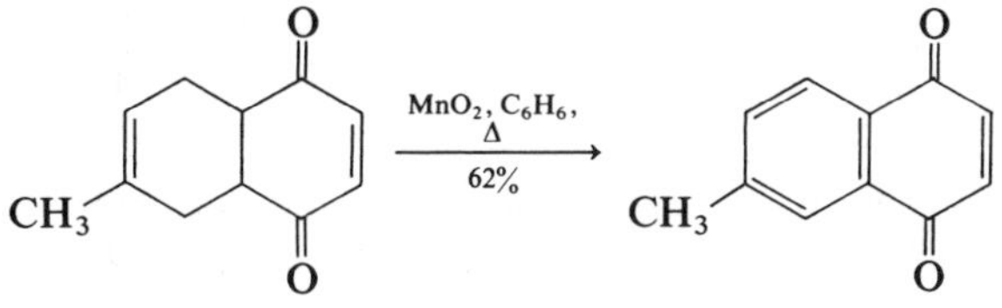

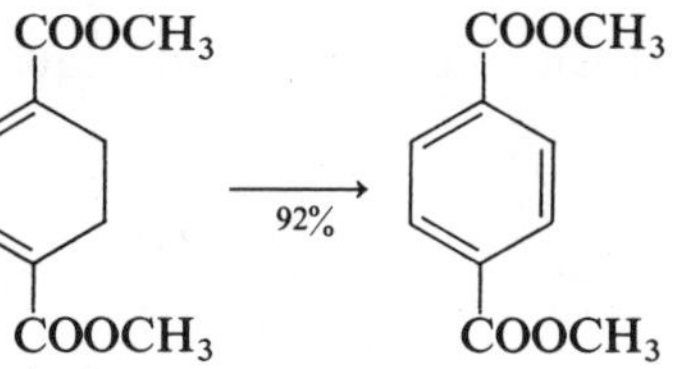

$\xrightarrow[62\%]{MnO_2,\ C_6H_6,\ \Delta}$

$\xrightarrow[92\%]{}$

1,4-*Benzoquinones.*[2] Manganese dioxide is often as efficient as the conventional, expensive oxidant silver oxide for conversion of 1,4-hydroquinones to the quinones, although it is ineffective for certain hydroquinones (2,3-dicyano- and 2,5-diformylhydroquinone, quinizarin). Activated MnO_2 is not necessary.

[1] S. Mashraqui and P. Keehn, *Syn. Comm.*, **12**, 637 (1982).
[2] J. M. Bruce, S. Fitzjohn, and R. T. Pardasani, *J. Chem. Res.* (*S*), 252 (1981).

Meldrum's acid, 8, 313–314; **10**, 252.

Heterocycles. Indoles can be prepared by reaction of phenylhydroxylamine with an acyl Meldrum's acid (**1**) in refluxing CH_3CN to give a product (**2**), which forms a 2-substituted indole (**3**) on treatment with another acyl Meldrum's acid (**1**).[1]

Example:

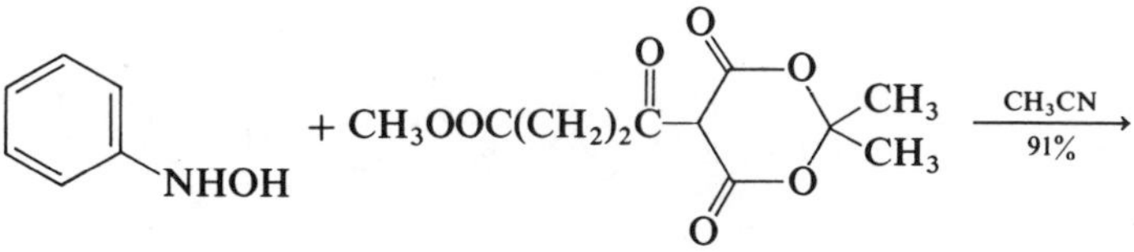

$C_6H_5NHOH + CH_3OOC(CH_2)_2CO$–(Meldrum's acid) **1** $\xrightarrow[91\%]{CH_3CN}$ **2** $\xrightarrow[58\%]{1,\ C_6H_5CH_3}$ **3**

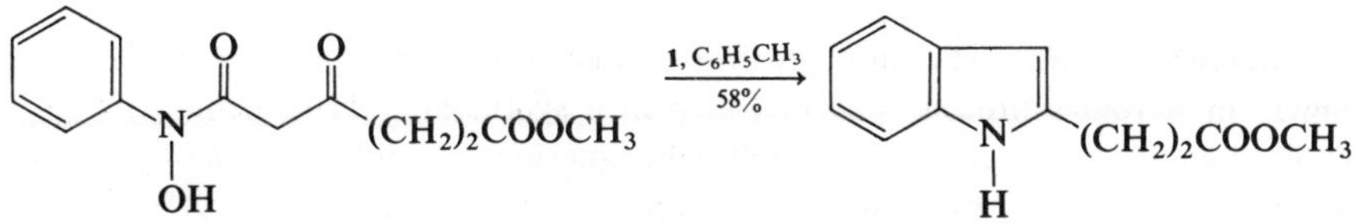

The initial products (**2**) when refluxed in benzene containing a trace of TsOH are converted into 5-substituted 2-phenylisoxazoline-3-ones (**4**).[2]

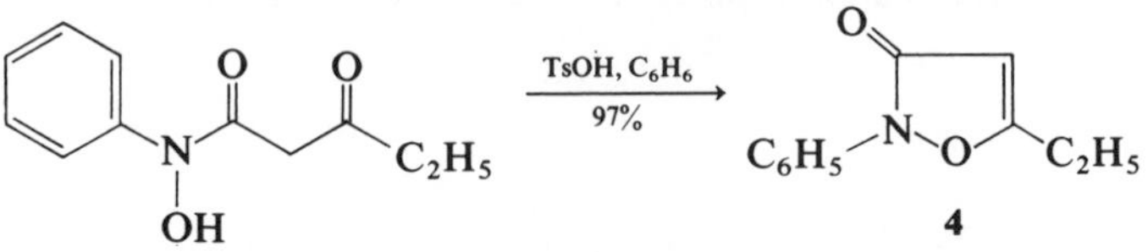

TsOH, C_6H_6, 97%

2b → **4**

[1] K. Mohri, Y. Oikawa, K. Hirao, and O. Yonemitsu, *Heterocycles*, **19**, 515 (1982).
[2] *Idem, ibid.*, **19**, 521 (1982).

(S)-(−)-Menthyl *p*-toluenesulfinate, m.p. 110°, $\alpha_D -202°$; **10**, 406–407. Review.[1]

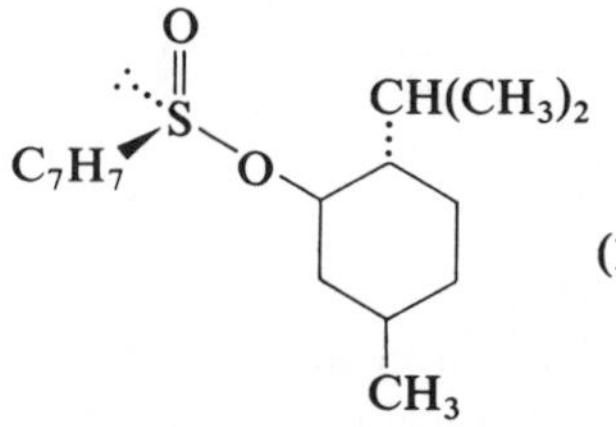

(**1**),

(S)*-2-(p-*Tolylsulfinyl*)-2-*alkenones. A general synthesis of these useful building blocks is shown in equation I for the chiral arylsulfinyl-2-cyclopentenone (**2**). These products are valuable in synthesis because they undergo conjugate addition with various organometallic reagents with high optical induction at the β-position.[2]

(I) 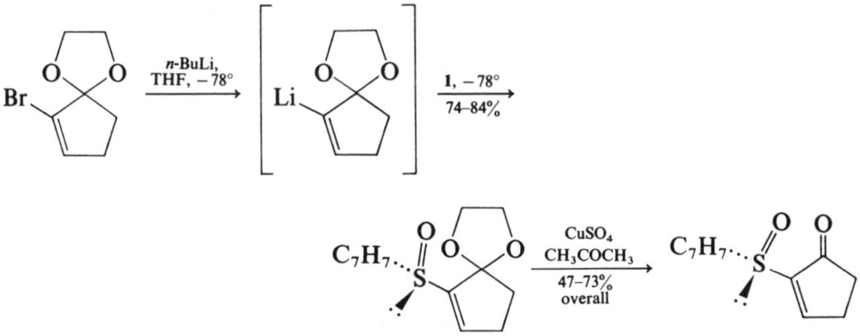 *n*-BuLi, THF, −78° → **1**, −78°, 74–84% → $CuSO_4$, CH_3COCH_3, 47–73% overall

(S)-**2**, α_D + 142°

Steroid synthesis. The sulfoxide (S)-(+)-**2** has been used as a chiral synthon for ring D in steroid synthesis. The first step in a synthesis of 11-ketoequilenin (**5**)[3] involves addition of 6-methoxy-2-naphthylmagnesium bromide followed by *in situ* methylation of the intermediate enolate ion to give **3** in greater than 98% optical purity. This product was converted into optically pure (S, S)-(+)-**4**, the racemate of which has been converted into (±)-11-ketoequilenin.

1) CH_3O-naphthyl-MgBr
2) CH_3I, HMPT
42%

(S)-(+)-**2** → **3** (>98% ee)

1) $(CH_3)_2CuLi$, THF
2) $BrCH_2COOCH_3$, HMPT
98%

(S,S)-(+)-**4** **5**

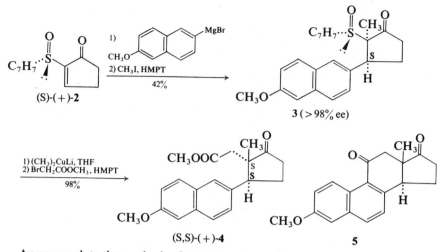

An approach to the synthesis of nonaromatic steroids involves conjugate addition of vinylmagnesium bromide to **2**. In this reaction, prior addition of zinc bromide to

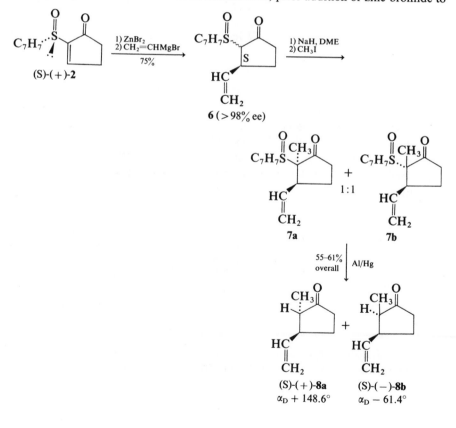

1) $ZnBr_2$
2) $CH_2{=}CHMgBr$
75%

(S)-(+)-**2** → **6** (>98% ee)

1) NaH, DME
2) CH_3I

7a + **7b** (1:1)

55–61% overall Al/Hg

(S)-(+)-**8a** $\alpha_D + 148.6°$

(S)-(−)-**8b** $\alpha_D - 61.4°$

form a chelate gives the desired adduct (**6**) in high optical yield. Methylation of **6** followed by reductive cleavage of the sulfonyl group results in the optically pure steroid intermediates **8a** and **8b** in 55–61% overall yield.[4]

Chiral 3-substituted cycloalkanones. More recently Posner *et al.*[5] have found that the enantioselectivity of the conjugate addition to **2** can be controlled by changes in the experimental conditions. In the absence of a chelating metal, dialkylmagnesium species add predominantly to **2** to give, after desulfonylation, (R)-3-alkylcycloalkanones (equation I). Zinc-chelated **2** on the other hand reacts with organometallic

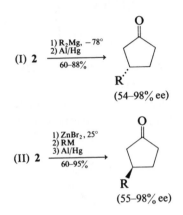

(I) **2** → 1) R_2Mg, −78° 2) Al/Hg; 60–88% → R (54–98% ee)

(II) **2** → 1) $ZnBr_2$, 25° 2) RM 3) Al/Hg; 60–95% → R (55–98% ee)

reagents to give, after desulfonylation, predominantly (S)-3-alkylcycloalkanones (equation II). The bulky alkyltriisopropoxytitanium reagents add in this way even in the absence of $ZnBr_2$ and with particularly high enantioselectivity (>85% ee). The report includes one example of an asymmetric synthesis of a quaternary carbon center (equation III).

(III) $C_7H_7\overset{*}{S}$(O) … CH_3 → 1) $ZnBr_2$ 2) $(p\text{-}CH_3C_6H_4)_2CuLi$ 3) Al/Hg; 50% → $p\text{-}CH_3C_6H_4$ … CH_3

Chiral α-sulfinylhydrazones.[6] The anions of N,N-dimethylhydrazones are converted into optically active α-sulfinylhydrazones **2** by treatment with **1** (equation I).

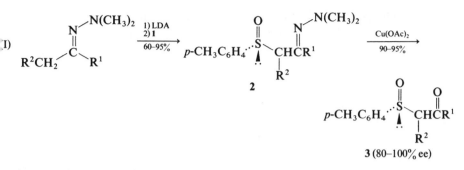

I) $R^2CH_2C(=NN(CH_3)_2)R^1$ $\xrightarrow[60\text{–}95\%]{1)\ LDA,\ 2)\ \mathbf{1}}$ p-$CH_3C_6H_4S(O)CH(R^2)C(=NN(CH_3)_2)R^1$ (**2**) $\xrightarrow[90\text{–}95\%]{Cu(OAc)_2}$ p-$CH_3C_6H_4S(O)CH(R^2)C(O)R^1$

3 (80–100% ee)

[1] G. Solladié, *Synthesis*, 185 (1981).
[2] G. H. Posner, J. P. Mallamo, and K. Miura, *Am. Soc.*, **103**, 2886 (1981); M. Hulce, J. P. Mallamo, L. L. Frye, T. M. Kogan, and G. H. Posner, *Org. Syn.* submitted (1982).
[3] G. H. Posner, M. J. Chapdelaine, and C. M. Lentz, *J. Org.*, **44**, 3661 (1979).
[4] G. H. Posner, M. Hulce, J. P. Mallamo, S. A. Drexler, and J. Clardy, *ibid.*, **46**, 5244 (1981).
[5] G. H. Posner, J. P. Mallamo, M. Hulce, and L. L. Frye, *Am. Soc.*, **104**, 4180 (1982); G. H. Posner, M. Hulce, J. P. Mallamo, and T. P. Kogan, *Org. Syn.*, submitted (1982).
[6] L. Banfi, L. Colombo, and C. Gennari, *Synthesis*, 829 (1982).

Mercury(II) acetate, 1, 644–652;**2**, 264—267; **3**, 194–196, **4**, 319–323; **5**, 424–427; **6**, 358–359; **7**, 222–223; **8**, 315–316; **9**, 291; **10**, 252–253.

Intramolecular aminomercuration.[1] The key step in a synthesis of (−)-deoxoprosophylline (**4**), a piperidine alkaloid, from (S)-serine (**1**) is aminomercuration of **2**. This step proceeds stereoselectively to give **3** with only traces of the C_8-epimer. Hydrolysis of the acetonide group furnishes **4**.

1 ($HOCH_2CH(NH_2)COOH$, S) $\xrightarrow{\text{Several steps}}$ **2** (acetonide, $(CH_3)_2C$, NH$_2$, $(CH_2)_{10}CH_3$) $\xrightarrow[76\%]{1)\ Hg(OAc)_2,\ CH_3OH;\ 2)\ NaBH_4}$ **3** (8, $(CH_2)_{11}CH_3$) $\xrightarrow[86\%]{HCl,\ CH_3OH}$ **4** (HO, HO, $(CH_2)_{11}CH_3$)

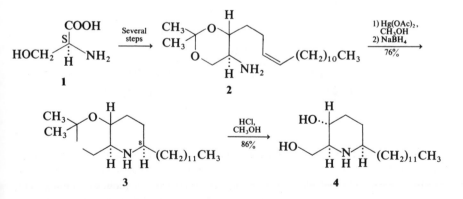

Reductive coupling of alkenes (**10**, 253–254). During the past few years Giese and coworkers[2] have reported that hydride reduction, preferably with sodium trimethoxyborohydride, of solvomercuration products of alkenes generates radicals that can be trapped by electrophilic alkenes such as acrylonitrile.

Examples:

$$C_6H_5CH{=}CH_2 \xrightarrow[2)\ NaCl]{1)\ Hg(OAc)_2,\ CH_3OH} [C_6H_5CH(OCH_3){-}CH_2HgCl] \xrightarrow[60\%]{NaBH(OCH_3)_3,\ CH_2{=}CHCN} C_6H_5CH(OCH_3)CH_2CH_2CH_2CN$$

$$\text{cyclohexene} \xrightarrow[NaCl]{Hg(OAc)_2,\ CH_3OH} [\text{2-(HgCl)-1-}OCH_3\text{-cyclohexane}] \xrightarrow[77\%]{NaBH(OCH_3)_3,\ CH_2{=}CHCN} \text{1-}OCH_3\text{-2-}CH_2CH_2CN\text{-cyclohexane}$$

cis/*trans* = 35:65

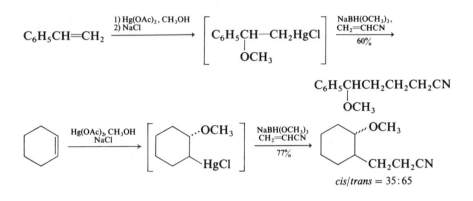

This reductive coupling reaction is a useful route to 5-substituted δ-lactones such as malyngolide (**1**), an antibiotic obtained from blue-green algae.[3]

$$H_2C{=}C(CH_2OTHP)(CH_2)_8CH_3 \xrightarrow[91\%]{1)\ Hg(OAc)_2,\ H_2O,\ THF;\ 2)\ NaBr} BrHgCH_2C(OH)(CH_2OTHP)(CH_2)_8CH_3 \xrightarrow[49\%]{NaBH(OCH_3)_3,\ CH_2Cl_2,\ CH_2{=}C(CH_3)CN}$$

$$CH_3CH(CN)CH_2CH_2C(OH)(CH_2OTHP)(CH_2)_8CH_3 \xrightarrow{TsOH,\ THF} \mathbf{1} + \text{diastereoisomer}\ (1:1)$$

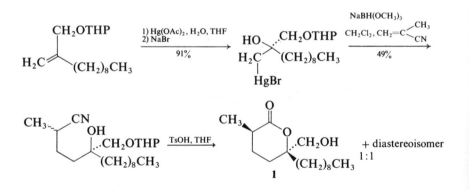

The Giese reaction has been used for reductive cyclization of α,β-enones containing an isolated terminal double bond. Solvomercuration can be carried out selectively on the latter group. On hydride reduction, the mercurial group is reduced preferentially and undergoes addition to the enone group. Experimentally, no products resulting from prior reduction of the enone are observed. Yields of cyclized products are usually satisfactory.[4]

Examples:

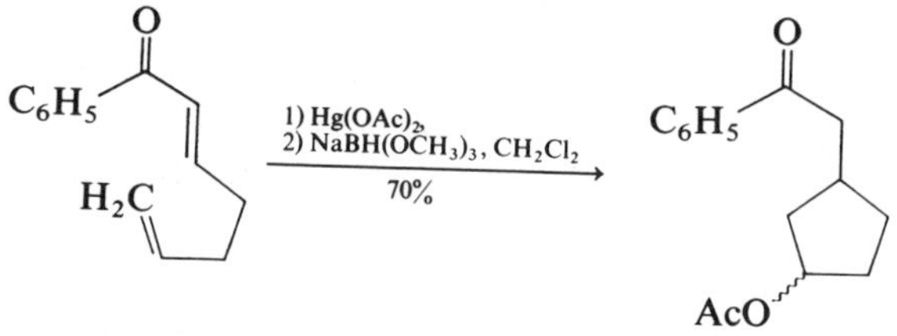

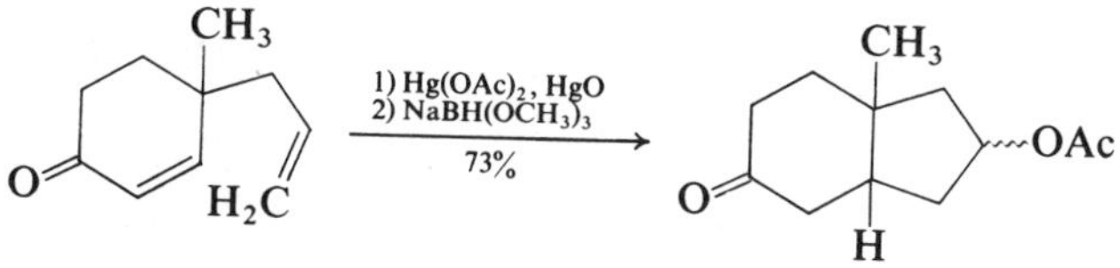

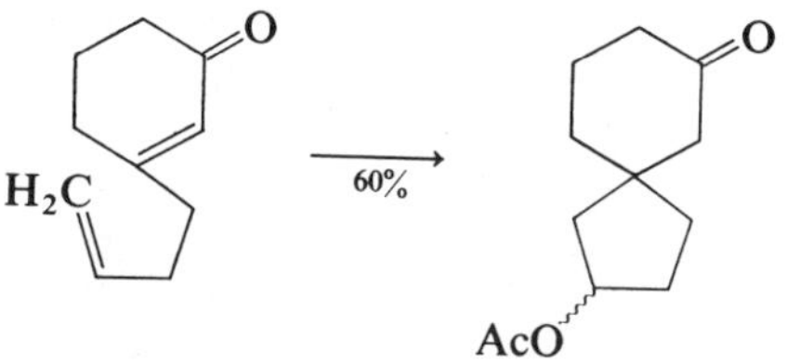

C_6H_5, H_2C — 1) $Hg(OAc)_2$, 2) $NaBH(OCH_3)_3$, CH_2Cl_2 — 70% — C_6H_5, AcO

CH_3, O, H_2C — 1) $Hg(OAc)_2$, HgO 2) $NaBH(OCH_3)_3$ — 73% — CH_3, OAc, O, H

O, H_2C — 60% — O, AcO

[1] Y. Saitoh, Y. Moriyama, H. Hirota, T. Takahashi, and Q. Khuong-Huu, *Bull. Chem. Soc. Japan*, **54**, 488 (1981).

[2] B. Giese, K. Heuck, and U. Lüning, *Tetrahedron Letters*, **22**, 2155 (1981) and references cited therein.

[3] A. P. Kozikowski, T. R. Nieduzak, and J. Scripto, *Organometallics*, **1**, 675 (1982).

[4] S. Danishefsky, S. Chackalamannil, and B.-J. Uang, *J. Org.*, **47**, 2231 (1982).

Mercury(II) nitrate, 7, 223; **10,** 254.

Amidomercuration.[1] The reaction of alkenes with anhydrous $Hg(NO_3)_2$ in the presence of primary amides in CH_2Cl_2 followed by demercuration ($NaBH_4$–NaOH–*n*-$BuNH_2$) results in N-substituted amides in 45–99% yield (equation I).

$$\text{(I)}\quad \begin{matrix} R^1 \\ R^2 \end{matrix}\!\!>C{=}CHR^3 + R^4CONH_2 \xrightarrow[45\text{–}99\%]{1)\,Hg(NO_3)_2\;\;2)\,NaBH_4} R^4CONH{-}\overset{\displaystyle R^1}{\underset{\displaystyle R^2}{C}}{-}CH_2R^3$$

Sulfonamidomercuration.[2] In the presence of $Hg(NO_3)_2$ (1 equiv.), *p*-toluenesulfonamide adds to alkenes to form adducts which are reduced by $NaBH_4$ in NaOH–*n*-$BuNH_2$ to give N-substituted sulfonamides in 30–75% yield. When applied to 1,4- or 1,5-dienes, this sequence leads to pyrrolidines. This reaction is an indirect way to add NH_3 to alkenes, since the N-alkylsulfonamides are reduced by Na–NH_3 to N-alkylamines.[3] $Hg(OAc)_2$ is not effective for this reaction.

Examples:

$$CH_3(CH_2)_3CH{=}CH_2 + TsNH_2 \xrightarrow[52\%]{1)\ Hg(NO_3)_2\ \ 2)\ NaBH_4} TsNHCH(CH_3)(CH_2)_3CH_3$$

$$CH_3CH{=}CHCH_2CH{=}CH_2 + TsNH_2 \xrightarrow[80\%]{} \text{2,5-dimethyl-N-tosylpyrrolidine}$$

$$\text{cycloocta-1,5-diene} + TsNH_2 \xrightarrow[73\%]{} \text{NTs bicyclic} + \text{NTs bicyclic}$$

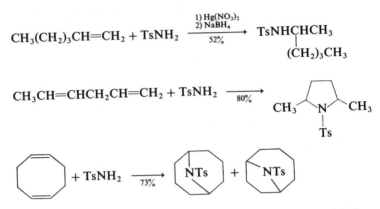

[1] J. Barluenga, C. Jiménez, C. Nájera, and M. Yus, *J.C.S. Chem. Comm.*, 670 (1981).
[2] J. Barluenga, C. Jiménez, C. Nájera, and M. Yus, *ibid.*, 1178 (1981).
[3] J. E. Bäckwell, K. Oshima, R. E. Palermo, and K. B. Sharpless, *J. Org.*, **44**, 1953 (1979).

Mercury(II) oxide–Boron trifluoride.

β-Keto esters. Enolates of aliphatic ketones condense selectively at the more substituted site with CS_2; β-keto dithioesters are obtained on methylation (CH_3I).[1] These products are converted into β-keto esters by HgO–BF_3 and CH_3OH in THF.[2]

Examples:

$$C_6H_5COCH(CH_3)_2 \xrightarrow[67\%]{1)\ t\text{-AmONa},\ CS_2\ \ 2)\ CH_3I} C_6H_5{-}CO{-}C(CH_3)_2(C({=}S)SCH_3) \xrightarrow[87\%]{HgO,\ BF_3,\ CH_3OH,\ THF} C_6H_5COC(CH_3)_2COOCH_3$$

$$\text{2,6-dimethylcyclohexanone} \xrightarrow[45\%\ \text{overall}]{} \text{2-}CH_3\text{-6-}CH_3\text{-6-}COOCH_3\text{-cyclohexanone}$$

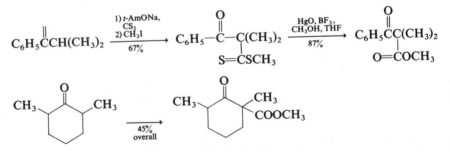

[1] A. Thiullier and J. Vaille, *Bull. Soc.*, 2187 (1962).
[2] A. S. Kende and D. A. Becker, *Syn. Comm.*, **12**, 829 (1982).

Mercury(II) oxide–Tetrafluoroboric acid, 9, 293; 10, 253–257.

Cycloamination of 1,3-dienes.[1] Linear and cyclic 1,3-dienes react with the reagent and primary aryl amines to form 1,4-cycloadducts.

Examples;

$$\text{cyclooctadiene} + 2C_6H_5NH_2 \xrightarrow{HgO/HBF_4,\ DMF,\ 80^\circ} \left[\text{NHC}_6\text{H}_5 \ldots \text{NHC}_6\text{H}_5 \right] \xrightarrow{45\%} C_6H_5N$$

$$CH_3CH{=}CHCH{=}CHCH_3 + C_6H_5NH_2 \xrightarrow{58\%} CH_3 \ldots N(C_6H_5) \ldots CH_3$$

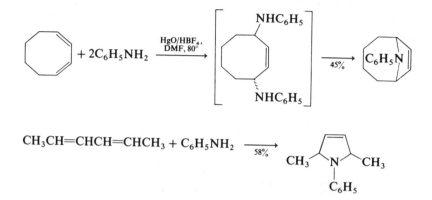

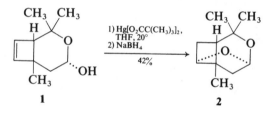

[1] J. Barluenga, J. Pérez-Prieto, and G. Asenio, *J.C.S. Chem. Comm.*, 1181 (1982).

Mercury(II) pivalate, $Hg[O_2CC(CH_3)_3]_2$.

Intramolecular oxymercuration.[1] Oxymercuration of **1** with mercury(II) pivalate followed by reduction of the intermediate mercury complex results in **2**, isolineatin.

$$\mathbf{1} \xrightarrow[42\%]{1)\ Hg[O_2CC(CH_3)_3]_2,\ THF,\ 20^\circ;\ 2)\ NaBH_4} \mathbf{2}$$

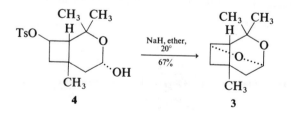

This structure had originally been assigned to lineatin, an aggregation pheromone of the female ambrosia beetle, which has since been shown to have structure **3**. The same laboratory has effected synthesis of **3** from **4**. Both **1** and **4** were obtained from a common intermediate in the total synthesis.

$$\mathbf{4} \xrightarrow[67\%]{NaH,\ ether,\ 20^\circ} \mathbf{3}$$

[1] J. D. White, M. A. Avery, and J. P. Carter, *Am. Soc.*, **104**, 5486 (1982).

Mercury(II) trifluoroacetate, 3, 195; **4,** 325; **6,** 360; **8,** 316–317; **9,** 294–296.

Cyclic enol ethers.[1] Reaction of the *cis*-acetylenic alcohol **(1)** with this Hg(II) salt in the presence of a base leads to **2**. The organomercury intermediate can be trapped by a N-halosuccinimide to give **3**.

$(CH_2)_3CH_3$, OH, CH_2 — $Hg(O_2CCF_3)_2$, $N(C_2H_5)_3$, 45% → H, O, $(CH_2)_3CH_3$, H, H

1 **2**

87% 1) $Hg(O_2CCF_3)_2$, $N(C_2H_5)_3$ 2) NIS

H, O, $(CH_2)_3CH_3$, I, H

3

$(CH_2)_3CH_3$, OH, CH_2 — 88% → H, O, $(CH_2)_3CH_3$, H

4 **5**

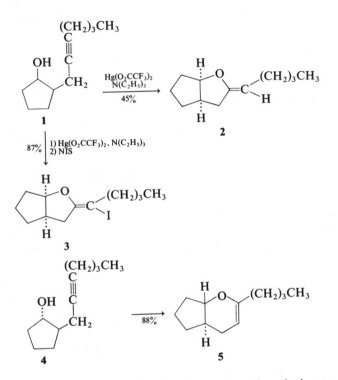

Cyclization of *trans*-isomer **(4)** proceeds more slowly and results in an endocyclic enol ether **5**.

Oxy-Cope rearrangement. Tertiary 1,5-hexadiene-3-ols undergo oxy-Cope rearrangement at room temperature on treatment with 1 equiv. of mercury(II) trifluoroacetate and subsequent demercuration with sodium borohydride.[2] The corresponding secondary alcohols undergo polymerization in the presence of this salt.

Examples:

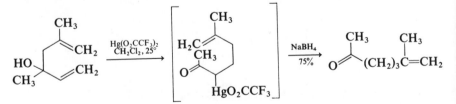

CH_3, CH_2, HO, CH_2, CH_3 — $Hg(O_2CCF_3)_2$, CH_2Cl_2, 25° → [CH_3, H_2C, CH_3, O, HgO_2CCF_3] — $NaBH_4$, 75% → CH_3, CH_3, O, $(CH_2)_3C{=}CH_2$

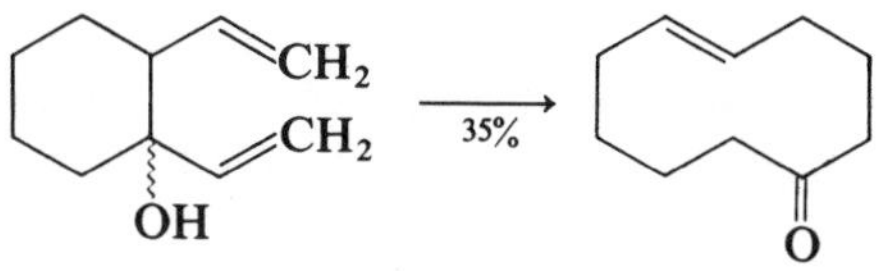

CH2 CH2 OH 35% O

α-Alkylidene-β-hydroxy-γ-methylenebutyrolactones.[3] This unit (**2**) is found in several species of the Lauracea family. In a recent synthesis of these lactones, the final step is enol lactonization of the acetylenic acids **1** with mercury(II) trifluoroacetate (**8**, 315). Despite the moderate yields, no other products could be isolated. The enol lactones (**2**) could also be obtained in comparable yield by treatment of **1** with aqueous sodium bicarbonate.

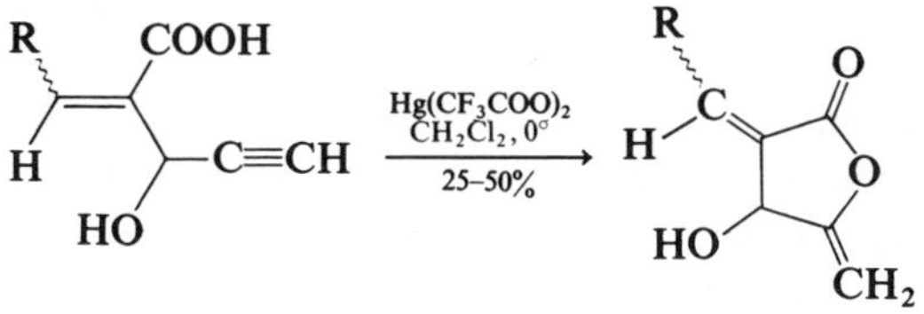

R COOH H C≡CH HO **1** $\xrightarrow[25\text{–}50\%]{Hg(CF_3COO)_2,\ CH_2Cl_2,\ 0^\circ}$ R C H O O HO CH2 **2**

[1] M. Riediker and J. Schwartz, *Am. Soc.*, **104**, 5842 (1982).
[2] N. Bluthe, M. Malacria, and J. Gore, *Tetrahedron Letters*, **23**, 4263 (1982).
[3] S. W. Rollinson, R. A. Amos, and J. A. Katzenellenbogan, *Am. Soc.*, **103**, 4114 (1981).

Methanesulfonic acid, 1, 666–667; **2**, 270; **4**, 326; **10**, 256.

Cyclodehydration. Neat, anhydrous CH_3SO_2OH is at least as useful as polyphosphoric acid for cyclization of 3-arylpropanoic acids and 4-arylbutanoic acids to indanones and tetralones.[1] However, it does not appear to be useful for intermolecular Friedel-Crafts condensation. Examples:

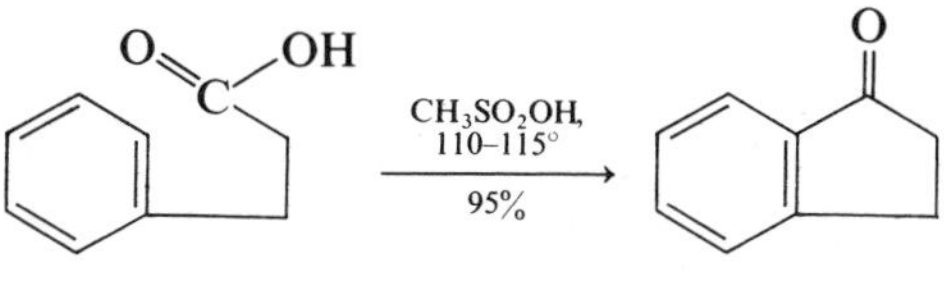

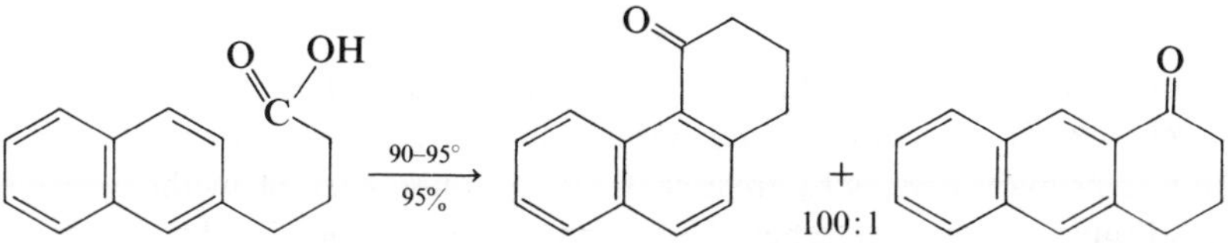

O C OH $\xrightarrow[95\%]{CH_3SO_2OH,\ 110\text{–}115^\circ}$ O

O OH C $\xrightarrow[95\%]{90\text{–}95^\circ}$ O + O 100:1

[1] V. Premasagar, V. A. Palaniswamy, and E. J. Eisenbraun, *J. Org.*, **46**, 2974 (1981).

Methanesulfonyl chloride, 1, 662–664; **2,** 268–269; **9,** 326–327; **5,** 425–436; **6,** 362–363.

RCOOH→RC≡N. A one-pot method for this conversion is outlined in equation (I). It is not applicable to acids substituted by NH_2, NHR, OH, or SH groups.[1]

$$\text{(I)}\ RCOOH + CH_3SO_2Cl \xrightarrow{Py} RC(=O){-}OSO_2CH_3 \xrightarrow{NH_3} RC(=O)NH_2 \xrightarrow[\text{65–80\% overall}]{CH_3SO_2Cl} RC{\equiv}N$$

[1] A. D. Dunn, M. T. Mills, and W. Henry, *Org. Prep. Proc. Int.*, **14**, 396 (1982).

Methanol–Sodium tetraborate.

7α-Methoxylation of cephalosporins.[1] Bromination of the cephalosporin derivative **1** followed by addition of methanol and sodium borate (**4**, 461) introduces a methoxy group at C_7 (**2**). The side chain is removed by successive treatment with PCl_5 and Girard's reagent T to give **3**.

$C_6H_5SO(CH_3)C{=}CHNH$–[cephem]–CH_2STet, $COOCH(C_6H_5)_2$ (**1**) $\xrightarrow{\text{1) } Br_2,\ CH_2Cl_2;\ \text{2) } CH_3OH,\ Na_2B_4O_7}$ **2** (7α-OCH_3)

Tet = 1-methyltetrazol-5-yl (N—N, N, CH_3)

2 $\xrightarrow{\text{1) } PCl_5;\ \text{2) Girard T}}$ H_2N, OCH_3, CH_2STet, $COOCH(C_6H_5)_2$ (**3**)

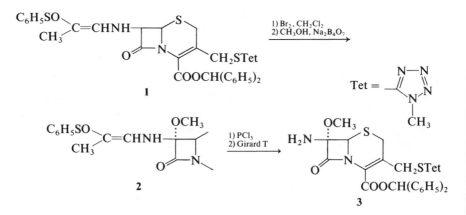

[1] K. Katano, K. Atsumi, K. Nishihata, F. Kai, E. Akita, and T. Niida, *Chem. Pharm. Bull. Japan*, **30**, 3054 (1982).

Methoxyamine, CH_3ONH_2. Mol. wt. 47.06, b.p. 49 – 50°. *Highly poisonous.*

RLi→RNH₂.[1] Various organolithium reagents are converted into the corresponding primary amines by treatment with 2 equiv. each of methoxyamine and methyllithium in hexane–ether. Yields are in the range 55–95%. Omission of methyllithium or substitution by *n*-butyllithium markedly reduces the yield. In fact, use of methoxyamine alone for amination of organometallics was first reported in

1938,[2] but received little attention because of the requirement for 2 equiv. of the substrate for satisfactory yields.

[1] P. Beak and B. J. Kokko, *J. Org.*, **47**, 2822 (1982).
[2] N. J. Sheverdina and Z. Kocheshkov, *J. Gen. Chem. U.S.S.R.*, **8**, 1825 (1938).

4(R)-Methoxycarbonyl-1,3-thiazolidine-2-thione, 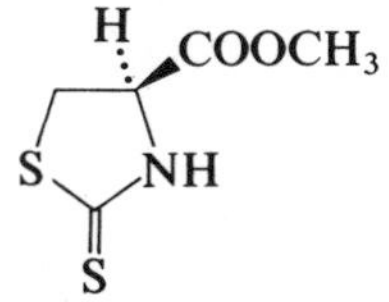 (**1**), $\alpha_D -67°$.

The thione is prepared by reaction of L-cysteine methyl ester hydrochloride and CS_2 in the presence of $N(C_2H_5)_3$.[1]

Chiral induction. Earlier investigations have shown that 3-acyl derivatives of the parent 1,3-thiazolidine-2-thione react with amines under mild conditions to form amides (equation I). Amides can also be prepared from amino alcohols and aminophenols.[2]

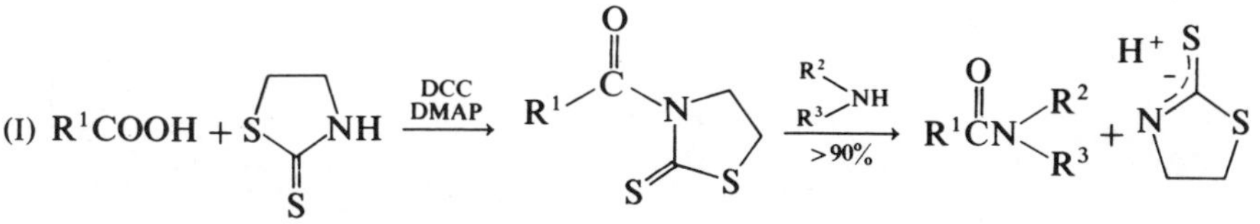

(I) R^1COOH + [thiazolidine-2-thione] $\xrightarrow{\text{DCC, DMAP}}$ $R^1C(O)$–N[thiazolidine-2-thione] $\xrightarrow[>90\%]{R^2R^3NH}$ $R^1C(O)NR^2R^3$ + H^+ [thiazolidine-2-thione anion]

The chiral 1,3-thiazolidine-2-thione (**1**) has been used to differentiate between two identical groups attached to a prochiral carbon atom, a distinction that has been limited to certain enzymes.[3] Thus, the two ligands in the optically active diamide **2**, prepared from 3-methylglutaric acid and **1**, differ in their reactivity with amines, particularly cyclic secondary amines (Figure I). Reaction with piperidine results

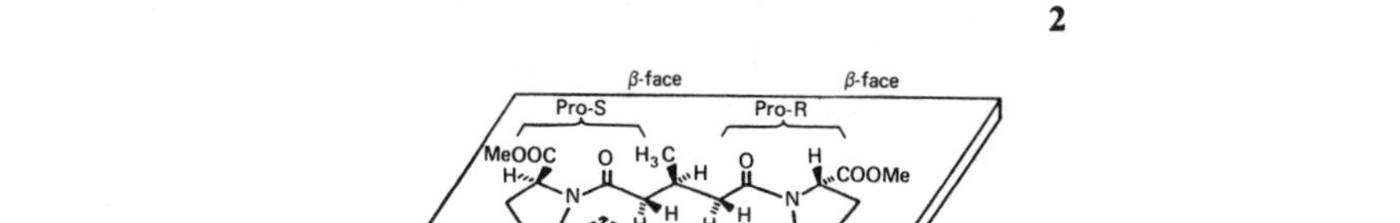

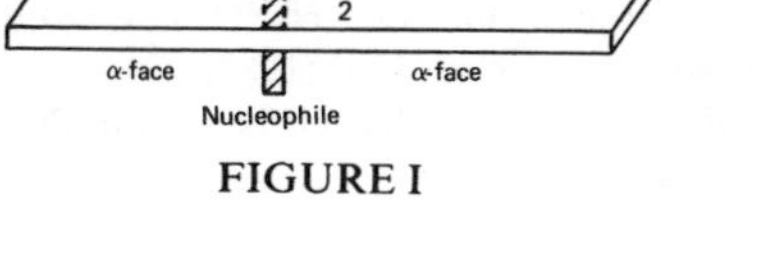

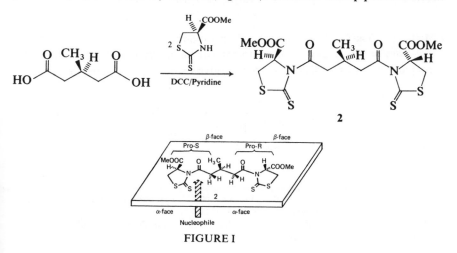

FIGURE I

mainly in attack at the pro-S ligand to give **3** as the major product. The structure was established by reaction with a second nucleophile to give **4**, which was converted into (−)-(3S)-3-methylvalerolactone (**6**). Compound **3** was also converted into a number of optically pure acyclic products in generally high yield. This new chemical differentiation between enantiotopic groups may have general applications.

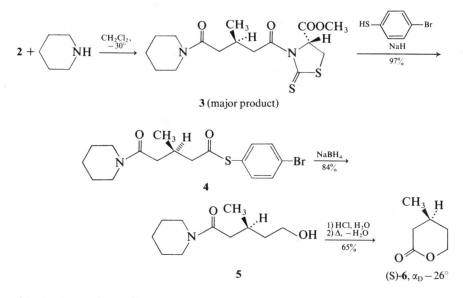

2 + piperidine $\xrightarrow{CH_2Cl_2,\ -30°}$ **3** (major product) $\xrightarrow[97\%]{HS\text{-}C_6H_4\text{-}Br,\ NaH}$

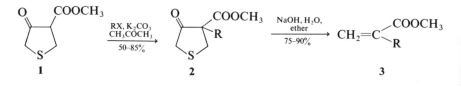

4 $\xrightarrow[84\%]{NaBH_4}$ **5** $\xrightarrow[65\%]{1)\ HCl,\ H_2O;\ 2)\ \Delta,\ -H_2O}$ (S)-**6**, $\alpha_D -26°$

[1] Y. Nagao, M. Yagi, T. Ikeda, and E. Fujita, *Tetrahedron Letters*, **23**, 205 (1982).
[2] E. Jujita, *Pure Appl. Chem.*, **53**, 1141 (1981).
[3] Y. Nagao, T. Ikeda, M. Yagi, E. Fujita, and M. Shiro, *Am. Soc.*, **104**, 2079 (1982).

4-Methoxycarbonylthiolane-3-one (1). Preparation.[1]

α-Substituted acrylates.[2] On treatment with base, alkyl derivatives of **1** undergo retrograde Dieckmann-Michael reactions to give α-substituted acrylates (**3**).

1 $\xrightarrow[50\text{–}85\%]{RX,\ K_2CO_3,\ CH_3COCH_3}$ **2** $\xrightarrow[75\text{–}90\%]{NaOH,\ H_2O,\ ether}$ $CH_2{=}C(R)COOCH_3$ (**3**)

[1] R. B. Woodward and R. H. Eastman, *Am. Soc.*, **68**, 2229 (1946); G. Yamada, T. Ishii, M. Kimura, and K. Hosaka, *Tetrahedron Letters*, 1353 (1981).
[2] P. G. Baraldi, A. Barco, S. Benetti, F. Moroder, G. P. Pollini, D. Simoni, and V. Zanirato, *J.C.S. Chem. Comm.*, 1265 (1982).

2-Methoxy-4-furyllithium (1). Preparation:

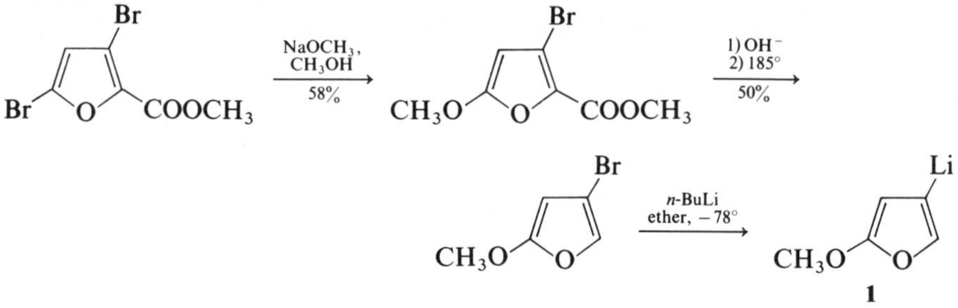

Br, Br, O, $COOCH_3$ — $NaOCH_3$, CH_3OH, 58% → Br, CH_3O, O, $COOCH_3$ — 1) OH^- 2) 185°, 50% →

Br, CH_3O, O — n-BuLi, ether, −78° → Li, CH_3O, O

1

Cardenolides. A short and efficient synthesis of digitoxigenin (**4**) utilizes as the key step the reaction of an α,β-unsaturated 17-keto steroid with **1**.

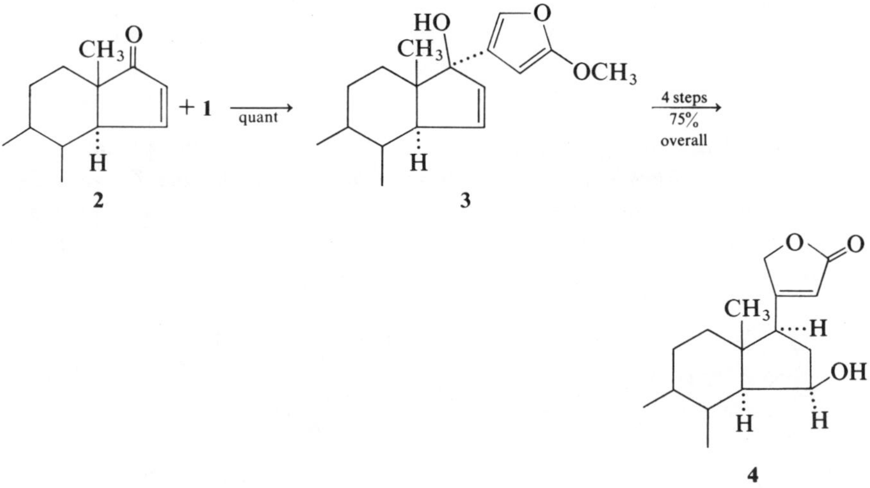

CH_3, O, H — **2** + **1** — quant → HO, CH_3, O, OCH_3, H — **3** — 4 steps, 75% overall →

O, O, CH_3, H, OH, H, H

4

[1] R. Marini-Bettolo, P. Flecker, T. Y. R. Tsai, and K. Wiesner, *Can. J. Chem.*, **59**, 1403 (1981).

(E)-1-(Methoxymethoxy)butene-2-yl(tri-n-butyl)tin (1).

Preparation:

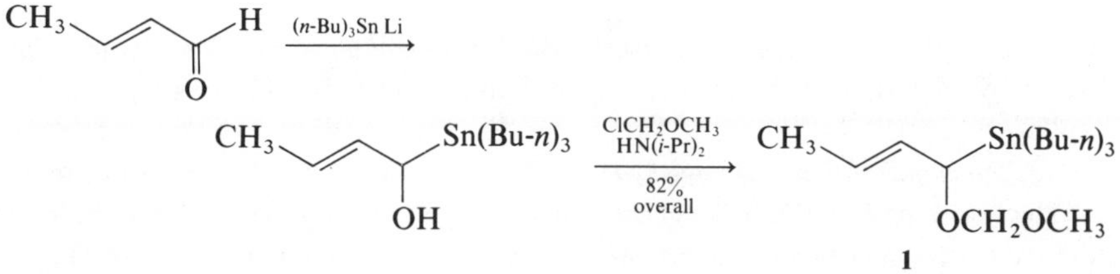

CH_3, H, O — $(n\text{-Bu})_3SnLi$ →

CH_3, $Sn(Bu\text{-}n)_3$, OH — $ClCH_2OCH_3$, $HN(i\text{-}Pr)_2$, 82% overall → CH_3, $Sn(Bu\text{-}n)_3$, OCH_2OCH_3

1

trans-4,5-*Disubstituted butyrolactones.*[1] This reagent (**1**) reacts with aldehydes to give stereoselectively *threo*-4-hydroxy-3-methyl-*cis*-1,2-enol ethers. The products can be converted into *trans*-4,5-disubstituted butyrolactones.

Example:

$C_6H_5CHO + \mathbf{1} \xrightarrow[79\%]{115°} C_6H_5CH(OH)CH(CH_3)CH{=}CH(OCH_2OCH_3) \xrightarrow{HCl, THF}$ [2-hydroxy-4-methyl-5-phenyltetrahydrofuran] $\xrightarrow[89\%]{PCC}$ *trans*-4-methyl-5-phenylbutyrolactone

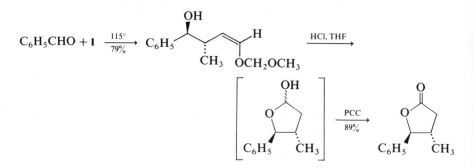

[1] A. J. Pratt and E. J. Thomas, *J.C.S. Chem. Comm.*, 1115 (1982).

(S)-(+)-2-Methoxymethylpyrrolidine, 10, 259–260. Preparation from proline.[1]

Asymmetric Michael addition to ω-nitrostyrenes.[2] The enamine (**2**) formed from cyclohexanone and this prolinol derivative reacts with 2-aryl-1-nitroethylenes (**3**) to form, after acid hydrolysis of the primary adduct, essentially only one (**4**) of the four possible γ-nitro ketones.

2 + $Ar{-}CH{=}CH{-}NO_2$ (**3**) $\xrightarrow[65–80\%]{1)\ Ether,\ -80°\quad 2)\ H_3O^+}$ (2S, 1′R)-**4** (> 90% de)

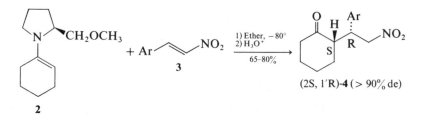

[1] D. Seebach, H.-O. Kalinowski, B. Bastani, G. Crass, H. Daum, H. Dorr, N. P. DuPreez, V. Ehrig, W. Langer, C. Nüssler, H.-A. Oei, and M. Schmidt. *Helv.*, **60**, 301 (1977).
[2] S. J. Blarer, W. B. Schweizer, and D. Seebach, *ibid.*, **65**, 1637 (1982).

(E,Z)-1-Methoxy-2-methyl-3-trimethylsilyloxy-1,3-pentadiene (1). The diene is prepared by silylation of the corresponding ketone with $ClSi(CH_3)_3$ using $N(CH_3)_3$–$ZnCl_2$.[1]

Cyclocondensations with aldehydes.[2] The diene **1** reacts with aldehydes to give products of type **2** and/or **3**, depending on the Lewis acid catalyst. With BF_3 etherate, the *trans*-adduct **3** predominates, whereas with $ZnCl_2$ catalysis the *cis*-adduct **2** is

1) Cat. 2) TFA, CCl_4

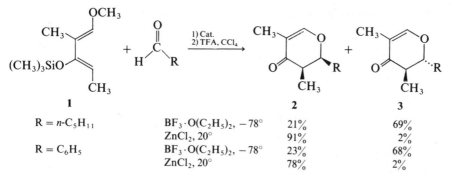

	Cat.	2	3
$R = n\text{-}C_5H_{11}$	$BF_3 \cdot O(C_2H_5)_2$, $-78°$	21%	69%
	$ZnCl_2$, 20°	91%	2%
$R = C_6H_5$	$BF_3 \cdot O(C_2H_5)_2$, $-78°$	23%	68%
	$ZnCl_2$, 20°	78%	2%

greatly favored. In the latter reaction, by workup under anhydrous conditions, the simple product (**4**) of pericyclic addition can be isolated in 41% yield.[3] Evidence supporting pericyclic cycloaddition has been obtained from reactions with other 1,3-dienes.

4 $\xrightarrow{\text{TFA}}$ **2** $(R = C_6H_5)$

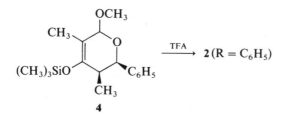

In contrast, acyclic intermediates (**5** and **6**) have been isolated in the reaction of **1** with C_6H_5CHO in the presence of BF_3 etherate. Evidently this cycloaddition operates through a *threo*-selective aldol-type process.

1 + C_6H_5CHO $\xrightarrow{46\%}$ **5** + **6** (2:1)

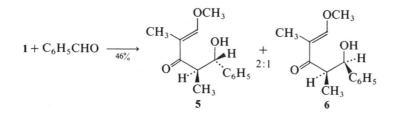

[1] S. Danishefsky, C.-F. Yan, R. K. Singh, R. B. Gammill, P. M. McCurry, Jr., N. Fritsch, and J. Clardy, *Am. Soc.*, **101**, 7001 (1979).

[2] S. Danishefsky, E. R. Larson, and D. Rakin, *ibid.*, **104**, 6457 (1982).

[3] E. R. Larson and S. Danishefsky, *ibid.*, **104**, 6458 (1982).

(Z)-2-Methoxy-1-(phenylthio)-1,3-butadiene (1). Mol. wt. 192.27, oil.

Preparation[1]:

$$LiCH(SC_6H_5)_2 \xrightarrow[48\text{–}52\%]{1)\ CH_2{=}CHCHO\quad 2)\ (CH_3O)_2SO_2} H_2C{=}CH{-}CH(OCH_3){-}CH(SC_6H_5)_2 \xrightarrow[55\text{–}64\%]{(CuOSO_2CF_3)_2\cdot C_6H_6} H_2C{=}CH{-}C(OCH_3){=}CH{-}SC_6H_5\ (\mathbf{1})$$

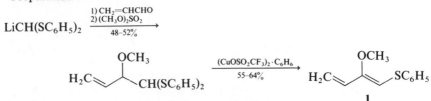

Diels-Alder reaction with α,β-enones.[2] The Diels-Alder reaction of **1** with α,β-enones catalyzed with a Lewis acid ($MgBr_2$ or $C_2H_5AlCl_2$) proceeds stereospecifically to give *endo*-adducts in which the phenylthio function is oriented *ortho* to the carbonyl group (7, 75–76). The related dienes **2** and **3** undergo cycloaddition with similar stereo- and regiospecificity.

Examples:

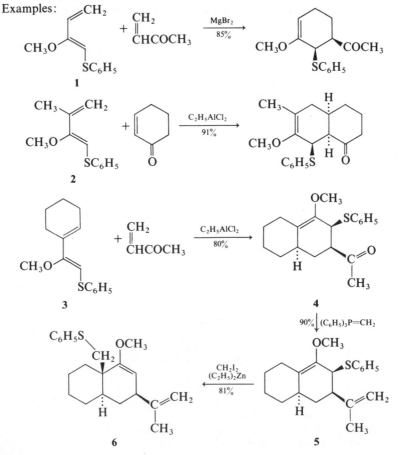

1 + $CH_2{=}CHCOCH_3$ $\xrightarrow[85\%]{MgBr_2}$ (adduct with CH_3O, SC_6H_5, $COCH_3$)

2 + cyclohexenone $\xrightarrow[91\%]{C_2H_5AlCl_2}$ (adduct with CH_3, CH_3O, C_6H_5S, H, H, O)

3 + $CH_2{=}CHCOCH_3$ $\xrightarrow[80\%]{C_2H_5AlCl_2}$ **4** (with OCH_3, SC_6H_5, $C{=}O$, CH_3, H)

4 $\xrightarrow[90\%]{(C_6H_5)_3P{=}CH_2}$ **5** (with OCH_3, SC_6H_5, $C{=}CH_2$, CH_3, H)

5 $\xrightarrow[81\%]{CH_2I_2,\ (C_2H_5)_2Zn}$ **6** (with C_6H_5S, CH_2, OCH_3, $C{=}CH_2$, CH_3, H)

Reaction of these adducts with a zinc carbenoid is accompanied by a [2,3]-sigmatropic rearrangement of the phenylthio group to generate a new carbon-carbon bond at the original allylic center. This reaction converts **5** stereospecifically into **6**, related to eudesmane sesquiterpenes.

[1] T. Cohen, R. J. Ruffner, D. W. Shull, E. R. Fogel, and J. R. Falck, *Org. Syn.*, **59**, 202 (1980).
[2] T. Cohen and Z. Kosarych, *J. Org.*, **47**, 4005 (1982).

2-Methoxypropene (Isopropenyl ether), 2, 230–231; **5,** 360.

Acetonation. Gelas and Horton[1] have reviewed the acetonation of diol groups in carbohydrates with this reagent. In contrast with acetonation with acetone and an acid catalyst, acetonation with 2-methoxypropene in DMF and a trace of TsOH proceeds under kinetic control, and the products often differ from those obtained by the classical method. The striking difference is best illustrated with D-glucose (equation I).[2] Usually both modes of acetonation obtain with 2,2-dimethoxypropane as reagent. Generally the anomeric group does not participate in reactions with 2-methoxypropene.

(I)

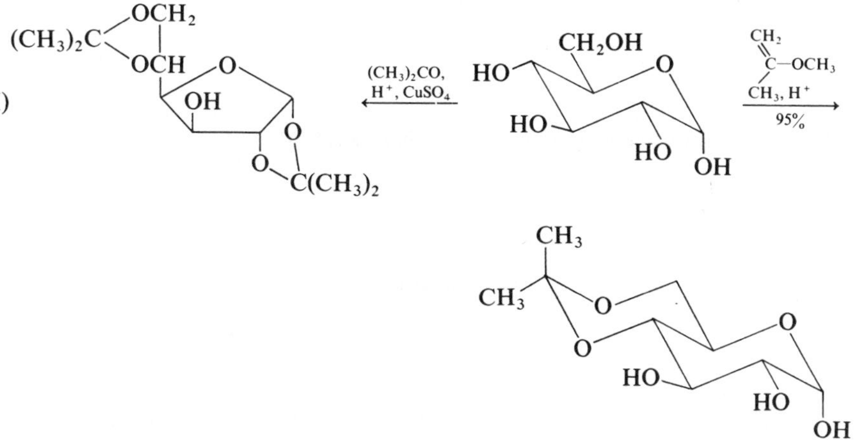

[1] G. Gelas and D. Horton, *Heterocycles*, **16**, 1587 (1981).
[2] E. Fanton, J. Gelas, and D. Horton, *J. Org.*, **46**, 4057 (1981).

(R)- and (S)-α-Methoxy-α-trifluoromethylphenylacetic acid (1, MTPA). Mol. wt. 234.17; (R)-(+)-, b.p. 116–118°/1.5 mm, $\alpha_D - 70°$; (S)-(−)-, b.p. 95–97°/0.05 mm, $\alpha_D - 72°$. Supplier: Aldrich.

Preparation:

$$C_6H_5\overset{\overset{\displaystyle O}{\|}}{C}CF_3 \xrightarrow[]{\substack{1)\ NaCN \\ 2)\ (CH_3)_2SO_4}} C_6H_5\overset{\overset{\displaystyle OCH_3}{|}}{\underset{\underset{\displaystyle CF_3}{|}}{C}}-CN \xrightarrow[\substack{63\% \\ overall}]{H_3O^+} C_6H_5\overset{\overset{\displaystyle OCH_3}{|}}{\underset{\underset{\displaystyle CF_3}{|}}{C}}-COOH$$

DL-**1**

Racemic **1** is resolved with (+)-α-phenylethylamine.

The reagent has been recommended for determination of the optical purity of alcohols and amines by use of fluorine NMR, which is simple and in an uncongested region.[1]

Resolution of bromohydrins.[2] Diastereoisomeric esters of bromohydrins with MTPA are readily separated by fractional crystallization and characterized by NMR. The optically pure bromo MTPA esters are convertible by known methods into chiral alcohols, diols, and epoxides, including arene oxides.

[1] J. A. Dale, D. L. Dulh, and H. S. Mosher, *J. Org.*, **34**, 2543 (1969).
[2] S. K. Balani, D. R. Boyd, E. S. Cassidy, R. M. E. Greene, K. M. McCombe, N. D. Sharma, and W. B. Jennings, *Tetrahedron Letters*, **22**, 3277 (1981).

4-Methoxy-2,3,6-trimethylbenzenesulfonyl chloride, (MtrCl).

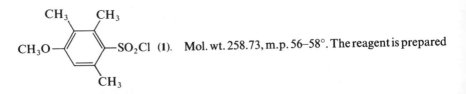

CH_3O–(2,3,6-trimethylphenyl)–SO_2Cl (**1**). Mol. wt. 258.73, m.p. 56–58°. The reagent is prepared by reaction of 2,3,5-trimethylanisole with chlorosulfonic acid (73% yield).[1]

Protection of amino groups. The Mtr group is a useful acid-labile protecting group for amines.[1,2] It is generally removed by TFA-thioanisole at 50°. When attached to the ε-amino group of lysine it is somewhat more resistant, but addition of methanesulfonic acid allows ready cleavage. Removal of the protecting group from an imidazole ring is accomplished with TFA–$S(CH_3)_2$. Deprotection of the Mtr group attached to an indole ring is possible with HF or CH_3SO_2OH, but not with TFA.

[1] M. Fujino, M. Wakimasu, and C. Kitada, *Chem. Pharm. Bull*, **29**, 2825 (1981).
[2] M. Wakimasu, C. Kitada, and M. Fujino, *ibid*, **30**, 2766 (1982); T. Fukuda, M. Wakimasu, S. Kobayashi, and M. Fujino, *ibid.*, **30**, 2825 (1982).

[Methoxy(trimethylsilyl)methyl]lithium, $(CH_3)_3Si\underset{\underset{Li}{|}}{C}HOCH_3$ (**1**). The reagent is prepared by treatment of (methoxymethyl)trimethylsilane with *sec*-butyllithium in THF at $-78°$ and then at $-30°$.

Carbonyl homologation. The reagent forms adducts with aldehydes and ketones, which do not undergo spontaneous elimination of $(CH_3)_3SiOLi$ (Peterson reaction). However, the alcohols are converted into enol ethers on treatment with KH in THF at 0°, or into aldehydes when treated with 90% formic acid.[1]

Example:

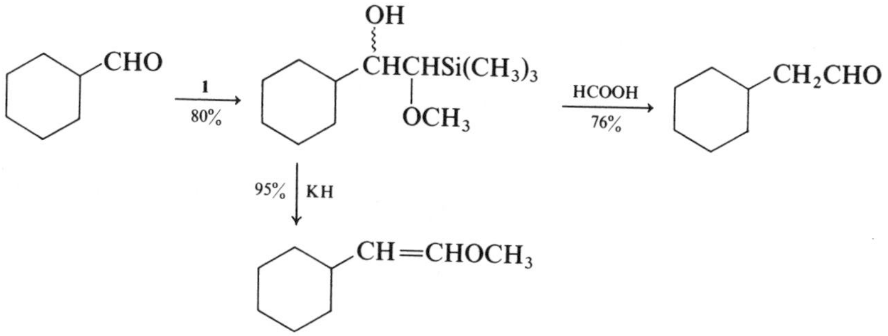

The major advantage of this reagent is its ability to react with carbonyl compounds that do not react with Wittig reagents. Thus it was used to convert the ketone group of **2** into the isomeric enol ethers (**4**), a reaction that failed with a number of other reagents.[2]

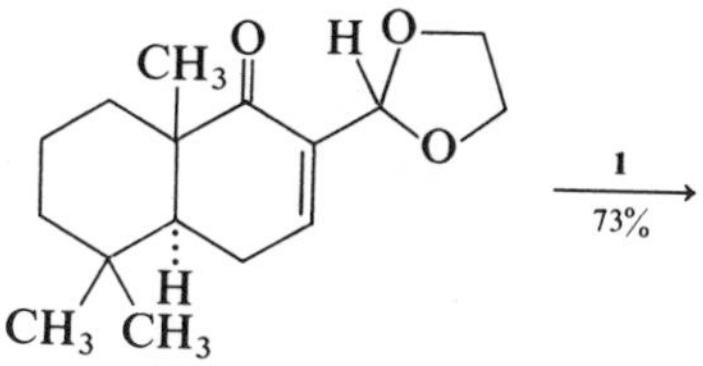

2

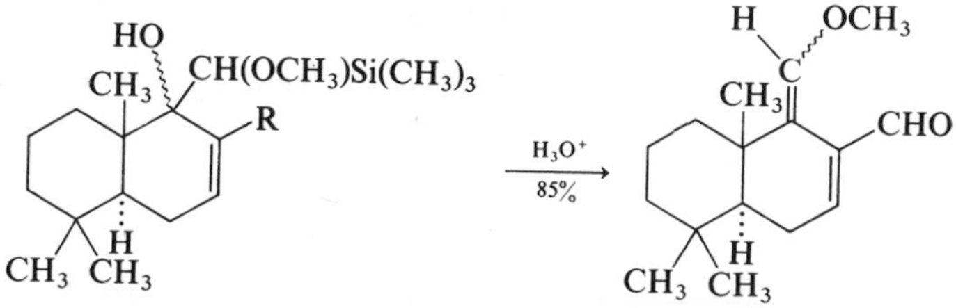

3 **4**

[1] P. Magnus and G. Roy, *Organometallics*, **1**, 553 (1982).
[2] A. S. Kende and T. J. Blacklock, *Tetrahedron Letters*, **21**, 3119 (1980).

(E)-1-Methoxy-3-trimethylsilyloxy-1,3-butadiene (1), 6, 370–372; **9,** 303–304, **10,** 260.

Cyclocondensation of dioxygenated* 1,3-*dienes with aldehydes. In the presence of $ZnCl_2$ in benzene or BF_3 etherate in ether, this diene **1** undergoes cyclization with a wide variety of aldehydes to afford 2,3-dihydro-γ-pyrones (equation I). The products

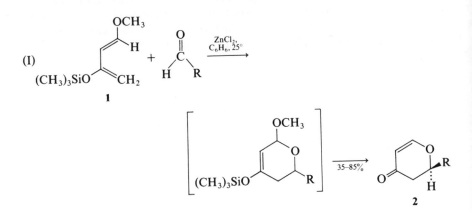

(I) $(CH_3)_3SiO$–diene (**1**, with OCH_3, H, CH_2) + H–C(=O)–R $\xrightarrow{ZnCl_2,\ C_6H_6,\ 25°}$ [cyclic adduct with OCH_3, $(CH_3)_3SiO$, R] $\xrightarrow{35\text{–}85\%}$ **2**

are useful precursors to various hexoses. For example, the dihydro-γ-pyrone **3**, formed by a similar cyclocondensation, is converted into the derivative **4** of talose by reduction (DIBAH), hydroxylation, and peracetylation.

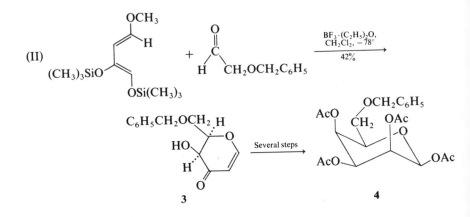

(II) $(CH_3)_3SiO$–diene (with OCH_3, H, $OSi(CH_3)_3$) + H–C(=O)–$CH_2OCH_2C_6H_5$ $\xrightarrow[42\%]{BF_3\cdot(C_2H_5)_2O,\ CH_2Cl_2,\ -78°}$ **3** ($C_6H_5CH_2OCH_2$, H, HO, H, O) $\xrightarrow{\text{Several steps}}$ **4** (AcO, $OCH_2C_6H_5$, CH_2, OAc, O, AcO, OAc)

Application of this condensation to aldehydes with an α-chiral center gives rise to two chiral centers, the relative stereochemistry of which can be related to the Prelog-Cram selectivity rule. Such a reaction was used for a short stereoselective synthesis of the Prelog-Djerassi lactone (**8**) as well as the C_6-epimer, starting with the reaction of diene **5** with **6** to give **7** and the C_4-epimer.[2]

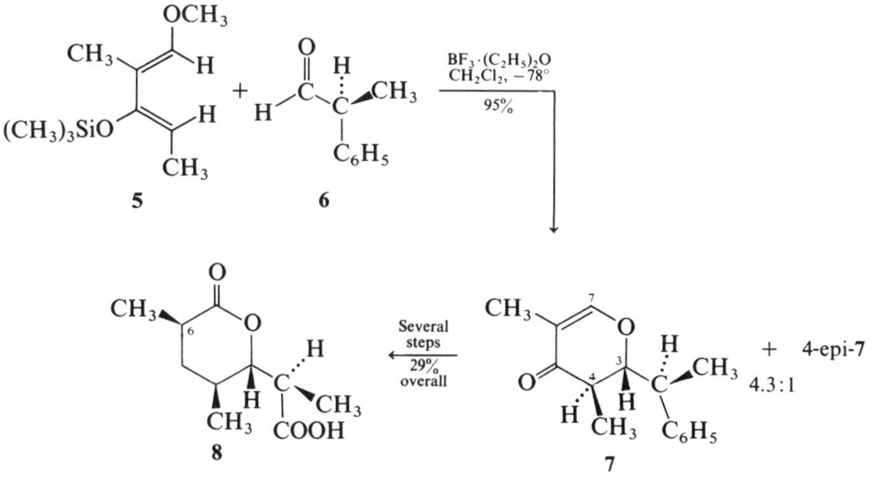

The cyclocondensation of the diene (**1**) with (R)-glyceraldehyde acetonide (**9**) results in high asymmetric induction at C_5 of the dihydropyrone (**10**). The configuration (S) was established by degradation to 2-deoxyribonolactone (**11**). The result is in accord with the Cram rule for addition to chiral carbonyl compounds. The paper also describes conversion of the pyrone (**10**) to chiral 2,4-dideoxy-D-glucose.

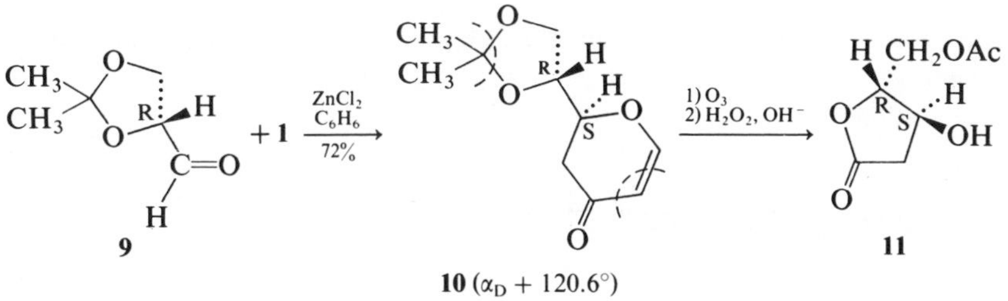

10 (α_D + 120.6°)

This methodology was used for a short stereoselective synthesis of the rare amino acid statine as the Boc derivative (**14**) by condensation of **1** with racemic N-Boc-leucinal (**12**) followed by oxidative degradation (O_3,H_2O_2–NaOH) of the major product (**13**).[3]

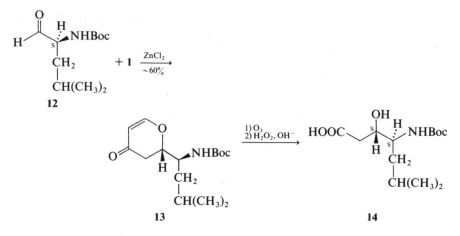

H–C(=O)–C(H)(NHBoc)(S)–CH_2–$CH(CH_3)_2$ **12** + **1** $\xrightarrow[\sim 60\%]{ZnCl_2}$ **13** $\xrightarrow{1)\ O_3;\ 2)\ H_2O_2,\ OH^-}$ **14**

NHBoc, H, CH_2, $CH(CH_3)_2$, HOOC, OH, S

Cyclocondensation with imines. The diene (**1**) undergoes a similar [2+4] cycloaddition with imines, even nonactivated imines. Use of a three- or fourfold excess of the diene improves the yield of 5,6-dihydro-γ-pyridones (equation I).[4]

(I) OCH_3, $(CH_3)_3SiO$, CH_2 + R^1–N=CH–R^2 $\xrightarrow[45–75\%]{ZnCl_2,\ THF}$ N–R^1, R^2, O

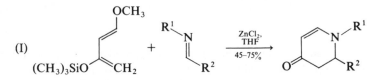

[1] S. Danishefsky, J. F. Kerwin, Jr., and S. Kobayashi, *Am. Soc.*, **104**, 358 (1982).
[2] S. Danishefsky, N. Kato, D. Askin, and J. F. Kerwin, Jr., *ibid.*, **104**, 360 (1982); for a related synthesis *see* R. H. Schlessinger and M. A. Poss, *ibid.*, **104**, 357 (1982).
[3] Danishefsky, S. Kobayashi, and J. F. Kerwin, Jr., *J. Org.*, **47**, 1981 (1982).
[4] J. F. Kerwin, Jr. and S. Danishefsky, *Tetrahedron Letters*, **23**, 3739 (1982).

Methyl 2-acetylacrylate (1). Mol. wt. 128.2, b.p. 82–86°/14 mm. Stable at −78° for short periods.

Preparation:

$CH_3C(=O)CH(CH_3)COOCH_3$ $\xrightarrow{1)\ C_6H_5SCl,\ CH_2Cl_2\ (100\%);\ 2)\ ClC_6H_4CO_3H,\ -40°(100\%)}$ $CH_3C(=O)C(CH_3)(S(=O)C_6H_5)COOCH_3$ $\xrightarrow[47\%]{60°,\ 0.1\ mm,\ sulfolane}$ $CH_3C(=O)C(=CH_2)COOCH_3$ **1**

Diels-Alder reaction with cyclopentadienes. An improved synthesis of a key intermediate (**6**) to gibberellic acid (**7**) begins with the cycloaddition of **1** to a 2:1 mixture of 2- and 1-(2-bromoallyl)cyclopentadiene (**2**) to give the adduct **3** in which the acetyl group has the *endo*-orientation. The silyl enol ether of **3** when heated undergoes a Cope rearrangement to give a *cis*-hydrindene (**4**), which was converted

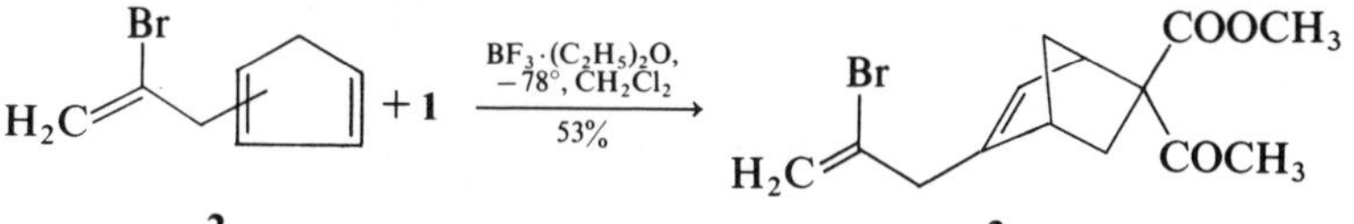

$BF_3 \cdot (C_2H_5)_2O$, $-78°$, CH_2Cl_2; 53%

1) $(CH_3)_3SiOTf$ (99%) 2) 160°

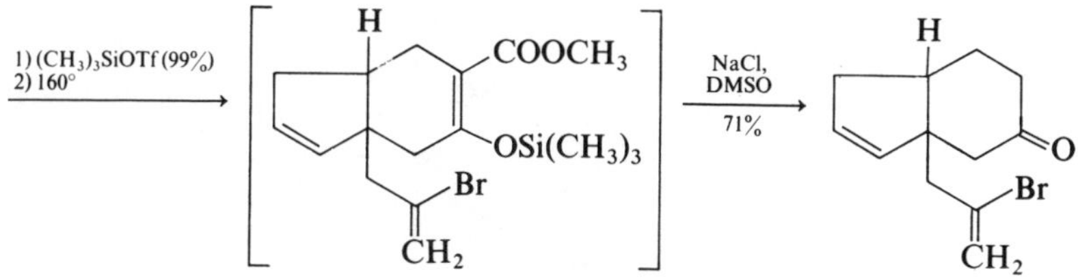

NaCl, DMSO; 71%

without purification into the ketone **5**. The ketone **5** was converted in five steps to **6**, which had been converted previously into gibberellic acid.[2]

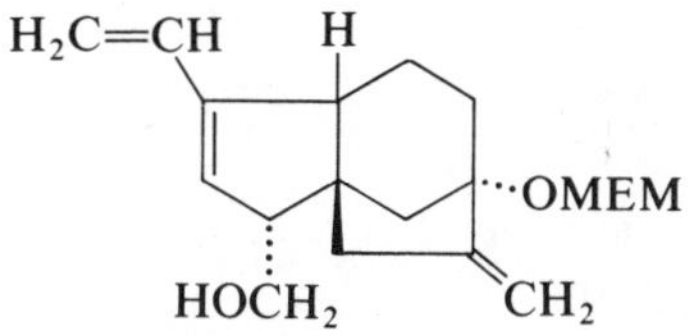

[1] E. J. Corey and J. E. Munroe, *Am. Soc.*, **104**, 6129 (1982).

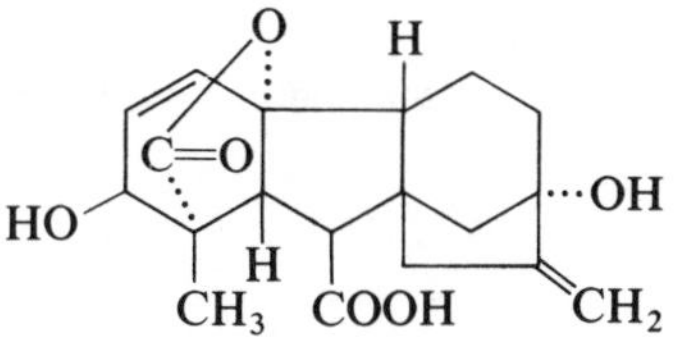

[2] E. J. Corey, R. L. Danheiser, S. Chandrasekaran, P. Siret, G. E. Keck, G.-L. Gras, *ibid.*, **100**, 8031 (1978); E. J. Corey, R. L. Danhesier, S. Chandrasekaran, G. E. Keck, S. D. Larsen, P. Siret, and G.-L. Gras, *ibid.*, **100**, 8034 (1978).

Methyl bis(methylthio)sulfonium hexachloroantimonate, 6, 375; **7**, 235. Supplier: Fluka.

***Cleavage of* 1,3-*Dithianes*.**[1] Conversion of 1,3-dithianes to the corresponding carbonyl compound can be effected by reaction with the reagent at $-78°$ in high yield (6 examples).

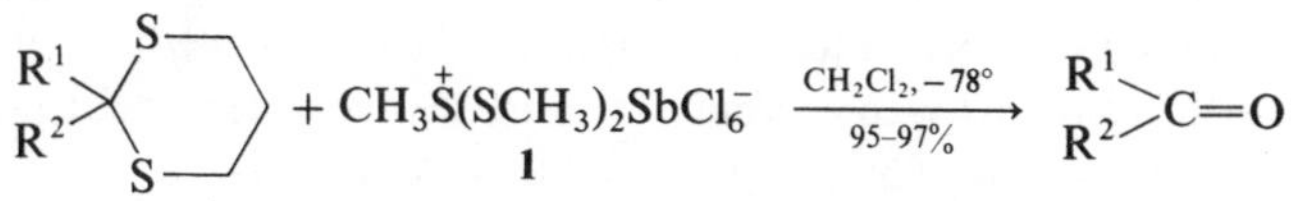

$$\text{R}^1\text{R}^2\text{C(S}(CH_2)_3\text{S)} + CH_3\overset{+}{S}(SCH_3)_2SbCl_6^- \ (\mathbf{1}) \xrightarrow[95\text{–}97\%]{CH_2Cl_2,\ -78°} \text{R}^1\text{R}^2\text{C=O}$$

The same reaction can also be carried out with dimethyl sulfoxide and hydrochloric acid in dioxane at 25°. The actual reagent is presumed to be chlorodimethylsulfonium ion. The method is not suitable for acid-sensitive substrates.

2,3-Dihydrobenzofuranes.[2] The reaction of **1** with 2-allylphenol at $-17°$ in CH_2Cl_2 results in the 2,3-dihydrobenzofurane **2** in 48% yield. At higher temperatures, thiomethylation is observed at the position *para* to the oxygen.

$$\text{2-allylphenol } (o\text{-}HOC_6H_4CH_2CH{=}CH_2) + (CH_3S)_2\overset{+}{S}CH_3SbCl_6^- \ (\mathbf{1}) \xrightarrow[-(CH_3S)_2]{CH_2Cl_2,\ -17°}$$

$$\left[o\text{-}HOC_6H_4CH_2\text{–}CH\text{–}CH_2 \ (\text{episulfonium } \overset{+}{S}\text{–}CH_3) \quad SbCl_6^- \right] (\mathbf{a}) \xrightarrow{48\%} \text{2,3-dihydrobenzofuran-2-yl–}CH_2SCH_3 \ (\mathbf{2})$$

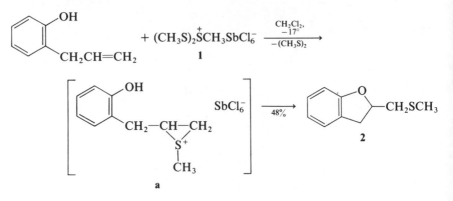

[1] M. Prato, U. Quintily, G. Scorrano, and A. Stararo, *Synthesis*, 679 (1982).
[2] G. Capozzi, V. Lucchini, F. Marcuzzi, and G. Modena, *J.C.S. Perkin I*, 3106 (1981).

N-Methyl-N-(*t*-butyldimethylsilyl)trifluoroacetamide,

$$CF_3CON\begin{matrix} CH_3 \\ SiMe_2\text{-}t\text{-}Bu \end{matrix} \quad (\mathbf{1}).$$

Mol. wt. 241.83, b.p. 168–170°. The reagent is prepared by reaction of N-methyltrifluoroacetamide with sodium hydride and then with t-$BuMe_2SiCl$.

t-Butyldimethylsilylation.[1] This amide, in the presence of 1% of *t*-butyldimethylsilyl chloride as catalyst, is an extremely reactive silylation reagent for alcohols, thiols, amines, and carboxylic acids. Complete silylation is generally effected within 5 minutes at 25° in acetonitrile. N-Methyl-N-(*t*-butyldimethylsilyl)acetamide is somewhat less reactive than **1**.

[1] T. P. Mawhinney and M. A. Madson, *J. Org.*, **47**, 3336 (1982).

Methyl chloroformate, 6, 376; 7, 236.

Enol carbonates.[1] Site-specific metal enolates, such as those generated by conjugate addition of cuprates to α,β-enones, can be trapped as the enol carbonates by a quench with methyl chloroformate. Unlike silyl enolates, enol carbonates are only weakly nucleophilic. Thus they are stable to ozone, peracids, and Wittig-like reagents. After operations on a remote double bond, they can be reconverted to the enolate of the original structural type by treatment with an alkyllithium for further transformations.

Example:

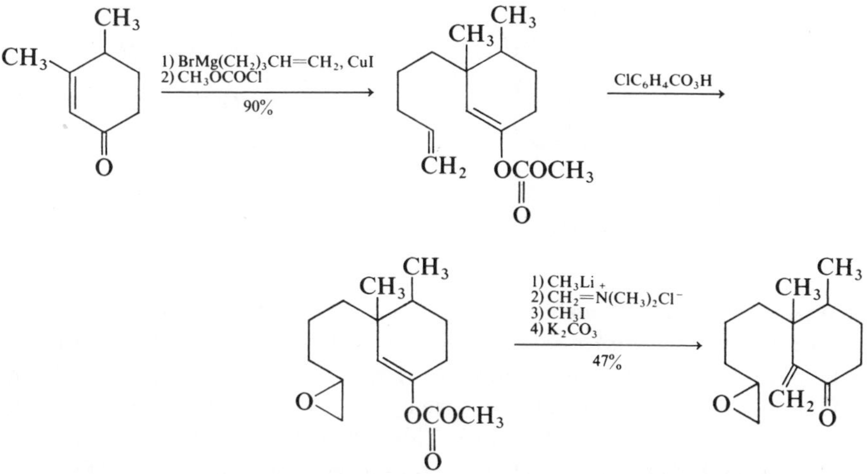
CH3
CH3
1) BrMg(CH2)3CH=CH2, CuI
2) CH3OCOCl
90%
O
CH3
CH3
CH2
OCOCH3
O
ClC6H4CO3H
CH3
CH3
O
OCOCH3
O
1) CH3Li
2) CH2=N(CH3)2Cl−
3) CH3I
4) K2CO3
47%
CH3
CH3
O
CH2
O

[1] S. Danishefsky, M. Kahn, and M. Silvestri, *Tetrahedron Letters*, **23**, 703 (1982).

L-N-Methyl-N-dodecylephedrinium bromide, 7, 238–239.

Correction.[1] Under basic conditions, this ephedrine derivative decomposes to the oxide **1**, which has a very high specific rotation. Contamination with this oxide is probably responsible for most or even all the optical activity observed previously on reduction of ketones with $NaBH_4$–NaOH in the presence of this phase-transfer catalyst.

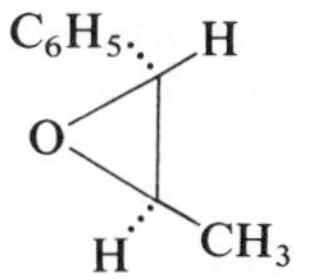
C6H5
H
O
H
CH3

1, $\alpha_D + 117.6°$

[1] E. V. Dehmlow, P. Singh, and J. Heider, *J. Chem. Res.*, 292 (1981).

Methylene bromide–Zinc–Titanium(IV) chloride, 8, 339.

Methylenation.[1] The species obtained from the reaction of $TiCl_4$ with zinc in CH_2Br_2 and THF is the reagent of choice for methylenation of the gibberellin ketones **1** and **2**, which undergo marked epimerization at the adjacent chiral center on methylenation with $(C_6H_5)_3P{=}CH_2$ or $(CH_3)_2S(O){=}CH_2$. Yields of 90% or more were realized in both cases with no evidence of epimerization.

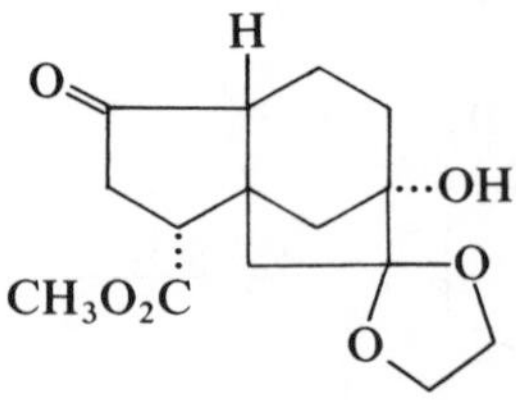

1

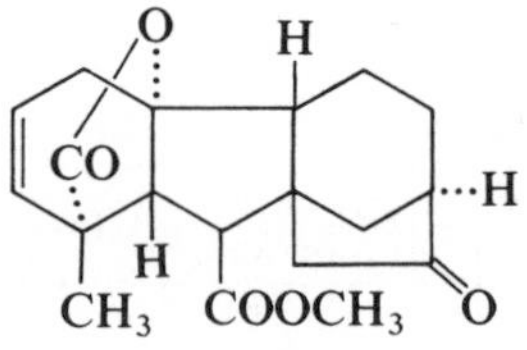

2

[1] L. Lombardo, *Tetrahedron Letters*, **23**, 4293 (1982).

Methylenetriphenylphosphorane, 1, 678; **6,** 380–381; **8,** 339–340; **9,** 307; **10,** 264–265.

α-Lithiomethylenetriphenylphosphorane, $(C_6H_5)_3P{=}CHLi$ **(1)**.[1] This reagent can be prepared by deprotonation of methylenetriphenylphosphorane with *t*-butyllithium in THF at $-78 \rightarrow -40°$ or by reaction of methyltriphenylphosphonium bromide with 2 equiv. *sec*-butyllithium ($-78° \rightarrow 20°$). This lithiated ylide **(1)** shows enhanced reactivity over the parent ylide. Thus it reacts with fenchone in the presence of HMPT to form an adduct that decomposes in the presence of excess *t*-butyl alcohol to give the corresponding methylene derivative **(2)** in 87% yield.

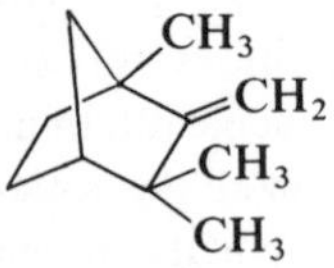

2

The reaction of **1** with 2 equiv. of an aldehyde produces a *trans*-allylic alcohol **(3)** by way of a *β*-oxido ylide **a** (equation I).

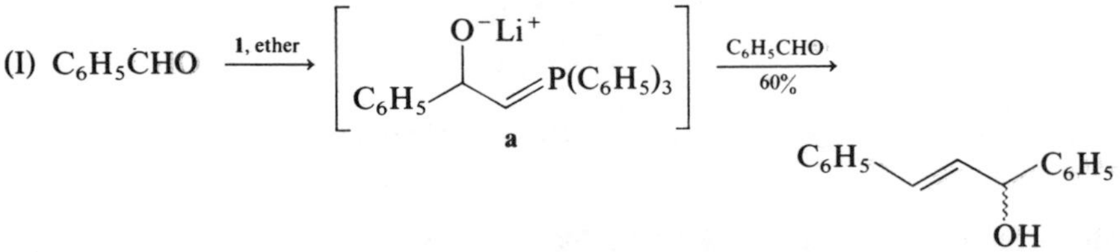

(I) C_6H_5CHO $\xrightarrow{\text{1, ether}}$ [a] $\xrightarrow[60\%]{C_6H_5CHO}$ $C_6H_5CH{=}CHCH(OH)C_6H_5$

Unlike Wittig reagents, **1** reacts with epoxides to form a *γ*-oxido ylide **(b)**, which reacts with an aldehyde to form a *trans*-homoallylic alcohol.

Examples:

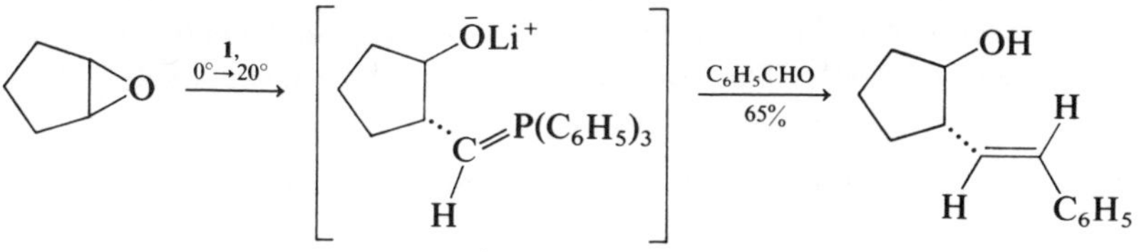

b

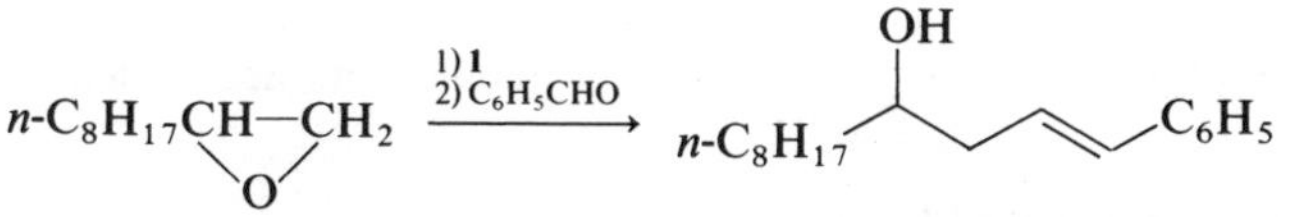

Chiral 2,5-dimethylvalerolactones.[2] A novel route to these lactones involves as the key step the Wittig reaction with the α-enoside **1**, available from D-glucose, to give the diene **2** in 72% yield. This unusual transformation presumably involves initial reaction with the enol acetate group to give an enone, which then gives the normal reaction with a second equiv. of the Wittig reagent. Hydrogenation of **2** results principally in the *trans*-isomer **3**. This is converted by several known reactions into the dimethylvalerolactone **4**, isomeric with **5**, the major component of the sex pheromone of the carpenter bee. The natural product would result from a similar synthesis with L-glucose as starting material.

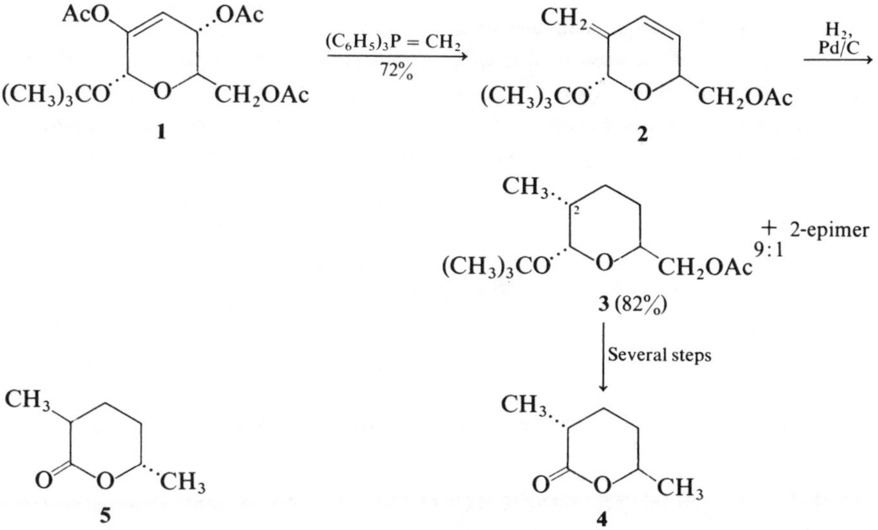

[1] E. J. Corey and J. Kang, *Am. Soc.*, **104**, 4724 (1982).

[2] S. Hanessian, G. Demailly, Y. Chapleur, and S. Leger, *J.C.S. Chem. Comm.*, 1125 (1981).

Methylketene methyl trimethylsilyl ketal (1), 9, 310.

Trimethylsilyl enol ethers.[1] Enolizable ketones are converted into silyl enol ethers by reaction with **1** and catalytic amounts of *n*-tetrabutylammonium fluoride in THF in yields of 70–85%. This silylation can be used for a one-pot ketalization and thioketalization of ketones (45–65% overall yield).

[1] Y. Kita, H. Yasuda, J. Haruta, J. Segawa, and Y. Tamura, *Synthesis*, 1089 (1982).

Methyl lithiodithioacetate, $H_2C{=}C(SLi)(SCH_3)$ (**1**). The anion is obtained by deprotonation of methyl dithioacetate with LDA in THF.

Conjugate addition. The reagent (**1**) undergoes selective 1,4-addition to enones to give potential precursors to 1,5-dicarbonyl compounds.[1]

Example:

$$CH_3\overset{O}{\overset{\|}{C}}CH{=}CHCH_3 \xrightarrow[74\%]{1,\ THF,\ -60^\circ} CH_3\overset{O}{\overset{\|}{C}}CH_2\overset{CH_3}{\overset{|}{C}}HCH_2\overset{S}{\overset{\|}{C}}SCH_3$$

[1] P. Metzner, *J.C.S. Chem. Comm.*, 335 (1982).

Methyllithium–Tetramethylethylenediamine, 9, 311–312.

Cyclization of dianions of bis(diphenyl thiocetals).[1] Treatment of **1** with 2–4 equiv. of methyllithium and of TMEDA results in formation of **2** (major product) and **3**. The paper presents evidence that these cyclizations involve an intermediate dianion, which loses a thiophenoxide ion to give an anionic sulfur-stabilized carbene.

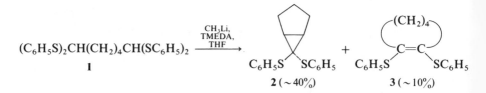

$$\underset{\mathbf{1}}{(C_6H_5S)_2CH(CH_2)_4CH(SC_6H_5)_2} \xrightarrow{CH_3Li,\ TMEDA,\ THF} \underset{\mathbf{2}\,(\sim 40\%)}{\text{bicyclo[3.1.0]hexane-6,6-}(SC_6H_5)_2} + \underset{\mathbf{3}\,(\sim 10\%)}{C_6H_5S\text{-}C{=}C\text{-}SC_6H_5\ \text{(ring closed by }(CH_2)_4)}$$

[1] T. Cohen, R. H. Ritter, and D. Ouellette, *Am. Soc.*, **104**, 7142 (1982).

Methyl methoxypropiolate, $CH_3OC{\equiv}CCO_2CH_3$ (**1**). The reagent is prepared (25–30% yield) by methoxycarbonylation of methoxyacetylene (hazardous).[1]

Methoxycarbonylketene equivalent. The reagent gives the adduct **2** on reaction with cyclopentadiene.

$$\text{cyclopentadiene} \xrightarrow{\mathbf{1},\ C_6H_6} \mathbf{2}\ (\text{bicyclic diene with } COOCH_3 \text{ and } OCH_3) \xrightarrow[\text{50\% overall}]{HCl,\ CH_3OH} \mathbf{3}\ (\text{keto ester, } COOCH_3,\ O)$$

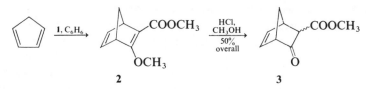

A safer and more useful equivalent of methoxycarbonylketene is methyl (phenylthio)propiolate (**4**).[2]

$$\text{cyclopentadiene} + \underset{\mathbf{4}}{C_6H_5SC{\equiv}CCOOCH_3} \xrightarrow{90\%} \mathbf{5}\ (COOCH_3,\ SC_6H_5)$$

$$\xrightarrow[CH_3CN]{HgCl_2,\ H_2O} \mathbf{3} \xrightarrow{\text{DABCO}} \text{bicyclic ketone } (=O)$$

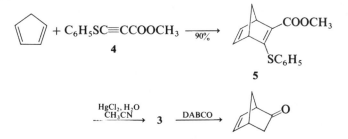

[1] B. R. Vogt, J. Bernstein, and F. L. Weisenborn, Fr. Pat. 2, 138, 112 (1973) [C.A., 79, 18452q (1973)].
[2] I. Gupta and P. Yates, *J.C.S. Chem. Comm.*, 1227 (1982).

Methyl O-methyllactate, $CH_3CH(OCH_3)COOCH_3$ (**1**). Mol. wt. 118.13, b.p. 131°. Preparation.[1]

Aldol condensation.[2] The enolate of **1** reacts with aldehydes to form preferentially *erythro*-α-methyl-α,β-dihydroxy carboxylic esters (equation I). The O-MEM ether of methyl lactate shows similar *erythro*-selectivity (*ca.* 6:1). In contrast, the anion of

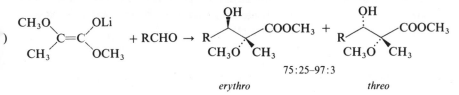

$$\text{(I)}\quad (CH_3O)(CH_3)C{=}C(OLi)(OCH_3) + RCHO \rightarrow \underset{erythro}{RCH(OH)C(OCH_3)(CH_3)COOCH_3} + \underset{threo}{RCH(OH)C(OCH_3)(CH_3)COOCH_3}$$

75:25–97:3

ester **2** reacts with very high *threo*-selectivity.

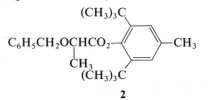

$$C_6H_5CH_2OCH(CH_3)CO_2\text{–}[2,6\text{-}((CH_3)_3C)_2\text{-}4\text{-}CH_3C_6H_2] \quad \mathbf{2}$$

[1] A. A. Petrov, B. V. Gantseva, and O. A. Kiseleva, *Zhur. Obschei. Khim.*, **23**, 737 (1953).
[2] C. H. Heathcock, J. P. Hagen, E. T. Jarvi, M. C. Pirrung, and S. D. Young, *Am. Soc.*, **103**, 4972 (1981).

N,N-Methylphenylaminotri-*n*-butylphosphonium iodide (1), 8, 345–346; **10**, 268–269.

γ-Quaternary methylation of a tertiary allyl alcohol.[1] In a synthesis of the antibiotic pleuromutilin (**2**), the final stages involved the conversion of the hindered C_{12}-ketone **3** into the quaternary methyl vinyl derivative (**6**) without elimination of the C_{11}-alkoxy group. The conversion was affected via the α,β-unsaturated aldehyde (**4**) obtained with (Z)-(2-ethoxyvinyl)lithium (**8**, 221–222) followed by DIBAH reduction to the corresponding allylic alcohol (**5**). The final step was γ-methylation of **5** by Murahashi's method (**8**, 345), which resulted in only the desired epimer **6**.

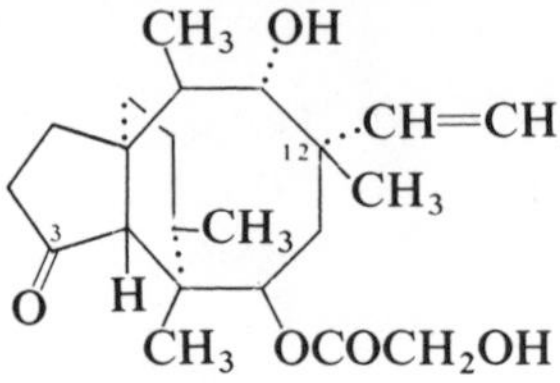

CH₃ OH; CH=CH₂; CH₃; CH₃; O; H; CH₃ OCOCH₂OH

2

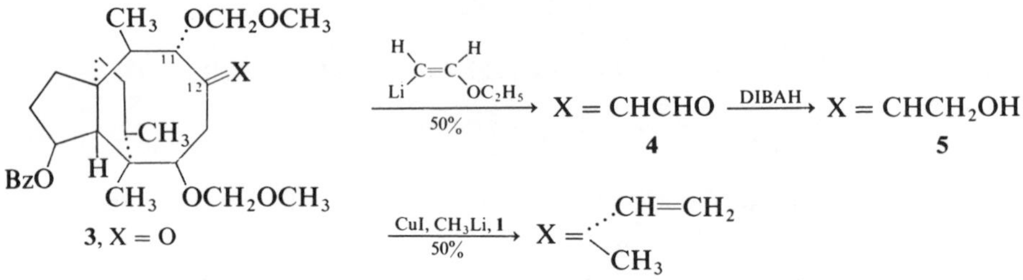

CH₃ OCH₂OCH₃; X; CH₃; BzO H; CH₃ OCH₂OCH₃

3, X = O

$\xrightarrow[50\%]{\text{(Z)-}LiCH=CHOC_2H_5}$ X = CHCHO (**4**) $\xrightarrow{\text{DIBAH}}$ X = $CHCH_2OH$ (**5**)

$\xrightarrow[50\%]{CuI,\ CH_3Li,\ \mathbf{1}}$ X = C(CH=CH₂)(CH₃) (**6**)

Allylsilanes. A general synthesis of allylsilanes involves alkylation at the γ-position of an allylic alcohol substituted at the γ-position by a trimethylsilyl group or at the α-position by a trimethylsilylmethyl group with an organolithium compound mediated by this phosphonium iodide (**1**).[2]

Examples;

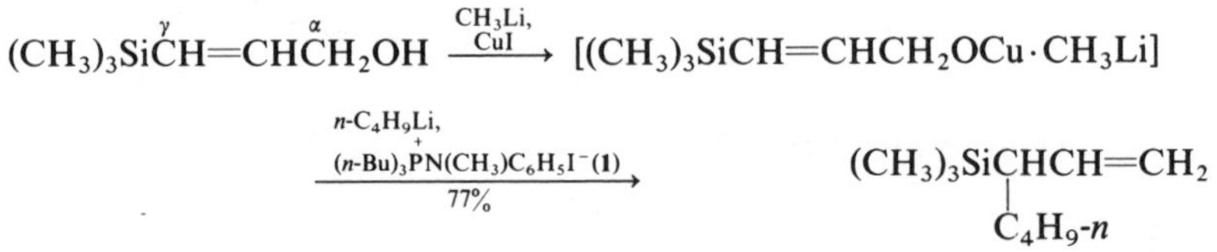

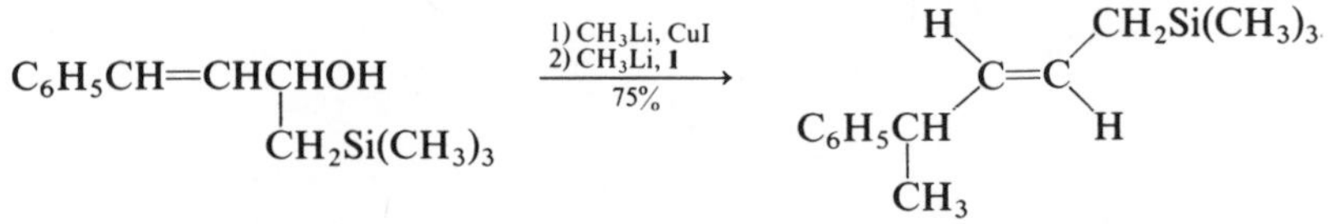

$(CH_3)_3Si\overset{\gamma}{C}H{=}CH\overset{\alpha}{C}H_2OH \xrightarrow[CuI]{CH_3Li,} [(CH_3)_3SiCH{=}CHCH_2OCu\cdot CH_3Li]$

$\xrightarrow[77\%]{n\text{-}C_4H_9Li,\ (n\text{-}Bu)_3\overset{+}{P}N(CH_3)C_6H_5I^-\ (\mathbf{1})} (CH_3)_3SiCH(C_4H_9\text{-}n)CH{=}CH_2$

$C_6H_5CH{=}CHCH(OH)CH_2Si(CH_3)_3 \xrightarrow[75\%]{1)\ CH_3Li,\ CuI;\ 2)\ CH_3Li,\ \mathbf{1}} C_6H_5CH(CH_3)CH{=}CHCH_2Si(CH_3)_3$ (H, $CH_2Si(CH_3)_3$ / $C_6H_5CH(CH_3)$, H)

[1] E. G. Gibbons, *Am. Soc.*, **104**, 1767 (1982).
[2] Y. Tanigawa, Y. Fuse, and S.-I. Murahashi, *Tetrahedron Letters*, **23**, 557 (1982).

Methyl(phenyl)selenoniomethanide (1). The selenonium ylide is generated *in situ* from dimethyl(phenyl)selenonium methyl sulfate[1] with NaH (equation I).

$$C_6H_5SeH + (CH_3)_2SO_4 \xrightarrow[78\%]{NaOH,\ H_2O} C_6H_5SeCH_3 \xrightarrow[92\%]{(CH_3)_2SO_4,\ 90^\circ} C_6H_5Se(CH_3)_2^+CH_3SO_4^-$$

$$\xrightarrow{NaH} C_6H_5Se(CH_3)=CH_2 \quad \mathbf{1}$$

Oxiranes.[2] The reagent reacts with ketones and aromatic aldehydes to form oxiranes in 75–95% yield (equation I).

$$\text{(I)}\quad R^1R^2C{=}O + \mathbf{1} \xrightarrow[75\text{–}95\%]{DMF,\ THF,\ 0\rightarrow 50^\circ} R^1R^2C(-O-)CH_2 + C_6H_5SeCH_3$$

[1] H. M. Gilow and G. L. Walker, *J. Org.*, **32**, 2580 (1967).
[2] K. Takaki, M. Yasumura, and K. Negoro, *Angew. Chem. Int. Ed.*, **20**, 671 (1981).

Methyl (phenylsulfinyl)acetate, $C_6H_5SOCH_2COOCH_3$ (1). Mol. wt. 198.24, m.p. 53–55°, b.p. 130–131°/0.5 mm. Supplier: Aldrich.

α,β-Unsaturated-δ-lactones.[1] This reagent undergoes conjugate addition to alkyl vinyl ketones in the presence of catalysts such as diisopropylamine, DBU, and various phosphines, particularly tri-*n*-butylphosphine. DMSO is the best solvent; DMF and CH_3CN are only slightly less effective. The products are precursors to α,β-unsaturated-δ-lactones.

Example:

$$CH_2{=}CHC(O)CH_2CH(CH_3)_2 \xrightarrow{\mathbf{1},\ P(C_4H_9)_3,\ DMSO} (CH_3)_2CHCH_2C(O)CH_2CH_2CH(SOC_6H_5)COOCH_3$$

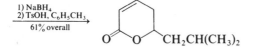

$$\xrightarrow[61\%\ \text{overall}]{1)\ NaBH_4;\ 2)\ TsOH,\ C_6H_5CH_3} \text{6-}(CH_2CH(CH_3)_2)\text{-5,6-dihydro-2H-pyran-2-one}$$

[1] T. Yoshida and S. Saito, *Chem. Letters*, 1587 (1982).

N-Methylphenylsulfonimidoylmethyllithium (1), 6, 395.

Ketone methylenation with optical resolution.[1] The key step in a synthesis of β-panasinsene (**5**), an important component of Ginseng, involves methylenation of the ketone **2** with the sulfoximine anion (**1**). Although **2** is inert to $(C_6H_5)_3P{=}CH_2$ under usual conditions, it reacts with **1** at $-78°$ to give **3** and **4** in a ratio $>30:1$. The major product (**3**) is reduced by Al/Hg and HOAc in THF to **5** in 67% overall yield.

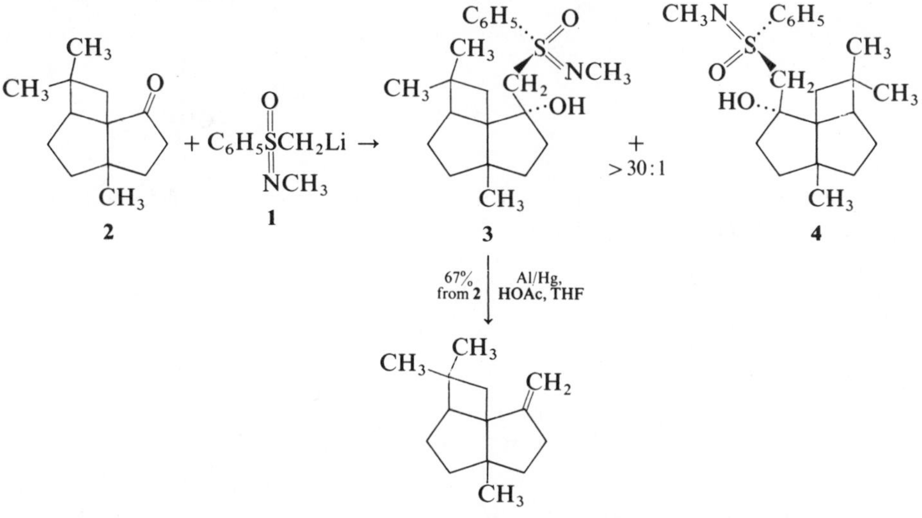

2 + $C_6H_5S(O)(NCH_3)CH_2Li$ (1) → 3 + 4 (>30:1); 3 → 5 (Al/Hg, HOAc, THF; 67% from 2)

Reaction of **2** with (S)-**1** results in (+)-**3**, $\alpha_D +11°$ (42% yield), and (+)-**4**, $\alpha_D +108°$ (33% yield). Reduction of (+)-**3** results in the natural (−)-**5** in 96% yield. Reduction of (+)-**4** results in (+)-**5** in 92% yield. Raney nickel desulfurization of (+)-**3** and (+)-**4** results in the corresponding enantiomeric tertiary carbinols in 78% and 94% yield, respectively.

[1] C. R. Johnson and N. A. Meanwell, *Am. Soc.*, **103**, 7667 (1981).

(Z)-3-Methyl-1-phenylthio-2-trimethylsilyloxy-1,3-butadiene (1).

Preparation:

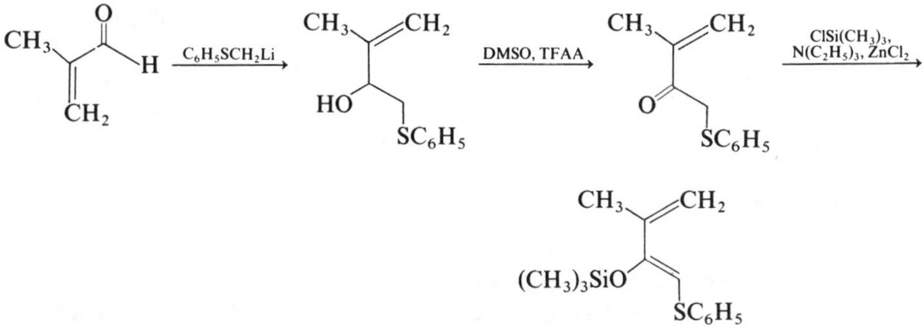

$CH_2{=}C(CH_3)CHO$ →($C_6H_5SCH_2Li$) $CH_2{=}C(CH_3)CH(OH)CH_2SC_6H_5$ →(DMSO, TFAA) $CH_2{=}C(CH_3)COCH_2SC_6H_5$ →($ClSi(CH_3)_3$, $N(C_2H_5)_3$, $ZnCl_2$) $CH_2{=}C(CH_3)C(OSi(CH_3)_3){=}CHSC_6H_5$ (**1**)

Diels-Alder reactions.[1] The diene undergoes uncatalyzed [4+2]cycloaddition with very reactive dienophiles, but generally a Lewis acid catalyst ($ZnCl_2$ or $C_2H_5AlCl_2$) is required. Thus, the catalyzed reaction of **1** with methyl acrylate proceeds at 0° to afford, after silyl ether cleavage, the cycloadducts **2** and **3**.

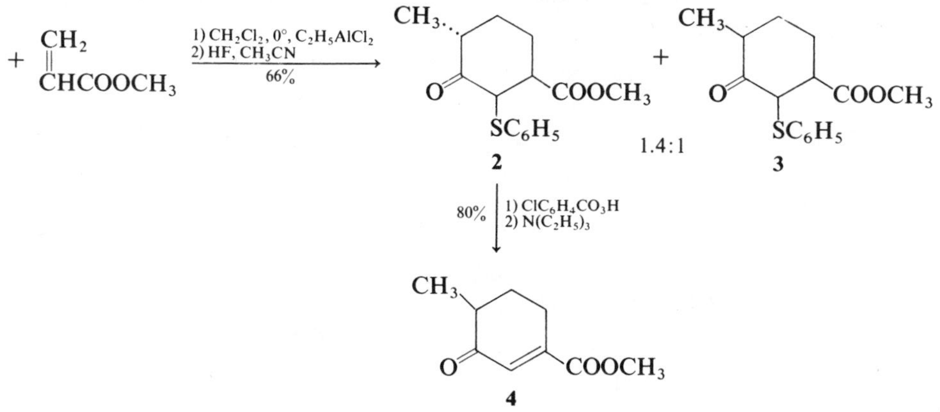

The catalyzed reaction of **1** with α,β-enals takes an unexpected course. The adduct with methacrolein, for example, obtained after silyl ether cleavage does not contain the expected aldehyde group, but has the bicyclic structure **5**. Apparently the initial product **a** undergoes an intramolecular aldol condensation to give **b** ,which undergoes silyl transfer to **c**, which is hydrolyzed by acid to the 7-hydroxy-bicyclo[2.2.1]heptanone **5**.

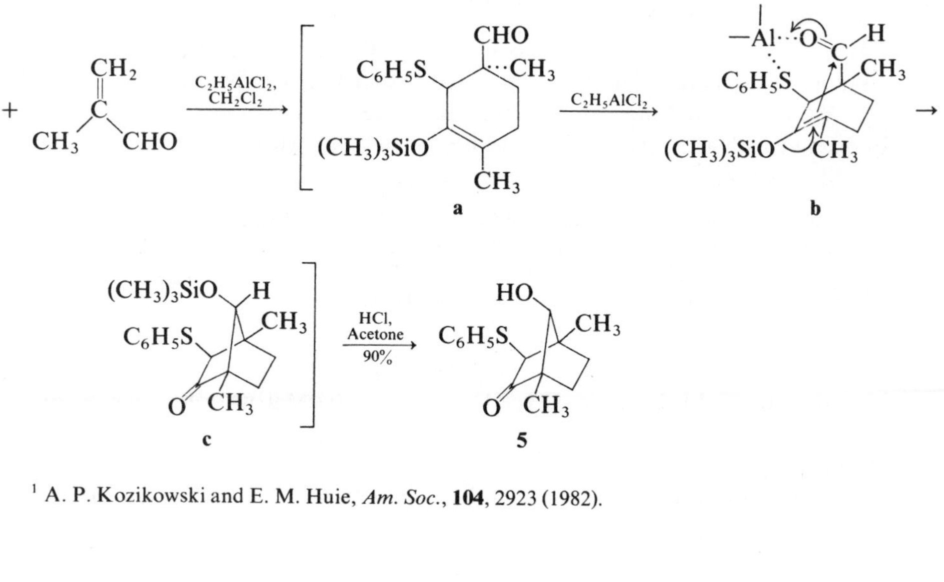

[1] A. P. Kozikowski and E. M. Huie, *Am. Soc.*, **104**, 2923 (1982).

N-Methyl-2-pyrrolidone (NMP), **1**, 696; **2**, 281; **9**, 316.

Alkylation of phenyl sulfones.[1] A convergent synthesis in which two synthons become attached through a double bond depends on the α-alkylation of the anion of a phenyl sulfone followed by a β-elimination. The synthesis of retinoic acid (**4**) is typical of the process. Potassium *t*-butoxide is used as base; the recommended solvent is THF/NMP = 50:10.

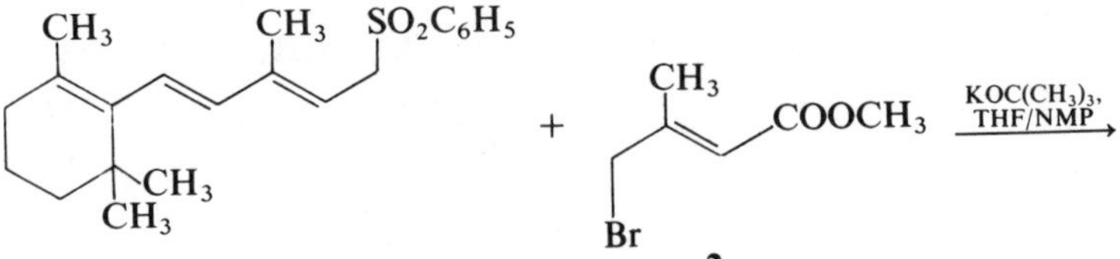

CH_3 CH_3 $SO_2C_6H_5$ CH_3 CH_3 **1** + CH_3 $COOCH_3$ Br **2** $\xrightarrow[\text{THF/NMP}]{\text{KOC(CH}_3)_3,}$

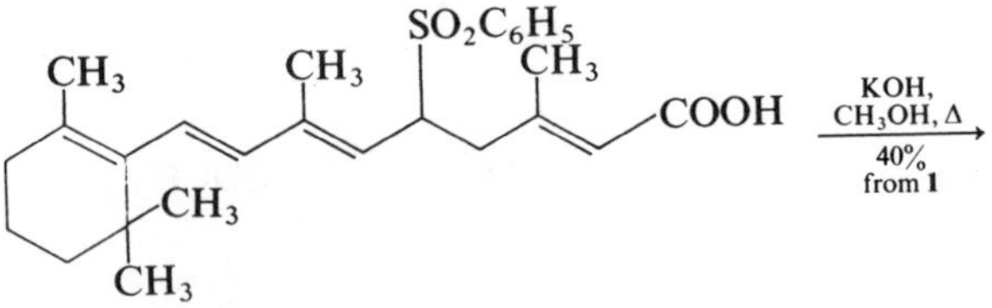

$SO_2C_6H_5$ CH_3 CH_3 CH_3 $COOH$ CH_3 CH_3 $\xrightarrow[\text{40\% from 1}]{\text{KOH, CH}_3\text{OH, }\Delta}$

3 (2 *cis*/all *trans* = 45/55)

CH_3 CH_3 CH_3 $COOH$ CH_3 CH_3

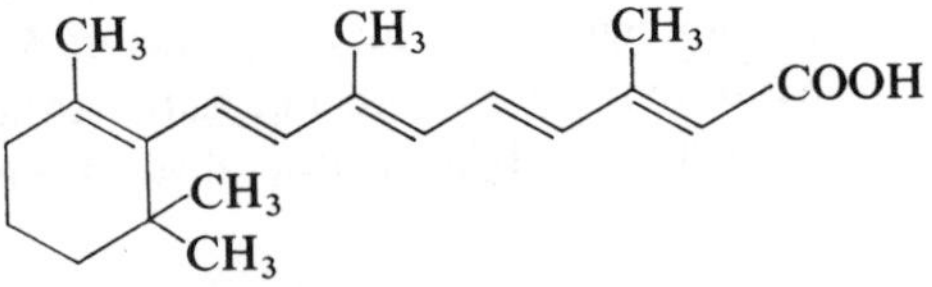

4 (all-*trans*)

—Cl→—Br. NMP or DMF is superior to THF or acetone for conversion of alkyl or allylic chlorides into bromides by halide exchange with NaBr.[2, 3]

Nitriles.[4] Aldehydes can be converted directly in satisfactory yield into nitriles by reaction with hydroxylamine hydrochloride at 90–140° when N-methylpyrrolidone is used as solvent.

[1] D. Arnould, M. Julia, P. Chabardès, and G. Farges, *Org. Syn.* submitted (1982).
[2] W. E. Willy, D. R. McKean, and B. A. Garcia, *Bull. Soc. Chem. Japan*, **49**, 1989 (1976).
[3] M. P. Cooke, Jr., and D. L. Burman, *J. Org.*, **47**, 4955 (1982).
[4] P. Andoye, A. Gaset, J.-P. Gorrichon, *Chimia*, **36**, 4 (1982).

Methyl 2-trimethylsilylacrylate, $CH_2{=}C\begin{smallmatrix} Si(CH_3)_3 \\ COOCH_3 \end{smallmatrix}$ (**1**). Mol. wt. 158.27, b.p. 53–56/18 mm. Preparation.[1]

Michael reactions. A synthesis of the 2-aminotetralin (**4**) uses **1** as the Michael acceptor in the key step to obtain **2**.[2] Methyl acrylate is not useful for this step mainly because of polymerization.

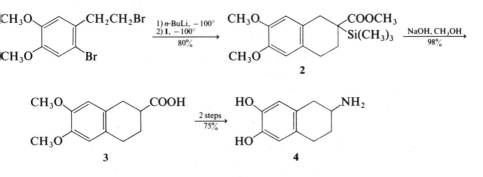

CH_3O, CH_3O, CH_2CH_2Br, Br $\xrightarrow[80\%]{1)\ n\text{-BuLi},\ -100°;\ 2)\ \mathbf{1},\ -100°}$ **2** (CH_3O, CH_3O, $COOCH_3$, $Si(CH_3)_3$) $\xrightarrow[98\%]{NaOH,\ CH_3OH}$ **3** (CH_3O, CH_3O, COOH) $\xrightarrow[75\%]{2\ steps}$ **4** (HO, HO, NH_2)

[1] A. Ottolenghi, M. Fridken, and A. Zilkha, *Can. J. Chem.*, **41**, 2977 (1963); R. F. Cunico, H. M. Lee, and J. Herbach, *J. Organometal. Chem.*, **52**, C7 (1973).

[2] A. P. S. Narula and D. I. Schuster, *Tetrahedron Letters*, **22**, 3707 (1981).

N-Methyl-N-trimethylsilymethyl-N′-*t*-butylformamidine (1), b.p. 78–80°/7 mm.

Preparation from dimethylformamide:

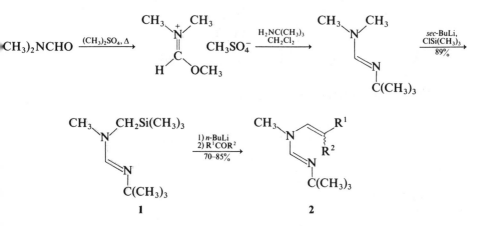

$(CH_3)_2NCHO \xrightarrow{(CH_3)_2SO_4,\ \Delta} (CH_3)_2\overset{+}{N}{=}CH(OCH_3)\ CH_3SO_4^- \xrightarrow[CH_2Cl_2]{H_2NC(CH_3)_3} (CH_3)_2N{-}CH{=}N{-}C(CH_3)_3 \xrightarrow[89\%]{sec\text{-BuLi},\ ClSi(CH_3)_3}$

1 ($CH_3(CH_2Si(CH_3)_3)N{-}CH{=}N{-}C(CH_3)_3$) $\xrightarrow[70\text{–}85\%]{1)\ n\text{-BuLi};\ 2)\ R^1COR^2}$ **2** ($CH_3N(CH{=}CR^1R^2){-}CH{=}N{-}C(CH_3)_3$)

Enamidines **(2)**.[1] Lithiation of **1** followed by treatment with aldehydes or ketones results in Peterson olefination to give a mixture of isomeric enamidines (**2**) in good yield. These enamidines can be used to convert the carbonyl compounds used in their preparation to homologated amines, aldehydes, and ketones. Conversion to a methylamine involves reduction with sodium borohydride (pH 6) to an aminal, which is then hydrolyzed by dilute acid. The sequence can be carried out from **1** without isolation of any intermediates (equation I).

$$\text{(I)}\ \mathbf{2} \xrightarrow{NaBH_4} CH_3N(CH_2CH(R^1)(R^2))CH_2NHC(CH_3)_3 \xrightarrow[\text{50–70\% overall}]{H_3O^+} CH_3\underset{H}{N}CH_2CH(R^1)(R^2)$$

Enamidines are cleaved to aldehydes by hydrazinolysis with 1,1-dimethylhydrazine and cleavage of the resultant hydrazone (equation II). Alternately, enamidines can be converted by aluminum amalgam to the corresponding N-methyl enamine, which is cleaved by dilute acid.

$$\text{(II)}\ \mathbf{2} \xrightarrow{H_2NN(CH_3)_2} \left[(CH_3)_2NN{=}CHCH(R^1)(R^2)\right] \xrightarrow[\text{55–85\%}]{Cu(OAc)_2,\ THF,\ H_2O} OHCCH(R^1)(R^2)$$

Enamidines can be converted to homologated ketones by alkylation of the anion (*t*-butyllithium) followed by hydrazinolysis and subsequent cleavage (equation III).

$$\text{(III)}\ \mathbf{2} \xrightarrow{1)\ t\text{-BuLi};\ 2)\ R^3I} \left[CH_3N(C(R^3){=}C(R^1)(R_2))CH{=}NC(CH_3)_3\right] \xrightarrow{\text{50–75\%}} R^3\overset{O}{\overset{\|}{C}}CH(R^1)(R^2)$$

[1] A. I. Meyers and G. E. Jagdmann, Jr., *Am. Soc.*, **104**, 877 (1982).

2-Methyl-2-trimethylsilyloxypentane-3-one.

$$C_2H_5\overset{O}{\overset{\|}{C}}-C(CH_3)_2-OSi(CH_3)_3 \quad (\mathbf{1}).$$

Mol. wt. 188.34, b.p. 71–75°/15 mm.

Preparation[1]:

$$C_2H_5CH(OH)CN + CH_2{=}CHOC_2H_5 \xrightarrow[\text{70–85\%}]{HCl} C_2H_5CH(OCH(CH_3)OC_2H_5)CN \xrightarrow[\text{45–55\%}]{1)\ LDA,\ THF;\ 2)\ CH_3COCH_3;\ 3)\ H^+}$$

$$C_2H_5\overset{O}{\overset{\|}{C}}-C(CH_3)_2-OH \xrightarrow[\text{75–80\%}]{CH_3C(OSi(CH_3)_3){=}NSi(CH_3)_3} \mathbf{1}$$

Stereoselective aldol condensations (**8**, 295; **9**, 280–281). The lithium enolate of **1**, which has the (Z)-configuration, reacts with aldehydes to form *erythro*-aldols with high stereoselectivity.[2] These aldols are cleaved directly to *erythro*-α-methyl-β-hydroxy acids by periodic acid.

Example:

$$\mathbf{1} \xrightarrow[93\%]{\text{1) LDA, THF, }-78^\circ\text{; 2) }(CH_3)_2CHCHO} (CH_3)_2CH\text{–}CH(OH)\text{–}CH(CH_3)\text{–}CO\text{–}C(CH_3)_2OSi(CH_3)_3 \xrightarrow[61\%]{H_5IO_6,\ \text{THF, }25^\circ} (CH_3)_2CH\text{–}CH(OH)\text{–}CH(CH_3)\text{–}COOH$$

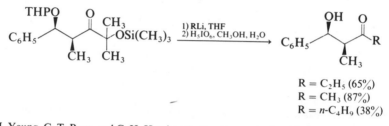

Conversion of the *erythro*-aldols to *erythro*-α-methyl-β-hydroxy aldehydes is effected by reduction with $LiAlH_4$ to a mixture of isomeric diols followed by glycol cleavage by sodium periodate in aqueous methanol.[3]

A general synthesis of *erythro*-α-methyl-β-hydroxy ketones involves protection of the aldol as THP ether followed by addition of an alkyllithium reagent to the carbonyl group. The resulting adduct is then cleaved by periodic acid to the desired ketone.[4]

Example:

$$C_6H_5\text{–}CH(OTHP)\text{–}CH(CH_3)\text{–}CO\text{–}C(CH_3)_2OSi(CH_3)_3 \xrightarrow{\text{1) RLi, THF; 2) }H_5IO_6,\ CH_3OH,\ H_2O} C_6H_5\text{–}CH(OH)\text{–}CH(CH_3)\text{–}CO\text{–}R$$

$R = C_2H_5$ (65%)
$R = CH_3$ (87%)
$R = n\text{-}C_4H_9$ (38%)

[1] S. I. Young, C. T. Buse, and C. H. Heathcock, *Org. Syn.*, submitted (1982).
[2] C. H. Heathcock, C. T. Buse, W. A. Kleschick, M. C. Pirrung, J. E. Sohn, and J. Lampe, *J. Org.*, **45**, 1066 (1980); B. Bal, C. T. Buse, and C. H. Heathcock, *Org. Syn.*, submitted (1982).
[3] C. H. Heathcock, S. D. Young, J. P. Hagen, M. C. Pirrung, C. T. White, and D. Van Derveen, *J. Org.*, **45**, 3846 (1980).
[4] C. T. White and C. H. Heathcock, *ibid*, **46**, 191 (1981).

Methyl α-trimethylsilylvinyl ketone, 5, 461–463.

Recent synthesis from acrolein:[1]

$$CH_2{=}C(Br)\text{–}CH(OC_2H_5)_2 \xrightarrow{\text{1) }n\text{-BuLi; 2) }ClSi(CH_3)_3} CH_2{=}C(Si(CH_3)_3)\text{–}CH(OC_2H_5)_2 \xrightarrow[73\%\ \text{overall}]{p\text{-TsOH, }(CH_3)_2C{=}O,\ H_2O}$$

$$CH_2{=}C(Si(CH_3)_3)\text{–}CHO \xrightarrow{CH_3MgI} \left[CH_2{=}C(Si(CH_3)_3)\text{–}CH(OH)CH_3\right] \xrightarrow[47\%]{PCC} CH_2{=}C(Si(CH_3)_3)\text{–}C(=O)\text{–}CH_3$$

[1] H. Okumoto and J. Tsuji, *Syn. Comm.*, **13**, 1015 (1982).

Molecular sieves, 1, 703–705; **2**, 286–287; **3**, 206; **4**, 345; **5**, 465; **6**, 411–412; **9**, 316–317; **10**, 273.

Removal of peroxides.[1] Self-indicating activated 4 Å molecular sieve (J. T. Baker, Merck) effectively removes peroxides from various ethers. The effectiveness is not a result of adsorption but of interaction with the impregnated indicator.

Intramolecular cyclization. Knoevenagel condensation of diphenylacetaldehyde with diethyl malonate (piperidinium acetate catalysis) results in only a trace of the expected product; when molecular sieves are present, the α-naphthol **1** is formed in 52% yield.[2] A similar reaction is observed with ethyl acetoacetate and ethyl benzoylacetate.

$$(C_6H_5)_2CHCHO + CH_2(COOC_2H_5)_2 \xrightarrow{\text{base}} [(C_6H_5)_2CHCH{=}C(COOC_2H_5)_2]$$

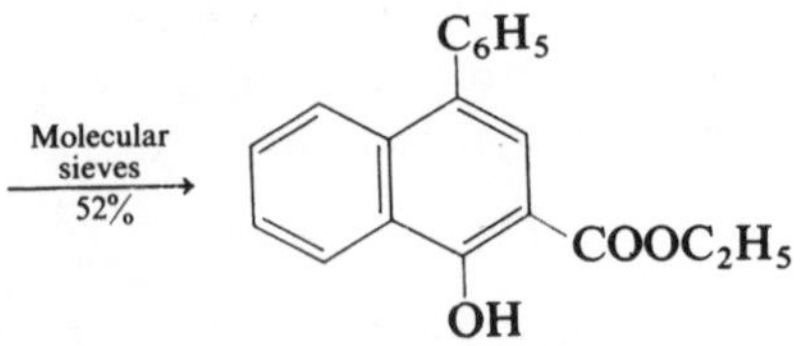

$\xrightarrow[\text{52\%}]{\text{Molecular sieves}}$ 4-C_6H_5-1-OH-naphthalene-2-$COOC_2H_5$ (**1**)

$$\underset{\mathbf{2}}{(C_6H_5)_2C{=}CHCH(COOC_2H_5)_2} \xrightarrow[\text{87\%}]{\text{molecular sieves, }\Delta} \mathbf{1}$$

The ester **2** is also converted into **1** in high yield when heated with molecular sieves.

[1] D. R. Burfield, *J. Org.*, **47**, 3821 (1982).
[2] G. A. Taylor, *J.C.S. Perkin I*, 3132 (1981).

Molybdenum carbonyl, 2, 287; **3**, 206–207; **4**, 346; **7**, 247–248; **9**, 317; **10**, 273–274.

Reductive cleavage of isoxazoles. In the presence of $Mo(CO)_6$ and water (1 equiv.), isoxazoles are cleaved to β-amino-α,β-enones in good yield.[1]

Example:

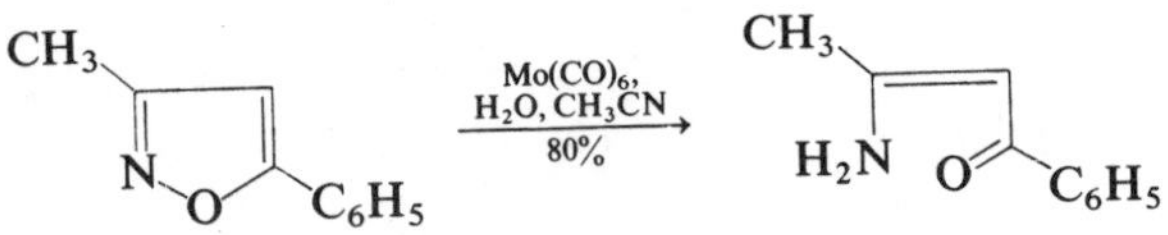

3-CH_3-5-C_6H_5-isoxazole $\xrightarrow[\text{80\%}]{Mo(CO)_6,\ H_2O,\ CH_3CN}$ $CH_3C(NH_2){=}CHCOC_6H_5$

[1] M. Nitta and T. Kobayashi, *J.C.S. Chem. Comm.*, 877 (1982).

Monoisopinocampheylborane (1), 8, 267; **9**, 317; **10**, 224. Several syntheses of this borane have been disclosed but the simplest, and one that furnishes material of ~100% ee, is formulated in equation (I).[1]

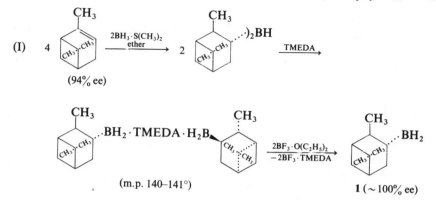

(I) 4 [α-pinene] (94% ee) $\xrightarrow[\text{ether}]{2BH_3\cdot S(CH_3)_2}$ 2 [diisopinocampheylborane] $\xrightarrow{\text{TMEDA}}$

[adduct]·BH$_2$·TMEDA·H$_2$B·[adduct] (m.p. 140–141°) $\xrightarrow[-2BF_3\cdot TMEDA]{2BF_3\cdot O(C_2H_5)_2}$ **1** (~100% ee)

Asymmetric hydroboration.[2] Extensive studies indicate that **1** is the reagent of choice for chiral hydroboration of *trans*-disubstituted alkenes and trisubstituted alkenes. The corresponding alcohols are obtained in 72–100% ee and all have the same absolute configuration. Surprisingly, this configuration is the opposite to that obtained by hydroboration with diisopinocampheylborane.

Chiral ketones.[3] Asymmetric hydroboration of a prochiral alkene with monoisocampheylborane followed by a second hydroboration of a nonprochiral alkene provides a chiral mixed trialkylborane. This product reacts with acetaldehyde with elimination of α-pinene to give a chiral borinic acid ester in 73–100% ee. Treatment of this intermediate with α,α-dichloromethyl methyl ether (**2**, 120; **5**, 200–203) and lithium triethylcarboxide followed by oxidation results in an optically active ketone in 60–90% ee.

Example:

$CH_3CH{=}CHCH_3$ (trans) $\xrightarrow[2)\ CH_3(CH_2)_2CH=CH_2]{1)\ \mathbf{1}}$ [Ipc]B(C_5H_{11}-*n*)CH(CH$_3$)CH$_2$CH$_3$ $\xrightarrow{CH_3CHO}$

n-C_5H_{11}B(OC$_2$H$_5$)CH(CH$_3$)CH$_2$CH$_3$ $\xrightarrow[66\%]{1)\ Cl_2CHOCH_3,\ LiOC(C_2H_5)_3;\ 2)\ H_2O_2,\ OH^-}$ *n*-C_5H_{11}COCH(CH$_3$)CH$_2$CH$_3$ (70% ee)

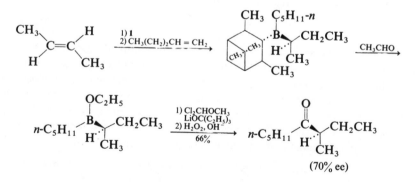

[1] H. C. Brown, A. K. Mandal, N. M. Yoon, B. Singaram, J. R. Schwier, and P. K. Jadhav, *J. Org.*, **47**, 5069 (1982).
[2] H. C. Brown, P. K. Jadhav, and A. K. Mandal, *ibid.*, **47**, 5074 (1982).
[3] H. C. Brown, P. K. Jadhav, and M. Desai, *Am. Soc.*, **104**, 6844 (1982).

Morpholine, 1, 705.

Decarboxylation of α-keto carboxylic acids.[1] The conversion of α-keto carboxylic acids into the morpholine enamine (TsOH) is accompanied by loss of CO_2 to form the enamine of an aldehyde in almost quantitative yield. The free aldehyde is liberated as usual by acidic hydrolysis.

I. K. Stamos, *Tetrahedron Letters*, **23**, 459 (1982).

Morpholine–Camphoric acid, 9, 317–318.

Intramolecular aldol condensation.[1] The opposite regioselectivity of this system (**1**) as compared to that of piperidinium acetate (**1**, 888–889) is also observed with dialdehydes of the general structure **2**.

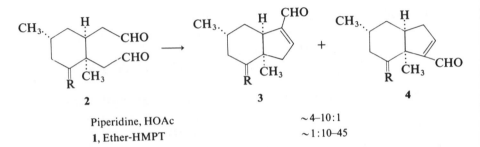

	2 → 3 + 4
Piperidine, HOAc	~4–10:1
1, Ether-HMPT	~1:10–45

[1] T. Harayama, M. Takatani, A. Yamanaka, H. Ikeda, M. Ono, and Y. Inubushi, *Chem. Pharm. Bull. Japan*, **29**, 766 (1981).

2-Morpholinoethyl isocyanide, O(CH$_2$CH$_2$)$_2$NCH$_2$CH$_2$NC (**1**). B.p. 72–73°/0.07 mm.

Supplier: Fluka.

Peptides. This isocyanide is preferable to simple aliphatic isocyanides (**9**, 82) for coupling of amino acid derivatives to peptides.[1] Addition of hydroxysuccinimide (**9**, 246) or 1-hydroxybenzotriazole (**6**, 288) suppresses racemization. It is important to allow the acid component and the additive to react with the isocyanide for a suitable period before addition of the amine component and triethylamine or DMAP. Coupling to di- and tripeptides in yields of 55–72% have been reported. The by-product, (N-morpholinoethyl)formamide, is removed by an acid wash.

[1] H. Aigner, G. Koch, and D. Marquarding, Symposium on Chemistry of Peptides and Proteins (Moscow), 56 (1980); D. Marquarding and H. Aignar, Ger. Pat. 2, 942,606.

N-Morpholinomethyldiphenylphosphine oxide (1), 9, 318–319.

α-Aminomethyl ketones.[1] Thermolysis of the adduct (**2**) of the anion of **1** with aldehydes results in loss of diphenylphosphine oxide and formation of α-aminomethyl ketones (**3**). Thermolysis is usually conducted in ethylene glycol at 140–150° for 3 hours.

$$\mathrm{RCHO} + (C_6H_5)_2\overset{O}{\overset{\|}{P}}\underset{Li}{\underset{|}{C}}H\text{—N}\langle\text{morpholine}\rangle\text{O} \longrightarrow (C_6H_5)_2\overset{O}{\overset{\|}{P}}\underset{R\text{—}CHOH}{\underset{|}{C}}H\text{—N}\langle\text{morpholine}\rangle\text{O}$$

2

$$\xrightarrow[75\text{–}90\%]{140\text{–}150^\circ} \mathrm{R}\overset{O}{\overset{\|}{C}}\text{—}CH_2N\langle\text{morpholine}\rangle\text{O}$$

3

[1] N. L. J. M. Broekhof and A. van der Gen, *Tetrahedron Letters*, **22**, 2799 (1981).

N

Nafion, 9, 320; **10,** 275–276.

Catalyst for photoreactions. Irradiation with short-wave light causes the triene **1** to isomerize to **2**.[1] When **1** is absorbed by Nafion-H the cation **3** is formed, and then irradiation at 380 nm results mainly in formation of the cyclopentadiene **5**.[2]

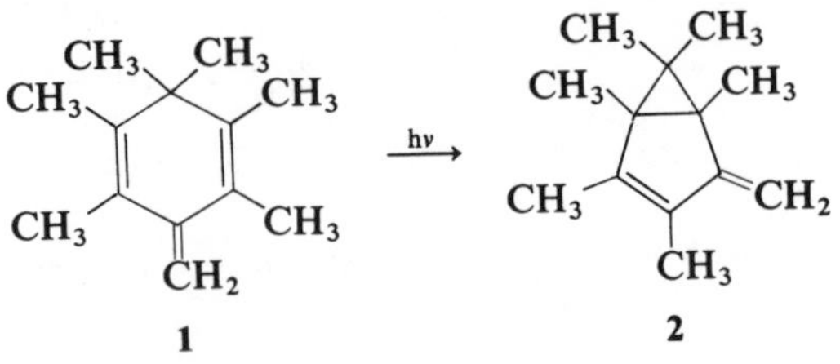

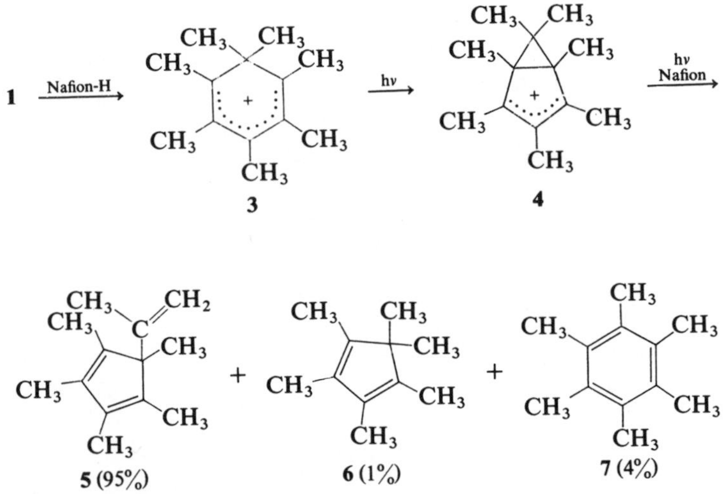

Methoxymethyl ethers. Nafion-H resin is a suitable catalyst for etherification of alcohols with dimethoxymethane.[3]

Rearrangement of α-ethynyl alcohols. Nafion-H is an efficient catalyst for rearrangement of α-ethynyl alcohols to α,β-enones.[4]

Examples:

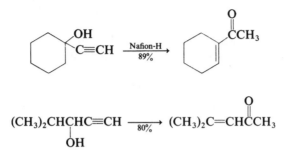

$$\text{1-ethynylcyclohexanol (C}_6\text{H}_{10}\text{(OH)C}\equiv\text{CH)} \xrightarrow[89\%]{\text{Nafion-H}} \text{1-acetylcyclohexene (C}_6\text{H}_9\text{C(=O)CH}_3\text{)}$$

$$(CH_3)_2CHCH(OH)C{\equiv}CH \xrightarrow[80\%]{} (CH_3)_2C{=}CHCCH_3 \ (\text{C=O})$$

Dehydration of* 1,4-*diols. Nafion-H can be used for dehydration of 1,4-diols to tetrahydrofuranes (90-95% yield). Yields of cyclic ethers from 1,2-, 1,5-, and 1,6-diols are lower (50–85%).[5]

[1] R. F. Childs and A. Mika-Gibala, *J. Org.*, **47**, 4204 (1982).
[2] H. Hart, J. D. DeVrieze, R. M. Lange, and A. J. Sheller, *J.C.S. Chem. Comm.*, 1650 (1968).
[3] G. A. Olah, A. Husain, B. G. B. Gupta, S. C. Narang, *Synthesis*, 471 (1981).
[4] G. A. Olah and A. P. Fung, *ibid.*, 473 (1981).
[5] G. A. Olah, A. P. Fung, and R. Malhotra, *ibid.*, 474 (1981).

Nickel on alumina.

Hydrogenation catalyst. The catalyst is prepared from $Ni(NO_3)_2 \cdot 6H_2O$ slurried with Al_2O_3 and then dried at 120°. The catalyst is activated before use by heating at 400° for 30 minutes.[1] The flow apparatus and general procedure for hydrogenation in the gas phase at atmospheric pressure have been described.[2]

This catalyst effects reduction of aromatic and tertiary nitriles to methyl groups in yields $>85\%$. Aliphatic nitriles, however, mainly undergo hydrogenolysis.[3]

The catalyst effects hydrogenolysis of large numbers of *t*-groups: OH, Br, NH_2, CH_2OH, COOH, $COOCH_3$, $OSi(CH_3)_3$. However, *t*-CH_2Br is converted into CH_3.

Unhindered ketones can be deoxygenated by hydrogenation with this catalyst.[4] Steric hindrance can prevent this reaction. Thus, 2-norbornanone does not react, although camphor is reduced. The alcohols corresponding to inactive ketones are reduced. Cyclopentanone is reduced to cyclopentane in 24% yield; cyclohexanone is reduced in 90% yield. When successful, this method is a useful one-step deoxygenation.

[1] W. F. Maier, K. Bergmann, W. Bleicher, and P. v. R. Schleyer, *Tetrahedron Letters*, **22**, 4227 (1981).
[2] P. Grubmuller, W. F. Maier, P. v. R. Schleyer, M. A. McKervey, and J. J. Rooney, *Ber.*, **113**, 1989 (1980).
[3] J. G. Andrade, W. F. Maier, L. Zapf, and P. v. R. Schleyer, *Synthesis*, 802 (1980).
[4] W. F. Maier, P. Grúuller, I. Thies, P. M. Stein, M. A. McKervey, and P. v. R. Schleyer, *Angew. Chem. Int. Ed.*, **18**, 939 (1979).

Nickel-Graphite. A highly dispersed nickel on graphite (Ni-Gr 1) is obtained by reduction of $NiBr_2 \cdot 2DME$ dissolved in THF–HMPT (15:1) with C_8K (2 equiv.) at 20° under argon. This material is slowly oxidized by air to give a less active metal (Ni–Gr 2), which can be stored for several months.

***Ni-Gr* 1.**[1] This nickel catalyst reduces disubstituted alkynes to *cis*-alkenes; the stereoselectivity is comparable to that observed with Lindlar catalyst or nickel boride (**3**, 208–210; **5**, 471–472).

***Ni-Gr* 2.**[2] Semihydrogenation of triple bonds proceeds more slowly and with somewhat less stereoselectivity with Ni–Gr 2 than with Ni–Gr 1. Ni–Gr 2 is particularly useful as a catalyst for hydrogenation of α,β-enones and β-diketones.

Examples:

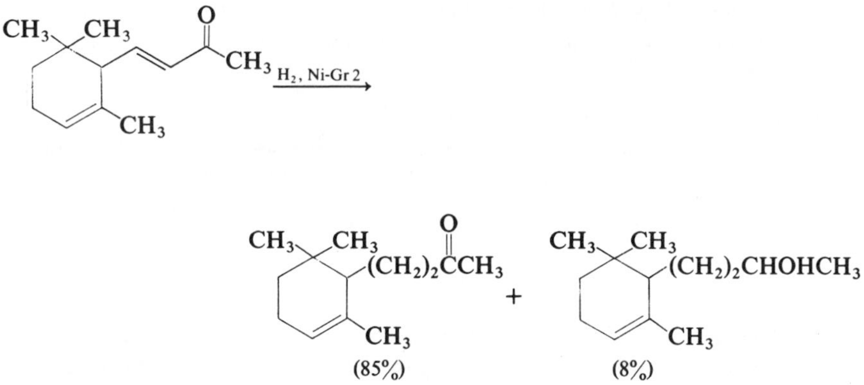

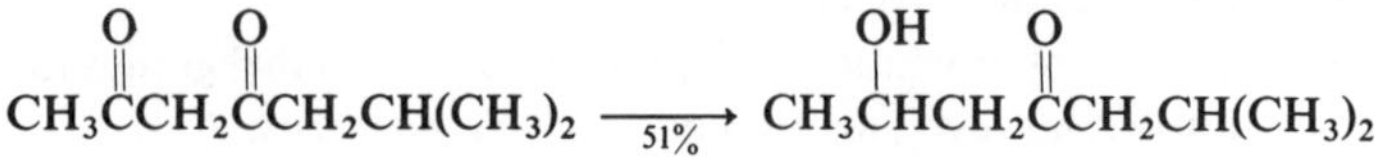

(structure: 2,6,6-trimethylcyclohex-2-enyl–CH=CH–C(=O)CH$_3$) $\xrightarrow{H_2,\ \text{Ni-Gr 2}}$

(2,6,6-trimethylcyclohex-2-enyl)–$(CH_2)_2CCH_3$ (C=O) (85%) + (2,6,6-trimethylcyclohex-2-enyl)–$(CH_2)_2CHOHCH_3$ (8%)

$$CH_3\overset{O}{\overset{\|}{C}}CH_2\overset{O}{\overset{\|}{C}}CH_2CH(CH_3)_2 \xrightarrow[51\%]{} CH_3\overset{OH}{\overset{|}{C}}HCH_2\overset{O}{\overset{\|}{C}}CH_2CH(CH_3)_2$$

[1] D. Savoia, E. Tagliavini, C. Trombini, and A. Umani-Ronchi, *J. Org.*, **46**, 5340 (1981).
[2] *Idem, ibid.*, **46**, 5344 (1982).

Nickel carbonyl, 1, 720–723; **2**, 290–293; **4**, 353–355; **5**, 472–474; **6**, 417–419; **7**, 250; **10**, 276.

α-Methylene-γ-lactones.[1] A recent synthesis of α-methylene-γ-lactones uses $Ni(CO)_4$ to promote addition of an allylic methanesulfonate ester group to an aldehyde (cyclization) and to effect carbonylation of a vinyl halide (**3**, 211–212). This strategy was used for a stereospecific synthesis of frullanolide (**2**, equation I). Thus treatment of **1** with $Ni(CO)_4$ results in formation of the desired lactone (**2**) and **3** in about equal amounts.

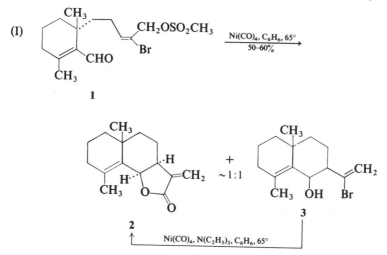

(I) [structure 1] $\xrightarrow[50\text{–}60\%]{Ni(CO)_4,\ C_6H_6,\ 65°}$ 2 + 3 (~1:1)

3 $\xrightarrow{Ni(CO)_4,\ N(C_2H_5)_3,\ C_6H_6,\ 65°}$ 2

Conjugate addition of acyl groups to naphthoquinones (**3**, 211).[2] The acyl nickel carbonylate anion (**1**) undergoes conjugate addition to the ethylene ketal **2** to form a lithium adduct that can be trapped by allyl iodide to give **3** in 81% yield. This product can be converted in a few steps to the hydroxyquinone **4**, a precursor of the naphthoquinone antibiotic deoxyfrenolicin (**5**).

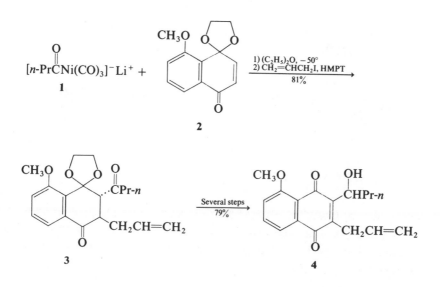

$[n\text{-PrC(=O)Ni(CO)}_3]^-Li^+$ (**1**) + **2** $\xrightarrow[81\%]{1)\ (C_2H_5)_2O,\ -50°;\ 2)\ CH_2{=}CHCH_2I,\ HMPT}$ **3**

3 $\xrightarrow[79\%]{\text{Several steps}}$ **4**

OH O Pr-*n*

O

CH_2COOH

O

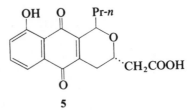

5

[1] M. F. Semmelhack and S. J. Brickner, *Am. Soc.*, **103**, 3945 (1981); *idem*, J. Org., **46**, 1723 (1981).

[2] M. F. Semmelhack, L. Keller, T. Sato, and E. Spiess, *ibid.*, **47**, 4382 (1982).

Nickel(II) chloride-Triphenylphosphine.

Desulfuration.[1] A secondary Grignard reagent in the presence of a catalytic amount of a 1:1 mixture of $NiCl_2$ and triphenylphosphine reduces alkylthio groups attached to a trigonal carbon center. The desulfuration coupled with the known substitution of alkylthio groups attached to a double bond with Grignard reagents catalyzed by $[(C_6H_5)_3P]_2NiCl_2$ or $(dppp)_2NiCl_2$ (**9**, 147–148) provides a general, regio- and stereoselective synthesis of disubstituted alkenes from 1,1-bis(alkylthio)alkenes, readily available from orthothioesters. An example is the transformation of 1,1-bis(ethylthio)-3-phenylpropene-1 (**1**) into either (E)- or (Z)-1-phenyl-1-pentene, (**3a**) or (**3b**), respectively.

$C_6H_5CH_2(H)C{=}C(SC_2H_5)_2$ (**1**) $\xrightarrow[85\%]{C_2H_5MgBr,\ (dppp)NiCl_2}$ $C_6H_5CH_2(H)C{=}C(SC_2H_5)(C_2H_5)$ (**2a**) $\xrightarrow[39\%]{(CH_3)_2CHMgBr,\ NiCl_2\text{-}PR_3}$ $C_6H_5CH_2(H)C{=}C(H)(C_2H)$ (**3a**)

1 $\xrightarrow[58\%]{(CH_3)_2CHMgBr,\ NiCl_2\text{-}P(C_6H_5)_3}$ $C_6H_5CH_2(H)C{=}C(SC_2H_5)(H)$ (**2b**) $\xrightarrow[50\%]{C_2H_5MgBr,\ (dppp)_2NiCl_2}$ $C_6H_5CH_2(H)C{=}C(C_2H_5)(H)$ (**3b**)

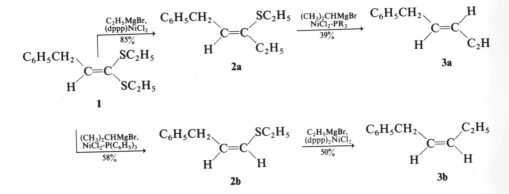

[1] E. Wenkert and T. W. Ferreira, *J.C.S. Chem. Comm.*, 840 (1982).

Nickel(II) 2-ethylhexanoate, $Ni(C_8H_{15}O_2)_2$ (1). Mol. wt. 345.13. Supplier: Alfa.

Selective oxidations with bromine. Bromine in combination with this Ni(II) alkanoate promotes the selective oxidation of primary, secondary 1,4-diols to γ-butyrolactones (equation I).[1]

(I) R^1–CH(OH)–$CH_2CH_2CH_2$–OH $\xrightarrow[60\text{–}70\%]{Br_2,\ Ni(OOCR^2)_2}$ R^1–(γ-lactone)

[1] M. P. Doyle and V. Bagheri, *J. Org.*, **46**, 4806 (1981).

p-Nitrobenzenesulfonyl 4-nitroimidazole, O_2N–C_6H_4–SO_2–N(4-nitroimidazolyl) (**1**).

Mol. wt. 298.2, m.p. 179–185° (dec).

The reagent is prepared by reaction of *p*-nitrobenzenesulfonyl chloride with 4-nitroimidazole (triethylamine).[1]

Polynucleotides. Triisopropylbenzenesulfonyl chloride (**1**, 1228–1229; **3**, 308) has been widely used as the condensing agent in triester condensation of nucleotide fragments. However, it can sulfonate the free 5′-hydroxy group of the incoming nucleotide. More recently, arylsulfonyltetrazoles (**7**, 13–14) have been used successfully, although they suffer from instability. A recent improvement is use of a mixture of triisopropylbenzenesulfonyl chloride and tetrazole.[2] The bis-nitroaromatic reagent **1** gives yields of polynucleotides comparable to those obtained with the combination reagent, with the only significant difference being a slower rate of reaction.[1]

[1] C. A. Leach, F. Waldmeier, and C. Tamm, *Helv.*, **64**, 2515 (1981).
[2] A. K. Seth and E. Jay, *Nucleic Acids Res.*, **8**, 5445 (1980).

5-Nitro-[3H]-1,2-benzoxathiole S,S-dioxide, (structure: O_2N-substituted benzene fused to CH_2–SO_2–O ring) (**1**). Mol. wt. 215.19, m.p. 148°. Preparation.[1]

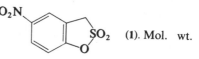

Peptide synthesis.[2] The sulfone can effect condensation of an amino group with salts of N-protected amino acids with only slight racemization. The by-product is water-soluble.

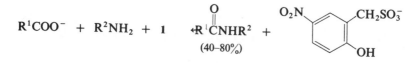

$$R^1COO^- + R^2NH_2 + \mathbf{1} \longrightarrow \underset{(40\text{–}80\%)}{R^1\overset{O}{\overset{\|}{C}}NHR^2} + O_2N\text{–}C_6H_3(CH_2SO_3^-)(OH)$$

[1] W. Marckwald and H. H. Frahne, *Ber.*, **31**, 1854 (1898); E. T. Kaiser and K.-W. Lo, *Am. Soc.*, **91**, 4912 (1969).
[2] M. Wakselman and F. Acher, *J.C.S. Chem. Comm.*, 632 (1981).

Nitroethane–Pyridinium chloride.

Aryl nitriles.[1] Aromatic aldehydes are converted directly into nitriles in 80% yield when refluxed with a slight excess of nitroethane and pyridinium chloride. The reaction involves an aldoxime as intermediate. Nitromethane or 1-nitropropane can replace nitroethane, but the reaction is then slower.

[1] D. Dauzonne, P. Demerseman, and R. Royer, *Synthesis*, 739 (1981).

p-Nitroperbenzoic acid, 1, 743; **9,** 324.

4-Vinylcyclopentenones. 1,2,4,6-Tetraenes are converted into 4-vinylcyclopentenones by epoxidation under the mild conditions of Anderson and Veysoglu (**5**, 120).[1] Acetoxymercuration is not useful for this cyclization.

Example:

$$CH_2{=}C{=}C(C_5H_{11}){-}CH{=}CHCH{=}CH_2 \xrightarrow[70\%]{p\text{-}NO_2C_6H_4CO_3H} \text{2-}C_5H_{11}\text{-4-}(CH{=}CH_2)\text{-cyclopent-2-enone}$$

[1] G. Balme, M. Malacria, and J Goré, *J. Chem. Research* (*M*), 2869 (1981).

o-Nitrophenylsulfenyl chloride, 1, 745.

Amino acid esters.[1] *o*-Nitrophenylsulfenyl N-protected amino acids can be esterified without racemization by the dicyclohexylcarbodiimide–4-dimethylaminopyridine procedure (**8**, 163). Urethane-protected amino acids undergo partial racemization on esterification.

[1] B. Neises, T. Andries, and W. Steglich, *J.C.S. Chem. Comm.*, 1132 (1982).

2-Nitro-3-pivaloyloxypropene, $CH_2{=}C(NO_2)CH_2OCOC(CH_3)_3$ (**1**).

Preparation[1]:

$$CH_3NO_2 \xrightarrow[2)\ (CH_3)_3CCOCl]{1)\ HCHO,\ NaOH} (CH_3)_3C\overset{O}{\overset{\|}{C}}OCH_2\overset{NO_2}{\overset{|}{C}}HCH_2O\overset{O}{\overset{\|}{C}}C(CH_3)_3 \xrightarrow{NaOCH_3} \mathbf{1}$$

Nitroallylation of nucleophiles.[1, 2] The most reactive nucleophiles (*e.g.*, alkyllithiums) do not react satisfactorily with 2-nitropropenes, but owing to the pivaloyloxy group, **1** reacts with a variety of nucleophiles to effect transfer of the 2-nitroallyl group selectively. The products can undergo Michael addition with another nucleophile to give unsymmetrical nitroalkanes (**2**). The nitro group in the final product can be transformed into other useful functional groups. A typical sequence is shown in Scheme (I).

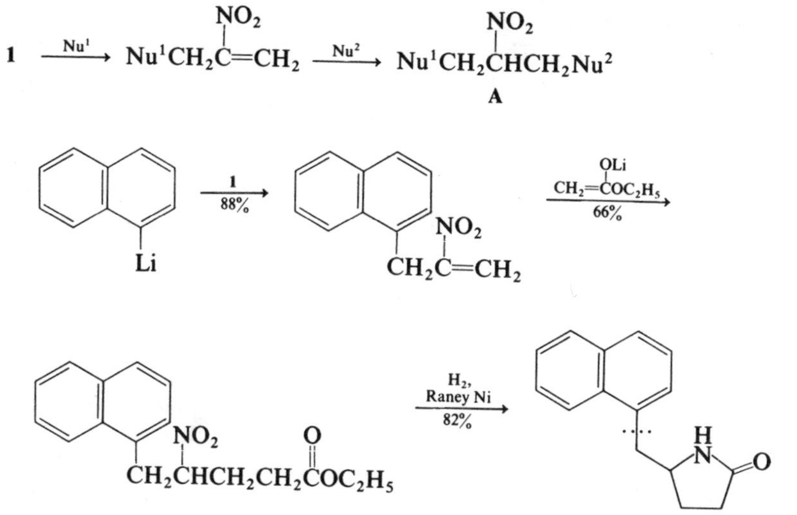

$$1 \xrightarrow{Nu^1} Nu^1CH_2C(NO_2){=}CH_2 \xrightarrow{Nu^2} Nu^1CH_2CH(NO_2)CH_2Nu^2 \quad \textbf{A}$$

Scheme I

[1] P. Knochel and D. Seebach, *Nouv. J. Chim.*, **5**, 75 (1981); *idem*, *Tetrahedron Letters*, **22**, 3223 (1981).
[2] *Idem*, *ibid.*, **23**, 3897 (1982).

(5-Nitropyridyl)diphenyl phosphinate,

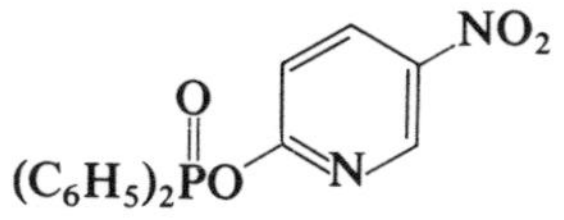

$(C_6H_5)_2P(O)O$–(5-nitro-2-pyridyl) (**1**).

Mol. wt. 340.27, m.p. 125–127°. The phosphinic ester is prepared by reaction of diphenylphosphinyl chloride and 2-hydroxy-5-nitropyridine in the presence of triethylamine.

Cyclic peptides. In continuing investigation of organophosphorus condensing reagents (**9**, 50–51), Mukaiyama's group[1] found that **1** is particularly promising for coupling to dipeptides at −10° without detectable racemization. A further advantage is that the by-product, 2-hydroxy-5-nitropyridine, is easily separated from the peptide. The reagent effected cyclization of a linear decapeptide to the corresponding cyclic decapeptide gramicidin S in 80% yield.

[1] T. Mukaiyama, K. Kamekawa, and Y. Watanabe, *Chem. Letters*, 1367 (1981).

Nitrosobenzene, 5, 478–479.

Diels-Alder reactions. The cycloaddition of nitrosobenzene with **1** has been used for synthesis of an amino sugar (**4**).[1]

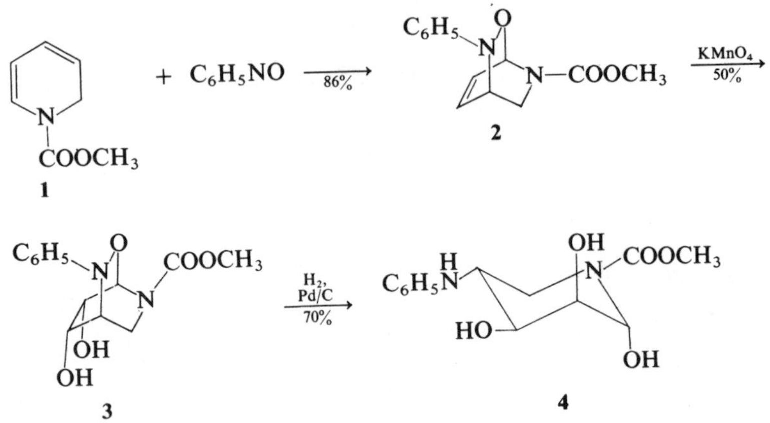

The reaction of nitrosobenzene with pyrane-2-thione (**5**) unexpectedly results in the adduct **6**, which may arise from an initial Diels-Alder reaction to give (**a**).[2]

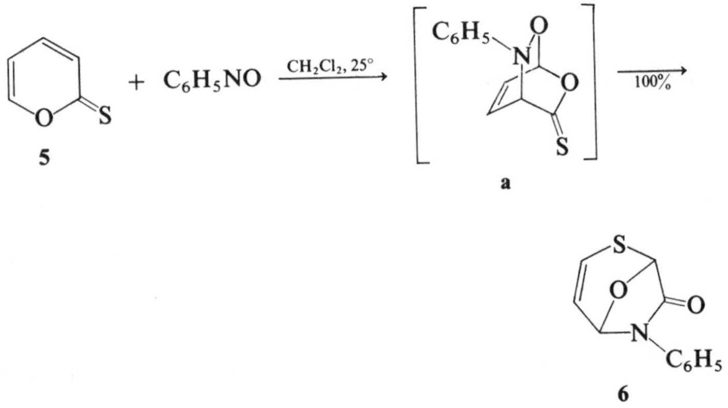

[1] J. Streith, G. Augelmann, H. Fritz, and H. Strub, *Tetrahedron Letters*, **23**, 1909 (1982).
[2] G. Augelmann, H. Fritz, G. Rihs, and J. Streith, *J.C.S. Chem. Comm.*, 1112 (1982).

Nitrosocarbonylmethane, CH_3CONO (**1**), **9,** 326.

Generation *in situ*:

$$CH_3CONHOH + \text{9,10-dimethylanthracene} + (n\text{-}C_3H_7)_4\overset{+}{N}IO_4^- \xrightarrow{95\%} \mathbf{2} \xrightarrow{\Delta} CH_3CONO\ (\mathbf{1})$$

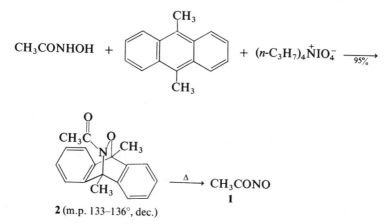

2 (m.p. 133–136°, dec.)

Intramolecular ene reactions.[1] In addition to the ene reactions of **1** with olefins (**9**, 326), suitable derivatives of **1** undergo intramolecular ene reactions to form bicyclic systems. The desired substrates are conveniently obtained by condensation of the enolate of **2** with an aldehyde followed by silylation, as formulated for a typical example in equation I.

(I) **2** $\xrightarrow[67\%]{\text{1) LDA, HMPT; 2) } CH_3CH{=}CHCHO\ (75\%)\text{; 3) } ClSi(CH_3)_2C(CH_3)_3\ (95\%)}$ [R_3SiO, HC, CH_3, H, $N{-}O$, O] $\xrightarrow[100\%]{\Delta}$ [R_3SiO, $CH{=}CH_2$, N, OH, O]

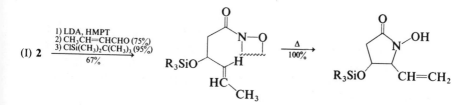

[1] G. E. Keck, R. R. Webb and J. B. Yates, *Tetrahedron*, **37**, 4007 (1981).

O

Organoaluminum compounds, 10, 281.

cis, vic-*Diols.* (Z)-γ-Alkoxyallyldiethylaluminum compounds (**1**), available *in situ* from allyl ethers, react with aldehydes exclusively at the γ-position to form monoprotected *cis, vic*-diols (**2**) with high diastereoselectivity (equation I). Addition of **1** to ketones is generally slower and is less diastereoselective.[1]

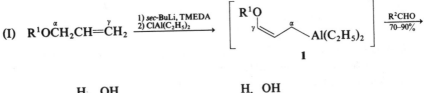

$$\text{(I)}\quad R^1O\overset{\alpha}{C}H_2CH{=}\overset{\gamma}{C}H_2 \xrightarrow[\text{2) ClAl}(C_2H_5)_2]{\text{1) }sec\text{-BuLi, TMEDA}} \left[\,R^1O\overset{\gamma}{C}H{=}CH\overset{\alpha}{C}H_2Al(C_2H_5)_2\,\right]\ \mathbf{1} \xrightarrow[70\text{–}90\%]{R^2CHO} \mathbf{2} + \mathbf{3}\ (9\text{–}11:1)$$

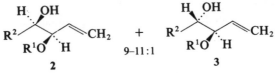

γ-Allyloxyaluminum ate complexes, $RO\overset{\gamma}{C}H{=}CH\overset{\alpha}{C}H_2\overset{+}{A}l(C_2H_5)_3Li^-$, also react with carbonyl compounds and reactive halides almost exclusively at the γ-position. The diastereoselectivity was not examined.[2] In contrast, the regioselectivity of reactions of carbanions of allylic esters with electrophiles can depend on the electrophiles, the gegenions, and other factors.

Reaction with epoxy alochols.[3] Epoxides formed from (E)-allylic alcohols react regioselectively with organoaluminum compounds to provide 1,2-diols with inversion at C_3 in high yield.

Example:

$$n\text{-Bu-epoxy alcohol (E)} \xrightarrow[95\%]{C_6H_5C{\equiv}CAl(C_2H_5)_2} n\text{-BuCH}(C{\equiv}CC_6H_5)CH(OH)CH_2OH$$

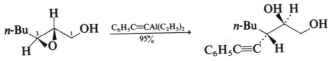

However, epoxides derived from (Z)-allylic alcohols react slowly with the same reagents and with lower regioselectivity.

Example:

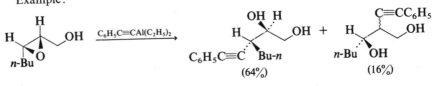

$$n\text{-Bu-epoxy alcohol (Z)} \xrightarrow{C_6H_5C{\equiv}CAl(C_2H_5)_2} n\text{-BuCH}(C{\equiv}CC_6H_5)CH(OH)CH_2OH\ (64\%) + n\text{-BuCH}(OH)CH(C{\equiv}CC_6H_5)CH_2OH\ (16\%)$$

[1] M. Koreeda and Y. Tanaka, *J.C.S. Chem. Comm.*, 845 (1982).
[2] Y. Yamamoto, H. Yatagai, and K. Maruyama, *J. Org.*, **45**, 195 (1980).
[3] T. Suzuki, H. Saimoto, H. Tomioka, K. Oshima, and H. Nozaki, *Tetrahedron Letters*, **23**, 3597 (1982).

Organocopper reagents, 9, 328–334; **10**, 282–290.

Hindered cuprates. A novel route to lithium dialkylcuprates containing a secondary or a tertiary alkyl group involves reaction of the tosylhydrazone of an aldehyde with a copper reagent, such as dilithium trimethylcuprate (**6**, 386; **7**, 115). An example is shown in equation (I) for preparation of a cuprate containing the *t*-butyl group.[1]

$$\text{(I)}\quad (CH_3)_3C\overset{N-NHTs}{\overset{\|}{C}}H \xrightarrow[-LiTs,\ -N_2]{(CH_3)_3CuLi_2} \left[(CH_3)_3C\overset{LiCuCH_3}{\overset{|}{C}}HCH_3\right] \xrightarrow[70\%]{C_6H_5COCl} C_6H_5\overset{O}{\overset{\|}{C}}\underset{CH_3}{\underset{|}{C}}HC(CH_3)_3$$

$$\left[\ \right] \xrightarrow[90\%]{CH_3I} (CH_3)_2CHC(CH_3)_3$$

$$\left[\ \right] \xrightarrow[94\%]{CH_2=C(CH_3)CHO} (CH_3)_3C\overset{CH_3}{\overset{|}{C}}HCH_2\overset{CH_3}{\overset{|}{C}}HCHO$$

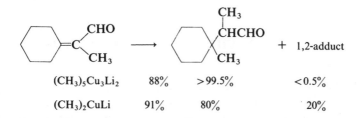

$(CH_3)_5Cu_3Li_2$. This organocuprate is obtained by addition of $CH_3Li \cdot LiCl$ (5 equiv.) to CuI (3 equiv.) in ether at 0°.[2, 3] It is superior to $(CH_3)_2CuLi$ for conjugate addition to α,β-enals, particularly in cases where a quaternary center is generated.[4, 5]

Example:

$$\text{cyclohexylidene}{=}C(CHO)CH_3 \longrightarrow 1\text{-}CH_3\text{-}1\text{-}(CH(CH_3)CHO)\text{cyclohexane} + \text{1,2-adduct}$$

$(CH_3)_5Cu_3Li_2$	88%	>99.5%	<0.5%
$(CH_3)_2CuLi$	91%	80%	20%

Esters or ketones. (S)-2-Pyridylthioates can be converted by reaction with lithium dialkylcuprates in ether at −78° into either esters or ketones. The former products are obtained without contamination by ketones if the reaction is conducted under oxygen; ketones are formed when the reaction is conducted under nitrogen. Yields of both products are generally in the range 70–95%.[6]

Examples:

$$\text{2-pyridyl-}SCOC_6H_5 + (CH_3)_2CuLi \xrightarrow[71\%]{O_2,\ ether} C_6H_5COOCH_3$$

$$\text{2-pyridyl-}SCOCH(CH_3)_2 + (n\text{-}C_4H_9)_2CuLi \xrightarrow[79\%]{N_2,\ ether} (CH_3)_2CHC(=O)C_4H_9\text{-}n$$

Mixed homocuprates.[7] Mixed cuprates (**1**) in which the nontransferable ligand is an α-sulfonyl carbanion are easily prepared from dimethyl sulfone or methyl phenyl sulfone (equation I), and are effective for conjugate addition to enones and for a synthesis of ketones from acid chlorides.

$$\text{(I)}\quad R^1SO_2CH_3 \xrightarrow{\text{1) } n\text{-BuLi, THF; 2) CuI}} R^1SO_2CH_2Cu \xrightarrow{RLi} \underset{\mathbf{1}}{(R^1SO_2CH_2CuR)^-Li^+}$$

$$R^1 = CH_3, C_6H_5$$

Examples:

$$\text{cyclohexenone} + [C_6H_5SO_2CH_2CuC(CH_3)_3]^-Li^+ \xrightarrow[97\%]{THF,\ 0^\circ} \text{3-}C(CH_3)_3\text{-cyclohexanone}$$

$$C_6H_5COCl + (CH_3SO_2CH_2CuCH_3)^-Li^+ \xrightarrow[90\%]{THF,\ -78^\circ} C_6H_5COCH_3$$

$R_2Cu(CN)Li_2$. These mixed cuprates are readily available by reaction of thoroughly dry CuCN with 2RLi in THF at $-78\rightarrow0^\circ$. Although displacement reactions of secondary halides with lithium dialkylcuprates proceed in low yield (because of reduction and/or elimination), these new mixed cuprates give high yields of coupled products in the case of secondary bromides and iodides. Mesylates are inert but tosylates react to some extent (5–60%). The R group can be CH_3, C_2H_5, *n*-propyl, and *n*-butyl, and even vinyl ($CH_2{=}CH$), but C_6H_5 is not useful because of very low yields. The halides can be cyclic or acyclic.[8]

Examples:

$$\text{C}_5\text{H}_9\text{—Br} + (n\text{-Bu})_2\text{Cu(CN)Li}_2 \xrightarrow[86\%]{\text{THF, }0^\circ} \text{C}_5\text{H}_9\text{—}n\text{-Bu}$$

$$\text{C}_6\text{H}_5(\text{CH}_2)_2\text{CH(I)}(\text{CH}_2)_3\text{CH}_3 + (\text{CH}_2\text{=CH})_2\text{Cu(CN)Li}_2 \xrightarrow[70\text{–}80\%]{} \text{C}_6\text{H}_5(\text{CH}_2)_2\text{CH(CH=CH}_2)(\text{CH}_2)_3\text{CH}_3$$

These mixed cuprates undergo conjugate addition to mono-, di-, and trisubstituted enones in good to excellent yield. Additives such as $S(CH_3)_2$ and R_3P, which are sometimes necessary in Michael additions of R_2CuLi, are not necessary.[9]

These cuprates react more readily than R_2CuLi reagents with epoxides. Reaction occurs cleanly at the less sterically hindered position with the expected inversion of configuration.[10] Yields are consistently higher than those obtained with RCu(CN)Li.[11]

Examples:

$$\text{C}_6\text{H}_5\text{-epoxide} + (n\text{-C}_4\text{H}_9)_2\text{Cu(CN)Li}_2 \longrightarrow \underset{(85\%)}{\text{C}_6\text{H}_5\text{CH(OH)}(\text{CH}_2)_4\text{CH}_3} + \underset{(8\%)}{\text{C}_6\text{H}_5\text{CH}(\text{C}_4\text{H}_9\text{-}n)\text{CH}_2\text{OH}}$$

$$\text{CH}_3\text{, O (epoxide), C(CH}_3)_3\text{-cyclohexane} + (\text{C}_2\text{H}_5)_2\text{Cu(CN)Li}_2 \xrightarrow[98\%]{} \text{CH}_3\text{, OH, C}_2\text{H}_5\text{, C(CH}_3)_3\text{-cyclohexane}$$

Lithium anilidocyanocuprates, $LiCu(CN)N(R^1)\text{-}C_6H_4R^2$ (**1**). *ortho*-Lithiated benzamides (**2**) undergo oxidative coupling with these cuprates (**1**) to form N-arylanthranilamides (**3**), which cyclize to acridones (**4**) in refluxing heptafluorobutyric acid. Chlorocuprates can be used instead of cyanocuprates, but yields are usually lower.[12]

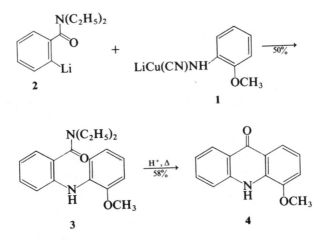

Example:

$$\underset{\mathbf{2}}{2\text{-}[(C_2H_5)_2NCO]C_6H_4Li} + \underset{\mathbf{1}}{LiCu(CN)NH\text{-}C_6H_4\text{-}2\text{-}OCH_3} \xrightarrow{50\%} \underset{\mathbf{3}}{2\text{-}[(C_2H_5)_2NCO]C_6H_4NH\text{-}C_6H_4\text{-}2\text{-}OCH_3} \xrightarrow[58\%]{H^+,\ \Delta} \underset{\mathbf{4}}{\text{4-methoxy-9(10H)-acridinone}}$$

Diphenylphosphidocopper(I), $(C_6H_5)_2PCu$ (**1**). The reagent is prepared from $(C_6H_5)_2PLi$ and $CuBr \cdot S(CH_3)_2$. Exposure of **1** to oxygen yields a new complex, $Cu[P(C_6H_5)_2]O_2$ (**2**).[13]

Diphenylphosphidocuprates can be prepared by addition of RLi to either **1** or **2**. These cuprates are more stable than previously known heterocuprates, being fairly stable even at 25° for 30 minutes. Comparably stable cuprates can be prepared by addition of RLi to (dicyclohexylamido)copper(I). Even so, these cuprates are highly active and undergo the usual reactions of cuprates, often in higher yields than those obtained with conventional heterocuprates.[14]

RCu · BF$_3$ (**8**, 334–335; **9**, 333). Yamamoto *et al.*,[15] have reviewed and extended their research on the $RCu \cdot BF_3$ system. This system is superior to R_2CuLi for conjugate addition to sterically hindered α,β-unsaturated ketones and esters, but is less useful in reactions with α,β-unsaturated acids and is unsatisfactory for additions to α,β-unsaturated nitriles. A typical example is formulated in equation (I).

$$\text{(I)}\quad (CH_3)_2C{=}C(COCH_3)CH_3 \longrightarrow (CH_3)_2C(n\text{-}C_4H_9)\text{—}CH(CH_3)COCH_3 + (CH_3)_2C{=}C(CH_3)C(OH)(CH_3)C_4H_9\text{-}n$$

Reagent	Conjugate adduct	Carbinol
$2\,n\text{-}C_4H_9Cu \cdot BF_3$	74%	7%
$(n\text{-}C_4H_9)_2CuLi$	16%	64%

In conjugate additions to cyclic enones, $RCu \cdot BF_3$ is superior to R_2CuLi in the case of cisoid enones, but less efficient in the case of transoid enones (equations II and III).

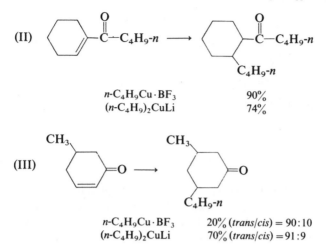

(II) 1-cyclohexenyl–$C(=O)$–C_4H_9-n ⟶ 2-(n-C_4H_9)cyclohexyl–$C(=O)$–C_4H_9-n

n-$C_4H_9Cu \cdot BF_3$	90%
$(n$-$C_4H_9)_2CuLi$	74%

(III) 5-CH_3-cyclohex-2-enone ⟶ 3-C_4H_9-n-5-CH_3-cyclohexanone

n-$C_4H_9Cu \cdot BF_3$	20% (*trans*/*cis*) = 90:10
$(n$-$C_4H_9)_2CuLi$	70% (*trans*/*cis*) = 91:9

$RCu \cdot BF_3$ differs markedly from R_2CuLi in conjugate addition to α,β-unsaturated esters. The latter reagent gives mainly the product of dialkylation via a 1,2-addition-conjugate addition sequence (equation IV).

(IV) $(CH_3)(H)C{=}C(CH_3)COOC_2H_5$ ⟶ $CH_3CH(C_4H_9\text{-}n)CH(CH_3)COOC_2H_5$ + $CH_3CH(C_4H_9\text{-}n)CH(CH_3)COC_4H_9\text{-}n$

$3n$-$C_4H_9Cu \cdot BF_3$	96%	1%
$2(n$-$C_4H_9)_2CuLi$	16%	57%

Stable heterocuprates. Bertz and Dabbagh[16] have examined the effect of various ligands on the thermal stability of n-butylcopper, which decomposes at temperatures greater than $-50°$ by elimination of CuH. Of the 14 butylcopper reagents of the type LiCuBuL examined, the most effective ligands favoring stability are diphenylphosphido and dicyclohexylphosphido, both bidentate. In addition to bridging, steric and electronic effects play a role.

4-Alkylpyridines. RLi and RMgX both add selectively to pyridines to form 1,2-dihydro-2-alkylpyridines. A number of organometallic reagents add to N-ethoxycarbonylpyridinium chloride (**1**) to give a mixture of 1,4- and 1,2-dihydropyridine derivatives (equation I). The most efficient and also selective reactions are shown by n-BuCu and n-BuCu $\cdot$ BF_3. The products are oxidized to the desired 4-alkylpyridines in 40–70% yield when stirred under a stream of oxygen.[17]

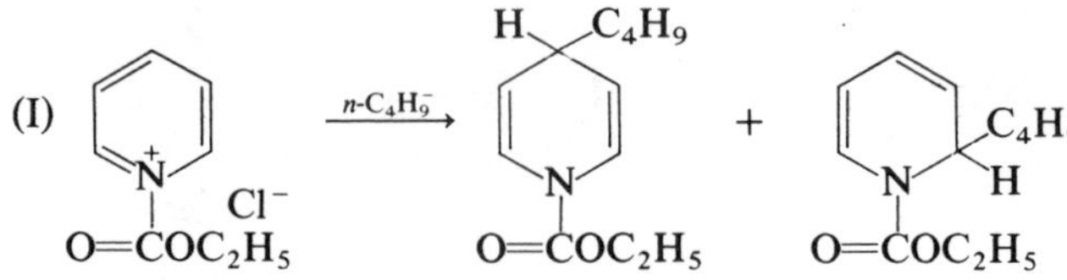

(I) [N-ethoxycarbonylpyridinium chloride] $\xrightarrow{n\text{-}C_4H_9^-}$ [4-butyl-1,4-dihydropyridine, N-$COOC_2H_5$] + [2-butyl-1,2-dihydropyridine, N-$COOC_2H_5$]

n-BuLi · B(C_4H_9-n)$_3$	42%	75:25
n-BuMgBr · B(C_4H_9-n)$_3$	58%	62:38
n-BuCu	78%	99.5:0.5
n-BuCu · BF_3	89%	99.5:0.5

Conjugate addition to enones.[18] The conjugate addition of organocopper reagents to α,β-enones is generally increased over 1,2-addition by generation of the copper species under sonication followed by *in situ* reaction with the enone. In a typical procedure, the organic bromide, lithium sand, and the soluble copper salt 1-pentynylcopper complexed with hexamethylphosphorous triamide[19] is sonicated in diethyl ether–THF. After consumption of the lithium, the enone is added and sonication is continued for a short period before work-up. The method is not useful in the case of methyl bromide.

Addition to $\alpha,\beta;\gamma,\delta$-dienyl sulfones. Lithium dialkylcuprates react with these diunsaturated sulfones by 1,6-addition, and the new double bond formed has the (Z)-geometry. The adducts on treatment with base extrude SO_2(Ramberg-Bäcklund reaction) to give a mixture of polyenes.[20]

Example:

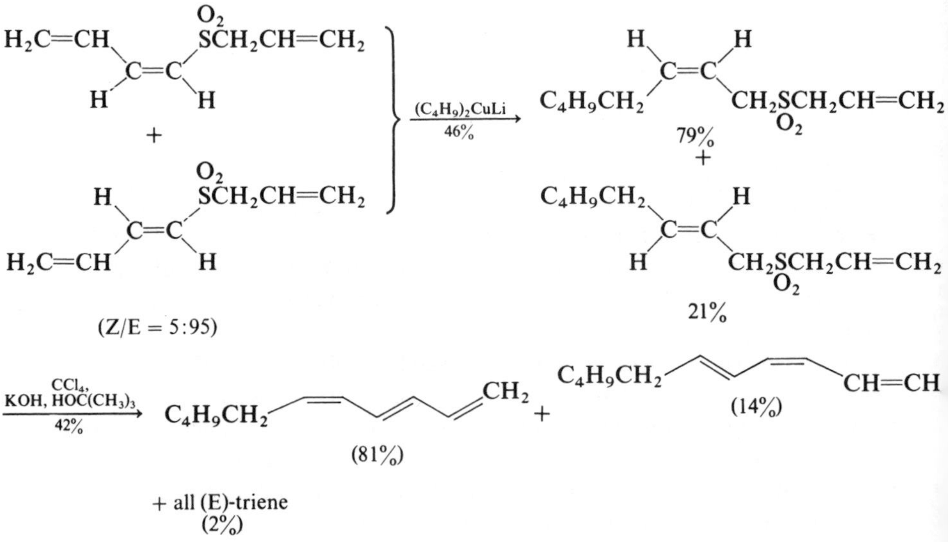

$H_2C{=}CH(H)C{=}C(H)SO_2CH_2CH{=}CH_2$ + $H_2C{=}CH(H)C{=}C(H)SO_2CH_2CH{=}CH_2$ (Z/E = 5:95) $\xrightarrow[46\%]{(C_4H_9)_2CuLi}$ (Z)-$C_4H_9CH_2CH{=}CHCH_2SO_2CH_2CH{=}CH_2$ (79%) + (E)-$C_4H_9CH_2CH{=}CHCH_2SO_2CH_2CH{=}CH_2$ (21%)

$\xrightarrow[42\%]{CCl_4,\ KOH,\ HOC(CH_3)_3}$ $C_4H_9CH_2$–triene–CH_2 (81%) + $C_4H_9CH_2$–triene–$CH{=}CH$ (14%)

+ all (E)-triene (2%)

Addition to β-acyloxy enol esters. Addition of lithium dimethylcuprate to chiral enol esters derived from carbohydrates provides access to C_6 chiral synthons with four alternating and/or consecutive methyl and hydroxyl groups.[21]

Example:

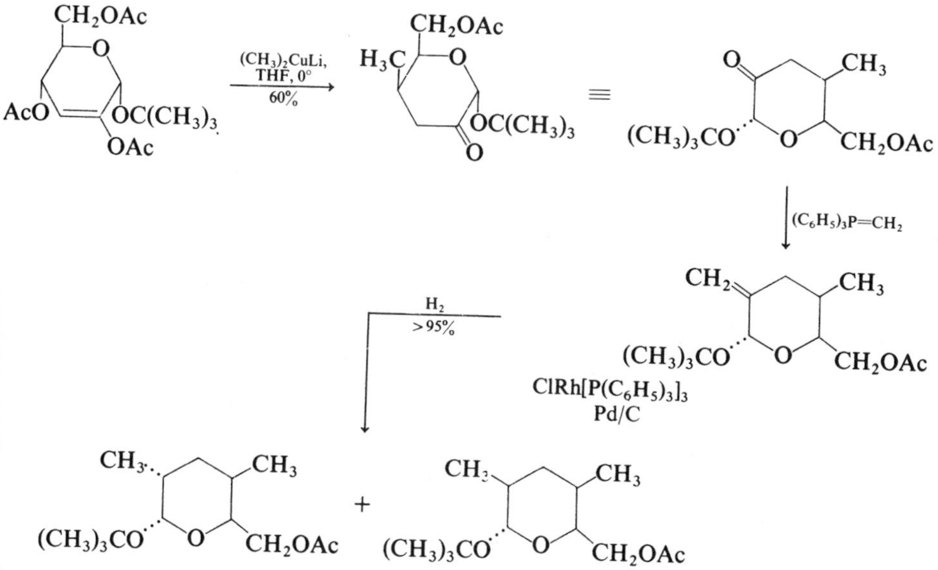

$(CH_3)_2CuLi$, THF, 0° 60%

$(C_6H_5)_3P{=}CH_2$

H_2 >95%

$ClRh[P(C_6H_5)_3]_3$

Pd/C

1:1

3:7

Addition to dimethyl acetylenedicarboxylate.[22] The copper reagent prepared by reaction of RMgBr and $CuBr \cdot S(CH_3)_2$ in THF adds stereoselectively to dimethyl acetylenedicarboxylate to give dimethyl 2-alkylmaleates in 70–95% yield (equation I).

$$\text{(I)}\quad CH_3OOCC{\equiv}CCOOCH_3 + RCu[S(CH_3)_2] \cdot MgBr \xrightarrow[70\text{–}95\%]{THF,\ -78^\circ}$$

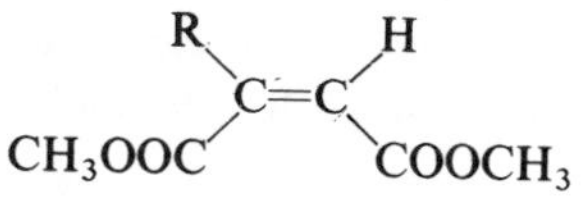

$$\underset{CH_3OOC}{\overset{R}{}}C{=}C\underset{COOCH_3}{\overset{H}{}}$$

1,5-*Addition to a vinylcyclopropyl ketone.*[23] Concluding steps in a recent synthesis of (±)-modhephene (**3**) required the introduction of a methyl group by a novel 1,5-addition of $(CH_3)_2CuLi$ to the vinylcyclopropyl ketone (**1**) to afford **2** after *in situ* enolate trapping with N,N-dimethylphosphoramidic dichloride.[24] The synthesis of **3** was completed by reduction of the phosphordiamidate group and hydrogenation.

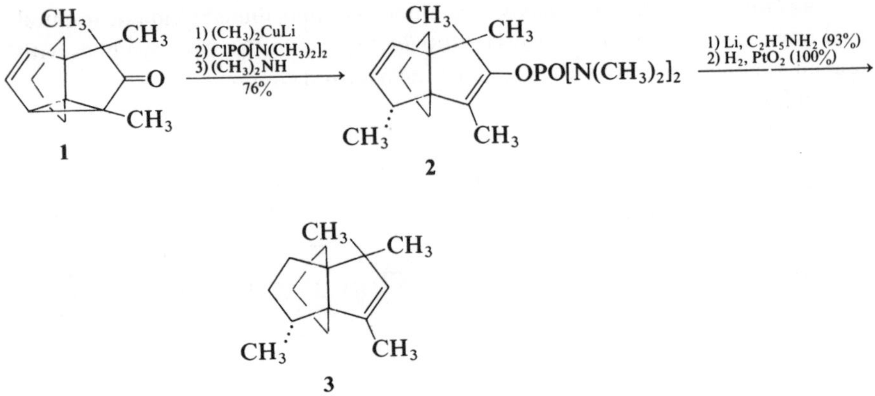

1) $(CH_3)_2CuLi$
2) $ClPO[N(CH_3)_2]_2$
3) $(CH_3)_2NH$
76%

1) Li, $C_2H_5NH_2$ (93%)
2) H_2, PtO_2 (100%)

Selective opening of allylic epoxides (**9**, 329–330). The reaction of the trimethylsilyl enol ether (**1**) of α, β-epoxycyclohexanone with lithium dimethylcuprate (or di-*n*-butylcuprate) in THF proceeds as expected to give **2**.[25, 26] However when the reaction is conducted in ether, **2** and **3** are formed in about equal amounts.[26]

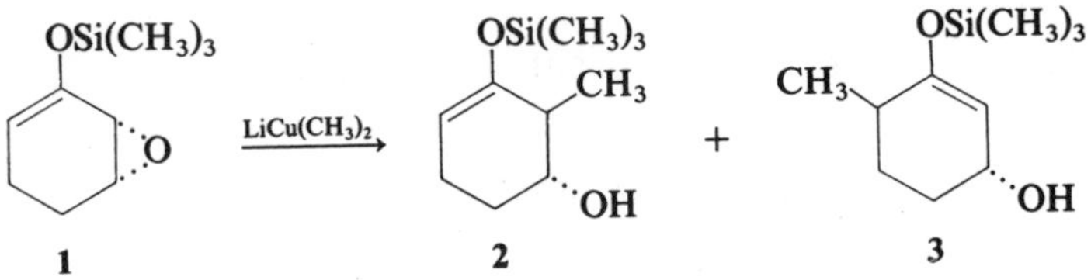

		2:3
THF, −20°	62%	100:0
$(C_2H_5)_2O$, −40°	81%	41:59

***Lithium (cyclopropyl)(phenylthio) cuprate* (1).**[27] This cuprate is prepared by addition of cyclopropyllithium to phenylthiocopper in THF at −78→−20°. It converts β-iodo enones (**2**) into β-cyclopropyl enones (**3**) in excellent yield, even when **2** is substituted at the α-position by a methyl group. The products are useful because they rearrange when heated to annelated cyclopentenes (**4** and/or **5**). The rearrangement is particularly efficient when R = H, and results in **5** as the major product (~85% yield).

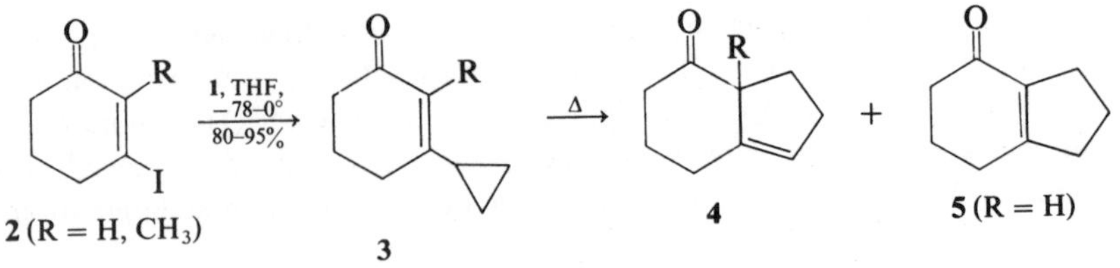

2 (R = H, CH_3) —(**1**, THF, −78–0°, 80–95%)→ **3** —(Δ)→ **4** + **5** (R = H)

This annelation provided a key step in a synthesis of the sesquiterpene zizaene (**6**).

H_2C H CH_3 CH_3

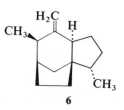

6

Lithium bis(N-methyl-N-phenylcarbamoyl)cuprate, $[C_6H_5N(CH_3)CO]_2CuLi$ (**1**).[28] The cuprate is prepared by reaction of CO with lithium N-methylanilide in the presence of 0.5 equiv. of CuI. The complex converts alkyl halides into amides in good yield. The reaction is not useful with aryl halides or benzyl halides.

1 S. H. Bertz, *Tetrahedron Letters*, **21**, 3151 (1980).
2 E. C. Ashby and J. J. Watkins, *Am. Soc.*, **99**, 5312 (1977).
3 E. C. Ashby, J. J. Liu, and J. J. Watkins, *J. Org.*, **42**, 1099 (1977).
4 D. L. J. Clive, V. Farina, and P. Beaulieu, *J.C.S. Chem. Comm.*, 643 (1981).
5 *Idem, J. Org.*, **47**, 2572 (1982).
6 S. Kim, J. I. Lee, and B. Y. Chung, *J.C.S. Chem. Comm.*, 1231 (1981).
7 C. R. Johnson and D. S. Dhanoa, *ibid.*, 358 (1982).
8 B. H. Lipshutz, R. S. Wilhelm, and D. M. Floyd, *Am. Soc.*, **103**, 7672 (1981).
9 B. H. Lipshutz, R. S. Wilhelm, and J. Kozlowski, *Tetrahedron Letters*, 3755 (1982).
10 R. D. Acker, *ibid.*, 3407 (1977); *ibid.*, 2399 (1978).
11 B. H. Lipshutz, J. Kozlowski, and R. S. Wilhelm, *Am. Soc.*, **104**, 2305 (1982).
12 M. Iwao, J. N. Reed, and V. Sniekus, *ibid.*, **104**, 5531 (1982).
13 K. Issleib and H. O. Frohlich, *Ber.*, **95**, 375 (1962).
14 S. H. Bertz, G. Dabbagh, and G. M. Villacorta, *Am. Soc.*, **104**, 5824 (1982).
15 Y. Yamamoto, S. Yamamoto, H. Yatagai, Y. Ishihara, and K. Maruyama, *J. Org.*, **47**, 119 (1982).
16 S. H. Bertz and G. Dabbagh, *J.C.S. Chem. Comm.*, 1030 (1982).
17 K. Akiba, Y. Iseki, and M. Wada, *Tetrahedron Letters*, **23**, 429 (1982).
18 J. L. Luche, C. Petrier, A. L. Gemal, and N. Zikra, *J. Org.*, **47**, 3805 (1982).
19 E. J. Corey and D. J. Beames, *Am. Soc.*, **94**, 7210 (1972).
20 F. Näf, R. Decorzant, and S. D. Escher, *Tetrahedron Letters*, **23**, 5043 (1982).
21 S. Hanessian, P. C. Tyler, and Y. Chapleur, *ibid.*, **22**, 4583 (1981).
22 H. Nishiyama, M. Sasaki, and K. Itoh, *Chem. Letters*, 905 (1981).
23 P. A. Wender and G. B. Dreyer, *Am. Soc.*, **104**, 5805 (1982).
24 H.-J. Liu, S. P. Lee, and W. H. Chan, *Can. J. Chem.*, **55**, 3797 (1977).
25 P. A. Wender, J. M. Erhardt, and L. J. Letendre, *Am. Soc.*, **103**, 2114 (1981).
26 M. F. Schlecht, *J.C.S. Chem. Comm.*, 1331 (1982).
27 E. Piers, J. Banville, C. K. Lau, and I. Nagakura, *Can. J. Chem.*, **60**, 2965 (1982).
28 Y. Wakita, T. Kobayashi, M. Maeda, and M. Kajima, *Chem. Pharm. Bull. Japan*, **30**, 3395 (1982).

Organopentafluorosilicates, K_2RSiF_5.

Review.[1] These compounds are usually prepared by hydrosilylation of alkenes and alkynes (equations I and II).

(I) $RCH{=}CH_2 + HSiCl_3 \xrightarrow{H_2PdCl_6} RCH_2CH_2SiCl_3 \xrightarrow{KF} K_2[RCH_2CH_2SiF_5]$

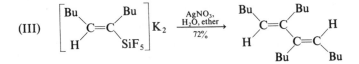

(II) $RC{\equiv}CH \rightarrow$ (R)(H)C=C(H)($SiCl_3$) $\rightarrow K_2$[(R)(H)C=C(H)(SiF_5)]

The C—Si bond of these products is cleaved by various electrophiles including X_2, NBS, CuX_2, and $ClC_6H_4CO_3H$. Ag(I) salts promote coupling of alkenylsilicates (equation III). Palladium salts promote various C—C bond forming reactions of alkenylsilicates.

(III) [(Bu)(H)C=C(Bu)(SiF_5)]K_2 $\xrightarrow[72\%]{AgNO_3,\ H_2O,\ ether}$ (Bu)(H)C=C(Bu)–C(Bu)=C(H)(Bu)

Example:

K_2[(C_6H_5)(H)C=C(H)(SiF_5)] + CO + CH_3OH $\xrightarrow[76\%]{PdCl_2,\ CH_3COONa}$ (C_6H_5)(H)C=C(H)($COOCH_3$)

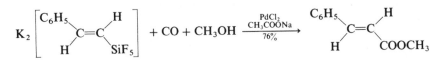

[1] M. Kumada, K. Tamao, and J.-I. Yoshida, *J. Organometal. Chem.*, **239**, 115 (1982).

Organotitanium reagents, 10, 138, 271, 422.

Review. Reetz[1] has reviewed newer published reactions of these reagents as well as numerous unpublished results from his own research. The review emphasizes the chemo-, diastereo-, and enantioselective reactions that can be achieved. Drawbacks are that secondary and tertiary alkyltitanium compounds generally are not available, owing to β-hydride elimination, and that ready ligand exchange makes the preparation of derivatives chiral at titanium difficult. The review includes useful suggestions for experimental techniques.

Chiral reagents for enantioselective addition to aldehydes. Seebach's group[2] has prepared a series of chiral reagents of the general type $RTi(OR^*)_3$ and $ArTi(OR^*)_3$ in which R* is derived from (−)-menthol, quinine, cinchonine, and (S)-1,1′-

(I) $ClTi[OCH(CH_3)_2]_3 + R^*OH \xrightarrow[-3(CH_3)_2CHOH]{C_6H_6,\ \Delta} ClTi(OR^*)_3 \xrightarrow{RLi\ or\ RMgX} RTi(OR^*)_3 \xrightarrow{R'CHO}$ R′–C*H(OH)–R

binaphthol. The highest enantioselectivities (as much as 88% ee in one case) in addition to aldehydes were observed with menthol and binaphthol derivatives. The enantioselectivity is strongly dependent on the substrate. Surprisingly, marked solvent as well as temperature effects are observed.

Alkyltris(dialkylamino)titanium derivatives, $R^1Ti(NR_2)_3$.[3] These reagents are readily available by reaction of halotris(dialkylamino)titanium with organolithium or -magnesium compounds.[4] The reagents, prepared *in situ* in ether, react with nonenolizable aldehydes to give tertiary amines (**1**), possibly via an iminium salt (equation I). With enolizable aldehydes and ketones, the reaction results in enamines after nonaqueous workup.

$$\text{(I)}\quad RC_6H_4CHO + CH_3Ti[N(C_2H_5)_2]_3 \xrightarrow[-60^\circ]{\text{ether,}} \left[RC_6H_4(H)C(OTi(CH_3)[N(C_2H_5)_2]_2)N(C_2H_5)_2 \right.\ (\mathbf{a})$$

$$\longrightarrow RC_6H_4(H)C{=}\overset{+}{N}(C_2H_5)_2 \left.\right]\ (\mathbf{b}) \xrightarrow{28\text{–}48\%} RC_6H_4(H)C(CH_3)N(C_2H_5)_2\ (\mathbf{1})$$

Chlorotitanium triisopropoxide, $ClTi[OCH(CH_3)_2]_3$ (**1**), b.p. 63–66°/0.1 mm, hygroscopic but stable under N_2 for months. The reagent is prepared by reaction of $TiCl_4$ with 3 equiv. of titanium(IV) isopropoxide. This reagent, or $BrTi[N(C_2H_5)_2]_3$, converts lithium enolates into titanium enolates, which show pronounced *erythro*-selectivity in reactions with aldehydes, irrespective of the geometry of the enolate.[5]

Examples:

$$\text{1-(OLi)cyclohexene} \xrightarrow{\mathbf{1}} \text{1-(OTi[OCH(CH}_3)_2]_3)\text{cyclohexene} \xrightarrow[\text{Pentane, }-120^\circ]{C_6H_5CHO} \text{2-(C}_6H_5\text{CH(OH))cyclohexanone } (\textit{erythro}) + \textit{threo}\text{-aldol}\quad 92:8$$

$$C_2H_5(OTi[N(C_2H_5)_2]_3)C{=}CHCH_3 + C_6H_5CHO \xrightarrow[-78^\circ]{\text{Pentane,}} C_2H_5COCH(CH_3)CH(OH)C_6H_5\ (\textit{erythro}) + \textit{threo}\text{-aldol}\quad 76:24$$

Z/E = 36:64

In contrast to titanium enolates of ketones, titanium enolates of aldehydes exhibit practically no stereoselectivity in aldol reactions. However, titanation of dimethylhydrazones of aldehydes with **1** results in substrates **(2)** that show high *erythro*-selectivity in aldol-type reactions with aldehydes (equation I). Bromotitanium tris(diethylamide) can be used in place of **1**, but is less efficient, as is Ti(IV) isopropoxide.[6]

Erythro-selectivity is also observed with titanated ketone hydrazones (equation II).

(I) $R^1CH_2CH{=}NN(CH_3)_2 \xrightarrow{\text{1) LDA; 2) 1}}$ $R^1(H)C{=}C(H)NN(CH_3)_2$ with $(i\text{-PrO})_3Ti$ on N (**2**) $\xrightarrow[\text{60–95\%}]{R^2CHO}$

3 (*erythro*) + **4** (*threo*), 90–98 : 10–2 [OH, $NN(CH_3)_2$, R^2, R^1, H]

(II) cyclohexanone $NN(CH_3)_2$ $\xrightarrow{\text{1) LDA; 2) 1; 3) } C_6H_5CHO}$ [$(CH_3)_2NN$, H, OH, C_6H_5] + [$(CH_3)_2NN$, H, OH, C_6H_5], ~98 : 2

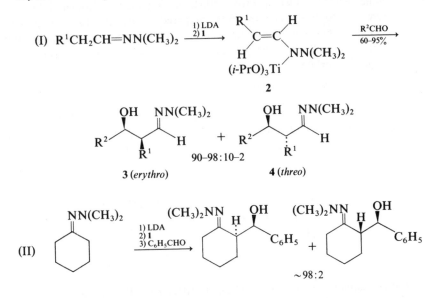

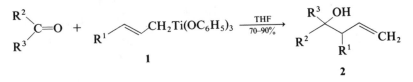

(2-Alkenyl)triphenoxytitanium derivatives.[7] Allylic organotitanium reagents of type **1** are readily available from allylic Grignard reagents (Rieke Mg) and chlorotriphenoxytitanium in THF. They add to ketones at low temperatures to furnish homoallylic alcohols **(2)** in 70–90% yield. The reaction is also diastereoselective, 55–97% ds, the selectivity being highest when the steric bulk of R^2 and R^3 is greatest. Tentatively, the majority of the adducts are assigned the *erythro*-configuration.

$R^2R^3C{=}O$ + $R^1CH{=}CHCH_2Ti(OC_6H_5)_3$ (**1**) $\xrightarrow[\text{70–90\%}]{THF}$ $R^2R^3C(OH)CH(R^1)CH{=}CH_2$ (**2**)

Allyltitanium ate complexes.[8] In contrast to organotitanium compounds, which are aldehyde selective, the chemoselectivity of allyltitanium ate complexes depends on the ligands attached to titanium. The most aldehyde-selective allyltitanium ate complex prepared to date is $CH_2{=}CHCH_2Ti[OCH(CH_3)_2]_4MgCl$ **(1)** which reacts with a mixture of an aldehyde and a ketone to give the corresponding adducts in the ratio 98:2. In contrast, $CH_2{=}CHCH_2Ti[N(CH_3)_2]_4Li^+$ (**2**) reacts with the same mixture to give the adducts in the ratio 2:98. The ate complex **1** also reacts selectively with the keto group of the keto ester **3** to give **4** in 81% isolated yield.

$$\underset{\mathbf{3}}{CH_3\overset{O}{\overset{\|}{C}}(CH_2)_2COOC_2H_5} \xrightarrow[81\%]{\mathbf{1},\ THF,\ -78^\circ} \underset{\mathbf{4}}{(HO)(CH_3)C(CH_2CH{=}CH_2)(CH_2)_2COOC_2H_5}$$

Crotyltitanium ate complexes can also be chemoselective; **5** is aldehyde selective (98:2), whereas **6** is ketone selective (>98%).

$$\underset{\mathbf{5}}{CH_3CH{=}CHCH_2\overset{-}{Ti}[OCH(CH_3)_2]_4\overset{+}{Mg}Cl} \qquad \underset{\mathbf{6}}{CH_3CH{=}CHCH_2\overset{-}{Ti}[N(CH_3)_2]_4\overset{+}{Mg}Cl}$$

(Alkylthio)allyltitanium reagents, $RSCH{=}CHCH_2TiL_n$ (**1**).[9] The reagents are prepared by deprotonation of allylic alkyl (aryl) sulfides with *sec*- or *t*-butyllithium followed by addition of $Ti(O\text{-}i\text{-}Pr)_4$ at -78°. They can react with carbonyl compounds at the α- or γ-position. α-Adducts predominate in reactions with α- and β-mono- and disubstituted sulfides, whereas γ-adducts predominate in reaction with γ-substituted sulfides. The α-adducts show high *erythro*-selectivity. The products are useful precursors to alkenyl oxiranes and to 2-(arylthio)-1,3-butadienes.

Examples:

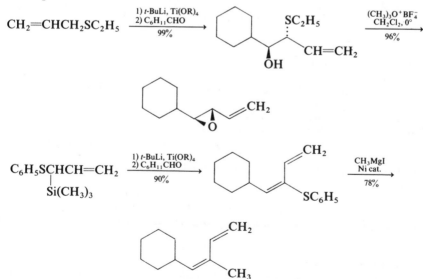

$$CH_2{=}CHCH_2SC_2H_5 \xrightarrow[99\%]{1)\ t\text{-BuLi, Ti(OR)}_4;\ 2)\ C_6H_{11}CHO} C_6H_{11}CH(OH)CH(SC_2H_5)CH{=}CH_2 \xrightarrow[96\%]{(CH_3)_3O^+BF_4^-,\ CH_2Cl_2,\ 0^\circ} C_6H_{11}\text{-(oxirane)-}CH{=}CH_2$$

$$C_6H_5S\underset{Si(CH_3)_3}{\underset{|}{C}}HCH{=}CH_2 \xrightarrow[90\%]{1)\ t\text{-BuLi, Ti(OR)}_4;\ 2)\ C_6H_{11}CHO} C_6H_{11}CH{=}C(SC_6H_5)CH{=}CH_2 \xrightarrow[78\%]{CH_3MgI,\ Ni\ cat.} C_6H_{11}CH{=}C(CH_3)CH{=}CH_2$$

Propargylic titanium reagents.[10] The propargylic titanium reagents **1** and **2**, prepared by treatment of the corresponding propargylic lithium compounds with titanium(IV) isopropoxide, show unusual regioselectivity and stereoselectivity in reactions with aldehydes. Thus **1** reacts with aldehydes to form only β-allenic alcohols (**3**), whereas **2** reacts with aldehydes to form only β-acetylenic alcohols (**4** and **5**), with the *threo*-isomer predominating. The IR spectrum of **1** is characteristic of an acetylenic structure, whereas that of **2** corresponds to an allenic structure. The different behavior of **1** and **2** is apparently general. The diastereoselectivity observed with **2** is associated in large part with the metal center, since it is markedly decreased when Ti is replaced by Li, Mg, or Zn. Replacement of the methyl group of **2** by a THP ether group does not alter the diastereoselectivity in the condensation reaction and thus provides a synthetically useful route to α,β-acetylenic diols.

$[(CH_3)_3SiC{\equiv}C{-}CH_2]^-\overset{+}{Ti}L_n$ + C_6H_{11}CHO $\xrightarrow{93\%}$ **3**

1 $[L = OCH(CH_3)_2]$

Si(CH$_3$)$_3$, OH, C, CH$_2$

3

$[(CH_3)_3SiC{\equiv}C{-}CHCH_3]^-\overset{+}{Ti}L_n$ + C_6H_{11}CHO $\xrightarrow{69\%}$

2

4 + **5** (89:11)

CH$_3$, C≡C, (CH$_3$)$_3$Si, OH — **4**

CH$_3$, C≡C, (CH$_3$)$_3$Si, OH — **5**

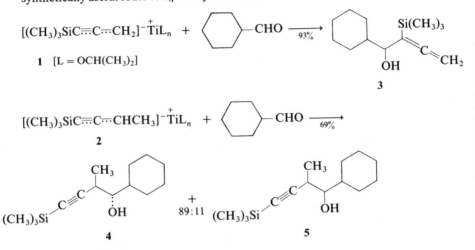

[1] M. Reetz, *Topics Curr. Chem.*, **106**, 3 (1982).
[2] B. Weidmann, L. Widler, A. G. Olivero, C. D. Maycock, and D. Seebach, *Helv.*, **64**, 357 (1981); A. G. Olivero, R. Weidmann, and D. Seebach, *ibid.*, **64**, 2485 (1981).
[3] D. Seebach and M. Schiess, *ibid.*, **65**, 2598 (1982).
[4] H. Bürger and H. J. Neese, *Chimia*, **24**, 209 (1970).
[5] M. T. Reetz and R. Peter, *Tetrahedron Letters*, **22**, 4691 (1981).
[6] M. T. Reetz, R. Steinbach, and K. Keeseler, *Angew. Chem. Int. Ed.*, **21**, 864 (1982); *Supplement*, 1899 (1982).
[7] L. Widler and D. Seebach, *Helv.*, **65**, 1085 (1982); D. Seebach and L. Widler, *ibid.*, **65**, 1972 (1982).
[8] M. T. Reetz and B. Wenderoth, *Tetrahedron Letters*, **23**, 5259 (1982).
[9] Y. Ikeda, K. Furuta, N. Meguriya, N. Ikeda, and H. Yamamoto, *Am. Soc.*, **104**, 7663 (1982).
[10] M. Ishiguro, N. Ikeda, and H. Yamamoto, *J. Org.*, **47**, 2225 (1982).

Organozirconium compounds.

Alkyl(tributoxy)zirconium reagents, $RZr(OBu)_3$ (**1**).[1] These reagents are prepared as shown in equation (I). They resemble the titanium anologs in preferential addition to aldehydes in the presence of ketones, but the selectivity is not as pronounced. They show a greater reactivity than the titanium analogs toward readily enolizable ketones. Vinylzirconium compounds are sufficiently stable to be used, whereas vinyltitanium reagents are unstable above −60°.

$$\text{(I)} \qquad Zr(OBu)_4 \xrightarrow{ZrCl_4} ClZr(OBu)_3 \xrightarrow{RLi} \mathbf{1}$$

[1] B. Weidman, C. D. Maycock, and D. Seebach, *Helv.*, **64**, 1552 (1981).

Oxalyl chloride, 1, 767–772; **2**, 301–302; **3**, 216–217; **4**, 361; **5**, 481–482; **6**, 424; **7**, 257–258; **8**, 365.

p-*Formylation of alkylbenzenes.*[1] Reaction of alkylbenzenes with oxalyl chloride and $AlCl_3$ (1 equiv.) in trichloroethylene at room temperature results in 4-alkylbenzoyl chlorides in 80–90% yield. The products are hydrogenated to aldehydes in the presence of 10% Pd/C and tributylamine. The intermediate Friedel-Crafts complex can be hydrogenated without isolation to give the aldehydes directly in somewhat lower yield.

[1] M. A. Osman, *Helv.*, **65**, 2448 (1982).

2-Oxazolidones, chiral, (1) and **(2)**. Evans *et al.*[1] have prepared these two chiral 2-oxazolidones by reaction of phosgene with (S)-valinol and (1S, 2R)-norephedrine, respectively, and used them as recyclable chiral auxiliaries for carboxylic acids in enantioselective reactions of the derived imides.

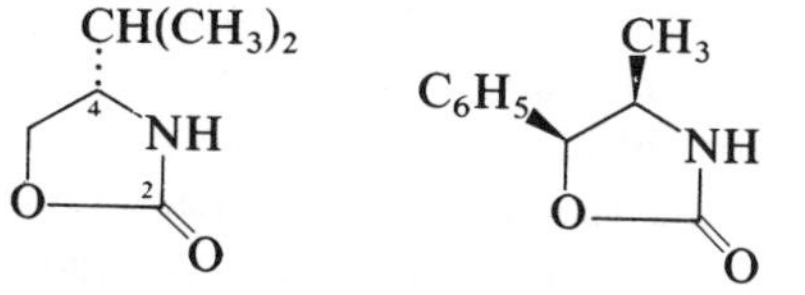

1, α_D + 14.8[2]

2, α_D + 163.7°

erythro *Chiral aldol condensations.*[1] The N-propionylamides **3** and **7** derived from **1** and **2**, respectively, are converted by di-*n*-butylboryl triflate (**9**, 140) exclusively into the (Z)-enolates, which react with aldehydes to form almost exclusively *erythro* adducts (**4** and **8**) but with the opposite sense of asymmetric induction (equations I and II). The lithium enolates of **3** and **7** exhibited rather low levels of stereoregulation. The adducts are readily hydrolyzed without racemization to give the chiral β-hydroxy acids (**6** and **9**) in high purity.

(I)

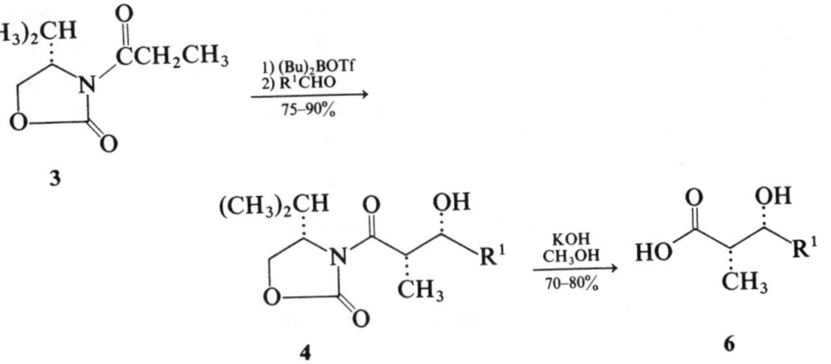

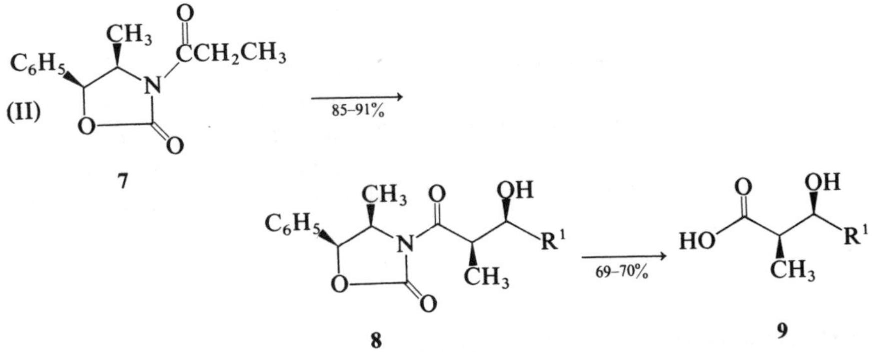

Asymmetric alkylation of imide enolates.[2] The sodium enolates of **3** and **7** are alkylated with marked but opposite diastereoselectivity by alkyl halides. The selectivity is improved by an increase in the size of the electrophile, with methylation being the least stereoselective process. The asymmetric induction results from formation of (Z)-enolates (chelation) with the diastereoselectivity determined by the chirality of the C_4-substituent on the oxazolidone ring (equations I and II). The products can be hydrolyzed to the free carboxylic acids or reduced by $LiAlH_4$ to the corresponding primary alcohols and the unreduced oxazolidone (**1** or **2**).

(I)

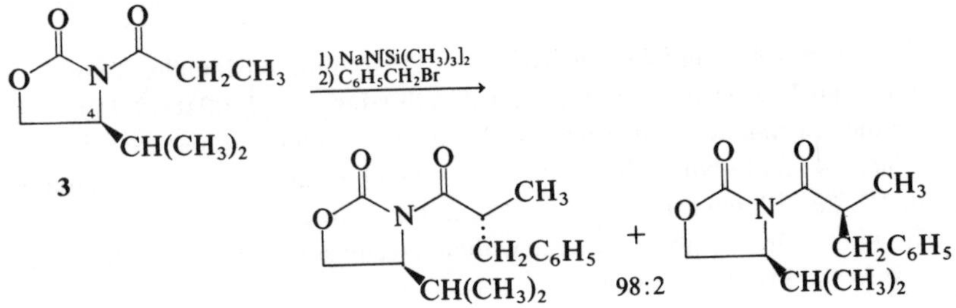

(II) 7 →

99:1

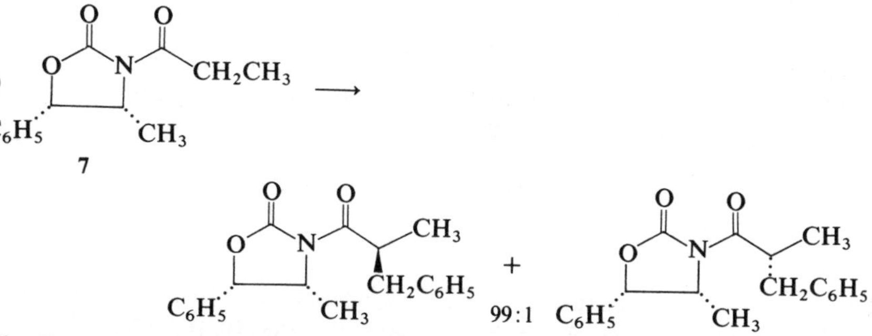

The diastereoselective alkylation and aldol reactions possible with the propionimide 7 were used to obtain the three contiguous asymmetric centers in a stereoselective synthesis[3] of (+)-Prelog-Djerassi lactone (**8**), a degradation product of methymycin.

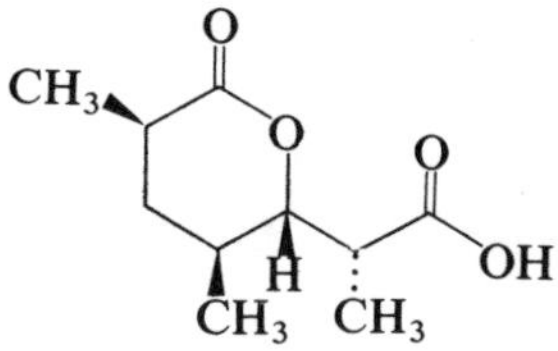

8

[1] D. A Evans, J. Bartroli, and T. L. Shih, *Am. Soc.*, **103**, 2127 (1981).
[2] D. A. Evans, M. D. Ennis, and D. J. Mathre, *ibid.*, **104**, 1737 (1982).
[3] D. A. Evans and J. Bartroli, *Tetrahedron Letters*, **23**, 807 (1982).

μ-Oxobis(chlorotriphenylbismuth), 9, 335.

17α-Methyl cardenolides. A synthesis of 17α-methyl cardenolides (**4**)[1] involves

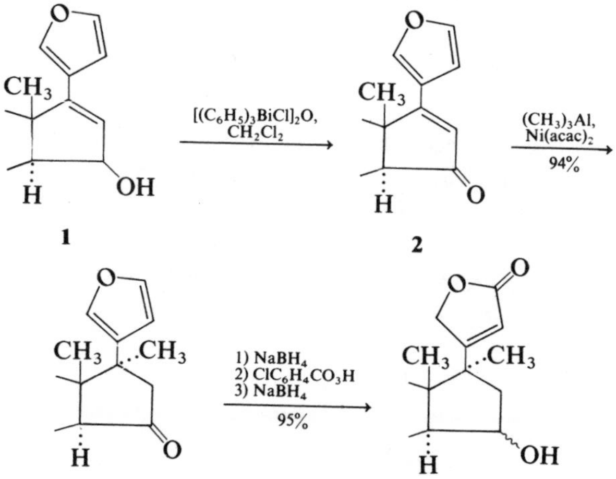

1 → $[(C_6H_5)_3BiCl]_2O$, CH_2Cl_2 → 2 → $(CH_3)_3Al$, $Ni(acac)_2$, 94% →

3 → 1) $NaBH_4$ 2) $ClC_6H_4CO_3H$ 3) $NaBH_4$, 95% → **4** (2 epimers)

as the first step oxidation of the furyl steroid (**1**) with this reagent. The resulting α,β-enone **2** could only be induced to undergo conjugate addition to **3** by treatment with $Al(CH_3)_3$ catalyzed by $Ni(acac)_2$ (**6**, 623–624). The product was converted in three steps into the cardenolides **4a** and **4b**.[2]

[1] K. S. Atwal, S. P. Sahoo, T. Y. R. Tsai, and K. Wiesner, *Heterocycles*, **19**, 641 (1982).
[2] T. Y. R. Tsai, A. Minta, and K. Wiesner, *ibid.*, **12**, 1397 (1979).

Oxodiperoxymolybdenum(pyridine)(hexamethylphosphoric triamide), $MoO_5 \cdot Py \cdot HMPT$ (MoOPH), **5**, 269; **7**, 136; **8**, 206–208; **9**, 197–198: **10**, 169–170.

Enolate hydroxylation.[1] Complete details are available for preparation of MoOPH (equation I). The reagent can be stored with protection from moisture and light in a refrigerator for several months. Partially decomposed reagent cannot be purified.

$$\text{(I)}\quad MoO_3 \xrightarrow[63\text{–}67\%]{\text{1) } 30\%\ H_2O_2,\ <40^\circ;\ \text{2) HMPT}} MoO_5\cdot H_2O\cdot HMPT \xrightarrow{-H_2O} [MoO_5\cdot HMPT] \xrightarrow[37\text{–}42\%\ \text{overall}]{Py} \underset{\mathbf{1}}{\text{MoOPH}}$$

Yields of α-hydroxy ketones by enolate hydroxylation with **1** are generally in the range 45–80%; the α-diketone is sometimes obtained in addition in ~5–26% yield. For unknown reasons, these hydroxylations often do not go to completion, and 5–15% of the carbonyl substrate is recovered. Yields are poor from hydroxylations of methyl ketones.

[1] E. Vedejs and S. Larsen, *Org. Syn.*, submitted (1982).

μ_3-Oxohexakis(μ-trimethylacetato)trimethanoltriiron(III) chloride, $\{Fe_3O[(CH_3)_3CCO_2]_6(CH_3OH)_3\}^+Cl^-$. The complex is prepared from $FeCl_3$ and sodium pivalate, and crystallized from $CHCl_3$–CH_3OH.[1]

Epoxidation of unsaturated acetates. The complex catalyzes epoxidation of unsaturated acetates by oxygen (1 molar equiv.).[2] The regioselectivity of epoxidation of monoterpene acetates is similar to that of peracids and opposite to that of *t*-BuOOH/VO(acac)$_2$.

Examples:

$$(CH_3)_2C{=}CHCH_2CH_2C(CH_3){=}CHCH_2OAc \xrightarrow[87\%]{O_2,\ \mathbf{1}} (CH_3)_2\overset{O}{C\!-\!C}HCH_2CH_2C(CH_3){=}CHCH_2OAc$$

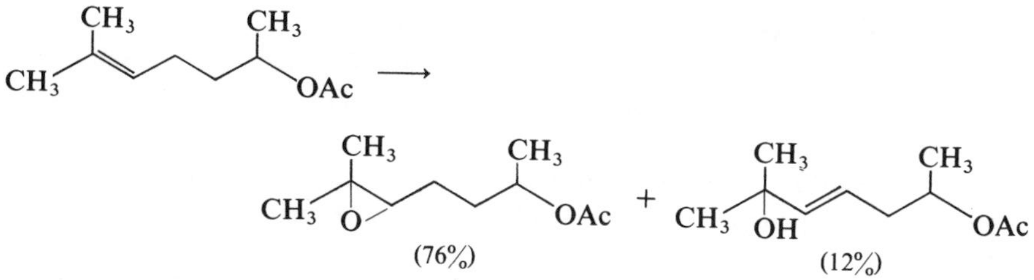

CH3 CH3 CH3 OAc ⟶

CH3 CH3 CH3 O OAc (76%) + CH3 CH3 CH3 OH OAc (12%)

[1] A. B. Blake and L. R. Fraser, *J.C.S. Dalton*, 193 (1975).
[2] S. Ito, K. Inoue, and M. Mastumoto, *Am. Soc.*, **104**, 6450 (1982).

Oxomethoxymolybdenum(V) 5,10,15,20-tetraphenylporphyrin, O=Mo(TPP)OCH_3. Preparation.[1, 2]

Stereoselective epoxidation.[2] O=Mo(TPP)OCH_3, O=Mo(TPP)Cl, and *cis*-O_2Mo(TPP) catalyze the epoxidation of cyclohexene with *t*-butyl hydroperoxide with formation of the insoluble $[OMo(TPP)]_2O$. These molybdenum complexes are comparable to $Mo(CO)_6$, but exhibit a longer induction period. The epoxidation of *cis*- and *trans*-2-hexene catalyzed with O-=Mo(TPP)Cl is highly stereospecific; the *cis*- and *trans*-2-epoxides are formed in 97% and 99% yield, respectively. The *cis*-2-hexene oxide is formed 3.5 times faster than the *trans*-isomer, an indication of efficient steric control because of the bulky ligand. Another difference is observed in epoxidation of isoprene with $Mo(CO)_6$ and O-=Mo(TPP)OCH_3 (equation I).

(I) 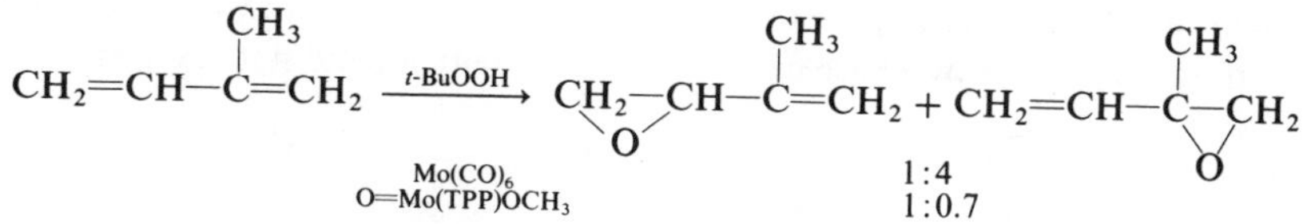 $CH_2{=}CH{-}C(CH_3){=}CH_2 \xrightarrow{t\text{-BuOOH}}$ $CH_2{-}CH{-}C(CH_3){=}CH_2$ (epoxide, O) + $CH_2{=}CH{-}C(CH_3){-}CH_2$ (epoxide, O)

Catalyst	Ratio
$Mo(CO)_6$	1:4
O=Mo(TPP)OCH_3	1:0.7

These epoxidations are the first examples of chemical models for oxidation with cytochrome-dependent monooxygenases.

[1] H. Ledon and B. Mentzen, *Inorg. Chem. Acta*, **31**, L393 (1978); H. J. Ledon, M. C. Bonnet, Y. Brigandat, and F. Varescon, *Inorg. Chem.* **19**, 3488 (1980).
[2] H. J. Ledon, P. Durbut, and F. Varescon, *Am. Soc.*, **103**, 3601 (1981).

2-Oxo-3-trimethylsilyltetrahydro-1,3-oxazole,

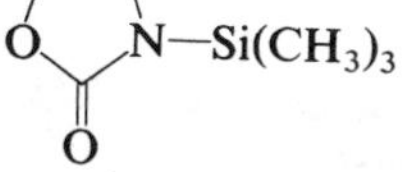

O N—$Si(CH_3)_3$ (**1**). O

Mol. wt. 145.28, b.p. 100°/6 mm. The reagent is prepared by reaction of 2-oxazolidinone with $ClSi(CH_3)_3$ and triethylamine (90% yield).

Trimethylsilyl esters.[1] Carboxylic acids are converted into the trimethylsilyl esters in generally high yield on reaction with **1** in CCl_4 even in the absence of a catalyst. The method is particularly useful for silylation of malonic acids without decarboxylation.

Silylation of* 1,3-*dicarbonyl compounds.[2] The reagent is particularly useful for preparation of trimethylsilyl enol ethers of 1,3-dicarbonyl compounds.

Examples:

$$CH_3\overset{O}{\overset{\|}{C}}CH_2\overset{O}{\overset{\|}{C}}OC_2H_5 + \mathbf{1} \xrightarrow{N(C_2H_5)_3,\ 20°} CH_3\overset{OSi(CH_3)_3}{\overset{|}{C}}{=}CHCOOC_2H_5 \ (90\%) + \text{2-oxazolidinone}$$

$$C_6H_5\overset{O}{\overset{\|}{C}}CH_2\overset{O}{\overset{\|}{C}}CH_3 + \mathbf{1} \xrightarrow[98\%]{} C_6H_5\overset{OSi(CH_3)_3}{\overset{|}{C}}{=}CHCOCH_3$$

[1] C. Palomo, *Synthesis*, 809 (1981).
[2] J. M. Aizpurua and C. Palomo, *ibid.*, 280 (1982).

Oxygen, 4, 362; **5**, 482–486; **6**, 426–430; **7**, 258–260; **8**, 366–367; **9**, 335–337; **10**, 293–294.

17α-*Hydroxypregnane-20-ones.* The aldehyde **1**, obtained in two steps from stigmasterol, can be cleaved to the lower homologous α-hydroxy (**2**) or α,β-unsaturated (**4**) pregnane-20-one by a base-catalyzed reaction with ground state oxygen.[1]

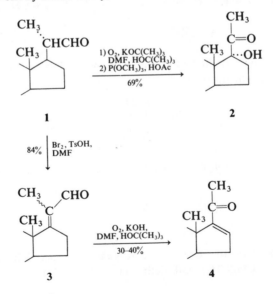

1 → **2**: 1) O_2, $KOC(CH_3)_3$, DMF, $HOC(CH_3)_3$; 2) $P(OCH_3)_3$, HOAc; 69%

1 → **3**: Br_2, TsOH, DMF; 84%

3 → **4**: O_2, KOH, DMF, $HOC(CH_3)_3$; 30–40%

Bishydroxylation of an α-tetralone. The potassium *t*-butoxide catalyzed oxygenation (**1**, 921–923; **2**, 341) of the α-tetralone **1**, gives minor amounts of the expected product **2**. The major product (**3**) results from further hydroxylation of **2**. The major product was shown to be a *cis*-1,3-diol. It is reduced by lithium borohydride selectively to the triol **4**. The triol has the same substitution pattern present in the A ring of certain anthracyclones.[2]

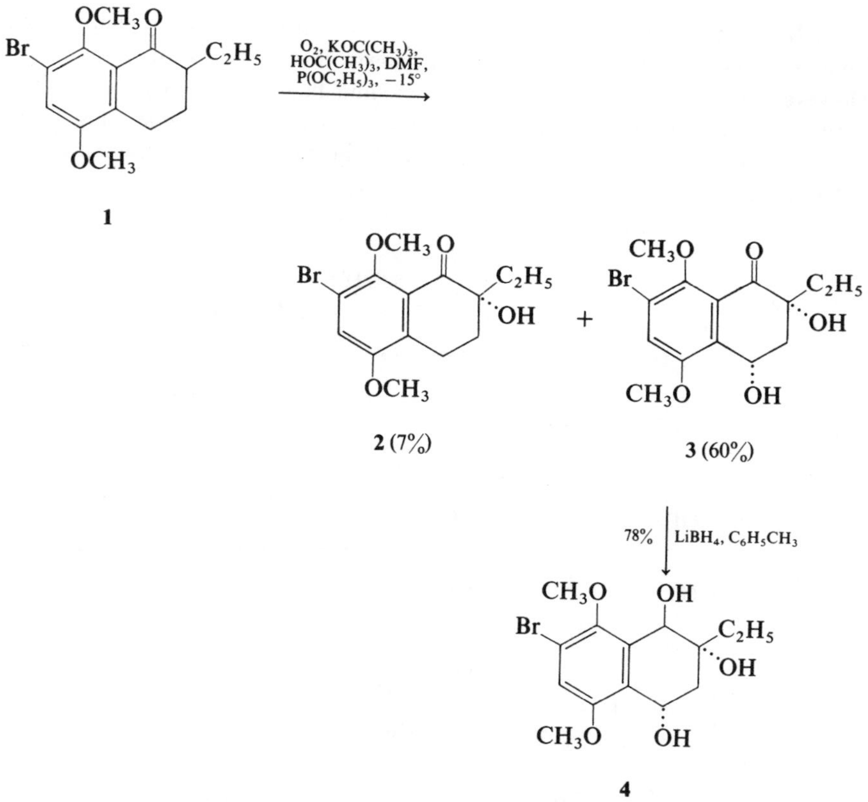

O_2, $KOC(CH_3)_3$, $HOC(CH_3)_3$, DMF, $P(OC_2H_5)_3$, $-15°$

1 → **2** (7%) + **3** (60%)

3 → **4**: 78%, $LiBH_4$, $C_6H_5CH_3$

[1] B. Biancini, V. Caciagli, F. Centini, G. Eletti-Bianchi, and L. Re, *Ann.*, 1829 (1982).
[2] C. E. Coburn, D. K. Anderson, and J. S. Swenton, *J.C.S. Chem. Comm.*, 987 (1982).

Oxygen, singlet, 4, 362–363; **5**, 486–491; **6**, 431–436; **7**, 261–269; **8**, 367–374; **9**, 338–341; **10**, 294–295.

(−)-α-Pinene→(−)-β-Pinene. A novel method for this transformation is photooxygenation of α-pinene (**1**) to give pinocarveol (**2**), which is reduced to β-pinene (**3**) by $LiAlH_4/TiCl_4$.[1]

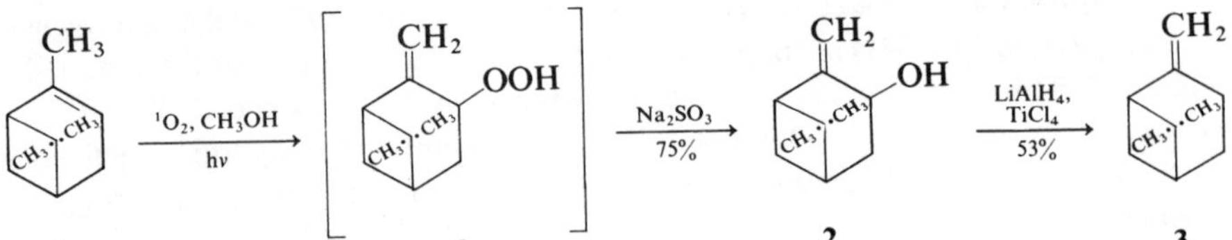

1 —($^1O_2$, CH_3OH, $h\nu$)→ [a] —(Na_2SO_3, 75%)→ 2 —($LiAlH_4$, $TiCl_4$, 53%)→ 3

Photooxidation of enols.[2] Although singlet oxygen adds to enol ethers, photooxidation of enols of ketones proceeds slowly, if at all, under usual conditions. However, in the presence of tetra-*n*-butylammonium fluoride (1 equiv.), β-diketones, α-diketones, and β-keto esters undergo photooxidation[3] at reasonable rates. In all cases, hydroperoxides are formed initially, which can undergo further reactions.

Examples:

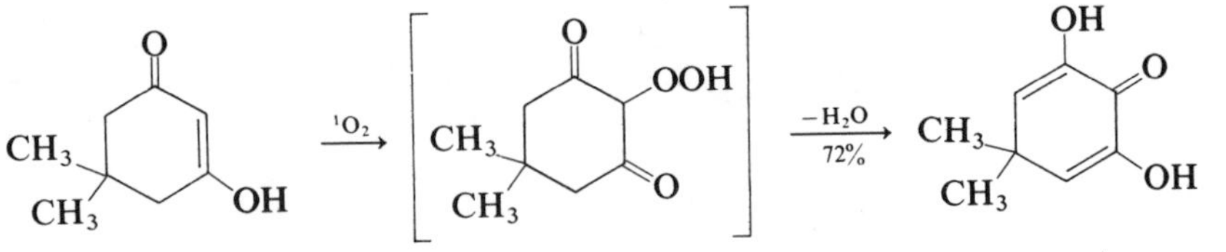

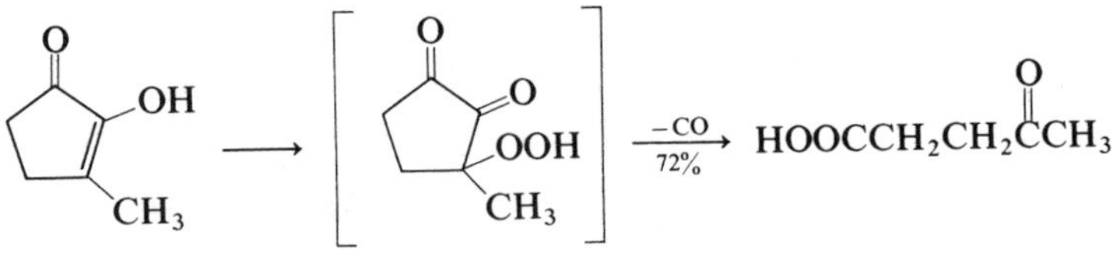

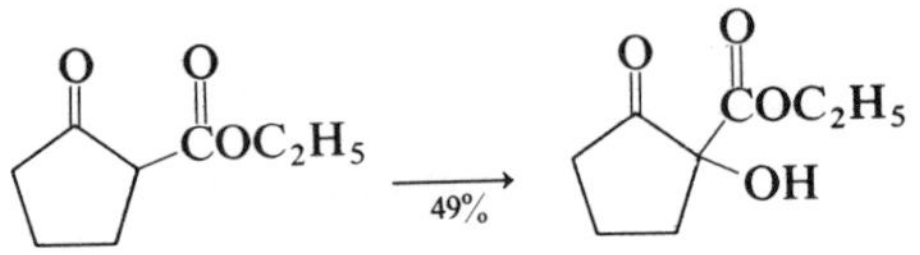

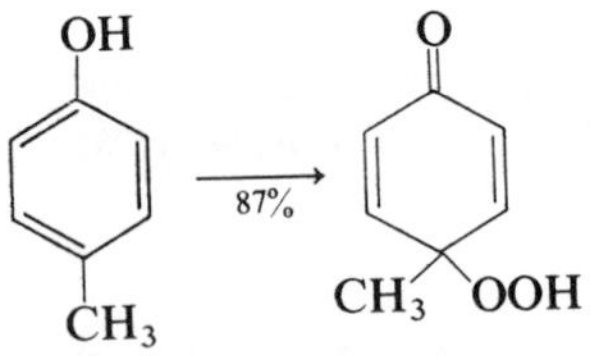

—($^1O_2$)→ [ ] —($-H_2O$, 72%)→

→ [ ] —($-CO$, 72%)→ $HOOCCH_2CH_2\overset{O}{\overset{\|}{C}}CH_3$

—(49%)→

—(87%)→

Review.[4] The varied uses of singlet oxygen in organic syntheses have been reviewed (70 references). The best known are the ene reaction of alkenes, which results in an allylic hydroperoxide, and the Diels-Alder reaction with 1,3-dienes to form 1,4-endoperoxides.

[1] Y.-F. Min, B.-W. Zhang, and Y. Cao, *Synthesis*, 875 (1982).
[2] H. H. Wasserman and J. E. Pickett, *Am. Soc.*, **104**, 4695 (1982).
[3] A polymer-bound Rose Bengal was used as sensitizer. This is available as Sensitox from Hydro Laboratories, New Brunswick, N.J.
[4] H. H. Wasserman and J. L. Ives, *Tetrahedron Rept.*, **109**, 1825 (1981).

Ozone, 1, 773–777; **4**, 363–364; **5**, 491–495; **6**, 436–441; **7**, 269–271; **8**, 374–377; **9**, 341–343; **10**, 295–296.

Isoquinolines. A one-pot synthesis of isoquinolines from indenes involves ozonation followed by treatment of the intermediate with ammonia (equation I).[1]

(I) R– (indene) $\xrightarrow[60\text{–}80\%]{1)\ O_3,\ CH_3OH,\ CH_2Cl_2;\ 2)\ (CH_3)_2S,\ NaHCO_3;\ 3)\ NH_4OH}$ R– (isoquinoline, N)

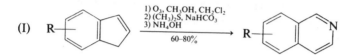

An alternative procedure[2] for the oxidation uses $NaIO_4$ (2 equiv.) and a catalytic amount of OsO_4 in $(CH_3)_3COH$ (equation II).

(II) R^1, R^2-indene $\xrightarrow[55\text{–}75\%]{1)\ NaIO_4,\ OsO_4;\ 2)\ NH_4OAc}$ R^1, R^2-isoquinoline (N)

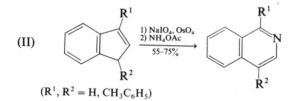

($R^1, R^2 = H, CH_3C_6H_5$)

1,2,3-*Triketones.* *vic*-Triketones (**2**) can be obtained by ozonolysis[3] of (dimethylsulfonio)diacylmethanides (**1**), prepared[4] as shown. A similar reaction has been observed with the methoxymethylene derivative (**3**) of Meldrum's acid.[5]

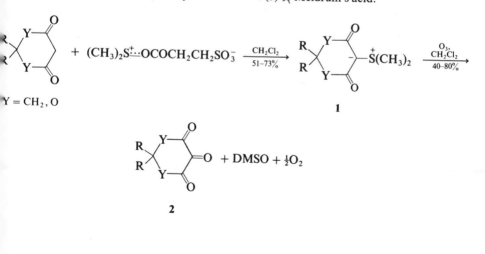

$$R_2C(Y\text{–}CO)_2CH_2 + (CH_3)_2\overset{+}{S}\text{–}OCOCH_2CH_2SO_3^- \xrightarrow[51\text{–}73\%]{CH_2Cl_2} R_2C(Y\text{–}CO)_2\overset{-}{C}\text{–}\overset{+}{S}(CH_3)_2 \xrightarrow[40\text{–}80\%]{O_3,\ CH_2Cl_2}$$

$Y = CH_2, O$ **1**

$$R_2C(Y\text{–}CO)_2C{=}O + DMSO + \tfrac{1}{2}O_2$$

2

1) O_3, CH_2Cl_2
2) PCl_3
61–65%

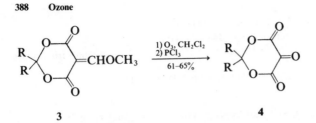

3

4

Ozonolysis of cycloalkenes.[6] Cycloalkenes are converted by ozonation in the presence of an alcohol (usually methanol) into an acyclic product with an aldehyde group and an α-alkoxy hydroperoxide group at the terminal positions. The products are usually difficult to purify, but they can be converted into useful products that retain differentiated terminal functionality.

Examples:

O_3, CH_3OH, $NaHCO_3$ → [CHO, CH(OCH₃)OOH] → Ac_2O, $N(C_2H_5)_3$, 96% → CHO, $COOCH_3$

O_3, CH_3OH, TsOH → [$CH(OCH_3)_2$, CH(OCH₃)OOH] → $NaHCO_3$, $S(CH_3)_2$, 92% → $CH(OCH_3)_2$, CHO

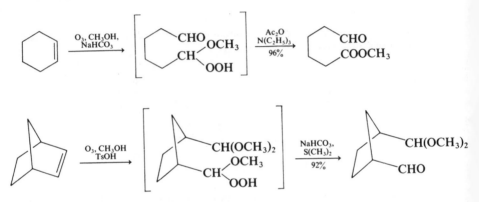

[1] R. B. Miller and G. M. Frincke, *J. Org.*, **45**, 5312 (1980); *Org. Syn.*, submitted (1982).
[2] D. S. Dime and S. McLean, *J. Org.*, **46**, 4999 (1981).
[3] K. Schank and C. Schuhknecht, *Ber.*, **115**, 3032 (1982).
[4] *Idem, Synthesis*, 678 (1978).
[5] *Idem, Ber.*, **115**, 2000 (1982).
[6] S. L. Schreiber, R. E. Claus, and J. Reagan, *Tetrahedron Letters*, **23**, 3867 (1982).

P

Palladium(II) acetate, 1, 778; **2**, 303; **4**, 365; **5**, 496–497; **6**, 442–443; **7**, 274–277; **8**, 378–382; **9**, 344–349; **10**, 297–298.

Bridged and spirocyclic bicycloalkenones. The conversion of silyl enol ethers to α,β-enones by $Pd(OAc)_2$ (**8**, 378) can result in cyclization to bicyclic systems when applied to silyl enol ethers of cyclohexanones bearing an alkenyl side chain α or γ to the carbonyl group. Although the factors favoring cyclization are not fully defined, this cyclization offers a route to a variety of bridged and spirocyclic systems.[1]

Examples:

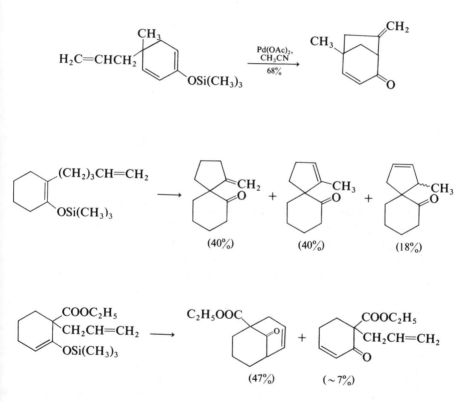

$H_2C{=}CHCH_2$, CH_3, $OSi(CH_3)_3$ → [$Pd(OAc)_2$, CH_3CN; 68%] → CH_3, CH_2, O

$(CH_2)_3CH{=}CH_2$, $OSi(CH_3)_3$ → CH_2, O (40%) + CH_3, O (40%) + CH_3, O (18%)

$COOC_2H_5$, $CH_2CH{=}CH_2$, $OSi(CH_3)_3$ → C_2H_5OOC, O (47%) + $COOC_2H_5$, $CH_2CH{=}CH_2$, O (~7%)

This methodology provides a remarkably short synthesis of the tetracylic antitumor agent quadrone (**7**) from (**1**) (Scheme I).[2]

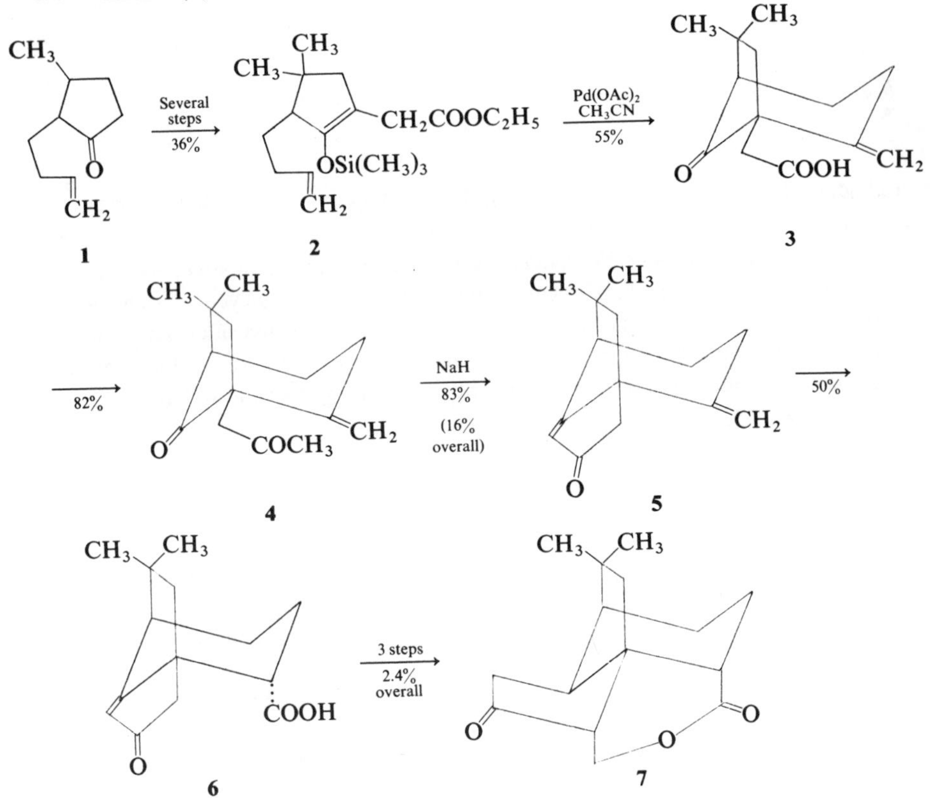

Scheme I

Ketones from halohydrins.[3] Palladium acetate complexed with a triarylphosphine, particularly tri-*o*-tolylphosphine, converts halohydrins into ketones in the presence of K_2CO_3. Yields are about 70–85% for substrates in which the halogen is secondary or tertiary, but less than 50% when the halogen is primary because of epoxide formation. The reaction is useful for conversion of alkenes to ketones in those instances in which halohydrins are formed regioselectively.

Example:

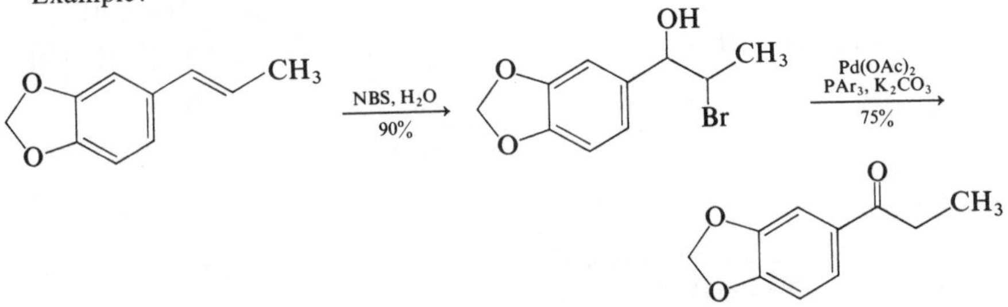

Arylation of quinones. Oxidation of quinones with $Pd(OAc)_2$ in HOAc in the presence of an arene results in coupling to form arylated quinones in moderate to high yield.[4]

Examples:

$$\text{1,4-naphthoquinone} + p\text{-}C_6H_4(CH_3)_2 \xrightarrow[78\%]{Pd(OAc)_2,\ HOAc} \text{2-(2,5-dimethylphenyl)-1,4-naphthoquinone}$$

$$\text{1,2-naphthoquinone} + p\text{-}C_6H_4Cl_2 \xrightarrow[27\%]{} \text{4-(2,5-dichlorophenyl)-1,2-naphthoquinone}$$

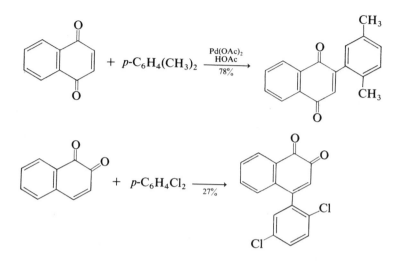

Cyclopropanation (**5**, 496; **6**, 442–443).[5] Nonactivated 1-alkenes react with CH_2N_2 in the presence of $Pd(OAc)_2$ to form cyclopropanes, generally in yields > 75%. Even selective cyclopropanation is possible (equation I).

$$\text{4-vinylcyclohexene } (CH{=}CH_2) + CH_2N_2 \xrightarrow[77\%]{Pd(OAc)_2,\ \text{ether, } 5^\circ} \text{4-cyclopropylcyclohexene}$$

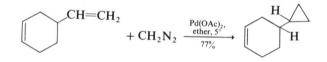

[1] A. S. Kende, B. Roth, and P. J. Sanfilippo, *Am.* Soc., **104**, 1784 (1982).
[2] A. S. Kende, B. Roth, P. J. Sanfilippo, and T. J. Blacklock, *ibid.*, **104**, 5808 (1982).
[3] J. Tsuji, H. Nagashima, and K. Sato, *Tetrahedron Letters*, **23**, 3085 (1982).
[4] T. Itahara, *J.C.S. Chem. Comm.*, 859 (1981).
[5] M. Suda, *Synthesis*, 714 (1981).

Palladium(II) acetate–1,2-Bis(diphenylphosphino)ethane (dppe).

α-Substituted-α,β-enones.[1] Allyl β-ketocarboxylates undergo decarboxylation and dehydrogenation in the presence of this Pd(II) catalyst to give mainly α-substituted-α,β-enones. This reaction is subject to solvent effects (equation I). In addition, a substituent in the α-position is essential for reasonable yields.

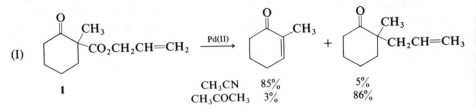

(I) 1 (2-methyl-2-(allyloxycarbonyl)cyclohexanone) $\xrightarrow{Pd(II)}$ 2-methyl-2-cyclohexenone + 2-methyl-2-allylcyclohexanone ($CH_2CH{=}CH_2$)

CH_3CN	85%	5%
CH_3COCH_3	3%	86%

Other palladium catalysts convert **1** into different products (equation II).

(II) 2-methyl-2-allylcyclohexanone (main product) $\xleftarrow{Pd\text{-}P(C_6H_5)_3}$ **1** $\xrightarrow[87\%]{PdCl_2[P(C_6H_5)_3]_2,\ HCOONH_4}$ 2-methylcyclohexanone

[1] I. Shimizu and J. Tsuji, *Am. Soc.*, **104**, 5844 (1982).

Palladium acetate–Triphenylphosphine, 5, 497–498; **9**, 349–350; **10**, 298–299.

Decarboxylation–carbonylation; β,γ-unsaturated esters. Allylic carbonates undergo decarboxylation-carbonylation in the presence of this palladium catalyst at 50° under 10 atms of CO to give β,γ-unsaturated esters in 70–95% yield.[1]

Example:

$$CH_3(CH_2)_3CH(OCO_2C_2H_5)CH{=}CH_2 + CO \xrightarrow[74\%]{Pd(OAc)_2,\ 2P(C_6H_5)_3} CH_3(CH_2)_3CH{=}CHCH_2COOC_2H_5$$

[1] J. Tsuji, K. Sato, and H. Okumoto, *Tetrahedron Letters*, **23**, 5189 (1982).

Palladium catalysts, 1, 778–782; **2**, 203; **4**, 368–369; **5**, 499; **6**, 445–446; **7**, 275–277; **8**, 382–383; **9**, 351–352; **10**, 299–300.

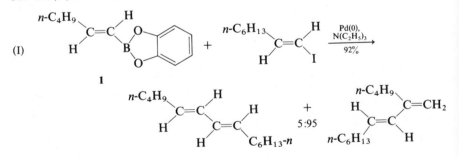

(I) (E)-$n\text{-}C_4H_9CH{=}CH\text{-}B$(catecholborane) (**1**) + (Z)-$n\text{-}C_6H_{13}CH{=}CHI$ $\xrightarrow[92\%]{Pd(0),\ N(C_2H_5)_3}$ $n\text{-}C_4H_9CH{=}CH\text{-}CH{=}CH\text{-}C_6H_{13}\text{-}n$ + $n\text{-}C_6H_{13}CH{=}CH\text{-}C(n\text{-}C_4H_9){=}CH_2$

5:95

$$\text{(II)}\quad \text{I} + \underset{\text{H}}{\overset{\text{C}_6\text{H}_5}{}}\text{C}=\text{C}\overset{\text{H}}{\underset{\text{Br}}{}} \xrightarrow[91\%]{\text{Pd[P(C}_6\text{H}_5)_3]_4,\ \text{NaOCH}_3} n\text{-C}_4\text{H}_9(\text{H})\text{C}=\text{C}(\text{H})\text{-}(\text{H})\text{C}=\text{C}(\text{H})\text{C}_6\text{H}_5$$

2-Alkyl-1,3-dienes. Palladium black, prepared from $Pd(OAc)_2$ and B_2H_6, is the most effective catalyst for head-to-tail cross-coupling of 1-alkenyl iodides with 1-alkenyl-1,3,2-benzodioxaboroles to give mainly 2-alkyl-1,3-dienes (equation I).[1]

$Pd[P(C_6H_5)_3]_4$ with a strong base has been used to prepare the minor products of this reaction (equation II).[2]

Catalytic reduction of alkynes to* cis*-alkenes. This reduction is not possible with 10% Pd/C alone because this metal is too reactive and the alkane is formed readily. The selective reaction is possible if the Pd/C is deactivated by either Hg(0) or Pb(0), obtained by reduction of metal acetate with $NaBH_4$. Sodium phosphinate, H_2PO_2Na, is the preferred hydride donor. Since this donor is not soluble in the organic solvents used, a phase-transfer catalyst, benzyltriethylammonium chloride, is added.[3]

Hydrodehalogenation. Aryl halides are reduced when heated in formic acid with catalytic amounts of Pd/C. Nitro group are also reduced, but carbonyl groups are not. Yields are 80–95%.[4]

Decarbonylation of acyl chlorides.[5] Rhodium-based catalysts are generally used for decarbonylation of aromatic acyl chlorides to aryl chlorides. This reaction can also be effected, often in high yield, at 265–360° with Pd/C, $PdCl_2$, or $Pd[P(C_6H_5)_3]_4$.

Palladium–Poly(ethylenimine). Preparation as beads or powder.[6] This material is nonpyrophoric and can be stored for many months. Supplier: Aldrich.

This Pd catalyst is recommended for deblocking of carbobenzoxy groups by transfer hydrogenolysis (formic acid donor).[7]

[1] N. Miyaura and A. Suzuki, *J. Organometal. Chem.*, **213**, C53 (1981).
[2] N. Miyaura, H. Suginome, and A. Suzuki, *Tetrahedron Letters*, **22**, 127 (1981).
[3] R. A. W. Johnstone and A. H. Wilby, *Tetrahedron*, **37**, 3667 (1981).
[4] P. N. Pandey and M. L. Purkayastha, *Synthesis*, 876 (1982).
[5] J. W. Verbicky, Jr., B. A. Dellacoletta, and L. Williams, *Tetrahedron Letters*, **23**, 371 (1982).
[6] W. E. Meyers and G. P. Royes, *Am. Soc.*, **99**, 6141 (1977).
[7] D. R. Coleman and G. P. Royes, *J. Org.*, **45**, 2268 (1980).

Palladium(II) chloride, 1, 782; **3**, 303–305; **4**, 367–370; **5**, 500–503; **6**, 447–450; **7**, 277; **8**, 384–385; **9**, 352; **10**, 300–302.

Carbonylation of (π-allyl)palladium dimers. Reaction of carbon monoxide in methanol with (π-allyl)palladium dimers requires high pressure and results in modest yields of products. However, addition of the sodium salt of a carboxylic acid permits carbonylation at 25° under low pressure of CO and results in high yields of β,γ-unsaturated esters. The reaction may involve formation of a (carbomethoxy)(π-allyl)palladium complex.[1]

Example:

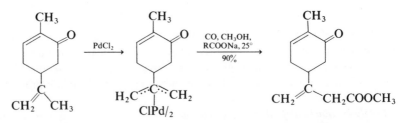

Coupling of π-(allylic)palladium complexes with an alkenylzirconium(IV) complex; 1,4-dienes.[2] A recent stereospecific synthesis of natural (20R)-cholestanone-3 (**4**) involves coupling of the alkenylzirconium complex **2** with the π-(allylic)palladium chloride dimer (**1**) of a (Z)-17(20)-pregnene. The major product is a 1,4-diene (**3**), formed by regioselective attack of **2** at C_{20}, the less hindered terminus of the allylic unit, and with inversion at C_{20}. Coupling of **2** with the π-(allylic)palladium chloride dimer of the (E)-isomer of **1** results in a 1,4-diene epimeric at C_{20} with **3**. Hydrogenation of the diene completes the synthesis of the desired natural cholestanone-3 (**4**).

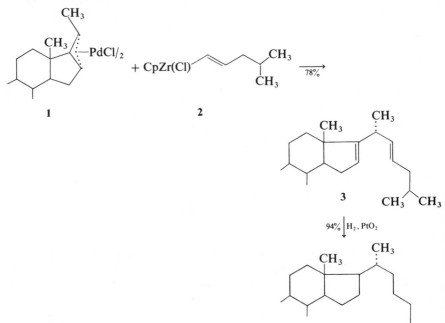

(Z)-Δ^3-Alkenyl sulfones. The π-allylpalladium complexes of 1,3-dienes react with sodium neophylsulfinate to give a palladium complex which on treatment with dimethylglyoxime gives (Z)-Δ^3-sulfones with high regio- and stereoselectivity.[3]

Example:

$$C_2H_5CH{=}CHCH{=}CH_2\ (E+Z)\ \xrightarrow[70\%]{1)\ PdCl_2,\ HOAc{-}H_2O;\ 2)\ NaSO_2CH_2C(CH_3)_2C_6H_5;\ 3)\ CH_3C(=NOH){-}C(=NOH)CH_3,\ CH_3OH}\ (Z)\text{-}C_2H_5CH{=}CH(CH_2)_2SO_2CH_2C(CH_3)_2C_6H_5$$

$$+\ CH_2{=}CHCH_2CH(C_2H_5)SO_2CH_2C(CH_3)_2{-}C_6H_5 \qquad 91:9$$

α,β-Unsaturated carbonyl compounds. Irradiation (366 nm) of an oxygenated solution of η^3-allylpalladium complexes leads to α,β-unsaturated carbonyl compounds in 30–70% yield.[4]

Examples:

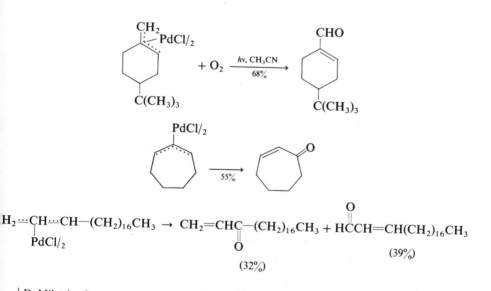

$$\text{(4-}tert\text{-butylcyclohexylidene)methyl-}\eta^3\text{-}PdCl/_2 + O_2 \xrightarrow[68\%]{h\nu,\ CH_3CN} \text{4-}C(CH_3)_3\text{-cyclohex-1-ene-1-CHO}$$

$$\eta^3\text{-cycloheptenyl-}PdCl/_2 \xrightarrow{55\%} \text{2-cycloheptenone}$$

$$H_2C{\cdots}CH{\cdots}CH{-}(CH_2)_{16}CH_3\ (PdCl/_2) \rightarrow CH_2{=}CHC(=O){-}(CH_2)_{16}CH_3\ (32\%) + HC(=O)CH{=}CH(CH_2)_{16}CH_3\ (39\%)$$

[1] D. Milstein, *Organometallics*, **1**, 888 (1982); *see also* D. E. James and J. K. Stille, *Am. Soc.*, **98**, 1810 (1976).
[2] J. S. Temple, M. Riediker, and J. Schwartz, *Am. Soc.*, **104**, 1310 (1982).
[3] Y. Tamaru, M. Kagotani, R. Suzuki, and Z. Yoshida, *J. Org.*, **46**, 3374 (1981).
[4] J. Muzart, P. Pale, and J.-P. Pete, *J.C.S. Chem. Comm.*, 668 (1981).

Palladium(II) chloride-Copper(I) chloride, 7, 278.

Oxidation of allylic and homallylic acetates (cf. **10**, 175–176).[1] This system is an efficient catalyst for oxygenation of terminal alkenes to methyl ketones (Wacker process). Similar oxidation of internal olefins is not useful because it is not regioselective. However, this catalyst effects oxygenation of allylic ethers and acetates regioselectively to give the corresponding β-alkoxy ketones in 40–75% yield. Under the same conditions, homoallylic acetates are oxidized to γ-acetoxy ketones as the major products.

Examples:

R ... OCH_3 $\xrightarrow[40-75\%]{O_2, PdCl_2, CuCl, DMF, H_2O}$ R ... OCH_3 (C=O)

$CH_3(CH_2)_4$ ... OAc $\xrightarrow[42\%]{}$ $CH_3(CH_2)_4$ ... OAc (C=O)

CH_3 ... OAc $\xrightarrow[80\%]{}$ CH_3 ... OAc (C=O) + CH_3 ... OAc (C=O)

91:9

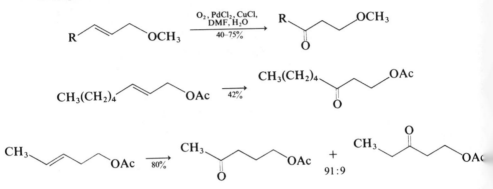

γ-Keto esters; 1,4-diketones. Oxygenation of β,γ-unsaturated esters or ketones with this catalyst system effects regioselective introduction of a keto group at the γ-position with formation of γ-keto esters or 1,4-diketones, respectively, in about 45–60% yield.[2]

Example:

$$C_2H_5CH{=}CHCH_2COOCH_3 \xrightarrow[52\%]{} C_2H_5\overset{\displaystyle O}{\overset{\|}{C}}CH_2CH_2COOCH_3$$

[1] J. Tsuji, H. Nagashima, and K. Hori, *Tetrahedron Letters*, **23**, 2679 (1982).
[2] H. Nagashima, K. Sakai, and J. Tsuji, *Chem. Letters*, 859 (1982).

Palladium(II) chloride–Silver(I) acetate.

Allylic acetoxylation.[1] The combination of *t*-butyl hydroperoxide and SeO_2 has been used for allylic hydroxylation of alkenes (**8**, 64–65), but this system is not useful for oxidation of cycloalkenes. Allylic acetoxylation of cycloalkenes is possible, but in modest yield, with $PdCl_2$ and AgOAc, which probably form a reactive species such as $[PdCl(OAc)]_n$. This system can be used in catalytic amounts in the presence of *t*-butyl hydroperoxide for a reoxidation step. The yield is improved by addition of TeO_2, which seems to accelerate the oxidation. The most satisfactory ratios of reagents for oxidation of 20 equiv. of the alkene are: $PdCl_2/AgOAc/t\text{-}BuOOH/TeO_2 = 1:2:10:1$. This system converts 1-alkenes into methyl ketones as the major product.

Examples:

$PdCl_2$, AgOAc, *t*-BuOOH, TeO_2, HOAc, Ac_2O

OAc (34%) + O (0.75%)

CH_2 → 26% → CH_2 ..OAc + CH_2OAc 63:37

CH_3 CH_3

[1] S. Uemura, S. Fukuzawa, A. Toshimitsu, and M. Okano, *Tetrahedron Letters*, **23**, 87 (1982).

Palladium(II) chloride–Thiourea, 6, 450–451; **7,** 278–279; **8,** 385.

α-Methylene-γ-butyrolactones (**7,** 278–279; **8,** 385).[1] The complete study of the Princeton group on synthesis of α-methylene lactones by cyclocarbonylation of ethynyl alcohols has been published. The original catalytic system, $PdCl_2$/thiourea, has the disadvantage that the only suitable solvent, acetone, does not give a homogeneous reaction mixture and, consequently, yields are not consistent. Moreover, the catalytic efficiency of $PdCl_2$ is low (about four turnovers). The most satisfactory catalytic system is $PdCl_2/SnCl_2/P(C_6H_5)_3$ or $P(n\text{-}Bu)_3$, which is soluble in CH_3CN. With this catalyst, high yields of a variety of α-methylene-γ-butyrolactones are often obtainable, particularly with low concentrations of the substrate alcohol to favor intramolecular cyclization.

[1] T. F. Murray, E. G. Samsel, V. Varma, and J. R. Norton, *Am. Soc.*, **103**, 7520 (1981).

Pentacarbonyl(methoxyphenylcarbene)chromium(0). (1). Mol. wt. 312.20, orange-red crystals, m.p. 46°.

Preparation[1]:

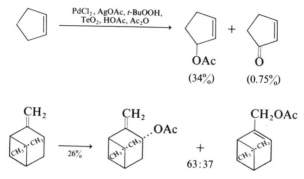

$$Cr(CO)_6 + C_6H_5Li \xrightarrow{\text{Ether}} (CO)_5CrC(OLi)(C_6H_5)\cdot(C_2H_5)_2O \xrightarrow[H_2O]{(CH_3)_3O^+BF_4^-}$$

$$(CO)_5Cr{=}C(OCH_3)(C_6H_5) + LiBF_4 + (C_2H_5)_2O$$

1 (69% overall)

α-Naphthoquinones. Dötz and coworkers[2] have shown that **1** reacts with alkynes to form (α-naphthol)chromium tricarbonyl complexes. A typical reaction is formulated in equation (I).

(I) $C_6H_5C{\equiv}CC_6H_5 + \mathbf{1} \xrightarrow[62\%]{(n\text{-}C_4H_9)_2O,\ 45°}$

OH, C_6H_5, C_6H_5, $(CO)_3Cr$, OCH_3 $\xrightarrow{70°}$ OH, C_6H_5, C_6H_5, $Cr(CO)_3$, OCH_3

2 (brick-red) **3**

This cyclization is general. The phenyl group of **1** can be replaced by various aryl groups, and a variety of alkynes can be used.

The complexes obtained in this way are converted to the corresponding α-naphthoquinones by CAN, I_2, $FeCl_3$, or $FeCl_3$–DMF in 65–70% yield.[3] The reaction is somewhat regioselective with unsymmetrical alkynes (equations II, III, and IV).

(II) $C_2H_5C{\equiv}CCH_3 + \mathbf{1} \xrightarrow[70\%]{1)\ THF,\ 45°;\ 2)\ CAN,\ CH_3OH}$ O, C_2H_5, CH_3, CH_3O, OCH_3 + (2:1) O, CH_3, C_2H_5, CH_3O, OCH_3

(III) $CH(CH_3)_2C{\equiv}CCH_3 + (CO)_5Cr{=}C(OCH_3)(C_6H_4OCH_3\text{-}o) \xrightarrow[61\%]{1)\ THF,\ 45°;\ 2)\ CAN,\ H_2O}$

O, $CH(CH_3)_2$, CH_3, CH_3O, O + (4.8:1) O, CH_3, $CH(CH_3)_2$, CH_3O, O

(IV) $CH_3CH_2CH_2C{\equiv}CH$ + $(CO)_5Cr{=}C(OCH_3)(C_6H_4OCH_3\text{-}m)$ $\xrightarrow[76\%]{1)\ THF,\ 45^\circ;\ 2)\ CAN}$

6-methoxy-2-propyl-1,4-naphthoquinone (CH_3O, $CH_2CH_2CH_3$) + 8-methoxy-2-propyl-1,4-naphthoquinone (CH_3O, $CH_2CH_2CH_3$)

10.5:1

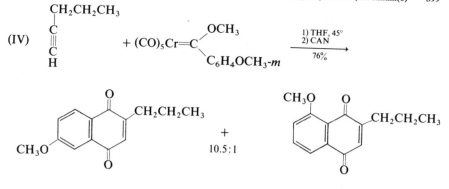

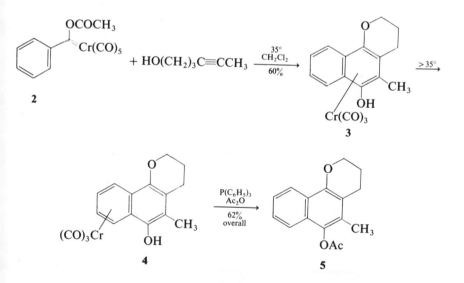

Intramolecular version. Because of the inherent lack of regiocontrol in the original version of this naphthoquinine synthesis, Semmelhack and Bozell[4] have developed an intramolecular cycloaddition variation using the acetoxycarbene complex **2**, prepared by reaction of phenyllithium with $Cr(CO)_6$ followed by acetylation. The (unstable) complex undergoes acyl transfer with an alkynol, which cyclizes at 35° with insertion of CO to give naphthalene complex (**3**) which rearranges to the more stable complex **4**. The naphthol **5** is obtained on removal of the $Cr(CO)_3$ group with $P(C_6H_5)_3$.

$C_6H_5C(OCOCH_3){=}Cr(CO)_5$ (**2**) + $HO(CH_2)_3C{\equiv}CCH_3$ $\xrightarrow[60\%]{35^\circ,\ CH_2Cl_2}$ **3** (CH_3, OH, $Cr(CO)_3$) $\xrightarrow{>35^\circ}$

4 ($(CO)_3Cr$, OH, CH_3) $\xrightarrow[62\%\ overall]{P(C_6H_5)_3,\ Ac_2O}$ **5** (OAc, CH_3)

A specific example of a regioselective synthesis of a 2,3-disubstituted naphthoquinone (**6**) by this procedure is shown in equation (I).

(I)

$\xrightarrow[86\%]{1}$ $\xrightarrow[65\%]{1)\ \Delta;\ 2)\ P(C_6H_5)_3;\ Ac_2O}$

$\xrightarrow{1)\ OH^-;\ 2)\ DDQ;\ 3)\ TsCl;\ NaI;\ Zn;\ 4)\ DDQ}$ **6**

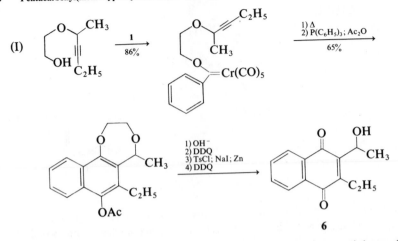

Reaction with ethyl propiolate.[5] The carbene reacts with ethyl propiolate to form isomeric vinyl ether derivatives (**2**) of diethyl malonate. The products are converted into the corresponding ketones on hydrolysis.

$$HC{\equiv}CCOOC_2H_5 + \mathbf{1} \xrightarrow[83\%]{C_2H_5OH} (C_6H_5)(CH_3O)C{=}CH{-}CH(COOC_2H_5)_2 \ (\mathbf{2a}) + (\mathbf{2b}) \quad 6:1$$

2a **2b**

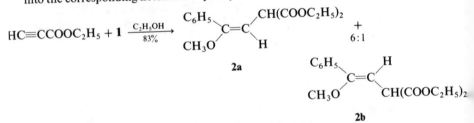

β-Lactams.[6] Sunlight photolysis of solutions of imines and chromium carbene complexes results in formation of β-lactams in fair to good yield.

Examples:

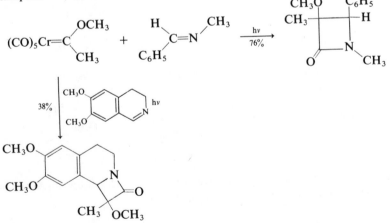

$$(CO)_5Cr{=}C(OCH_3)(CH_3) + C_6H_5CH{=}NCH_3 \xrightarrow[76\%]{h\nu}$$

$\xrightarrow[38\%]{h\nu}$

$$(CO)_5Cr{=}C(OCH_3)(CH_3) + \text{thiazoline} \xrightarrow[81\%]{h\nu} \text{CH}_3\text{O, CH}_3\text{, H-substituted bicyclic }\beta\text{-lactam (S, N)}$$

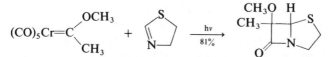

[1] E. O. Fisher, B. Heckl, K. H. Dötz, J. Müller, and H. Werner, *J. Organometal. Chem.*, **16**, P29 (1969); E. O. Fischer, C. G. Kreiter, H. J. Kollmeier, J. Müller, and R. D. Fisher, *ibid.*, **28**, 237 (1971).

[2] K. H. Dötz, *Angew. Chem. Int. Ed.*, **14**, 644 (1975); K. H. Dötz, R. Dietz, A. von Imhof, H. Lorenz, and G. Huttner, *Ber.*, **109**, 2033 (1976); K. H. Dötz and R. Dietz, *ibid.*, **110**, 1555 (1977); *idem, ibid.*, **111**, 2517 (1978); K. H. Dötz and B. Fügen-Köster, *ibid.*, **113**, 1449 (1980).

[3] W. D. Wulff, P.-C. Tang, and J. S. McCallum, *Am. Soc.*, **103**, 7677 (1981).

[4] M. F. Semmelhack and J. J. Bozell, *Tetrahedron Letters*, **23**, 2931 (1982).

[5] A. Yamashita and T. A. Scahill, *ibid.*, **23**, 3765 (1982).

[6] M. A. McGuire and L. S. Hegedus, *Am. Soc.*, **104**, 5538 (1982).

Pentacarbonyl(trimethylsilyl)manganese (1). Mol. wt. 268.18, m.p. 66.5°.

Preparation[1]:

$$[Mn(CO)_5]_2 \xrightarrow{\text{K(}sec\text{-Bu)}_3\text{BH, THF, }0^\circ} K[Mn(CO)_5] \xrightarrow[\text{60–80\% overall}]{BrSi(CH_3)_3} (CO)_5MnSi(CH_3)_3$$

1

Vinyl ethers.[2] Methyl enol ethers are usually prepared by treatment of dimethyl ketals with HCl at 100–200°. The same reaction is effected with **1** at 50° (equation I). Reaction of **1** with dimethyl acetals is less useful.

$$\text{(I)}\quad R^1(R^2CH_2)C(OCH_3)_2 + \mathbf{1} \xrightarrow{CH_3CN,\ 50^\circ} R^2CH{=}C(OCH_3)R^1 + CH_3OSi(CH_3)_3 + (CO)_5MnH$$

Other examples[3]:

$$\text{cyclohexanone ethylene ketal} \xrightarrow[87\%]{\mathbf{1},\ CH_3CN} \text{1-}[O(CH_2)_2OSi(CH_3)_3]\text{cyclohexene}$$

$$\text{cyclohexanone bis(}CH_2{=}CHCH_2O\text{) ketal} \xrightarrow[42\%]{} \text{1-}(OCH_2CH{=}CH_2)\text{cyclohexene}$$

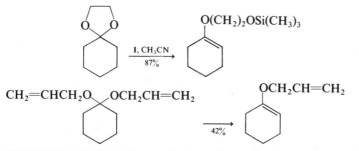

[1] J. A. Gladysz, G. M. Williams, W. Tam, D. L. Johnson, D. W. Parker, and J. C. Selover, *Inorg. Chem.*, **18**, 553 (1979).

[2] M. Marsi and J. A. Gladysz, *Tetrahedron Letters*, **23**, 631 (1982).

[3] *Idem, Organometallics*, **1**, 1467 (1982).

Peracetic acid, 1, 785–791; **2,** 307–309; **3,** 219; **4,** 372; **5,** 505–506; **6,** 452–453; **8,** 386.

Δ^3-Butenolides.[1] 5-Alkyl-2-trimethylsilylfuranes (**1**), readily obtained as shown,[2] are oxidized in moderate to high yield to Δ^3-butenolides by 40% peracetic acid in buffered CH_2Cl_2. The oxidation probably proceeds through **a** and **b**. The trimethylsilyl group is essential for this oxidation.

(CH3)3Si–furan → [1) n-BuLi, 2) RX] → (CH3)3Si–furan–R (**1**) → [CH3CO3H, CH2Cl2] → [(CH3)3Si epoxide–R (**a**) → (CH3)3Si–butenolide–R (**b**)] → [CH3COOH] → butenolide–R (**2**)

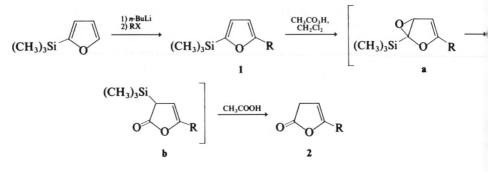

[1] I. Kuwajima and H. Urabe, *Tetrahedron Letters*, **22**, 5191 (1981).
[2] H. W. Gschwend and H. R. Rodriguez, *Org. React.*, **26**, 1 (1979).

Perchloric acid, 1, 796–802; **2,** 309–310; **3,** 220; **5,** 506–507; **6,** 453–459; **7,** 279–280.

Rearrangement of enynols to dienones.[1] The acetylenic alcohol **1** is stable to 3% H_2SO_4 in ether but rearranges in 10% $HClO_4$ in THF to the enynol **2** and the dienone **3** in about equal amounts. The latter product becomes the major product (85%) when 35% $HClO_4$–THF is used. This rearrangement to a dienone does not occur in the absence of the C_8-methyl group.

1 (BzlO, HO, C≡CH, CH3, CH3, H, CH3) → [H⁺] → **2** (BzlO, C≡CH, CH3, CH3, OH, H, CH3) → [H⁺] → **3** (BzlO, CH=CH2, CH3, CH3, O, H, CH3)

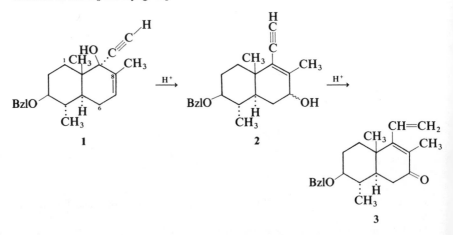

[1] T. Leapheart and P. Magnus, *J.C.S. Chem. Comm.*, 1391 (1982).

Perfluoroallyl fluorosulfate (1). Mol. wt. 230.10, b.p. 63–64°.
Preparation:

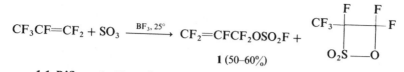

$$CF_3CF{=}CF_2 + SO_3 \xrightarrow{BF_3,\ 25^\circ} \underset{\mathbf{1}\ (50\text{–}60\%)}{CF_2{=}CFCF_2OSO_2F} + \begin{matrix} & F & F & \\ CF_3- & C & C & -F \\ & | & | & \\ & O_2S & -O & \end{matrix}$$

1,1-Difluoro-1-alkenes.[1] Examples of synthesis of these 1-alkenes are shown in equations (I) and (II).

$$\text{(I)}\quad (CF_3)_2C{=}O + KF \longrightarrow (CF_3)_2CFOK \xrightarrow{\mathbf{1}} (CF_3)_2CFOCF_2CF{=}CF_2$$

$$\text{(II)}\quad FSO_2CF_2COF + KF \longrightarrow FSO_2CF_2CF_2OK \xrightarrow{\mathbf{1}} FSO_2CF_2CF_2OCF_2CF{=}CF_2$$

[1] C. G. Krespan and D. C. England, *Am. Soc.*, **103**, 5598 (1981).

Phase-transfer catalysts, 8, 387–391; **9**, 356–361; **10**, 305–306.

Amino acid synthesis (**8**, 389). Alkylation of the aldimine (**1**) from glycine ethyl ester and *p*-chlorobenzaldehyde under phase-transfer conditions offers a general route to amino acids. Either liquid-liquid phase-transfer or solid-liquid phase-transfer catalytic conditions are satisfactory with active halides, but alkylation with allylic halides and less active alkyl halides is best effected under ion-pair extraction conditions (**6**, 41), with 1 equiv. of tetra-*n*-butylammonium hydrogen sulfate (76–95% yields).[1]

$$\underset{\mathbf{1}}{p\text{-}ClC_6H_4CH{=}NCH_2COOC_2H_5} + RX \xrightarrow{76\text{–}95\%} \underset{\displaystyle NH_2}{RCHCOOH}$$

α-Methyl-substituted phenylalanine and tyrosine can be prepared by appropriate benzylation of the aldimine (**2**) of alanine ethyl ester or by methylation of the aldimine (**3**) of the ethyl ester of phenylalanine and of O-benzyltyrosine, both under phase-transfer conditions in 80–95% yield.[2]

$$\underset{\mathbf{2}}{\begin{matrix} CH_3CHCOOC_2H_5 \\ | \\ N{=}CHC_6H_4Cl\text{-}p \end{matrix}} \qquad \underset{\mathbf{3}}{\begin{matrix} RC_6H_4CH_2CHCOOC_2H_5 \\ | \\ N{=}CHC_6H_4Cl\text{-}p \end{matrix}}$$

Glycosylation.[3] The α-anomer of ara-7-deazaguanosine (**4**) has been prepared by glycosylation with tetrabutylammonium hydrogen sulfate as catalyst of the pyrimidine **1** with 2,3,5-tri-O-benzyl-1-bromoarabinofuranose (**2**). The catalyst equilibrates **2** to a mixture of the α- and β-anomers; the more reactive β-anomer reacts with **1** by a S_N2-displacement to give mainly **3**.

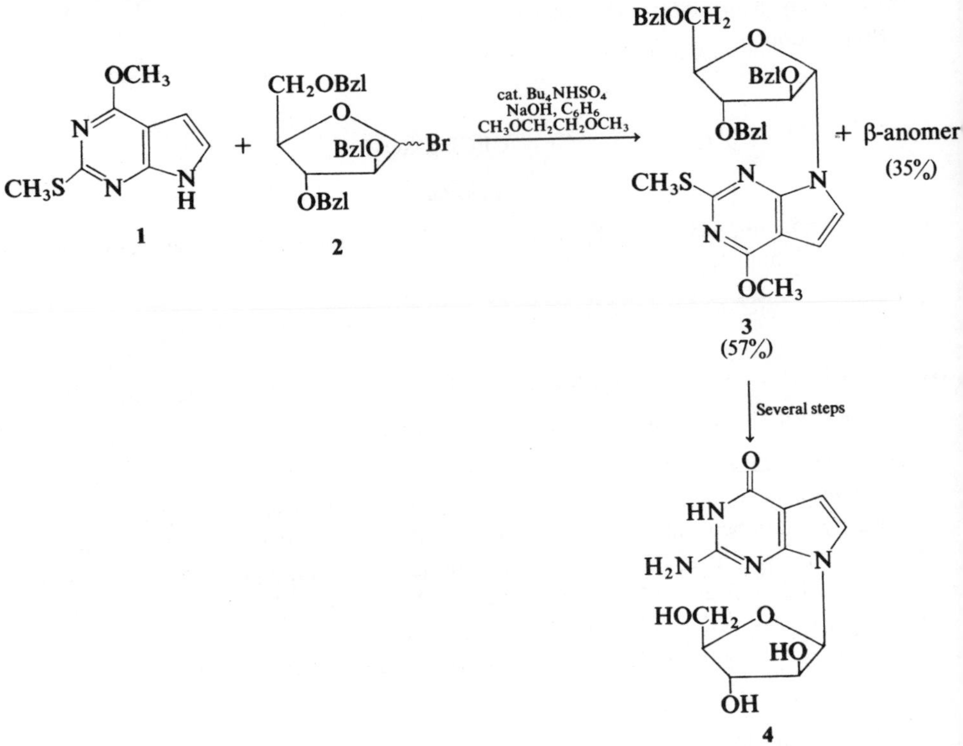

$$\mathbf{1} + \mathbf{2} \xrightarrow[\text{CH}_3\text{OCH}_2\text{CH}_2\text{OCH}_3]{\text{cat. Bu}_4\text{NHSO}_4,\ \text{NaOH, C}_6\text{H}_6} \mathbf{3}\ (57\%) + \beta\text{-anomer}\ (35\%)$$

$\mathbf{3}$ $\xrightarrow{\text{Several steps}}$ $\mathbf{4}$

Thioacetals.[4] Thioacetals can be prepared in 60–95% yield by reaction of a thiol with a *gem*-dichloride with K_2CO_3 as base and Aliquat 336 as catalyst in H_2O—$C_6H_5CH_3$.

Examples:

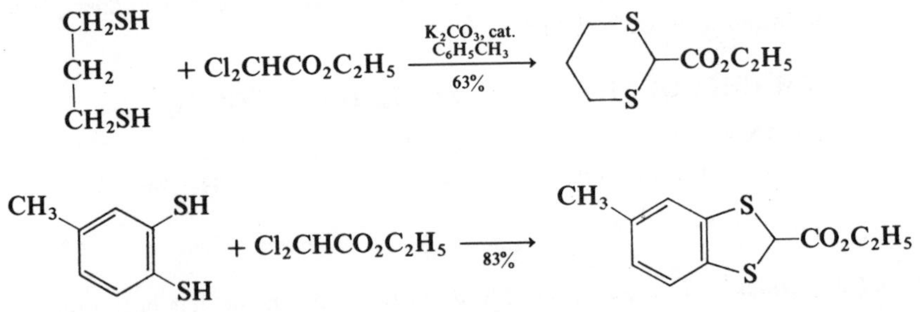

$$HSCH_2CH_2CH_2SH + Cl_2CHCO_2C_2H_5 \xrightarrow[63\%]{K_2CO_3,\ \text{cat.},\ C_6H_5CH_3} \text{2-}(CO_2C_2H_5)\text{-1,3-dithiane}$$

$$\text{CH}_3\text{-benzene-1,2-dithiol} + Cl_2CHCO_2C_2H_5 \xrightarrow{83\%} \text{CH}_3\text{-benzo-1,3-dithiole-2-}CO_2C_2H_5$$

These thioacetals can be deprotonated by K_2CO_3 in the presence of Aliquat 336; and the anions obtained react with various electrophiles including allylic halides. In contrast to reactions with NaH as the base, reaction with an allylic bromide results in allylic transposition.

Example:

$$\text{(1,3-dithian-2-yl)}-CO_2C_2H_5 + CH_3CH{=}CHCH_2Br \xrightarrow[68\%]{K_2CO_3,\ \text{cat., toluene}}$$

$$\text{2-}CO_2C_2H_5\text{-2-}[CH(CH_3)CH{=}CH_2]\text{-1,3-dithiane} \quad + \quad \text{2-}CO_2C_2H_5\text{-2-}(CH_2CH{=}CHCH_3)\text{-1,3-dithiane} \qquad 95:5$$

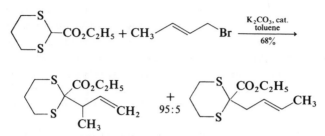

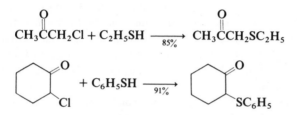

α-Alkylthio carbonyl compounds. These compounds are obtained in high yield by reaction of α-chloro carbonyl compounds with RSH or ArSH in aqueous toluene with Na_2CO_3 as base and Aliquat 336 as catalyst.[5]

Examples:

$$CH_3\overset{O}{\overset{\|}{C}}CH_2Cl + C_2H_5SH \xrightarrow[85\%]{} CH_3\overset{O}{\overset{\|}{C}}CH_2SC_2H_5$$

$$\text{2-chlorocyclohexanone} + C_6H_5SH \xrightarrow[91\%]{} \text{2-}(SC_6H_5)\text{cyclohexanone}$$

Solid-liquid phase-transfer catalyzed alkylation of anions. Aliquat 336 is a particularly effective catalyst for alkylation of carboxylate anions (equation I). The combination of Bu_4NBr and TiO_2 is slightly less effective (93% yield).[6]

$$\text{(I)}\quad CH_3COOK + CH_3(CH_2)_7Br \xrightarrow[98\%]{\text{cat., }25^\circ} CH_3(CH_2)_7OCOCH_3$$

Alkylation of alkynes.[7] The reaction of alkynes with carbonyl compounds in toluene–aqueous NaOH is markedly accelerated by catalytic amounts of tetra-*n*-butylammonium bromide.

Example:

$$(CH_3)_2\overset{OH}{\overset{|}{C}}-C{\equiv}CH + (CH_3)_2C{=}O \xrightarrow{NaOH,\ C_6H_5CH_3,\ H_2O} (CH_3)_2\overset{HO}{\overset{|}{C}}-C{\equiv}C-\overset{OH}{\overset{|}{C}}(CH_3)_2$$

no cat., 2 hours	7%
$(C_4H_9)_4NBr$, 30 minutes	100%

N-Alkylation of amides and sulfonamides.[8] Sodium hydride has been recommended for deprotonation of amides that are only weakly acidic. Actually benzamide, $C_6H_5CONH_2$, can be monalkylated in 50% aqueous NaOH–C_6H_6 (refluxing) in the presence of tetra-*n*-butylammonium hydrogen sulfate as phase-transfer catalyst. A second alkyl group cannot be introduced under these conditions. To effect dialkylation a solid-liquid two-phase system composed of $NaOH/K_2CO_3$ suspended in refluxing benzene or toluene with the same catalyst is necessary. In practice, the alkylation is limited to primary alkyl halides. The solid-liquid system is also necessary for monoalkylation for propanamide, $C_2H_5CONH_2$, a weaker acid than benzamide.

Sulfonamides, which are relatively strong NH-acids, can be alkylated under liquid-liquid or solid-liquid conditions; by either method dialkylation obtains with primary alkyl halides.

Dichlorocarbene cyclopropanations.[9] Tetramethylammonium chloride is apparently the most selective catalyst for monoaddition of Cl_2C: to dienes yet reported. The nature of the diene and the concentration of the catalyst seem to be more significant than the type of catalyst.

Alkynes.[10] Alkynes can be prepared by double dehydrobromination of *vic*-dibromides in petroleum ether by use of powdered KOH and catalytic amounts of a phase-transfer catalyst. Tetraoctylammonium bromide, 18-crown-6, or even Aliquat 336 are much more effective than more hydrophilic quarternary ammonium salts, such as tetrabutylammonium hydrogen sulfate, previously used (7, 354–355).[11] Isolated yields are 80–98%. Yields are generally lower when *vic*-dichlorides are used as the starting material.

Aldol condensations.[12] Phase-transfer condensation of aldehydes with α,β-unsaturated carboxylic acid derivatives is possible with anhydrous K_2CO_3 as base and Aliquat 336 as catalyst.

Example:

RCHO + CH3–CH=CH–C(=O)–N(piperidine) —(K_2CO_3, cat., $C_6H_5CH_3$, 90°)→ R–CH=CH–CH=CH–C(=O)–N(piperidine)

—(ArCHO, K_2CO_3, cat.; 62%)→ Ar–CH=CH–CH=CH–CH=CH–C(=O)–N(piperidine)

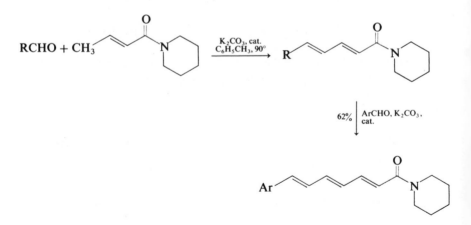

β-Lactams.[13] A convenient route to β-lactams is shown in equation (I). The second step requires high dilution for satisfactory yields.

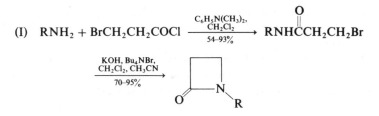

$$\text{(I)}\quad RNH_2 + BrCH_2CH_2COCl \xrightarrow[54\text{–}93\%]{C_6H_5N(CH_3)_2,\ CH_2Cl_2} RNH\overset{O}{\overset{\|}{C}}CH_2CH_2Br$$

$$\xrightarrow[70\text{–}95\%]{KOH,\ Bu_4NBr,\ CH_2Cl_2,\ CH_3CN} \text{(N-R azetidin-2-one)}$$

[1] L. Ghosez, J. -P. Antoine, E. Deffense, M. Navarro, V. Libert, M. J. O'Donnell, W. A. Bruder, K. Willey, and K. Wojciechowski, *Tetrahedron Letters*, **23**, 4255 (1982).
[2] M. J. O'Donnell, B. LeClef, D. B. Rusterholz, L. Ghosez, J. -P. Antoine, and M. Navarro, *ibid*, **23**, 4259 (1982).
[3] F. Seela and H.-D. Winkeler, *Ann.*, 1634 (1982).
[4] M. Lissel, *ibid.*, 1589 (1982).
[5] M. Lissel, *J. Chem. Res.* (*M*), 2946 (1982).
[6] J. Barry, G. Bram, G. Decodts, P. Pigeon, and J. Sansoulet, *Tetrahedron Letters*, **23**, 5407 (1982).
[7] E. V. Dehmlow and A. R. Sharmout, *Ann.*, 1750 (1982).
[8] T. Gajda and A. Zwierzak, *Synthesis*, 1005 (1981).
[9] E. V. Dehmlow and M. Prashad, *J. Chem. Res.* (*S*), 354 (1982).
[10] E. V. Dehmlow and M. Lissel, *Tetrahedron*, **37**, 1653 (1981).
[11] A. Le Coq and A. Gorgues, *Org. Syn.*, **59**, 10 (1980).
[12] E. V. Dehmlow and A. R. Shamout, *J. Chem. Res.* (*M*), 1178 (1981).
[13] H. Takahata, Y. Ohnishi, H. Takehara, K. Tsuitani, and T. Yamazaki, *Chem. Pharm. Bull. Japan*, **29**, 1063 (1981).

Se-Phenyl benzeneselenosulfonate (1). Mol. wt. 297.23, m.p. 55°, *cf.*, **10**, 315.

Preparation:

$$C_6H_5SeO_2H + 2C_6H_5SO_2H \xrightarrow[61\%]{C_2H_5OH,\ 0^\circ} \underset{\mathbf{1}}{C_6H_5SO_2SeC_6H_5}$$

Photoaddition to alkenes. On irradiation **1** dissociates to $C_6H_5Se\cdot$ and $C_6H_5SO_2\cdot$; irradiation in the presence of an alkene initiates a chain reaction leading to β-phenylseleno sulfones in good yield.

Examples:

$$\text{cyclohexene} + \mathbf{1} \xrightarrow[80\%]{h\nu} \text{1-}SO_2C_6H_5\text{-2-}SeC_6H_5\text{-cyclohexane}$$

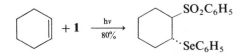

$$RCH{=}CH_2 + 1 \xrightarrow[65\text{–}75\%]{} R\underset{SeC_6H_5}{\underset{|}{C}}HCH_2SO_2C_6H_5$$

This reaction can also be induced thermally (**10**, 315). The adducts are useful precursors to α,β-unsaturated sulfones, obtained on oxidation with H_2O_2 or $ClC_6H_4CO_3H$.[1]

Addition to allenes.[2] Se-Phenyl areneselenosulfonates also undergo free-radical addition to allenes, with the arylsulfonyl group adding to the central carbon, and the C_6H_5Se group becoming attached to the less substituted terminal carbon. The products when oxidized with H_2O_2 undergo [2, 3] sigmatropic rearrangement to a selenenate, which can be converted into a β-arylsulfonyl-substituted allylic alcohol.

Example:

$$CH_3CH{=}C{=}C(CH_3)_2 + p\text{-}CH_3C_6H_4SO_2SeC_6H_5 \xrightarrow[CCl_4]{h\nu,}$$

$$\underset{C_6H_5Se}{CH_3\overset{|}{C}H}-\underset{Ts}{\overset{|}{C}}{=}C(CH_3)_2 \xrightarrow[\substack{98\% \\ \text{overall}}]{\substack{1)\ H_2O_2,\ THF \\ 2)\ N(C_2H_5)_3}} CH_3CH{=}\underset{Ts}{C}-\overset{OH}{C}(CH_3)_2$$

Addition to alkynes. Se-Phenyl areneselenosulfonates undergo *anti*-addition to alkynes at elevated temperatures to afford β-(phenylseleno)vinyl sulfones (equations I, II). The products can undergo selenoxide elimination to alkynes.[3]

$$\text{(I)}\quad TsSeC_6H_5 + R^1C{\equiv}CR^2 \xrightarrow[50\text{–}90\%]{C_6H_6,\ \Delta} \begin{matrix} R^1 & & SeC_6H_5 \\ & C{=}C & \\ Ts & & R^2 \end{matrix}$$

$$\text{(II)}\quad TsSeC_6H_5 + HC{\equiv}CC_6H_5 \xrightarrow[93\%]{} \begin{matrix} H & & SeC_6H_5 \\ & C{=}C & \\ Ts & & C_6H_5 \end{matrix} \xrightarrow[88\%]{m\text{-}ClC_6H_4CO_3H} TsC{\equiv}CC_6H_5$$

[1] R. A. Gancarz and J. L. Kice, *J. Org.*, **46**, 4899 (1981).
[2] Y. -H. Kang and J. L. Kice, *Tetrahedron Letters*, **23**, 5373 (1982).
[3] T. G. Back and S. Collins, *ibid.*, **22**, 5111 (1981).

Phenylboronic acid, 1, 833–834; **2**, 317; **3** 221–222; **5**, 513–514; **7**, 284; **9**, 23–24.

1,3-cis-*Diols.*[1] A recent synthesis of 4-dimethoxyanthracyclinone (**6**) obtained the desired *cis*-relationship of the 7- and 9-hydroxy groups by reaction of **1**, consisting of a mixture of 1,3-diols with phenylboronic acid. The corresponding quinone (**3**) is converted into **5** by a Diels-Alder reaction with the *o*-quinonedimethide from **4**. Final steps include deprotection with BCl_3 and 2-methylpentane-2,4-diol.

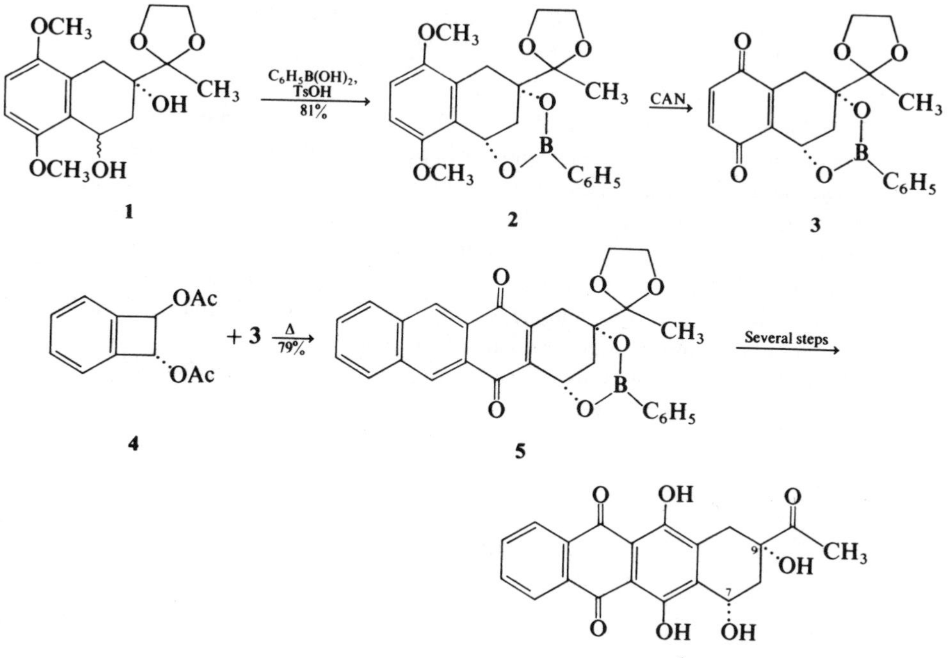

α,β*-Dihydroxy ketones.*[2] The α-hydroxy ketone **1** is converted on reaction with phenylboronic acid into a cyclic vinyloxyborane (**a**). On addition of an aldehyde, a 1,3,2-dioxaborole (**b**) is formed stereoselectively. Hydrolysis of **b** with aqueous H_2O_2 affords as the major product an α,β-dihydroxy ketone with the *anti*-configuration of the diol group (three examples).

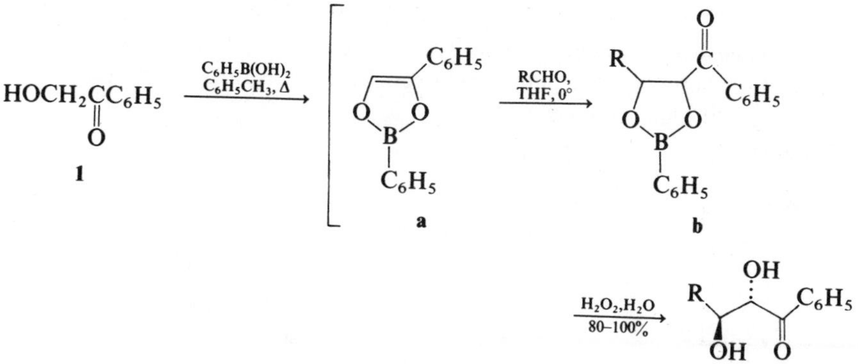

(diastereomer ratio ~9:1)

[1] M. J. Broadhurst, C. H. Hassall, and G. J. Thomas, *J.C.S. Perkin I*, 2239 (1982).
[2] T. Mukaiyama and M. Yamaguchi, *Chem. Letters*, 509 (1982).

Phenyl chlorosulfate, $C_6H_5SO_3Cl$. Mol. wt. 192.62, b.p. 61–65°/50 mm. The reagent is prepared by reaction of sodium phenoxide in benzene with sulfuryl chloride.

Sugar sulfates.[1] Chlorosulfonic acid is usually used for sulfation of sugars, but the products are often difficult to purify. A new method uses phenyl chlorosulfate. The products, phenyl sulfates, are stable to trifluoroacetic acid, acetic anhydride–acetic acid, sodium methoxide, and fluoride ion. The phenyl group is removed by hydrogenolysis (PtO_2, K_2CO_3 buffer), during which the phenyl group is initially reduced to a cyclohexyl group, which is then cleaved by base. 1,2;5,6-Di-O-isopropylidene-α-D-glucofuranose (**1**) is converted into the 3-phenyl sulfate (**2**) in 75% yield by consecutive reaction with NaH and then $C_6H_5SO_3Cl$.

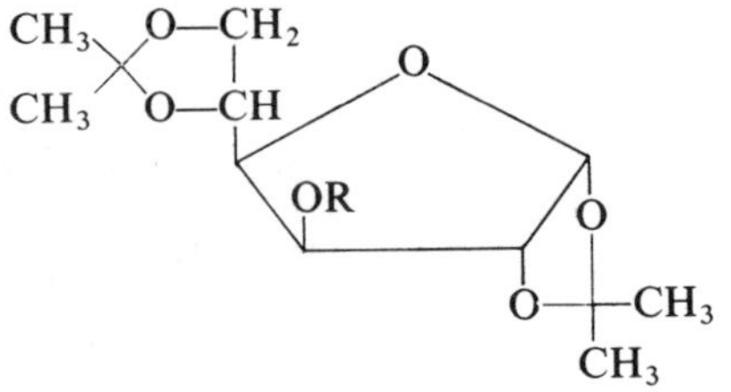

1, R = H
2, R = $SO_3C_6H_5$

[1] C. L. Penney and A. S. Perlin, *Carbohydrate Res.*, **93**, 241 (1981).

Phenyldichlorophosphate–Dimethyl formamide, $C_6H_5O\overset{O}{\underset{Cl}{P}}{-}OCH{=}\overset{+}{N}(CH_3)_2Cl^-$ **(1).**

α-Phthalimido-β-lactams. A new synthesis of these β-lactams (**4**) uses the complex **1** to phosphorylate a phthalimidoacetic acid (**2**) to give **3**, which reacts with imines to give the final products.[1]

Example:

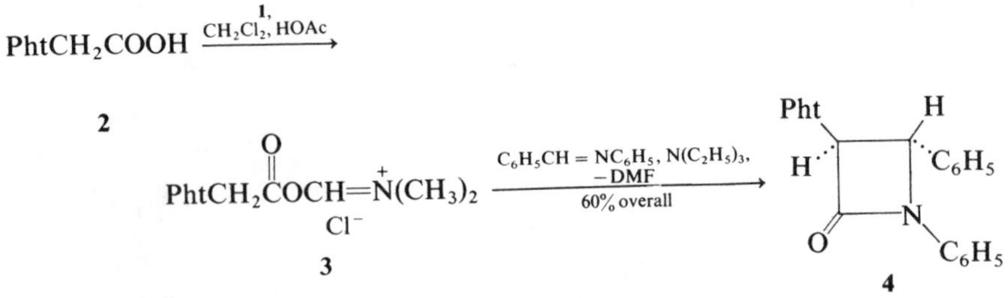

[1] A. Arrieta, J. M. Aizpurua, and C. Palomo, *Syn. Comm.*, **12**, 967 (1982).

α-Phenylethylamine (α-Methylbenzylamine), 1, 838; **2,** 271–273; **3,** 199–200; **5,** 441; **6,** 457; **7,** 393–395; **9,** 363–364.

cis-2-*Alkylcyclohexanamines.*[1] An asymmetric synthesis of these amines is formulated in equation (I). The key step is the hydrogenation of the imine to give optically pure *cis*-products.

$$\text{(I)}\quad \text{2-}CH_3\text{-cyclohexanone}{=}O + H_2N{-}\overset{H}{\underset{CH_3}{C^*}}{-}C_6H_5 \longrightarrow \text{2-}CH_3\text{-cyclohexylidene}{=}N{-}\overset{H}{\underset{CH_3}{C^*}}{-}C_6H_5 \xrightarrow[\text{88\% overall}]{\text{1) } H_2\text{, Raney Ni; 2) HCl}}$$

(+) or (−)

$$\text{2-}CH_3\text{-cyclohexyl}^{**}{-}NH{-}\overset{H}{\underset{CH_3}{C^*}}{-}C_6H_5\cdot HCl \xrightarrow{H_2\text{, Pd/C}} \text{2-}CH_3\text{-cyclohexyl}^{**}{-}NH_2\cdot HCl \xrightarrow{OH^-} \text{2-}CH_3\text{-cyclohexyl}^{**}{-}NH_2$$

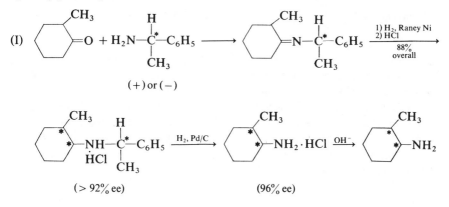

(> 92% ee) (96% ee)

A. W. Frahm and G. Knupp, *Tetrahedron Letters*, **22**, 2633 (1981).

Phenyliodine (III) diacetate (Iodosobenzene diacetate), 1, 508–509, **3,** 166; **4,** 266; **10,** 214–215.

Review.[1] The reactions of $C_6H_5I(OAc)_2$ are mainly oxidations in which iodine(III) is reduced to iodine(I) and closely resemble those of lead tetraacetate, such as cleavage of glycols. For this reaction, phenyliodine(III) bis(trifluoroacetate) is the preferred reagent (**6,** 301).

α-Methoxy-α-arylacetic acid esters. Treatment of methyl or ethyl α-arylacetates with $C_6H_5I(OAc)_2$ and sodium methoxide in CH_3OH results in α-methoxylation. The reaction is also observed, but in lower yields, with alkyl-substituted acetates (equation I).[2]

$$\text{(I)}\quad ArCH_2COOCH_3 \xrightarrow[70\text{–}80\%]{C_6H_5I(OAc)_2,\ NaOCH_3,\ CH_3OH} Ar\overset{OCH_3}{CH}COOCH_3$$

Dimethyl acetals.[3] Reaction of arylaldazines with $C_6H_5I(OAc)_2$ and $NaOCH_3$ in methanol gives the dimethyl acetal of the aldehyde in moderate to high yield. Yields are low with azines of aliphatic aldehydes.

$$\text{(I)}\quad (ArCH{=}N{-})_2 \xrightarrow{C_6H_5I(OAc)_2,\ NaOCH_3,\ CH_3OH} \underset{(50\text{–}95\%)}{ArCH(OCH_3)_2} + C_6H_5I + NaOAc + N_2$$

[1] A. Varvoglis, *Chem. Soc. Rev.*, **10**, 377 (1981).

[2] R. M. Moriarty and H. Hu, *Tetrahedron Letters*, **22**, 2747 (1981).
[3] *Idem, ibid.*, **23**, 1537 (1982).

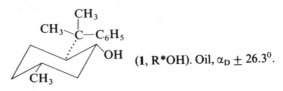

8-Phenylmenthol, (**1**, R*OH). Oil, $\alpha_D \pm 26.3^0$.

The (+)- or (−)-alcohol is prepared by CuI-catalyzed conjugate addition of C_6H_5MgBr to (−) or (+)-pulegone, respectively, followed by reduction of the carbonyl group with sodium–isopropanol.[1]

Asymmetric Diels-Alder reactions. The acrylate of (+)-**1** is markedly superior to (−)-menthyl acrylate in effecting asymmetric induction in Diels-Alder reaction with cyclopentadiene (Lewis acid catalysis, equation I). A chiral intermediate in Corey's prostaglandin synthesis was obtained by the reaction of (+)-8-phenylmenthyl acrylate with a cyclopentadiene derivative.[1]

(I) cyclopentadiene + $CH_2{=}CHCOR^*$ (with C=O) $\xrightarrow{SnCl_4 \text{ or } TiCl_4}$ H, COOR*

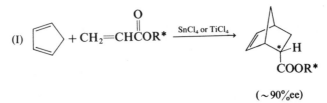

(~90%ee)

The reagent has one handicap; only (+)-pulegone is readily available. Consequently, Oppolzer and coworkers[2] have examined a number of chiral alcohols, derived from various monoterpenes, in which the hydroxyl group is similarly shielded. The *endo*, *cis*-diphenylmethylisoborneol (**2**), obtained from (+)-camphor, and the enantiomer, obtained from (−)-camphor, are almost as effective as **1** as chiral adjuncts in this Diels-Alder reaction.

CH_3 CH_3 $CH(C_6H_5)_2$ CH_3 OH

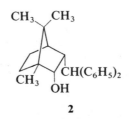

2

Asymmetric Michael addition to enoates.[3] The known conjugate addition of $RCu{\cdot}BF_3$ (**8**, 324–325) to α,β-unsaturated carbonyl compounds[4] proceeds with high chiral induction to *trans*-(−)-8-phenylmenthyl crotonate (**2**). Addition to the isomeric *cis*-crotonate (**5**) is less selective, but results mainly in the enantiomeric acid (**6**).

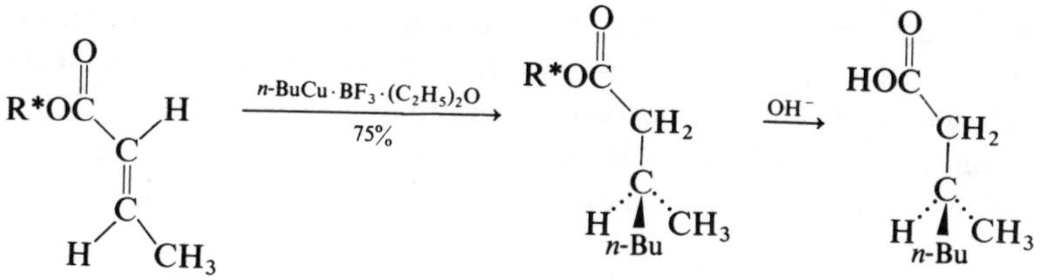

$$\mathrm{R^{*}OC(=O)CH{=}CHCH_3}\ \xrightarrow[75\%]{n\text{-BuCu}\cdot\mathrm{BF_3}\cdot(\mathrm{C_2H_5})_2\mathrm{O}}\ \mathrm{R^{*}OC(=O)CH_2C(H)(CH_3)(}n\text{-Bu)}\ \xrightarrow{\mathrm{OH^-}}\ \mathrm{HOC(=O)CH_2C(H)(CH_3)(}n\text{-Bu)}$$

2 [R* = (−)-**1**] **4** (99% ee)

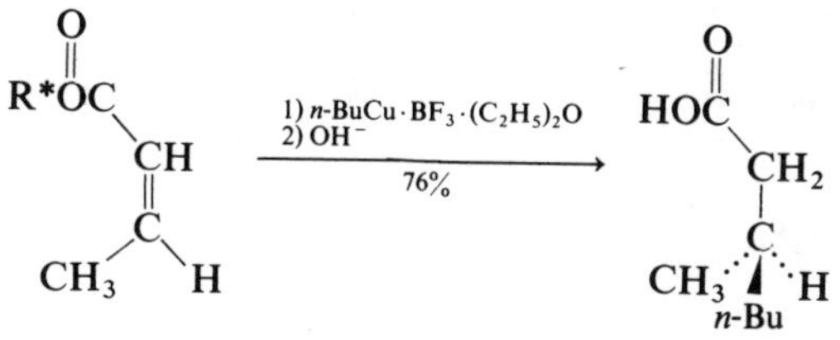

$$\mathrm{R^{*}OC(=O)CH{=}CHCH_3}\ \xrightarrow[76\%]{1)\ n\text{-BuCu}\cdot\mathrm{BF_3}\cdot(\mathrm{C_2H_5})_2\mathrm{O};\ 2)\ \mathrm{OH^-}}\ \mathrm{HOC(=O)CH_2C(CH_3)(H)(}n\text{-Bu)}$$

5 **6** (70% ee)

Asymmetric ene reactions.[5] A key step in an asymmetric synthesis of the amino diacid (+)-α-allokainic acid (**5**) involves an intramolecular ene reaction of the (Z)-diene **2**, in which the enophile is a (Z)-(−)-8-phenylmenthyl acrylate unit. In the presence of $(C_2H_5)_2AlCl$, (Z)-**2** cyclizes to the two pyrrolidines **3** and **4** in the ratio of

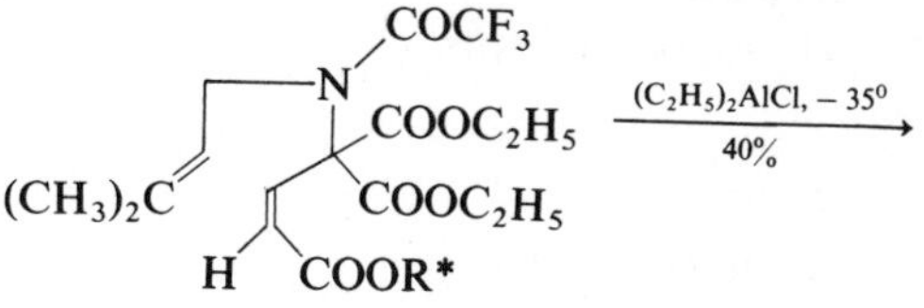

$$(Z)\text{-}\mathbf{2}\ \xrightarrow[40\%]{(\mathrm{C_2H_5})_2\mathrm{AlCl},\ -35^\circ}\ \mathbf{3}\ +\ \mathbf{4}\quad (95:5)$$

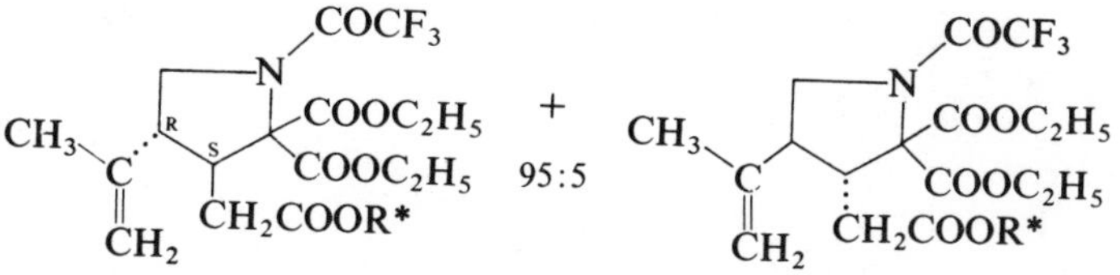

(Z)-**2**: $(CH_3)_2C{=}CHCH_2N(COCF_3)C(COOC_2H_5)_2CH{=}CHCOOR^*$

3, **4**: N-COCF₃ pyrrolidines bearing $CH_3C({=}CH_2)$, $(COOC_2H_5)_2$ and CH_2COOR^* substituents (R, S in **3**)

$$\mathbf{3}\ \xrightarrow[73\%]{1)\ \mathrm{NaOH}\ (-\mathrm{CO_2})}\ \mathbf{5}$$

5: NH pyrrolidine bearing $CH_3C({=}CH_2)$, COOH and CH_2COOH

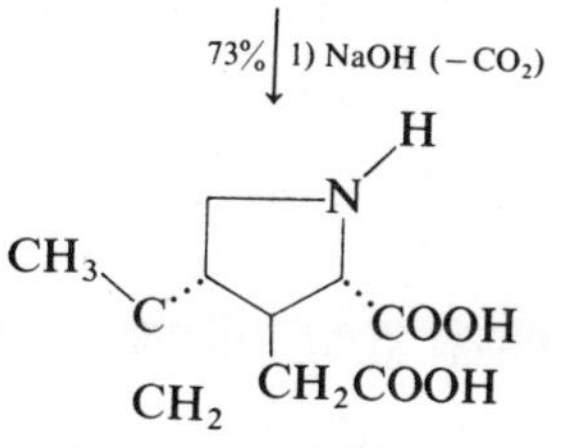

95:5. The same cyclization of (E)-**2** also results in **3** and **4**, but in the ratio 15:85. Saponification of **3** followed by acid-catalyzed decarboxylation provides the natural product **5**. Very slight optical induction is observed when R = menthyl.

Asymmetric reactions of a chiral glyoxylate ester. Grignard reagents add to the glyoxylate ester (**2**) of (−)-**1** with unusually high asymmetric induction. The chirality

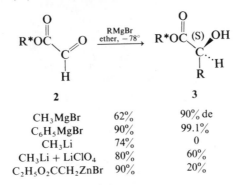

$$\text{R*OC(=O)CH=O}\ (\mathbf{2}) \xrightarrow[\text{ether, } -78^\circ]{\text{RMgBr}} \text{R*OC(=O)}\overset{(S)}{\text{C}}\text{H(OH)R}\ (\mathbf{3})$$

		3
CH_3MgBr	62%	90% de
C_6H_5MgBr	90%	99.1%
CH_3Li	74%	0
$CH_3Li + LiClO_4$	80%	60%
$C_2H_5O_2CCH_2ZnBr$	90%	20%

(S) is that expected from addition to the front face of **2** with a *syn*-orientation of the carbonyl groups. The asymmetric induction is low with lithium and zinc reagents.[6]

The glyoxylate **2** also undergoes highly diastereoselective Lewis acid-catalyzed ene reactions with 1-hexene and *trans*-2-butene (equations II and III). Again, the major product has the (S)-configuration at C_2. The degree and sense of induction at C_3 in the ene reaction with 2-butene is not certain.[7]

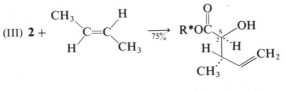

$$\text{(II)}\ \mathbf{2} + C_4H_9CH{=}CH_2 \xrightarrow[92\%]{SnCl_4} \text{R*OC(=O)}\overset{S}{\text{C}}\text{H(OH)CH}_2\text{CH}{=}\text{CH(CH}_2)_2\text{CH}_3$$

4 (97.6% de)

$$\text{(III)}\ \mathbf{2} + \text{CH}_3\text{CH}{=}\text{CHCH}_3\ (trans) \xrightarrow{75\%} \text{R*OC(=O)}\overset{S}{\text{C}}\text{H(OH)CH(CH}_3)\text{CH}{=}\text{CH}_2$$

5 (93% de at C_2)

[1] E. J. Corey and H. E. Ensley, *Am. Soc.*, **97**, 6908 (1975).

[2] W. Oppolzer, M. Kurth, D. Reichlin, C. Chapuis, M. Mohnhaupt, and F. Moffatt, *Helv.*, **64**, 2802 (1981).

[3] W. Oppolzer, M. Kurth, D. Reichlin, C. Chapius, M. Mohnhaupt, and F. Moffatt, *ibid.*, **64**, 2802 (1981).

[4] Y. Yamato and K. Maruyama, *Am. Soc.*, **100**, 3240 (1978).
[5] W. Oppolzer, C. Robbiani, and K. Battig, *Helv.*, **63**, 2015 (1980).
[6] J. K. Whitesell, A. Bhattacharya, and K. Henke, *J.C.S. Chem. Comm.*, 988 (1982).
[7] J. K. Whitesell, A. Bhattacharya, D. A. Aguilar, and K. Henke, *ibid.*, 989 (1982).

Phenylmercuric hydroxide, C_6H_5HgOH. Mol. wt. 294.70, m.p. 190° (dec.), severe poison. Supplier: Aldrich.

Hydration of terminal alkynes.[1] The reagent does not react with internal alkynes, but it is useful for conversion of terminal, nonconjugated alkynes to methyl ketones in 50–65% yield.[1]

Examples:

$$CH_3(CH_2)_4C{\equiv}CH \xrightarrow[CHCl_3, 60°]{C_6H_5HgOH,} \left[CH_3(CH_2)_4C{\equiv}CHgC_6H_5 \right] \xrightarrow[56\%]{H_2O, 60°} CH_3(CH_2)_4COCH_3$$

OH / C≡CH (cyclohexane) —65%→ OH / COCH₃ (cyclohexane)

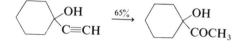

[1] V. Janout and S. L. Regen, *J. Org.*, **47**, 3331 (1982).

Phenylmercuric perchlorate, $C_6H_5HgClO_4$(**1**). The reagent is toxic but not explosive. For the reaction described below it should be freshly prepared *in situ* by reaction of C_6H_5HgCl and $AgClO_4$ in THF at 25° (15 minutes).

Cyclization of piperazinediones.[1] An efficient synthesis of the unusual ring system (**3**) of the antibiotic bicyclomycin is based on treatment of a protected piperazinedione (**2**) with 2.0 equiv. of **1** at 25°. The reagent effects deprotection of the silyl group and removal of the pyridylthio residue prior to cyclization to give **3** in 90–99% yield in 2–3 minutes. The same overall transformation can be effected by treatment first with the HF·pyridine complex and then with $AgClO_4$ in yields of 60–

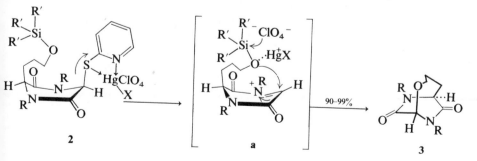

R′ R′ R′ Si O O H R N S N HgClO₄ X H RN O

2

R′ R′ ClO₄⁻ R′ Si O HgX O H R N H N R O

a

90–99%

R O N H O O H N R

3

93%. Or the cyclization can be induced with camphorsulfonic acid at 80° (48–77% yield).

[1] R. M. Williams, O. P. Anderson, R. W. Armstrong, J. Josey, H. Meyers, and C. Eriksson, *Am. Soc.*, **104**, 6092 (1982).

Phenyl N-phenylphosphoramidochloridate, $C_6H_5OP(=O)(NHC_6H_5)Cl$ **(1).**

Mol. wt. 298.64, m.p. 129–130°. The reagent is prepared by reaction of phenyl phosphorodichloridate with aniline in benzene.

Amides.[1] Amides can be prepared in one step from a carboxylic acid and an amine by reaction with **1** (1 equiv.) and triethylamine (2 equiv.) in CH_2Cl_2 at 25°. Presumably a mixed anhydride is first formed, which reacts subsequently with the amine. A variation involves conversion of the carboxylic acid into the carboxylic anhydride by reaction with **1** (0.5 equiv.) and triethylamine; subsequent addition of the amine results in the formation of the amide. Yields are generally greater than 90%.

[1] R. Mestres and C. Palomo, *Synthesis*, 218 (1981); *idem, ibid.*, 288 (1982).

Phenyl selenocyanate, C_6H_5SeCN, **8**, 119–120; **9**, 34. A simple preparation involves the reaction of C_6H_5SeCl with $(CH_3)_3SiCN$ in THF or CH_2Cl_2 (96% yield).

Alkynyl selenides.[1] Terminal alkynes react with C_6H_5SeCN in CH_2Cl_2 in the presence of CuCN or CuBr and triethylamine to give alkynyl selenides in yields usually >90%. The synthesis is compatible with hydroxyl and ethoxyl carbonyl groups.

Cyanoselenenylation.[2] In the presence of a Lewis acid, particularly $SnCl_4$, phenyl selenocyanate adds to alkenes to form *trans*-adducts in high yield. Two regioisomers are formed from unsymmetrical alkenes.

Examples:

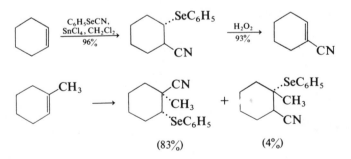

Cyclohexene $\xrightarrow[96\%]{C_6H_5SeCN,\ SnCl_4,\ CH_2Cl_2}$ trans-1-SeC_6H_5-2-CN-cyclohexane $\xrightarrow[93\%]{H_2O_2}$ 1-cyanocyclohexene (CN)

1-Methylcyclohexene (CH_3) $\longrightarrow$ 1-CN-1-CH_3-2-SeC_6H_5-cyclohexane (83%) + 1-SeC_6H_5-1-CH_3-2-CN-cyclohexane (4%)

α-Keto nitriles.[3] The reagent adds to ketene ketals regioselectively to give ketals of α-keto nitriles, which on oxidation yield the corresponding β,γ-unsaturated nitriles.

Example:

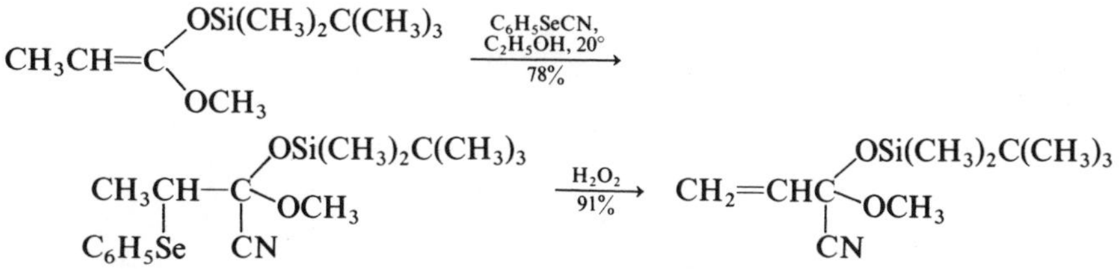

$$CH_3CH{=}C(OSi(CH_3)_2C(CH_3)_3)(OCH_3) \xrightarrow[78\%]{C_6H_5SeCN,\ C_2H_5OH,\ 20^\circ}$$

$$CH_3CH(SeC_6H_5){-}C(CN)(OSi(CH_3)_2C(CH_3)_3)(OCH_3) \xrightarrow[91\%]{H_2O_2} CH_2{=}CHC(CN)(OSi(CH_3)_2C(CH_3)_3)(OCH_3)$$

[1] S. Tomoda, Y. Takeuchi, and Y. Nomura, *Chem. Letters*, 1069 (1981).
[2] *Idem, J.C.S. Chem. Comm.*, 871 (1982).
[3] *Idem, Chem. Letters*, 1733 (1982).

N-Phenylselenophthalimide (1), 9, 366–367; **10,** 310–312.

(cis-6-*Methyltetrahydropyran-2-yl)acetic acid* (2).[1] A short synthesis of this natural product is outlined in equation (I).

(I) $H_2C{=}CHCH_2CH_2C(=O)CH_2COOCH_2C_6H_5 \xrightarrow[57\%]{1,\ SnCl_4}$ $C_6H_5SeCH_2$-dihydropyran-CH_2CO_2Bzl + (4:1) $C_6H_5SeCH_2$-tetrahydropyranylidene-$CHCO_2Bzl$ $\xrightarrow[47\%]{1)\ Bu_3SnH;\ 2)\ H_2,\ Raney\ Ni}$ CH_3-tetrahydropyran-CH_2COOH (**2**)

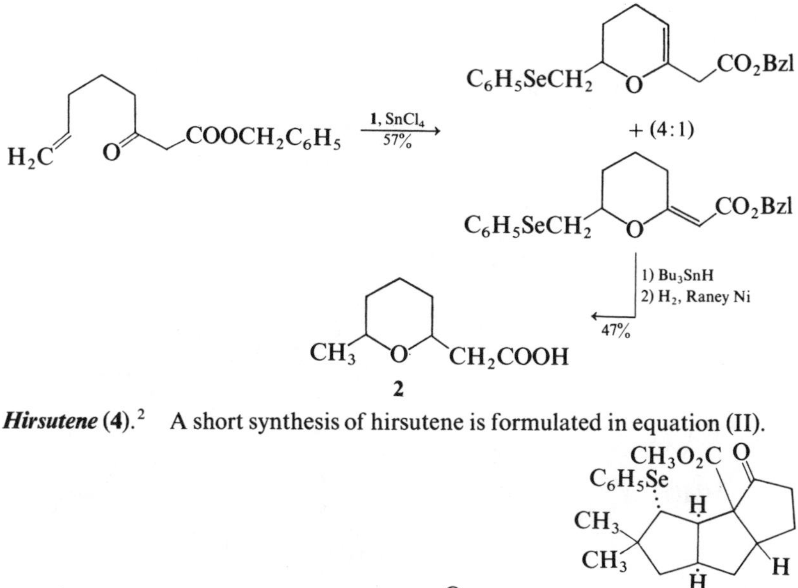

***Hirsutene* (4).**[2] A short synthesis of hirsutene is formulated in equation (II).

(II) CH_3OOC, CH_3, CH_3 (cyclopentenyl-methyl-cyclopentanone) $\xrightarrow{1,\ SnCl_4}$

CH_3O_2C, O, C_6H_5Se, H, CH_3, CH_3, H, H

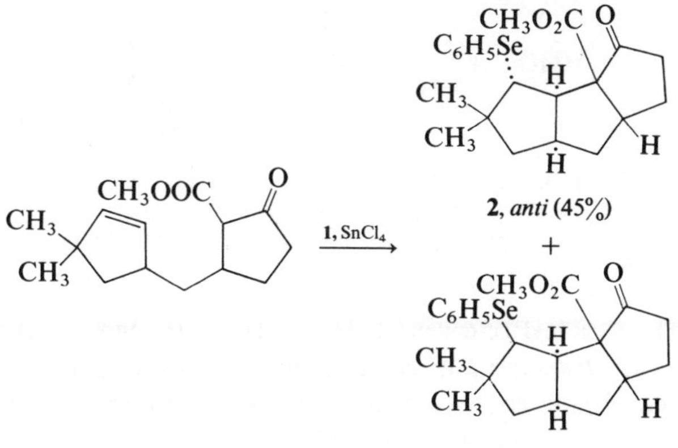

2, *anti* (45%)

+

CH_3O_2C, O, C_6H_5Se, H, CH_3, CH_3, H, H

3, *syn* (45%)

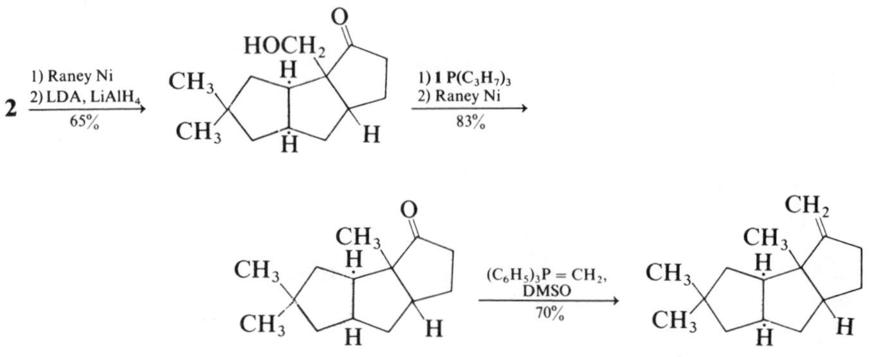

2 —1) Raney Ni 2) LDA, $LiAlH_4$, 65%→ [HOCH₂-ketone] —1) 1 $P(C_3H_7)_3$ 2) Raney Ni, 83%→ [ketone] —$(C_6H_5)_3P=CH_2$, DMSO, 70%→ (4)

Spiro ketals. Spiro ketals can be prepared using organoselenium mediated cyclizations.[3]

Examples:

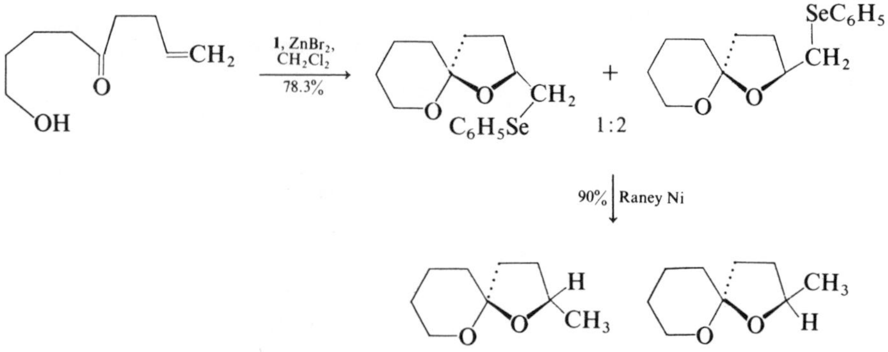

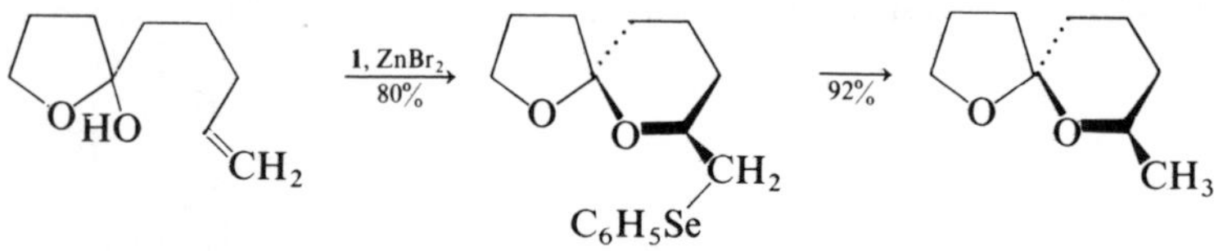

—1, $ZnBr_2$, CH_2Cl_2, 78.3%→ 1:2 —90% Raney Ni→

—1, $ZnBr_2$, 80%→ —92%→

[1] S. V. Ley, B. Lygo, H. Molines, and J. A. Morton, *J.C.S. Chem. Comm.*, 1251 (1982).
[2] S. V. Ley and P. J. Murray, *ibid.*, 1252 (1982).
[3] S. V. Ley and B. Lygo, *Tetrahedron Letters*, **23**, 4625 (1982).

Phenylsulfinylacetonitrile, $C_6H_5SOCH_2CN$ (**1**). Mol. wt. 165.21, m.p. 63–64°.

γ-Hydroxy-α,β-unsaturated nitriles. This reagent reacts with an aldehyde or ketone in the presence of piperidine to give γ-hydroxy-α,β-unsaturated nitriles.[1] These products are converted to furanes by reduction with DIBAH followed by acid hydrolysis.[2] An example is the synthesis of the sesquiterpene dendrolasin (**2**) from geranylacetone.

$$(CH_3)_2C{=}CH(CH_2)_2C(CH_3){=}CH(CH_2)_2COCH_3 \xrightarrow[85\%]{\mathbf{1},\ \text{base}}$$

$$(CH_3)_2C{=}CH(CH_2)_2C(CH_3){=}CH(CH_2)_2C(CH_2OH){=}CH(CN) \xrightarrow[75\%]{\text{DIBAH}}$$

$$(CH_3)_2C{=}CH(CH_2)_2C(CH_3){=}CH(CH_2)_2\text{-(3-furyl)}$$

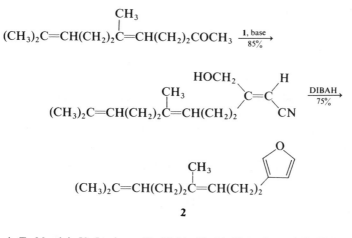

2

[1] J. Nokami, T. Mandai, Y. Imakura, K. Nishiuchi, M. Kawada, and S. Wakabayashi, *Tetrahedron Letters*, **22**, 4489 (1981).

[2] T. Mandai, S. Hashio, J. Goto, and M. Kawada, *ibid.*, **22**, 2187 (1981).

2-Phenylsulfonylethylchloroformate, $C_6H_5SO_2CH_2CH_2OCOCl$ (**1**). The chloroformate is prepared by reaction of 2-phenylsulfonylethanol with phosgene in toluene solution. The reagent is unstable on distillation *in vacuo*.

Protection of alcohols.[1] Alcohols can be converted into the 2-phenylsulfonylethylcarbonyl derivatives by reaction with **1** in pyridine at 20° (75–100% yield). The protecting group can be removed by treatment with $N(C_2H_5)_3$ in pyridine at 20°, conditions that do not affect benzyl, levulinyl, and several base-labile protecting groups. The PSEC group is stable to 80% HOAc at 20°.

[1] N. Balgobin, S. Josephson, and J. B. Chattopadhyaya, *Tetrahedron Letters*, **22**, 3667 (1981).

(Phenylsulfonyl)nitromethane, $C_6H_5SO_2CH_2NO_2$ (**1**). Mol. wt. 201.21, m.p. 78–79°. Supplier: Alfa.

α-Nitrosulfones.[1] The sodium salt of **1** undergoes C-alkylation with primary iodides and benzylic bromides or iodides (50–75% yield). Desulfonylation of the products (**2**) with sodium amalgam is not satisfactory, but can be effected in 60–70% yield with 1-benzyl-1,4-dihydronicotinamide (**6**, 36–37; **7**, 15–16) under irradiation. The overall result is equivalent to C-alkylation of nitromethane (equation I).

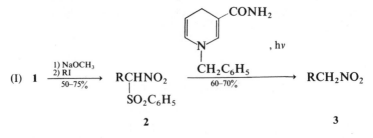

$$(I)\quad \mathbf{1} \xrightarrow[50\text{–}75\%]{1)\ NaOCH_3;\ 2)\ RI} \underset{\mathbf{2}}{RCH(SO_2C_6H_5)NO_2} \xrightarrow[60\text{–}70\%]{\text{1-benzyl-1,4-dihydronicotinamide},\ h\nu} \underset{\mathbf{3}}{RCH_2NO_2}$$

Reaction of the α-nitrosulfones (**2**) with aqueous $TiCl_3$ results unexpectedly in nitriles, RC≡N, in about 75% yield. α-Nitrosulfones are oxidized in alkaline solution by excess $KMnO_4$ to carboxylic acids, RCOOH, in 80–90% yield, probably via the acyl sulfone.

[1] P. A. Wade, H. R. Hinney, N. V. Amin, P. D. Vail, S. D. Morrow, S. A. Hardinger, and M. S. Saft, *J. Org.*, **46**, 765 (1981).

1-Phenyl-2-tetrazoline-5-thione, (**1**). Mol. wt. 178.22, m.p. 150°. The thione is prepared by reaction of phenyl isothiocyanate with sodium azide (96%).[1]

Macrolides.[2] On treatment of **1** with *t*-butyl isocyanide, either **2a** or **2b** is formed, which on treatment with a carboxylic acid forms a mixture of the interconvertible **3a** and **3b**. When the R group of **3** bears a terminal hydroxyl group, lactonization occurs (with release of **1**) very readily, even without silver ion catalysis (**6**, 246). Work-up is particularly simple if the phenyl group is substituted by an amino group, which permits recovery of the thione reagent by a wash with dilute acid.

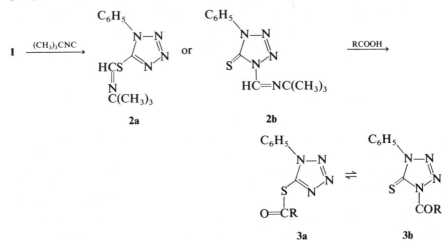

$$\mathbf{1} \xrightarrow{(CH_3)_3CNC} \underset{\mathbf{2a}}{\text{1-}C_6H_5\text{-tetrazol-5-yl–S–CH=NC(CH_3)_3}} \text{ or } \underset{\mathbf{2b}}{\text{1-}C_6H_5\text{-4-(HC=NC(CH_3)_3)-tetrazoline-5-thione}} \xrightarrow{RCOOH}$$

$$\underset{\mathbf{3a}}{\text{1-}C_6H_5\text{-tetrazol-5-yl–S–C(=O)R}} \rightleftharpoons \underset{\mathbf{3b}}{\text{1-}C_6H_5\text{-4-COR-tetrazoline-5-thione}}$$

Dilution is not required, and 16-, 17-, and 20-membered lactones are formed in > 80% yield. A 13-membered lactone is obtained in yields of 50–55% yield.

Example:

COOH, OH $\xrightarrow[80\%]{2}$ (lactone) + 1

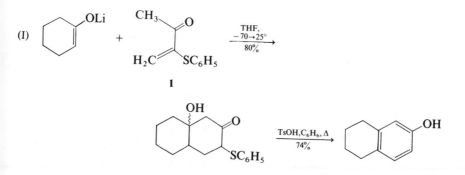

[1] E. Lieber and J. Ramachandran, *Can. J. Chem.*, **37**, 101 (1959).
[2] U. Schmidt and M. Dietsche, *Angew. Chem. Int. Ed.*, **20**, 771 (1981).

3-Phenylthio-3-butene-2-one (1). Mol. wt. 209.14, pale yellow oil.

Preparation[1]:

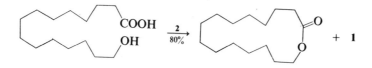

$$CH_3COCH_2COCOOC_2H_5 \xrightarrow{CH_2O} \text{(3-acetyltetrahydrofuran-2,3-dione)} \xrightarrow[90\%]{1)\,KHCO_3;\ 2)\,C_6H_5SCl} CH_3\overset{O}{\overset{\|}{C}}-\underset{CH_2}{\underset{\|}{C}}-SC_6H_5$$

1

Phenol annelation.[2] This modified methyl vinyl ketone can be used for synthesis of 5,6,7,8-tetrahydro-2-naphthol or 5-indanol by reaction with the lithium enolate of cyclohexanone or cyclopentanone, respectively. The former reaction is formulated in equation (I).

(I) (lithium cyclohexenolate, OLi) + $CH_3C(=O)C(=CH_2)SC_6H_5$ (**1**) $\xrightarrow[80\%]{THF,\ -70\rightarrow25°}$ (OH, O, SC_6H_5 bicyclic ketol) $\xrightarrow[74\%]{TsOH, C_6H_6, \Delta}$ (5,6,7,8-tetrahydro-2-naphthol, OH)

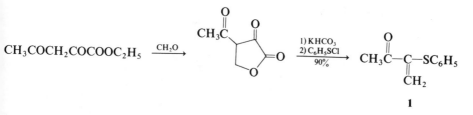

[1] G. M. Ksander, J. E. McMurry, and M. Johnson, *J. Org.*, **42**, 1180 (1977).
[2] K. Takaki, M. Okada, M. Yamada, and K. Negoro, *J.C.S. Chem. Comm.*, 1183 (1980).

2-Phenylthiocyclobutanone (1), Mol. wt. 178.045, yellow oil.

An earlier preparation of **1** (**8**, 311–312) has been improved (equation I).

(I) $(C_6H_5S)_2CH\,CH(SC_6H_5)_2$ (cyclobutane) $\xrightarrow[72\%]{sec\text{-BuLi, TMEDA}}$ 1,2-bis(phenylthio)cyclobutene (C_6H_5S, SC_6H_5) $\xrightarrow[85\%]{TiCl_4, CuCl_2}$ **1** (C_6H_5S, O)

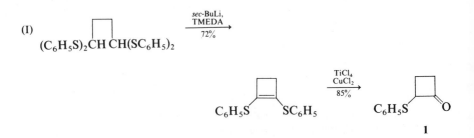

Cyclobutanone equivalent.[1] Use of lithium 1-dimethylaminonaphthalenide (**10**, 244) or Raney nickel for desulfuration permits use of **1** as a surrogate for cyclobutanone, which can not be monoalkylated satisfactorily. Some synthetic uses are formulated in equations (II) and (III).

(II) **1** $\xrightarrow[65\%]{1)\ KOC(CH_3)_3;\ 2)\ n\text{-BuI}}$ 2-*n*-butyl-2-phenylthiocyclobutanone (Bu-*n*, SC_6H_5) $\xrightarrow[89\%]{LDMAN}$ 2-*n*-butylcyclobutanone (Bu-*n*)

(III) **1** $\xrightarrow{C_6H_{11}MgBr,\ ether}$ 1-cyclohexyl-2-phenylthiocyclobutanol (OH, SC_6H_5) $\xrightarrow[87\%]{Raney\ Ni}$ 1-cyclohexylcyclobutanol (OH)

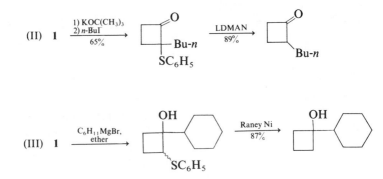

[1] T. Cohen, D. Ouellete, K. Pushpananda, A. Senaratne, and L. -C. Yu, *Tetrahedron Letters*, **22**, 3377 (1981).

Phenylthiomethyllithium. $C_6H_5SCH_2Li$ is prepared by reaction of *n*-BuLi with thioanisole in THF containing DABCO.[1]

Butenolide annelation.[2] A new method was used to add the butenolide ring of the novel sesquiterpene colorata-4(13), 8-dienolide (**5**) to the ketone **1**. Reaction of $C_6H_5SCH_2Li$ with **2** followed by hydrolysis gives **3** in 65% yield, but the hydrolysis at 20° requires about 4 weeks to prevent isomerization of the double bond. Oxidation of **3** followed by dehydration gives the thiophenylfurane **4**. A similar slow hydrolysis (1 week) of **4** gives the desired **5**.

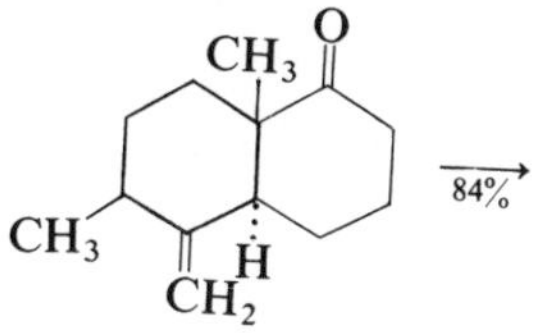

1 $\xrightarrow{84\%}$

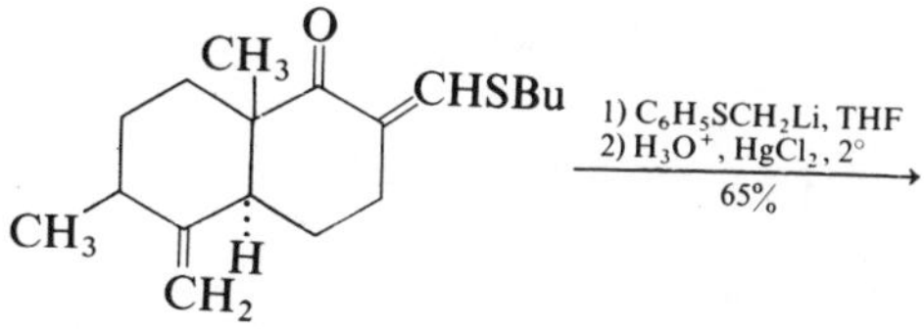

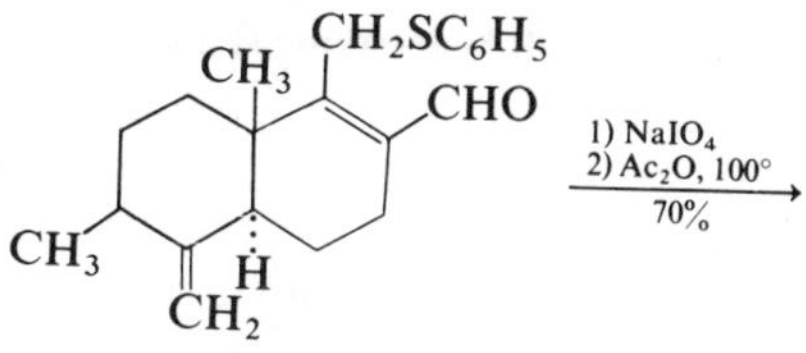

2 $\xrightarrow[65\%]{1)\ C_6H_5SCH_2Li,\ THF;\ 2)\ H_3O^+,\ HgCl_2,\ 2^\circ}$ 3 $\xrightarrow[70\%]{1)\ NaIO_4;\ 2)\ Ac_2O,\ 100^\circ}$

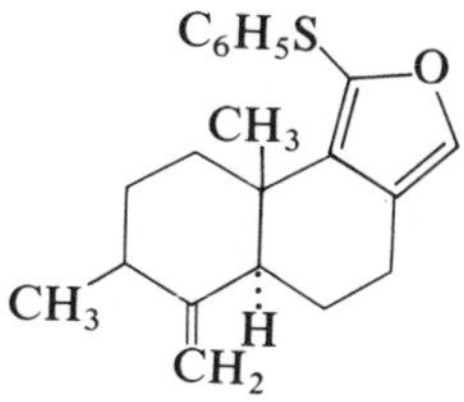

4 $\xrightarrow[60\%]{HgCl_2,\ H_3O^+}$ 5

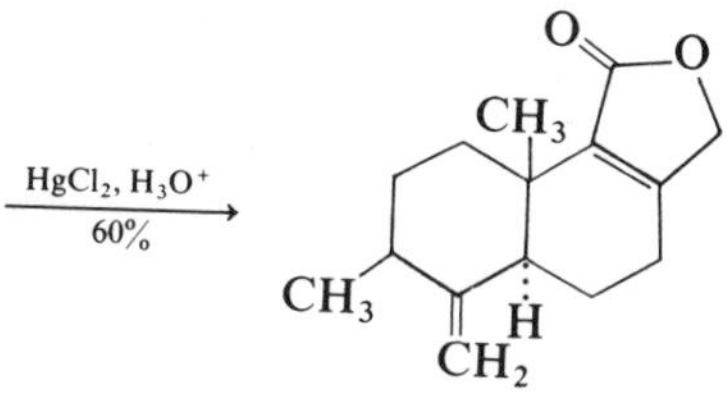

[1] E. J. Corey and D. Seebach, *J. Org.*, **31**, 4097 (1966).
[2] A. de Groot, M. P. Broekhuysen, L. L. Doddema, M. C. Vollering, and J. M. M. Westerbeek, *Tetrahedron Letters*, **23**, 4831 (1982).

Phenylthiomethyl(trimethyl)silane (1). 10, 313–314.

Reaction with various electrophiles.[1] The anion (**2**, *n*-BuLi) of **1** in the presence of TMEDA forms adducts with esters, which on chromatography produce α-phenylthio ketones (equation I). Other useful reactions are formulated in equations (II) and (III).

$$\text{(I)}\quad C_6H_5S\overset{Li}{C}HSi(CH_3)_3\ (\mathbf{2}) \xrightarrow{RCO_2R^1,\ TMEDA} \left[\begin{matrix} C_6H_5S \\ (CH_3)_3Si \end{matrix} CH-\overset{O}{\overset{\|}{C}}R \right] \xrightarrow[50\text{–}80\%]{SiO_2} R\overset{O}{\overset{\|}{C}}CH_2SC_6H_5$$

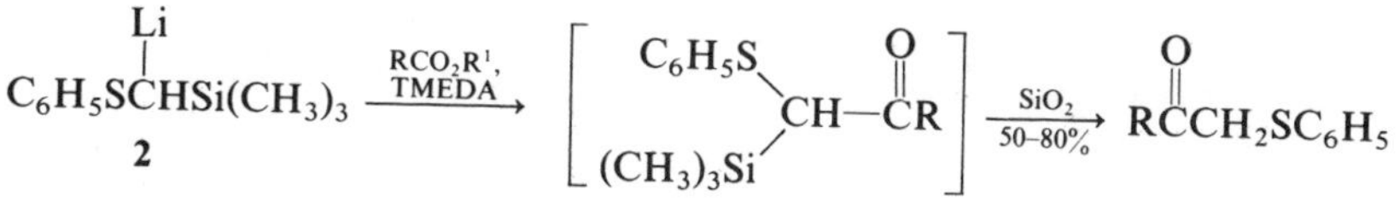

$$\text{(II)}\quad \begin{matrix} R^1 \\ R^2 \end{matrix} C=O \xrightarrow[65\text{–}80\%]{\mathbf{2}} \begin{matrix} R^1 \\ R^2 \end{matrix} C=CHSC_6H_5$$

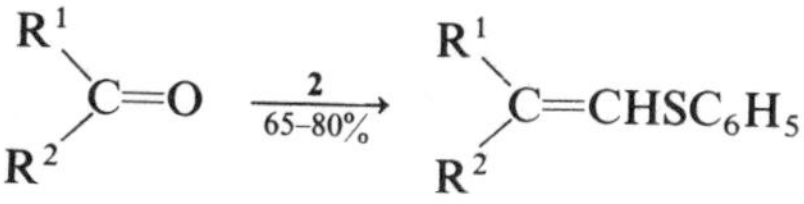

(III) 2-Cyclohexenone $\xrightarrow[\sim 100\%]{2,\ DME}$ 3-[CH(SC_6H_5)($Si(CH_3)_3$)]cyclohexanone

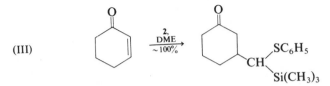

[1] D. J. Ager, *Tetrahedron Letters*, **22**, 2803 (1981).

Phenylthio(triphenylstannyl)methyllithium, $(C_6H_5)_3SnCHLiSC_6H_5$ (**1**).

Preparation:

$$(C_6H_5)_3SnCl \xrightarrow[56\%]{LiCH_2SC_6H_5} (C_6H_5)_3SnCH_2SC_6H_5 \xrightarrow[\sim 100\%]{LDA} \mathbf{1}$$

Carbonyl olefination.[1] The reaction of **1** with benzaldehyde results in a 1:1 separable mixture of the *threo*- and *erythro*-adducts (**2a** and **2b**, respectively). The adducts undergo stereospecific *syn*-elimination when heated to give β-phenylthiostyrene (**3**). The (E)-isomer (**3a**) is formed from **2a**, and the (Z)-isomer (**3b**) is formed from **2b**. On the other hand, *anti*-elimination obtains on treatment of **2** with perchloric acid in methanol. This carbonyl olefination has one advantage over the Peterson reaction in that intermediate adducts can be isolated and converted as desired to an (E)- or a (Z)-olefin.

$$C_6H_5CHO \xrightarrow[91\%]{1} (C_6H_5)_3Sn\text{–}CH(SC_6H_5)\text{–}CH(OH)C_6H_5\ (\mathbf{2a}) + (C_6H_5)_3Sn\text{–}CH(SC_6H_5)\text{–}CH(OH)C_6H_5\ (\mathbf{2b})\quad 1:1$$

$$\mathbf{2a} \xrightarrow[93\%]{110^\circ} \mathbf{3a}\ [(E)\text{-}C_6H_5SCH{=}CHC_6H_5];\qquad \mathbf{2a} \xrightarrow[100\%]{H^+} \mathbf{3b}\ [(Z)\text{-}C_6H_5SCH{=}CHC_6H_5]$$

$$\mathbf{2b} \xrightarrow[90\%]{110^\circ} \mathbf{3b};\qquad \mathbf{2b} \xrightarrow[95\%]{H^+} \mathbf{3a}$$

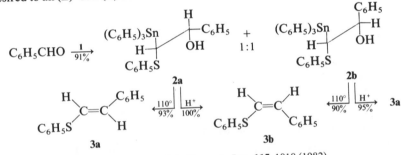

[1] T. Kauffmann, R. Kriegesmann, and A. Hamsan, *Ber.*, **115**, 1818 (1982).

1-(Phenylthio)vinyldiphenylphosphine oxide (**1**). Mol. wt. 336.39, oil.

Preparation:

$$(C_6H_5)_2P(O)CH_2SC_6H_5 \xrightarrow[60\%]{1)\ n\text{-BuLi, TMEDA};\ 2)\ CH_3I} (C_6H_5)_2P(O)CH(CH_3)SC_6H_5 \xrightarrow{NCS}$$

$$\left[(C_6H_5)_2P(O)\text{–}C(Cl)(CH_3)\text{–}SC_6H_5\right] \xrightarrow[62\text{–}66\%]{ZnBr_2\ \text{or}\ SiO_2} (C_6H_5)_2P(O)\text{–}C(SC_6H_5){=}CH_2\ (\mathbf{1})$$

Ketone synthesis.[1] Addition of an alkyl- or aryllithium to **1** forms an anion (**a**) that undergoes a Wittig-Horner reaction with an aldehyde to form a vinyl sulfide (**2**),

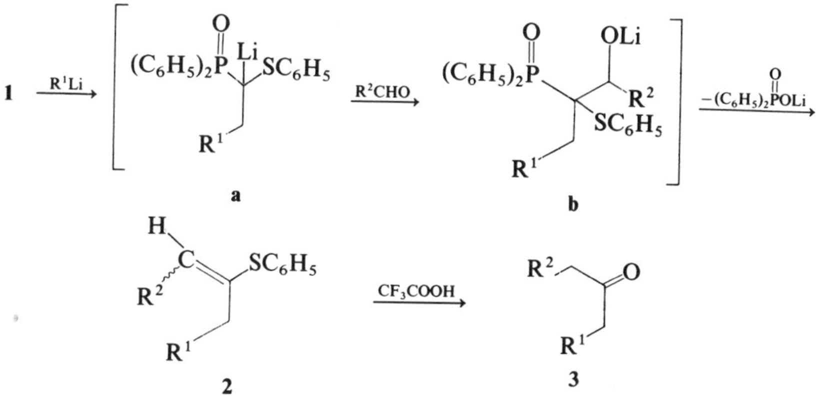

which is converted by acid hydrolysis to a ketone (**3**). Overall yields are moderate to high.

[1] S. Warren and A. T. Zaslona, *Tetrahedron Letters*, **23**, 4167 (1982).

1-Phenylthiovinyl(triphenyl)phosphonium iodide, 9, 502.

Cyclopentanones. Hewson *et al.*[1] have used the related reagent, 1-methylthiovinyl(triphenyl)phosphonium chloride (**1**) for a synthesis of prostaglandin D_1 methyl ester (**8**). Thus reaction of the diketodithiane **2** with **1** in the presence of NaH gives **3**, which is readily converted into the enone **4**. Addition of the cuprate reagent **5** to **4** shows unexpected selectivity in favor of the natural *trans*-arrangement of the side chains, perhaps because of the spiro dithiane unit.

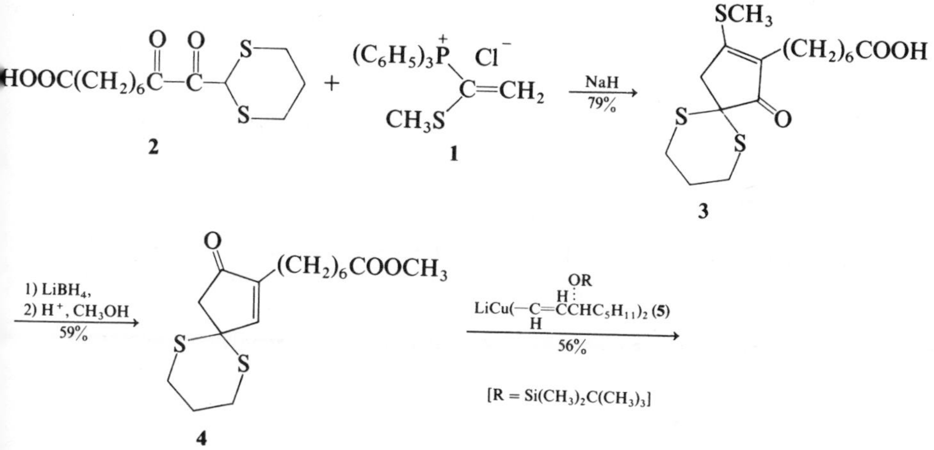

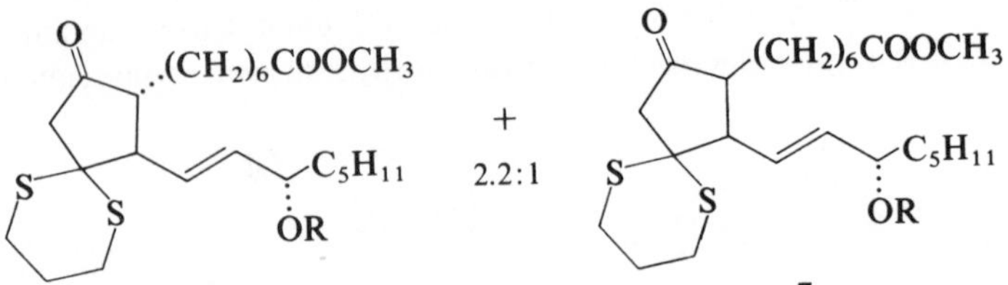

6 + 2.2:1 7

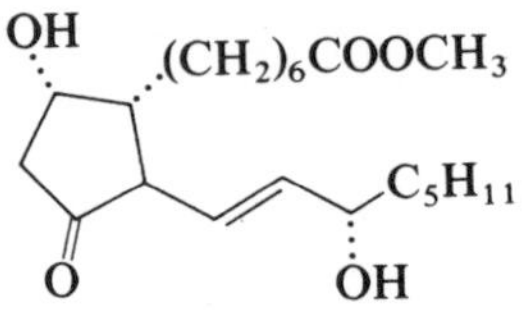

8

[1] A. G. Cameron, A. T. Hewson, and A. H. Wadsworth, *Tetrahedron Letters*, **23**, 561 (1982).

Phenyltrimethylammonium perbromide, 1, 855; **2,** 328; **4,** 386–387; **5,** 531–532.

α-Bromoketals.[1] These useful precursors to α,β-enones can be prepared by reaction of a ketone with this reagent and ethylene glycol in THF at 28°. The reagent preferentially attacks the more substituted α-position.

Example:

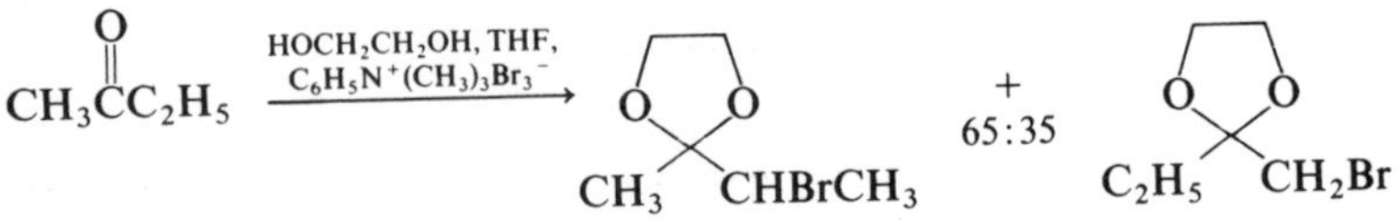

$CH_3COC_2H_5 \xrightarrow[C_6H_5N^+(CH_3)_3Br_3^-]{HOCH_2CH_2OH,\ THF,}$ (ketal: CH_3, $CHBrCH_3$) + (ketal: C_2H_5, CH_2Br) 65:35

[1] S. Visweswariah, G. Prakash, V. Bhushan, and S. Chandrasekaran, *Synthesis*, 309 (1982).

Phenyl 2-(trimethylsilyl)ethynyl sulfone, $C_6H_5SO_2C{\equiv}CSi(CH_3)_3$ **(1)**. Mol. wt. 238.38, m.p. 64–65°. Prepared by reaction of bis(trimethylsilyl)acetylene with $C_6H_5SO_2Cl$–$AlCl_3$ in CH_2Cl_2 (65% yield).[1]

α-Vinylation of carbonyl compounds. The anion of carbonyl compounds (NaH, *t*-BuOK, or KF) undergoes Michael addition to **1**; vinyl sulfones (**2**) are formed on treatment of the adducts with acetic acid. Desulfuration of **2** can be effected with Al/Hg.[2]

Example:

$$C_2H_5CH(COOC_2H_5)_2 + \mathbf{1} \xrightarrow[80\%]{1)\ NaH,\ THF;\ 2)\ HOAc} C_6H_5SO_2CH{=}CHC(C_2H_5)(COOC_2H_5)_2$$

2 (Z, 100%)

$$\xrightarrow[81\%]{Al/Hg,\ H_2O,\ THF,\ 0\rightarrow 20^\circ} CH_2{=}CHC(C_2H_5)(COOC_2H_5)_2$$

[1] S. N. Bhattacharya, B. M. Josiah, and D. R. M. Walton, *Organometal. Chem. Syn.*, **1**, 145 (1971).
[2] T. Ohnuma, N. Hata, H. Fujiwara, and Y. Ban, *J. Org.*, **47**, 4713 (1982).

Phenyl vinyl sulfoxide, 8, 399–400.

The presently preferred preparation of the sulfoxide and the sulfone is shown in equation (**I**).[1]

$$\text{(I)}\quad C_6H_5SH \xrightarrow{1)\ NaOC_2H_5,\ C_2H_5OH;\ 2)\ BrCH_2CH_2Br} [C_6H_5SCH_2CH_2Br] \xrightarrow[79\text{–}88\%]{NaOC_2H_5} C_6H_5SCH{=}CH_2$$

$$C_6H_5S(O)CH{=}CH_2 \xleftarrow[90\text{–}95\%]{ClC_6H_4CO_3H,\ CH_2Cl_2} C_6H_5SCH{=}CH_2 \xrightarrow[74\text{–}78\%]{H_2O_2,\ HOAc} C_6H_5SO_2CH{=}CH_2$$

b.p. 98°/0.6 mm (m.p. 64–65°)

The intermediate bromosulfide is a potent vesicant that causes severe blistering.

[1] L. A. Paquette and R. V. C. Carr, *Org. Syn.*, submitted (1981).

Phosphorus(V) oxide–Hexamethyldisiloxane (Trimethylsilylpolyphosphate, PPSE), P_4O_{10}–$(CH_3)_3SiOSi(CH_3)_3$.

Cyclodehydration.[1] This reagent is superior to polyphosphate ester (PPE) for synthesis of benzimadazoles, indoles, and isoquinolines.

Example:

$$o\text{-}C_6H_4(NH_2)(YH) + CH_3COOH \xrightarrow[65\text{–}95\%]{PPSE,\ ClCH_2CH_2Cl} \text{2-methylbenzazole (N, Y ring)–}CH_3$$

Y = NH,O,S

Beckmann rearrangement.[2] PPSE is comparable to PPE for Beckmann rearrangement of oximes to amides (equation II).

$$\text{(II)} \quad R^1R^2C{=}NOH \xrightarrow[\text{50–95\%}]{\text{PPSE, } CH_2Cl_2 \text{ or } C_6H_6} R^2CONHR$$

[1] K. Yamamoto and H. Watanabe, *Chem. Letters*, 1225 (1982).
[2] T. Imamoto, H. Yokoyama, and M. Yokoyama, *Tetrahedron Letters*, **22**, 1803 (1981).

Phosphorus(V) oxide–Methanesulfonic acid, 5, 535; **7**, 291–292.

Cyclodehydration.[1] This reagent is clearly the most satisfactory Lewis acid for cyclization of **1** to the tricyclic **2**, a model for tricyclic diterpenes. The by-product **3** is also converted to **2** by further treatment with the same reagent. Stannic chloride is almost as effective, but PPA is definitely inferior. The *trans* product, **2**, is the product of kinetic control but is also more stable thermodynamically than *cis*-**2**.

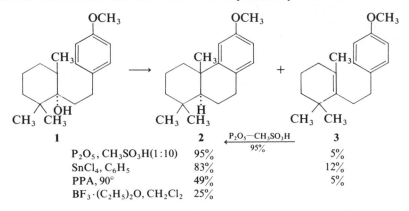

OCH₃ OCH₃ OCH₃
CH₃ OH CH₃ CH₃ → CH₃ H CH₃ CH₃ + CH₃ CH₃ CH₃

1 → **2** $\xleftarrow[95\%]{P_2O_5{-}CH_3SO_3H}$ **3**

	2	3
P_2O_5, CH_3SO_3H(1:10)	95%	5%
$SnCl_4$, C_6H_5	83%	12%
PPA, 90°	49%	5%
$BF_3\cdot(C_2H_5)_2O$, CH_2Cl_2	25%	

[1] B. W. Axon, B. R. Davis, and P. D. Woodgate, *J.C.S. Perkin I*, 2956 (1981).

Phosphorus(V) sulfide, 1, 870–871; **3**, 226–228; **4**, 389; **5**, 534–535; **8**, 401; **9**, 374; **10**, 320.

Reduction of sulfonic acids.[1] P_4S_{10} in sulfolane reduces sulfonic acids (and sulfinic acids) directly to polysulfides in good yield. The products are reduced to thiols by $LiAlH_4$ or $NaBH_4$ (equation I).

$$\text{(I)} \quad RSO_3H \xrightarrow[\text{70–90\%}]{P_4S_{10}} R(S)_nR \xrightarrow[\text{60–90\%}]{LiAlH_4} RSH$$

Thioamides.[2] The preparation of thioamides from amides and P_4S_{10} is improved by use of an ultrasonic laboratory cleaner. Lower temperatures and shorter reaction times can be used; P_4S_{10} is not required in large excess. Yields by use of ultrasound are ~ 80–95%.

[1] S. Oae and H. Togo, *Tetrahedron Letters*, **23**, 4701 (1982).
[2] S. Raucher and P. Klein, *J. Org.*, **46**, 3558 (1981).

Phosphoryl chloride, 1, 876–882; **2**, 330–331; **3**, 228; **4**, 390; **5**, 535–537; **7**, 292–293; **8**, 401–402; **9**, 374–375.

Arenesulfonyl chlorides.[1] Sodium arenesulfonates are converted into the sulfonyl chlorides by reaction with $POCl_3$ and sulfolane in acetonitrile in 85–95% yields. Yields are poor in the absence of sulfolane. The rate can be accelerated by catalytic amounts of DMF.

Benzoxazoles.[2] Benzoxazoles (**2**) are obtained in generally good yield by Beckmann rearrangement of the oxime of 2′-hydroxyacetophenones (**1**) using $POCl_3$–DMA in CH_3CN at 30° (equation I). The method is applicable to synthesis of hydroxy-substituted benzoxazoles, which are usually difficult to prepare.

$$\text{(I)}\quad \mathbf{1},\ R = \text{alkyl, OH} \xrightarrow[60\text{–}90\%]{POCl_3,\ CH_3CON(CH_3)_2} \mathbf{2}$$

1, R = alkyl, OH **2**

[1] S. Fujita, *Synthesis*, 423 (1982).
[2] S. Fujita, K. Koyama, and Y. Inagaki, *ibid.*, 68 (1982).

B-3-Pinanyl-9-borabicyclo [3.3.1] nonane, 8, 403; **9**, 320–321; **10**, 320–321. The reagent is available from Aldrich as "Alpine Borane".

Asymmetric reductions of prochiral ketones.[1] Reduction with this reagent proceeds at a higher rate and with higher enantioselectivity in the absence of a solvent. Neat reagent also effects reduction of α,β-unsaturated ketones in good to high optical yields (60–90% ee). Of greater interest, even purely aliphatic ketones can be reduced stereoselectively with yields of 40–60% ee. The optical induction increases as the steric inequality of the two alkyl groups increases. The asymmetric reduction of the α-keto ester ethyl pyruvate to ethyl lactate in 76% ee is possible with this organoborane.

Asymmetric reduction of α,β-acetylenic ketones.[2] Detailed directions for reduction of 1-octyne-3-one to (R)-1-octyne-3-ol with this reagent are available. This method has been used to reduce 10 other acetylenic ketones in >70% ee.

Chiral γ-lactones (**10**, 321). Midland[3] has extended his synthesis of chiral γ-lactones to an efficient synthesis of the sex pheromone (**2**) of the Japanese beetle. The overall optical yield of (**2**) from optically pure α-pinene is 97%.

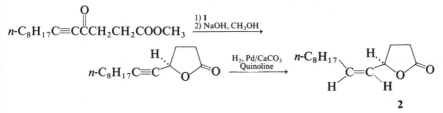

$$n\text{-}C_8H_{17}C{\equiv}C\overset{O}{\overset{\|}{C}}CH_2CH_2COOCH_3 \xrightarrow{1)\ \mathbf{1}\quad 2)\ NaOH,\ CH_3OH}$$

$$n\text{-}C_8H_{17}C{\equiv}C\text{-lactone} \xrightarrow[\text{Quinoline}]{H_2,\ Pd/CaCO_3} n\text{-}C_8H_{17}CH{=}CH\text{-lactone}\ (\mathbf{2})$$

2

[1] H. C. Brown and G. G. Pai, *J. Org.*, **47**, 1606 (1982).
[2] M. M. Midland and R. S. Graham, *Org. Syn.*, submitted (1982).
[3] M. M. Midland and N. H. Nguyen, *J. Org.*, **46**, 4107 (1981).

Platinum *t*-butyl peroxide trifluoroacetate, $\{(CF_3CO_2)_2Pt[OOC(CH_3)_3](CH_3)_3COH\}_2$ (**1**). Mol. wt. 1168.72, orange powder. This peroxidic complex is prepared by the reaction of *t*-butyl hydroperoxide with (norbornadiene)$Pt(CF_3CO_2)_2$.

***Methyl ketones from* 1-*alkenes*.**[1] This complex selectively oxidizes 1-alkenes to methyl ketones in high yield at 70°.

[1] J.-M. Bregeault and H. Mimoun, *Nouv. J. Chem.*, **5**, 287 (1981).

Polyphosphate ester (PPE), **1**, 892–894; **2**, 333–334; **3**, 229–231; **4**, 394–395; **5**, 539–540; **6**, 474; **9**, 376–377.

Cyclodehydration.[1] A key step in a recent synthesis of cleavamine (**3**) involves the cyclization of **1** to **2**, which was accomplished with PPE. Cyclization with polyphosphoric acid gave **2** in 12% yield. This cyclization with PPE is unsuccessful with substrates lacking the $\Delta^{4,5}$-double bond.

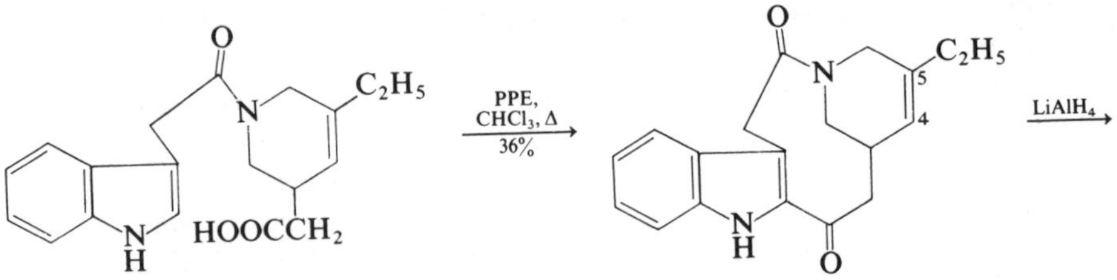

PPE, $CHCl_3$, Δ, 36%; $LiAlH_4$

1   **2**

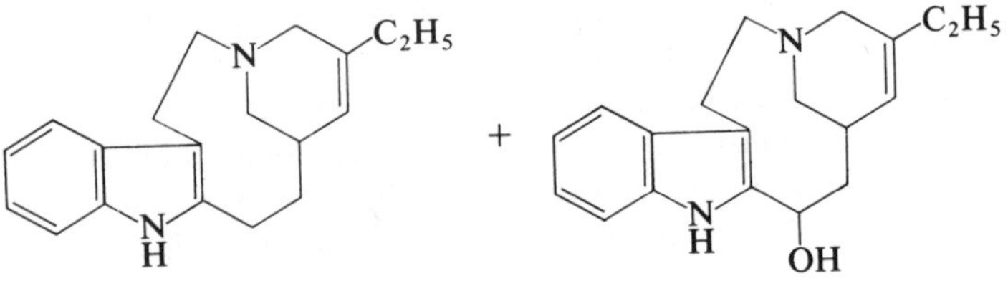

3, 19% + **4**, 24%

Thiol esters.[2] Condensation of carboxylic acids and thiols to form thiol esters can be effected with PPE in yields of 75–100%.

[1] T. Imanishi, A. Nakai, N. Yagi, and M. Hanaoka, *Chem. Pharm. Bull. Japan*, **29**, 901 (1981).
[2] T. Imamoto, M. Kodera, and M. Yokoyama, *Synthesis*, 134 (1982).

Potassium–Crown ether, 9, 387; **10**, 322.

Reductive cleavage of alkyl fluorides.[1] Potassium or Na/K and dicyclohexyl-18-crown-6 in toluene is the most effective of known systems for reductive cleavage of unactivated C—F bonds. Addition of a proton source is not necessary and actually is deleterious.

Reductive cleavage of sulfonamides.[2] Sodium naphthalenide can be used for this reaction, but it does not cleave mesyl amides of primary and dialkyl amines. A wide variety of sulfonamides is cleaved with potassium and dicyclohexyl-18-crown-6.

Examples:

(cyclohexane ring)NMs —[1) K, crown ether, diglyme; 2) HCl; 76%]→ (cyclohexane ring)NH·HCl

$(C_6H_{11})(C_{10}H_{21})NTs$ —[1) K, crown ether; 2) Ac_2O, Py; 57%]→ $(C_6H_{11})(C_{10}H_{21})NAc$

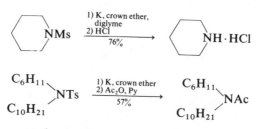

This system, as expected, reductively cleaves *p*-toluenesulfonates in good yield (76% for cholestanyl *p*-toluenesulfonate).

Sulfonamides can also be cleaved by K in diglyme without a crown ether, but with a proton source (isopropanol). They are also cleaved by Na/K alloy and isopropanol in toluene (**1**, 1102–1103), but explosions have been reported for this alloy.

[1] T. Ohsawa, T. Takagaki, A. Haneda, and T. Oishi, *Tetrahedron Letters*, **22**, 2583 (1981).
[2] T. Ohsawa, T. Takagaki, F. Ikehara, Y. Takahashi, and T. Oishi, *Chem. Pharm. Bull. Japan*, **30**, 3178 (1982).

Potassium 3-aminopropylamide (KAPA), **6**, 476; **7**, 296; **8**, 406–407; **9**, 378–379.

Isomerization of chiral propargyl alcohols.[1] The isomerization of chiral alcohols of the type $R\overset{*}{C}HOHC{\equiv}C(CH_2)_nCH_3$ to terminal acetylenic alcohols, $R\overset{*}{C}HOH(CH_2)_{n+1}C{\equiv}CH$, in the presence of KAPA occurs with no significant loss of enantiomeric purity. Evidently, formation of the alkoxide suppresses racemization. Retention of configuration is observed even when the triple bond moves through several methylene groups.

[1] M. M. Midland, R. L. Hatterman, C. A. Brown, and A. Yamaichi, *Tetrahedron Letters*, **22**, 4171 (1981).

Potassium bromate, $KBrO_3$. Mol. wt. 167.00. Supplier: Alfa.

Bromination of nitrobenzene.[1] Bromination of nitrobenzene can be effected by addition of $KBrO_3$ (slight excess) to the arene in aqueous H_2SO_4 (1:1 v/v) at a rate such that the temperature does not exceed 35°. The 3-bromonitrobenzene thus obtained (76–91% yield) is contaminated with a slight amount of dibromonitrobenzenes. The same procedure when applied to benzoic acid and phthalic acid gives the *meta*-bromo derivatives in 50–60% yield.

[1] J. J. Harrison, J. P. Pellegrini, and C. M. Selwitz, *J. Org.*, **46**, 2169 (1981); *Org. Syn.* submitted (1981).

Potassium *t*-butoxide, 1, 911–927; **2**, 338–339; **3**, 233–234; **4**, 399–405; **5**, 544–553; **6**, 477–479; **7**, 296–298; **8**, 407–408; **9**, 380–381; **10**, 323.

Dialkylaminoallenes.[1] $KOC(CH_3)_3$ complexed with $HOC(CH_3)_3$ in HMPT is the most useful base for isomerization of dialkyl-2-propynylamines (**1**) to dialkylaminoallenes (**2**).

$$HN(CH_3)_2 + HC{\equiv}CCH_2Br \xrightarrow[>85\%]{} \underset{\mathbf{1}}{HC{\equiv}CCH_2N(CH_3)_2} \xrightarrow[\text{HMPT, }23^\circ]{KOC(CH_3)_3\text{–}HOC(CH_3)_3}$$

$$\underset{\mathbf{2}}{H_2C{=}C{=}CHN(CH_3)_2} + \underset{\mathbf{3}}{CH_3C{\equiv}CN(CH_3)_2} \quad 96:4$$

Wittig reactions. Dehmlow and Barahona-Naranjo[2] have examined use of phase-transfer catalysts in addition to a base in Wittig reactions and conclude that best results are obtained with solid potassium *t*-butoxide in benzene without an extra catalyst. Solid NaOH is superior to aqueous NaOH. KF and K_2CO_3 are not generally useful.

[1] H. D. Verkruÿsse, H. T. T. Bos, L. J. de Noten, and L. Brandsma, *Rec. trav.*, **100**, 244 (1981).
[2] E. V. Dehmlow and S. Barahona-Naranjo, *J. Chem. Res.* (*M*), 1748 (1981).

Potassium *t*-butoxide–Dimethyl sulfoxide, 6, 479; **7**, 298.

Protiodesilylation of α- and β-hydroxysilanes. Both α- and β-hydroxysilanes undergo protiodesilylation when treated with potassium *t*-butoxide and DMSO containing a proton source (H_2O). Both reactions take place with retention of configuration at carbon.[1]

Examples:

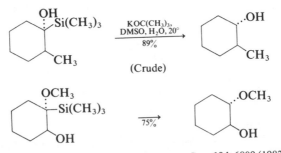

1-(trimethylsilyl)-2-methylcyclohexanol (OH, $Si(CH_3)_3$, CH_3) $\xrightarrow[89\%]{KOC(CH_3)_3,\ DMSO,\ H_2O,\ 20^\circ}$ 2-methylcyclohexanol (OH, CH_3) (Crude)

1-methoxy-1-(trimethylsilyl)-2-hydroxycyclohexane (OCH_3, $Si(CH_3)_3$, OH) $\xrightarrow[75\%]{}$ 2-methoxycyclohexanol (OCH_3, OH)

[1] P. F. Hudrlik, A. M. Hudrlik, and A. K. Kulkarni, *Am. Soc.*, **104**, 6809 (1982).

Potassium carbonate, 5, 552–553; **8,** 408; **9,** 382–383; **10,** 323.

Wittig reactions.[1] Potassium carbonate can function as a solid-liquid transfer catalyst for Wittig reactions of aliphatic or aromatic aldehydes with alkyltriphenylphosphonium bromides in dioxane containing a trace of water. K_2CO_3 is superior to NaOH because it does not catalyze aldol condensation. As expected, the (Z)/(E) ratio is ~ 70 : 30.

[1] Y. Le Bigot, M. Delmas, and A. Gaset, *Syn. Comm.*, **12**, 107 (1982).

Potassium cyanide, 5, 553; **7,** 299; **8,** 409; **10,** 324.

α,β-Unsaturated nitriles from vinyl sulfones.[1] KCN in combination with dicyclohexyl-18-crown-6 and methylene blue (radical inhibitor) reacts with vinyl sulfones in refluxing *t*-butyl alcohol to form α,β-unsaturated nitriles (equation I, II).

$$\text{(I)}\quad RCH{=}CHSO_2C_6H_5 \xrightarrow[\sim 70\%]{KCN} RC(CN){=}CH_2$$

(II) 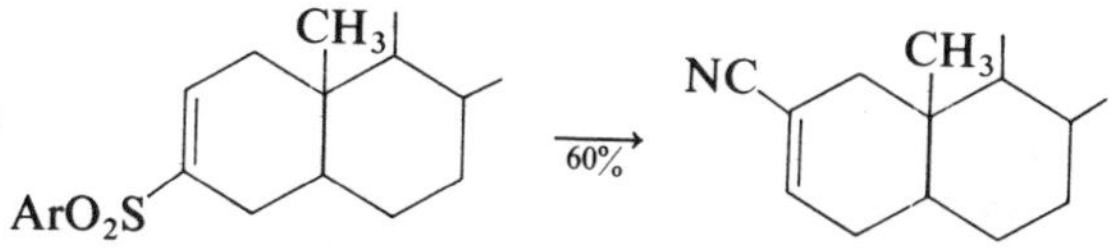 [ArO₂S-substituted methyl octalin → NC-substituted methyl octalin, 60%]

This reaction can be used to construct a trisubstituted olefin, as shown in a new synthesis of nuciferal (**1**).

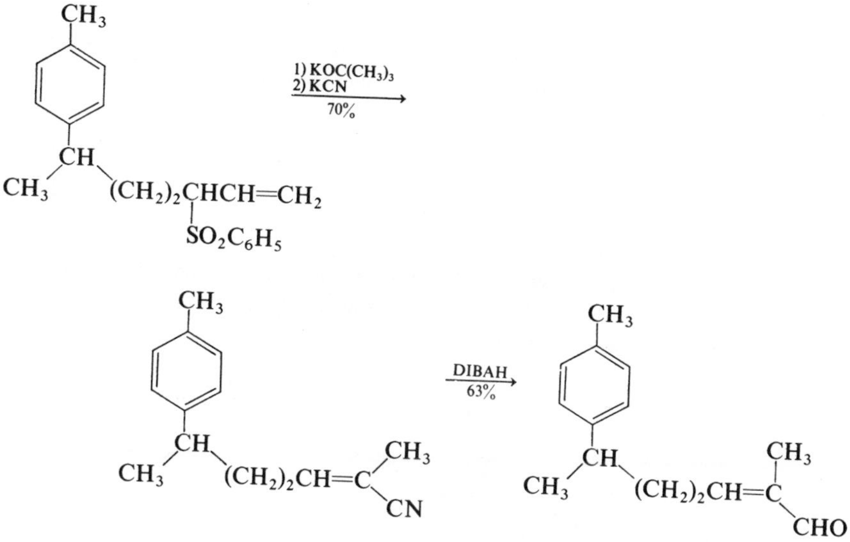

$$p\text{-}CH_3C_6H_4CH(CH_3)(CH_2)_2CH(SO_2C_6H_5)CH{=}CH_2 \xrightarrow[70\%]{1)\ KOC(CH_3)_3\quad 2)\ KCN}$$

$$p\text{-}CH_3C_6H_4CH(CH_3)(CH_2)_2CH{=}C(CH_3)CN \xrightarrow[63\%]{DIBAH} p\text{-}CH_3C_6H_4CH(CH_3)(CH_2)_2CH{=}C(CH_3)CHO$$

1

Aroyl cyanides.[2] Aroyl cyanides can be prepared by reaction of an aroyl chloride with KCN in acetonitrile. Addition of a trace of water markedly accelerates the rate and improves the yield (by as much as 100%). No other additives show this effect.

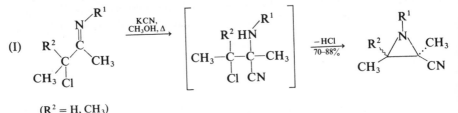

$$\text{(I)}\quad \text{ArC(=O)Cl} + \text{KCN} \xrightarrow[75\text{–}90\%]{CH_3CN,\ H_2O,\ 80^\circ} \text{ArC(=O)CN}$$

α-Cyanoaziridines.[3] α-Cyanoaziridines can be obtained by reaction of KCN with α-chloroketimines in methanol. The reaction involves addition followed by 1,3-dehydrochlorination (equation I).

$$\text{(I)}\quad R^2(CH_3)C(Cl)\text{–}C(=NR^1)CH_3 \xrightarrow{KCN,\ CH_3OH,\ \Delta} \left[CH_3\text{–}C(R^2)(Cl)\text{–}C(NHR^1)(CN)\text{–}CH_3 \right] \xrightarrow[70\text{–}88\%]{-HCl} \text{aziridine } (N\text{–}R^1;\ R^2,\ CH_3;\ CH_3,\ CN)$$

$(R^2 = H, CH_3)$

[1] D. F. Taber and S. A. Saleh, *J. Org.*, **46**, 4817 (1981).
[2] M. Tanaka and M. Koyanagi, *Synthesis*, 973 (1981).
[3] N. DeKimpe, L. Moëns, R. Verhé, L. DeBuyck, and N. Schamp, *J.C.S. Chem. Comm.*, 19 (1982).

Potassium fluoride, 1, 933–935; **2**, 346; **5**, 555–556; **6**, 481–482; **8**, 410–412; **10**, 325–326.

Alkylation of phenols and alcohols.[1] Both KF and CsF impregnated on alumina are highly effective as catalysts for alkylation of phenols and alcohols. Acetonitrile or DME are superior solvents for this reaction.

[1] T. Ando, J. Yamawaki, T. Kawate, S. Sumi, and T. Hanafusa, *Bull. Chem. Soc. Japan*, **55**, 2504 (1982).

Potassium fluoride–Oxygen.

Oxidation of cyclic β-keto esters. These substrates were oxidized by KF and O_2 in DMSO to α-hydroxy-β-keto esters in fair yields, which can be improved by addition of triethyl phosphite and 18-crown-6.[1]

Examples:

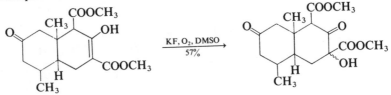

$$\text{(enol ester: } CH_3,\ COOCH_3,\ OH,\ COOCH_3,\ H,\ CH_3) \xrightarrow[57\%]{KF,\ O_2,\ DMSO} \text{(α-hydroxy-β-keto ester: } CH_3,\ COOCH_3,\ =O,\ COOCH_3,\ OH,\ H,\ CH_3)$$

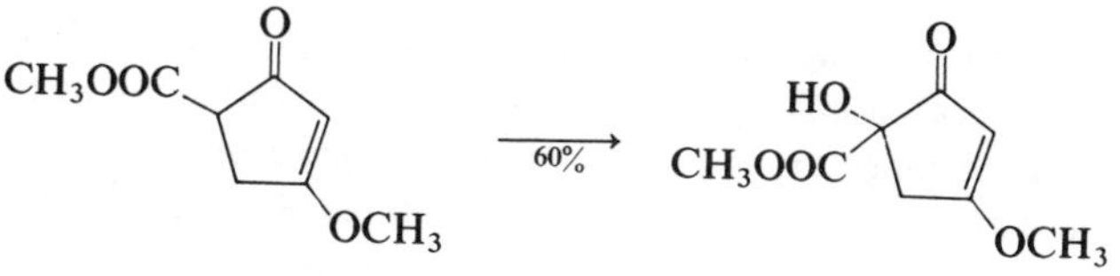

[1] H. Irie, J. Katakawa, M. Tomita, and Y. Mizuno, *Chem. Letters*, 637 (1981).

Potassium hydride, 1, 935; **2,** 346; **4,** 409; **5,** 557; **6,** 482–483; **7,** 302–303; **8,** 412–415; **9,** 386–387; **10,** 327–328.

Desilylation of α-hydroxy silanes.[1] Acylsilanes can serve as the equivalent of aldehydes with several advantages. They are less prone to self-condensation and the $Si(CH_3)_3$ group can exert useful steric effects. Thus the acylsilane **2** reacts with 3-methylpentadienyllithium (**1**) to give only the desired 1,3-diene **3**, whereas the corresponding aldehyde condenses with **1** to form mainly a 1,4-dienol. After an intramolecular Diels-Alder reaction to produce **4**, the trimethylsilyl group can be removed by a Brook rearrangement with KH in HMPT to give **5**. Alternatively, the trimethylsilyl group can be removed prior to the Diels-Alder step, in which case **7**, the epimer of **5**, is obtained as the major product.

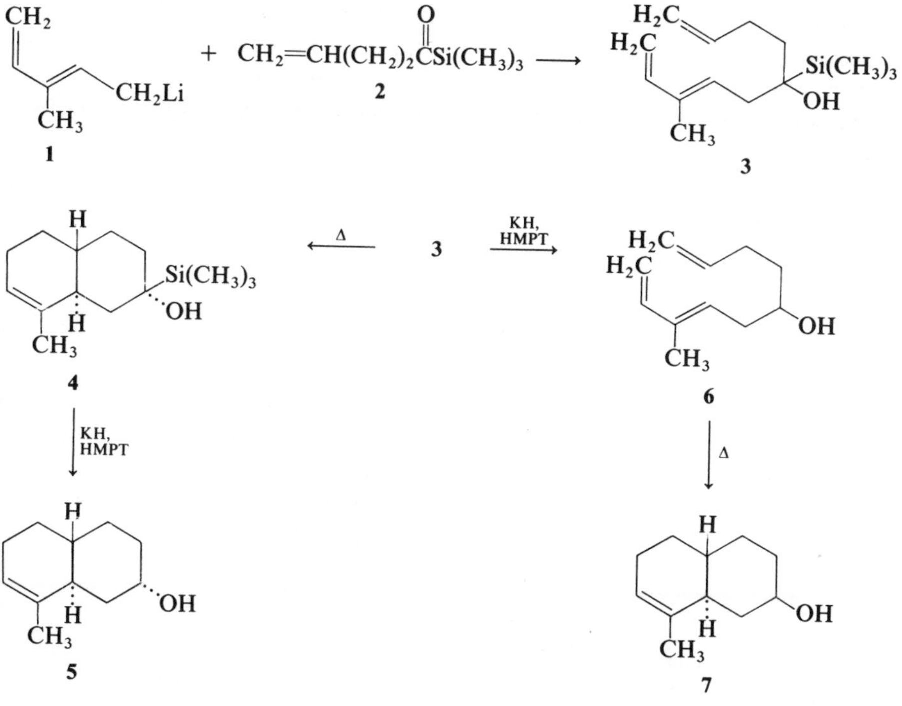

Acceleration of Claisen rearrangements.[2] The Claisen rearrangement of an allyl vinyl ether is markedly accelerated by a stabilized α-sulfonyl carbanion at the 2-position. Thus **1** and **2** rearrange to the γ,δ-unsaturated ketone **3** in the presence of potassium hydride and 18-crown-6 at moderate temperatures. Rates can be further enhanced by addition of HMPT. Substitution of methyl groups on either the allyl or vinyl units does not affect the regioselectivity but can accelerate the rate of rearrangement.

$C_6H_5SO_2$ ... H_2C^1 ... H_2C **1**; $C_6H_5SO_2$ ... H_3C ... H_2C **2** $\xrightarrow[\text{76–78\%}]{\text{KH, 18-Cr-6; THF, HMPT, 50°}}$ $C_6H_5SO_2CH_2\overset{O}{\overset{\|}{C}}CH_2CH_2CH{=}CH_2$ **3**

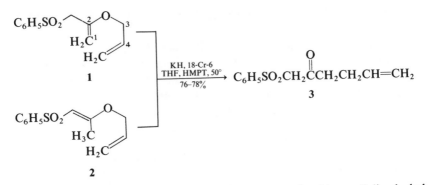

Fragmentation of homoallylic alkoxides.[3] Thermolysis of homoallylic alcohols requires temperatures >200°. However, fragmentation of bicyclo[2.2.2]-5-octene-2-alkoxides (**1**) prepared as shown in equation (I), occurs at 90–120° and results mainly in unsaturated ketones (**2**) formed by cleavage of the allylic (C_1–C_2) bond. Although the *exo*-alkoxide is cleaved noticeably faster than the *endo*-epimer, both alkoxides give the same product.

(I) bicyclo[2.2.2]octenone $\xrightarrow[\text{ether}]{\text{RMgX,}}$ alcohol (R, OH) (2 isomers) $\xrightarrow{\text{KH, HMPT}}$ [alkoxide (R, OK) **1**] $\xrightarrow[\text{50–75\%}]{\text{90–120°}}$ cyclohexenyl–$CH_2\overset{O}{\overset{\|}{C}}R$ **2**

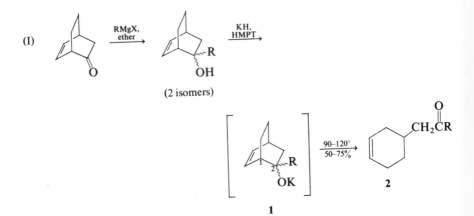

A new synthesis of propenyl ketones involves reaction of an ester with 2 equiv. of allylmagnesium chloride (or methallylmagnesium chloride) to form the tertiary bishomoallylic alcohol. The corresponding potassium alkoxide, formed with KH in HMPT or with $KOC(CH_3)_3$ in DMF, fragments at 80° with loss of propene to give a mixture of β,γ-and α,β-unsaturated ketones.[4]

Examples:

$$\begin{matrix} C_2H_5 \\ CH_3 \end{matrix}\!\!>\!CHCOOCH_3 \xrightarrow[80\%]{2\,CH_2{=}CHCH_2MgCl,\ THF}$$

$$\begin{matrix} C_2H_5 \\ CH_3 \end{matrix}\!\!>\!CHC(OH)(CH_2CH{=}CH_2)-CH_2CH{=}CH_2 \xrightarrow[-CH_3CH{=}CH_2]{1)\ KH,\ HMPT\quad 2)\ 80^\circ} \left[\begin{matrix} C_2H_5 \\ CH_3 \end{matrix}\!\!>\!CHC(OK){=}CHCH{=}CH_2\right] \xrightarrow[79\%]{H_3O^+}$$

$$\begin{matrix} C_2H_5 \\ CH_3 \end{matrix}\!\!>\!CHC(=O)CH_2CH{=}CH_2 + \begin{matrix} C_2H_5 \\ CH_3 \end{matrix}\!\!>\!CHC(=O)CH{=}CHCH_3 \qquad 4:1$$

(cyclohexene ring with CH_3, CH_3, $COOCH_3$, CH_3) $\xrightarrow{82\%}$ (cyclohexene ring with CH_3, CH_3, $C(OH)(CH_2CH{=}CH_2)_2$, CH_3) $\xrightarrow{61\%}$ (cyclohexene ring with CH_3, CH_3, $C(=O)CH{=}CHCH_3$, CH_3)

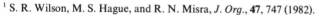

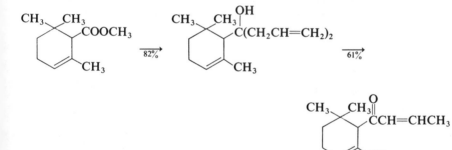

[1] S. R. Wilson, M. S. Hague, and R. N. Misra, *J. Org.*, **47**, 747 (1982).
[2] S. E. Denmark and M. A. Harmata, *Am. Soc.*, **104**, 4972 (1982).
[3] R. L. Snowden and K. H. Schulte-Elte, *Helv.*, **64**, 2193 (1981).
[4] R. L. Snowden, B. L. Muller, and K. H. Schulte-Elte, *Tetrahedron Letters*, **23**, 335 (1982).

Potassium hydride–*sec*-Butyllithium/Tetramethylethylenediamine.

α,β-Dideprotonation of γ,δ-unsaturated ketones. This combination of bases, first used to effect this reaction in the case of allylacetophenone (**8**, 7–8), has been shown to be generally useful for deprotonation of γ,δ-enones. Use of potassium enolates is essential, since lithium enolates are not deprotonated by *sec*-BuLi/TMEDA or *t*-BuLi/TMEDA. The dianions react with benzophenone exclusively at the δ-position.[1]

Examples:

$$(CH_3)_2C(OH)COCH_2CH_2CH{=}CH_2 \xrightarrow[\;]{1)\ 2KH,\ THF;\ 2)\ sec\text{-}BuLi,\ 3TMEDA} \left[(CH_3)_2C(OK){-}C(OK){=}CHCH{\cdots}\bar{C}H{\cdots}CH_2\ \ Li^+\right]$$

$$\xrightarrow[48\%]{(C_6H_5)_2C{=}O} (CH_3)_2C(OH)COCH_2CH{=}CHCH_2C(OH)(C_6H_5)_2$$

$$(CH_3)_3CCOCH_2CH_2CH{=}CH_2 \xrightarrow[80\%]{1)\ KH,\ THF,\ 20°;\ 2)\ sec\text{-}BuLi,\ TMEDA;\ 3)\ (C_6H_5)_2C{=}O} (CH_3)_3CCOCH_2CH{=}CHCH_2C(OH)(C_6H_5)_2$$

[1] D. Seebach and M. Pohnmakotr, *Tetrahedron*, **37**, 4047 (1981).

Potassium hydride–18-Crown-6, 9, 387.

Bicyclo[5.4.0.]undecanes. An attractive route to this ring system (**2**), characteristic of some sesquiterpenes, involves as the final step an oxy-Cope rearrangement of

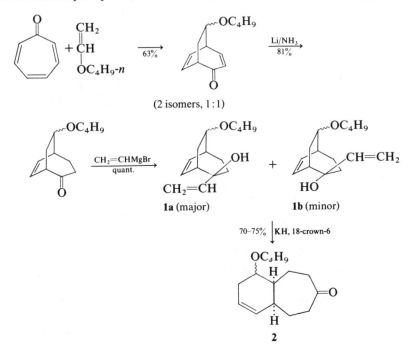

Tropone + $CH_2{=}CHOC_4H_9$-*n* $\xrightarrow{63\%}$ bicyclic dienone with OC_4H_9 (2 isomers, 1:1) $\xrightarrow[81\%]{Li/NH_3}$ bicyclic enone with OC_4H_9 $\xrightarrow[quant.]{CH_2{=}CHMgBr}$ **1a** (major) (OH, $CH_2{=}CH$) + **1b** (minor) (HO, $CH{=}CH_2$)

$\xrightarrow[70\text{–}75\%]{KH,\ 18\text{-crown-6}}$ **2** (OC_4H_9, H, H, =O)

substrates such as **1**, obtained in three steps from tropone. The rearrangement results in a *cis*-ring fusion exclusively. It is initiated by KH in refluxing THF containing 18-crown-6 (essential).[1]

[1] J. H. Rigby, J.-M. Sage, and J. Raggon, *J. Org.*, **47**, 4815 (1982).

Potassium hydroxide, 5, 557–560; **6**, 486; **7**, 303–304; **8**, 415–416; **9**, 387.

Cyclopentenones. A recent synthesis of (±)-albene (**5**), a tricyclic natural product, from the Diels-Alder adduct (**1**) of cyclopentadiene and 2,3-dimethylmaleic anhydride required transformation of a succinic anhydride group to a cyclopentenone (**4**). This was accomplished via a protected keto aldehyde **3** in four steps as formulated in 51% overall yield.[1]

1) CH_3Li (95%) 2) DHP, TsOH (78%); DIBAH 92%; KOH, CH_3OH 71% from **2**, 51% from **1**; 1) DIBAH 2) Ac_2O, NaOAc 3) $Li/C_2H_5NH_2$ 53%

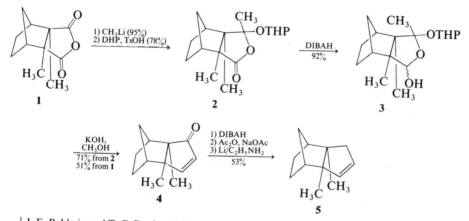

[1] J. E. Baldwin and T. C. Barden, *J. Org.*, **46**, 2442 (1981).

Potassium hydroxide–Dimethyl sulfoxide, 9, 388.

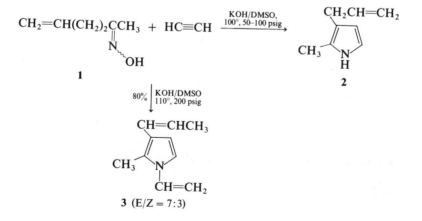

$CH_2{=}CH(CH_2)_2C({=}NOH)CH_3$ (**1**) + $HC{\equiv}CH$ $\xrightarrow{\text{KOH/DMSO, 100°, 50–100 psig}}$ 2-methyl-3-($CH_2CH{=}CH_2$)pyrrole (**2**)

1 $\xrightarrow{\text{KOH/DMSO, 110°, 200 psig; 80\%}}$ **3** (E/Z = 7:3) (1-$CH{=}CH_2$, 2-CH_3, 3-$CH{=}CHCH_3$ pyrrole)

Trofimov reaction.[1] The reaction of ketoximes with acetylene in the super base KOH/DMSO forms N-vinylpyrroles. An extension is the reaction of the unsaturated ketoxime **1** under the same conditions to form either **2** or **3** as the major product.[2]

[1] B. A. Trofimov, *Russian Chem. Rev.*, **50**, 248 (1981).
[2] A. I. Mikhaleva, M. V. Sigalov, and G. A. Kalabin, *Tetrahedron Letters*, **23**, 5063 (1982).

Potassium iodide–Dimethyl sulfoxide.

Aryl iodides.[1] DMSO is a particularly useful solvent for decomposition of aryl diazonium salts. Decomposition in the presence of KI (and I_2 to minimize formation of iodinated biphenyls) results in aryl iodides in >85% yield.

[1] A. Citterio and A. Arnoldi, *Syn. Comm.*, **11**, 639 (1981).

Potassium permanganate, 1, 942–952; **2**, 348; **4**, 412–413; **5**, 562–563; **8**, 416–417; **9**, 388–391; **10**, 330.

Oxidation of nitroalkane salts (cf., **10**, 330). Straight-chain primary nitroalkanes can be oxidized to aldehydes by prior conversion to the alkali metal salt. The reaction is even possible in the presence of a double bond.[1]

Examples:

$$CH_3(CH_2)_6CH_2NO_2 \xrightarrow[85\%]{\text{1) NaH, }(CH_3)_3COH \quad \text{2) } KMnO_4,\ B(OH)_3} CH_3(CH_2)_6CHO$$

$$CH_2{=}CH(CH_2)_8CH_2NO_2 \xrightarrow[59\%]{} CH_2{=}CH(CH_2)_8CHO$$

Oxidative expansion of a β-lactam.[2] Treatment of the penicillinate **1** with $KMnO_4$ in aqueous acetic acid (conditions usually resulting in oxidation of the —S— to —SO_2—) provides **2** in 56% yield.

$$\mathbf{1} \xrightarrow[56\%]{KMnO_4,\ CH_3COOH,\ H_2O} \mathbf{2}$$

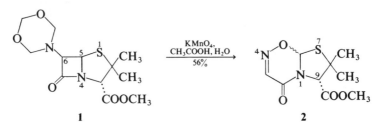

Nef transformation.[3] A versatile route to 1,4-diketones involves Michael addition of a nitroalkane to an α,β-enone, which can be effected in ~80% yield when catalyzed by fluoride ion (**9**, 446–447). The resulting γ-nitro ketone is oxidized efficiently by a stoichiometric amount of $KMnO_4$ impregnated on silica gel to a 1,4-

diketone. Overall isolated yields for this two-step reaction (equation I) are as high as 70%.

$$\text{(I)}\quad R^1\overset{O}{\overset{\|}{C}}\underset{R^2}{\underset{|}{C}}H{=}CHR^3 + RCH_2NO_2 \xrightarrow[\text{18-Crown-6}]{\text{KF,}} R^1\overset{O}{\overset{\|}{C}}\underset{R^2}{\underset{|}{C}}H{-}\overset{R^3}{\overset{|}{C}}H\underset{R}{\underset{|}{C}}HNO_2 \xrightarrow[\text{SiO}_2]{\text{KMnO}_4,} R^1\overset{O}{\overset{\|}{C}}\underset{R^2}{\underset{|}{C}}H{-}\overset{R^3}{\overset{|}{C}}H\underset{O}{\underset{\|}{C}}R$$

$KMnO_4/CuSO_4 \cdot 5H_2O$ (**9**, 389).[4] $KMnO_4$ supported on $CuSO_4 \cdot 5H_2O$ oxidizes secondary allylic alcohols to α,β-unsaturated ketones in 80–90% yield, and thus is competitive with the classical reagent, activated MnO_2, for this oxidation. Conjugation of the unsaturation with a phenyl group (e.g., $C_6H_5CH{=}CHCHOHCH_3$) facilitates the oxidation. Nonallylic secondary alcohols are more difficult to oxidize than the corresponding allylic alcohols. Bentonite, mainly $Al_2O_3 \cdot 4SiO_2 \cdot H_2O$, is somewhat less effective than $CuSO_4 \cdot 5H_2O$.

[1] N. Kornblum, A. S. Erickson, W. J. Kelly, and B. Henggeler, *J. Org.*, **47**, 4534 (1982).
[2] J. Steele and R. J. Stoodley, *J.C.S. Chem. Comm.*, 1215 (1982).
[3] J. H. Clark and D. G. Cork, *ibid.*, 635 (1982).
[4] N. A. Noureldin and D. G. Lee, *Tetrahedron Letters*, **22**, 4889 (1981).

Potassium peroxodisulfate, 1, 952–954; **8**, 417; **10**, 331.

Oxidation of benzylic groups.[1] $K_2S_2O_8$ in the presence of catalytic amounts of $CuSO_4$ is effective for oxidation of benzylic compounds to carbonyl compounds.

Examples:

$$p\text{-}CH_3OC_6H_4CH_2CH_3 \xrightarrow[\text{86\%}]{K_2S_2O_8,\ CuSO_4,\ CH_3CN} p\text{-}CH_3OC_6H_4COCH_3$$

$$o\text{-}CH_3OC_6H_4CH_3 \xrightarrow{64\%} o\text{-}CH_3OC_6H_4CHO$$

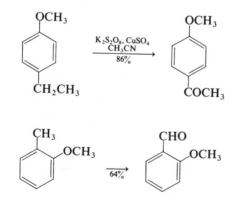

[1] M. V. Bhatt and P. T. Perumal, *Tetrahedron Letters*, **22**, 2605 (1981).

Potassium peroxomonosulfate (Oxone), $2KHSO_5 \cdot KHSO_4 \cdot K_2SO_4$ **(1)**, **10**, 328. This persulfate was first prepared by DuPont chemists.[1]

γ-Hydroxy-α,β-unsaturated ketones.[2] This reagent oxidizes the methyl ether **(2a)** or the acetate **(2b)** of the dienol to the axial γ-hydroxy enone **3** in higher yield than other reagents previously used for similar transformations: *m*-$ClC_6H_4CO_3H$, $C_6H_5I{=}O$, or $O_2/h\nu$.

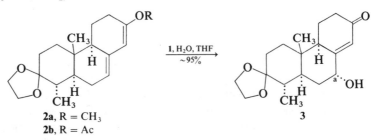

OR CH3 H H CH3 — **1**, H_2O, THF, ~95% → O CH3 H a OH H CH3

2a, R = CH_3 **3**
2b, R = Ac

Epoxidation. Oxone decomposes in the presence of a ketone (such as acetone) to form a species, possibly a dioxirane (**a**), which can epoxidize alkenes in high yield in reactions generally conducted in CH_2Cl_2-H_2O with a phase-transfer catalyst. An added ketone is not necessary for efficient epoxidation of an unsaturated ketone. The method is particularly useful for preparation of epoxides that are unstable to heat or acids and bases.[3] The acetone–Oxone system is comparable to *m*-chloroperbenzoic acid in the stereoselectivity of epoxidation of allylic alcohols. It is also similar to the peracid in preferential attack of the double bond in geraniol (dienol) that is further removed from the hydroxyl group.[4]

R, R, C, O, O

a

[1] R. J. Kennedy and A. M. Stock, *J. Org.*, **25**, 1901 (1960).
[2] S. N. Suryawanshi and P. L. Fuchs, *Tetrahedron Letters*, **22**, 4201 (1981).
[3] R. Curci, M. Fiorentino, L. Troisi, J. O. Edwards, and R. H. Pater, *J. Org.*, **45**, 4758 (1980).
[4] G. Cicala, R. Curci, M. Fiorentino, and O. Laricchiuta, *ibid.*, **47**, 2670 (1982).

Potassium superoxide, 6, 488–490; **7,** 304–307; **8,** 417–419; **9,** 391–393.

Dihydroisoquinolines. The N-chlorination-dehydrochlorination sequence with KO_2 (**8**, 417) can be used to convert 1,2,3,4-tetrahydroisoquinoline (**1**) to 3,4-dihydroisoquinoline (**2**) and 1-substituted tetrahydroisoquinolines (**3**).[1]

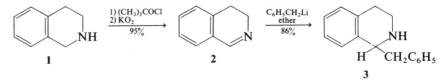

NH — 1) $(CH_3)_3COCl$, 2) KO_2, 95% → N — $C_6H_5CH_2Li$, ether, 86% → NH, H, $CH_2C_6H_5$

1 **2** **3**

Dehydrohalogenation.[2] KO_2 in DMF effects dehydrohalogenation of dichlorovinyl groups to chloroethynyl groups and also of other polychloro groups. This reaction is of considerable interest as a means of detoxifying several polychlorinated pesticides, such as *cis*-permethrin (**1**) and DDT (**3**).

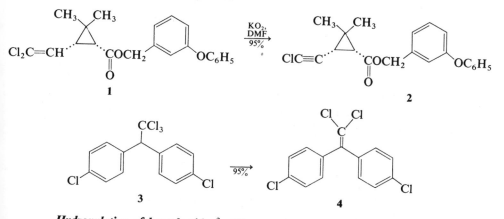

$$\text{Cl}_2\text{C}{=}\text{CH}\text{—}[\text{(CH}_3)_2\text{-cyclopropane}]\text{—COOCH}_2\text{C}_6\text{H}_4\text{OC}_6\text{H}_5\ (\mathbf{1}) \xrightarrow[95\%]{KO_2,\ DMF} \text{ClC}{\equiv}\text{C}\text{—}[\text{(CH}_3)_2\text{-cyclopropane}]\text{—COOCH}_2\text{C}_6\text{H}_4\text{OC}_6\text{H}_5\ (\mathbf{2})$$

$$(p\text{-ClC}_6\text{H}_4)_2\text{CHCCl}_3\ (\mathbf{3}) \xrightarrow{95\%} (p\text{-ClC}_6\text{H}_4)_2\text{C}{=}\text{CCl}_2\ (\mathbf{4})$$

Hydroxylation of benzylamides.[3] The reaction of N-benzylamides with KO_2 and 18-crown-6 results in *o*- and *p*-hydroxylation (equation I).

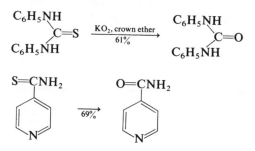

$$\text{(I)}\quad \text{C}_6\text{H}_5\text{CH}_2\text{NHCOC}_6\text{H}_5 \xrightarrow{KO_2} p\text{-HOC}_6\text{H}_4\text{CH}_2\text{NHCOC}_6\text{H}_5\ (58\%) + o\text{-HOC}_6\text{H}_4\text{CH}_2\text{NHCOC}_6\text{H}_5\ (4\%)$$

Desulfuration.[4] Thioureas and thioamides are converted into ureas and amides in 60–80% yield by reaction with KO_2–crown ether in DMF at 25°.

Examples:

$$(\text{C}_6\text{H}_5\text{NH})_2\text{C}{=}\text{S} \xrightarrow[61\%]{KO_2,\ \text{crown ether}} (\text{C}_6\text{H}_5\text{NH})_2\text{C}{=}\text{O}$$

$$4\text{-(S}{=}\text{CNH}_2\text{)pyridine} \xrightarrow{69\%} 4\text{-(O}{=}\text{CNH}_2\text{)pyridine}$$

The same conversion is possible with thiouracils and thiouridines (equation I).

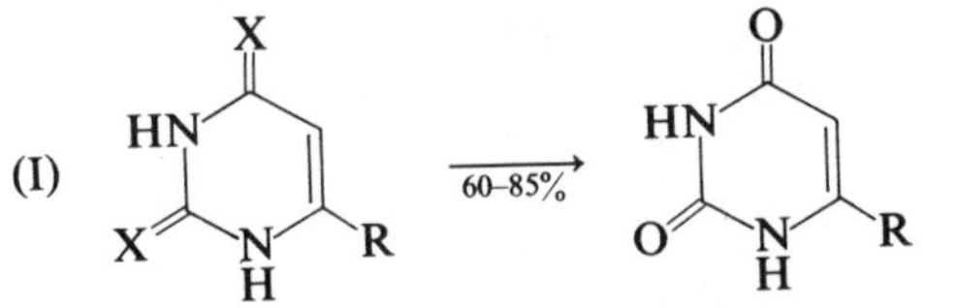

(I) [HN–pyrimidine with X at C4, X at C2, N–H, R] $\xrightarrow{60-85\%}$ [uracil with O, O, R]

($X = O$ or S,
$R = H, CH_3, C_6H_5$)

Oxidation of indoles.[5] Indoles react with KO_2 in conjunction with 18-crown-6 or benzyl(triethyl)ammonium chloride in THF to give products (**1**, **2**, **3**) that arise from an indoxyl hydroperoxide intermediate (**a**).

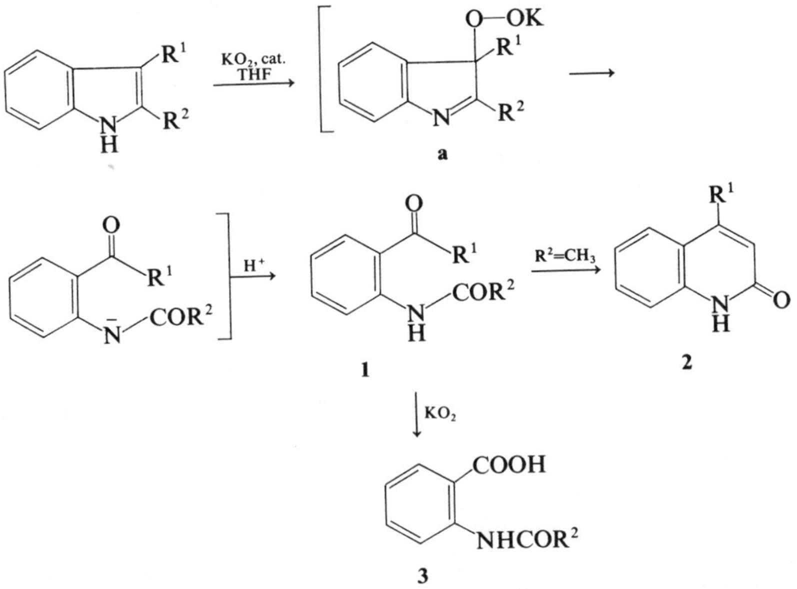

KO_2, cat. THF; H$^+$; R^2=CH$_3$; KO_2

a — **1** — **2** — **3**

COOH; NHCOR^2; COR^2; R^1; R^2; O—OK

Oxidation of 5α- and 5β-3-ketosteroids.[6] The oxidation of 3-ketosteroids (**1** and **3**) involves the enol. The 5α-ketones are converted into pseudo acids (**2**); the 5β-ketones are cleaved at C_3–C_4 to give mainly keto acids.

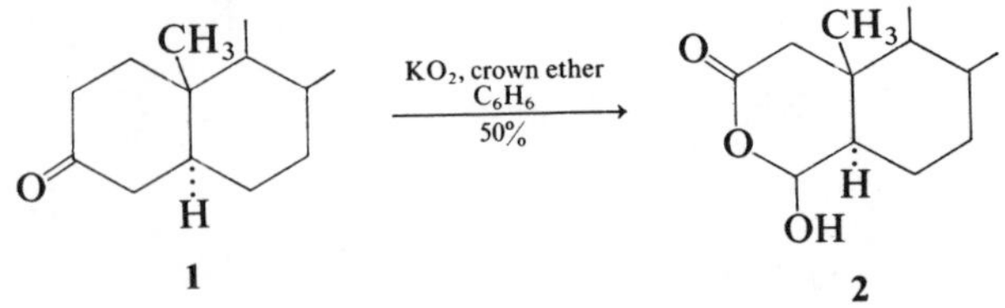

KO_2, crown ether, C_6H_6, 50%

1 → **2**

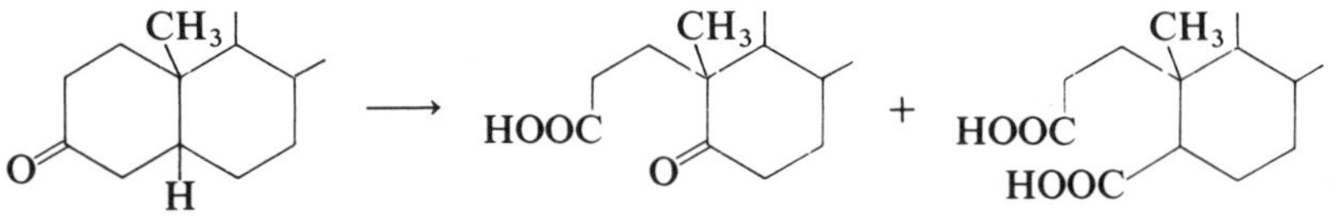

3 → 4 (45%) + 5 (7%)

Inversion of a neopentylic hydroxyl group.[7] Although S_N2 reactions at a neopentylic position proceed with difficulty, the alcohol **1** was epimerized by reaction of the mesylate with KO_2 and dibenzo-18-crown-6 in DMSO-DME in high yield.

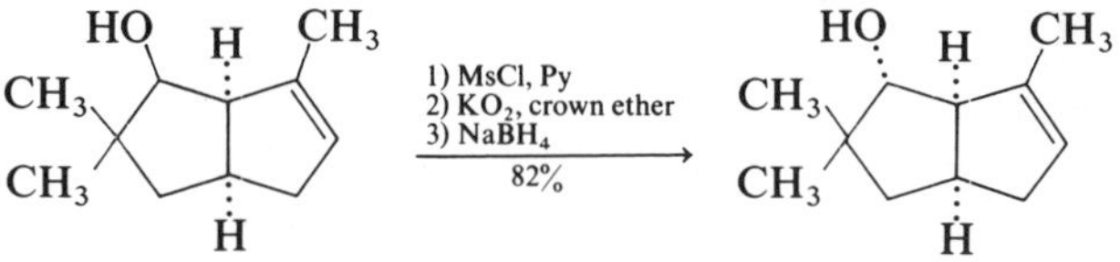

1) MsCl, Py
2) KO_2, crown ether
3) $NaBH_4$
82%

1 → **2**

Inversion of 7α-hydroxycholanic acids.[8] Chenodeoxycholic acid (**1**, 3α, 7α) has been converted to ursodeoxycholic acid (**3**, 3α, 7β) by treatment of the 3-cathylate-7-mesylate (**2**) of **1** with KO_2–crown ether in DMSO.

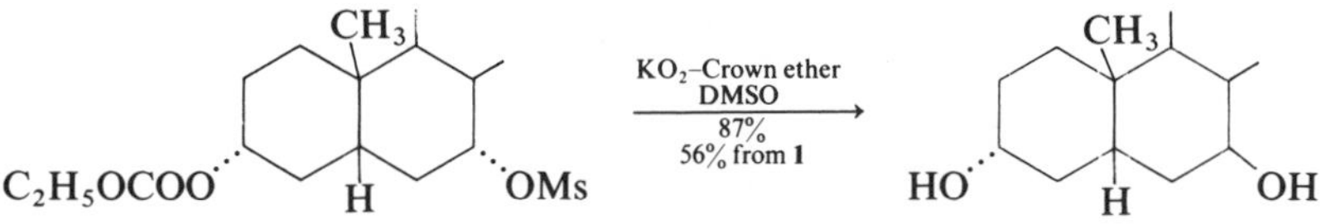

KO_2–Crown ether
DMSO
87%
56% from 1

2 → **3**

[1] F. E. Scully, Jr. and J. J. Schlager, *Heterocycles*, **19**, 653 (1982).
[2] P. Dureja, J. E. Casida, and L. O. Ruzo, *Tetrahedron Letters*, **23**, 5003 (1982).
[3] G. Galliani and B. Rindone, *Tetrahedron*, **37**, 2313 (1981).
[4] E. Katori, T. Nagano, T. Kunieda, and M. Hirobe, *Chem. Pharm. Bull.*, **29**, 3075 (1981).
[5] E. Balogh-Hergovich and G. Speier, *Tetrahedron Letters*, **23**, 4473 (1982).
[6] E. Alvarez, C. Betancor, R. Freire, A. Martin, and E. Suárez, *ibid.*, **22**, 4335 (1981).
[7] T. Ito, N. Tomiyoshi, K. Nakamura, S. Azuma, M. Izawa, F. Maruyama, M. Yanagiya, H. Shirahama, and T. Matsumoto, *ibid.*, **23**, 1721 (1982).
[8] T. Iida, H. R. Taneja, and F. C. Chang, *Lipids*, **16**, 863 (1981).

Potassium tri-*sec*-butylborohydride, 6, 490–492; **7**, 307.

Stereoselective reduction of 3-ketogibberellin acids.[1] The 3-ketogibberellin acid **1** is reduced by this borohydride almost entirely to the 3β-alcohol. Reduction with lithium tri-*sec*-butylborohydride proceeds with the opposite stereoselectivity. The difference is considered to be the result of differences in size between potassium and lithium borate complexes with the carboxylic acid group.

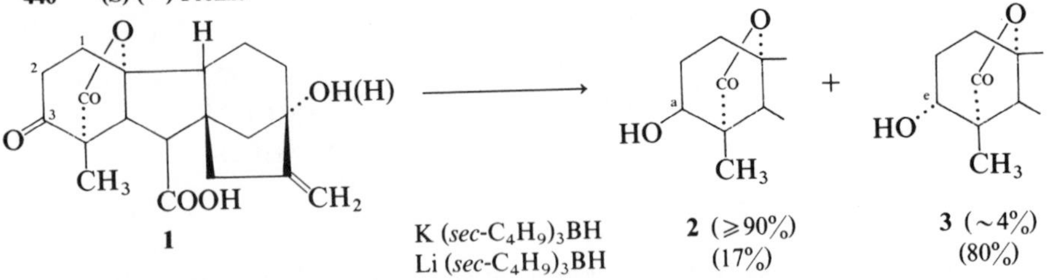

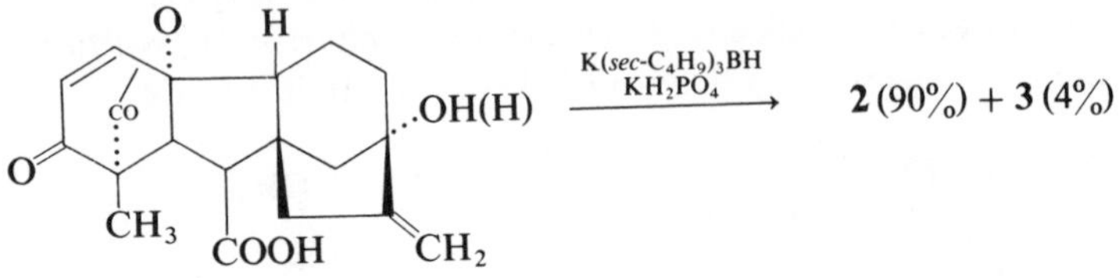

K $(sec\text{-}C_4H_9)_3BH$ — 2 (≥90%) + 3 (~4%)
Li $(sec\text{-}C_4H_9)_3BH$ — 2 (17%) + 3 (80%)

4 → $K(sec\text{-}C_4H_9)_3BH$, KH_2PO_4 → 2 (90%) + 3 (4%)

The 1,2-dehydro ketone **4** can be reduced selectively directly to **2**, but in this case a proton source is necessary to decompose an intermediate enol borate formed by 1,4-addition of hydride. Potassium dihydrogen phosphate serves this purpose and also functions as a buffer.

Reduction of steroid ketones.[2] Steroid 3-ketones can be reduced selectively by potassium tri-*sec*-butylborohydride in the presence of 17- and 20-keto groups to the 3-axial alcohol (3α-OH for 5α-steroids; 3β-OH for 5β-steroids).

[1] R. A. Bell and J. V. Turner, *Tetrahedron Letters*, **22**, 4871 (1981).
[2] G. Göndös and J. C. Orr, *J.C.S. Chem. Comm.*, 1239 (1982).

(S)-(−)-Proline, 6, 492–493; **7,** 307; **8,** 421–424; **10,** 331–332.

Asymmetric bromolactonization **(8,** 421–423**);** ***α,β-epoxy aldehydes.***[1] (2R,3S)-Epoxy aldehydes (**2**) can be prepared in high optical yield by an extension of the asymmetric bromolactonization of N-(α,β-unsaturated) acylprolines (**1**), as shown in the example.

Example:

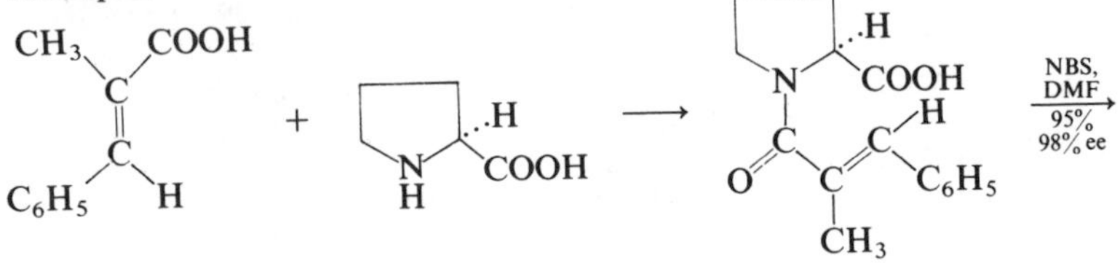

→ **1** (NBS, DMF; 95%, 98% ee) → (NaOCH₃, 90%) → (SMEAH, 72%, 98% ee) → **2**

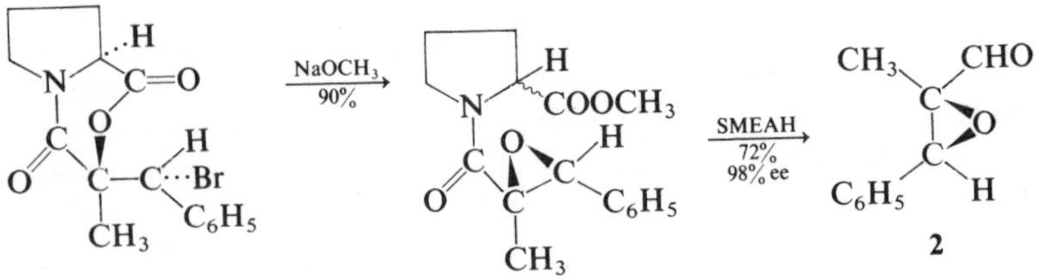

Asymmetric reduction of imines. Japanese chemists[2] have prepared a number of chiral sodium triacyloxyborohydrides from N-acyl derivatives of natural α-amino acids. The most effective are derived from (S)-proline. A particularly useful reducing agent is **1**, derived from N-benzyloxycarbonyl-(S)-proline (equation I).

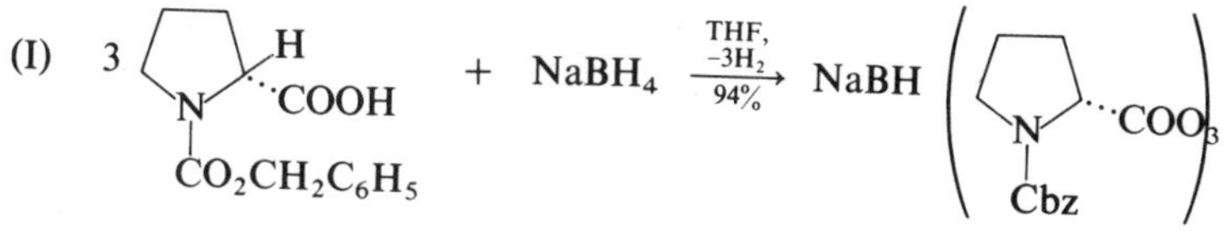

(I) 3 [N-($CO_2CH_2C_6H_5$)-(S)-proline, H, COOH] + $NaBH_4$ $\xrightarrow[94\%]{THF, -3H_2}$ NaBH(N-Cbz-prolyl-COO)$_3$

1, m.p. 55–65° dec.

This reagent has been used to reduce several 3,4-dihydroisoquinoline derivatives to optically active alkaloids in 60–85% ee.

Example:

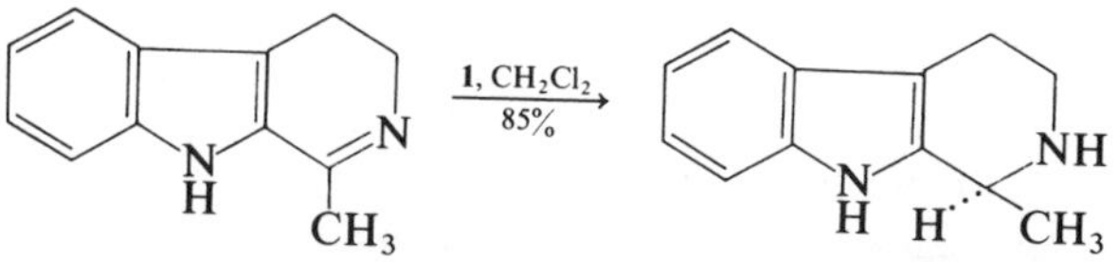

[1-methyl-3,4-dihydro-β-carboline] $\xrightarrow[85\%]{1,\ CH_2Cl_2}$ [1-methyl-1,2,3,4-tetrahydro-β-carboline, NH, H, CH_3] (79% ee)

[1] M. Hayashi, S. Terashima, and K. Koga, *Tetrahedron*, **37**, 2797 (1981).
[2] K. Yamada, M. Takeda, and T. Iwakuma, *Tetrahedron Letters*, **22**, 3869 (1981).

2-(E)-1-Propenyl-1,3-dithiane, [structure: CH_3CH=C(CH_3)-dithianyl] (**1**). This reagent is prepared by reaction of (E)-2-methyl-2-butenal with 1,3-propanedithiol (BF_3 etherate, HOAc, $HCCl_3$).

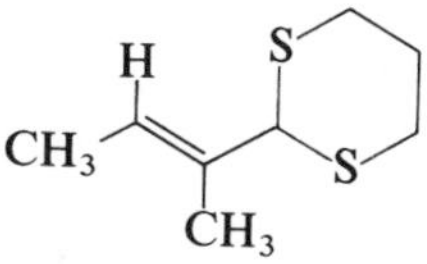

Pseudoguaianolides.[1] The anion of **1** can undergo 1,2- and 1,4-addition to enones. In general, 1,4-addition is favored by use of Li^+ as the counter ion. α-1,4-

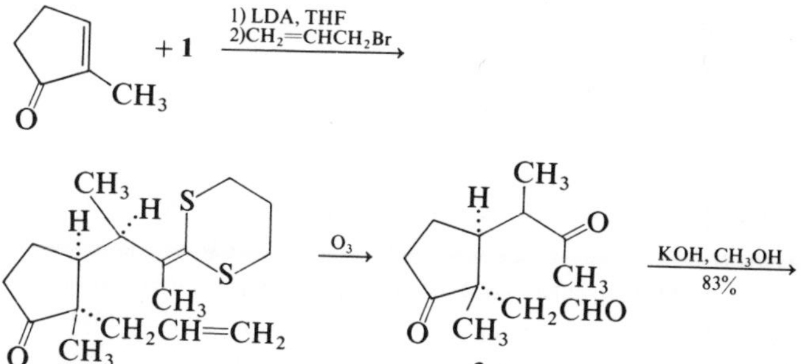

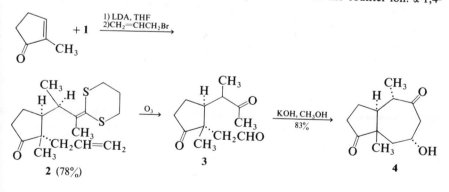

[2-methylcyclopentenone] + **1** $\xrightarrow{1)\ LDA,\ THF;\ 2)\ CH_2{=}CHCH_2Br}$

2 (78%) $\xrightarrow{O_3}$ **3** $\xrightarrow[83\%]{KOH,\ CH_3OH}$ **4**

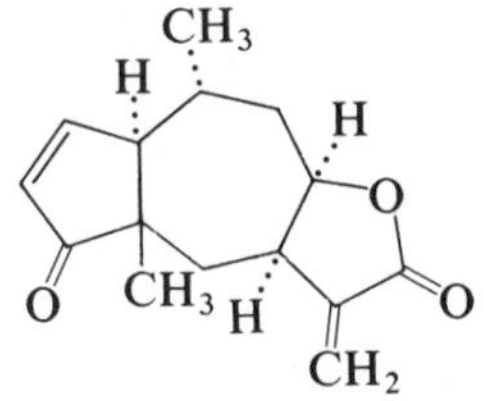

generally enhanced over γ-1,4-addition by addition of HMPT. A useful pseudoguaianolide synthesis involves the conjugate addition of the lithium anion of **1** to 2-methylcyclopentenone followed by monoallylation to give **2** in about 78% yield. The product was ozonized to give **3**, which undergoes aldolization to **4** in base. This product is a useful precursor to aromatin (**5**).

[1] F. E. Ziegler, J.-M. Fang, and C. C. Tam, *Am. Soc.*, **104**, 7174 (1982).

Pyridine, 1, 958–963; **2**, 349–351; **4**, 914–915; **6**, 497.

Dehydrochlorination*; *4-alkylidene-2-butenolides.[1] 2,2-Dichloro-4-alkylbutanolides (**1**), (**9**, 491–492), are converted into 4-alkylidene-2-butenolides (**2**) on didehydrochlorination with pyridine at reflux.

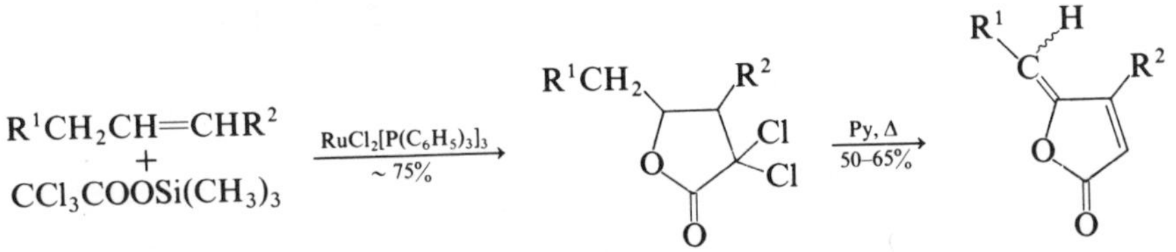

$R^1CH_2CH{=}CHR^2$ + $CCl_3COOSi(CH_3)_3$ $\xrightarrow[\sim 75\%]{RuCl_2[P(C_6H_5)_3]_3}$ **1** $\xrightarrow[50–65\%]{Py, \Delta}$ **2** (Z/E ~ 85:15)

Borane scavenger.[2] Sodium borohydride in DMF can be used to reduce an acid chloride to an aldehyde in >70% yield if a molar excess of pyridine is present as a borane scavenger. This methodology is now preferred to an earlier method from the same laboratory (**10**, 358) in which the reduction is limited by a quench with ethyl vinyl ether and propionic acid.

[1] T. Nakano and Y. Nagai, *J.C.S. Chem. Comm.*, 815 (1981).
[2] J. H. Babler, *Syn. Comm.*, **12**, 839 (1982).

4-Pyridinecarboxaldehyde, 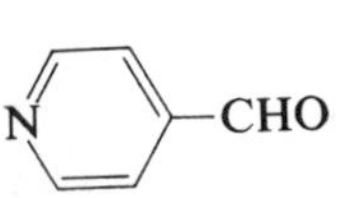 (**1**). Mol. wt. 107.11, b.p. 77–78°/12 mm. Supplier: Aldrich.

Conversion of primary amines to carbonyl compounds.[1] 4-Pyridinecarboxaldehyde is recommended for biomimetric transamination leading to carbonyl compounds. It is particularly useful for conversion of α-amino acids into α-keto carboxylic acids.

$$p\text{-}ClC_6H_4CH_2NH_2 + \mathbf{1} \xrightarrow{\text{DMF, DBU}} \left[p\text{-}ClC_6H_4CH{=}N{-}CH{=}\langle\quad NH \right] \xrightarrow[83\%]{H_3O^+} p\text{-}ClC_6H_4CHO$$

$$(CH_3)_2CHCH_2CH(NH_2)COOH \xrightarrow[85\%]{} (CH_3)_2CHCH_2C(=O)COOH$$

[1] S. Ohta and M. Okamoto, *Synthesis*, 756 (1982).

Pyridinium (*d*)-camphor-10-sulfonate, (**1**).

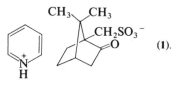

Mol. wt. 311.40, m.p. 153–155.° The salt is obtained from reaction of pyridine and the free acid at 25° (80% yield).

Asymmetric cyclization to a pyrrolizidine.[1] Treatment of the secondary amine **2** with methanolic HCl followed by reduction of the Mannich base **3** gives the alkaloid (±)-trachelanthamidine (**4**) in about 40% yield. Use of the chiral salt **1** results in asymmetric cyclization to give (+)-**4** in 33% enantiomeric excess. Unfortunately the chemical yield is low.

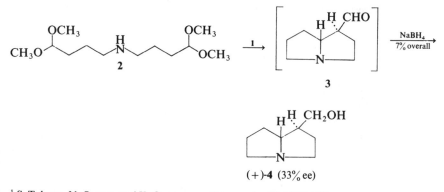

$$\mathbf{2} \xrightarrow{\mathbf{1}} [\mathbf{3}] \xrightarrow[\text{7\% overall}]{NaBH_4} (+)\text{-}\mathbf{4}$$

(+)-**4** (33% ee)

[1] S. Takano, N. Ogawa, and K. Ogasawara, *Heterocycles*, **16**, 915 (1981).

Pyridinium chloride, 1, 964–966; **2**, 352–353; **3**, 239–240; **4**, 415–418; **5**, 566–567; **6**, 497–498; **7**, 308; **8**, 424–425; **10**, 333.

Ester hydrolysis.[1] Pyridinium chloride selectively demethylates methyl esters of *o*-substituted aromatic carboxylic acids.

Example:

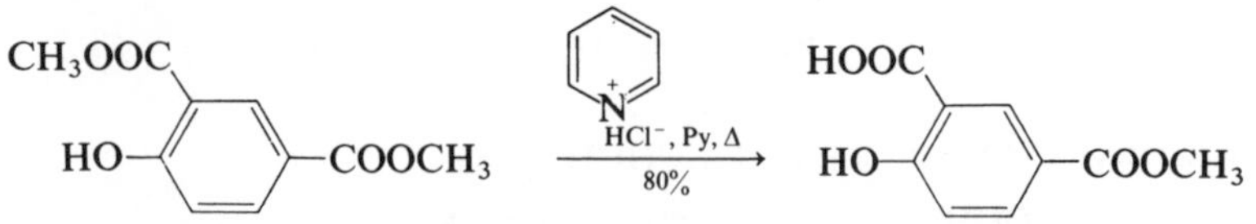

Aromatization and cyclodeamination.[2] The final steps in a synthesis of the antibiotic resistomycin (**2**) from **1** involved desilylative aromatization, demethylation, and cyclization. These were originally carried out sequentially with *p*-toluenesulfonic acid, boron trichloride, and concentrated sulfuric acid in 65% overall yield. Surprisingly, reaction of **1** with pyridinium chloride results directly in **2** in 84% yield.

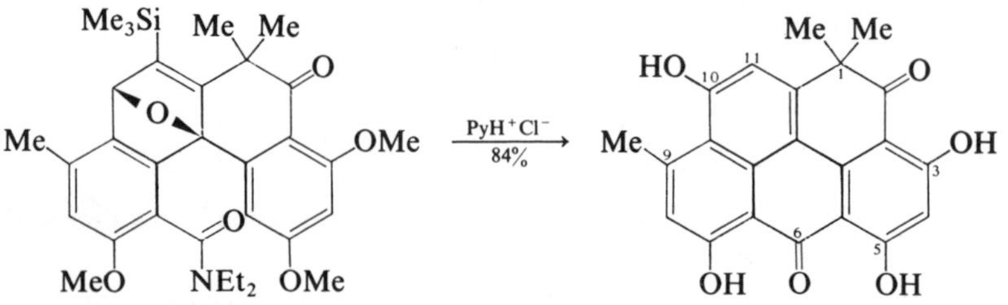

1 **2**

[1] I. G. C. Coutts, M. Edwards, and D. J. Richards, *Synthesis*, 487 (1981).
[2] B. A. Keay and R. Rodrigo, *Am. Soc.*, **104**, 4725 (1982).

Pyridinium chlorochromate (PCC), **6**, 498–499; **7**, 308–309; **8**, 425–427; **9**, 397–399; **10**, 334–335.

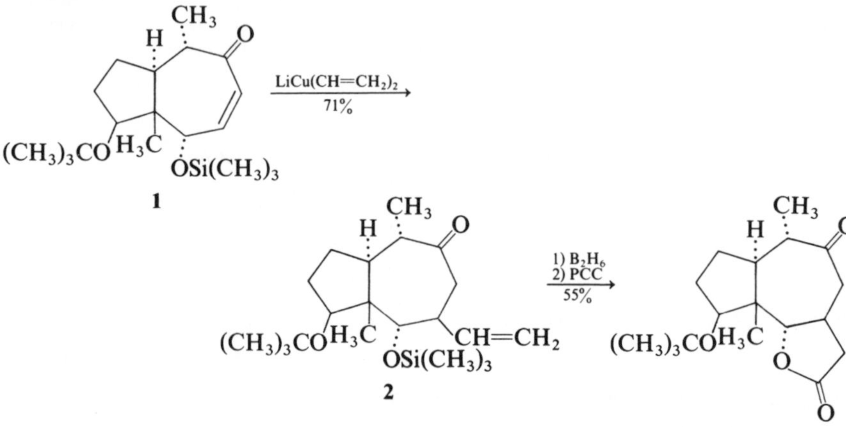

3

γ-Lactone annelation. A new method for γ-lactone annelation involves conjugate addition of lithium divinylcuprate to the γ-trimethylsilyloxycycloheptenone **1**, from the side opposite to the bulky $(CH_3)_3SiO$-substituent. The product (**2**) is converted in a one-pot reaction to the lactone **3** by hydroboration followed by oxidation with pyridinium chlorochromate to a transient aldehyde, which interacts with the adjacent oxygen function to form the lactone **3**, probably via a lactol.[1]

Cleavage of vic-diols. The reagent cleaves *vic*-diols to aldehydes and ketones. Examples:

$$C_6H_5CH(OH)\!-\!CH(OH)C_6H_5 \xrightarrow[18\%]{PCC} 2\,C_6H_5CHO$$

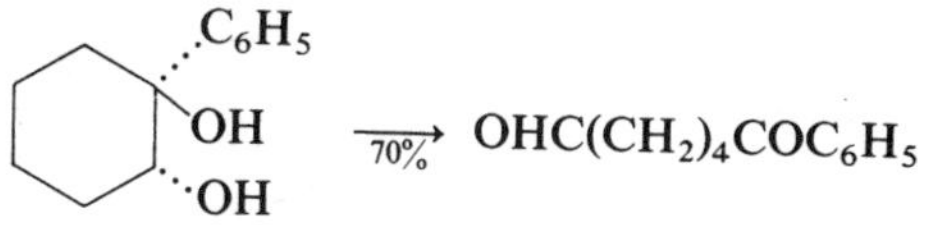

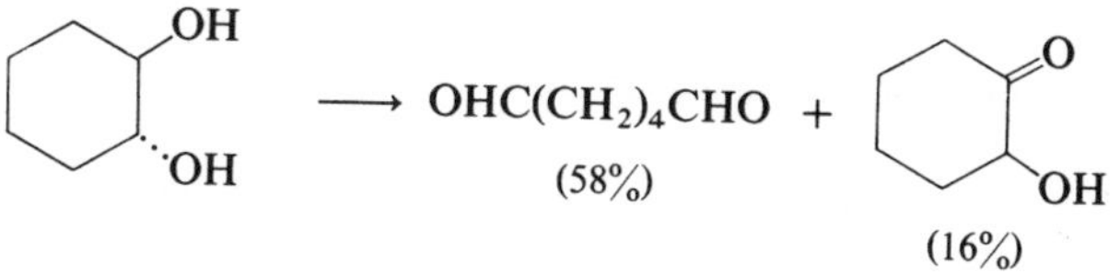

$$\xrightarrow{70\%} OHC(CH_2)_4COC_6H_5$$

$$\longrightarrow OHC(CH_2)_4CHO \; (58\%) \; + \; (16\%)$$

Benzopinacol is not oxidized even under more drastic conditions.

Oxidative desilylation.[3] Silyl ethers (*t*-butyldimethylsilyl and trimethylsilyl) of hydroquinones are converted to quinones in 60–90% yield by PCC. Electrochemical oxidation is also possible.

Furanes. (Z)-2-Butene-1,4-diols are converted in one step into furanes by reaction with PCC in CH_2Cl_2 at 25°. Activated MnO_2 is somewhat less efficient for the reaction, but is superior to PCC for oxidation of substrates containing an allylsilane unit.[4]

Example:

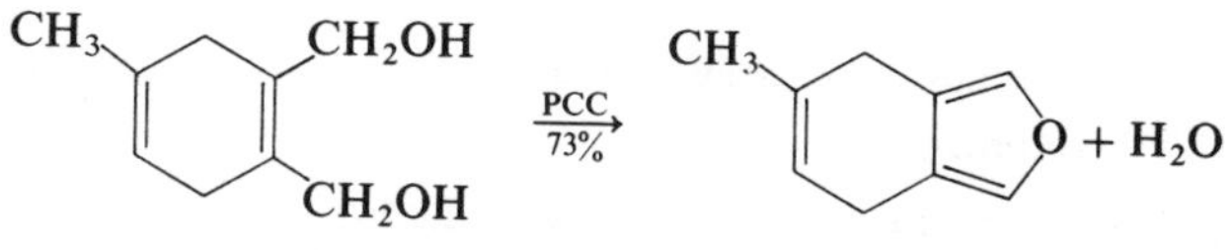

$$\xrightarrow[73\%]{PCC} \; + H_2O$$

Oxidation of alkylfuranes.[5] Furanes substituted by alkyl groups at C_2 and C_5 are oxidatively cleaved by PCC to *trans*-enediones, which are useful precursors to cyclopentenones. The *cis*-isomers are formed originally and are then isomerized by the acidic reagent. 2-Alkylfuranes are also oxidized in this manner.

Examples:

2-Methyl-5-(n-dodecyl)furan $\xrightarrow[90\%]{PCC, CH_2Cl_2}$ $CH_3COCH{=}CHCOC_{12}H_{25}$-n $\xrightarrow[80\%]{NaOH, CH_3OH}$ 3-(n-undecyl)-2-methyl-4-methoxycyclopent-2-enone ($C_{11}H_{23}$-n, CH_3O, CH_3)

2-(n-dodecyl)furan $\xrightarrow{60\%}$ $HCOCH{=}CHCOC_{12}H_{25}$-n

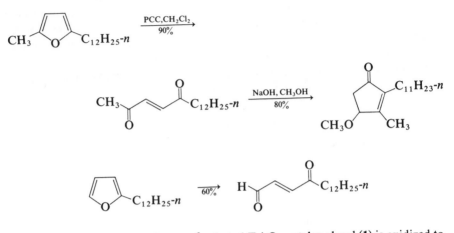

Oxidation of glucals to lactones.[6] 3, 4, 6-Tri-O-acetyl-D-glucal (**1**) is oxidized to the unsaturated lactone **2** in 78% yield by PCC. In contrast, tri-O-benzyl-D-glucal (**3**) is oxidized without elimination to the saturated lactone (**4**).

1, R = Ac (CH_2OR, OR, RO) $\xrightarrow[78\%]{PCC, ClCH_2CH_2Cl}$ **2** (CH_2OAc, AcO, =O)

3, R = Bzl $\xrightarrow{60\%}$ **4** (CH_2OR, OR, RO, =O)

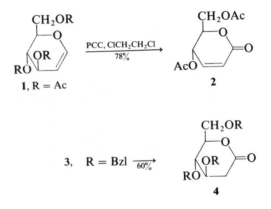

Review.[7] The various applications of this oxidant in organic synthesis have been reviewed (60 references).

[1] P. T. Lansbury and T. E. Nickson, *Tetrahedron Letters*, **23**, 2627 (1982).
[2] A. Cisneros, S. Fernández, and J. E. Hernández, *Syn. Comm.*, **12**, 833 (1982).
[3] J. P. Willis, K. A. Z. Gogins, and L. L. Miller, *J. Org.*, **46**, 3215 (1981).
[4] H. Nishiyama, M. Sasaki, and K. Itoh, *Chem. Letters*, 1363 (1981).
[5] G. Piancatelli, A. Scettri, and M. D'Auria, *Tetrahedron*, **36**, 661, 1877 (1980).
[6] P. Rollin and P. Sinaÿ, *Carbohydrate Res.*, **98**, 139 (1981).
[7] G. Piancatelli, A. Scettri, and M. D'Auria, *Synthesis*, 245 (1982).

Pyridinium dichromate (PDC), **9**, 399; **10**, 335–336.

Oxidation of hemiacetals.[1] Pyridinium dichromate in DMF effects a novel oxidative cyclization of the aldehyde **1** to the macrocyclic lactone verrucarin T (**2a**). The conversion is believed to involve cyclization to a hemiacetal (**2b**) followed by oxidation to **2a**. Manganese dioxide does not oxidize **1**, and a variety of other oxidants attack the dienic ester side chain.

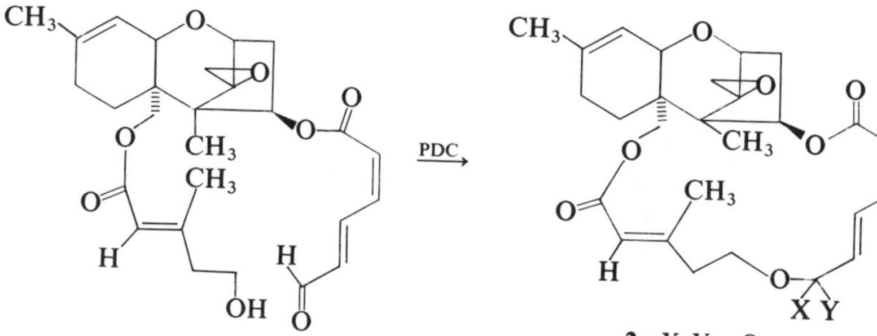

1 $\xrightarrow{PDC}$ **2a**, X, Y = O; **2b**, X = H; Y = OH

Oxidative cleavage of oximes.[2] PDC is recommended for regeneration of ketones and aromatic aldehydes from the oximes (80–100% conversion). The rate of deoximation is increased by addition of 3Å molecular sieves, but the yield of the carbonyl compound is lowered.

[1] R. Esmond, B. Fraser-Reid, and B. B. Jarvis, *J. Org.*, **47**, 3358 (1982).
[2] S. Satish and N. Kalyanam, *Chem. Ind.*, 809 (1981).

Pyridinium fluorochromate, $C_5H_5NHCrO_3F$. Mol. wt. 199.1, m.p. 106–108°, soluble in H_2O, DMF, CH_3CN, acetone, CH_2Cl_2. The reagent is prepared by addition of pyridine to a solution of CrO_3 in 40% HF (93–94% yield).

Oxidation.[1] The reagent is comparable to PCC for oxidation of primary and secondary alcohols to carbonyl compounds in high yield. It is somewhat less acidic than PCC.

It has been used to effect oxidation of anthracene and phenanthrene to quinones in >70% yields.

[1] M. N. Bhattacharjee, M. K. Chaudhuri, H. S. Dasgupta, and D. T. Khathing, *Synthesis*, 588 (1982).

Pyridinium poly(hydrogen fluoride), 5, 528–539; **6**, 473–474; **7**, 294; **9**, 399–400.

vic-***Fluoro amines.*** Two laboratories[1, 2] have reported the synthesis of *vic*-fluoro amines by reaction of aziridines with Olah's reagent. French chemists[1] have reported an extensive study of this reaction with three different reagents: HF, $(HF)_n \cdot C_5H_5N$, and a modified Olah's reagent, $N(C_2H_5)_3 \cdot HF$. This last reagent is prepared by addition in controlled amounts of triethylamine to commercial preparations of $(HF)_n \cdot C_5H_5N$ in CH_2Cl_2. It is particularly useful for cleavage of N-activated aziridines, particularly N-(*t*-butoxycarbonyl)aziridines, which undergo exclusive *trans*-addition of HF (equation I).

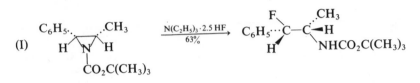

(I) C_6H_5, H, CH_3, H aziridine, N–$CO_2C(CH_3)_3$ $\xrightarrow[63\%]{N(C_2H_5)_3 \cdot 2.5\,HF}$ C_6H_5(F)(H)C–C(CH_3)(H)$NHCO_2C(CH_3)_3$

β-Fluoro-α-keto esters. Cleavage of glycidic esters with this reagent followed by Jones oxidation gives fluoropyruvic esters in 60–85% yield.[3]

Examples:

(cyclohexane spiro-epoxide)$CHCOOC_2H_5$ $\xrightarrow{HF,\ Py}$ [(1-fluorocyclohexyl)CH(OH)$COOC_2H_5$] $\xrightarrow[66\%]{Jones\ oxid.}$ (1-fluorocyclohexyl)$C(=O)COOC_2H_5$

$C_6H_5(CH_3)C$–$CHCOOC_2H_5$ (epoxide, O) $\xrightarrow{80\%}$ $C_6H_5(CH_3)C(F)COCOOC_2H_5$

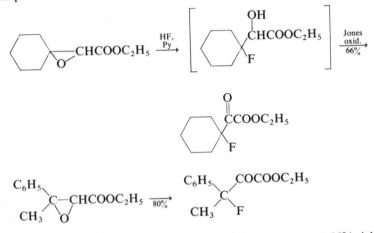

α-Fluoro carboxylic acids.[4] These acids can be prepared in 40–85% yield by *in situ* diazotization ($NaNO_2$) of α-amino acids in 48:52 (w/w) HF-pyridine. Use of the more acidic 70:30 (w/w) of HF-pyridine usually employed for fluorination results in rearrangement of an alkyl, aryl, or hydroxyl group during diazotization.

1 G. M. Alvernhe, C. M. Ennakoua, S. M. Lacombe, and A. J. Laurent, *J. Org.*, **46**, 4938 (1981).
2 T. N. Wade, *ibid.*, **45**, 5328 (1980).
3 A. Ourari, R. Condom, and R. Guedj, *Can. J. Chem.*, **60**, 2707 (1982).
4 G. A. Olah, G. K. S. Prakash, and Y. L. Chao, *Helv.*, **64**, 2528 (1981).

Pyridinium poly(hydrogen fluoride)–N-Bromosuccinimide.

Vinyl fluorides.[1] This combination results in formation of a reagent "BrF" (**1**), which reacts with alkenes by *trans*-addition to form bromofluorides **2**. On elimination of HBr (KOH), the adducts form vinyl fluorides stereospecifically.

$$\underset{(Z)}{R^1(H)C{=}C(H)R^2} + \mathbf{1} \xrightarrow{45\text{–}60\%} \underset{threo\text{-}\mathbf{2}}{Br{-}C(R^1)(H){-}C(H)(R^2){-}F} \xrightarrow[45\text{–}70\%]{KOH} \underset{(Z)\text{-}\mathbf{3}}{R^1(H)C{=}C(F)R^2}$$

$$\underset{(E)}{R^1(H)C{=}C(H)R^2} + \mathbf{1} \xrightarrow{45\text{–}60\%} \underset{erythro\text{-}\mathbf{2}}{Br{-}C(R^1)(H){-}C(R^2)(H){-}F} \xrightarrow[40\text{–}65\%]{KOH} \underset{(E)\text{-}\mathbf{3}}{R^1(H)C{=}C(R^2)F}$$

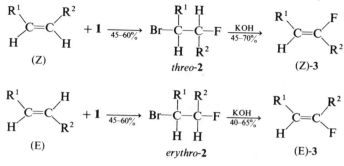

[1] G. Boche and U. Fährmann, *Ber.*, **114**, 4005 (1981).

2-Pyridylselenenyl bromide, 2-pyridyl–SeBr (**1**). The reagent is prepared *in situ* by reaction of 2,2′-dipyridyl diselenide with bromine.

α,β-Enones and -enals.[1] Aldehydes and ketones are converted into α-pyridylseleno carbonyl derivatives by reaction with **1** in the presence of hydrogen chloride. These compounds undergo selenoxide elimination on oxidation to give α,β-unsaturated compounds. Yields are generally higher than those obtained by selenoxide elimination of α-phenylselenyl aldehydes and ketones. Oxidation of α-pyridylseleno ketones is best effected with ozone at −78° followed by addition to refluxing carbon tetrachloride. Oxidative elimination of α-pyridylseleno aldehydes is effected with sodium periodate at 23°.

Examples:

$$C_6H_5\overset{O}{\overset{\|}{C}}CH_2CH_3 \xrightarrow[78\%]{1,\ C_2H_5OH,\ HCl} C_6H_5\overset{O}{\overset{\|}{C}}\underset{SePy}{\underset{|}{C}}HCH_3 \xrightarrow[97\%]{1)\ O_3,\ -78^\circ;\ 2)\ CCl_4,\ \Delta} C_6H_5\overset{O}{\overset{\|}{C}}CH{=}CH_2$$

$$C_6H_5CH_2CH_2CHO \xrightarrow[69\%]{} C_6H_5CH_2\underset{SePy}{\underset{|}{C}}HCHO \xrightarrow[83\%]{NaIO_4,\ 23^\circ} \underset{C_6H_5}{H}{>}C{=}C{<}\underset{H}{CHO}$$

The selenoxide elimination of primary alkyl 2-pyridyl selenides to give terminal alkenes also proceeds in higher yield than the corresponding reaction with alkyl phenyl selenides.[2]

Example:

$$PySe{-}CH_2CH_2C_{16}H_{33}\text{-}n \xrightarrow[82\%]{H_2O_2,\ 32^\circ} CH_2{=}CHC_{16}H_{33}\text{-}n$$

[1] A. Toshimitsu, H. Owada, S. Uemura, and M. Okano, *Tetrahedron Letters*, **23**, 2105 (1982).
[2] *Idem*, *ibid.*, **21**, 5037 (1980).

Q

Quinine, 6, 501; **7**, 311; **8**, 430–431; **9**, 403; **10**, 338.

Asymmetric phenylthiolation.[1] The reaction of **1**, probably via **a**, with thiophenol in benzene containing 1.2 equiv. of cinchonidine results in optically active **2** (54% ee). The same reaction conducted on the acetate corresponding to **1** results in (+)-**2** in 38% optical yield. Recrystallization of the impure (±)-**2** from C_6H_6–C_6H_{12} gives **2** in low optical purity, but optically pure (+)-**2**, $\alpha_D + 105°$, can be obtained from the mother liquor in 28% yield.

Pure (+)-**2** was used for a fairly efficient synthesis of (+)-thienamycin.

$SO_2C_6H_5$ ... **1** → [**a**] $\xrightarrow[96\%]{C_6H_5SH,\ C_6H_6,\ \text{cat.*},\ 35°}$ SC_6H_5 ... (+)-**2** (54% ee)

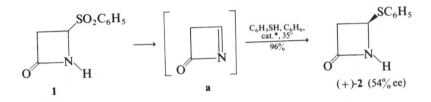

[1] M. Shibasaki, A. Nishida, and S. Ikegami, *J.C.S. Chem. Comm.*, 1324 (1982).

R

Raney nickel, 1, 723–731; **2**, 293–294; **5**, 570–571; **6**, 502; **7**, 312; **8**, 433; **9**, 405–406; **10**, 339–340.

β-Hydroxy ketones. Catalytic hydrogenation of isoxazolines has been used for some time to obtain β-amino alcohols. Partial reduction to a β-hydroxy ketone (aldol) via a β-hydroxy imine is now possible with Raney nickel as catalyst and buffered aqueous methanol as solvent. Since the heterocycle is usually obtained by a [3+2] dipolar cycloaddition of nitrile oxides to olefins, which is regioselective and stereospecific, either *threo*- or *erythro*-alkyl-β-hydroxy ketones can be obtained on reduction, depending on the stereochemistry of the starting olefin. Epimerization is largely suppressed by addition of methyl borate as buffer[1] or of $AlCl_3$ or HCl[2] to catalyze hydrolysis of the imino group. The complete sequence for conversion of *cis*- and *trans*-2-butene into epimeric α-methyl-β-hydroxy ketones is shown in equations (I) and (II).

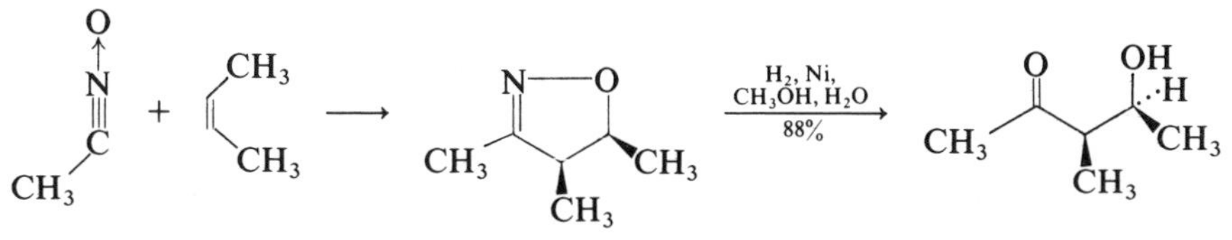

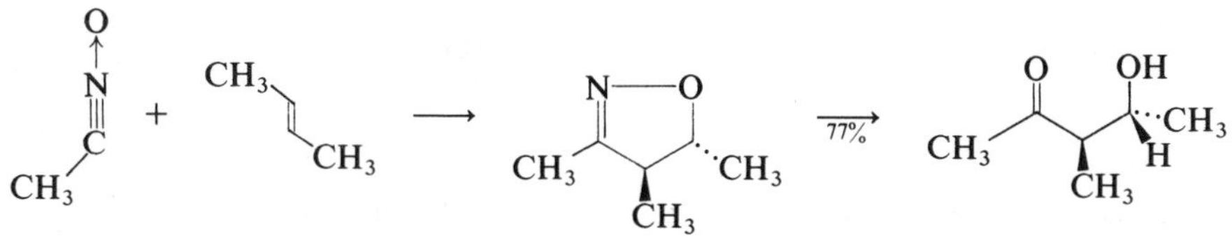

H_2, Ni, CH_3OH, H_2O 88%

77%

This reduction permits use of the isoxazoline ring as a masked α,β-enone, as in the synthesis of the α-methylene ketone **1**, convertible into sarkomycin (**2**).[3]

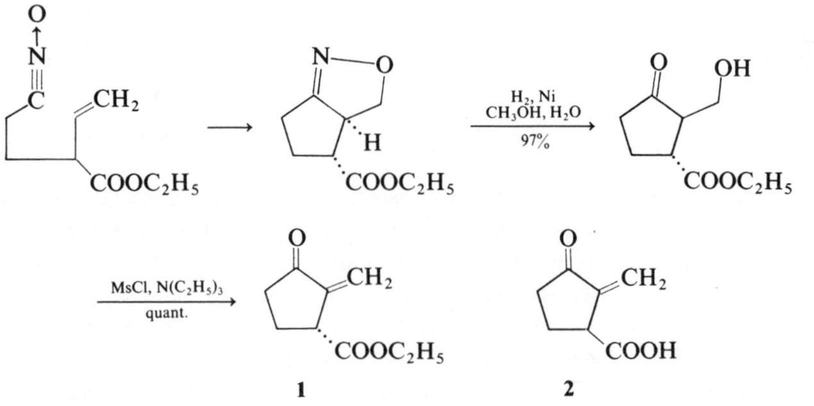

H_2, Ni
CH_3OH, H_2O
97%

MsCl, $N(C_2H_5)_3$
quant.

1

2

This reduction has also been used to effect ring annelation by an intramolecular [3 + 2] cycloaddition of a nitrile oxide to a cyclohexene.[4]

Example:

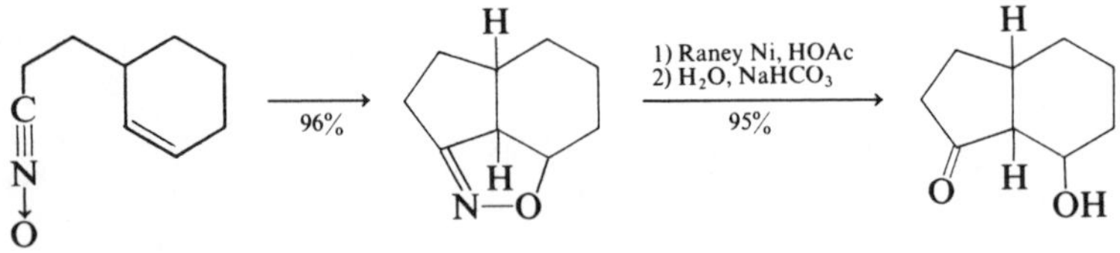

96%

1) Raney Ni, HOAc
2) H_2O, $NaHCO_3$
95%

[1] D. P. Curran, *Am. Soc.*, **104**, 4024 (1982).
[2] A. P. Kozikowski and M. Adamczyk, *Tetrahedron Letters*, **23**, 3123 (1982).
[3] A. P. Kozikowski and P. D. Stein, *Am. Soc.*, **104**, 4023 (1982).
[4] R. H. Wollenberg and J. E. Goldstein, *Synthesis*, 757 (1980).

Rhodium(II) carboxylates, 5, 571–572; **7**, 313; **8**, 434–435; **9**, 406–408; **10**, 339–340.

Intramolecular cycloaddition of a diazo ketone to a cyclopropene.[1] Rhodium(II) acetate is markedly superior to copper or copper(II) sulfate as the catalyst for cyclopropanation of 1,4-diacetoxy-2-butyne with *t*-butyl diazoacetate. The product (**1**) was converted by known steps into the diazo ketone **2**. In the presence of rhodium(II) acetate, **2** undergoes intramolecular cycloaddition to the cyclopropene double bond to give the highly strained tricyclic pentanone derivative **3** in 30% yield. Copper catalysts are less efficient for this conversion.

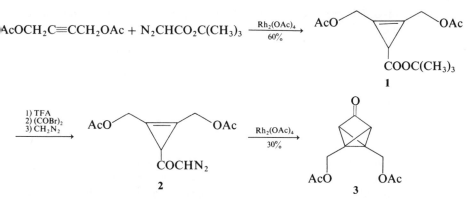

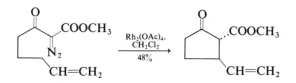

$$AcOCH_2C{\equiv}CCH_2OAc + N_2CHCO_2C(CH_3)_3 \xrightarrow[60\%]{Rh_2(OAc)_4} \mathbf{1}$$

1 (cyclopropene bearing two CH_2OAc groups and $COOC(CH_3)_3$)

$$\xrightarrow{1)\ TFA;\ 2)\ (COBr)_2;\ 3)\ CH_2N_2} \mathbf{2} \xrightarrow[30\%]{Rh_2(OAc)_4} \mathbf{3}$$

2 (AcO, OAc, $COCHN_2$) **3** (O, AcO, OAc)

Cyclopentanones.[2] 2-Carboalkoxycyclopentanones can be obtained by cyclization of α-diazo-β-keto esters catalyzed by $Rh_2(OAc)_4$. Allylic C—H insertion can be favored over cyclopropanation.

Example:

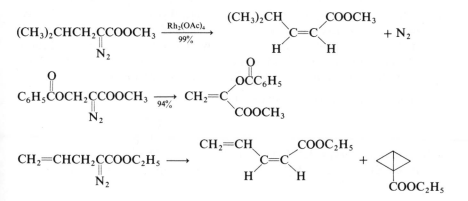

$$\xrightarrow[48\%]{Rh_2(OAc)_4,\ CH_2Cl_2}$$

(O, $COOCH_3$, N_2, $CH{=}CH_2$ → O, $COOCH_3$, $CH{=}CH_2$)

cis-α,β-*Unsaturated esters.*[3] The decomposition of α-diazo esters in the presence of $Rh_2(OAc)_4$ in benzene at room temperature results in formation of *cis*-α,β-unsaturated esters.

Examples:

$$(CH_3)_2CHCH_2C(=N_2)COOCH_3 \xrightarrow[99\%]{Rh_2(OAc)_4} (CH_3)_2CH(H)C{=}C(H)COOCH_3 + N_2$$

$$C_6H_5C(=O)OCH_2C(=N_2)COOCH_3 \xrightarrow[94\%]{} CH_2{=}C(OC(=O)C_6H_5)COOCH_3$$

$$CH_2{=}CHCH_2C(=N_2)COOC_2H_5 \longrightarrow CH_2{=}CH(H)C{=}C(H)COOC_2H_5 + \text{(bicyclobutane)}\ COOC_2H_5$$

The α-diazo-β-hydroxy ester **1** rearranges under the same conditions to the formylacetic ester **2**.

$$\underset{\mathbf{1}}{\mathrm{HOCH_2\underset{\underset{N_2}{\|}}{C}COOC_2H_5}} \xrightarrow[\text{quant.}]{\mathrm{Rh_2(OAc)_4}} \underset{\mathbf{2}}{\mathrm{OHCCH_2COOC_2H_5}} + \mathrm{N_2}$$

Unsaturated ethers.[4] The efficient insertion of carboalkoxycarbenes into the O—H bond of alcohols catalyzed by Rh(II) acetate (**5**, 571–572) extends to reactions with unsaturated alcohols. For this reaction copper(II) triflate is usually comparable to rhodium(II) alkanoates. Insertion predominates over cyclopropanation in the case of ethylenic alcohols. In reactions with acetylenic alcohols, cyclopropenation can predominate over insertion because of steric effects, as in reactions of $HC{\equiv}CC(CH_3)_2OH$ where the insertion/addition ratio is 36:56.

[1] H. Irngartinger, A. Goldman, R. Schappert, P. Garner, and P. Dowd, *J.C.S. Chem. Comm.*, 455 (1981); P. Dowd, P. Garner, R. Schappert, H. Irngartinger and A. Goldman, *J. Org.*, **47**, 4240 (1982).
[2] D. F. Taber and E. H. Petty, *ibid.*, **47**, 4808 (1982).
[3] N. Ikota, N. Takamura, S. D. Young, and B. Ganem, *Tetrahedron Letters*, **22**, 4163 (1981).
[4] A. F. Noels, A. Demonceau, N. Petinot, A. J. Hubert, and P. Teyssié, *Tetrahedron*, **38**, 2733 (1982).

Rhodium catalysts, 1, 982–983; **4**, 418–419; **6**, 503; **8**, 433–434.

Hydrogenation of β-acyloxy-α,β-unsaturated esters and ketones.[1] Attempted hydrogenation of these substrates is difficult because of lack of reactivity of the double bond and susceptibility to hydrogenolysis. Thus hydrogenation of ethyl 3-acetoxycrotonate with PtO_2 as catalyst gives only ethyl butyrate. This reduction can be effected, however, with 5% rhodium on carbon in quantitative yield. The hydrogenation was shown to be *syn*-selective in the case of ethyl (E)- and (Z)-2-deuterio-3-acetoxybutyrate. The method is general for a variety of β-acyloxy-α,β-unsaturated esters and ketones (>90% yield). Although hydrogenation of ethyl β-aminocrotonate fails, the corresponding trifluoroacetate can be hydrogenated in 95% yield with this catalyst.

[1] J. D. Rozzell, *Tetrahedron Letters*, **23**, 1767 (1982).

Rhodium(III) chloride, 4, 421; **8**, 437–438; **10**, 343.

Resorcinols. $RhCl_3 \cdot 3H_2O$ catalyzes isomerization of ω-alkenyl-1,3-cyclohexanedione derivatives to resorcinols.[1]

Examples:

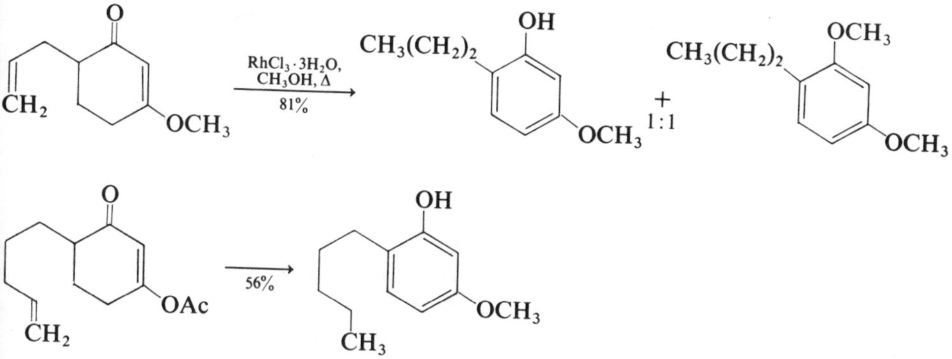

$\xrightarrow[81\%]{RhCl_3 \cdot 3H_2O,\ CH_3OH,\ \Delta}$ 1:1

$\xrightarrow{56\%}$

[1] I. S. Blagbrough, G. Pattenden, and R. A. Raphael, *Tetrahedron Letters*, **23**, 4843 (1982).

Rhodium(III) porphyrins, 1–5.

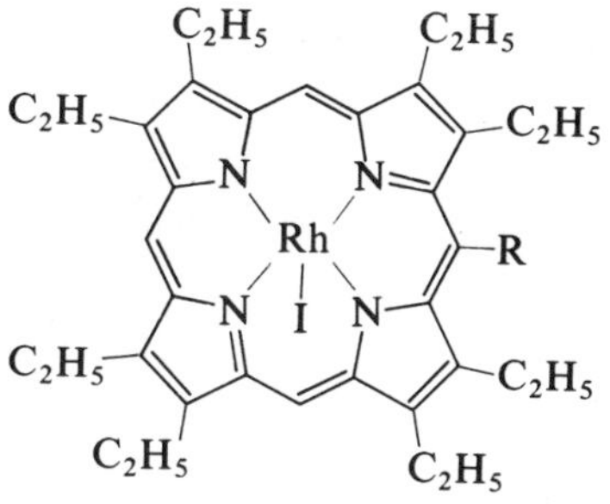

1, R = H
2, R = CN

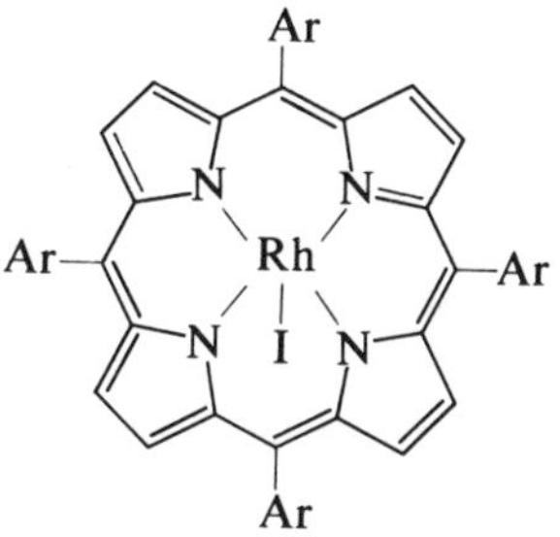

3, Ar = C_6H_5
4, Ar = $C_6H_4CH_3$-*o*
5, Ar = $C_6H_2(CH_3)_3$-2, 4, 6

Ethoxycarbonylcarbene cyclopropanation. These Rh(III) porphyrins catalyze the decomposition of ethyl diazoacetate and the transfer of the resulting carbene to alkenes to give ethyl cyclopropylcarboxylates.[1] Previously, the highest *syn/anti* selectivity has been observed with rhodium(II) pivalate (**9**, 406). The highest *syn/anti* ratios in cyclopropanation of *cis*-alkenes are observed with **5** as catalyst and to a lesser extent with **3**, both being markedly more *syn*-selective than Rh(II) pivalate. However, only slight selectivity is observed with *trans*-alkenes. *anti*-Preference is also observed with arenes such as phenanthrene. The stereochemical selectivity is thus largely a consequence of steric effects. Metalloporphyrin-catalyzed epoxidation of alkenes also shows a marked *cis*-selectivity.[2]

[1] H. J. Callot, F. Metz, and C. Piechocki, *Tetrahedron*, **38**, 2365 (1982).
[2] J. T. Groves, R. C. Haushalter, M. Nakamura, T. E. Nemo, and B. J. Evans, *Am. Soc.*, **103**, 2884 (1981), and references cited therein.

Ruthenium(III) chloride, 4, 421; **8**, 437–438; **10**, 343.

Allyl methyl ethers.[1] Allylic alcohols can be converted into allyl methyl ethers when heated in methanol in the presence of $RuCl_3 \cdot 3H_2O$. The etherification usually produces the more thermodynamically stable product of rearrangement. $RuCl_2[P(C_6H_5)_3]_3$ is inactive; $FeCl_3$, $RhCl_3$, and $PdCl_2$ are only slightly active.

Examples:

$$C_6H_5CH{=}CHCH(OH)CH_3 + CH_3OH \xrightarrow[93\%]{RuCl_3 \cdot 3H_2O,\ 60^\circ} CH_3CH(OCH_3)CH{=}CHC_6H_5$$

$$CH_2{=}CHCH(C_6H_5)OH \longrightarrow \underset{(64\%)}{CH_3OCH_2CH{=}CHC_6H_5} + \underset{(7\%)}{CH_2{=}CHCH(C_6H_5)OCH_3}$$

[1] S. Ito and M. Matsumoto, *Syn. Comm.*, **12**, 807 (1982).

Ruthenium tetroxide, 1, 986–989; **2**, 357–359; **3**, 293–294; **4**, 420–421; **6**, 504–506; **7**, 315; **8**, 438; **10**, 343.

Catalyzed oxidations.[1] In catalytic procedures with RuO_4, periodate or hypochlorite are generally used as the stoichiometric oxidants. The addition of acetonitrile, which is inert to oxidation but an effective ligand for lower valent transition metals, results in much higher yields. A third solvent, chloroform, also plays a significant part. The ruthenium tetroxide is generated *in situ* from $RuCl_3 \cdot (H_2O)_n$ or RuO_2 with sodium or potassium metaperiodate; sodium hypochlorite is less effective.

This system is highly efficient for cleavage of alkenes to carboxylic acids (75–95% yield). Examples are given of some other oxidations now possible with RuO_4.[2]

Examples:

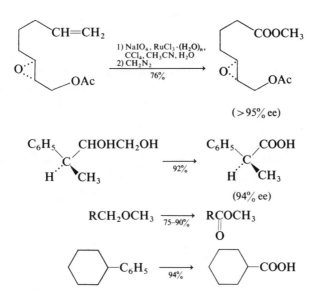

$$\text{(epoxide)}\ \mathrm{CH{=}CH_2 \ldots OAc} \xrightarrow[76\%]{1)\ NaIO_4,\ RuCl_3\cdot(H_2O)_n,\ CCl_4,\ CH_3CN,\ H_2O;\ 2)\ CH_2N_2} \text{(epoxide)}\ \mathrm{COOCH_3 \ldots OAc}$$

(>95% ee)

$$\mathrm{C_6H_5(H)(CH_3)C{-}CHOHCH_2OH} \xrightarrow{92\%} \mathrm{C_6H_5(H)(CH_3)C{-}COOH}$$

(94% ee)

$$\mathrm{RCH_2OCH_3} \xrightarrow{75\text{–}90\%} \mathrm{RC({=}O)OCH_3}$$

$$\mathrm{C_6H_{11}{-}C_6H_5} \xrightarrow{94\%} \mathrm{C_6H_{11}{-}COOH}$$

[1] P. H. J. Carlsen, T. Katsuki, V. S. Martin, K. B. Sharpless, *J. Org.*, **46**, 3936 (1981).
[2] B. E. Rossiter, T. Katsuki, and K. B. Sharpless, *Am. Soc.*, **103**, 464 (1981).

S

Samarium(II) iodide, 8, 439; **10**, 344.

Alkylation of aldehydes.[1] SmI_2 (stoichiometric amount) effects alkylation of aliphatic aldehydes with allylic and benzylic halides. The reaction is not applicable to aromatic aldehydes, which are rapidly converted to pinacols in the presence of SmI_2.

Examples:

$$n\text{-}C_7H_{15}CHO + CH_3CH{=}CHCH_2Br \xrightarrow{SmI_2,\ THF} C_7H_{15}CH(OH)CH_2CH{=}CHCH_3 \ (60\%) + C_7H_{15}CH(OH)CH(CH_3)CH{=}CH_2 \ (27\%)$$

$$n\text{-}C_7H_{15}CHO + HC{\equiv}CCH_2Br \longrightarrow C_7H_{15}CH(OH)CH_2C{\equiv}CH \ (72\%) + C_7H_{15}CH(OH)CH{=}C{=}CH \ (16\%)$$

Reductive cleavage of isoxazoles.[2] SmI_2 (2 equiv.) in the presence of CH_3OH as a proton source is an efficient reagent for reductive cleavage of the O—N bond in isoxazoles.

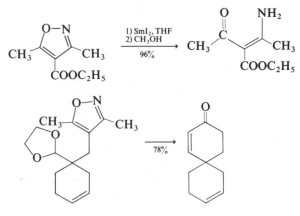

$$\text{3,5-dimethyl-4-(ethoxycarbonyl)isoxazole} \xrightarrow[96\%]{1)\ SmI_2,\ THF;\ 2)\ CH_3OH} CH_3C(O)C(COOC_2H_5){=}C(NH_2)CH_3$$

$$\text{isoxazole acetal (spiro-cyclohexene)} \xrightarrow{78\%} \text{spiro[5.5]undeca-dienone}$$

α-Diketones.[3] The reaction of an acid chloride (1 equiv.) with SmI_2 (2 equiv.) in THF results in coupling to an α-diketone as the major product. Yields are 50–80%, being the highest in the case of aliphatic acid chlorides. Trimethylacetyl chloride is converted mainly into the α-ketol, $(CH_3)_3CCOCHOHC(CH_3)_3$ (46% yield).

Selective reduction. SmI_2 is reported to effect cleavage of a 2-chloroethyl carbamate that could not be effected with Zn–HOAc or $CrCl_2$–HCl.[4]

[1] J. Souppe, J. L. Namy, and H. B. Kagan, *Tetrahedron Letters*, **23**, 3497 (1982).
[2] N. R. Natale, *ibid.*, **23**, 5009 (1982).
[3] P. Girard, R. Couffignal, and H. B. Kagan, *ibid.*, **22**, 3959 (1981).
[4] T. P. Ananthanarayan, T. Gallagher, and P. Magnus, *J.C.S. Chem. Comm.*, 709 (1982).

Selenium, 1, 990–992; **4,** 222; **5,** 575; **6,** 507–509; **8,** 439; **10,** 345.

Carbonylation of o-hydroxyacetophenone.[1] The reaction of *o*-hydroxyacetophenone (**1**) with Se (1 equiv.) and CO in the presence of DBU gives 4-hydroxycoumarin (**2**) in quantitative yield. Selenium can be used in catalytic amounts if nitrobenzene is added to oxidize H_2Se to Se.

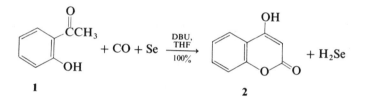

1 + CO + Se —(DBU, THF; 100%)→ **2** + H_2Se

Unsaturated β-dicarbonyl compounds.[2] An attractive route to this system involves selenylation of the enolate of β-dicarbonyl compounds in the presence of 3–6 equiv. of HMPT, followed by methylation. The resulting methyl selenide is then oxidized by H_2O_2 or *m*-$ClC_6H_4CO_3H$ in CH_2Cl_2–H_2O.

Examples:

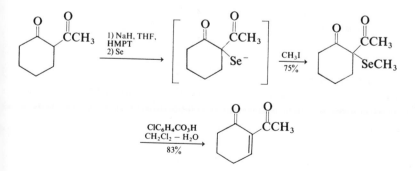

2-acetylcyclohexanone —(1) NaH, THF, HMPT; 2) Se)→ [2-acetyl-2-Se⁻-cyclohexanone] —(CH_3I; 75%)→ 2-acetyl-2-($SeCH_3$)cyclohexanone

—($ClC_6H_4CO_3H$, CH_2Cl_2 – H_2O; 83%)→ 2-acetyl-2-cyclohexenone

$$C_2H_5O\overset{O}{\overset{\|}{C}}-\underset{C_2H_5}{\underset{|}{CH}}-\overset{O}{\overset{\|}{C}}OCH_3 \xrightarrow[97\%]{} C_2H_5O_2C-\overset{SeCH_3}{\overset{|}{\underset{C_2H_5}{\underset{|}{C}}}}-CO_2CH_3 \xrightarrow[96\%]{H_2O_2,\ CH_2Cl_2,\ H_2O} C_2H_5O\overset{O}{\overset{\|}{C}}-\underset{CHCH_3}{\underset{\|}{C}}-\overset{O}{\overset{\|}{C}}OCH$$

[1] A. Ogawa, K. Kondo, S. Murai, and N. Sonoda, *J.C.S. Chem. Comm.*, 1283 (1982).
[2] D. Liotta, M. Saindane, C. Barnum, H. Ensley, and P. Balakrishnan, *Tetrahedron Letters*, **22**, 3043 (1981).

Silica, 6, 510; **9**, 910; **10**, 346–347.

Beckmann rearrangement.[1] The *p*-toluenesulfonates of an oxime rearrange to amides in 45–90% yield when stirred at room temperature with SiO_2.

Azo coupling. Silica gel facilitates solid-liquid interfacial azo coupling. The yield is increased by the presence of 4% H_2O.[2]

Example:

$$p\text{-}O_2NC_6H_4\overset{+}{N}\equiv NBF_4^- + \text{(9-ethylcarbazole)} \xrightarrow[\sim 95\%]{SiO_2\ (H_2O)} \text{3-}(N{=}NC_6H_4NO_2\text{-}p)\text{-9-ethylcarbazole}$$

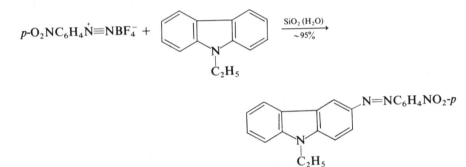

Acetylation of phenols.[3] Phenols adsorbed on silica (or Al_2O_3, ZnO, Celite) are acetylated by ketene at room temperature without an acid catalyst. Yields are almost quantitative.

[1] A. Costa, R. Mestres, and J. M. Riego, *Syn. Comm.*, **12**, 1003 (1982).
[2] S. Tamagaki, K. Suzuki, and W. Tagaki, *Chem. Letters*, 1237 (1982).
[3] T. Chihara, S. Teratani, and H. Ogawa, *J.C.S. Chem. Comm.*, 1120 (1981).

Silicon(IV) chloride–Sodium iodide.

Cleavage of ethers.[1] The actual reagent formed from these two reagents is believed to be iodotrichlorosilane and is similar to iodotrimethylsilane for cleavage of ethers but somewhat more regioselective. It also cleaves acetals and ketals quantitatively to the carbonyl compound at room temperature.

[1] M. V. Bhatt and S. S. El-Morey, *Synthesis*, 1048 (1982).

Silver hexafluoroantimonate, 5, 577; **7**, 320.

Remote hydroxylation.[1] Dehalogenation of α-bromo ketosteroids by $AgSbF_6$ generates by a series of hydride shifts cyclic oxonium salts, which are converted on hydrolysis into hydroxylated steroids.

Examples:

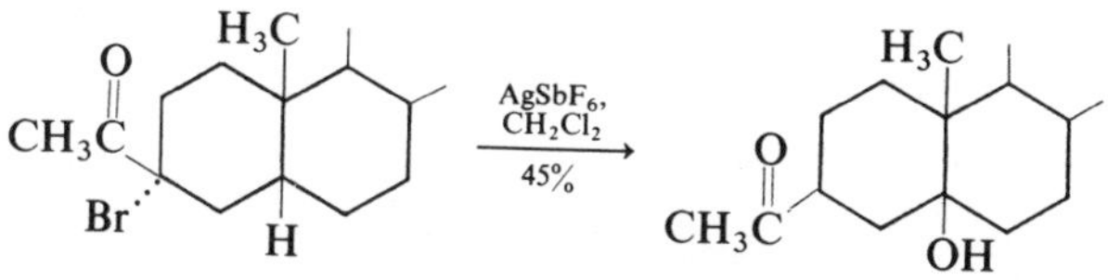

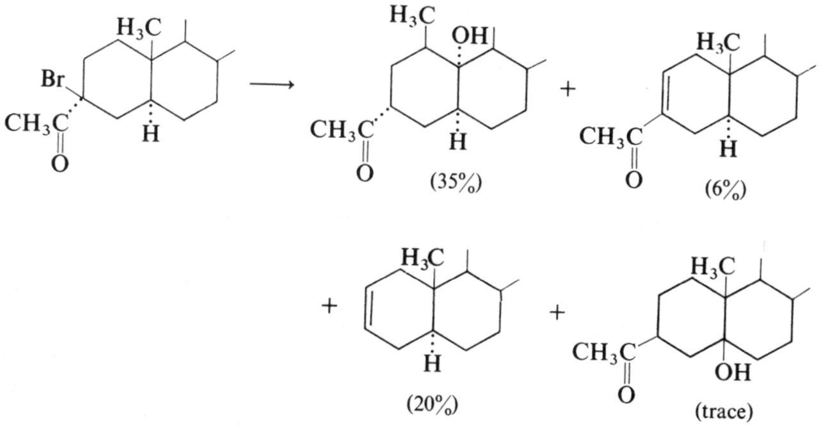

$AgSbF_6$, CH_2Cl_2 45%

(35%) (6%)

(20%) (trace)

[1] J.-P. Bégué, *J. Org.*, **47**, 4268 (1982).

Silver imidazolate, 10, 349–350.

Glycosidation. The silver imidazolate-$HgCl_2$ system favors formation of α-glucosides from reaction of 2,3,4,6-tetra-O-benzyl-α-D-glucopyranosyl chlorides and -α-D-gallactopyranosyl chlorides with sterically hindered alcohols. Addition of tetra-*n*-butylammonium chloride (1.4 equiv.) enhances the rate.[1]

The same system improves the yields of β-D-mannopyranosides in reaction of 2,3;4,6-di-O-cyclohexylidine-α-D-mannopyranosyl chloride, particularly with carbohydrates with a single free secondary hydroxyl group.[2]

[1] P. J. Garegg, C. Ortega, and B. Samuelsson, *Acta. Chem. Scand.*, **B35**, 161 (1981).
[2] P. J. Garegg, R. Johansson, and B. Samuelsson, *ibid.*, **B35**, 635 (1981).

Silver nitrite–Mercury(II) chloride.

Nitroselenenylation.[1] A variation of Corey and Estreicher's nitromercuration of alkenes as a route to 1-nitroalkenes (**9**, 292–293) is based on nitroselenenylation of alkenes (equation I). The yield of **4** is somewhat lower than that previously obtained by nitromercuration, but has the advantage of use of a nonaqueous medium.

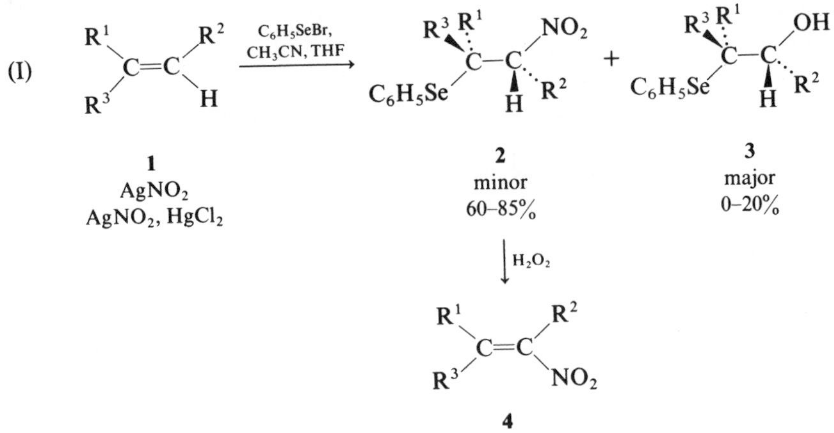

(I) $R^1R^3C{=}CHR^2$ (**1**) $\xrightarrow{C_6H_5SeBr,\ CH_3CN,\ THF}$ $C_6H_5Se{-}CR^1R^3{-}CH(NO_2)R^2$ (**2**, minor) + $C_6H_5Se{-}CR^1R^3{-}CH(OH)R^2$ (**3**, major)

	2	**3**
	minor	major
	60–85%	0–20%
$AgNO_2$		
$AgNO_2$, $HgCl_2$		

2 $\xrightarrow{H_2O_2}$ $R^1R^3C{=}CR^2NO_2$ (**4**)

[1] T. Hayama, S. Tomoda, Y. Takeuchi, and Y. Nomura, *Tetrahedron Letters*, **23**, 4733 (1982).

Silver(I) oxide, 1, 1011; **2**, 368; **3**, 252–254; **4**, 930–931; **5**, 583–585; **6**, 515–518; **7**, 321–322; **8**, 442–443; **10**, 350–352.

CHO→COOH. Oxidation of the complement inhibitor K-76 (**1**) in 1 *N* NaOH with excess Ag_2O cleanly affords a monocarboxylic acid, shown to be **2** by lactonization experiments. Thus the formyl group *meta* to the phenolic hydroxyl is the more reactive one.[1]

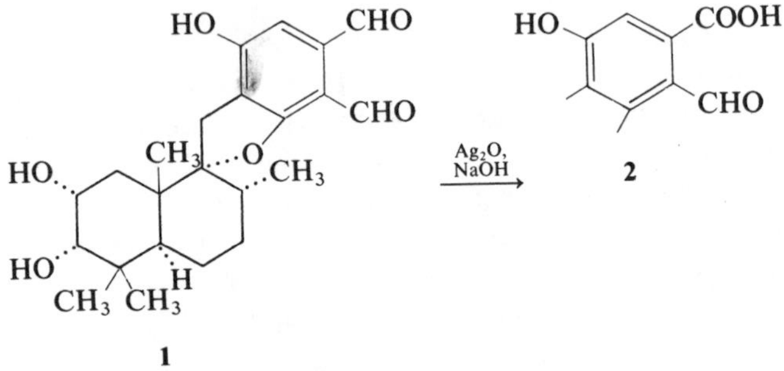

1 $\xrightarrow{Ag_2O,\ NaOH}$ **2**

Alkaline silver oxide is the most satisfactory reagent for oxidation of the aldehyde **3** to the acid **4**.[2] This reagent has been recommended for oxidation of aryl aldehydes (**1**, 1012–1013), but only moderate yields have been reported for oxidation of α,β-enals with Ag_2O.[3] Apparently, the successful results in the oxidation of **3** are the result of the activating effect of the keto group.

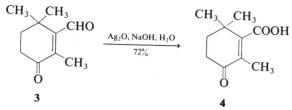

CH_3 CH_3 CHO CH_3 O **3** — Ag_2O, NaOH, H_2O; 72% → CH_3 CH_3 COOH CH_3 O **4**

[1] E. J. Corey and J. Das, *Am. Soc.*, **104**, 5551 (1982).
[2] A. B. Pepperman, *J. Org.*, **46** 5039 (1981).
[3] S. C. Thomason and D. G. Kubler, *J. Chem. Ed.*, **45**, 546 (1968).

Silver(II) oxide, 1, 1011–1015; **2**, 368; **3**, 252–254; **4**, 430–431; **5**, 583–585; **6**, 575–578; **7**, 321–322; **9**, 412–413; **10**, 352–354.

Hydroxyquinones. A substituted methyl sesamol (**1**) is oxidized by AgO to a hydroxyquinone (**2**).[1]

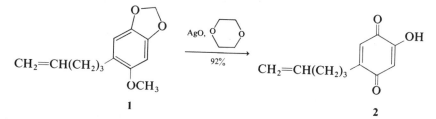

$CH_2{=}CH(CH_2)_3$ … OCH_3 **1** — AgO, dioxane; 92% → $CH_2{=}CH(CH_2)_3$ … OH **2**

[1] G. A. Kraus and K. Neuenschwander, *Syn. Comm.*, **10**, 9 (1980).

Silver perchlorate, 2, 369–370; **4**, 432–435; **5**, 585–587; **6**, 518–519; **7**, 322–323; **9**, 413–414; **10**, 354–355.

***Hydration of* 2-*butyne*-1,4-*diols* (1).** Treatment of **1** with acetic anhydride and pyridine (25°) results in acetylation of the less hindered hydroxyl group. When treated with $AgClO_4$, the monoacetate (**2**) is converted into the enol acetate of a dihydro-3(2H)-furanone (**3**, equation I). The regioisomeric dihydro-3(2H)-furanone is formed by hydration of **1** with $Hg(OAc)_2$–Nafion–H preferentially (equation II).[1]

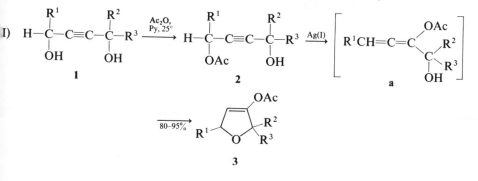

I) $H{-}C(R^1)(OH){-}C{\equiv}C{-}C(R^2)(OH){-}R^3$ (**1**) $\xrightarrow{Ac_2O,\ Py,\ 25^\circ}$ $H{-}C(R^1)(OAc){-}C{\equiv}C{-}C(R^2)(OH){-}R^3$ (**2**) $\xrightarrow{Ag(I)}$ [$R^1CH{=}C{=}C(OAc){-}C(R^2)(R^3)OH$] (**a**) $\xrightarrow{80\text{–}95\%}$ **3** (OAc, R^1, R^2, R^3, O)

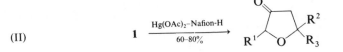

(II) $\mathbf{1} \xrightarrow[60\text{–}80\%]{Hg(OAc)_2\text{–Nafion-H}}$ [tetrahydrofuran-3-one with R^1, R^2, R_3]

[3+4] *Cycloadditions.* Silver perchlorate converts the 1,1-dimethyl-2-(trimethylsiloxy)allyl chloride (**1**) and related allylic chlorides into an oxyallyl cation (**a**), which reacts with 1,3-dienes to form seven-membered ketones, often in high yield, particularly in reactions conducted in nitromethane.

Examples:

$$(CH_3)_2C(Cl)\text{–}C(OSi(CH_3)_3)\text{=}CH_2 \ (\mathbf{1}) \xrightarrow{AgClO_4,\ CH_3NO_2} \left[(CH_3)_2C\cdots\overset{+}{C}(OSi(CH_3)_3)\cdots CH_2\ \ ClO_4^-\right] (\mathbf{a}) \xrightarrow[92\%]{\text{furan}} \mathbf{2}$$

$$(CH_3)_2C(Cl)\text{–}C(OSi(CH_3)_3)\text{=}C(H)Cl \ (\mathbf{3}) \xrightarrow[89\%]{Ag^+,\ \text{furan}} \mathbf{4}$$

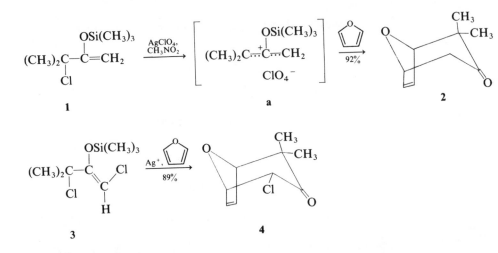

The course of the reaction, however, is markedly dependent on the solvent. Thus, when the reaction of **3** with furane is carried out in the THF/ether, the main product (**5**) results from substitution. Marked differences in regio- and stereoselectivities are also observed in the two solvents.[2]

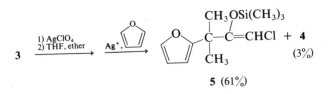

$$\mathbf{3} \xrightarrow{1)\ AgClO_4\ \ 2)\ THF,\ ether} \xrightarrow{Ag^+,\ \text{furan}} \text{2-furyl–}C(CH_3)_2\text{–}C(OSi(CH_3)_3)\text{=CHCl}\ \ \mathbf{5}\ (61\%) + \mathbf{4}\ (3\%)$$

[1] H. Saimoto, T. Hiyama, and H. Nozaki, *Am. Soc.*, **103**, 4975 (1981).

[2] N. Shimizu, M. Tanaka, and Y. Tsuno, *ibid.*, **104**, 1330 (1982).

Silver tetrafluoroborate, 1, 1015–1018; **2**, 365–366; **3**, 250–251; **4**, 428–429; **5**, 587–588; **6**, 519–520; **8**, 443–444; **9**, 414.

Methyl α-arylalkanoates. Treatment of methyl ketals of α-halogenoalkyl aryl ketones, prepared *in situ*, with a number of silver salts, particularly silver tetrafluoroborate or hexafluoroantimonate, in methanol induces a rearrangement to methyl α-arylalkanoates in yields generally around 98%. The rate is influenced by substituents on the aryl group; the oxygens of the ketal group greatly facilitate this reaction (equation I).[1]

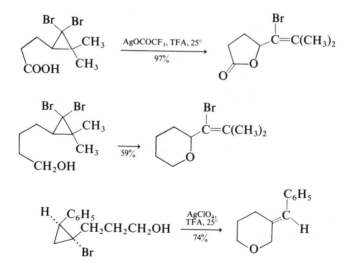

(I) ArC(=O)CHBr(Cl)(R) $\xrightarrow{HC(OCH_3)_3,\ H^+}$ [ArC(OCH$_3$)$_2$–CHBr(Cl)(R)] $\xrightarrow[\sim 98\%]{Ag(I),\ CH_3OH}$ ArCH(R)COOCH$_3$

[1] C. Giordano, G. Castaldi, F. Casagrande, and A. Belli, *J.C.S. Perkin I*, 2575 (1982).

Silver(I) trifluoroacetate, 1, 1018–1019; **7**, 323–324; **8**, 444–445; **10**, 355.

Cleavage-cationic cyclization of bromocyclopropanes. Bromocyclopropanes substituted by a suitably situated hydroxyl or carboxyl group undergo cleavage-cyclization at 25° in the presence of 1–1.5 equiv. of $AgOCOCF_3$ or $Hg(OAc)_2$. Vinyl lactones, tetrahydropyranes, or tetrahydrofuranes can be prepared in this way.[1]

Examples:

(Br, Br, CH$_3$, CH$_3$-cyclopropane with COOH side chain) $\xrightarrow[97\%]{AgOCOCF_3,\ TFA,\ 25°}$ (γ-lactone)–C(Br)=C(CH$_3$)$_2$

(Br, Br, CH$_3$, CH$_3$-cyclopropane with CH$_2$OH side chain) $\xrightarrow{59\%}$ (tetrahydropyran)–C(Br)=C(CH$_3$)$_2$

(H, C$_6$H$_5$, Br-cyclopropane)–CH$_2$CH$_2$CH$_2$OH $\xrightarrow[74\%]{AgClO_4,\ TFA,\ 25°}$ (tetrahydropyran)=C(C$_6$H$_5$)H

[1] R. L. Danheiser, J. M. Morin, Jr., M. Yu, and A. Basak, *Tetrahedron Letters*, **22**, 4205 (1981).

Simmons-Smith reagent, 1, 1019–1922; **2**, 371–372; **3**, 255–258; **4**, 436–437; **5**, 588–589; **6**, 521–523; **8**, 445; **9**, 415.

Cyclopropanation.[1] Ultrasound irradiation facilitates Simmons-Smith cyclopropanation. Ordinary zinc can be used. The reaction is faster and yields are considerably improved. 1,2-Dimethoxyethane is recommended as solvent in place of the usual solvent (ether).

[1] O. Repič and S. Vogt, *Tetrahedron Letters*, **23**, 2729 (1982).

Sodium–Ammonia, 1, 1041; **2**, 374–376; **3**, 259; **4**, 438; **5**, 589–591; **6**, 523; **7**, 324–325; **9**, 415; **10**, 355–356.

Metal-ammonia reduction of ketones. Swiss chemists[1] have reported a detailed study of the mechanism of this reaction, using for the most part (+)-[3,3-D_2] camphor as substrate. The conclusions drawn have some useful practical applications. The choice of metal (Li, Na, or K) has little effect on the course of reduction to the thermodynamically more stable diastereoisomeric alcohol. The most important conclusion is that pinacol reduction can be suppressed completely in Na–NH_3 reductions by use of ammonium chloride as the proton source, a finding first reported by Murphy and Sullivan.[2] This salt also partially suppresses pinacol formation in Li–NH_3 reductions. It also suppresses reduction of enolates, and thus should decrease racemization in reduction of chiral ketones.

[1] V. Rautenstrauch, B. Willholm, W. Thommen, and U. Burger, *Helv.*, **64**, 2109 (1981).
[2] W. S. Murphy and D. F. Sullivan, *Tetrahedron Letters*, 3707 (1971).

Sodium–Ethanol, 9, 916.

Desulfonylation of α-sulfonylacetates. Desulfonylation of these substrates is not satisfactory by the usual methods (Zn-HOAc, Raney Ni–C_2H_5OH, sodium amalgam), but can be conducted in 70–75% yield with sodium–ethanol in THF (modified Bouveault-Blanc reduction). The substrates (**2**) are obtained by Wolff rearrangement of α-acyl-α-benzyl-sulfonyldiazomethanes (**1**).[1]

$$\mathrm{RCOCl} \xrightarrow{C_6H_5CH_2SO_2CHN_2} \underset{\mathbf{1}}{\mathrm{RCO\underset{\large SO_2CH_2C_6H_5}{\overset{|}{C}}{=}N_2}} \xrightarrow[60\text{–}85\%]{H_2O,\ C_6H_5CH_3}$$

$$\underset{\mathbf{2}}{\mathrm{R\underset{\large SO_2CH_2C_6H_5}{\overset{|}{C}}HCOOH}} \xrightarrow[70\text{–}75\%]{Na,\ C_2H_5OH,\ THF} \underset{\mathbf{3}}{\mathrm{RCH_2COOH}}$$

[1] Y.-C. Kuo, T. Aoyama, and T. Shioiri, *Chem. Pharm. Bull.*, 2787 (1982).

Sodium–Hexamethylphosphoramide, 9, 416.

Selective cleavage of methoxythioanisoles.[1] *o*-, *m*-, and *p*-Methoxythioanisoles are selectively cleaved by sodium in HMPT to methoxythiophenols by cleavage of the CH_3—S bond. Cleavage of the C—O bond occurs to a minor extent (less than 5%) in the case of *o*- and *m*-methoxythioanisoles. In contrast, sodium isopropanethiolate, $(CH_3)_2CHSNa$,[2] cleaves the same substrates selectively to the sodium salts of (methylthio)phenols. Again, this process is not completely selective in the case of *o*- and *m*-methoxythioanisoles. Sodium isopropanethiolate also effects dealkylation of aryl alkyl sulfides when HMPT is used as solvent, and this reaction provides a convenient synthesis of aromatic thiols.[3]

The sodium salts of (methylthio)phenols, $Ar(ONa)SCH_3$ are also cleaved by sodium and HMPT. Whereas $C_6H_4(OCH_3)_2$ is cleaved by sodium only to $C_6H_4(OCH_3)ONa$, $C_6H_4(SCH_3)_2$ can be cleaved by sodium to $C_6H_4(SNa)_2$.

Reductive cleavage of 2H-thiopyrane **(1)**.[4] Treatment of **1** with 2 equiv. of sodium in HMPT generates a dianion (**a**) by a selective cleavage of the S—CH_2 bond. Protonoloysis followed by methylation yields the pure (Z,Z)-diunsaturated methyl sulfide **2** with retention of configuration of the double bonds.

$$\mathbf{1} \xrightarrow[\text{THF}]{\text{2Na, HMPT,}} \left[\text{CH}_2\text{Na},\ \text{SNa} \right] \mathbf{a} \xrightarrow[\text{2) CH}_3\text{I}]{\text{1) (CH}_3)_3\text{COH}} \mathbf{2}\ (\text{CH}_3,\ \text{SCH}_3)$$

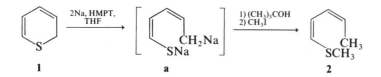

Reductive cleavage of the potassium anions (**3**) of (methylthio)pentadienes with Li in NH_3 followed by methylation gives **4** with retention of configuration of the C_1—C_2 bond, but the C_3—C_4 bond is always (Z) regardless of the configuration in **3**.

$$\underset{\substack{\mathbf{3},\ 1Z,3Z \\ \mathbf{3},\ 1E,3E}}{CH_3S\overset{(Z\ or\ E)}{CH{=}CH}{-}\overset{(Z\ or\ E)}{CH{=}CH}{-}CH_2K} \xrightarrow[\text{2) CH}_3\text{I}]{\text{1) 2 Li, NH}_3} \underset{\substack{\mathbf{4},\ 1Z,3Z = \mathbf{2} \\ \mathbf{4},\ 1E,3Z}}{CH_3S\overset{(Z\ or\ E)}{CH{=}CH}{-}\overset{(Z)}{CH{=}CH}{-}CH_3}$$

[1] L. Testaferri, M. Tiecco, M. Tingoli, D. Chianelli, and F. Maiolo, *Tetrahedron*, **38**, 2721 (1982).
[2] L. Testaferri, M. Tingoli, and M. Tiecco, *J. Org.*, **45**, 4376 (1980).
[3] *Idem, Tetrahedron Letters*, **21**, 3099 (1980).
[4] R. Gräfing, A. V. E. George, and L. Brandsma, *Rec. trav.*, **101**, 346 (1982).

Sodium amalgam, 1, 1030–1033; **2**, 373; **3**, 259; **7**, 326–327; **9**, 416–417; **10**, 355–356.

Alkene synthesis. Some years ago Julia and Paris[1] reported an alkene synthesis by reaction of β-alkoxy sulfones with sodium amalgam (equation I).

(I) $R^1C(SO_2C_6H_5)(R^2)—C(OR)(R^4)(R^3) \xrightarrow[55–85\%]{Na/Hg,\ 25°,\ C_2H_5OH} R^1(R^2)C{=}C(R^3)R^4$

($R = Ac, SO_2CH_3,$ *p*-Ts)

Various aspects of this reaction have been summarized by Kocienski.[2] It is particularly useful for synthesis of disubstituted alkenes and conjugated dienes and trienes. It fails with some trisubstituted alkenes and most tetrasubstituted alkenes because the precursors are unstable. One advantage of this route is that *trans*-alkenes are formed preferentially or exclusively. Yields are highest when the eliminated groups can adopt a *trans*-coplanar arrangement.

Reductive elimination with sodium amalgam was used in a recent synthesis of diumycinol (**4**) to introduce two of the hindered double bonds. Thus, this reaction was successful in converting **1** to **2** when alternative methods failed and in introducing the *trans*-double bond of **4** more stereoselectively than other methods tried.[3]

1 $\xrightarrow[75\%]{Na(Hg),\ THF,\ CH_3OH}$ **2**

3 $\xrightarrow[33\%]{1)\ Na(Hg)\ \ 2)\ Na\text{-}NH_3}$ **4**

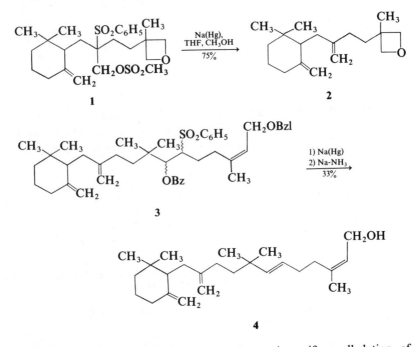

4-*Alkyl-2-cyclohexenones*.[4] A method for regiospecific γ-alkylation of 2-cyclohexenones involves alkylation of the ketal sulfone **1**, prepared as indicated, to give **2**. Conversion of **2** into the cyclohexenones **4** and **5** involves desulfonylation and

$(CH_3)_3SiO$ CH_2 $CHOCH_3$

1) $CH_2=CHSO_2C_6H_5$, Δ
2) $HOCH_2CH_2OH$, TsOH
85%

$SO_2C_6H_5$
1

1) NaH, DMF
2) RX
75–95%

R $SO_2C_6H_5$
2

Na(Hg), CH_3OH
Na_2HPO_4
90–100%

R
3a

+ ~65:35

R H
3b

TsOH,
C_2H_5OH
72–93%

R H
4

+ ~70:30

R
5

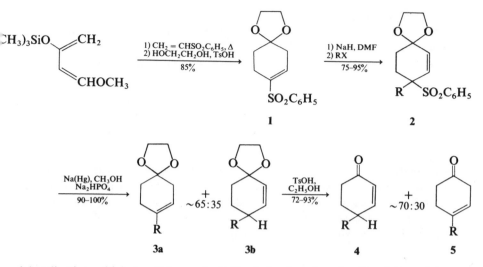

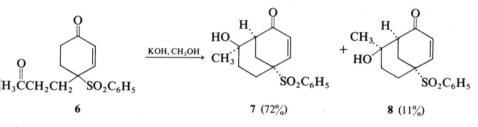

deketalization, which should be conducted in that order, since desulfonylation of the α,β-unsaturated ketones with sodium amalgam is not efficient.

This method was used to prepare **6**, an intermediate to the bicyclo keto alcohols **7** and **8**.

$H_3CCH_2CH_2$ $SO_2C_6H_5$
6

KOH, CH_3OH

H HO CH_3 $SO_2C_6H_5$
7 (72%)

+

H CH_3 HO $SO_2C_6H_5$
8 (11%)

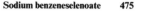

[1] M. Julia and J. -M. Paris, *Tetrahedron Letters*, 4933 (1973).
[2] P. J. Kocienski, *Chem. Ind.*, 548 (1981).
[3] P. Kocienski and M. Todd, *J.C.S. Chem. Comm.*, 1078 (1982).
[4] L. A. Paquette and W. A. Kinney, *Tetrahedron Letters*, **23**, 131 (1982).

Sodium benzeneselenoate, 5, 273; **6,** 548–549; **8,** 447–448; **10,** 356.

Cleavage of esters and lactones **(8,** 447–448).[1] S_N2-Cleavage of esters and lactones with $NaSeC_6H_5$ in HMPT–THF is considered to be the most effective known method. The order of reactivity of related reagents is $NaSeC_6H_5$–18-crown-6–THF > $LiSeC_6H_5$–HMPT–THF > $LiSeC_6H_5$–THF. Methyl esters are cleaved more readily than ethyl esters; benzyl, isoamyl, and isopropyl esters are cleaved in high yield, but more slowly. Methyl esters are cleaved easily, irrespective of steric hindrance. Amides are completely inert.

[1] D. Liotta, U. Sunay, H. Santiesteban, and W. Markiewicz, *J. Org.*, **46**, 2605 (1981).

Sodium bicarbonate, 5, 595.

Nitrosoalkenes.[1] A general route to nitrosoalkenes involves dehydrohalogenation of β-chloro oximes in CH_3OH or C_2H_5OH with $NaHCO_3$ (equation I). Formation of a blue or green color is evidence for formation of the nitrosoalkenes, but few are

(I) $R^2R^3C(Cl)\text{—}C(=NOH)R^1 \xrightarrow{NaHCO_3} R^2R^3C\text{=}C(R^1)N\text{=}O$

sufficiently stable for isolation. They can be stored for a few weeks at $-20°$, but are usually generated for cycloaddition reactions in the presence of the 1,3-diene. The most common reaction is [4+2] cycloaddition involving the N=O bond to form epoxyepimines (equation II). However, nitrosoalkenes that are not substituted in the β-position form oxazines as the adducts (equation III).

(II) cyclopentadiene + O=NCH=C(Cl)C(CH_3)$_3$ ⟶ [bicyclic adduct N—CH=C(Cl)C(CH_3)$_3$] $\xrightarrow{90\%}$ epoxy aziridine N—CH=C(Cl)C(CH_3)$_3$

(III) cyclopentadiene + O=NC(CH$_2$Cl)=CH$_2$ $\xrightarrow{72\%}$ bicyclic oxazine (H, O, N, H, CH$_2$Cl)

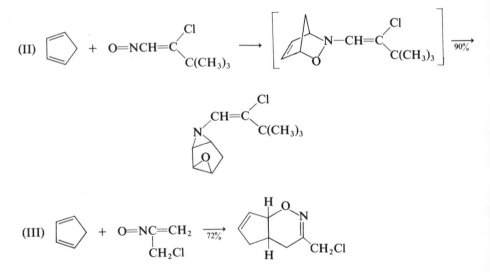

[1] E. Francotte, R. Merényi, B. Vandenbulcke-Coyette, and H. -G. Viehe, *Helv.* **64**, 1208 (1981).

Sodium bis(2-methoxyethoxy)aluminium hydride (SMEAH), **3**, 260–261; **4**, 441–442; **5**, 596; **6**, 528–529; **7**, 327–329; **8**, 998–999; **9**, 418–420; **10**, 357.

Reduction of α,β-epoxy alcohols. Reduction of 3-substituted-2,3-epoxy alcohols by SMEAH, particularly in THF, gives essentially only the 1,3-diol. However, a substituent at C_2 reverses the selectivity and decreases the reaction rate. The hydroxyl group also plays a role; the corresponding benzyl ethers show only a slight preference for reduction at C_3.[1]

Examples:

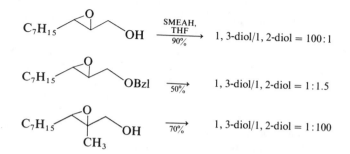

$$C_7H_{15}\text{-epoxide-}CH_2OH \xrightarrow[90\%]{\text{SMEAH, THF}} \text{1,3-diol/1,2-diol} = 100:1$$

$$C_7H_{15}\text{-epoxide-}CH_2OBzl \xrightarrow{50\%} \text{1,3-diol/1,2-diol} = 1:1.5$$

$$C_7H_{15}\text{-epoxide}(CH_3)\text{-}CH_2OH \xrightarrow{70\%} \text{1,3-diol/1,2-diol} = 1:100$$

[1] S. M. Viti, *Tetrahedron Letters*, **23**, 4541 (1982).

Sodium borohydride, 1, 1049–1055; **2**, 377–378; **3**, 262–264; **4**, 443–444; **5**, 597–601; **6**, 530–539; **7**, 329–331; **8**, 449–451; **9**, 420–421; **10**, 357–359.

Reductive amination; amides and lactams.[1] The α-nitro ketone **1**, prepared as shown, on reduction with $NaBH_4$ in the presence of ammonia gas is converted into the amide **2** in 30% yield.

$$C_2H_5C(=O)CH(NO_2)CH_3 + CH_2{=}CHCHO \xrightarrow[70\%]{P(C_6H_5)_3} C_2H_5C(=O)C(NO_2)(CH_3)(CH_2)_2CHO \; (\mathbf{1}) \xrightarrow[30\%]{\text{1) } NH_3\text{, 2) } NaBH_4\text{, HOAc}} C_2H_5C(=O)NH(CH_2)_3CH(NO_2)CH_3 \; (\mathbf{2})$$

A similar strategy was used to convert α-nitrocyclododecanone (**3**) into the lactam (**5**).

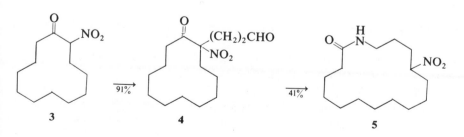

$$\mathbf{3} \text{ (2-nitrocyclododecanone)} \xrightarrow{91\%} \mathbf{4} \text{ (2-(CH}_2)_2\text{CHO-2-NO}_2\text{-cyclododecanone)} \xrightarrow{41\%} \mathbf{5} \text{ (NO}_2\text{-substituted macrocyclic lactam, O=C–NH)}$$

Reduction of halides and sulfonates.[2] Sodium borohydride effects this reaction under phase-transfer conditions (aqueous $C_6H_5CH_3$ or CH_2Cl_2). The more lipophilic salts hexadecyltributylphosphonium bromide or tetraoctylammonium bromide are the most effective catalysts.

$RN_3 \rightarrow RNH_2$.[3] The reduction of azides to amines normally proceeds in poor yield, but can proceed in 79–92% yield if carried out with $NaBH_4$ in toluene–water in the presence of hexadecyltributylphosphonium bromide as phase-transfer catalyst.

Reduction of epoxides.[4] Epoxides can be reduced by slow addition of methanol to a refluxing mixture of the epoxide and $NaBH_4$ in *t*-butyl alcohol. The more substituted alcohol is formed preferentially. Presumably the actual reducing agent is $NaBH_{4-n}(OCH_3)_n$. Amide, nitro, and nitrile groups are not reduced by the procedure.

Deoxygenation of naphthalene-1,4-endoxides. A new method for conversion of these Diels-Alder adducts of benzynes with furanes to naphthalenes consists in reduction with sodium borohydride in trifluoroacetic acid.[5] Excess acid is used when the substrate bears methyl groups at the bridgeheads. Substrates lacking such groups tend to undergo acid-catalyzed rearrangement to naphthols, but are reduced satisfactorily with THF as solvent and a limited amount of acid.

Examples:

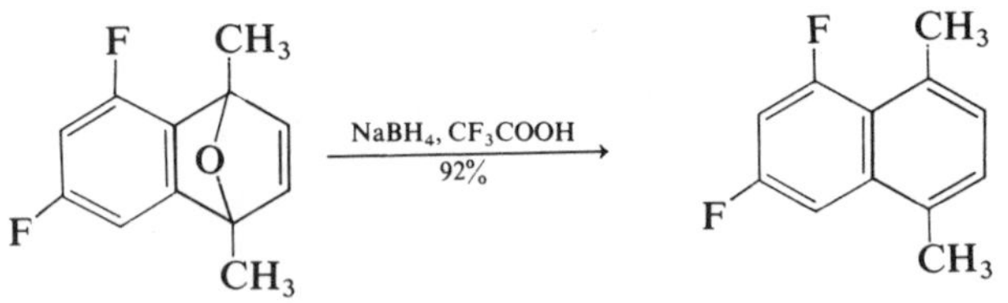

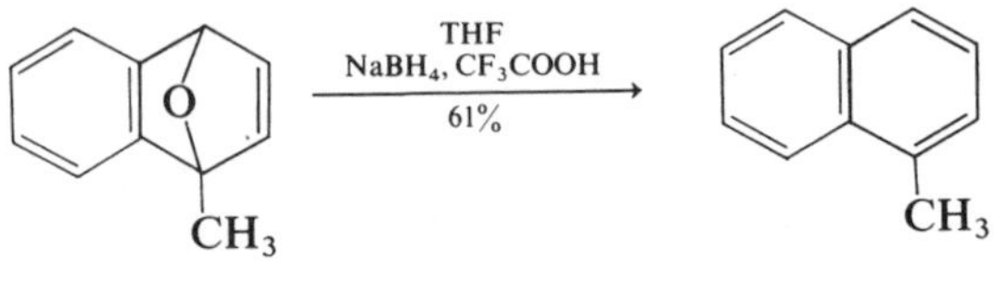

$NaBH_4$, CF_3COOH; 92%

$NaBH_4$, CF_3COOH, THF; 61%

Other recent methods for this conversion have been summarized by Hart and Nwokogu.[6]

Reduction of thiol esters. Thiol esters, R^1COSR^2, are reduced to primary alcohols, R^1CH_2OH, in high yield by $NaBH_4$ in C_2H_5OH without reduction of nitrile, ester, and amide substituents.[7]

Thiol esters can be prepared directly from carboxylic acids by reaction with thiols in the presence of phenyl dichlorophosphate as the activating agent (equation I).[8]

$$\text{(I)}\quad R^1COOH \xrightarrow[\text{Py}]{C_6H_5OPOCl_2,} \left[R^1\overset{O}{\overset{\|}{C}}O\overset{O}{\overset{\|}{\underset{OC_6H_5}{\underset{|}{P}}}}Cl \right] \xrightarrow[80\text{–}100\%]{R^2SH} R^1\overset{O}{\overset{\|}{C}}SR^2$$

Selective reduction of esters.[9] Esters can be reduced to primary alcohols by slow addition of methanol to a solution of the ester and sodium borohydride in refluxing *t*-butyl alcohol (78–90% yield). Under these conditions, amide, nitrile, and carboxylic acid groups are not affected.

[1] R. Wälchli and M. Hesse, *Helv.*, **65**, 2299 (1980).
[2] F. Rolla, *J. Org.*, **46**, 3909 (1981).
[3] *Idem, ibid.*, **47**, 4327 (1982).
[4] K. Soai, A. Ookawa, H. Oyamada, and M. Takase, *Heterocycles*, **19**, 1371 (1982).
[5] G. W. Gribble, W. J. Kelly, and M. P. Sibi, *Synthesis*, 143 (1982).
[6] H. Hart and G. Nwokogu, *J. Org.*, **46**, 1251 (1981).
[7] H. -J. Liu, R. B. Bukownik, and P. R. Pednekar, *Syn. Comm.*, **11**, 599 (1981).
[8] H. J. Liu and S. I. Sabesan, *Can. J. Chem.*, **58**, 2645 (1980).
[9] K. Soai, *Syn. Comm.*, **12**, 463 (1982).

Sodium borohydride–Cobalt(II) chloride, 10, 359–360.

Deoxygenation of sulfoxides.[1] Sulfoxides are reduced to sulfides by $NaBH_4$ (excess) and 1 equiv. of $CoCl_2 \cdot 6H_2O$ in ethanol in 45–95% yield. Sulfones are inert to this reagent.

[1] S. -K. Chung and G. Han, *Syn. Comm.*, **12**, 903 (1982).

Sodium borohydride–Methanesulfonic acid.

Reduction of amides.[1] Sodium borohydride combined with methanesulfonic acid in DMSO reduces amides to the corresponding amines in 60–90% isolated yield. The system also reduces acids and esters to primary alcohols. These reductions have been conducted with lithium aluminum hydride and with borane–tetrahydrofurane (**5**, 48),[2] but with somewhat different selectivities. This new reagent, however, appears to be less hazardous than the latter reagent.

[1] S. R. Wann, P. T. Thorsen, and M. M. Kreevoy, *J. Org.*, **46**, 2579 (1981).
[2] C. F. Lane, *Chem. Rev.*, **76**, 773 (1976).

Sodium borohydride–Palladium (II) chloride.

Reductions.[1] This combination of $NaBH_4$–$PdCl_2$ (5 : 1) reduces diaryl ketones or benzylic alcohols to the hydrocarbon (THF/H_2O, 7°) in ~ 70–90% yield. This system (CH_3OH, 20°) reduces chloroarenes to arenes (50–70% yield). The system also reduces 17-keto steroids to 17β-ols in high yield.

2,2-Dimethylchromenes.[2] 2,2-Dimethylchromanones (**1**) are converted to the corresponding chromenes **2** in good yield by reduction with $NaBH_4$–$PdCl_2$ followed by dehydration. $NaBH_4$ alone is less satisfactory. The carbonyl group is reduced by the modified reagent more easily than an ester group or an acetyl group.

1) $NaBH_4 - PdCl_2(\sim 9:1)$
2) $KHSO_4, C_6H_5CH_3, \Delta$
75–95%

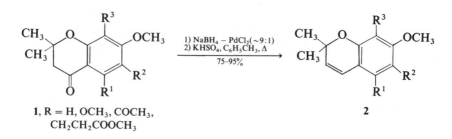

1, R = H, OCH_3, $COCH_3$, $CH_2CH_2COOCH_3$

2

[1] T. Satoh, N. Mitsuo, M. Nishiki, K. Nanba, and S. Suzuki, *Chem. Letters*, 1029 (1981).
[2] M. Tsukayama, T. Sakamoto, T. Horie, M. Masumura, and M. Nakayama, *Heterocycles*, **16**, 955 (1981).

Sodium borohydride–Rhodium(III) chloride.

Reduction of arenes. A number of phenyl-substituted compounds can be reduced to the corresponding cyclohexyl derivatives by $NaBH_4$ that has been incubated with $RhCl_3$ in ethanol at 0–30°.[1]

Examples:

$C_6H_5CH_2CH(NHCOCH_3)COOH$ → (1) $RhCl_3$, C_2H_5OH, 30°; 2) $NaBH_4$; 99%) cyclohexyl-$CH_2CH(NHCOCH_3)COOH$

4-pyridyl-$COOC_2H_5$ → (60%) HN(piperidin-4-yl)-$COOC_2H_5$

[1] M. Nishiki, H. Miyataka, Y. Niino, N. Mitsuo, and T. Satoh, *Tetrahedron Letters*, **23**, 198 (1982).

Sodium borohydride–Tin(II) chloride.

Reduction of aryl nitro compounds.[1] $NaBH_4$–$SnCl_2 \cdot 2H_2O$ (0.5 : 5 equiv.) in C_2H_5OH selectively reduces aromatic nitro groups to amino groups in the presence of C=O, COOR, C≡N, and C=C groups. Yields are 70–95%, being highest for nitro compounds containing *p*-electron-attracting substituents.

[1] T. Satoh, N. Mitsuo, M. Nishiki, Y. Inoue, and Y. Ooi, *Chem. Pharm. Bull.*, **29**, 1443 (1981).

Sodium chlorite, 5, 603.

Oxidation of α,β-enals. Sodium chlorite is superior to other reagents for oxidation of α,β-unsaturated aldehydes to the corresponding carboxylic acid, particularly of aldehydes containing an α-methylene group. In addition, the oxidation is stereospecific. 2-Methyl-2-butene[2] was used as the chlorine scavenger.

Examples:

$$C_6H_5CH{=}CHCHO \xrightarrow[95\%]{NaClO_2} C_6H_5CH{=}CHCOOH$$

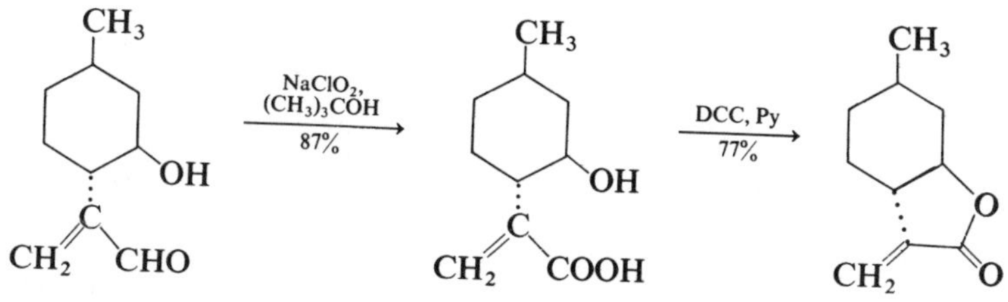

[1] B. S. Bal, W. E. Childers, Jr., and H. W. Pinnick, *Tetrahedron*, **37**, 2091 (1981); B. S. Bal and H. W. Pinnick, *Heterocycles*, **16**, 2091 (1981).
[2] G. A. Kraus and M. J. Taschner, *J. Org.*, **45**, 1175 (1980).

Sodium cyanide–Chlorotrimethylsilane–Sodium iodide.

Nitriles.[1] Alcohols are converted directly into nitriles by reaction with NaCN and $ClSi(CH_3)_3$ and a catalytic amount of NaI. $DMF/CH_3CN(50/50/v/v)$ is the most satisfactory solvent. Presumably, $ISi(CH_3)_3$ is formed *in situ*. This reaction is applicable to alcohols of all types; it proceeds with inversion in the case of 3β-cholestanol but with retention in the case of cyclohexanol.

Examples:

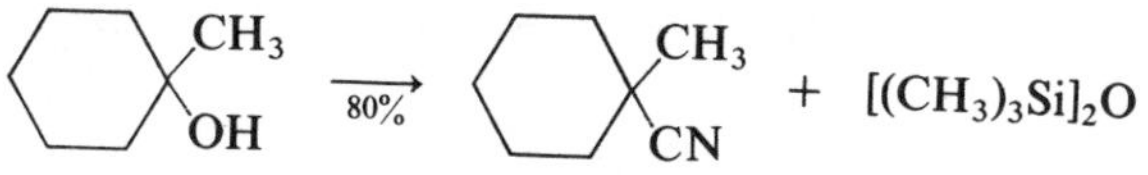

$$C_6H_5CH_2OH \xrightarrow[98\%]{} C_6H_5CH_2CN$$

[1] R. Davis and K. G. Untch, *J. Org.*, **46**, 2985 (1981).

Sodium cyanoborohydride, 4, 448–451; **5,** 607–609; **6,** 537–538; **7,** 334–335; **8,** 454; **9,** 424–426; **10,** 360–361.

Reduction of acetals (*cf.*, **8,** 454). 4,6-O-Benzylidene-hexopyranosides (**1**) are reduced by $NaBH_3CN$ and hydrogen chloride in the presence of 3 Å molecular sieves to 6-benzyl ethers with a free C_4-hydroxyl group (**2**).[1]

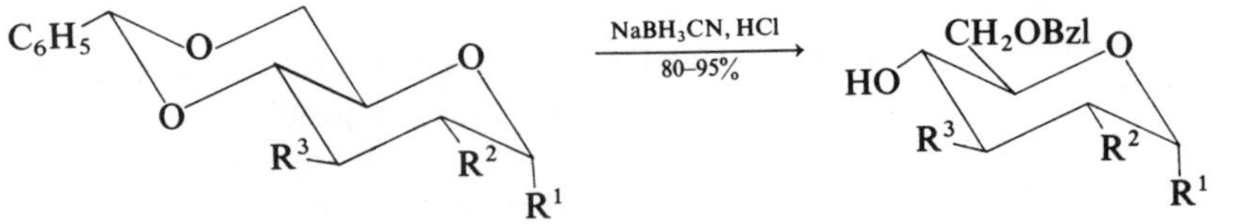

1 ($R^1 = R^3 = OCH_3$, OBz, OBzl)

2

Further example[2]:

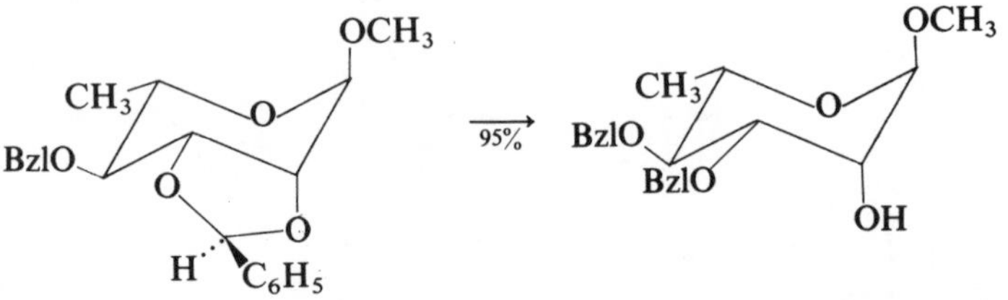

The 2-propenylidene acetals of hexopyranosides, such as **3**, are selectively reduced by acidic sodium cyanoborohydride to 6-O-2-propenyl ethers (eg., **4**).[3]

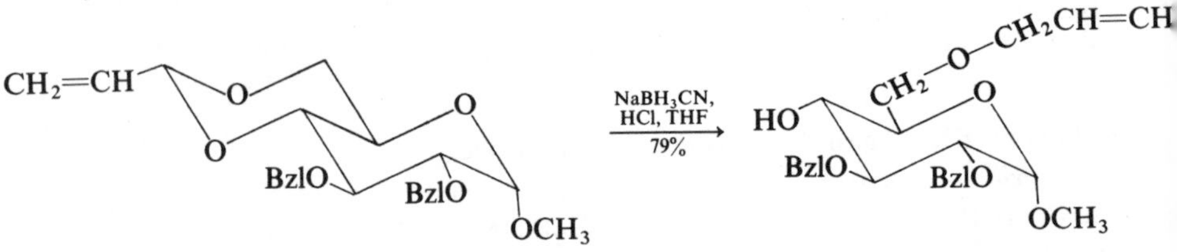

3

4

Reduction of O-acyl oximes. This reaction has been conducted with borane–pyridine and HOAc[4] (**1**, 963–964; **8**, 50–51; **9**, 59) and with either $NaCNBH_3$/HOAc or $(C_2H_5)_3SiH/CF_3COOH$.[5] Yields are somewhat higher with $NaCNBH_3$ than with $(C_2H_5)_3SiH$; surprisingly these reagents show opposite stereoselectivity (equation I).

(I)

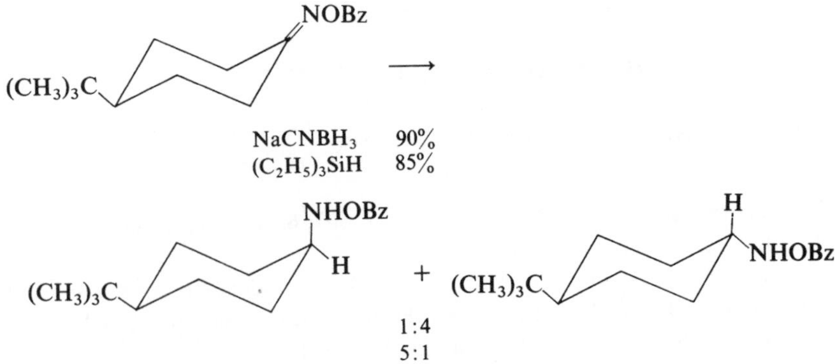

[1] P. J. Garegg and H. Hultberg, *Carbohydrate Research*, **93**, C10 (1981).
[2] P. J. Garegg, H. H. Hultberg, and S. Wallin, *ibid.*, **108**, 97 (1982).
[3] P. J. Garegg, H. Hultberg, and S. Oscarson, *J.C.S. Perkin I*, 2395 (1982).
[4] M. Kawase and Y. Kikugawa, *ibid.*, 643 (1979).
[5] D. D. Sternbach and W. C. L. Jamison, *Tetrahedron Letters*, **22**, 3331 (1981).

Sodium 9-cyano-9-hydrido-9-borabicyclo [3.3.1] nonane (9BBN–NaCN) (**1**). Mol. wt. 171.03, tacky noncrystalline solid. The reagent is prepared by reaction of NaCN with 9-BBN in THF.

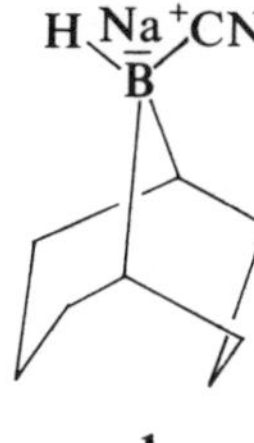

1

Reduction of alkyl halides and sulfonate esters.[1] This reagent in HMPT is somewhat more effective than $NaBH_3CN$ for reduction of primary alkyl iodides and allylic and benzylic iodides. It is also preferred for reduction in benzene or CH_2Cl_2.

Reduction of N-benzylimines.[2] Borch reduction of imines (**4**, 450) to amines is not stereoselective, although reduction to equatorial amines is somewhat favored. Reductions with **1** and a number of dialkylcyanoborohydrides are much more stereoselective. Thus substituted cycloalkanone imines are reduced mainly to axial amines. However, highest selectivities are obtained with lithium tri-*sec*-butylborohydride (**4**, 312).

[1] R. A. Hutchins, D. Kandasamy, C. A. Maryanoff, D. Masilamani, and B. E. Maryanoff, *J. Org.*, **42**, 82 (1977).
[2] J. E. Wrobel and B. Ganem, *Tetrahedron Letters*, **22**, 3447 (1981).

Sodium dicarbonylcyclopentadienylferrate, 5, 610; **6**, 538–539; **8**, 454–455; **9**, 426; **10**, 362.

α-Methylene-γ-lactones.[1] Reaction of NaFp with α-bromopyruvate diethyl ketal results in the stable complex **1**, which on reaction with HPF_6 is converted into the yellow crystalline, unstable complex **2**. The cation **2** serves an α-acrylic ester cation

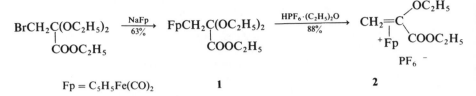

$$BrCH_2C(OC_2H_5)_2(COOC_2H_5) \xrightarrow[63\%]{NaFp} FpCH_2C(OC_2H_5)_2(COOC_2H_5) \; (\mathbf{1}) \xrightarrow[88\%]{HPF_6 \cdot (C_2H_5)_2O} [CH_2{=}C(OC_2H_5)(COOC_2H_5)\,{}^{+}Fp]\,PF_6^- \; (\mathbf{2})$$

$Fp = C_5H_5Fe(CO)_2$

equivalent in the conversion of the enolate of cyclohexanone to *cis*- and *trans*-2-methylene-γ-lactones (**5** and **6**).

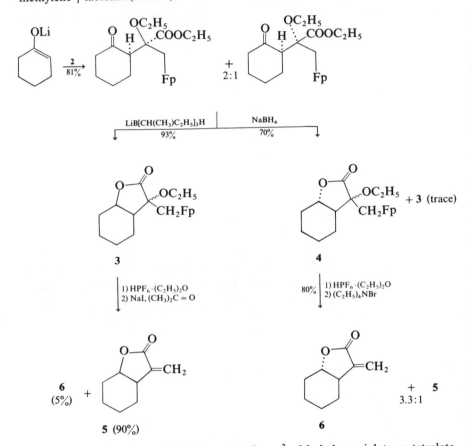

OLi; $\xrightarrow[81\%]{\mathbf{2}}$; O, H, OC_2H_5, $COOC_2H_5$, Fp + O, H, OC_2H_5, $COOC_2H_5$, Fp (2:1)

$LiB[CH(CH_3)C_2H_5]_3H$ (93%) | $NaBH_4$ (70%)

3: O, O, OC_2H_5, CH_2Fp

4: O, O, OC_2H_5, CH_2Fp + **3** (trace)

3 → 1) $HPF_6 \cdot (C_2H_5)_2O$ 2) NaI, $(CH_3)_2C{=}O$ → **6** (5%) + **5** (90%): O, O, $=CH_2$

4 → 80% 1) $HPF_6 \cdot (C_2H_5)_2O$ 2) $(C_2H_5)_4NBr$ → **6** + **5** (3.3:1): O, O, $=CH_2$

Condensation of propiolic esters with alkenes.[2] Methyl propiolate or tetrolate, when activated by complexation with the $C_5H_5Fe(CO)_2$ cation (Fp^+), condense with alkenes to form cyclobutenes, 1,3-dienes, and lactones. The type of products formed depend on the structure of the alkene. 1,2-Disubstituted alkenes yield cyclobutenes and 1,3-dienes mainly, whereas 1,1-disubstituted or trisubstituted alkenes form mainly lactones.

Examples:

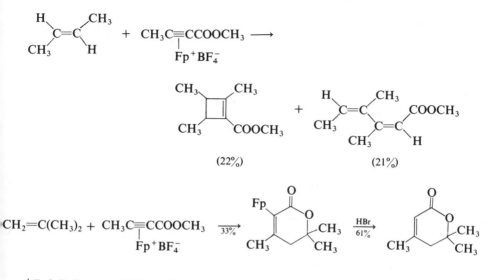

$$\underset{CH_3}{\overset{H}{>}}C{=}C\underset{H}{\overset{CH_3}{<}} + \underset{\underset{Fp^+BF_4^-}{|}}{CH_3C{\equiv}CCOOCH_3} \longrightarrow$$

$$\text{(1,2-dimethyl-3-methyl-4-carbomethoxycyclobutene: } CH_3, CH_3, CH_3, COOCH_3\text{)} \; (22\%) \quad + \quad \underset{CH_3}{\overset{H}{>}}C{=}C(CH_3){-}C(CH_3){=}C\underset{H}{\overset{COOCH_3}{<}} \; (21\%)$$

$$CH_2{=}C(CH_3)_2 + \underset{\underset{Fp^+BF_4^-}{|}}{CH_3C{\equiv}CCOOCH_3} \xrightarrow{33\%} \text{(3-Fp-4,6,6-trimethyl-5,6-dihydro-2-pyranone)} \xrightarrow[61\%]{HBr} \text{(4,6,6-trimethyl-5,6-dihydro-2-pyranone)}$$

[1] T. C. T. Chang and M. Rosenblum, *J. Org.*, **46**, 4626 (1981).
[2] M. Rosenblum and D. Scheck, *Organometallics*, **1**, 374 (1982).

Sodium O, O-diethyl phosphorotelluroate, 8, 455; **10**, 362.

Dehalogenation of α-halo ketones. α-Chloro or α-bromo ketones are dehalogenated in moderate to good yield by reaction with the reagent in ethanol at 25–80°. A possible pathway is formulated in equation (I).[1]

$$\text{(I)}\quad R\overset{O}{\overset{\|}{C}}CH_2Br \xrightarrow[-\,(C_2H_5O)_3PO]{(C_2H_5O)_2\overset{O}{\overset{\|}{P}}TeNa,\; C_2H_5OH} \left[R\overset{O}{\overset{\|}{C}}CH_2TeNa \xrightarrow[-\,Te]{} R\overset{O^-Na^+}{\overset{|}{C}}{=}CH_2 \right] \xrightarrow{H_2O} R\overset{O}{\overset{\|}{C}}CH_3$$

[1] D. L. J. Clive and P. L. Beaulieu, *J. Org.*, **47**, 1124 (1982).

Sodium dithionite, 1, 1081–1083; **5**, 615–617; **7**, 456–458; **9**, 426–427; **10**, 363–364.

Claisen rearrangement of allyloxyanthraquinones. 1-Allyloxyanthraquinones rearrange in high yield to 1-hydroxy-2-allylanthraquinones when heated in DMF–H_2O containing 1.3–1.8 equiv. of sodium dithionite. Rearrangement of 1,4-bisallyloxyanthraquinones is slow under these conditions, but the double rearrangement is effected readily if sodium hydroxide (4 equiv.) is present in addition to the reducing agent. These rearrangements involve the semiquinone anion or the hydroquinone dianion.[1]

Example:

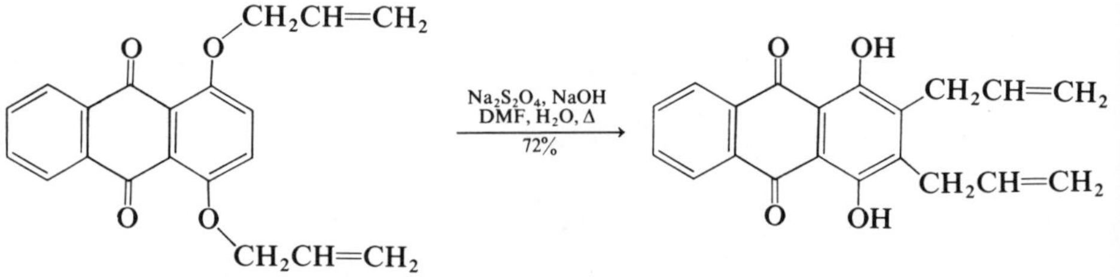

$$\xrightarrow[72\%]{Na_2S_2O_4,\ NaOH,\ DMF,\ H_2O,\ \Delta}$$

Dehalogenation of α-bromo or α-chloro ketones.[2] This reaction can be effected with $Na_2S_2O_4$ in aqueous DMF at 25–90° in yields of ~ 50–95%. The rate can be enhanced by addition of $NaHCO_3$.

Reduction of conjugated diunsaturated acids.[3] α,β; γ,δ-Diunsaturated acids are reduced by sodium dithionite in an alkaline medium (NaOH, $NaHCO_3$) to a mixture of (Z)- and (E)-β,γ-unsaturated acids (40–75% yield). A similar reduction of the diunsaturated esters is possible under phase-transfer conditions (Adogen 464, C_6H_6–H_2O).

[1] K. Boddy, P. J. Boniface, R. C. Cambie, P. A. Craw, D. S. Larsen, H. McDonald, P. S. Rutledge, and P. D. Woodgate, *Tetrahedron Letters*, **23**, 4407 (1982).
[2] S. -K. Chung and Q. -Y. Hu, *Syn. Comm.*, **12**, 261 (1982).
[3] F. Camps, J. Coll, A. Guerrero, J. Guitart, and M. Riba, *Chem. Letters*, 715 (1982).

Sodium hydride–Sodium *t*-amyl oxide–Zinc chloride, $NaH–C_2H_5C(CH_3)_2ONa–ZnCl_2$ (**1**).

Reduction of carbonyl groups.[1] This reagent effects highly selective 1,2-reduction of α,β-enones in THF.

Reductive silylation of carbonyl compounds.[2] This reagent in the presence of a threefold excess of chlorotrimethylsilane reduces saturated and α,β-unsaturated aldehydes and ketones, even highly enolizable ones, to trimethylsilyl ethers. Indeed this reaction is usually superior, in respect to yield, to reduction with **1** alone.

Examples:

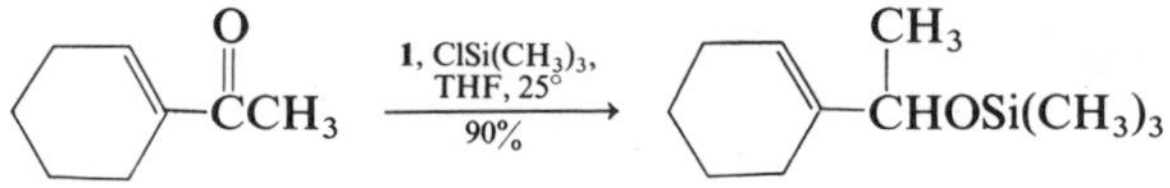

$$\xrightarrow[90\%]{1,\ ClSi(CH_3)_3,\ THF,\ 25^\circ}$$

$$CH_3(CH_2)_5CHO \xrightarrow[93\%]{} CH_3(CH_2)_5CH_2OSi(CH_3)_3$$

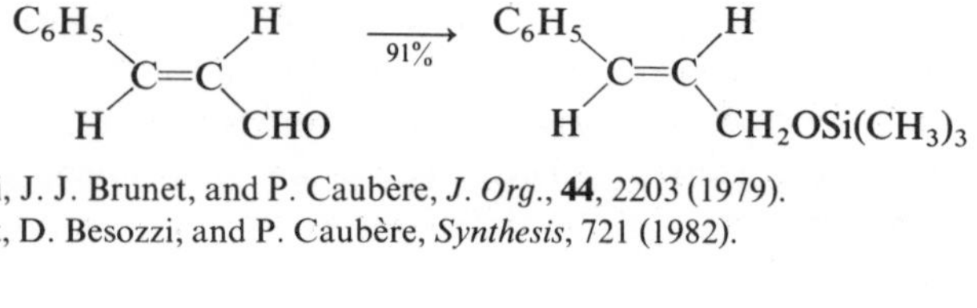

$$\xrightarrow[91\%]{}$$

[1] L. Mordenti, J. J. Brunet, and P. Caubère, *J. Org.*, **44**, 2203 (1979).
[2] J. -J. Brunet, D. Besozzi, and P. Caubère, *Synthesis*, 721 (1982).

Sodium hydrogen sulfide, NaSH, **9**, 429. Anhydrous reagent is prepared by reaction of sodium ethoxide and hydrogen sulfide in ethanol. The yield is essentially quantitative.[1]

N-Alkylthiolactams.[2] N-Alkyllactams can be converted into N-alkylthiolactams by the same method previously used to convert lactones into thiolactones (**10**, 417). However, the method is not applicable to thionation of acyclic N,N-disubstituted amides [RC(S)OR is obtained instead].

Example:

$$\text{N-(CH}_2)_3\text{F-2-piperidone} \xrightarrow{(C_2H_5)_3O^+BF_4^-} \left[\text{N}^+\text{-(CH}_2)_3\text{F}=\text{C(OC}_2\text{H}_5)\ BF_4^-\right] \xrightarrow[55\%]{\text{NaSH, } CH_3COCH_3} \text{N-(CH}_2)_3\text{F-2-piperidinethione}$$

[1] R. E. Eibeck, *Inorg. Syn.*, **7**, 128 (1963).
[2] J. J. Bodine and M. K. Kaloustian, *Syn. Comm.*, **12**, 787 (1982).

Sodium hydrogen sulfite, 1, 1047–1049.

Selective protection of an aldehyde.[1] Selective reduction of 4-acetylbenzaldehyde (**1**) to 4-(1-hydroxyethyl)benzaldehyde (**3**) is possible by conversion to the sodium bisulfate adduct (**2**), which is absorbed on silica to increase the stability. The dried adduct is then reduced with diborane. Use of alumina as support is less satisfactory.

$$\underset{\mathbf{1}}{CH_3C(O)C_6H_4CHO} \xrightarrow{\text{NaHSO}_3,\ H_2O} \underset{\mathbf{2}}{CH_3C(O)C_6H_4CH(OH)SO_3Na} \xrightarrow{\text{Dried on } SiO_2}$$

$$\xrightarrow[\text{quant.}]{B_2H_6} \underset{\mathbf{3}\ (70\text{–}93\%)}{CH_3CH(OH)C_6H_4CHO} + \underset{<2\%}{\left\{\begin{matrix} CH_3C(O)C_6H_4CH_2OH\text{-}p \\ CH_3CH(OH)C_6H_4CH_2OH\text{-}p \end{matrix}\right\}}$$

[1] T. Chihara, T. Wakabayashi, and K. Taya, *Chem. Letters*, 1657 (1981).

Sodium hypochlorite, 1, 1084–1087; **2**, 67; **3**, 45, 243; **4**, 456; **5**, 617; **6**, 543; **7**, 337–338; **8**, 461–463; **9**, 430; **10**, 365.

Oxidation of alcohols (**10**, 365). Selective oxidation of secondary alcohols is possible with 1 equiv. of NaOCl in acetic acid at 25°.

Examples:

$$HOCH_2CH(OH)(CH_2)_5CH_3 \xrightarrow[73\%]{NaOCl,\ CH_3COOH} HOCH_2C(=O)(CH_2)_5CH_3$$

$$\text{2-(hydroxymethyl)cyclohexanol} \xrightarrow[90\%]{} \text{2-(hydroxymethyl)cyclohexanone}$$

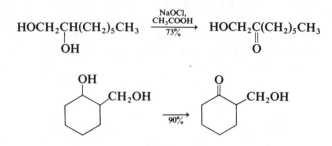

Aldehydes can be oxidized to methyl esters by NaOCl in CH_3OH containing 1 equiv. of CH_3COOH. Presumably, a methyl hemiacetal is an intermediate.

Examples:

$$CH_3C(=O)(CH_2)_4CHO \xrightarrow[90\%]{NaOCl,\ CH_3OH,\ CH_3COOH} CH_3C(=O)(CH_2)_4COOCH_3$$

$$m\text{-}O_2NC_6H_4CHO \xrightarrow[73\%]{} m\text{-}O_2NC_6H_4COOCH_3$$

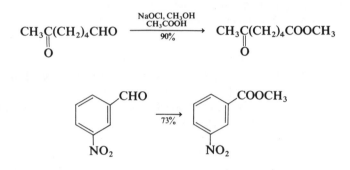

[1] R. V. Stevens, K. T. Chapman, C. A. Stubbs, W. W. Tam, and K. F. Albizati, *Tetrahedron Letters*, **23**, 4647 (1982).

Sodium iodide–Chloramine-T.

Alkyl iodides.[1] In the presence of a mild oxidizing agent such as chloramine-T, sodium iodide reacts with trialkylboranes to form alkyl iodides. The same system converts triphenylborane into iodobenzene in high yield (only one of the phenyl groups is utilized). With trialkylboranes, two of the three groups are used when the borane is prepared from a terminal alkene, but only one group is used in reactions of trialkylboranes prepared from internal alkenes. This system is superior to iodine monochloride, which requires a preparation step.[2] An example is synthesis of 19-iodononadecanic acid (equation I).[3] Alkyl bromides can be prepared in the same way

$$\text{(I)}\quad CH_2{=}CH(CH_2)_{15}CH_2COOH \xrightarrow{2\,(C_6H_{11})_2BH} \xrightarrow[89\%]{NaI,\ NaOAc,\ Chloramine\text{-}T} ICH_2(CH_2)_{16}CH_2COO\ldots$$

using sodium bromide and the oxidant. This methodology can be used to prepare primary halides in the carbohydrate field.[4]

[1] G. W. Kabalka and E. E. Gooch, *J. Org.*, **46**, 2582 (1981).
[2] G. W. Kabalka, K. A. R. Sastry, H. C. Hsu, and M. D. Hylarides, *ibid.*, **46**, 3113 (1981).
[3] G. W. Kabalka and E. E. Gooch, *ibid*, **45**, 3578 (1980).
[4] L. D. Hall and J.-R. Neeser, *Can. J. Chem.*, **60**, 2082 (1982).

Sodium iodide–Pivaloyl chloride.

Cleavage of ethers. A few reports have mentioned that acyl iodides can cleave ethers in the absence of a Lewis acid. Since acyl iodides are not readily available, Oku *et al.*[1] have used NaI and an acyl chloride as a possible equivalent. In any case, the system does cleave both cyclic and acyclic ethers selectively at the less substituted α-C—O bond. Although any acyl chloride can be used, use of pivaloyl chloride is particularly attractive because the resulting pivaloyl esters are readily hydrolyzed. Indeed this system is particularly useful for deprotection of methyl ethers.

Example:

$$\text{2-methyltetrahydrofuran} \xrightarrow[91\%]{\text{NaI, (CH}_3)_3\text{CCOCl, CH}_3\text{CN, 25°}} (CH_3)_3CCOOCH(CH_3)(CH_2)_3I + (CH_3)_3CCOO(CH_2)_3CH(CH_3)I$$

97:3

[1] A. Oku, T. Harada, and K. Kita, *Tetrahedron Letters*, **22**, 681 (1982).

Sodium methoxide, 1, 1091–1094; **2**, 385–386; **3**, 259–260; **4**, 457–459; **5**, 617–620; **8**, 963; **9**, 430–431.

Secondary amines.[1] Aminosilanes react with primary alkyl halides in the presence of sodium methoxide to form N-alkylamines in 45–85% yield (equation I).

$$\text{(I)}\quad RNHSi(CH_3)_3 + NaOCH_3 \xrightarrow{\text{THF}} RNHNa + CH_3OSi(CH_3)_3$$

$$RNHNa \xrightarrow[45\text{–}85\%]{R^1X} RNHR^1$$

[1] W. Ando and H. Tsumaki, *Chem. Letters*, 693 (1981).

Sodium methylsulfinylmethylide (Dimsylsodium), 1, 310–313; **3**, 166–169; **3**, 123–124; **4**, 195–196; **5**, 621; **6**, 546–547; **7**, 338–339; **9**, 431.

Long-chain alkanes.[1] Triphenylphosphonium salts such as **3**, prepared as shown from bromo acetals **2**, when treated with dimsylsodium form the long-chain bromo acetal **4**. The terminal bromine atom can be displaced by lithium triethylborohydride.

This methodology has been used to prepare a C_{104}-alkane, the longest alkane yet reported.

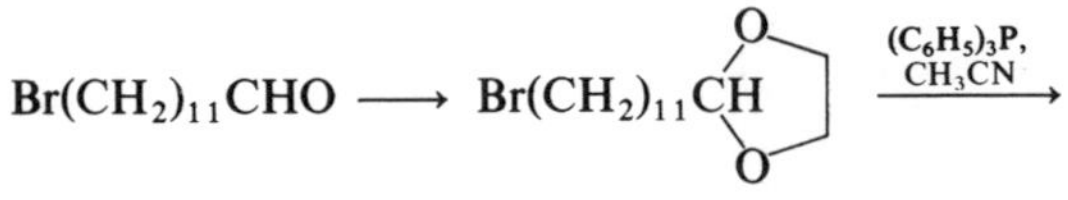

$Br(CH_2)_{11}CHO$ ⟶ $Br(CH_2)_{11}CH$(OCH₂CH₂O) —$(C_6H_5)_3P$, CH_3CN⟶

1 **2**

$(C_6H_5)_3\overset{+}{P}(CH_2)_{11}CH$(OCH₂CH₂O) Br^- —1) $NaCH_2SOCH_3$, 2) **1**⟶ $Br(CH_2)_{11}CH{=}CH(CH_2)_{10}CH$(OCH₂CH₂O)

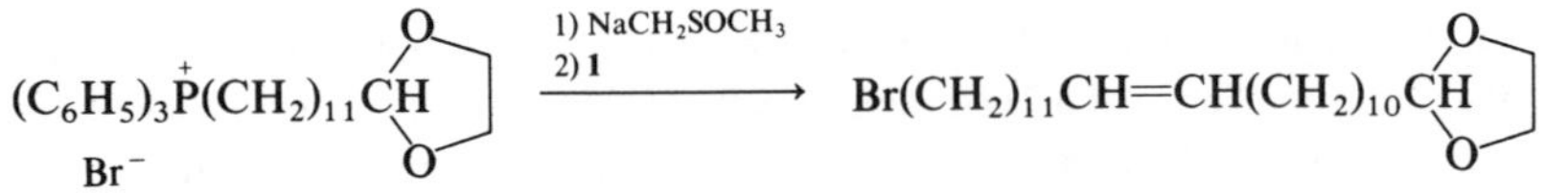

3 **4** (Z)

[1] O. I. Paynter, D. J. Simmonds, and M. C. Whiting, *J.C.S. Chem. Comm.*, 1165 (1982).

Sodium naphthalenide, 1, 711–712; **2**, 289; **4**, 349–350; **5**, 468–470; **7**, 340; **8**, 464; **9**, 431.

Reductive cyclization of γ-ethynyl ketones to allylic alcohols. This reaction was first reported by Stork *et al.*, who used an alkali metal and liquid ammonia.[1] The main by-products, at least in cyclization to A-norsteroids, result from overreduction. This side reaction can be prevented by use of sodium naphthalenide in THF or DME.[2]

Examples:

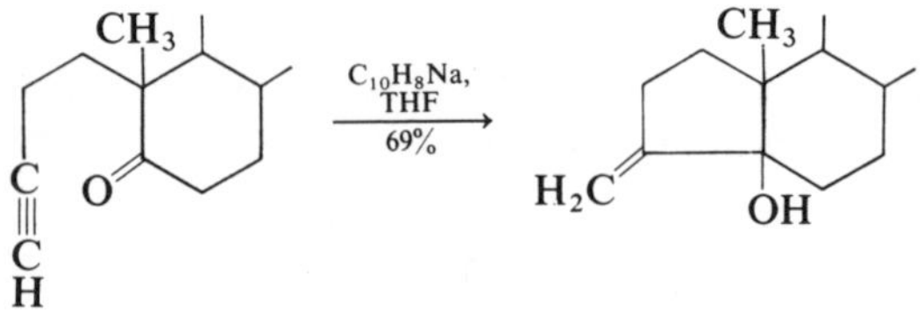

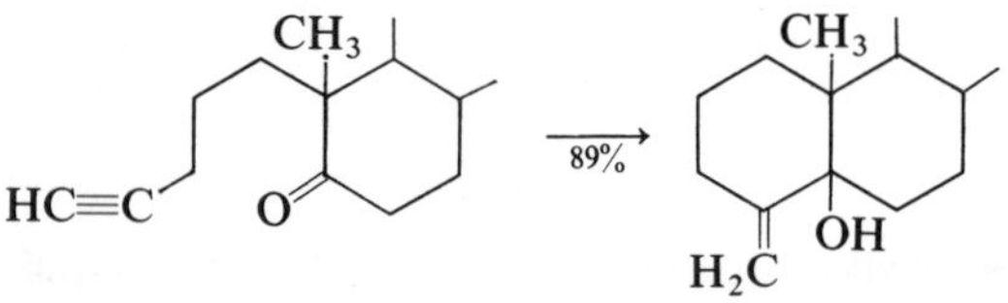

$C_{10}H_8Na$, THF, 69%

CH₃ ... C≡CH, O ⟶ CH₃, H₂C, OH

89%

CH₃, HC≡C, O ⟶ CH₃, H₂C, OH

This cyclization was used in a synthesis of the tricyclic bisallylic diol (**3**) from the keto-alkyne (**1**).[3]

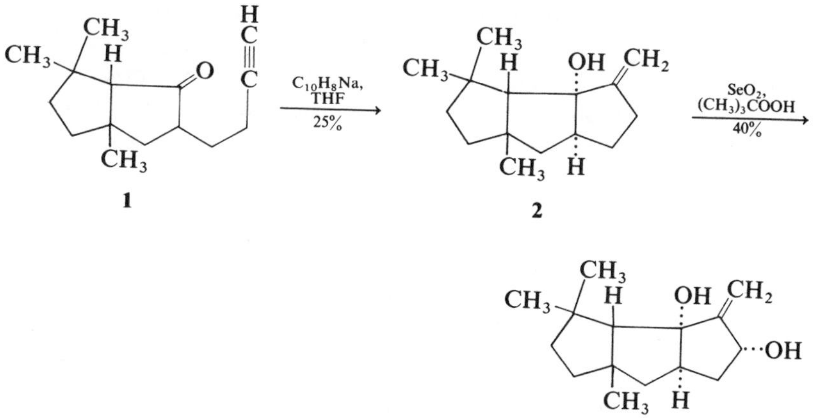

$$CH_3\text{-}C(CH_3)\ldots \xrightarrow[25\%]{C_{10}H_8Na,\ THF} \ldots \xrightarrow[40\%]{SeO_2,\ (CH_3)_3COOH} \ldots$$

1 2 3

[1] G. Stork, S. Malhotra, H. Thompson, and M. Uchibayashi, *Am. Soc.*, **87**, 1148 (1965).
[2] S. K. Pradhan, T. V. Radhakrishnan, and R. Subramanian, *J. Org.*, **41**, 1943 (1976).
[3] G. Pattenden and S. J. Teague, *Tetrahedron Letters*, **23**, 5471 (1982).

Sodium nitrite, 1, 1097–1101; **2**, 386–387; **4**, 459–460; **6**, 547; **9**, 432.

Tertiary amines.[1] A new preparation of tertiary amines involves alkylation of 1,1-dimethylhydrazine[2] (**9**, 184–185 and earlier volumes) followed by deamination of the resulting hydrazinium salt with nitrous acid in 4 *N* HCl.

Example:

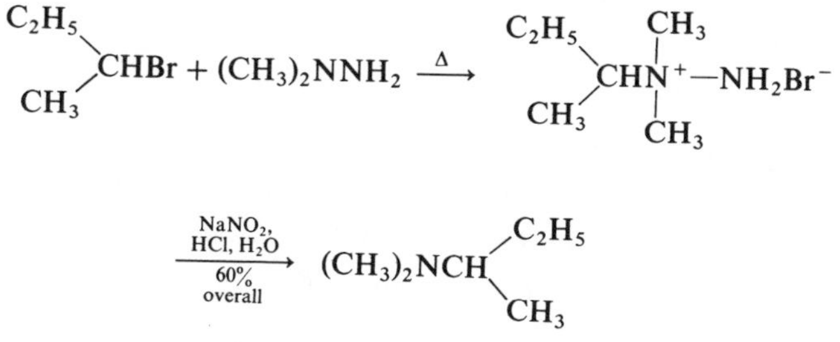

$$(C_2H_5)(CH_3)CHBr + (CH_3)_2NNH_2 \xrightarrow{\Delta} (C_2H_5)(CH_3)CHN^+(CH_3)_2\text{—}NH_2Br^-$$

$$\xrightarrow[60\%\ \text{overall}]{NaNO_2,\ HCl,\ H_2O} (CH_3)_2NCH(C_2H_5)(CH_3)$$

$>C{=}S \rightarrow >C{=}O$. Secondary and tertiary thioamides are converted into the corresponding amides by reaction with $NaNO_2$ in an acidic medium. The conversion is considered to involve an S-nitroso intermediate.[3]

[1] R. F. Smith and K. J. Coffman, *Syn. Comm.*, **12**, 801 (1982).
[2] Potentially carcinogenic. N-Nitrosodimethylamine is a known carcinogen.
[3] K. A. Jørgensen, A.-B. A. G. Ghattas, and S.-O. Lawesson, *Tetrahedron*, **38**, 1163 (1982).

Sodium periodate, 1, 809–815; **2,** 311–313; **4,** 373–374; **5,** 507–508; **6,** 455.

Supported reagent. $NaIO_4$ supported on silica gel has the significant advantage that it functions in nonpolar solvents such as CH_2Cl_2, benzene, or even ether. It oxidizes a variety of hydroquinones to quinones in high yield within a few hours. It is comparable to Ag_2CO_3 on Celite for this oxidation.[1]

Further oxidations:

$$\text{cyclohexane-1,2-diol (OH, OH)} \xrightarrow[81\%]{NaIO_4 - SiO_2} OHC(CH_2)_4CHO$$

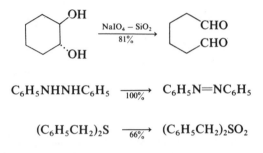

$$C_6H_5NHNHC_6H_5 \xrightarrow[100\%]{} C_6H_5N{=}NC_6H_5$$

$$(C_6H_5CH_2)_2S \xrightarrow[66\%]{} (C_6H_5CH_2)_2SO_2$$

[1] D. N. Gupta, P. Hodge, and J. E. Davies, *J.C.S. Perkin I*, 2970 (1981).

Sodium peroxide, 7, 341.

Δ^4-3,6-Diketosteroids.[1] Δ^4-3-Ketosteroids are oxidized to Δ^4-3,6-diketosteroids by aqueous Na_2O_2 in 45–65% isolated yield. Use of Li_2O_2 results only in epoxidation in low yield. Isolated carbonyl and hydroxyl groups are stable to Na_2O_2.

[1] H. L. Holland, U. Daum, and E. Riemland, *Tetrahedron Letters*, **22,** 5127 (1981).

Sodium sulfide, 3, 269–272; **5,** 623–624.

Dehalogenation.[1] *vic*-Dibromides and *vic*-dichlorides are converted into alkenes by reaction with $Na_2S \cdot 9H_2O$ in DMF at 25° in 70–95% isolated yield.

[1] K. Fukunaga and H. Yamaguchi, *Synthesis*, 879 (1981).

Sodium thioethoxide, C_2H_5SNa, **3,** 115.

Sulfenylation of active methylene compounds. Sulfenylation of active methylene compounds by the sulfenyl transfer reagents **1** ordinarily gives a mixture of mono- and bisulfenylated products in which the latter predominate.[1] However, the bisulfenylated products are reduced cleanly by sodium thioethoxide to monosulfenylated products.[2]

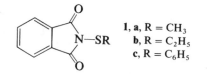

1, a, R = CH_3
b, R = C_2H_5
c, R = C_6H_5

$$CH_2(COOC_2H_5)_2 \xrightarrow[94\%]{\textbf{1c},\ N(C_2H_5)_3,\ CH_2Cl_2} (C_6H_5S)_2C(COOC_2H_5)_2 \xrightarrow[72\%]{C_2H_5SNa,} C_6H_5SCH(COOC_2H_5)_2$$

$$C_6H_5COCH_2SO_2CH_3 \xrightarrow[87\%]{\textbf{1b}} C_6H_5COC(SC_2H_5)_2SO_2CH_3 \xrightarrow[80\%]{} C_6H_5COCH(SC_2H_5)SO_2CH_3$$

Displacement of fluorine in arenes.[3] Attempted demethylation of **1** with C_2H_5SNa results also in displacement of the fluorine by SC_2H_5 to give **2**. The fluorine is also partially displaced by Br on demethylation with BBr_3 in CH_2Cl_2. Demethylation of **1** with NaI and $(CH_3)_3SiCl$ is not successful.

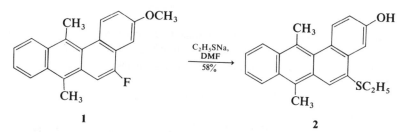

$$\mathbf{1} \xrightarrow[58\%]{C_2H_5SNa,\ DMF} \mathbf{2}$$

[1] D. N. Harpp and T. Z. Back, *J. Org.*, **41**, 2498 (1976).
[2] J. S. Grossert and P. K. Dubey, *J.C.S. Chem. Comm.*, 1183 (1982).
[3] Y. M. Sheikh, N. Ekwuribe, B. Dhawan, and D. T. Witiak, *J. Org.*, **47**, 4341 (1982).

Sodium trithiocarbonate, Na_2CS_3, **10**, 369–370. An aqueous solution of Na_2CS_3 can be prepared by reaction of $Na_2S{\cdot}9H_2O$ with CS_2 at 45–50°. The solution is stable for about 1 week.

—Cl→—SH.[1] Na_2CS_3 is superior to ammonium hydrogen sulfide for direct substitution of the Cl group of **1** to give Captropril (**2**), an orally active antihypertensive agent; the optical activity of **1** is retained.

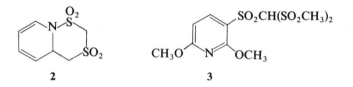

$ClCH_2CH(CH_3)CON$(pyrrolidine-COOH) **1** $\xrightarrow[H_2O,\ 85\%]{Na_2CS_3,\ Na_2CO_3,}$ $HSCH_2CH(CH_3)CON$(pyrrolidine-COOH) **2** (74%)

[1] M. Shimazaki, J. Hasegawa, K. Kan, K. Nomura, Y. Nose, H. Kondo, T. Ohashi, and K. Watanabe, *Chem. Pharm. Bull.*, **30**, 3134 (1982).

Sulfene, CH_2=SO_2 (**1**). Sulfene is generated by reaction of methanesulfonyl chloride and triethylamine. In the absence of a trapping agent, it forms a tetramer, ie salt.[1]

Adducts with pyridines. Generation of (**1**) in pyridine results in formation of the novel heterocycle **2**. Sulfene forms a 3:1 adduct (**3**) with 2,6-dimethoxypyridine in > 80% yield.[2]

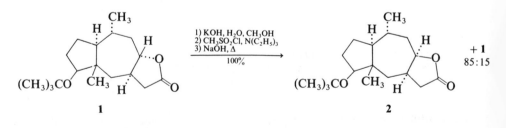

2 (structure with N, O_2S, SO_2)

3 (CH_3O, N, OCH_3, $SO_2CH(SO_2CH_3)_2$)

γ-Lactone inversion.[3] The *trans*-γ-lactone **1** has been converted into the *cis*-isomer (**2**) by treatment of the dry potassium salt of the hydroxy acid derived from **1** with sulfene, followed by relactonization. The Mitsunobu inversion failed in this case.

1 $\xrightarrow[100\%]{1)\ KOH,\ H_2O,\ CH_3OH;\ 2)\ CH_3SO_2Cl,\ N(C_2H_5)_3;\ 3)\ NaOH,\ \Delta}$ **2** + **1** (85:15)

(Structures: $(CH_3)_3CO$, CH_3, H, O labels on each lactone.)

[1] J. S. Grossert, M. M. Bharadwaj, R. F. Langler, T. S. Cameron, and R. E. Cordes, *Can. J. Chem.*, **56**, 1183 (1978).

[2] J. S. Grossert, J. Hoyle, M. M. Bharadwaj, and T. S. Cameron, *J.C.S. Chem. Comm.*, 1175 (1982).

[3] P. T. Lansbury and J. P. Vacca, *Tetrahedron Letters*, **38**, 2797 (1982).

Sulfur dioxide, 1, 1122; **2,** 392; **4,** 469; **5,** 633; **6,** 558; **7,** 346–347; **8,** 464–466; **9,** 440–441.

Homoadditions.[1] Under carefully controlled conditions, SO_2 adds to norbornadiene (**1**) to give the sulfone **2** in almost quantitative yield. To date, no other homoconjugated dienes are known to undergo this type of addition.

$$\mathbf{1} + SO_2 \xrightarrow[\sim 100\%]{20^\circ} \mathbf{2}$$

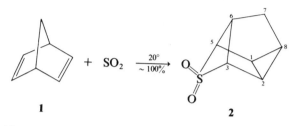

Sulfur dioxide cycloadds to quadricyclane **3** to give the unstable β-sultine **4** as the major product.[2] It also causes isomerization of **3** to **1** (a valence tautomer).

$$\mathbf{3} \xrightarrow{SO_2} \mathbf{4} + \mathbf{1}$$

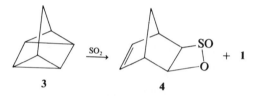

Thioacetals.[3] SO_2 serves as a catalyst for thioacetalization of aliphatic and aryl aldehydes and ketones. Yields are comparable to those obtained with HCl or TsOH as acid catalysts.

[1] O. De Lucchi and V. Lucchini, *J.C.S. Chem. Comm.*, 1105 (1982).
[2] *Idem, ibid.*, 464 (1982).
[3] B. Burczyk and Z. Kortylewicz, *Synthesis*, 831 (1982).

Sulfur dioxide–Copper(I) chloride.

Arenesulfonyl chlorides.[1] Aryldiazonium chlorides are converted into arenesulfonyl chlorides by reaction with SO_2 and CuCl. Yields from the corresponding aniline are 30–90%, being highest with anilines substituted with electron-withdrawing substituents.

[1] R. V. Hoffman, *Org. Syn.*, **60**, 121 (1981).

Sulfur monochloride, 1, 1122–1123; **6,** 560; **9,** 442.

Thiobenzaldehydes.[1] The first stable thiobenzaldehyde, 2,4,6-tri-*t*-butylthiobenzaldehyde (**2**), has been prepared by oxidative sulfuration of the hydrazone **1** with S_2Cl_2. It has also been prepared by treatment of 2,4,6-tri-*t*-butylphenyllithium (**3**)

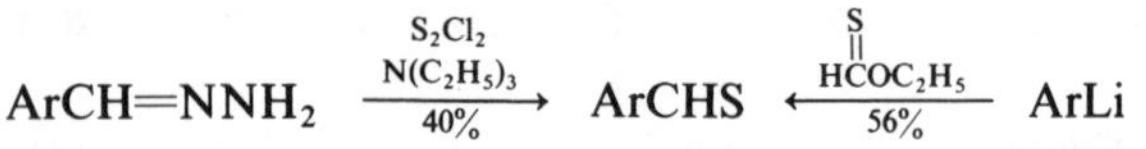

$$\underset{\mathbf{1}}{ArCH{=}NNH_2} \xrightarrow[40\%]{S_2Cl_2,\ N(C_2H_5)_3} \underset{\mathbf{2}}{ArCHS} \xleftarrow[56\%]{HC(=S)OC_2H_5} \underset{\mathbf{3}}{ArLi}$$

$Ar = 2,4,6\text{-}t\text{-}Bu_3C_6H_2$

with O-ethyl thioformate in somewhat higher yield. The thioaldehyde (**2**) can be stored at 25° for over a year in the absence of air; in the presence of oxygen, it is converted to the corresponding aldehyde in quantitative yield.

[1] R. Okazaki, A. Ishii, N. Fukuda, H. Oyama, and H. Inamoto, *J.C.S. Chem. Comm.*, 1187 (1982).

N,N′-Sulfuryldiimidazole, (Im–SO_2–Im) (**1**). Mol. wt. 198.21, m.p. 141°.

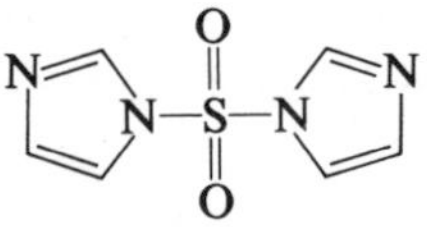

The reagent is prepared by reaction of sulfuryl chloride with imidazole. It is hydrolyzed by acid only at elevated temperatures, and does not react with carboxylic acids to form imidazolides.[1]

Imidazolylsulfonates.[2] Alcohols can be converted into the imidazolylsulfonates by reaction of the alkoxides (NaH, DMF) with **1**. Another route involves conversion to the chlorosulfate (sulfuryl chloride) followed by reaction with **1**. The imidazylate group is an efficient leaving group in reactions with nucleophiles. Displacement reactions with even secondary imidazylates can proceed in high yield. Imidazylates are also useful in elimination reactions (second example).

Examples:

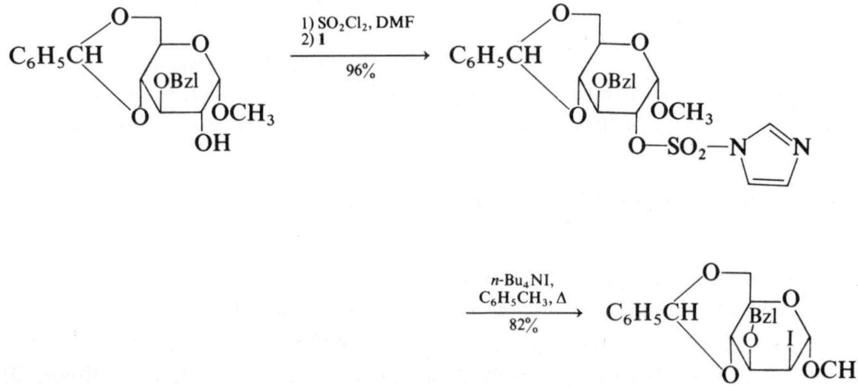

C_6H_5CH (O,O-acetal), OBzl, OCH_3, OH $\xrightarrow[96\%]{1)\ SO_2Cl_2,\ DMF;\ 2)\ \mathbf{1}}$ C_6H_5CH (O,O-acetal), OBzl, OCH_3, $O{-}SO_2{-}N$(imidazole)

$\xrightarrow[82\%]{n\text{-}Bu_4NI,\ C_6H_5CH_3,\ \Delta}$ C_6H_5CH (O,O-acetal), OBzl, I, OCH_3

CH$_3$ HO H 1) NaH, **1** 2) n-Bu_4NI, $C_6H_5CH_3$, Δ 84% CH$_3$ + ~1:1 CH$_3$

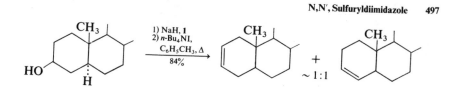

[1] H. A. Staab, *Angew. Chem. Int. Ed.*, **1**, 365 (1962).
[2] S. Hanessian and J. -M. Vatèle, *Tetrahedron Letters*, **22**, 3579 (1981).

T

Tellurium, Te. Atomic weight 127.60.

Carbonylation of amines.[1] The carbonylation of amines catalyzed by selenium to give ureas has been known for some time (**4**, 422). Tellurium can also be used as a catalyst, and in this case oxygen is not required to regenerate the catalyst (equation I).

$$\text{(I)}\quad 2RNH_2 + CO \xrightarrow{Te} (RNH)_2CO + H_2$$

[1] N. Kambe, K. Kondo, H. Ishii, and N. Sonoda, *Bull. Chem. Soc. Japan*, **54**, 1460 (1981).

2,4,4,6-Tetrabromocyclohexadienone (1), 4, 476–477; **5**, 643–644; **7**, 351–352; **10**, 377.

Demethylation. Cleavage of the methyl ether **2** could not be achieved directly in the desired sense but was accomplished indirectly by treatment with **1**, which affords the cyclic bromo ether (**3**). Reduction of **3** gives the desired hydroxy diene **4**.[1]

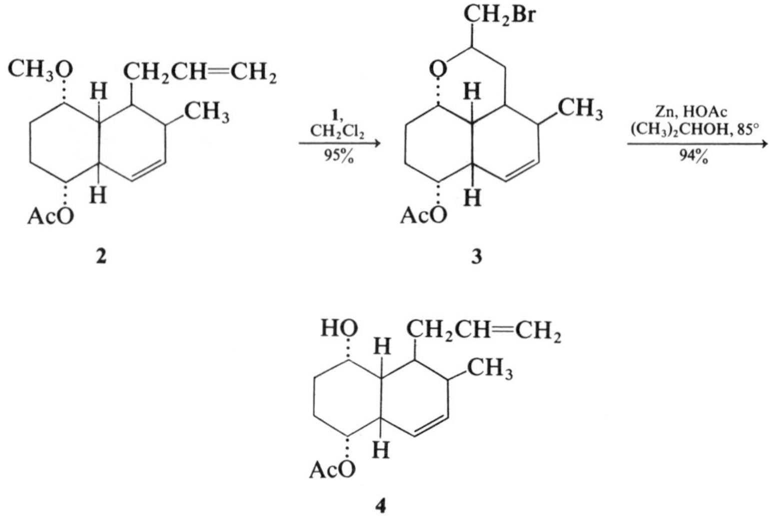

[1] N. N. Girotra and N. L. Wendler, *Tetrahedron Letters*, **23**, 5501 (1982).

Tetra-*n*-butylammonium borohydride, 6, 564–565; **7,** 352–353; **10,** 378.

Selective reductions. Raber *et al.*[1] have compared the ability of tetra-*n*-butylammonium borohydride, tetraethylammonium borohydride, and sodium borohydride to effect selective reduction of aldehydes in the presence of ketones, and conclude that no one of these reagents is generally effective for this purpose and that the three reagents are generally similar.

[1] D. J. Raber, W. C. Guida, and D. C. Shoenberger, *Tetrahedron Letters*, **22**, 5107 (1981).

Tetra-*n*-butylammonium cyanoborohydride, 5, 645.

Reductive amination.[1] The limited solubility of $NaBH_3CN$ restricts its use to aprotic, ethereal solvents. Tetrabutylammonium cyanoborohydride or $NaBH_3CN$ in combination with Aliquat 336 can be used in most common organic solvents. Either reagent is useful for reductive amination of aldehydes or ketones.

Example:

$(C_4H_9)_4NBH_3CN$, CH_2Cl_2, HCl; 94%

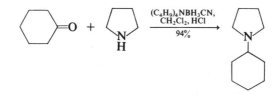

[1] R. O. Hutchins and M. Markowitz, *J. Org.*, **46**, 3571 (1981).

Tetra-*n*-butylammonium fluoride, 4, 477–478; **5,** 645; **7,** 353–354; **8,** 467–468; **9,** 444–446; **10,** 378–381.

erythro-*Nitroaldols* (10, 381).[1] Experimental details are available for the highly *erythro*-selective condensation of aldehydes with *t*-butyldimethylsilyl esters of 1-*aci*-nitroalkanes[2] catalyzed by tetra-*n*-butylammonium fluoride (equation I). Diastereo-

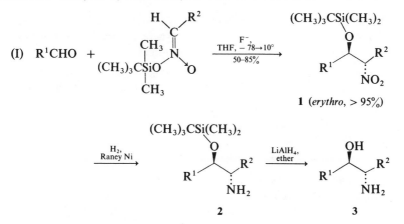

(I) R^1CHO + $(CH_3)_3CSi(CH_3)_2ON(O)=CHR^2$ —[F^-, THF, $-78\rightarrow10°$; 50–85%]→ **1** (*erythro*, > 95%)

—[H_2, Raney Ni]→ **2** —[$LiAlH_4$, ether]→ **3**

selectivity is observed consistently only when the catalyst is freshly dried over molecular sieves in a THF solution. Tetra-*n*-butylammonium fluoride dried with heat, absorbed on alumina, or bound to polymer effects the condensation, but with little or no diastereoselectivity. Trimethylsilyl *aci*-nitronates show no diastereoselectivity in this reaction, even when conducted at −100°.

These *erythro*-nitroaldols are attractive precursors to 1,2-amino alcohols **(3)**. The conversion can be effected with retention of configuration by reduction to **2** (H_2 and neutral Raney nickel) followed by silyl ether cleavage with lithium aluminum hydride in ether. A one-step reduction of **1** to a 1,2-amino alcohol can be effected with lithium aluminum hydride, but with loss of configurational purity and some C—C bond cleavage.

β-Hydroxy macrolides. The *β*-hydroxy macrolide **3**, 17-O-methyllythridine, was prepared in two steps from **1**. The hydroxyl group was converted into the trimethylsilylacetate **2** in 60% yield by reaction with trimethylsilylketene (**6**, 635). On treatment of **2** with tetra-*n*-butylammonium fluoride, the silylacetate group undergoes intramolecular addition to the aldehyde group to give **3** in 35% yield. No detectable trace of the 13-epimer is formed.[3]

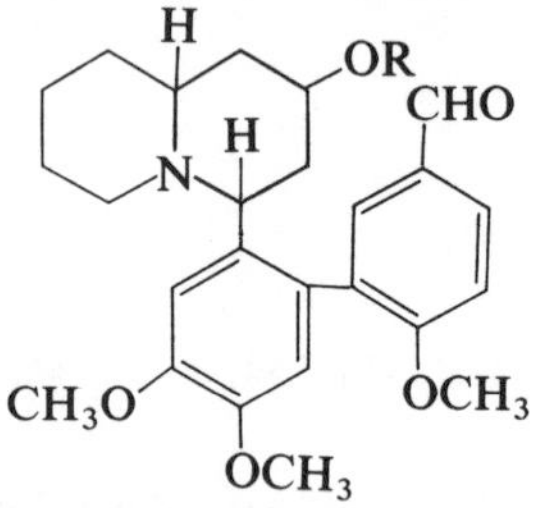

H OR CHO H N CH_3O OCH_3 OCH_3

1, R = H
2, R = $COCH_2Si(CH_3)_3$

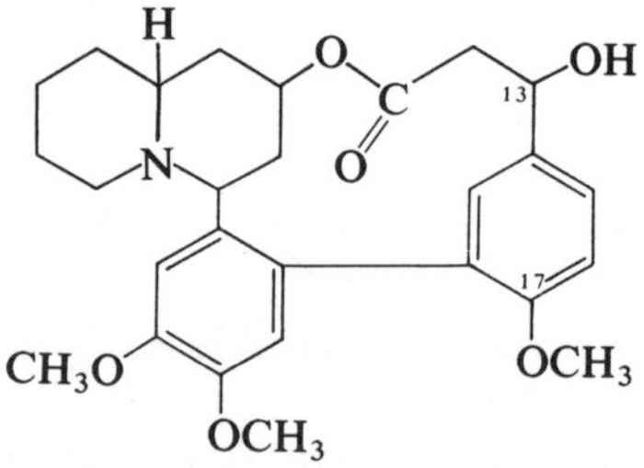

H O C O OH 13 N 17 CH_3O OCH_3 OCH_3

3

Michael addition. Fluoride ion can catalyze Michael additions.[4] In reactions that are slow owing to steric hindrance, or where elimination or fragmentation competes, use of high pressure[5] may prove expeditious.

[1] D. Seebach, A. K. Beck, T. Mukhopadhyay, and E. Thomas, *Helv.*, **65**, 1101 (1982).
[2] E. W. Colvin, A. K. Beck, B. Bastini, D. Seebach, Y. Kai, and J. D. Dunitz, *ibid.*, **63**, 697 (1980).
[3] D. E. Seitz, R. A. Milius, and J. Quick, *Tetrahedron Letters*, **23**, 1439 (1982).
[4] K. Matsumoto, *Angew. Chem. Int. Ed.*, **20**, 770 (1981).
[5] *Idem*, *ibid.*, **19**, 1013 (1980).

Tetra-*n*-butylammonium hydroxide, 5, 645–646.

Dehydration of* vic*-dihydrodiols. *vic*-Dihydrodiols of polycyclic aromatic hydrocarbons are converted into phenols by reaction with this base in CH_3OH (no dehydration occurs in CH_3CN, CH_2Cl_2, or THF).[1] The same phenol is formed from both *cis*- and *trans*-diols, and this phenol is generally different from that obtained by acid-catalyzed dehydration (H_3PO_4 in CH_3OH). The method fails with *vic*-diols that cannot be converted to phenols.

[1] D. W. McCourt, P. P. Roller, and H. V. Gelboin, *J. Org.*, **46**, 4157 (1981).

Tetra-*n*-butylammonium iodide, 5, 646–647; **6,** 566–567; **7,** 355; **9,** 447.

Monoetherification of polyols.[1,2] Monobenzylation and monoallylation of polyols can be conducted conveniently under mild conditions by conversion to the stannylene derivative (dimeric) by di-*n*-butyltin oxide (**5**, 189; **9**, 141). The stannylene is then treated with benzyl bromide or allyl bromide and tetra-*n*-butylammonium iodide (1 equiv.) in benzene. The same conditions can be used to prepare monomethoxymethyl ethers. Quaternary ammonium bromides are less efficient catalysts than the iodides. These salts also accelerate reaction of stannylenes with acid anhydrides. The mechanism for this activation is not clear; it may involve coordination of I^- to tin.

Examples:

OH, CH_2OH, O, BzlO, OBzl, OBzl $\xrightarrow[60\%]{\text{1) } Bu_2SnO, C_6H_6 \text{ 2) } C_6H_5CH_2Br, Bu_4NI}$ OH, CH_2OBzl, O, BzlO, OBzl, OBzl

OH, CH_2OBzl, O, HO, OBzl, OBzl $\xrightarrow[92\%]{\text{1) } Bu_2SnO, C_6H_6 \text{ 2) } CH_3OCH_2Cl, Bu_4NI}$ OH, CH_2OBzl, O, CH_3OCH_2O, OBzl, OBzl

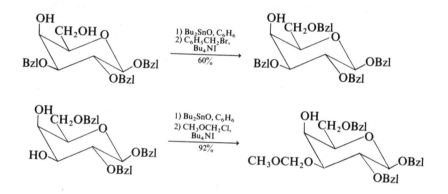

[1] J. Alais and A. Veyrières, *J.C.S. Perkin I*, 377 (1981).
[2] S. David, A. Thieffry, and A. Veyrières, *ibid.*, 1796 (1981).

Tetra-*n*-butylammonium octahydrotriborate (1). Mol. wt. 261.33, m.p. 210–212° (dec.), soluble in CH_2Cl_2 or CH_3CN, slightly soluble benzene, ether, water.

Preparation:

$$3NaBH_4 \xrightarrow[-\,2NaI,\,-\,2H_2]{I_2} NaB_3H_8 \xrightarrow[50\%]{n\text{-}Bu_4NI} n\text{-}Bu_4\overset{+}{N}B_3H_8^-$$

1

Reduction of carbonyl compounds.[2] The reagent reduces aromatic and aliphatic ketones, aldehydes, and acid chlorides to alcohols in moderate to excellent yield in chloroform solution. The order of substrate reactivity is RCOCl > RCHO > R_2CO. Only three of the eight hydrogens are transferred under the reaction conditions.

[1] G. E. Ryschkewitsch and K. C. Nainan, *Inorg. Syn.*, **15**, 111 (1974).
[2] W. H. Tamblyn, D. H. Weingold, E. D. Snell, and R. E. Waltermire, *Tetrahedron Letters*, **23**, 3337 (1982).

Tetraethyl dimethylaminomethylenediphosphonate, 3, 277–278. The reagent can be prepared in 90% yield by reaction of the Vilsmeier reagent (from DMF and oxalyl chloride) with triethyl phosphite.

The reagent is generally useful for conversion of aromatic, heteroaromatic, and aliphatic aldehydes into the homologous carboxylic acid in satisfactory yield.[1]

[1] C. R. Degenhardt, *Syn. Comm.*, **12**, 415 (1982).

5,6,8,9-Tetrahydro-7-phenyldibenzo[*c,h*]xanthylium tetrafluoroborate (1). Mol. wt. 464.38, m.p. 265°.

C_6H_5

BF_4^-

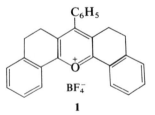

1

The reagent is prepared by reaction of 2-benzylidene-α-tetralone with α-tetralone in the presence of BF_3 etherate (42% yield).[1]

Dehydroamination.[2] The salt **1** reacts with primary amines, RCH_2NH_2, to form salts that are converted to alkenes at 150°. However, reaction at 20° of **1** with amines of the type $R^1CH_2CH(R^2)NH_2$ forms alkenes directly. For example, cyclopentylamine is converted in this way into cyclopentene (63% yield). In the case of open-chain amines, the ratio of alkenes formed is similar to that obtained by solvolysis of sulfonates.

[1] A. R. Katritzky and S. S. Thind, *J.C.S. Perkin I*, 1895 (1980).
[2] A. R. Katritzky and J. M. Lloyd, *ibid.*, 2347 (1982).

Tetrakis(triphenylphosphite)nickel(0), $Ni[P(OC_6H_5)_3]_4$ (**1**), **6**, 570; **7**, 357; **9**, 450.

Addition of HCN to —C≡C—; nitriles.[1] In the presence of this nickel(0) catalyst, HCN adds stereospecifically to both terminal and disubstituted alkynes. The addition is also regioselective.

Examples:

$$C_6H_5C\equiv CH + HCN \xrightarrow[51\%]{\mathbf{1},\ P(OC_6H_5)_3,\ C_6H_6,\ 75^\circ} (C_6H_5)(H)C{=}C(H)(CN)$$

$$C_6H_5C\equiv CC_6H_5 + HCN \xrightarrow[93\%]{} (C_6H_5)(H)C{=}C(C_6H_5)(CN)$$

$$(CH_3)_3CC\equiv CCH_3 \xrightarrow[80\%]{} (CH_3)_3C(H)C{=}C(CH_3)(CN) + (CH_3)_3C(NC)C{=}C(CH_3)(H) \quad 9:1$$

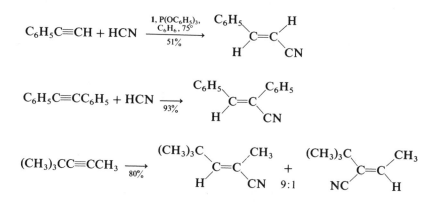

Reductive dimerization of propargyl chlorides.[2] Treatment of propargyl chlorides with this nickel(0) complex effects cyclodimerization to 3,4-bis(alkylidene)-cyclobutenes.

Example:

$$(CH_3)_2C(Cl){-}C\equiv CCH_3 \xrightarrow[65\text{–}75\%]{Ni(0),\ C_6H_6,\ 0^\circ} \text{1,2-di-}CH_3\text{-3,4-bis}(=C(CH_3)_2)\text{cyclobutene}$$

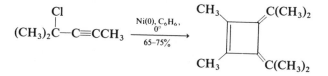

[1] W. R. Jackson and C. G. Lovel, *J.C.S. Chem. Comm.*, 1231 (1982).
[2] D. J. Pasto and D. K. Mitra, *J. Org.*, **47**, 1381 (1982).

Tetrakis(triphenylphosphine)palladium(0), 6, 571–573; **7**, 357–358; **8**, 472–476; **9**, 451–458; **10**, 384–391.

Decarboxylation of β-acetoxy carboxylic acids (**10**, 387). This reaction has been used in a $C_{10}+C_{10}$ synthesis of vitamin A ethyl ester (**3**) in 35–40% overall yield from 2,2,6-trimethylcyclohexanone (precursor to **1**).[1]

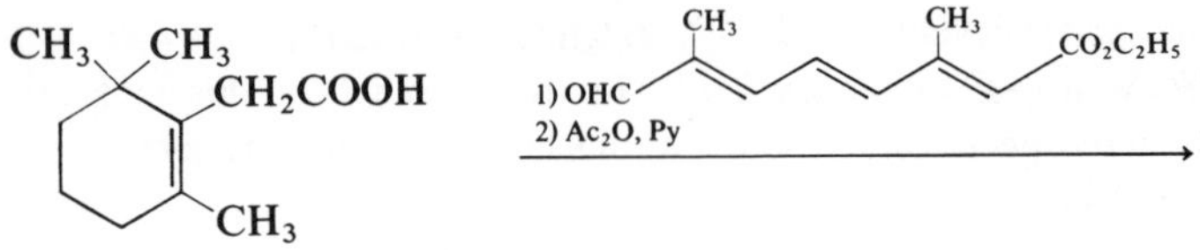

Cleavage of endoperoxides. 1,4-Endoperoxides in the presence of this Pd complex (5–10 mole %) are converted into three types of products, all involving cleavage of the O—O bond. Relative amounts depend on the particular ring system.[2]

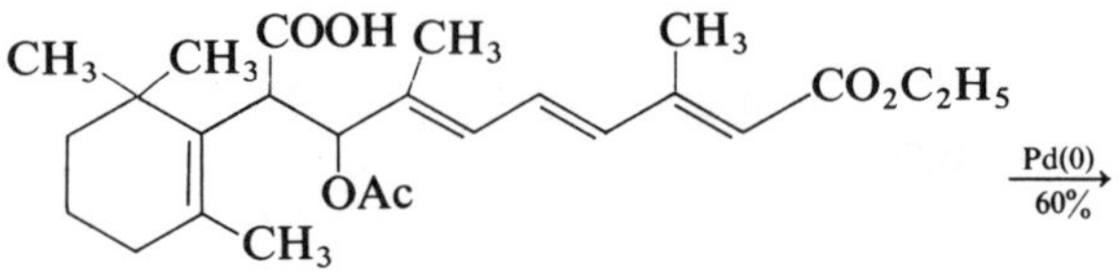

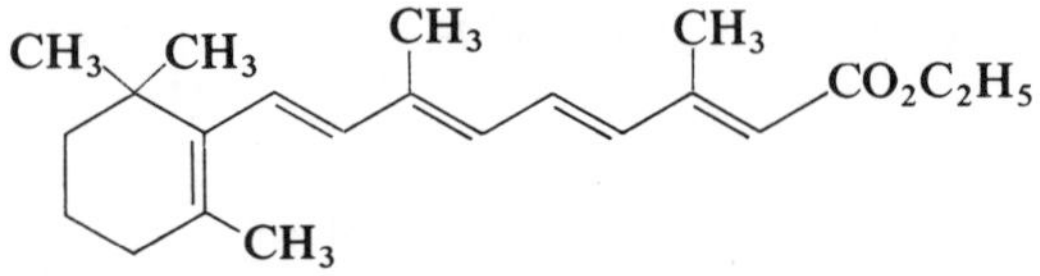

Examples:

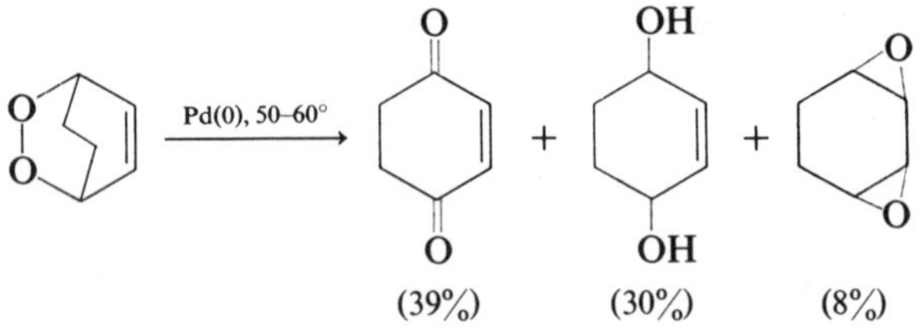

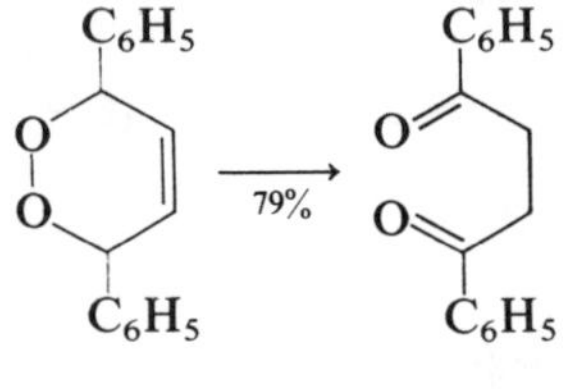

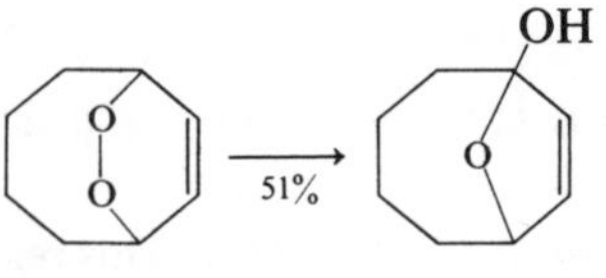

(39%) (30%) (8%)

This reaction converts PGH_2methyl ester (**1**) into primary prostaglandins (equation I).[3]

(I)

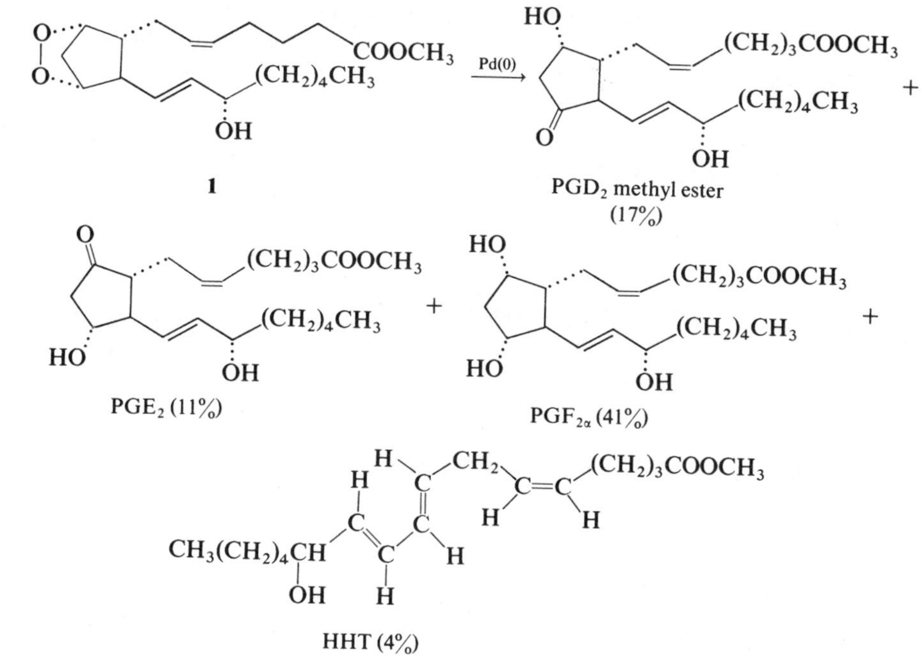

1 → Pd(0) → PGD$_2$ methyl ester (17%) + PGE$_2$ (11%) + PGF$_{2\alpha}$ (41%) + HHT (4%)

$(CH_2)_4CH_3$, COOCH$_3$, OH; HO, $(CH_2)_3COOCH_3$; $CH_3(CH_2)_4CH(OH)$

Alkylation of dihydropyranyl acetates. The Pd(0)-catalyzed alkylation of allylic acetates by stabilized carbanions originally reported by Trost and Verhoeven (**8**, 475; **9**, 451–457)[4] has been extended to alkylation of dihydropyranyl acetates as a route to natural C-glycopyranosides.[5] Again the displacement is regioselective and results in net retention of configuration in reactions with most carbanions. However, alkylation with phenylzinc chloride or vinylzinc chloride occurs with inversion.

Examples:

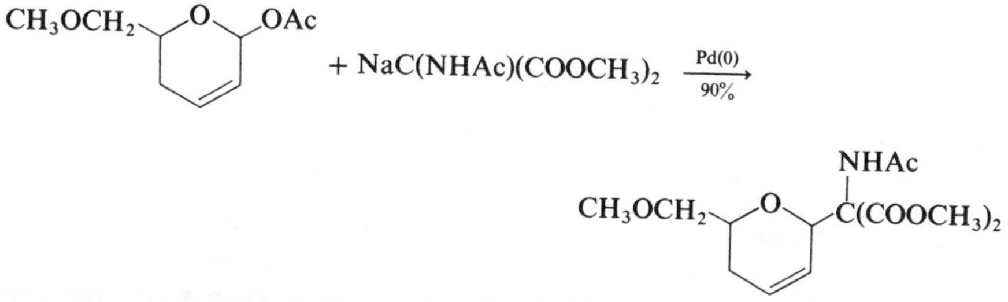

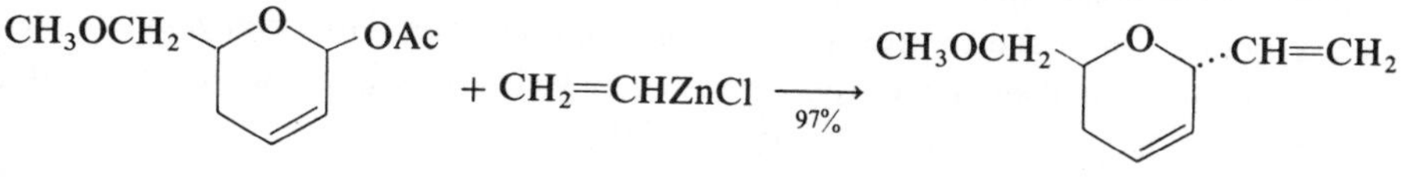

CH_3OCH_2–(dihydropyranyl)–OAc + NaC(NHAc)(COOCH$_3$)$_2$ $\xrightarrow[90\%]{Pd(0)}$ CH_3OCH_2–(dihydropyranyl)–C(NHAc)(COOCH$_3$)$_2$

CH_3OCH_2–(dihydropyranyl)–OAc + $CH_2{=}CHZnCl$ $\xrightarrow[97\%]{}$ CH_3OCH_2–(dihydropyranyl)–$CH{=}CH_2$

Alkylation and amination of allylic phosphates.[6] Allyl diethyl phosphates undergo efficient allylic alkylation and amination in the presence of this Pd(0) catalyst.

Examples:

$$\underset{(Z)}{CH_2OAc}\text{-}CH{=}CH\text{-}CH_2OP(O)(OC_2H_5)_2 + NaCH(COOCH_3)_2 \xrightarrow[83\%]{Pd(0),\ THF} AcO\text{-}CH_2CH{=}CHCH_2\text{-}CH(COOCH_3)_2 \quad (E/Z = 87/13)$$

$$\underset{(E\ or\ Z)}{C_6H_5CH{=}CHCH_2OP(O)(OC_2H_5)_2} + HN(C_2H_5)_2 \xrightarrow[\sim 65\%]{} C_6H_5CH{=}CHCH_2N(C_2H_5)_2 \quad (E/Z = 100:0)$$

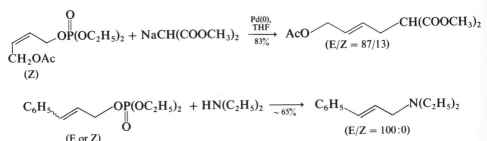

Sequential alkylation–amination is also possible.

Example:

$$CH_2OAc\text{-}CH{=}CH\text{-}CH_2OP(O)(OC_2H_5) \xrightarrow[79\%]{1)\ NaCH(COOCH_3)_2,\ Pd(0);\ 2)\ HN(CH_3)_2} (CH_3)_2N\text{-}CH_2CH{=}CHCH_2\text{-}CH(COOCH_3)_2$$

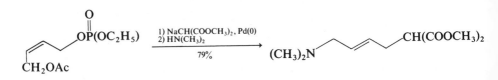

1,4-Dienes.[7] Palladium catalyzed allylation of alkenylmetals (**10**, 388) proceeds with almost complete inversion of configuration of the allylic center (equation I).

(I) $$\underset{CH_3}{\overset{n\text{-}Bu}{}}C{=}C\underset{Al(CH_3)_2}{\overset{H}{}} + \text{(bicyclic lactone)} \xrightarrow[90\%]{Pd(0)} \underset{CH_3}{\overset{n\text{-}Bu}{}}C{=}C\overset{H}{}\text{-(cyclohexenyl-COOH)}$$

HOOC (98% pure)

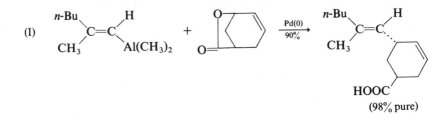

α-Allylated ketones.[8] Although allylation of alkali metal enolates of ketones with allylic acetates catalyzed by $Pd[P(C_6H_5)_3]_4$ is generally unsatisfactory, various allylic electrophiles undergo facile palladium-catalyzed reaction with potassium enoxyborates, prepared by treatment of potassium enolates with triethylborane (**9**, 482). The α-allylated ketones are obtained in high yield and with essentially complete retention of the enolate regiochemistry and of the allyl geometry.

Example:

$$CH_3\text{-cyclohexenyl-}OB(C_2H_5)_3K + AcO\text{-}CH_2CH{=}C(CH_3)CH_2CH_2CH{=}C(CH_3)CH_3 \xrightarrow[81\%]{Pd(0)} 2\text{-}CH_3\text{-}6\text{-}[CH_2CH{=}C(CH_3)CH_2CH_2CH{=}C(CH_3)CH_3]\text{cyclohexanone}$$

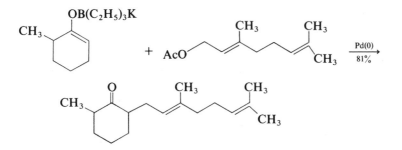

Alkylation of allyl nitroalkanes.[9] Cyclic allylic nitro compounds react with stabilized carbanions or amines in the presence of Pd(0) at the allylic position without allylic transposition. The reactive intermediate is presumably an allylpalladium(II) species.

Example:

$$1\text{-}(CH_2NO_2)\text{cyclopentene} + NaCH(COOCH_3)_2 \xrightarrow[70\%]{Pd(0),\ P(C_6H_5)_3,\ DMF} 1\text{-}[CH_2CH(COOCH_3)_2]\text{cyclopentene}$$

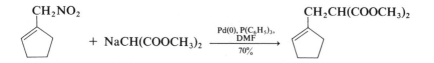

Allylation of nucleophiles. π-Allyl palladium complexes (**a**) of α,β-unsaturated epoxides react with active methylene compounds to give 1,4-adducts [allylic alcohols, equation (I)].[10]

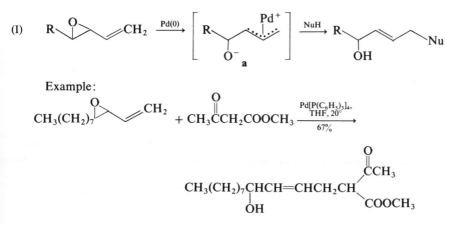

$$\text{(I)}\quad R\text{-epoxide-}CH{=}CH_2 \xrightarrow{Pd(0)} \left[R\text{-}CH(O^-)\text{-}\pi\text{-allyl-}Pd^+ \right]_{\mathbf{a}} \xrightarrow{NuH} R\text{-}CH(OH)CH{=}CHCH_2Nu$$

Example:

$$CH_3(CH_2)_7\text{-epoxide-}CH{=}CH_2 + CH_3\overset{O}{\overset{\|}{C}}CH_2COOCH_3 \xrightarrow[67\%]{Pd[P(C_6H_5)_3]_4,\ THF,\ 20^\circ}$$

$$CH_3(CH_2)_7\underset{OH}{\underset{|}{C}}HCH{=}CHCH_2CH(\overset{O}{\overset{\|}{C}}CH_3)(COOCH_3)$$

Allylic carbonates show high reactivity in a related palladium catalyzed decarboxylative allylation of active methylene compounds (equation II). In this case,

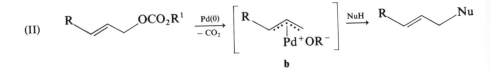

(II) R–CH=CH–CH$_2$–OCO$_2$R^1 $\xrightarrow[-\,CO_2]{Pd(0)}$ [R–(π-allyl)Pd$^+$OR$^-$] **b** $\xrightarrow{NuH}$ R–CH=CH–CH$_2$–Nu

tris(dibenzylideneacetone)dipalladium [$Pd_2(dba)_3 \cdot CHCl_3$] proved more effective than $Pd[P(C_6H_5)_3]_4$.[11]

Example:

$$CH_2{=}C(CH_3)CH_2OCO_2CH_3 + C_2H_5C(O)CH(CH_3)COOCH_3 \xrightarrow[92\%]{Pd_2(dba)_3 \cdot CHCl_3,\ P(C_6H_5)_3,\ THF} CH_2{=}C(CH_3)CH_2C(CH_3)(COOCH_3)COC_2H_5$$

An asymmetric intramolecular cyclization by this reaction in the presence of $Pd(OAc)_2$ and a chiral phosphine[12] proceeds in 48% optical yield (equation III).[13]

(III) $CH_3OCO_2CH_2CH{=}CH(CH_2)_3COCH_2COOCH_3$ $\xrightarrow{Pd(OAc)_2,\ (SR)\text{–}BPPFA^*,\ C_6H_6,\ 30°}$ [2-CO$_2$CH$_3$-3-(CH=CH$_2$)cyclohexanone] $\xrightarrow{83\%}$ 3-(CH=CH$_2$)cyclohexanone

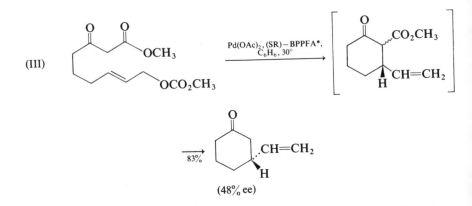

(48% ee)

Alkylation of vinyl epoxides. Although Pd(0) catalyzes the rearrangement of vinyl epoxides (**9**, 452–453), alkylation of cyclic or acylic vinyl epoxides with dimethyl malonate under neutral conditions is possible with the same catalyst or with bis[1,2-bis(diphenylphosphino)ethane]palladium. The rearrangement and alkylation proceed with different regio- and stereoselectivity.[19]

Examples:

$$\text{cyclopentenone} \xleftarrow[77^\circ]{Pd(0),\ CH_2Cl_2} \text{cyclopentadiene monoepoxide} \xrightarrow[57\%]{CH_2(COOCH_3)_2,\ Pd(0),\ THF} \text{cis-4-}CH(COOCH_3)_2\text{-2-cyclopentenol}$$

$$\text{epoxide} + O{=}C(CH_2COOCH_3)_2 \xrightarrow[79\%]{Pd(0)} \text{product with } CH_2CHCCH_2COOCH_3 \text{ (}COOCH_3\text{, }C{=}O\text{), } CH_3,\ CH_3,\ OH$$

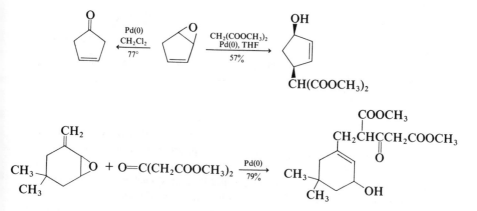

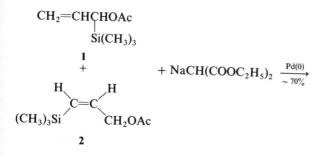

Vinylsilanes. Reaction of either 1- or 3-trimethylsilylallyl acetate (**1** or **2**) with several nucleophiles in the presence of Pd(0) affords vinylsilanes.[15]

Examples:

$$CH_2{=}CHCH(OAc)Si(CH_3)_3\ (\mathbf{1}) \text{ or } (CH_3)_3SiCH{=}CHCH_2OAc\ (\mathbf{2}) + NaCH(COOC_2H_5)_2 \xrightarrow[\sim 70\%]{Pd(0)}$$

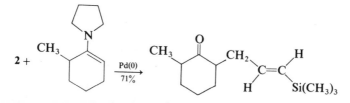

$$(CH_3)_3SiCH{=}CHCH_2CH(COOC_2H_5)_2$$

$$E/Z = 25:75$$

$$\mathbf{2} + \text{1-pyrrolidino-6-methylcyclohexene} \xrightarrow[71\%]{Pd(0)} \text{2-methyl-6-}[CH_2CH{=}CHSi(CH_3)_3]\text{cyclohexanone}$$

Nucleophilic substitution reactions with α-acetoxy-β,γ-unsaturated nitriles.[15] This reaction can be effected with Pd(0) as catalyst; it proceeds regioselectively with rearrangement to give γ-substituted-α,β-unsaturated nitriles.

Example:

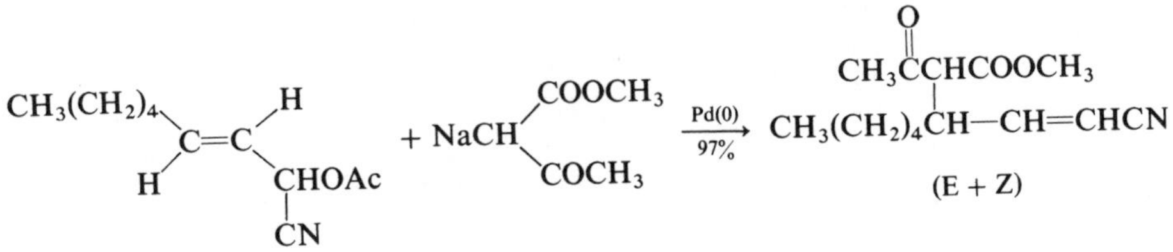

Macrocyclization. Trost and Warner[17] have effected efficient cyclization to ten- and fifteen-membered rings by use of Pd(0) supported on a polystyrene bearing phosphine ligands. Precursors with an epoxy vinyl terminal group proved particularly suitable. Thus, **1** in the presence of such a Pd(0) catalyst cyclizes to **2** and **3** in 71% yield. The products are isomeric at the double bond, since both are oxidized to the ketone **4**. The same catalyst system converted **1** (n = 9) into two macrocyclic isomers in 66% yield. Both cyclizations are concentration dependent. The temperature is also critical, the reaction being particularly clean at 65°.

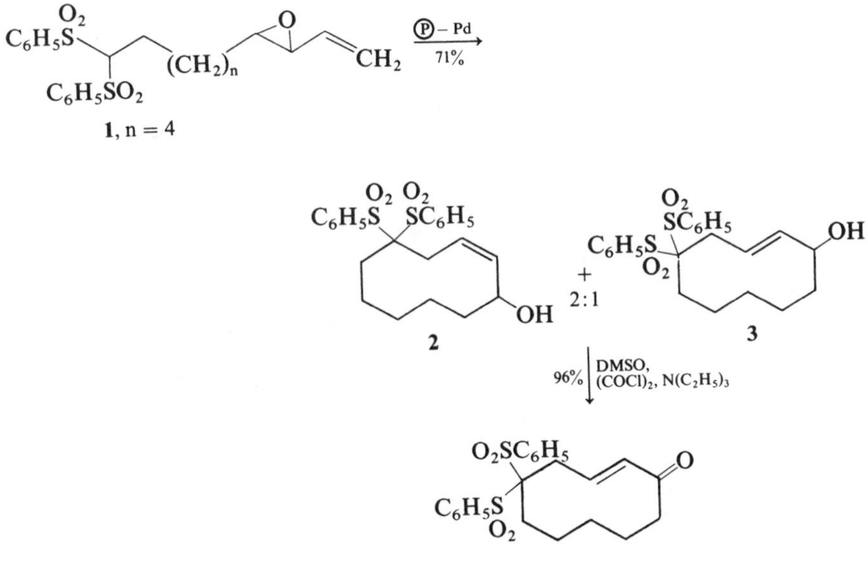

1, n = 4

2

3

4

Macroheterocyclization.[18] Pd(0) can catalyze macrocyclizations which form a C—N bond. This reaction was used in a synthesis of the spermidine alkaloid inandenine-12-one (**1**).

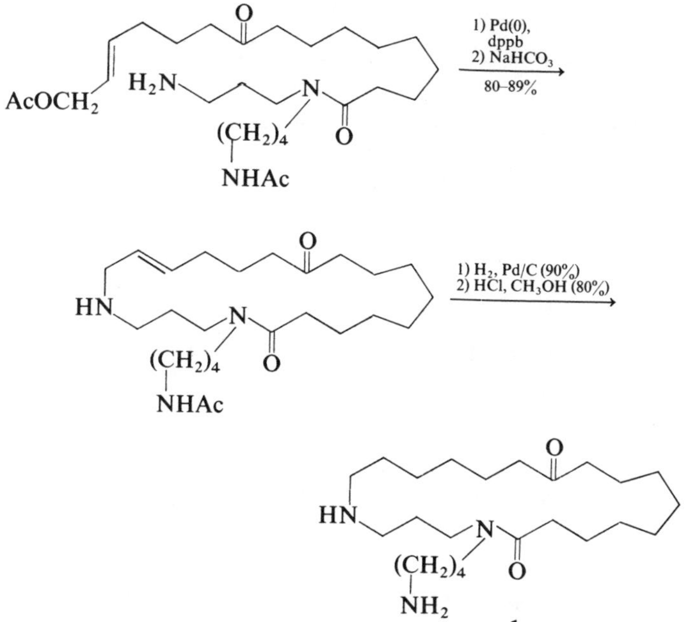

$$\text{AcOCH}_2\text{-CH=CH-(macrocyclic keto lactam with } H_2N\text{ chain, N-(CH}_2)_4\text{NHAc)} \xrightarrow[\text{80–89\%}]{\text{1) Pd(0), dppb; 2) NaHCO}_3}$$

$$\text{(macrocyclic amine, HN, N-(CH}_2)_4\text{NHAc)} \xrightarrow{\text{1) H}_2\text{, Pd/C (90\%); 2) HCl, CH}_3\text{OH (80\%)}}$$

$$\text{(saturated macrocycle, HN, N-(CH}_2)_4\text{NH}_2)$$

1

β-Keto esters.[9] Acyl chlorides condense with the Reformatsky reagent in the presence of $Pd[P(C_6H_5)_3]_4$ or, preferably, a Pd(0) catalyst obtained by reduction of $Cl_2Pd[P(C_6H_5)_3]_2$ with DIBAH,[10] to give β-keto esters in yields of 25–90%. The yields are highest with aromatic or heterocyclic acyl chlorides (equation I).

$$\text{(I)}\quad RCOCl + BrZnCH_2COOC_2H_5 \xrightarrow[\text{25–90\%}]{\text{Pd(0), DME}} RCOCH_2COOC_2H_5$$

Double silylation. Chloromethyldisilanes, methoxymethyldisilanes, and hexamethyldisilane add to allene and 1,2-butadiene in the presence of $Pd[P(C_6H_5)_3]_4$ to form compounds containing vinylsilane and allysilane groups.[21]

Examples:

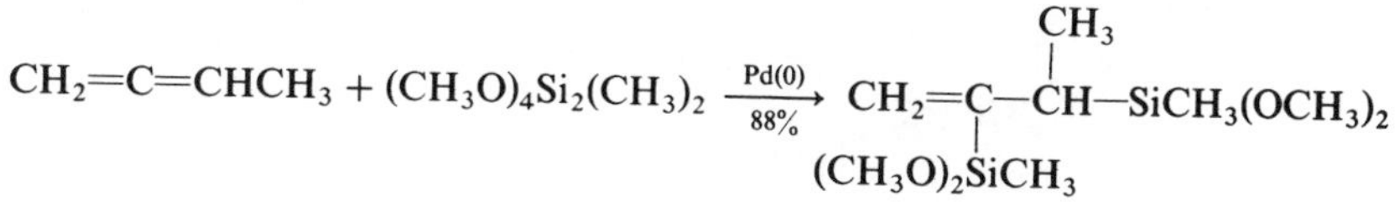

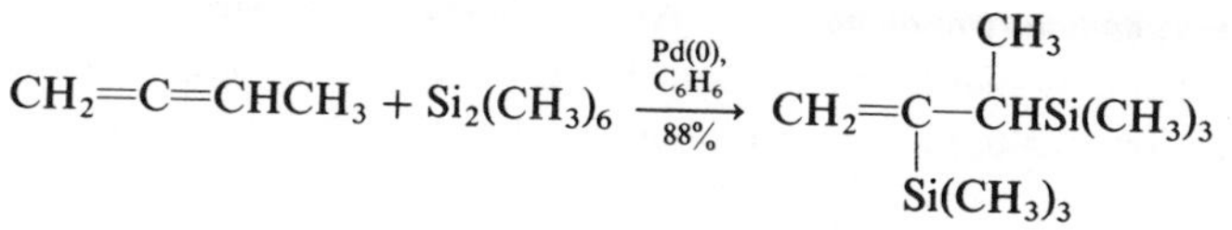

$$CH_2{=}C{=}CHCH_3 + (CH_3O)_4Si_2(CH_3)_2 \xrightarrow[88\%]{\text{Pd(0)}} CH_2{=}C(Si(OCH_3)_2CH_3){-}CH(CH_3){-}SiCH_3(OCH_3)_2$$

$$CH_2{=}C{=}CHCH_3 + Si_2(CH_3)_6 \xrightarrow[88\%]{\text{Pd(0), }C_6H_6} CH_2{=}C(Si(CH_3)_3){-}CH(CH_3)Si(CH_3)_3$$

$$CH_2{=}C{=}CH_2 + Cl_3Si_2(CH_3)_3 \xrightarrow[53\%]{} CH_2{=}C(SiCH_3Cl_2){-}CH_2Si(CH_3)Cl_2$$

Allylic sulfones (*cf.*, **9**, 132). In the presence of this Pd(0) complex, allylic acetates condense with sodium *p*-tolylsulfinate to form allylic sulfones (predominately E). The more substituted isomer is formed initially, but rearranges after extended reaction time to the more stable allylic sulfone.[22]

Example:

$$(CH_3)_2C{=}CH(CH_2)_2C(CH_3)(OAc)CH{=}CH_2 + TsNa \xrightarrow[78\%]{Pd(0)} (CH_3)_2C{=}CH(CH_2)_2C(CH_3)(Ts)CH{=}CH_2$$

$$\xrightarrow[84\%]{25^\circ} (CH_3)_2C{=}CH(CH_2)_2C(CH_3){=}CHCH_2Ts \quad (E/Z) = 87:13$$

***2,3-Disubstituted bicyclo*[2.2.1]*heptanes*.**[23] Bicyclo[2.2.1]heptene is converted into 2,3-disubstituted derivatives when treated with an aryl or vinyl bromide and an alkyne in the presence of potassium acetate (1 equiv.) and a catalytic amount of $Pd[P(C_6H_5)_3]_4$.

Example:

$$\text{bicyclo[2.2.1]heptene} \xrightarrow{C_6H_5Br,\ CH_3COOK,\ Pd(0)} \left[\text{2-}C_6H_5\text{-3-PdOAc-norbornane} \xrightarrow{C_6H_5C{\equiv}CH} \text{2-}C_6H_5\text{-3-}(CH{=}C(C_6H_5)PdOAc)\text{-norbornane} \right] \xrightarrow[42\%]{-Pd(0),\ -CH_3COOH} \text{2-}C_6H_5\text{-3-}(C{\equiv}CC_6H_5)\text{-norbornane}$$

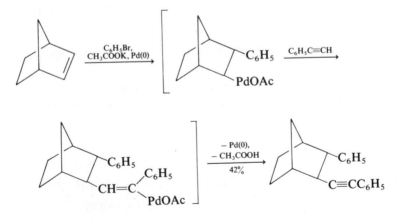

***Desulfonylation; homoallylic alcohols*.**[24] Allyl sulfones (**2**), prepared as shown, undergo desulfonylation by sodium borohydride in the presence of catalytic amounts of Pd(0). This reaction was used to prepare homoallylic alcohols (**3**) from allylic *p*-tolyl sulfones (**1**).

Example:

$$C_6H_5CH{=}CHCH_2Ts \xrightarrow[84\%]{1)\ n\text{-BuLi};\ 2)\ C_6H_5CHO} C_6H_5CH{=}CHCH(Ts)CH(OH)C_6H_5$$

1 → 2

$$\xrightarrow[75\%]{NaBH_4,\ (CH_3)_2CHOH,\ Pd(0)} C_6H_5CH{=}CHCH_2CH(OH)C_6H_5$$

3

Sequential π-allylpalladium alkylation. Pd(0)-catalyzed dialkylation of the allylic-homoallylic cyclopentenyl diester **1** with diethyl sodiomalonate gives only **2** (> 90% yield).

1 (OAc, OAc cyclopentene) $\xrightarrow{NaCH(COOC_2H_5)_2,\ Pd(0)}$ [$(C_2H_5O_2C)_2CH$–cyclopentenyl–Pd] $\xrightarrow[>90\%]{NaCH(COOC_2H_5)_2}$ 2: $(C_2H_5OOC)_2CH$, $CH(COOC_2H_5)_2$ cyclopentene

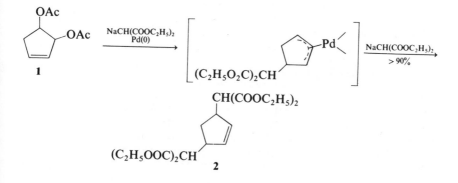

Use of a mixed diester permits sequential addition of two different nucleophiles (equation I).[25]

(I) AcO···(cyclopentene)–$OCOC(CH_3)_3$ $\xrightarrow[54\%]{1)\ NaCH(COOCH_3)_2,\ Pd(0);\ 2)\ NaCH(COOC_2H_5)_2,\ Pd(0)}$ $(CH_3OOC)_2CH$···(cyclopentene)–$CH(COOC_2H_5)_2$

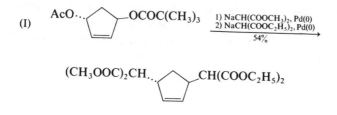

Biaryls.[26] In the presence of this Pd(0), catalyst, phenylboric acid couples with iodo- or bromoarenes under basic conditions to give biphenyls in 40–95% isolated yield. Since arylboric acids are generally accessible, this route probably is widely applicable (equation I).

$$\text{(I)}\quad C_6H_5B(OH)_2 + BrC_6H_4Z \xrightarrow[40\text{–}95\%]{\text{Pd(0)},\ Na_2CO_3} C_6H_5{-}C_6H_4Z$$

[1] B. M. Trost and J. M. D. Fortunak, *Tetrahedron Letters*, **22**, 3459 (1981).
[2] M. Suzuki, Y. Oda, and R. Noyori, *ibid.*, **22**, 4413 (1981).
[3] M. Suzuki, R. Noyori, and N. Hamanaka, *Am. Soc.*, **103**, 5606 (1981).
[4] B. M. Trost and T. R. Verhoeven, *ibid.*, **102**, 4730 (1980).
[5] L. V. Dunkerton, A. J. Serino, *J. Org.*, **47**, 2812 (1982).
[6] Y. Tanigawa, K. Nishimura, A. Kawasaki, and S.-I. Murahashi, *Tetrahedron Letters*, **23**, 5549 (1982).
[7] H. Matsushita and E. Negishi, *J.C.S. Chem. Comm.*, 160 (1982).
[8] E. Negishi, H. Matsushita, S. Chatterjee, and R. A. John, *J. Org.*, **47**, 3188 (1982).
[9] R. Tamura and L. S. Hegedus, *Am. Soc.*, **104**, 3727 (1982).
[10] T. Tsuji, H. Kataoka, and Y. Kobayashi, *Tetrahedron Letters*, **22**, 2572 (1981).
[11] J. Tsuji, I. Shimizu, I. Minami, and Y. Ohashi, *ibid.*, **23**, 4809 (1982).
[12] (S,R)-N,N-Dimethyl-1-[1′,2-bis(diphenylphosphine)ferrocenyl]ethylamine.
[13] K. Yamamoto and J. Tsuji, *Tetrahedron Letters*, **23**, 3089 (1982).
[14] B. M. Trost and G. A. Molander, *Am. Soc.*, **103**, 5969 (1981).
[15] T. Hirao, J. Enda, Y. Ohshiro, and T. Agawa, *Tetrahedron Letters*, **22**, 3079 (1981).
[16] T. Tsuji, H. Ueno, Y. Kobayashi, and H. Okumoto, *ibid.*, **22**, 2573 (1981).
[17] B. M. Trost and R. W. Warner, *Am. Soc.*, **104**, 6112 (1982).
[18] B. M. Trost and J. Cossy, *ibid.*, **104**, 6881 (1982).
[19] T. Sato, T. Itoh, and T. Fujisawa, *Chem. Letters*, 1559 (1982).
[20] E. Negishi, A. O. King, and N. Okukado, *J. Org.*, **42**, 1821 (1977).
[21] H. Watanabe, M. Saito, N. Sutou, and Y. Nagai, *J.C.S. Chem. Comm.*, 617 (1981).
[22] K. Inomata, T. Yamamoto, and H. Kotake, *Chem. Letters*, 1357 (1981).
[23] M. Catellani and G. P. Chiusoli, *Tetrahedron Letters*, **23**, 4517 (1982).
[24] H. Kotake, T. Yamamoto, and H. Kinoshita, *Chem. Letters*, 1331 (1982).
[25] R. S. Valpey, D. J. Miller, J. M. Estes, and S. A. Godleski, *J. Org.*, **47**, 4717 (1982).
[26] N. Miyaura, T. Yanagi, and A. Suzuki, *Syn. Comm.*, **11**, 513 (1981).

Tetrakis(triphenylphosphine)palladium(0)–Zinc.

Reductive coupling; benzyl ketones.[1] Benzyl bromides and acyl chlorides undergo reductive coupling when treated with zinc (2 equiv.) and catalytic amounts of $Pd[P(C_6H_5)_3]_4$ or $Cl_2Pd[P(C_6H_5)_3]_2$ to form benzyl ketones. The highest yields are obtained with aryl acyl chlorides. $ArCH_2ZnBr$ is considered to be an intermediate.

[1] T. Sato, K. Naruse, M. Enokiya, and T. Fujisawa, *Chem. Letters*, 1135 (1981).

Tetramethylammonium hydroxide, $(CH_3)_4\overset{+}{N}OH^-$, Mol. wt. 91.15.

1,2-*Dioxetanes.*[1] The epoxides (**1**) of allylic hydroperoxides are converted into 1,2-dioxetanes (**2**) on treatment with this base in aqueous ether. Alkaline hydroxides are unsatisfactory. The main limitation is the availability of the allylic hydroperoxides.

$$\text{R}^1\text{R}^2\text{C(OOH)C(R}^3\text{)=CH}_2 \xrightarrow{p\text{-NO}_2\text{C}_6\text{H}_4\text{CO}_3\text{H}} \mathbf{1} \xrightarrow[\text{65–80\% overall}]{(\text{CH}_3)_4\text{NOH, H}_2\text{O, ether}} \mathbf{2}$$

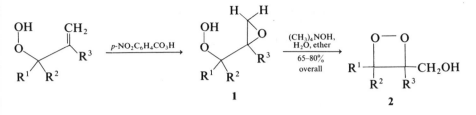

[1] D. Leclerq, J.-P. Bats, P. Picard, and J. Moulines, *Synthesis*, 778 (1982).

Tetraphenylbismuth trifluoroacetate, 10, 393.

O- and C-Phenylation. The reagent can be used to effect either O- or C-phenylation of enolic substrates, depending on conditions. O-Phenylation is favored under neutral conditions; under strongly basic conditions, C-phenylation is favored. Addition of Cl_3CCOOH significantly improves yields of phenyl ethers.[1]

Example:

$$\text{2-naphthyl-OC}_6\text{H}_5 \xleftarrow[90\%]{(\text{C}_6\text{H}_5)_4\text{BiOCOCF}_3,\ \text{Cl}_3\text{CCOOH}} \text{2-naphthol} \xrightarrow[90\%]{1)\ (\text{CH}_3)_2\text{NC(=NC(CH}_3)_3)\text{N(CH}_3)_2\quad 2)\ (\text{C}_6\text{H}_5)_4\text{BiOCOCF}_3} \text{1-C}_6\text{H}_5\text{-2-naphthol}$$

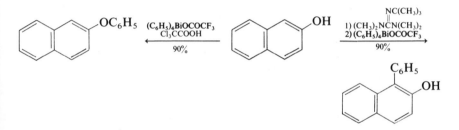

[1] D. H. R. Barton, B. Charpiot, and W. B. Motherwell, *Tetrahedron Letters*, **23**, 3365 (1982).

Thallium(I) carbonate, Tl_2CO_3. Mol. wt. 468.75, m.p. 273.° Supplier: Alfa.

Modified Hunsdiecker reaction. Primary carboxylic acids are converted into thallium(I) carboxylates by reaction with Tl_2CO_3. The salts are not isolated, but treated with Br_2 (1.5 equiv.) to give primary alkyl bromides (equation I).[1]

$$2\text{CH}_3(\text{CH}_2)_n\text{COOH} + \text{Tl}_2\text{CO}_3 \xrightarrow{-\text{H}_2\text{CO}_3} \left[\ 2\text{RCO}_2\text{Tl}\ \right] \xrightarrow[80\text{–}85\%]{\text{Br}_2} 2\text{RBr}$$

$n = 10\text{–}16$

[1] R. C. Cambie, R. C. Hayward, J. L. Jurlina, P. S. Rutledge, and P. D. Woodgate, *J.C.S. Perkin I*, 2608 (1981).

Thallium(III) trifluoroacetate, 3, 286–289; **4**, 496–501; **5**, 658–659; **7**, 365; **8**, 478–481; **9**, 462–464; **10**, 397.

Arene allylation.[1] This reaction can be carried out in ~40–70% yield (isolated) by reaction of the arene (large excess) with an allylsilane, allylgermane, or allylstannane and $Tl(OCOCF_3)_3$ to form an allyl cation. Thallium(III) nitrate can also be used.

Example:

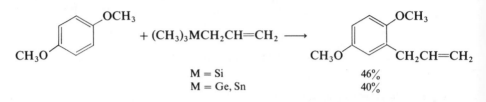

$$p\text{-}CH_3OC_6H_4OCH_3 + (CH_3)_3MCH_2CH{=}CH_2 \longrightarrow 2,5\text{-}(CH_3O)_2C_6H_3CH_2CH{=}CH_2$$

M = Si 46%

M = Ge, Sn 40%

Macrolides. Oxidation of the α,β-unsaturated ester **1** with TTFA in TFA results in the 8-membered lactone **2** in 36% yield.[2]

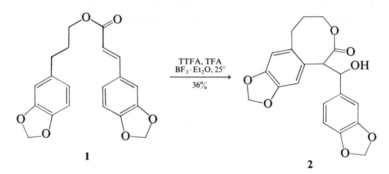

TTFA, TFA, $BF_3\cdot Et_2O$, 25°, 36%

1 → **2**

[1] M. Ochiai, M. Arimoto, and E. Fujita, *Tetrahedron Letters*, **22**, 4491 (1981).
[2] S. Nishiyama and S. Yamamura, *Chem. Letters*, 1511 (1981).

Thallium(III) trifluoroacetate–Palladium(II) acetate.

Biaryls.[1] TTFA oxidizes benzene at 80° to biphenyl in 14% yield; when a catalytic amount of $Pd(OAc)_2$ is added, the yield of biphenyl is increased to as much as 90% with no precipitation of Pd(0).

Arenes substituted with OCH_3, CH_3, F, and Cl are converted in satisfactory yield to 4,4′-substituted biphenyls in the same way.

[1] A. D. Ryabov, S. A. Deiko, A. K. Yatsimirsky, and I. V. Berezin, *Tetrahedron Letters*, **22**, 3793 (1981).

Thexylborane, 1, 276; **2**, 148–149; **4**, 175–176; **5**, 232–233; **6**, 207–208; **10**, 397–398.

Cyclic hydroboration (**10**, 397). Two laboratories[1,2] have reported stereoselective syntheses of the Prelog-Djerassi lactonic acid (**4**), both of which involve cyclic hydroboration of a slightly different 1,5-diene. One of these syntheses is outlined in equation I. Although two isomers of **2** are formed on hydroboration, the lactone **3** (2 isomers), obtained on oxidation, is readily equilibrated to give only one isomer. The remaining steps to **4** are known reactions. The sequence can be carried out with an enantiomerically pure diene.

(I) **1** $\xrightarrow[92\%]{1)\ B_2H_6\quad 2)\ H_2O_2,\ \bar{O}H}$ **2** (2 isomers 1:1) $\xrightarrow[>90\%]{Ag_2CO_3\text{–Celite}}$ (**3**, 2 isomers) $\longrightarrow$ **4**

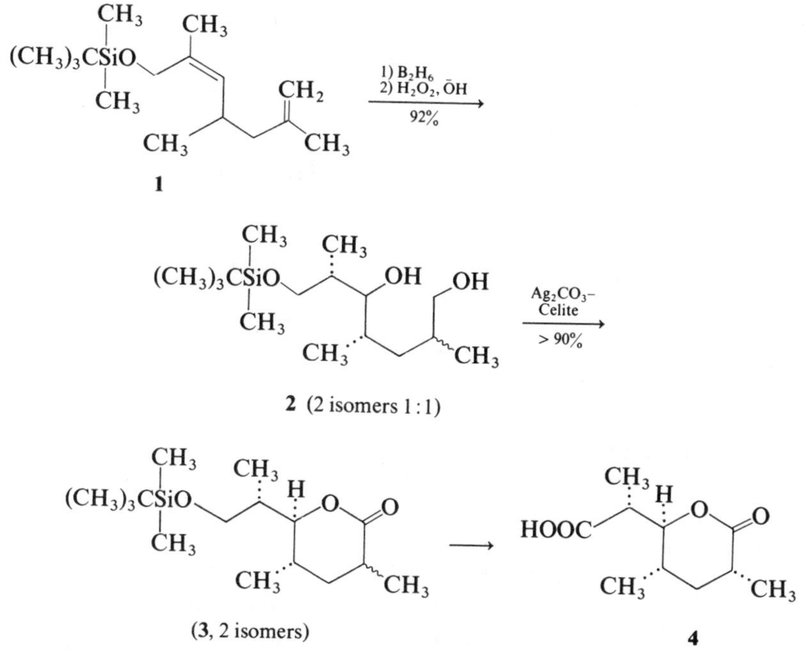

Acyclic diastereoselective hydroboration.[3] Hydroboration-oxidation of terminal alkenes substituted at C_4 by a large and a medium alkyl group can proceed asymmetrically with $BH_3 \cdot S(CH_3)_2$ or, even more selectively, with thexylborane.

Example:

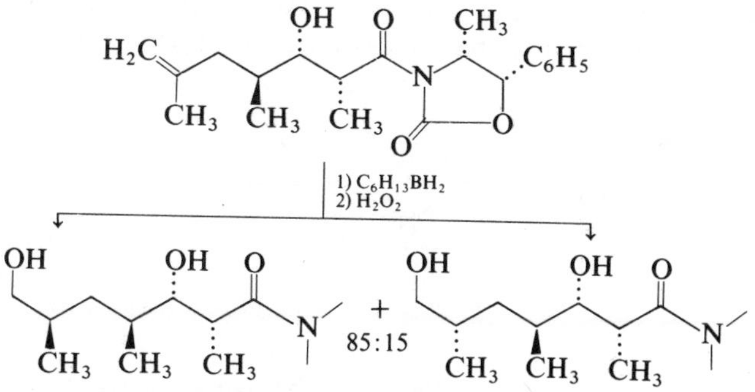

1) $C_6H_{13}BH_2$ 2) H_2O_2 → 85:15

This 1,3-asymmetric induction was first observed in a synthesis of (+)-Prelog-Djerassi lactonic acid (**1**), partially formulated in equation (I).[4]

(I) $H_2C{=}C(CH_3)CH_2CH(CH_3)CH(OSi(CH_3)_3)CH(CH_3)CON(CH_3)_2$ $\xrightarrow[79\%]{1)\ C_6H_{13}BH_2;\ 2)\ H_2O_2}$ $HOCH_2CH(CH_3)CH_2CH(CH_3)CH(OSi(CH_3)_3)CH(CH_3)CON(CH_3)_2$

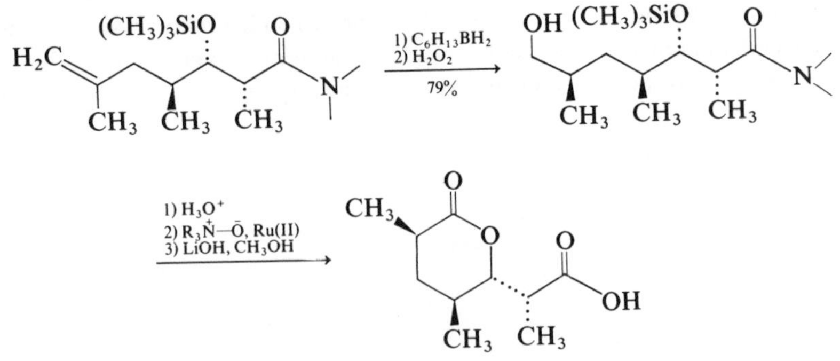

$\xrightarrow{1)\ H_3O^+;\ 2)\ R_3\overset{+}{N}-\bar{O},\ Ru(II);\ 3)\ LiOH,\ CH_3OH}$ **1**

[1] W. C. Still and K. R. Shaw, *Tetrahedron Letters*, **22**, 3725 (1981).
[2] D. J. Morgans, Jr., *ibid.*, **22**, 3721 (1981).
[3] D. A. Evans, J. Bartroli, and T. Godel, *ibid.*, **23**, 4577 (1982).
[4] D. A. Evans and J. Bartroli, *ibid.*, **23**, 807 (1982).

1,3-Thiazolidine-2-thione, 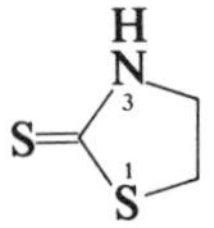 (**1**). Mol. wt. 119.21, m.p. 107°.

The thione is obtained on reaction of CS_2 with ethanolamine, β-bromoethylamine, or ethyleneimine.

Reduction of carboxylic acids to aldehydes.[1, 2] 3-Acylthiazolidine-2-thiones (**2**) can be prepared by reaction of 1,3-thiazolidine-2-thione (**1**) with carboxylic acids directly (using 2-chloro-1-methylpyridinium iodide, **8**, 95–96) or with acid chlorides (triethylamine or DCC). They can also be prepared from reaction of the thallium salt of **1** with an acid chloride. Yields by all four methods are 70–95%. The amides are reduced to aldehydes by either DIBAH or, generally in higher yield, by $NaBH_4$ (90–98% yield).

Amides.[3] Reaction of 3-acylthiazolidine-2-thiones (**2**) with amines gives amides in high yields. The reactivity of amines is $RNH_2 > ArNH_2$ and $RCH_2NH_2 > R_2CHNH_2 \gg R_3CNH_2$. An example is the synthesis of maytenine (**3**).

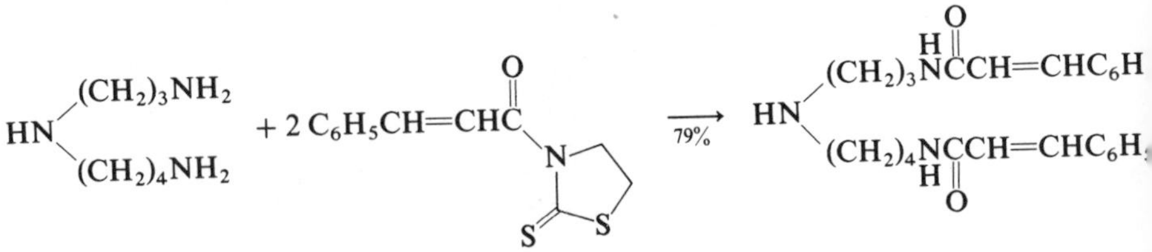

$HN[(CH_2)_3NH_2][(CH_2)_4NH_2] + 2\ C_6H_5CH{=}CHC(O){-}N$ (thiazolidine-2-thione) (**2**) $\xrightarrow{79\%}$ $HN[(CH_2)_3NHCOCH{=}CHC_6H_5][(CH_2)_4NHCOCH{=}CHC_6H_5]$ (**3**)

Review. Newer uses of **1** in synthesis have been outlined in a lecture.[4]

[1] T. Izawa and T. Mukaiyama, *Chem. Letters*, 1443 (1977); E. Fujita, Y. Nagao, K. Seno, S. Takao, T. Miyasaka, M. Kumura, and W. H. Watson, *J.C.S. Perkin I*, 914 (1980).

[2] Y. Nagao, K. Kawabata, K. Seno, and E. Fujita, *ibid.*, 2470 (1980).
[3] Y. Nagao, K. Seno, K. Kawabata, T. Miyasaka, S. Takao, and E. Fujita, *Tetrahedron Letters*, **21**, 841 (1980).
[4] E. Fujita, *Pure Appl. Chem.*, **53**, 1141 (1981).

Thiourea, 1, 1164–1167; **2,** 412–413; **3,** 290–291; **6,** 586.

Thiol synthesis.[1] A powerful flavor-impact constituent of grapefruit juice is 1-*p*-menthene-8-thiol (**3**). It has been synthesized by reaction of thiourea with the epoxide **1** to give **2** followed by reduction with $LiAlH_4$.

CH_3 CH_3 $S{=}C(NH_2)_2,$ H_2SO_4 Na_2CO_3 $\frac{1}{2}SO_4^{2-}$ O OH CH_3 CH_3 $CH_2SC(NH_2){=}\overset{+}{N}H_2$

1 **a**

CH_3 CH_3 $LiAlH_4,$ THF 41% overall S SH CH_3 CH_3 CH_3

2 **3**

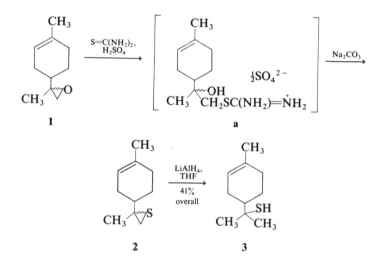

Ozonation.[2] Thiourea can be used to reduce the α-alkoxyhydroperoxides obtained on ozonation of alkenes in methanol to aldehydes and the dimethyl acetal.

[1] E. Demole, P. Enggist, and G. Ohloff, *Helv.*, **65**, 1785 (1982).
[2] D. Gupta, R. Soman, and S. Dev, *Tetrahedron*, **38**, 3013 (1982).

Tin(0), At. wt. 118.69.

Homoallylic alcohols. In the presence of metallic tin, both allyl iodide and bromide react with carbonyl compounds at room temperature to form homoallylic alcohols in 75–90% yield. Presumably a diallyltin dihalide is formed initially.[1]

Examples:

$$CH_3CH{=}CHCH_2Br + C_6H_5CHO \xrightarrow[76\%]{Sn(0),\ THF} CH_2{=}CHCH(CH_3)CH(OH)C_6H_5$$

$$Br_2Sn(CH_2CH{=}CH_2)_2 + C_6H_5CH_2COCH_3 \xrightarrow[75\%]{} CH_2{=}CHCH_2C(OH)(CH_3)CH_2C_6H_5$$

erythro-*Selective aldol reactions.*[2] Metallic tin converts α-bromo ketones regioselectively into a divalent tin enolate, which reacts with aldehydes and ketones to form aldols in high yield. High *erythro*-selectivity is attained in the reactions of α-substituted α-bromo ketones.

Example:

$$C_6H_5COCH(CH_3)Br + Sn \xrightarrow{DMF,\ CH_2Cl_2} \left[C_6H_5C(OSnBr){=}CHCH_3 \right] \xrightarrow[93\%]{C_6H_5CHO,\ -78^\circ}$$

$$C_6H_5COCH(CH_3)CH(OH)C_6H_5\ (erythro) + C_6H_5COCH(CH_3)CH(OH)C_6H_5\ (threo) \quad 93:7$$

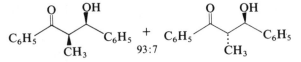

In contrast, *threo*-selectivity has been observed in the reaction of tin(IV) enolates of two ketones with benzaldehyde at $-78°$.[3]

α,β-Dihydroxy ketones. Treatment of an α-diketone or α-keto aldehyde (**1**) with activated tin(0), prepared by reaction of $SnCl_2$ and potassium, results in a tin(II) enediolate (**a**), which reacts with an aldehyde to form a mixture of isomeric α,β-dihydroxy ketones **2** and **3** in a ratio of 2–1:1. The ratio of **2** to **3** increases to 3–4:1 when hexafluorobenzene is added.[4]

$$R^1COCOR^2 \xrightarrow{Sn(0),\ THF,\ 25^\circ} \left[R^1C(O){=}C(O)R^2 \cdot Sn \right] \xrightarrow[70\text{–}95\%]{R^3CHO}$$

1 (R^2 = H, CH_3, C_2H_5) **a**

$$R^3CH(OH)C(R^1)(OH)COR^2\ (\mathbf{2}) + R^3CH(OH)C(R^1)(OH)COR^2\ (\mathbf{3}) \quad 2\text{–}1:1$$

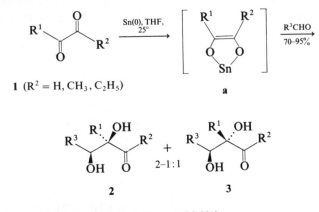

[1] T. Mukaiyama and T. Harada, *Chem. Letters*, 1527 (1981).
[2] T. Harada and T. Mukaiyama, *ibid.*, 467 (1982).

[3] S. Shenoi and J. K. Stille, *Tetrahedron Letters*, **23**, 627 (1982).
[4] T. Mukaiyama, J. Kato, and M. Yamaguchi, *Chem. Letters*, 1291 (1982).

Tin(II) chloride, $SnCl_2$ (**1**). Mol. wt. 189.60, m.p. 247°.

Rearrangement of 2-hydroxy-3-trimethylsilylpropyl selenides (**2**). These selenides (**2**), prepared as shown from α-phenylseleno aldehydes (**1**), rearrange in the presence of $SnCl_2$, usually to allylic selenides.[1]

$$\underset{\mathbf{1}}{(R^1)(R^2)C(SeC_6H_5)CHO} \xrightarrow[70\text{–}90\%]{(CH_3)_3SiCH_2Li,\ \text{ether},\ -78^\circ} \underset{\mathbf{2}}{C_6H_5Se\text{–}C(R^1)(R^2)\text{–}CH(OH)CH_2Si(CH_3)_3}$$

$$\underset{\mathbf{2a}}{C_6H_5SeCH(CH_3)CH(OH)CH_2Si(CH_3)_3} \xrightarrow[63\%]{SnCl_2,\ CH_2Cl_2,\ 20^\circ} CH_3CH{=}CHCH_2SeC_6H_5$$

$$\underset{\mathbf{2b}}{C_6H_5SeC(C_2H_5)_2\text{–}CH(OH)CH_2Si(CH_3)_3} \xrightarrow{SnCl_2} \underset{\mathbf{3}\ (46\%)}{(C_2H_5)_2C(CH_2Si(CH_3)_3)\text{–}CHO} + \underset{(20\%)}{(C_2H_5)_2C{=}CHCH_2SeC_6H_5}$$

$$\mathbf{2b} \xrightarrow[83\%]{AgNO_3} \mathbf{3}$$

α-Hydroxyallenes.[2] Propargyl iodides form an adduct with $SnCl_2$, which reacts with aldehydes to form a mixture of α-hydroxyallenes and β-hydroxyacetylenes in which the former product usually predominates. 1,3-Dimethyl-2-imidazolidone (DMI) is the preferred aprotic solvent.

Examples:

$$CH_3C{\equiv}CCH_2I + C_6H_5CHO \xrightarrow[79\%]{SnCl_2} CH_2{=}C{=}C(CH_3)\text{–}CH(OH)C_6H_5 + CH_3C{\equiv}CCH_2CH(OH)C_6H_5 \quad 97:3$$

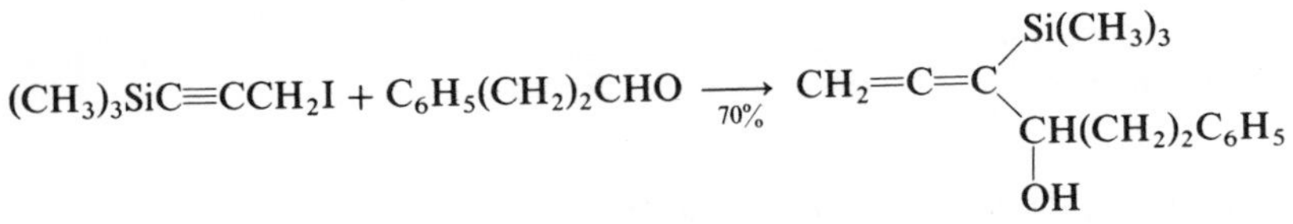

$$(CH_3)_3SiC{\equiv}CCH_2I + C_6H_5(CH_2)_2CHO \xrightarrow[70\%]{} CH_2{=}C{=}C(Si(CH_3)_3)CH(OH)(CH_2)_2C_6H_5$$

[1] H. Nishiyama, T. Kitajima, A. Yamamoto, and K. Itoh, *J.C.S. Chem. Comm.*, 1232 (1982).
[2] T. Mukaiyama and T. Harada, *Chem. Letters*, 621 (1981).

Tin(IV) chloride, 1, 1111–1113; **3**, 269; **5**, 627–631; **6**, 553–554; **7**, 342–345; **9**, 436–438; **10**, 370–373.

Intramolecular α-t-alkylation of ketones. $SnCl_4$ is more effective than other Lewis acids for cyclization of unsaturated ketones involving α-*t*-alkylation of the carbonyl group.[1]

Examples:

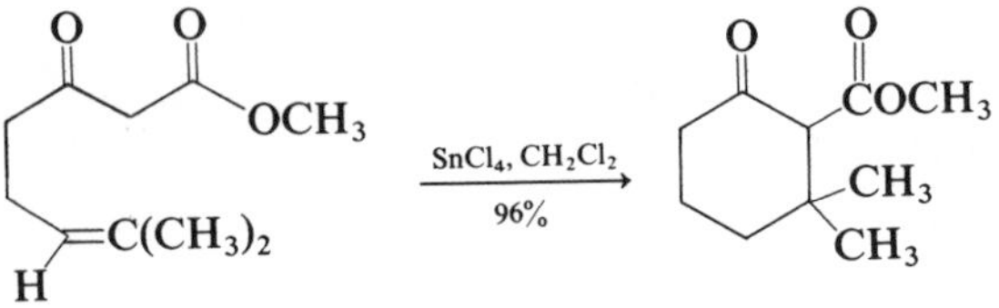

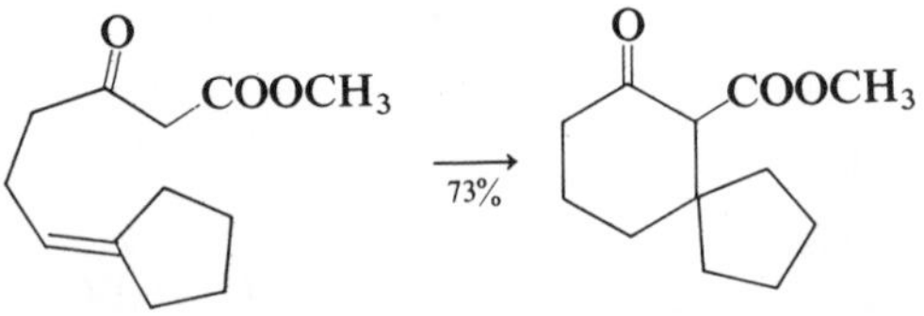

$$(CH_3)_2C{=}CH(CH_2)_2COCH_2COOCH_3 \xrightarrow[96\%]{SnCl_4,\ CH_2Cl_2} \text{2-}(COOCH_3)\text{-3,3-}(CH_3)_2\text{-cyclohexanone}$$

$$\text{(cyclopentylidene)}CH(CH_2)_2COCH_2COOCH_3 \xrightarrow[73\%]{} \text{spiro[4.5]decan-6-one-7-}COOCH_3$$

Cyclization of chiral imines.[2] Asymmetric induction to the extent of 36–65% is observed on cyclization of chiral imines derived from (R)- and (S)-α-phenylethylamine.

Example:

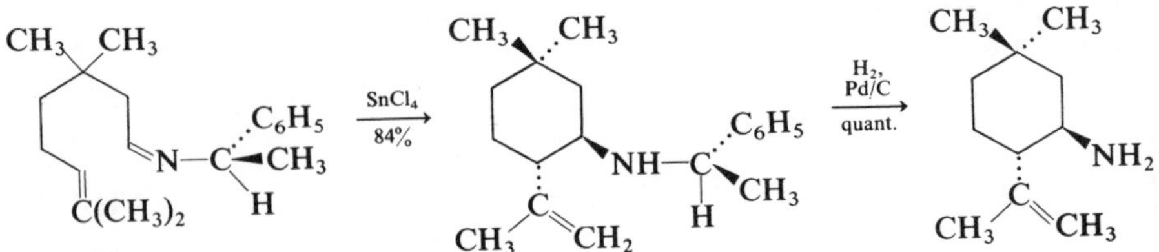

$$(CH_3)_2C{=}CH(CH_2)_2C(CH_3)_2CH{=}N{-}CH(CH_3)C_6H_5\ \text{(R)} \xrightarrow[84\%]{SnCl_4} \text{cyclohexyl}{-}NH{-}CH(CH_3)C_6H_5 \xrightarrow[\text{quant.}]{H_2,\ Pd/C} \text{3,3-dimethyl-6-isopropenylcyclohexylamine}\ (36\%\ ee)$$

Intramolecular ene reactions; α-hydroxy-δ-lactones. The intramolecular ene reaction of **1** catalyzed by $SnCl_4$ results in the α-hydroxy-δ-lactone **2** in 85% yield. This reaction was developed on connection with a projected synthesis of the sesquiterpene toxin anisatin (**3**).[3]

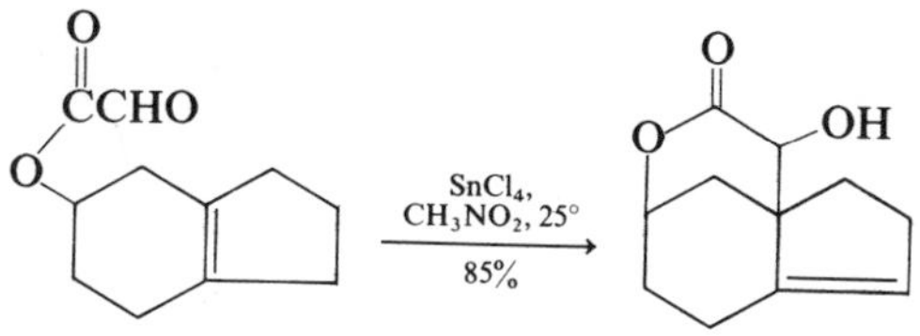
O
CCHO
O
SnCl4,
CH3NO2, 25°
85%
O
O
OH

1 **2**

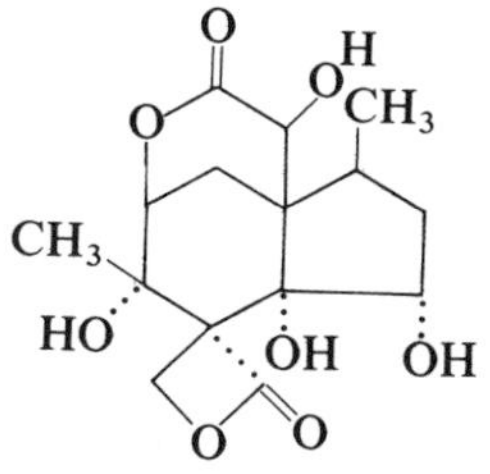
O
H
O
CH3
O
CH3
HO
OH
OH
O
O

3

Glycosylation of N-heterocycles. A few years ago Niedballa and Vorbrüggen[4] reported that pyrimidine nucleosides can be prepared by reaction of 2,4-trimethylsilyl derivatives of uracils or cytosines with 1-alkoxy or 1-acyloxy derivatives of protected sugars in the presence of Friedel-Crafts catalysts, particularly $SnCl_4$ (equation I).

(I)

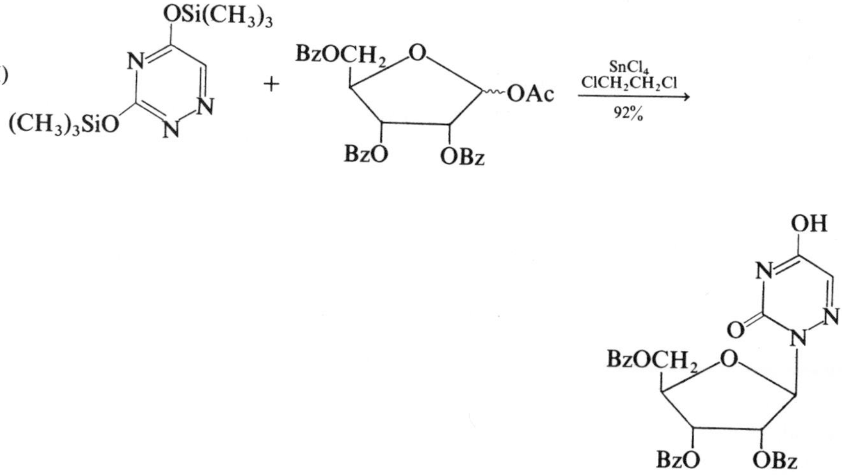
OSi(CH3)3
N
N
N
(CH3)3SiO
+
BzOCH2
O
OAc
BzO
OBz
SnCl4
ClCH2CH2Cl
92%
OH
N
N
N
O
BzOCH2
O
BzO
OBz

The same procedure has been applied to synthesis of pentostatin-like nucleosides (equation II).[5]

(II) [silylated imidazodiazepinone, $NSi(CH_3)_3$, N–$Si(CH_3)_3$] + [$BzOCH_2$ furanose with $\sim$Cl, BzO, OBz] $\xrightarrow[51\%]{SnCl_4,\ CH_3CN}$ [$BzOCH_2$ furanosyl nucleoside with BzO, OBz; imidazodiazepinone with NH]

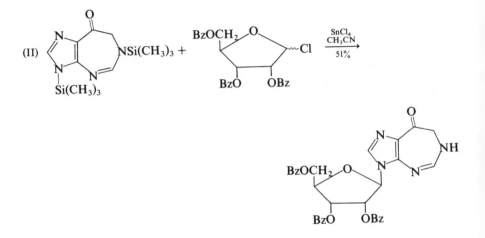

Silylation is essential in each case; the concentration of the catalyst is not critical.

[1] M. T. Reetz, I. Chatziiosifidis, and K. Schwellnus, *Angew. Chem. Int. Ed.*, **20**, 687 (1981).
[2] G. Demailly and G. Solladie, *J. Org.*, **46**, 3102 (1981).
[3] D. L. Lindner, J. B. Doherty, G. Shoham, and R. B. Woodward, *Tetrahedron Letters*, **23**, 5111 (1982).
[4] A. Niedballa and H. Vorbrüggen, *Angew. Chem. Int. Ed.*, **9**, 461 (1970).
[5] H. D. H. Showalter and S. R. Putt, *Tetrahedron Letters*, **22**, 3155 (1981).

Tin(II) fluoride. Mol. wt. 194.68, m.p. 219°, b.p. 850°. Supplier: Alfa.

(E)-α,β-Epoxy phenyl ketones.[1] The ketones can be formed by an aldol-type reaction of α,α-dibromoalkyl phenyl ketones with aldehydes in the presence of SnF_2 (equation I).

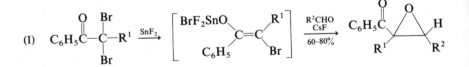

(I) $C_6H_5C(=O)-CBr_2-R^1 \xrightarrow{SnF_2} [BrF_2SnO(C_6H_5)C=C(R^1)Br] \xrightarrow[60-80\%]{R^2CHO,\ CsF}$ $C_6H_5C(=O)$–epoxide (R^1; H, R^2)

Homoallylic alcohols.[2] SnF_2 and allyl iodide form an allyltin difluoroiodide, which reacts with carbonyl compounds in 1,3-dimethyl-2-imidazolidinone to form homoallylic alcohols (equation I).

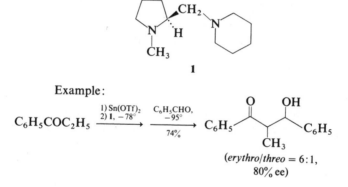

$$\text{(I)}\quad CH_2{=}CHCH_2I + \overset{R^1}{\underset{R^2}{>}}C{=}O \xrightarrow[75\text{–}95\%]{SnF_2} CH_2{=}CHCH_2\underset{R^2}{\overset{R^1}{|}}C{-}OH$$

[1] S. Shoda and T. Mukaiyama, *Chem. Letters*, 723 (1981).
[2] T. Mukaiyama, T. Harada, and S. Shoda, *ibid.*, 1507 (1980).

Tin(II) trifluoromethanesulfonate, $Sn(OSO_2CF_3)_2$. The triflate is obtained by reaction of $SnCl_2$ with triflic acid. It is a white powder that can be stored for some time under nitrogen.[1]

Enanantioselective aldol reactions. Divalent tin enolates of aldehydes and aryl ketones generated with tin(II) triflate undergo aldol condensation with aldehydes to form aldols.[2] The reaction is highly enantioselective if conducted in the presence of chiral diamines derived from (S)-proline, such as **1**.[3]

CH₂, N, H, N, CH₃

1

Example:

$$C_6H_5COC_2H_5 \xrightarrow{\text{1) Sn(OTf)}_2\ \text{2) } \mathbf{1},\ -78^\circ} \xrightarrow[74\%]{C_6H_5CHO,\ -95^\circ} C_6H_5C(=O)CH(CH_3)CH(OH)C_6H_5$$

(*erythro*/*threo* = 6:1, 80% ee)

Divalent tin enolates of ketones also react with another ketone to form cross-aldols in good to excellent yield. Enhanced *threo*-selectivity is observed in the case of aryl ketones.[4]

Example:

$$C_6H_5\overset{O}{\overset{\|}{C}}CH_2OCOC_6H_5 \xrightarrow[85\%]{\text{1) } \mathbf{1}\ \text{2) } C_6H_5COCH_3} C_6H_5\overset{O}{\overset{\|}{C}}\underset{OBz}{\underset{|}{C}H}\overset{OH}{\overset{|}{C}}(C_6H_5){-}CH_3$$

(*erythro*/*threo* = 30:70)

cis-α,β-*Epoxy ketones.* α-Bromoacetone and an aldehyde undergo an aldol-type reaction in the presence of $Sn(OTf)_2$; in THF the *syn*/*anti* ratio is ~80:20. Although the *syn*-product isomerizes very readily to the *anti*-isomer, KF/dicyclohexyl–18-crown-16 effects closure to the epoxy ketone with minimal isomerization.[5]

Example:

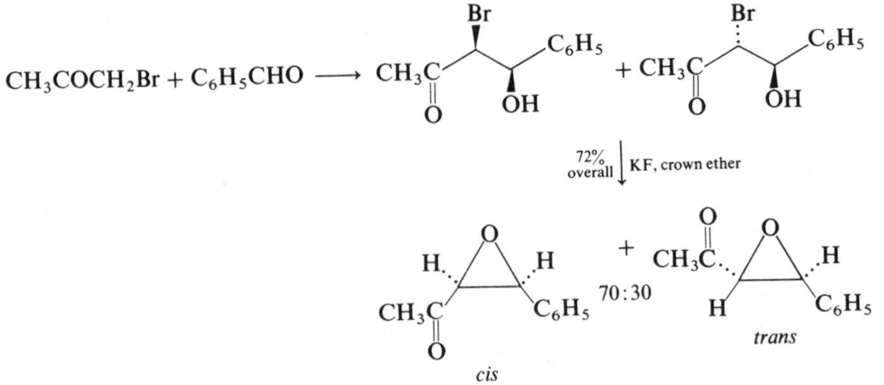

$$CH_3COCH_2Br + C_6H_5CHO \longrightarrow CH_3C(=O)CHBrCH(OH)C_6H_5 + CH_3C(=O)CHBrCH(OH)C_6H_5$$

72% overall ↓ KF, crown ether

cis 70:30 trans

[1] R. J. Batchelor, J. N. R. Ruddick, J. R. Sams, and F. Aubke, *Inorg. Chem.*, **16**, 1414 (1977).
[2] T. Mukaiyama, R. W. Stevens, and N. Iwasawa, *Chem. Letters*, 353 (1982).
[3] N. Iwasawa and T. Mukaiyama, *ibid.*, 1441 (1982).
[4] R. W. Stevens, N. Iwasawa, and T. Mukaiyama, *ibid.*, 1459 (1982).
[5] T. Mukaiyama, T. Haga, and N. Iwasawa, *ibid.*, 1601 (1982).

Titanium(0), 7, 368–369; **8**, 841–842; **9**, 466–467.

Cycloalkenes (**8**, 483). The last step in a synthesis of flexibilene (**2**), a 15-membered-ring diterpene, involved cyclization of the keto-aldehyde **1**, catalyzed with the active titanium metal prepared by reduction of $TiCl_3$ with Zn–Cu in DME. A mixture of two cyclized products is formed, in which **2** is the major constituent.[1]

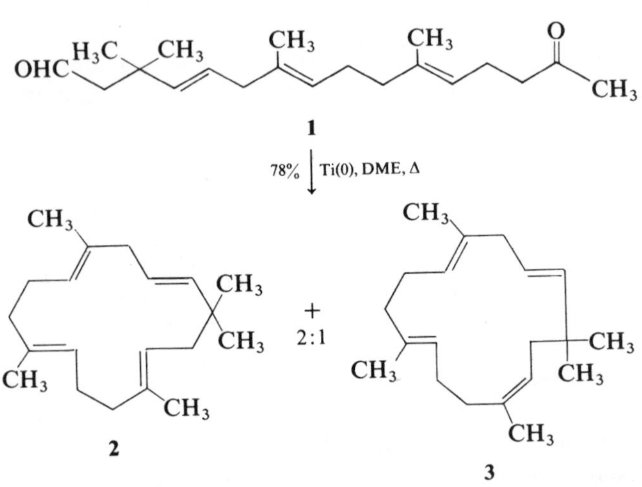

1

78% ↓ Ti(0), DME, Δ

2 + **3** (2:1)

1,3-Dienes. Low-valent titanium ($TiCl_3/LiAlH_4 = 2:1$) reacts with 2-ene-1,4-diols to give 1,3-dienes by a 1,4-reductive elimination.[2]

Examples:

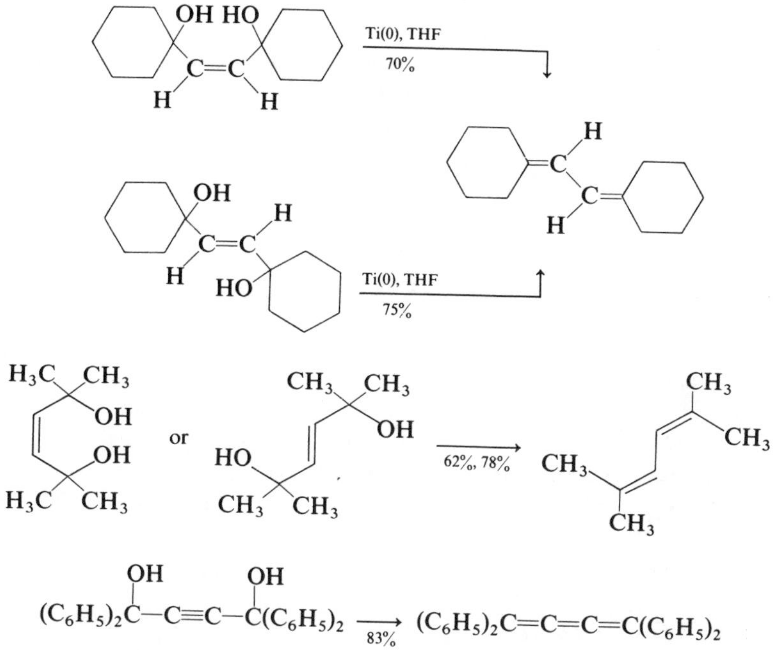

Ti(0), THF 70%

Ti(0), THF 75%

or

62%, 78%

$$(C_6H_5)_2C(OH)-C\equiv C-C(OH)(C_6H_5)_2 \xrightarrow{83\%} (C_6H_5)_2C=C=C=C(C_6H_5)_2$$

Methylenation of ketones.[3] A new method for this reaction is outlined in the example. The titanium metal used for the second step is prepared by treatment of $TiCl_3$ with potassium in refluxing THF.

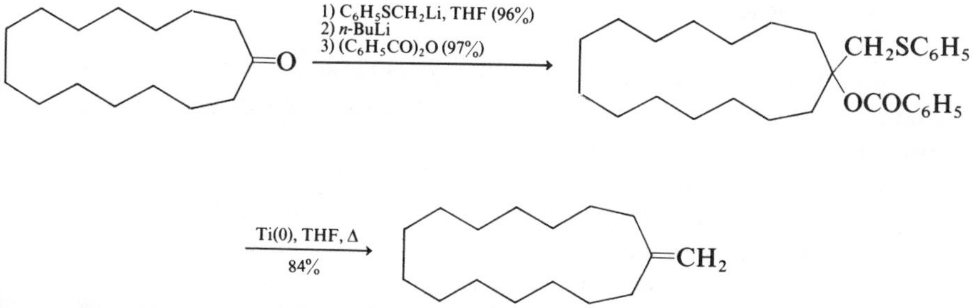

1) $C_6H_5SCH_2Li$, THF (96%)
2) n-BuLi
3) $(C_6H_5CO)_2O$ (97%)

$CH_2SC_6H_5$ $OCOC_6H_5$

Ti(0), THF, Δ 84%

$=CH_2$

[1] J. McMurry, J. R. Matz, K. L. Kees, and P. A. Bock, *Tetrahedron Letters*, **23**, 1777 (1982).
[2] H. M. Walborsky and H. H. Wüst, *Am. Soc.*, **104**, 5807 (1982).
[3] S. C. Welch and J.-P. Loh, *J. Org.*, **46**, 4072 (1981).

Titanium(IV) alkoxides, $Ti(OR)_4$. Supplier: Alfa and Fluka supply the ethoxide and isopropoxide.

Transesterifications. Use of titanates for transesterification has been described in the patent literature for some time and has now been recommended for research purposes by Seebach *et al.*[1] In the presence of catalytic amounts of a tetraalkyl titanate, esters undergo transesterification with the alcohol solvent. Acyl-protected alcohols are deprotected in the process, but a wide variety of functional groups are unaffected, such as C=O, OH, $OSi(CH_3)_2C(CH_3)_3$, acetonide, lactam. One advantage is that completely dry alcohols are not required.

Examples:

$$N{\equiv}CCH_2\overset{O}{\overset{\|}{C}}OCH_2CH_2OCH_3 + (CH_3)_2CHOH \xrightarrow[77\%]{Ti(OC_2H_5)_4,\ 82°} N{\equiv}CCH_2\overset{O}{\overset{\|}{C}}OCH(CH_3)_2$$

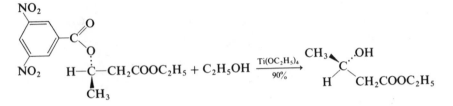

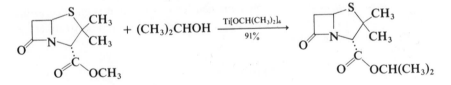

$$3,5\text{-}(NO_2)_2C_6H_3\overset{O}{\overset{\|}{C}}O\text{—}\underset{CH_3}{\overset{H}{C}}\text{—}CH_2COOC_2H_5 + C_2H_5OH \xrightarrow[90\%]{Ti(OC_2H_5)_4} CH_3(H)C(OH)CH_2COOC_2H_5$$

$$CH_3\overset{O}{\overset{\|}{C}}CH_2\overset{O}{\overset{\|}{C}}OC(CH_3)_3 + C_6H_5CH_2OH \xrightarrow[56\%]{} CH_3\overset{O}{\overset{\|}{C}}CH_2\overset{O}{\overset{\|}{C}}OCH_2C_6H_5$$

$$\text{(penam methyl ester, } 2,2\text{-}(CH_3)_2\text{, }COOCH_3) + (CH_3)_2CHOH \xrightarrow[91\%]{Ti[OCH(CH_3)_2]_4} \text{(penam ester, } 2,2\text{-}(CH_3)_2\text{, }COOCH(CH_3)_2)$$

Methyl esters cannot be obtained by this method because $Ti(OCH_3)_4$ has limited solubility in methanol. However, ethyl esters undergo slow transesterification to give methyl esters when refluxed in methyl propionate in the presence of catalytic amounts of $Ti(OC_2H_5)_4$. Presumably a mixed alkoxide $Ti(OC_2H_5)_{4-n}$—$(OCH_3)_n$ is the actual catalyst. Another expedient is to use a catalyst, "GlyTi," obtained from a 2:1 mixture of $Ti(OC_2H_5)_4$ and $HO(CH_2)_2OH$, which catalyzes slow conversion of ethyl esters into methyl esters in refluxing methanol. The two modifications are comparable with respect to rate of interconversion and yields.[2]

Protected dipeptide esters can be transesterified in the presence of $Ti(O\text{-}i\text{-}Pr)_4$ and molecular sieves (4 Å) without racemization in 70–85% yield. Benzyl esters can be easily converted into methyl esters in this way.[3]

[1] D. Seebach, E. Hungerbuhler, R. Naef, and P. Schnurrenberger, *Synthesis*, 138 (1982).
[2] P. Schnurrenberger, M. F. Züger, and D. Seebach, *Helv.*, **65**, 1197 (1982).
[3] H. Rehwinkel and W. Steglich, *Synthesis*, 826 (1982).

Titanium(III) chloride, 2, 415; **4,** 506–508; **5,** 669–671; **7,** 369; **8,** 482–483; **9,** 467; **10,** 400.

Reductive arylation of enones. $TiCl_3$ induces homolytic dediazotization of diazonium salts to produce aryl radicals, which arylate electron-deficient alkenes.[1]

Example:

$$p\text{-}Cl-C_6H_4-N_2^+Cl^- + CH_2{=}CHCOCH_3 \xrightarrow[75\text{–}82\%]{TiCl_3,\ DMF,\ H_2O,\ 0^\circ} p\text{-}Cl-C_6H_4-CH_2CH_2COCH_3$$

[1] A. Citterio and E. Vismara, *Synthesis*, **291**, 751 (1980); A. Citterio, *Org. Syn.*, submitted (1981).

Titanium(III) chloride–Lithium aluminum hydride, 6, 588–589; **7,** 369–370; **9,** 468; **10,** 401.

McMurry reductive coupling (**6**, 589; **9**, 368–369). Recent studies[1] confirm that the active species obtained by reduction of $TiCl_3$ with Li, K, or Mg is actually Ti(0), but suggest that the active species obtained by reduction of $TiCl_3$ with $LiAlH_4$ is Ti(I). The optimum ratio of $TiCl_3$ to $LiAlH_4$ for the coupling is 1:0.5. A slight excess is required because of impurities in $TiCl_3$. If the $TiCl_3$ is completely reduced prior to coupling, the method used is not important. THF is the most satisfactory solvent for the coupling.

[1] R. Dams, M. Malinowski, I. Westdorp, and H. Y. Giese, *J. Org.*, **47**, 248 (1982).

Titanium(IV) chloride, 1, 1169–1171; **2,** 414–415; **3,** 291; **9,** 507–508; **5,** 671–672; **6,** 590–596; **7,** 370–372; **8,** 483–486; **9,** 468–470; **10,** 401–403.

Conjugate allylation with allylsilanes. This Lewis acid allows conjugate addition of allylsilanes to enones with regiospecific transposition of the allyl group. Acyclic and cyclic enones and even fused cyclic enones react satisfactorily.[1]

Examples:

$$C_6H_5CH{=}CHCOCH_3 + (CH_3)_3SiCH_2CH{=}CH_2 \xrightarrow[80\text{–}87\%]{TiCl_4,\ CH_2Cl_2,\ -40^\circ} C_6H_5CH(CH_2CH{=}CH_2)CH_2COCH_3$$

$$\text{2-cyclopentenone} + (CH_3)_3SiCH_2C(CH_3){=}CH_2 \xrightarrow{70\%} \text{3-}(CH_2C(CH_3){=}CH_2)\text{cyclopentanone}$$

$$\text{(octalone)} + (CH_3)_3SiCH_2CH{=}CH_2 \xrightarrow[85\%]{} \text{(angular } CH_2CH{=}CH_2\text{ decalone, H)}$$

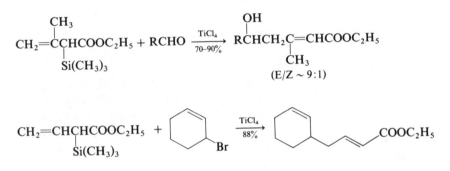

γ-Substituted-α,β-unsaturated esters. Regioselective γ-substitution of α,β-unsaturated esters can be effected indirectly by reaction of α-trimethylsilyl-β,γ-unsaturated esters with electrophiles promoted by $TiCl_4$ or trimethylsilyl triflate. In this case, the allylsilane group controls electrophilic substitution at the γ-position. Suitable electrophiles are carbonyl compounds, acetals, and acid chlorides,[2] and also secondary allylic ethers or halides.[3]

Examples:

$$CH_2{=}C(CH_3)CH(Si(CH_3)_3)COOC_2H_5 + RCHO \xrightarrow[70\text{–}90\%]{TiCl_4} RCH(OH)CH_2C(CH_3){=}CHCOOC_2H_5 \quad (E/Z \sim 9:1)$$

$$CH_2{=}CHCH(Si(CH_3)_3)COOC_2H_5 + \text{3-bromocyclohexene (Br)} \xrightarrow[88\%]{TiCl_4} \text{cyclohexenyl-}CH_2CH{=}CHCOOC_2H_5$$

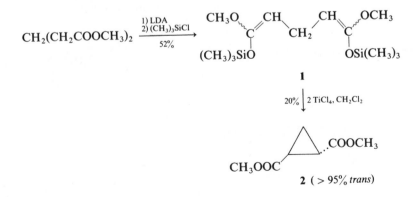

trans-*Cyclopropane-1,2-dicarboxylates.*[4] The isomeric mixture of bisketene silyl acetals **1**, derived from dimethyl glutarate, undergoes oxidative intramolecular cyclization in the presence of $TiCl_4$ to the *trans*-isomer of dimethyl cyclopropanedicarboxylate (**2**). The same selectivity is observed with unsymmetrically substituted derivatives of **1**, apparently because of thermodynamic control.

$$CH_2(CH_2COOCH_3)_2 \xrightarrow[52\%]{1)\ LDA\quad 2)\ (CH_3)_3SiCl} (CH_3O)((CH_3)_3SiO)C{=}CH\text{–}CH_2\text{–}CH{=}C(OCH_3)(OSi(CH_3)_3)$$

1

$$\mathbf{1} \xrightarrow[20\%]{2\ TiCl_4,\ CH_2Cl_2} \text{cyclopropane-1,2-}(COOCH_3)_2$$

2 (> 95% *trans*)

Butenolides. 4-Substituted butenolides can be prepared by reaction of 2-acetoxyfurane (**1**) with aldehydes and related electrophiles in the presence of $TiCl_4$.[5]

Example:

$$\mathbf{1}\ (\text{2-acetoxyfurane}) + CH_3(CH_2)_6CHO \xrightarrow[60\%]{TiCl_4,\ CH_2Cl_2,\ -78^\circ} CH_3(CH_2)_6CH(OH)\text{-butenolide}$$

$$\mathbf{1} \xrightarrow[89\%]{CH(OCH_3)_3,\ TiCl_4} (CH_3O)_2CH\text{-butenolide}$$

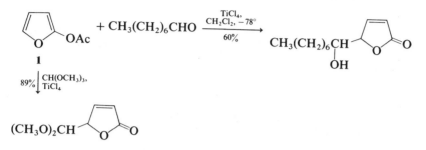

Biphenyls. A novel route to unsymmetrical biphenyls involves addition of allyltrimethylsilyllithium (**1, 8**, 273–274) to a keto acetal such as **2** to produce a vinylsilane (**3**). Treatment of **3** with $TiCl_4$ in CH_2Cl_2–ether at $-78 \rightarrow 0^\circ$ results in annelation to **4**.[6]

$$\mathbf{2}\ (p\text{-}CH_3C_6H_4COCH_2CH(OCH_3)_2) \xrightarrow{LiCH_2CH{=}CHSi(CH_3)_3\ (\mathbf{1})} \mathbf{3}\ (p\text{-}CH_3C_6H_4C(OH)(CH_2CH{=}CHSi(CH_3)_3)CH_2CH(OCH_3)_2)$$

$$\xrightarrow{TiCl_4} \mathbf{4}\ (59\%)\ (p\text{-}CH_3C_6H_4C_6H_5) + (CH_3)_3SiOH + 2\,CH_3OH$$

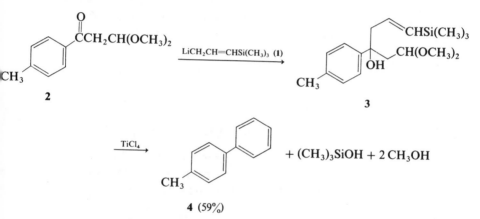

Diels-Alder catalyst. Unlike other Lewis acids, $TiCl_4$ favors formation of the *exo*-isomer in the Diels-Alder reaction of cyclopentadiene and mesityl oxide (equation I).[7]

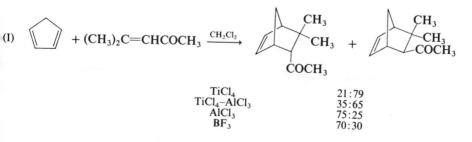

(I) $$\text{cyclopentadiene} + (CH_3)_2C{=}CHCOCH_3 \xrightarrow{CH_2Cl_2} \text{endo-}COCH_3 \text{ adduct} + \text{exo-}COCH_3 \text{ adduct}$$

$TiCl_4$	21:79
$TiCl_4$–$AlCl_3$	35:65
$AlCl_3$	75:25
BF_3	70:30

Selective cleavage of 2-methoxyethyl hemiacetals. 2-Methoxyethyl hemiacetals react with allyltrimethylsilane in the presence of $TiCl_4$ (1 equiv.) to give a homoallyl ether with selective cleavage of the 2-methoxyethoxy group (*cf.*, **7**, 370–371).[8]

Examples:

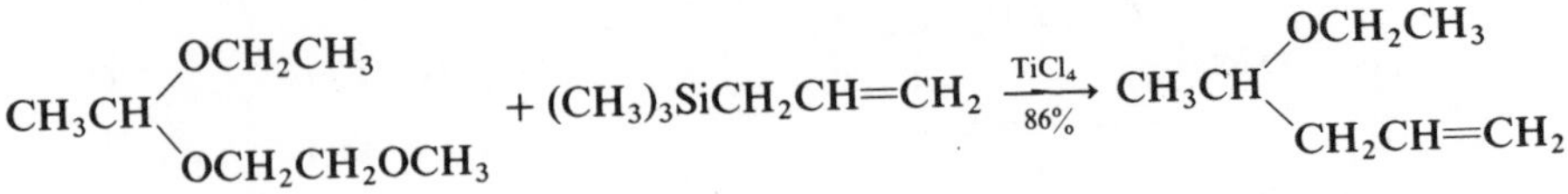

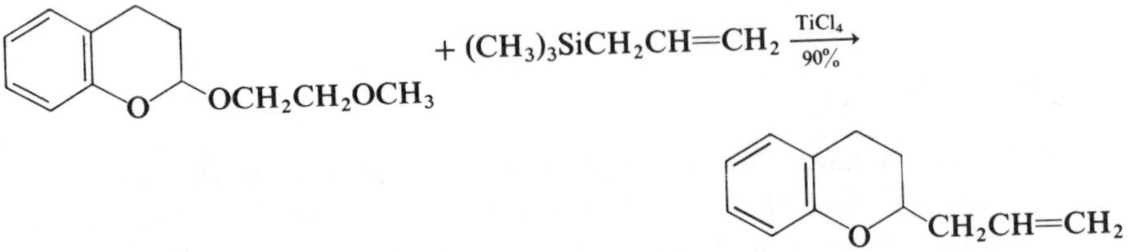

$$CH_3CH(OCH_2CH_3)(OCH_2CH_2OCH_3) + (CH_3)_3SiCH_2CH{=}CH_2 \xrightarrow[86\%]{TiCl_4} CH_3CH(OCH_2CH_3)(CH_2CH{=}CH_2)$$

$$\text{2-(OCH}_2\text{CH}_2\text{OCH}_3\text{)chroman} + (CH_3)_3SiCH_2CH{=}CH_2 \xrightarrow[90\%]{TiCl_4} \text{2-(CH}_2\text{CH=CH}_2\text{)chroman}$$

This cleavage reaction can be used to effect intramolecular homologated cyclization of hydroxy allylsilanes by conversion to the 2-methoxyethoxymethyl (MEM) ether (**7**, 228–229) followed by cleavage with $TiCl_4$ to the species $\overset{+}{O}{=}CH_2$.

Example:

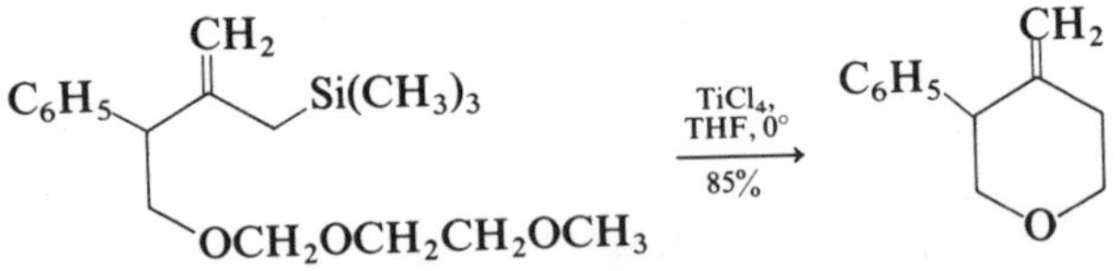

$$C_6H_5CH(CH_2OCH_2OCH_2CH_2OCH_3)CH_2C(=CH_2)CH_2Si(CH_3)_3 \xrightarrow[85\%]{TiCl_4,\ THF,\ 0^\circ} \text{3-}C_6H_5\text{-4-methylenetetrahydropyran}$$

Debenzylation–Dehydration.[9] The alcohol **1** is converted on treatment with $TiCl_4$ in good yield into the alkaloid norruspoline (**2**) by dehydration and debenzylation. Reaction of **1** with $SiCl_4$ effects only dehydration in about 30% yield.

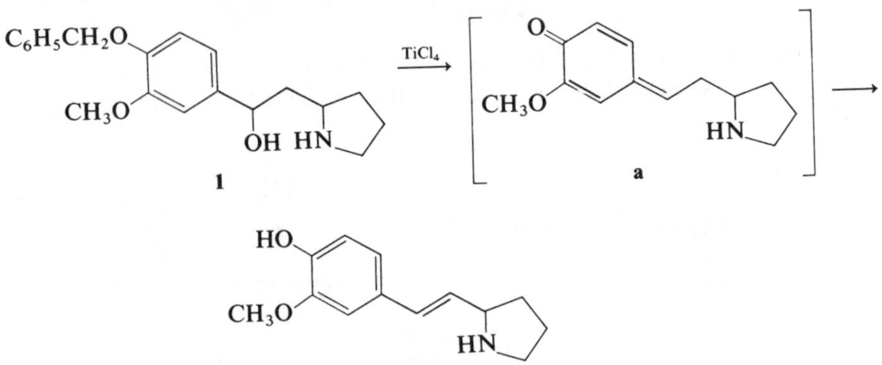

$$\mathbf{1}\ (C_6H_5CH_2O,\ CH_3O\text{-aryl}\ CH(OH)CH_2\text{-2-pyrrolidinyl}) \xrightarrow{TiCl_4} [\mathbf{a}\ (\text{quinone methide},\ CH_3O,\ HN)] \longrightarrow \mathbf{2}\ (HO,\ CH_3O\text{-aryl}\ CH{=}CH\text{-2-pyrrolidinyl})$$

Claisen rearrangement (**7**, 372). Treatment of allyl naphthyl ethers with $TiCl_4$ in CH_2Cl_2 at 0^0 results in naphthodihydrofuranes.[10]

Examples:

$OCH_2CH{=}CH_2$ $\xrightarrow[84\%]{TiCl_4,\ CH_2Cl_2}$ CH_3 O

$OCH_2CH{=}CHCH_3$ $\xrightarrow[56\%]{}$ CH_3 O CH_3

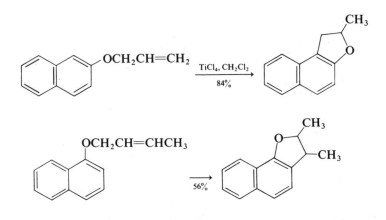

Nortricyclyl chloride (**3**).[11] The simplest synthesis of **3** is the two-step sequence formulated in equation (I).

1 $\xrightarrow[70\%]{HCl}$ **2** Cl + **3** Cl (80:20)

2 $\xrightarrow[75\%]{TiCl_4}$ **3**

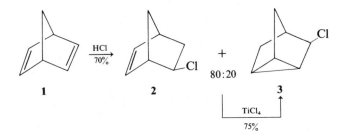

[1] H. Sakurai, A. Hosomi, and J. Hayashi, *Org. Syn.*, submitted (1981); A. Hosomi and H. Sakurai, *Am. Soc.*, **99**, 1673 (1977).
[2] P. Albaugh-Robertson and J. A. Katzenellenbogen, *Tetrahedron Letters*, **23**, 723 (1982).
[3] Y. Morizawa, S. Kanemoto, K. Oshima, and H. Nozaki, *ibid.*, **23**, 2953 (1982).
[4] T. H. Chan and I. H. M. Wallace, *Tetrahedron Letters*, **23**, 799 (1982).
[5] T. Shono, Y. Matsumura, and S. Yamane, *ibid.*, 3269 (1981).
[6] M. A. Tius, *ibid.*, **22**, 3335 (1981).
[7] J. Bachner, U. Huber, and G. Buchbauer, *Monats.*, **112**, 679 (1981).
[8] H. Nishiyama and K. Itoh, *J. Org.*, **47**, 2496 (1982).
[9] R. B. Herbert and P. C. Wormald, *J. Chem. Research* (*M*), 3001 (1982).
[10] M. R. Saidi, *Heterocycles*, **19**, 1473 (1982).
[11] M. T. Reetz and M. Sauerwald, *Ber.*, **114**, 2355 (1981).

Titanium(IV) chloride–Lithium aluminum hydride, 6, 596; **7,** 372–373; **8,** 486–487.

Deoxygenation of the adducts of furanes and alkynes. This low-valent form of Ti deoxygenates the Diels-Alder adducts of furanes and alkynes to form aromatic systems.[1]

Examples:

CH_3 COOCH$_3$ C≡C COOCH$_3$ $\xrightarrow[71\%]{100°}$ CH$_3$ COOCH$_3$ O COOCH$_3$ $\xrightarrow[77\%]{TiCl_4–LiAlH_4,\ N(C_2H_5)_3,\ 65°}$ CH$_3$ COOCH$_3$ COOCH$_3$

$\xrightarrow{57\%}$

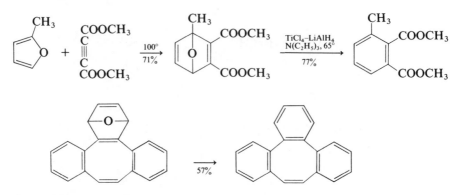

[1] Y. De Xing and N. Z. Huang, *J. Org.*, **47**, 140 (1982).

Titanium (IV) chloride–Magnesium amalgam.

Mixed coupling of fluorenone and α, β-enones. The major products of this reaction arise by conjugate addition of the fluorenone dianion to the enone.[1]

Examples:

1 =O + $(CH_3)_2C{=}CHCOCH_3$ $\xrightarrow[70\%]{TiCl_4,\ Mg/Hg}$ O OH CH$_3$ CH$_3$ CH$_3$

1 + CH$_3$ O C CH$_3$ CH$_2$ $\xrightarrow{70\%}$ HO CH$_3$ O C CH$_3$ CH$_2$

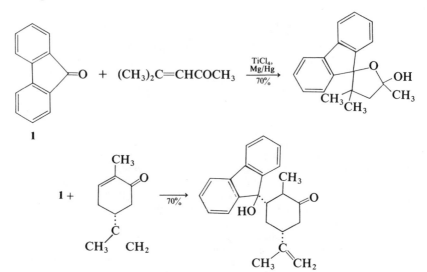

[1] J.-M- Pons and M. Santelli, *Tetrahedron Letters*, **23**, 4937 (1982).

p-Toluenesulfinylimidazolide, $CH_3-C_6H_4-SO-N(C_3H_3N)$ (**1**). The reagent is prepared *in situ* by reaction of *p*-toluenesulfinic acid and N, N′-carbonyldiimidazole in $ClCH_2CH_2Cl$.

Sugar tosylates.[1] Preparation of sugar tosylates of protected sugars by the usual method (TsCl, Py) is often slow and difficult. An alternative is reaction with **1**, which proceeds at 25° in 5–150 minutes to give *p*-toluenesulfinic esters in yields generally > 80%. These products are oxidized in high yield to tosylates by *m*-chloroperbenzoic acid. The method is not useful in the case of anomeric tosylates, which decompose as formed.

Example:

$$\text{HO-sugar}(CH_3, OBzl, OCH_3, OBzl) \xrightarrow[91\%]{\mathbf{1}} p\text{-}CH_3C_6H_4S(=O)\text{-O-sugar} \xrightarrow[90\%]{ClC_6H_4CO_3H} \text{TsO-sugar}$$

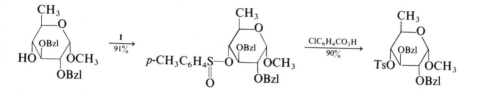

[1] H. Redlich and W.-U. Meyer, *Ann.*, 1354 (1981).

p-Toluenesulfonic acid, 1, 1172–1178; **4,** 508–510; **5,** 673–675; **7,** 374–375; **8,** 488–489; **9,** 971–972.

Acetal-acetal interchange.[1] Formaldehyde and acetaldehyde acetals can be prepared from methoxymethyl ethers and ethoxyethyl ethers, respectively, by reaction with TsOH in refluxing benzene.

Example:

$$CH_3(CH_2)_5OMOM \xrightarrow[78\%]{TsOH,\ C_6H_6,\ \Delta} [CH_3(CH_2)_5O]_2CH_2$$

[1] B. C. Barot and H. W. Pinnick, *J. Org.*, **46**, 2981 (1981).

p-Toluenesulfonyl azide, 1, 1178–1179; **2,** 415–417; **3,** 291–292; **4,** 510; **5,** 675; **6,** 597; **9,** 472.

Diazo transfer.[1] Tosyl azide is the most commonly used reagent for diazo transfer, even though it is now recognized as a potentially dangerous explosive. Merck chemists have examined a number of other organic azides; of these the safest is *p*-dodecylbenzenesulfonyl azide, prepared from a commercial sulfonic acid mixture.

Another relatively safe azide is naphthalene-2-sulfonyl azide. Either reagent, however, should be handled cautiously.

[1] G. H. Hazen, L. M. Weinstock, R. Connell, and F. W. Bollinger. *Syn. Comm.*, **11**, 947 (1981).

***p*-Toluenesulfonyl chloride, 1**, 1178–1179; **2**, 915–917; **3**, 291–292; **4**, 510; **5**, 675; **6**, 597; **9**, 472.

3-Methylene-2-azetidinones. A versatile route to these β-lactams starts with the 2,4,6-triisopropylbenzenesulfonylhydrazone (**1**) of an α-keto amide, which can be converted as shown into acrylamides (**2**).[1] Treatment of **2** with *n*-butyllithium (2 equiv.) and then TsCl results in the *p*-toluenesulfonates **3**, which can be isolated. They slowly cyclize to **4** on standing at 25°. The cyclization is more rapid in the presence of NaH.[2]

$$\underset{\mathbf{1}}{CH_3C(=N\text{-}NHSO_2C_6H_2(i\text{-}Pr)_3)CONHR^1} \xrightarrow{n\text{-BuLi}} \underset{\mathbf{a}}{\left[CH_2{=}C{=}C(NR^1Li)(OLi)\right]} \xrightarrow[60\text{–}80\%]{R^2CHO} \underset{\mathbf{2}}{H_2C{=}C(CONHR^1)CH(OH)R^2}$$

$$\xrightarrow[2)\ TsCl,\ THF]{1)\ n\text{-BuLi}} \underset{\mathbf{3}}{H_2C{=}C(CONHR^1)CH(OTs)R^2} \xrightarrow[\text{50–70\% from } \mathbf{2}]{NaH,\ 25^\circ} \underset{\mathbf{4}}{\text{3-methylene-4-}R^2\text{-1-}R^1\text{-2-azetidinone}}$$

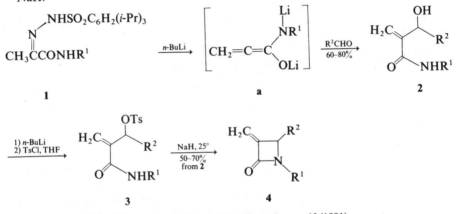

[1] R. M. Adlington and A. G. M. Barrett, *J. C. S. Chem. Comm.*, 65 (1981).
[2] R. M. Adlington, A. G. M. Barrett, P. Quayle, A. Walker, and M. J. Betts, *ibid.*, 404 (1981).

***p*-Toluenesulfonyl cyanide**, $p\text{-}CH_3C_6H_4SO_2CN$ (**1**). Mol. wt. 181.22, m.p. 49–50°. The reagent is prepared in high yield by reaction of sodium *p*-toluenesulfinate with cyanogen chloride.[1]

β-Keto nitriles.[2] β-Keto nitriles can be obtained in fair to good yield by reaction of TsCN with kinetic enolates (LDA) of ketones. Yields are poor, however, in the cyanation of methyl ketones.

Example:

$$\text{2-}(CH_2C_6H_5)\text{cyclohexanone} \xrightarrow[67\%]{1)\ LDA,\ THF,\ -78^\circ;\ 2)\ \mathbf{1},\ -78^\circ} \text{2-}(CH_2C_6H_5)\text{-6-CN-cyclohexanone}$$

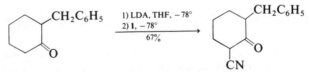

[1] J. M. Cox and R. Ghosh, *Tetrahedron Letters*, 3351 (1969).
[2] D. Kahne and D. B. Collum, *ibid.*, **22**, 5011 (1981).

***p*-Toluenesulfonylhydrazine, 1**, 1185–1187; **2**, 419–423; **3**, 293; **4**, 571–572; **5**, 678–681; **6**, 598–600; **7**, 375–376; **8**, 489–493; **9**, 472–473.

α-Phenylthio and α-phenylseleno aldehydes and ketones.[1] The tosylhydrazones of α-halo aldehydes or ketones react readily with thiophenol or benzeneselenol and an excess of triethylamine at −78° to give α-phenylthio or -seleno substituted tosylhydrazones. This conversion probably proceeds via a tosylazoalkene.

Example:

$$CH_3C(CH_3)(Br){-}CH{=}NNHTs \xrightarrow[75\%]{C_6H_5SH,\ N(C_2H_5)_3} CH_3C(CH_3)(SC_6H_5){-}CH{=}NNHTs$$

$$\xrightarrow[93\%]{BF_3\cdot(C_2H_5)_2O,\ (CH_3)_2C{=}O,\ H_2O} (CH_3)_2C(SC_6H_5){-}CHO$$

1,2-Carbonyl shift via allylic sulfides.[2] A method for transposition of the carbonyl group of α, β-enones to either the α- or α′- position is illustrated in equation (I) for the tosylhydrazone of 3-methyl-2-cyclohexenone.

(I)

NNHTs / CH_3 $\xrightarrow{1)\ 2\ n\text{-}BuLi;\ 2)\ C_6H_5SSC_6H_5}$ C_6H_5S, NNHTs, CH_3 $\xrightarrow{NaBH_4,\ HOAc}$ C_6H_5S, CH_3 (**1**)

$\xrightarrow{ClC_6H_4CO_3H}$ $C_6H_5S(=O)$, CH_3 (**2**) $\xrightarrow{1)\ LiN(C_2H_5)_2;\ 2)\ C_6H_5SSC_6H_5}$ C_6H_5S, OH, CH_3

2 $\xrightarrow[\text{71\% from 1}]{C_6H_5SH,\ KOC(CH_3)_3}$ α′, OH, CH_3 (**3**)

$\xrightarrow[\text{60\% from 1}]{HgCl_2,\ H_2O,\ CH_3CN}$ O, α, CH_3 (**4**)

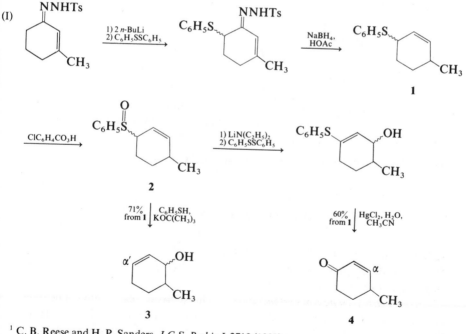

[1] C. B. Reese and H. P. Sanders, *J.C.S. Perkin I*, 2719 (1982).
[2] T. Mimura and T. Nakai, *Chem. Letters*, 1579 (1981).

(R)-(+)-(p-Tolylsulfinyl)acetic acid, 10, 405–407.

***t-Butyl 2-(p-tolylsulfinyl)propionate* (2).**[1] Methylation of (R)-*t*-butyl 2-(*p*-tolylsulfinyl)acetate **(1)** can be effected only with the lithium bases and only with CH_3I, but with poor diastereoselectivity. Aldol-type condensation of the anion of the

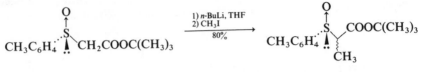

$CH_3C_6H_4S(O)CH_2COOC(CH_3)_3$ (R)-**1** —[1) *n*-BuLi, THF; 2) CH_3I; 80%]→ $CH_3C_6H_4S(O)CH(CH_3)COOC(CH_3)_3$ **2** (α_D + 148°, 2 isomers)

—[1) $(CH_3)_3CMgBr$; 2) *n*-$C_7H_{15}CHO$; 90%]→ **3** —[Al/Hg; 95%]→ $C_7H_{15}CH(OH)CH(CH_3)COOC(CH_3)_3$

—[1) $LiAlH_4$; 2) TsCl, Py; 3) $LiAlH_4$; 95%]→ $C_7H_{15}CH(OH)CH(CH_3)CH_3$

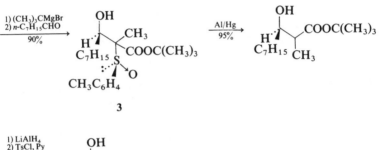

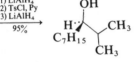

(R)-(+)-**5** (80% ee)

diastereoisomeric **2** is limited to aldehydes, because of steric hindrance. The asymmetric induction in the case of an aliphatic aldehyde was determined by conversion to (R)-(+)-**5** and found to be 80%.

[1] G. Solladié, F. Matloubi-Moghadam, C. Luttmann, and C. Miskowski, *Helv.* **65**, 1602 (1982).

(R)-(+)-*p*-Tolyl vinyl sulfoxide, *p*-$CH_3C_6H_4$—S(=O)—CH=CH_2 **(1)**.

Mol. wt. 164. 23, b.p. 110°/1.0 mm, α_D + 421°.

The sulfoxide is prepared by reaction of (—)-menthyl *p*-toluenesulfinate with vinylmagnesium bromide (88% yield).[1]

Asymmetric* [1,3] *dipolar cycloaddition.[2] This sulfoxide reacts with C-phenyl-N-methylnitrone **(2)** to form optically active 4-(*p*-tolylsulfinyl)-isoxazolidines (**3a** and **3b**), which differ only in chirality of the S=O group. The optical purity was shown to be no less than 80% by conversion to the amino alcohol **4**. The absolute configuration

(R) of **4** was established by synthesis of (S)-**4** from (S)-β-phenyl-β-alanine. Similar results were observed on addition of **1** to C, N-diphenylnitrone.

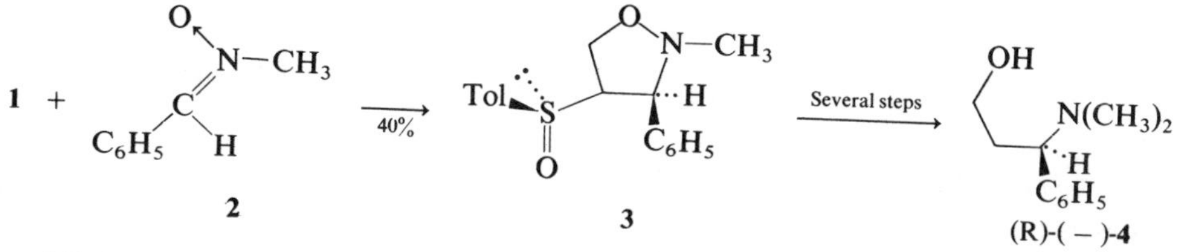

[1] D. J. Abbott, S. Colonna, and C. J. M. Stirling, *J.C.S. Perkin I*, 492 (1976).
[2] T. Koizumi, H. Hirai, and E. Yoshii, *J. Org.*, **47**, 4004 (1982).

Tosylmethylisocyanide (TosMIC), **4**, 514–516; **5**, 684–685; **6**, 600; **7**, 377; **8**, 493–494; **10**, 409.

Cycloalkylation of a* 1, ω*-dibromide.[1] A small number of sesquiterpene lactones contain the *cis, cis*-1,6-cyclodecadiene ring system. An interesting approach to the ketone **(3)** containing this ring system is reaction of the rather unstable dibromide **1** with TosMIC and KH in THF under high dilution to form **2**, which is hydrolyzed by acid to the ketone **(3)**.

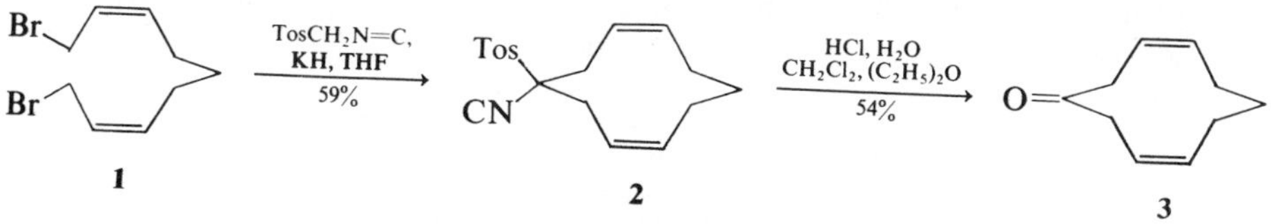

Use of methyl methylthiomethyl sulfoxide (**6**, 391–392) in place of TosMIC results in a difficultly separable mixture that fails to produce the dienone **3** on hydrolysis.

[1] M. S. Frazza and B. W. Roberts, *Tetrahedron Letters*, **22**, 4193 (1981).

Trialkylaluminum.

Beckmann rearrangement-alkylation sequence.[1] Oxime sulfonates react with AlR_3 to form rearranged imines, which are reduced to amines by DIBAH (equation I). The reaction involves selective migration of the R group *anti* to the oxime sulfonate.

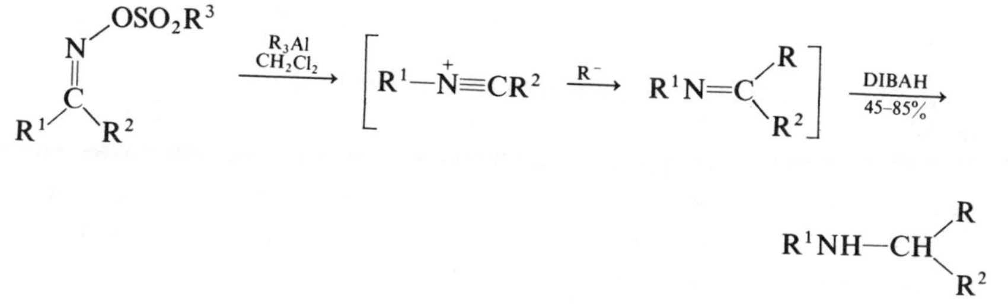

Example:

CH_3 (2-methylcyclohexanone NOTs) $\xrightarrow[48\%]{1)\ (CH_3CH_2CH_2)_3Al;\ 2)\ DIBAH}$ CH_3–(azepane, N–H)–$CH_2CH_2CH_3$

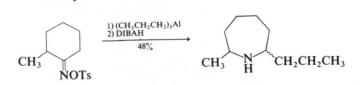

This sequence was used for an efficient synthesis of *dl*-pumiliotoxin C (**2**), a frog toxin, from **1**.

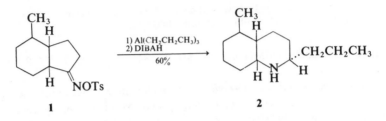

1 $\xrightarrow[60\%]{1)\ Al(CH_2CH_2CH_3)_3;\ 2)\ DIBAH}$ **2** (CH_3, H, $CH_2CH_2CH_3$, H, H H)

Selective cleavage of epoxides. The α- and β-benzyloxy epoxides **1** and **3** (either *cis* or *trans*) undergo almost exclusive β-addition with trimethylaluminum catalyzed by *n*-butyllithium or lithium methoxide to give the alcohols **2** and **4**, respectively. The regioisomeric alcohols are formed only in traces. CH_3MgBr, CH_3Li, or $(CH_3)_2Mg$ are much less selective.[2]

1 (CH_3–epoxide–CH_2OBzl) $\xrightarrow[76\%]{(CH_3)_3Al,\ n\text{-}BuLi,\ toluene}$ **2** ($(CH_3)_2CH–CH(OH)–CH_2OBzl$)

3 (CH_3–epoxide–CH_2CH_2OBzl) $\xrightarrow[85\%]{(CH_3)_3Al,\ CH_3OLi}$ **4** ($(CH_3)_2CH–CH(OH)–CH_2CH_2OBzl$)

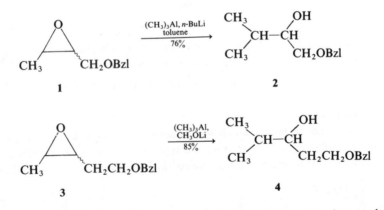

Another regioselective addition to an epoxide was used as one step in a synthesis of the *t*-butyldiphenylsilyl ether (**7**) of verrucarinic acid from **5**.[3] The diol was converted into the optically active epoxy alcohol by the Sharpless method (**10**, 64–65) and then oxidized to the epoxy acid **6** by the new ruthenium-catalyzed oxidation of Sharpless *et al.* (this volume). This epoxy acid undergoes almost exclusive β-addition with trimethylaluminum to give the desired product **7**.

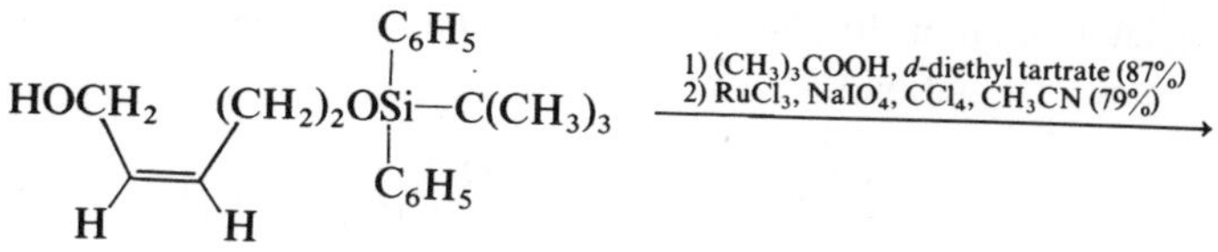

$$HOCH_2CH{=}CH(CH_2)_2OSi(C_6H_5)_2C(CH_3)_3 \;(\mathbf{5}) \xrightarrow[\text{2) } RuCl_3,\, NaIO_4,\, CCl_4,\, CH_3CN\ (79\%)]{\text{1) } (CH_3)_3COOH,\ d\text{-diethyl tartrate } (87\%)}$$

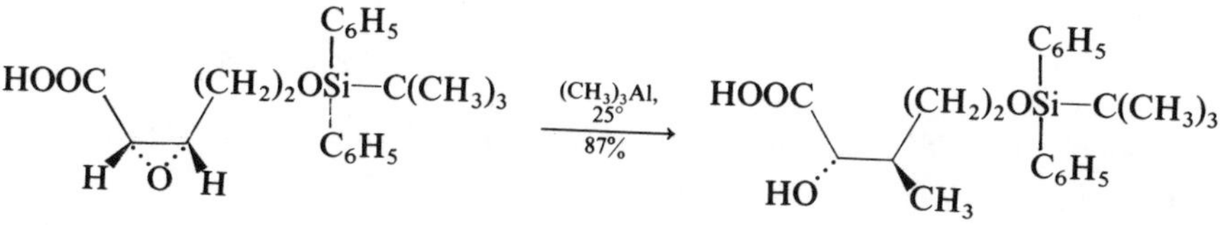

$$\mathbf{6} \xrightarrow[87\%]{(CH_3)_3Al,\ 25^\circ} \mathbf{7}$$

The successful synthesis of optically active **7** then led to the first synthesis of verrucarin A (**8**), a macrotrilactone with significant cytostatic activity. The synthesis involved esterification of the primary alcohol of verrucarol (the tricyclic fragment) with the acetate of **7** (DCC, 4-pyrrolinopyridine) and then with a protected derivative of (E, Z)-muconic acid. After deprotection (Bu_4NF), lactonization was effected by the Mitsunobu procedure (7, 405–406).

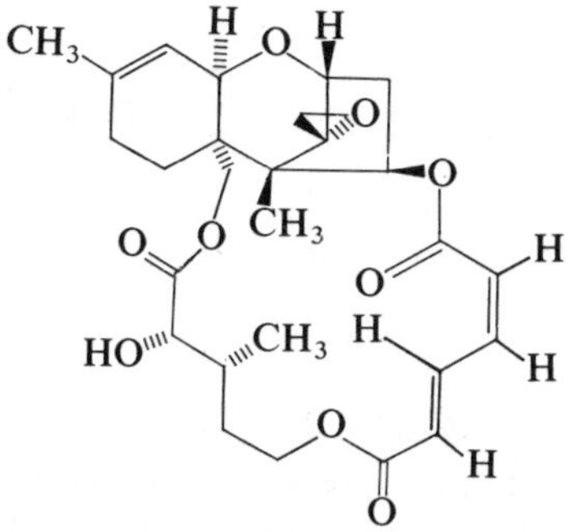

8

Aliphatic Claisen rearrangement.[4] Allyl vinyl ethers undergo [3, 3] sigmatropic rearrangement in the presence of $(CH_3)_3Al$ or $(C_2H_5)_3Al$ with substitution of CH_3 or C_2H_5 on the aldehydic carbon atom (equation I).

(I) 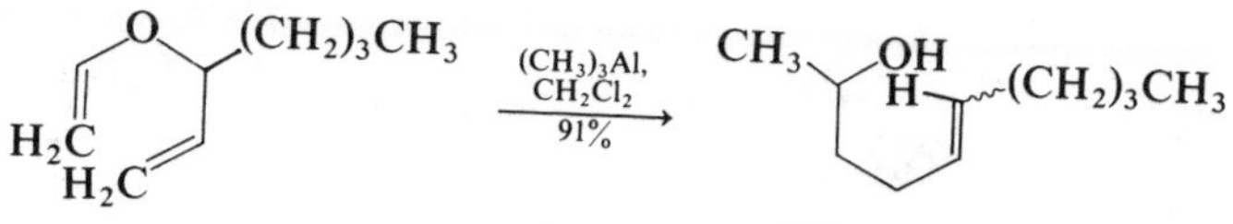 $$\xrightarrow[91\%]{(CH_3)_3Al,\ CH_2Cl_2}$$

E/Z = 47:53

Use of triisobutylaluminum or diisobutylaluminum chloride results in a primary alcohol (substitution by H, equation II).

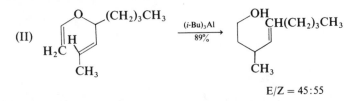

(II) [scheme: allyl vinyl ether with $(CH_2)_3CH_3$, H_2C, H, CH_3 $\xrightarrow[89\%]{(i\text{-}Bu)_3Al}$ OH, $CH(CH_2)_3CH_3$, CH_3]

E/Z = 45:55

Use of $(C_2H_5)_2AlSC_6H_5$ results in rearrangement to the aldehyde rather than the primary alcohol observed in equation (II).

[1] K. Hattori, Y. Matsumura, T. Miyazaki, K. Maruoka, and H. Yamamoto, *Am. Soc.*, **103**, 7368 (1981).

[2] A. Pfaltz and A. Mattenberger, *Angew. Chem. Int. Ed.*, **21**, 71 (1982).

[3] W. C. Still and H. Ohmizu, *J. Org.*, **46**, 5242 (1981).

[4] K. Takai, I. Mori, K. Oshima, and H. Nozaki, *Tetrahedron Letters*, **23**, 3985 (1981).

Triaryl phosphates, $O{=}P(OAr)_3$.

Aryl ethers.[1] Activated halogen compounds react with triaryl phosphates in the presence of KOH and DMF to form aryl ethers in 50–98% yield.

Example:

$$O_2N{-}C_6H_4{-}Cl + O{=}P(OC_6H_4CH_3\text{-}p)_3 \xrightarrow[98\%]{KOH,\ DMF} O_2N{-}C_6H_4{-}O{-}C_6H_4{-}CH_3$$

[1] A. Ohta, Y. Iwasaki, Y. Akita, *Synthesis*, 828 (1982).

Tri-*n*-butylcrotyltin, 10, 411.

Stereoselective reactions with aldehydes. The related (E)-2-pentenyltin reagent **2** also reacts with aldehydes to form predominantly *erythro* adducts such as **3**. This reaction was used for a stereoselective synthesis of an aggregation pheromone (**4**) of the European elm bark beetle.[1]

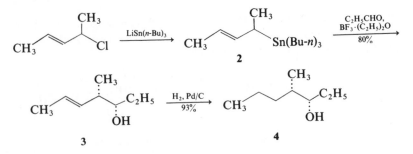

[scheme: $CH_3CH{=}CHCH(CH_3)Cl \xrightarrow{LiSn(n\text{-}Bu)_3}$ $CH_3CH{=}CHCH(CH_3)Sn(Bu\text{-}n)_3$ (**2**) $\xrightarrow[80\%]{C_2H_5CHO,\ BF_3\cdot(C_2H_5)_2O}$ $CH_3CH{=}CHCH(CH_3)CH(OH)C_2H_5$ (**3**) $\xrightarrow[93\%]{H_2,\ Pd/C}$ $CH_3(CH_2)_3CH(CH_3)CH(OH)C_2H_5$ (**4**)]

In contrast, (E)-cinnamyltin reagents **5** and **6** react with aldehydes under Lewis acid catalysis to form *threo* adducts with high stereoselectivity. These reactions presumably involve the conventional chairlike cyclic transition state.[2]

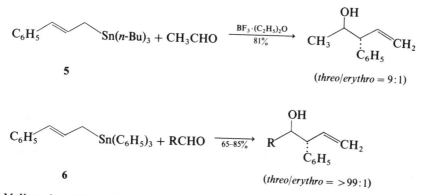

$$\underset{\mathbf{5}}{C_6H_5CH{=}CHCH_2Sn(n\text{-}Bu)_3} + CH_3CHO \xrightarrow[81\%]{BF_3\cdot(C_2H_5)_2O} CH_3CH(OH)CH(C_6H_5)CH{=}CH_2$$

(*threo*/*erythro* = 9:1)

$$\underset{\mathbf{6}}{C_6H_5CH{=}CHCH_2Sn(C_6H_5)_3} + RCHO \xrightarrow[65\text{–}85\%]{} RCH(OH)CH(C_6H_5)CH{=}CH_2$$

(*threo*/*erythro* = >99:1)

[1] M. Koreeda and Y. Tanaka, *Chem. Letters*, 1297 (1982).
[2] *Idem, ibid.*, 1299 (1982).

Tributylfluorophosphonium bromide, $(n\text{-}C_4H_9)_3\overset{+}{P}FBr^-$ (**1**). The salt is obtained by ligand exchange between $(n\text{-}C_4H_9)_3\overset{+}{P}BrBr^-$ and $(n\text{-}C_4H_9)_3PF_2$ in CH_2Cl_2 at 25°.[1]

Alkyl bromides.[2] The reagent converts alkyl trimethylsilyl ethers into alkyl bromides at room temperature in 60–85% yield (equation I).

$$\text{(I)}\quad ROSi(CH_3)_3 + (n\text{-}C_4H_9)_3\overset{+}{P}FBr^- \xrightarrow[-(CH_3)_3SiF]{} \left[(n\text{-}C_4H_9)_3\overset{+}{P}OR\ \ Br^-\right]$$

$$\longrightarrow \underset{(60\text{–}85\%)}{RBr} + (n\text{-}C_4H_9)_3P{=}O$$

[1] R. Bartsch, O. Stelzer, and R. Schmutzler, *Z. Naturforsch.* [*b*], **36**, 1349 (1981).
[2] *Idem, Synthesis*, 326 (1982).

Tri-*n*-butyl(iodomethyl)tin, 9, 475–476.

1, 3-*Dienes from allylic 2-pyridyl sulfides.* Treatment of the carbanion of the 2-pyridyl sulfide **2** with tri-*n*-butyl(iodomethyl)tin (**1**) results in formation of 1-phenylbutadiene (**3**) directly.

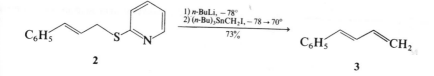

$$\underset{\mathbf{2}}{C_6H_5CH{=}CHCH_2S\text{-}(2\text{-pyridyl})} \xrightarrow[73\%]{1)\ n\text{-BuLi},\ -78^\circ;\ 2)\ (n\text{-Bu})_3SnCH_2I,\ -78\rightarrow70^\circ} \underset{\mathbf{3}}{C_6H_5CH{=}CHCH{=}CH_2}$$

Allylic sulfones undergo the same reaction, as formulated for the conversion of geranyl phenyl sulfone (**4**) into **5**.[1]

$$\underset{\mathbf{4}}{(CH_3)_2C{=}CHCH_2CH_2C(CH_3){=}CHCH_2SO_2C_6H_5} \xrightarrow[78\%]{1)\ n\text{-BuLi};\ 2)\ \mathbf{1}} \underset{\mathbf{5}}{(CH_3)_2C{=}CHCH_2CH_2C(CH_3){=}CHCH{=}CH_2}$$

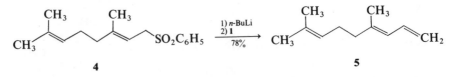

[1] M. Ochiai, S. Toda, K. Sumi, and E. Fujita, *Tetrahedron Letters*, **23**, 2205 (1982).

Tri-*n*-butylstannylcopper, n-Bu_3SnCu (**1**). This copper reagent is prepared by addition of CuI to n-Bu_3SnLi (**8**, 495–497) at $-78°$.

β-Tri-n-butylstannylacrylates. These vinylstannanes are available by reaction of **1** with β-halo or β-tosyloxy acrylates.[1]

Examples:

$$ClCH{=}CHCOOCH_3 \xrightarrow[58\%]{\mathbf{1},\ -78°} n\text{-}Bu_3SnCH{=}CHCOOCH_3$$

$$TsOCH{=}CHCOOC_2H_5 \xrightarrow[59\%]{\mathbf{1},\ -78\rightarrow 25°} n\text{-}BuSnCH{=}CHCOOC_2H_5$$

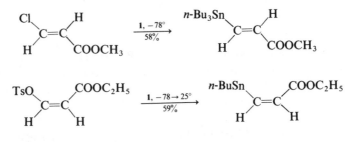

[1] D. E. Seitz and S.-H. Lee, *Tetrahedron Letters*, **22**, 4909 (1981).

Tri-*n*-butyltin fluoride (1). Mol. wt. 309.04. Supplier: Alfa.

Arylation of silyl enol ethers.[1] Silyl enol ethers of methyl *n*- or *sec*-alkyl ketones undergo arylation when treated with an aryl bromide and tri-*n*-butyltin fluoride and a

$$\text{(I)}\quad R{-}C(OSi(CH_3)_3){=}CH_2 + Bu_3SnF \xrightarrow{-(CH_3)_3SiF} \left[R{-}C(OSnBu_3){=}CH_2\right] \xrightarrow[45\text{–}85\%]{ArBr,\ PdCl_2[PAr_3]_2} RC(=O)CH_2Ar$$

palladium catalyst (equation I). Of a number of trialkyl or triaryl fluorides, only Bu_3SnF gave satisfactory results.

This arylation fails with methyl *t*-alkyl ketones or with ethyl alkyl ketones.

[1] I. Kuwajima and H. Urabe, *Am. Soc.*, **104**, 6831 (1982).

Tri-*n*-butyltin hydride, 1, 1192–1193; **2**, 424; **3**, 294; **4**, 518–520; **6**, 604; **7**, 379–380; **8**, 497–498; **9**, 379–380; **10**, 411–413.

Cyclization of vinyl radicals.[1] Vinyl radicals formed by reduction of vinyl halides with tri-*n*-butyltin hydride (AIBN, hν) can undergo intramolecular cyclization to a suitably situated double bond. When either a five- or a six-membered ring can be formed, the former ring is favoured. When cyclization could lead to a six- or seven-membered ring, only the former is formed, as in the cyclization of **1** to **2**. The cyclization of **3** to **4** is particularly interesting because the exocyclic double bond is formed at a predictable and useful position, and because a quaternary center is formed without difficulty.

CH_3, OH, I, CH_3, $C(CH_3)_2$ — $\xrightarrow[72\%]{(C_4H_9)_3SnH,\ h\nu}$ — CH_3, OH, CH_3, $HC(CH_3)_2$

1 → **2** (2 isomers, 1 : 1)

CH_3, Br, CH_2, OH, CN — $\xrightarrow{70\%}$ — CH_2, CH_3, OH, CN

3 → **4**

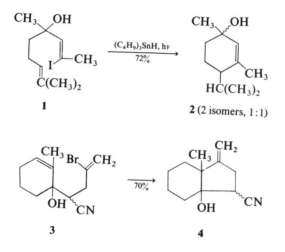

Radical cyclization; γ-butyrolactones.[2] On treatment of the bromoacetals **1** with Bu_3SnH (AIBN), the 2-alkoxytetrahydrofuranes (**2**) are formed in modest yield. The yields can be increased to 85–95% by use of a polymer-supported organotin.[3] The

(I) H, Br, R, CH_2, BuO, O — $\xrightarrow[\sim 60\text{–}65\%]{Bu_3SnH}$ — R, CH_3, BuO, O — $\xrightarrow[80\text{–}90\%]{1)\ \text{Jones oxid.}\ 2)\ \Delta}$ — O, O, R, CH_3

1 → **2** → **3**

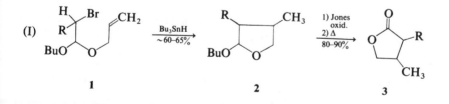

products **2** are converted into γ-butyrolactones by Jones oxidation followed by the usual workup and distillation.

High stereoselectivity can be attained by the reaction (equation II).

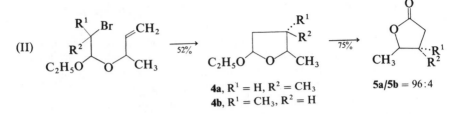

(II) [scheme] $\xrightarrow{52\%}$ **4a**, $R^1 = H$, $R^2 = CH_3$; **4b**, $R^1 = CH_3$, $R^2 = H$ $\xrightarrow{75\%}$ **5a/5b** = 96:4

Indolizidinones; pyrrolizidones.[4] Reaction of the thiophenoxylactam **1** with *n*-Bu_3SnH (AIBN) results to a minor extent in reduction. The major products result from cyclization of an intermediate α-acylamino radical to a pyrrolizidinone **2** and an indolizidinone **3**, in a 2:1 ratio. This ratio of *exo* and *endo* cyclization is dependent on

1 $\xrightarrow{n\text{-}Bu_3SnH,\ AIBN,\ C_6H_6}$ **2** + **3** (2:1)

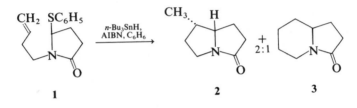

the alkyl substitution of the γ, δ-double bond. One advantage of this method is that both *exo* and *endo* cyclization proceed with high selectivity at the radical center (equation I).

(I) [scheme] $\xrightarrow{n\text{-}Bu_3SnH,\ AIBN}$ (51%) + (31%)

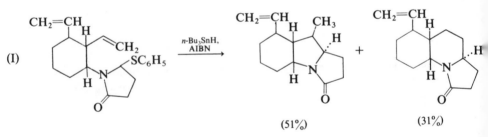

Bicyclic β-lactams.[5] This radical cyclization has been used for cyclization of the 4-allyloxy-2-azetidinone (**1**) to the bicyclic β-lactam **2**. In this case, only *endo* addition is observed.

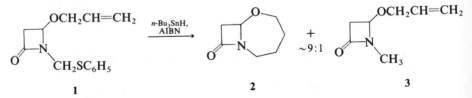

1 $\xrightarrow{n\text{-}Bu_3SnH,\ AIBN}$ **2** + **3** (~9:1)

Carbacephams.[6] This radical cyclization has been used for a synthesis of a carbacepham ring system (equation I).

(I) [Structure: azetidinone with SC_6H_5, N-substituent bearing $COOCH_3$ and allyl chain ending in CH_2] $\xrightarrow[\text{AIBN}]{Bu_3SnH}$ [bicyclic carbacepham with H and $COOCH_3$] (43%) + [azetidinone N-substituted with $CH(COOCH_3)CH_2CH{=}CH_2$] (trace)

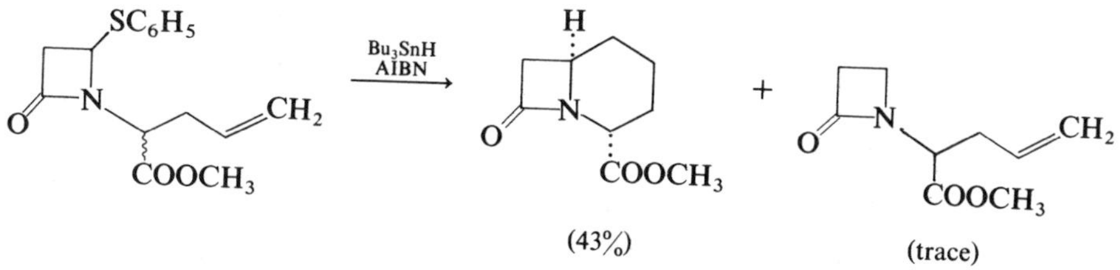

Reduction of nitro groups.[7,8] Trialkyltin hydrides under free-radical conditions replace tertiary nitro groups by hydrogen in yields > 75%. Some secondary nitro groups, when substituted by electron-accepting groups, can also be reduced satisfactorily. Yields are comparable to those obtained with sodium thiomethoxide, the reagent used previously for this reduction.[9]

This reaction can be used to effect conjugate addition of an alkyl group to α,β-unsaturated sulfoxides, which is generally unsatisfactory with RMgBr or R_2CuLi. However, nitroalkanes undergo this 1,4-addition in the presence of a base, particularly DBU, and the product can be selectively reduced to the desired sulfoxide. The same conditions can be used to effect conjugate addition to α,β-unsaturated ketones and esters.[10]

Example:

$$CH_2{=}CHSOC_6H_5 + (CH_3)_2CHNO_2 \xrightarrow[95\%]{\text{DBU, } CH_3CN} (CH_3)_2C(NO_2)CH_2CH_2SOC_6H_5$$

$$\xrightarrow[93\%]{Bu_3SnH,\ \text{AIBN, } C_6H_6} (CH_3)_2CHCH_2CH_2SOC_6H_5$$

Alkenes from* vic*-dinitro compounds.[11] *vic*-Dinitro compounds and β-nitro sulfones are converted into alkenes when treated with this reagent (2 equiv.) in benzene at 80°. The rate is accelerated by AIBN. Yields are 70–95%. Both (E)- and (Z)-alkenes are usually formed in about equivalent amounts.

Cyclohexanones. The ability of tri-*n*-butyltin hydride to effect denitration (**10**, 413) permits use of α-nitroalkenes for construction of cyclohexanones by a Diels-Alder reaction. Only one regioisomer (as a mixture of stereoisomers) is formed in the example cited, owing to the reinforcing effect of the nitro group in control of the direction of the cycloaddition.[12]

Example:

$$(CH_3)_3SiO(CH_2{=})C\text{–}CH{=}CH(OCH_3) + (H)(C_2H_5)C{=}C(CH_3)(NO_2) \xrightarrow[71\%]{1)\ C_6H_6,\ \Delta;\ 2)\ H_3O^+}$$

$$\text{3-CH}_3\text{O-5-CH}_3\text{-5-NO}_2\text{-6-C}_2\text{H}_5\text{-cyclohexanone} \xrightarrow[82\%]{(C_4H_9)_3SnH} \text{3-CH}_3\text{O-5-CH}_3\text{-6-C}_2\text{H}_5\text{-cyclohexanone}$$

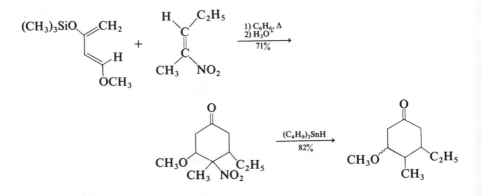

Reduction of acyl chlorides to aldehydes.[13] Tri-*n*-butyltin hydride reduces acyl chlorides to a mixture of aldehydes and esters. In the presence of a soluble Pd catalyst {usually $Pd[P(C_6H_5)_3]_4$}, only aldehydes obtain. This reduction is very general, and yields are usually > 80%. Reduction of double bonds is a minor competing reaction with α, β-unsaturated substrates (< 5% reduction). Of other reducible groups, only an allylic bromide group competes with the COCl group.

Stereoselective debromination.[14] A key step in a notably efficient synthesis of shikimic acid (**3**) from 1,4-dihydrobenzoic acid (see also, **8**, 84–86) is the stereoselective debromination of **1** with Bu_3SnH to give **2** in 72% yield. When Bu_3SnD is used for this reduction, 6β-deuterioshikimic acid is obtained after alkaline hydrolysis with

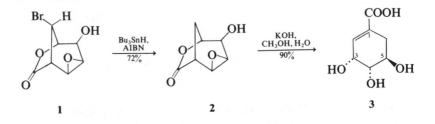

$$\mathbf{1} \xrightarrow[72\%]{Bu_3SnH,\ AIBN} \mathbf{2} \xrightarrow[90\%]{KOH,\ CH_3OH,\ H_2O} \mathbf{3}$$

high stereoselectivity. This retention of configuration is unusual in free-radical reactions and probably results from thermodynamic effects. Actually a related *anti*-bromolactone is reduced by Bu_3SnD with complete inversion of configuration.

Conjugate reduction of α,β-enals and -enones. Tri-*n*-butyltin hydride in the presence of tetrakis(triphenylphosphine)palladium effects conjugate reduction of α, β-unsaturated aldehydes and ketones in the presence of a proton source (water, acetic acid). Yields are improved by addition of a radical scavenger.[15] Double bonds bearing two or more substituents are not reduced readily unless catalytic amounts of zinc chloride are added.[16]

Wharton reaction (**1**, 439; **8**, 245). This reaction can also be conducted under free-radical conditions of reduction of thiocarbonylimidazolide derivatives of α,β-epoxy alcohols with tri-*n*-butyltin hydride.[17] In some cases, inverse addition is necessary to prevent further rearrangement of the allylic alcohol initially formed. Yields are around 50–60%. The modification is not useful in the carbohydrate field, where very complex product mixtures are formed.

[Scheme: epoxy alcohol (with CH_3, OH, and $C(=CH_2)CH_3$ substituents) → (thiocarbonyldiimidazole, 91%) → thiocarbonylimidazolide $OC(=S)$–N(imidazole) → $(C_4H_9)_3SnH$, AIBN, C_6H_6, 65% → allylic alcohol (HO, CH_3, $C(=CH_2)CH_3$)]

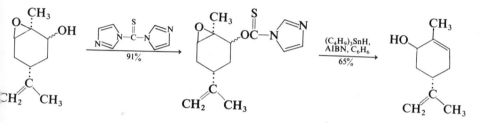

Alkyl hydroperoxides. Peroxymercuration of alkenes proceeds cleanly and in good yield. These products previously were reduced to the desired alkyl hydroperoxides with alkaline sodium borohydride. This reduction proceeds in reasonable yield in the case of terminal alkenes, but scission to form epoxides of the original alkene predominates among products from nonterminal alkenes. This difficulty is now overcome by use of tri-*n*-butyltin hydride for reduction.[18]

Example:

$$\underset{(Z)}{CH_3CH{=}CHCH_3} \xrightarrow[2)\ KBr]{1)\ HgX_2,\ (CH_3)_3COOH}$$

$$CH_3CH(OOC(CH_3)_3){-}CH(HgBr)CH_3 \xrightarrow[63\%]{(C_4H_9)_3SnH} CH_3CH(OOC(CH_3)_3){-}CH_2CH_3$$

The overall yields compare favourably with those obtained by *t*-butyl perhydrolysis of alkyl bromides assisted by CF_3COOAg (**7**, 175).[19]

2-Alkyl-1,3-butadienes.[20] A radical reaction of tri-*n*-butyltin hydride with α-(hydroxymethyl)allyl tolyl sulfones (**1**) results in stereoselective formation of allyltin intermediates, which fragment to 2-substituted 1,3-butadienes when heated.

Example:

$$\underset{\mathbf{1}}{CH_3C_6H_4SO_2(HOCH_2)C{=}C(CH{=}CH_2)(CH_2)_7CH_3} \xrightarrow[\text{AIBN}]{Bu_3SnH,\ 80^\circ} \underset{(Z)}{CH_3(CH_2)_7(HOCH_2)C{=}C(H)CH_2SnBu_3} + \text{(E)-isomer} \quad 96{:}4$$

$$(Z) \xrightarrow[92\%]{160^\circ} CH_3(CH_2)_7C({=}CH_2)CH{=}CH_2$$

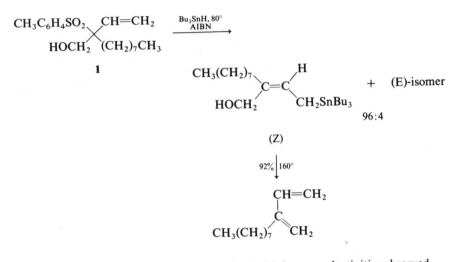

The hydroxymethyl group is important for the high stereoselectivities observed.

Deoxy sugars. Barton and Motherwell[21] have reviewed the reduction of a number of thiocarbonyl esters of alcohols by tri-*n*-butyltin hydride to give deoxygenated products. Rasmussen[22] recommends imidazolylthiocarbonyl derivatives for this deoxygenation. They are obtained in high yield with N, N′-thiocarbonyldiimidazole, and can be stored when protected from light and air for several weeks. The reduction of these derivatives of secondary alcohols of sugars generally proceeds in 55–90% yield (equation I).

$$\text{(I)}\quad R^1R^2CHOH + (C_3H_3N_2)\text{—}C(=S)\text{—}(C_3H_3N_2) \xrightarrow[80\text{–}95\%]{ClCH_2CH_2Cl} R^1R^2CHOC(=S)\text{—}(C_3H_3N_2) \xrightarrow[55\text{–}90\%]{Bu_3SnH} R^1R^2CH_2$$

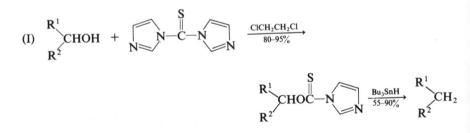

[1] G. Stork and N. H. Baine, *Am. Soc.*, **104**, 2321 (1982).
[2] K. Saigo, A. Morikawa, and T. Mukaiyama, *Bull. Chem. Soc. Japan*, **49**, 1656 (1976); J. M. Berge and S. M. Roberts, *Synthesis*, 471 (1979).
[3] Y. Ueno, K. Chino, M. Watanabe, O. Moriya, and M. Okawara, *Am. Soc.*, **104**, 5564 (1982).
[4] D. J. Hart and Y.-M. Tsai, *Am. Soc.*, **104**, 1430 (1982).
[5] M. D. Bachi and C. Hoornaert, *Tetrahedron Letters*, **22**, 2693 (1981).
[6] T. Kametani and T. Honda, *Heterocycles*, **19**, 1861 (1982).

[7] N. Ono, H. Miyake, R. Tamura, and A. Kaji, *Tetrahedron Letters*, **22**, 1705 (1980).
[8] D. D. Tanner, E. V. Blackburn, and G. E. Diaz, *Am. Soc.*, **103**, 1557 (1981).
[9] N. Kornblum, S. C. Carlson, and R. G. Smith, *ibid.*, **101**, 647 (1979).
[10] N. Ono, H. Miyake, A. Kamimura. N. Tsukui, and A. Kaji, *Tetrahedron Letters*, **23**, 2957 (1982).
[11] N. Ono, H. Miyake, R. Tamura, I. Hamamoto, and A. Kaji, *Chem. Letters*, 1139 (1981).
[12] N. Ono, H. Miyake, and A. Kaji, *J.C.S. Chem. Comm.*, 33 (1982).
[13] P. Four and F. Guibe, *J. Org.*, **46**, 4439 (1981).
[14] K. E. Coblens, V. B. Muralidharan, and B. Ganem, *J. Org.*, **47**, 5041 (1982).
[15] E. Keinan and P. A. Glieze, *Tetrahedron Letters*, **23**, 477 (1982).
[16] P. Four and F. Guibe, *ibid.*, **23**, 1825 (1982).
[17] D. H. R. Barton, R. S. H. Motherwell, and W. B. Motherwell, *J.C.S. Perkin I*, 2363 (1981).
[18] A. J. Bloodworth, J. A. Khan, and M. E. Loveitt, *ibid*, 621 (1981), and references cited therein.
[19] A. J. Bloodworth and J. L. Courtneidge, *J.C.S. Chem. Comm.*, 1117 (1981).
[20] Y. Ueno, H. Sano, S. Aoki, and M. Okawara, *Tetrahedron Letters*, **22**, 2675 (1981).
[21] D. H. R. Barton and W. B. Motherwell, *Pure Appl. Chem.* **53**, 15 (1981).
[22] J. R. Rasmussen, *J. Org.*, **45**, 2725 (1980); J. R. Rasmussen, C. J. Slinger, R. J. Kordish, and D. D. Newman-Evans, *ibid.*, **46**, 4843 (1981).

Tri-*n*-butyltinlithium, 8, 495–497, **10**, 413–414.

$RCH_2CHO \rightarrow RC{\equiv}CH$.[1] The first step in this process is addition of tri-*n*-butyltinlithium followed by conversion to an α-bromoalkylstannane (**10**, 413–414). Remaining steps are dehydrobromination with DBU and lead tetraacetate oxidation

$$\text{(I)}\quad C_6H_5(CH_2)_4CHO \xrightarrow{n\text{-}Bu_3SnLi} \left[C_6H_5(CH_2)_4\overset{\displaystyle OH}{\overset{|}{C}}HSnBu_3 \right] \xrightarrow[80\%]{P(C_6H_5)_3,\ CBr_4}$$

$$C_6H_5(CH_2)_4\overset{\displaystyle Br}{\overset{|}{C}}HSnBu_3 \xrightarrow[90\%]{DBU} \underset{E/Z\,=\,8:1}{C_6H_5(CH_2)_3CH{=}CHSnBu_3}$$

$$\xrightarrow[65\%]{Pb(OAc)_4,\ CH_3CN} C_6H_5(CH_2)_3C{\equiv}CH$$

(**5**, 43–44). The method is outlined in equation (I) and is generally applicable.

Aldehydes with an α-substituent can be converted into pure l-alkenylstannanes (equation II).

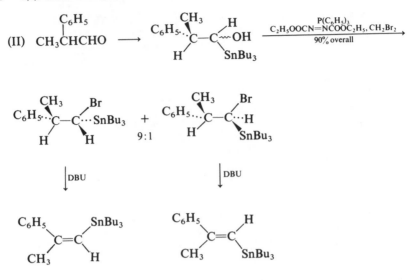

$$\text{(II)}\quad CH_3CH(C_6H_5)CHO \longrightarrow C_6H_5(CH_3)(H)C{-}C(H)(OH)(SnBu_3) \xrightarrow[\text{90\% overall}]{P(C_6H_5)_3,\ C_2H_5OOCN{=}NCOOC_2H_5,\ CH_2Br_2}$$

$$C_6H_5(CH_3)(H)C{-}C(Br)(SnBu_3)(H) \quad + \quad C_6H_5(CH_3)(H)C{-}C(Br)(H)(SnBu_3)$$

$$9:1$$

$$\downarrow \text{DBU} \qquad\qquad \downarrow \text{DBU}$$

$$(C_6H_5)(CH_3)C{=}C(SnBu_3)(H) \qquad\qquad (C_6H_5)(CH_3)C{=}C(H)(SnBu_3)$$

[1] M. Shibasaki, Y. Torisawa, and S. Ikegami, *Tetrahedron Letters*, **23**, 4607 (1982).

2,4,6-Trichlorobenzoyl chloride, 9, 478–479.

Thiol esters.[1] These esters can be prepared by reaction of mixed anhydrides prepared from this chloride and a carboxylic acid with thiols in the presence of 4-dimethylaminopyridine (equation I).

$$\text{(I)}\quad R^1SH + 2,4,6\text{-}Cl_3C_6H_2COOCOR^2 \xrightarrow[75\text{–}85\%]{\text{DMAP},\ CH_2Cl_2} R^2COSR^1$$

[1] Y. Kawanami, Y. Dainobu, J. Inanaga, T. Katsuki, and M. Yamaguchi, *Bull. Chem. Soc. Japan*, **54**, 943 (1981).

Trichloroethylene, 9, 479–480; **10,** 414–415.

Dichlorovinylation. Enolate dichlorovinylation with trichloroethylene involves proton elimination to form dichloroacetylene as the actual electrophile.[1] Dichloroacetylene can be prepared by reaction of trichloroethylene with $LiN[Si(CH_3)_3]_2$ followed by distillation. In the absence of a proton donor, it reacts with a variety of tertiary enolates to give α-chloroethynyl ketones and esters in 43–90% yields.

Examples:

$$\text{2,6-dimethylcyclohex-2-enone} \xrightarrow[64\%]{\text{1) LDA, HMPT, THF; 2) ClC}\equiv\text{CCl}} \text{6-(chloroethynyl)-2,6-dimethylcyclohex-2-enone}$$

$$\text{cyclohexyl-}COOCH_3 \xrightarrow{90\%} \text{1-(C}\equiv\text{CCl)cyclohexyl-}COOCH_3$$

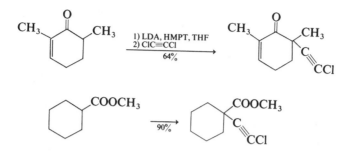

Similar adducts can be obtained from $C_6H_5C{\equiv}CCl$ and $C_6H_5SC{\equiv}CCl$. Reaction with enolates not at tertiary centers results in isomerization of the primary adducts to allenes.

The chloroethynyl group can be reduced to an ethynyl group by copper powder in THF/HOAc or to a vinyl group by Lindlar catalyst.[2]

[1] A. S. Kende and P. Fludzinski, *Tetrahedron Letters*, **23**, 2369 (1982).
[2] *Idem, Tetrahedron Letters*, **23**, 2373 (1982).

Trichloro(methyl)silane, CH_3SiCl_3 (**1**). Mol. wt. 149.50, b.p. 66.4°. Supplier: Alfa, Petrarch.

Cleavage of ethers with* 1 *and NaI.[1] This combination has certain advantages over $ClSi(CH_3)_3$–NaI for cleavage of ethers. It cleaves methyl (also trityl and tetrahydropyranyl) ethers to give quantitative yields of alcohols when the alkyl groups are primary or secondary. However, when the alkyl group of the methyl ethers is tertiary, the tertiary iodide is the main product. Ethyl ethers are cleaved, in varying amounts, to alcohols and iodides. The new procedure is also useful for selective cleavage of ethers in the presence of hydroxyl or ester groups.

[1] G. A. Olah, A. Husain, B. G. B. Gupta, and S. C. Narang, *Angew. Chem. Int. Ed.*, **20**, 690 (1981).

Trichlorosilane–Triethylamine.

Isocyanides. Isocyanates, RN=C=O, and alkyl carbamates, $R^1NHCOOR^2$, are reduced to isocyanides, RN≡C, by treatment with trichlorosilane and triethylamine in CH_2Cl_2 in yields of ~40–70%.[1] *t*-Butyldiphenylsilyllithium[2] also effects reduction of isocyanates to isocyanides, but this reagent is less convenient to use.

Trichlorosilane in combination with a tertiary amine has been used previously to reduce polyhalo compounds.[3] Both reductions are believed to involve the trichlorosilyl anion, which is isoelectronic with phosphines.

[1] J. E. Baldwin, J. C. Bottaro, P. D. Riordan, and A. E. Derome, *J.C.S. Chem. Comm.*, 942 (1982).
[2] H. Gilman, D. J. Peterson, and D. Wittenberg, *Chem. Ind.*, 1479 (1958).
[3] R. A. Benkeser and W. E. Smith, *Am. Soc.*, **90**, 5307 (1968).

Triethoxysilane, $HSi(OC_2H_5)_3$. The silane is prepared industrially by ethanolysis of trichlorosilane.

Reduction of esters.[1] In the presence of cesium fluoride, triethoxysilane (or diethoxymethylsilane) reduces esters to the corresponding primary alcohols in 65–90% yield. (equation I).

$$\text{(I)}\quad R^1C(=O)OR^2 \xrightarrow{2HSi(OC_2H_5)_3,\ CsF} [R^1CH_2OSi(OC_2H_5)_3 + R^2OSi(OC_2H_5)_3] \xrightarrow{H_2O} R^1CH_2OH + R^2OH + (C_2H_5O)_3SiOSi(OC_2H_5)_3$$

Selective reduction of carbonyl groups.[2] Heterogeneous reduction of carbonyl groups by triethoxysilane with CsF (or KF) as catalyst is highly selective, the reactivity order being aldehyde > ketone > ester. Double bonds, nitro, halo, and amido groups are inert to this reagent.

Examples:

$$CH_3\overset{O}{\overset{\|}{C}}NHC_6H_4CHO \xrightarrow[80\%]{(C_2H_5O)_3SiH,\ CsF} CH_3\overset{O}{\overset{\|}{C}}NHC_6H_4CH_2OH$$

$$C_6H_5\overset{O}{\overset{\|}{C}}(CH_2)_2COOCH_3 \xrightarrow{85\%} \text{5-phenyl-}\gamma\text{-butyrolactone } (C_6H_5\text{-substituted five-membered lactone})$$

$$\begin{matrix} CH_2\overset{O}{\overset{\|}{C}}COOCH(CH_3)_2 \\ | \\ CH_2COOCH(CH_3)_2 \end{matrix} \xrightarrow{70\%} \begin{matrix} CH_2\overset{OH}{\overset{|}{C}}HCOOCH(CH_3)_2 \\ | \\ CH_2COOCH(CH_3)_2 \end{matrix}$$

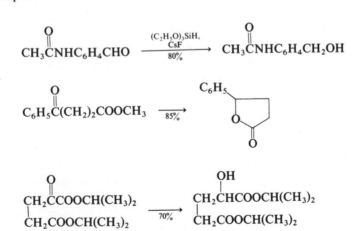

[1] J. Boyer, R. J. P. Corriu, R. Perz, M. Poirier, and C. Reye, *Synthesis*, 558 (1981).
[2] J. Boyer, R. J. P. Corriu, R. Perz, and C. Reye, *Tetrahedron*, **37**, 2165 (1981).

Triethyl orthoformate, 1, 1204–1209; **2**, 430; **4**, 527; **5**, 690–691; **6**, 610–611; **7**, 385–386.

Alkenes from* 1,2-*diols (**7**, 385–386). An efficient synthesis of (S)-(—)-γ-methoxymethyl-α, β-butenolide (**3**) from (+)-5-O-methyl-D-ribonolactone (**1**) involves conversion to the cyclic orthoformate **2** followed by pyrolysis to give **3** in 66.5% overall yield.[1] In this case, the Corey-Winter reaction and the Hanessian route (**8**, 192) were

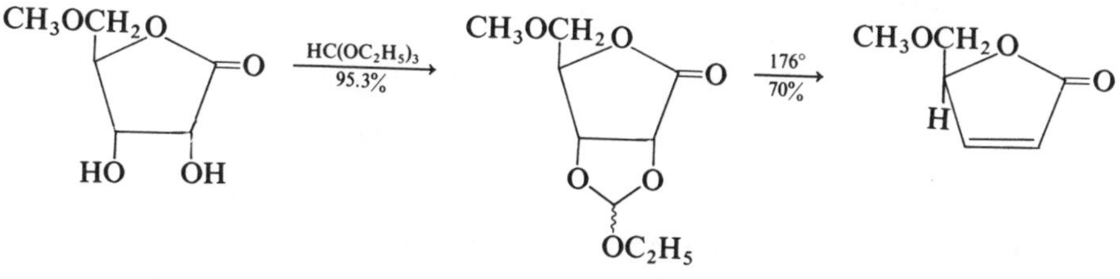

CH_3OCH_2 ... $\xrightarrow[95.3\%]{HC(OC_2H_5)_3}$ ... $\xrightarrow[70\%]{176°}$ ...

HO OH; OC_2H_5

1 **2** **3**

not useful. Application of this method to D-ribonolactone did not give **3**, but traces of the anhydro derivative **4** (protoanemonim).

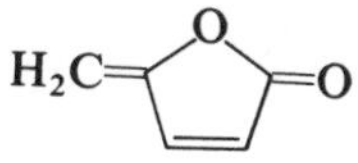

H_2C=

4

[1] P. Camps, J. Cardellach, J. Font, R. M. Ortuno, and O. Ponsati, *Tetrahedron*, **38**, 2395 (1982).

Triethyl phosphite, 1, 1212–1216; **2**, 432–433; **3**, 304; **4**, 529–530; **5**, 693; **6**, 612; **7**, 387; **8**, 501.

Reduction of* gem-*dibromides.[1] Triethyl phosphite and triethylamine reduces *gem*-dibromocyclopropanes and *gem*-dibromoalkenes to the corresponding monobromides.

Examples:

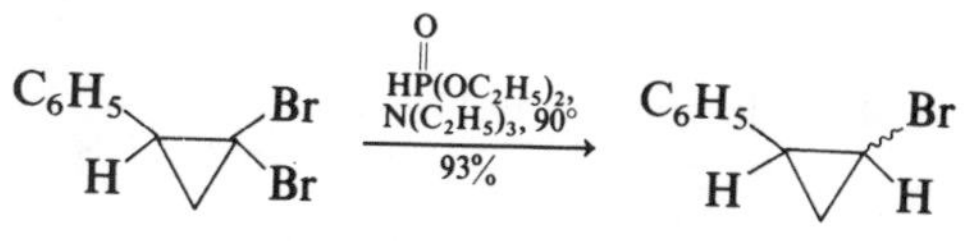

C_6H_5, H, Br, Br $\xrightarrow[93\%]{HP(O)(OC_2H_5)_2,\ N(C_2H_5)_3,\ 90°}$ C_6H_5, H, Br, H

E/Z = 75:25

$C_6H_5(H)C{=}CBr_2 \xrightarrow[96\%]{25°} C_6H_5(H)C{=}C(H)Br$

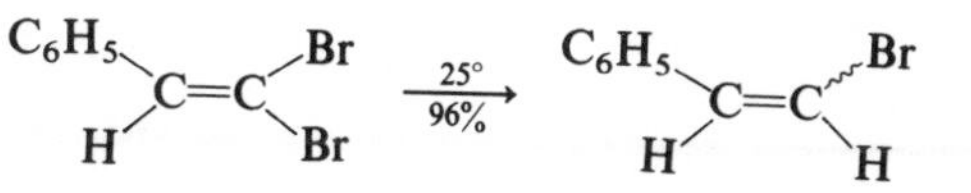

E/Z = 94:6

[1] T. Hirao, T. Masunaga, Y. Ohshiro, and T. Agawa, *J. Org.*, **46**, 3745 (1981).

3-Triethylsilyloxypentadienyllithium (1).

Preparation:

$$CH_2{=}CHCH(OH)CH{=}CH_2 + (C_2H_5)_3SiCl \xrightarrow[88\%]{\text{imidazole, DMF}} CH_2{=}CHCH(OSi(C_2H_5)_3)CH{=}CH_2$$

$$\xrightarrow[\text{THF, }-78^\circ]{sec\text{-BuLi}} [CH_2{\cdots}CH{\cdots}\overset{\alpha}{C}(OSi(C_2H_5)_3){\cdots}CH{\cdots}\overset{\gamma}{C}H_2]^- Li^+$$

1

The corresponding 3-trimethylsilyl ether has been prepared but it and the products formed by alkylation are less stable.

Electrophilic substitutions.[1] The anion **1** reacts with water and R_3SiCl exclusively at the γ-position. Alkylation is less selective; both positions are attacked, but the γ-position is usually preferred. The selectivity is not altered by addition of TMEDA or HMPT or by change of the counterion to K^+ or Zn^{2+}. The γ-silylated products provide a route to vinyl ketones (equation I).

$$\text{(I)}\quad (CH_3)_2C{=}CHCH_2Br + \mathbf{1} \xrightarrow{76\%} CH_2{=}CHC(CH_2CH{=}C(CH_3)_2)(OSi(C_2H_5)_3)CH{=}CH_2 \; + \; (8:92)$$

$$CH_2{=}CHC(=O)CH_2CH_2CH_2CH{=}C(CH_3)_2 \xleftarrow[74\%]{\text{KF, }CH_3OH} CH_2{=}CHC(OSi(C_2H_5)_3){=}CHCH_2CH_2CH{=}C(CH_3)_2$$

The γ-products can also serve as substrates for intramolecular Diels-Alder reactions (equation II).

$$\text{(II)}\quad CH_2{=}CH(CH_2)_2Br + \mathbf{1} \xrightarrow{80\%} \alpha\text{-product} \; + \; (39:61)\; CH_2{=}CHC(OSi(C_2H_5)_3){=}CHCH_2CH_2CH{=}CH_2$$

$$\xrightarrow[84\%]{C_6H_5CH_3,\ 160^\circ} \text{bicyclic silyl enol ether } (OSi(C_2H_5)_3) \xrightarrow[81\%]{\text{KF, }CH_3OH} \text{cis-hydrindanone (H, H)}$$

Aldehydes and ketones react with **1** mainly at the γ-position as a result of kinetic control.

1-(*Methylthio*)-3-*triethylsilyloxypentadienyllithium*.[2] Reaction of **1** with dimethyl disulfide gives **2** in 81% yield (stable oil). The anion of **2** is alkylated exclusively at the γ-position. The products (**3**) have two latent enone groups, which can be liberated separately by F^- and oxidation/sulfoxide elimination.

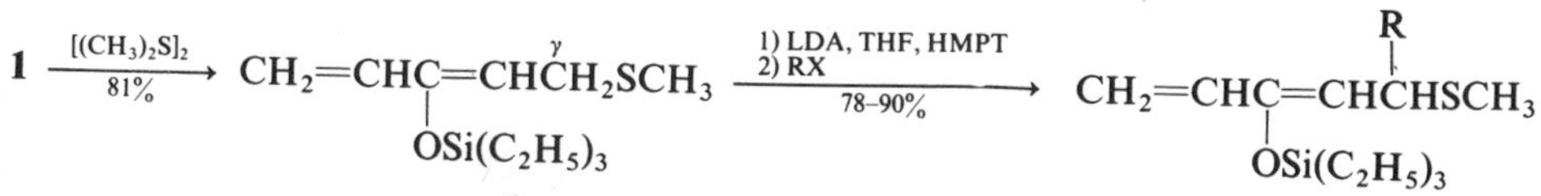

$$\mathbf{1} \xrightarrow[81\%]{[(CH_3)_2S]_2} \underset{\mathbf{2}}{CH_2{=}CHC(OSi(C_2H_5)_3){=}CH\overset{\gamma}{C}H_2SCH_3} \xrightarrow[78\text{–}90\%]{1)\ LDA,\ THF,\ HMPT;\ 2)\ RX} \underset{\mathbf{3}}{CH_2{=}CHC(OSi(C_2H_5)_3){=}CHCH(R)SCH_3}$$

[1] W. Oppolzer, R. L. Snowden, and D. P. Simmons, *Helv.*, **64**, 2002 (1981).
[2] W. Oppolzer, R. L. Snowden, and P. H. Briner, *ibid.*, **64**, 2022 (1981).

Trifluoroacetic acid, 1, 1219–1221; **2**, 433–434; **3**, 305–308; **4**, 530–532; **5**, 695–700; **6**, 613–615; **7**, 388–389; **8**, 503; **9**, 483; **10**, 418.

Diazo ketone cyclization. Some years ago Mander and his group[1] demonstrated that the protonated diazomethylcarbonyl group can initiate cyclizations in unsaturated systems. In the case of phenolic diazo ketones, formation of spirodienones can predominate over competing side reactions (dienone-phenol rearrangement). Tetrafluoroboric acid or boron trifluoride etherate can be used, but trifluoroacetic acid is usually the acid of choice.

The diazo ketone methodology has provided a general route to gibberellins. Thus treatment of the diazo ketone **1**, obtained from 1,7-dimethoxynaphthalene, with TFA resulted in the tricyclic dienedione **2**, which was converted in a few steps to **3**, which contains the B, C, and D rings of gibberellic acid (**4**).[2]

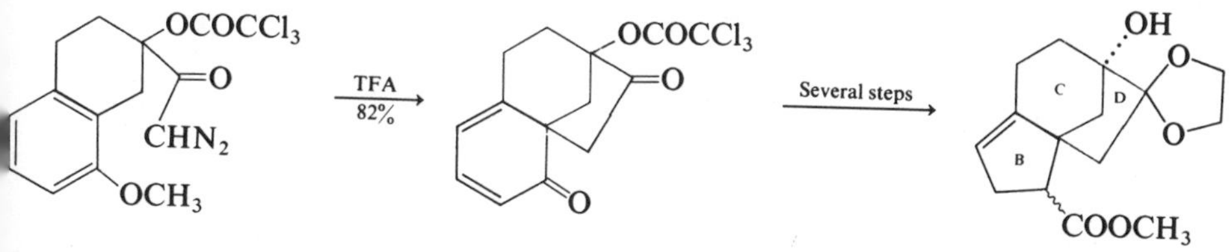

OCOCCl$_3$ =O CHN$_2$ OCH$_3$ **1** $\xrightarrow[82\%]{TFA}$ OCOCCl$_3$ =O O **2** $\xrightarrow{\text{Several steps}}$ OH O O C D B COOCH$_3$ **3**

O H CO HO H CH$_3$ COOH OH =CH$_2$

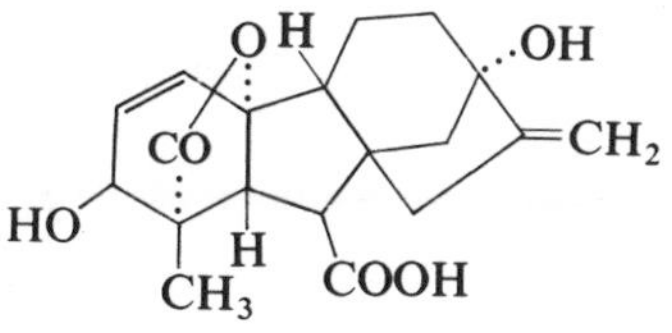

4

This cyclization also is useful in an efficient route to the ring system **5** of the natural product aphidicolin (**6**).[3]

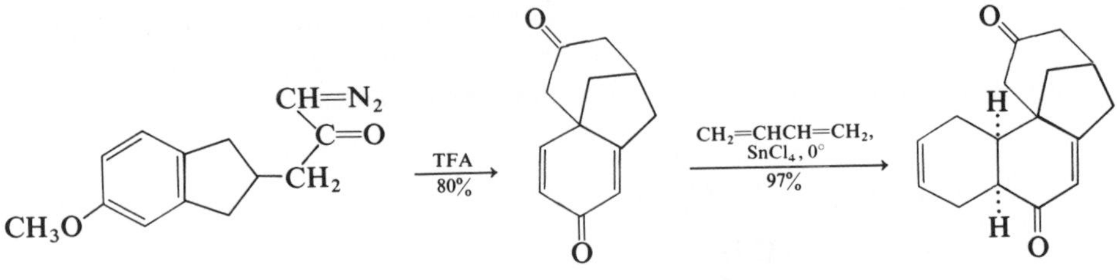

TFA 80%

$CH_2{=}CHCH{=}CH_2$, $SnCl_4$, 0° 97%

5

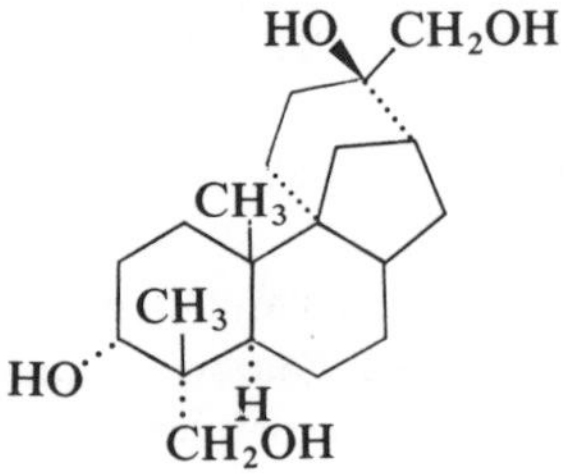

6

Claisen rearrangements (**4**, 532).[4] The 5-allyloxy-2-hydroxybenzoate **1** when dissolved in TFA rearranges in about 20 hours to **2** and **3**, probably via Claisen

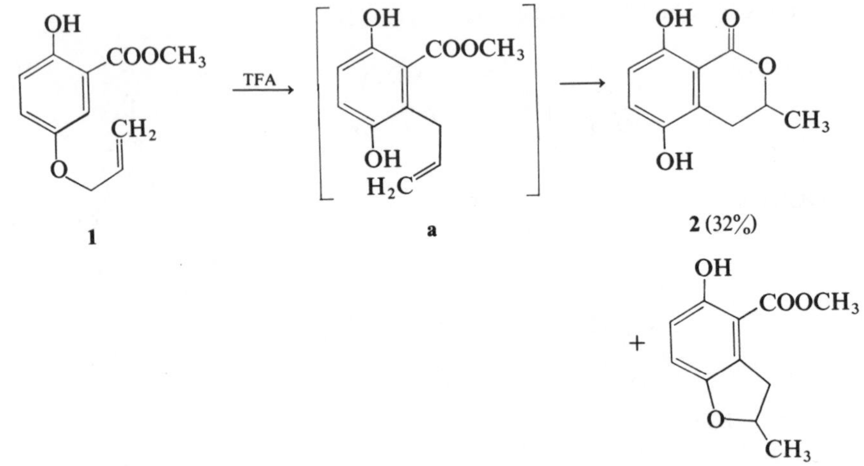

TFA

1

a

2 (32%)

3 (21%)

rearrangement to **a**. The dihydroisocoumarin (**2**) has been converted into (±)-mellein (**4**), which has pheromonal activity in the carpenter ant.

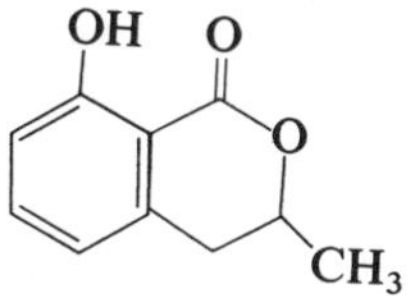

OH O

O

CH_3

4

Cope rearrangements. 2-Acyl-l,5-dienes, such as **1**, undergo smooth Cope rearrangement in the presence of TFA (1 equiv.).[5] The dienone-phenol rearrangement[6] of substrates such as **3** in the presence of TFA can be regarded as Cope rearrangements of 1-acyl-1,5-dienes.

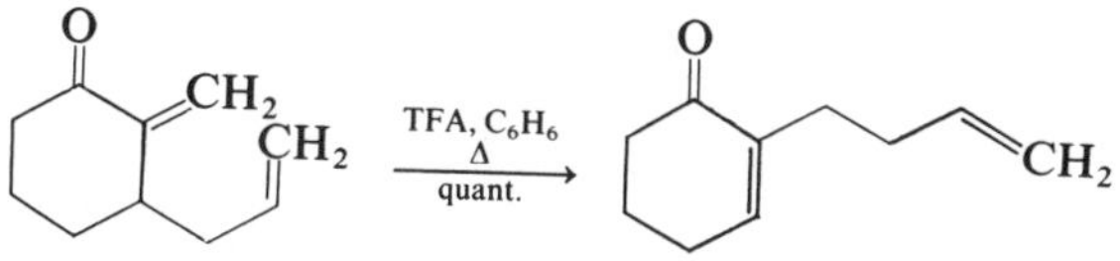

O CH_2 CH_2 — TFA, C_6H_6, Δ, quant. → O CH_2

1 **2**

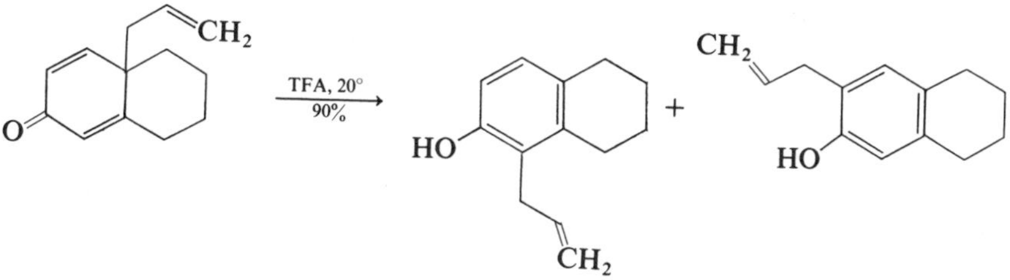

CH_2 O — TFA, 20°, 90% → HO CH_2 + CH_2 HO

3 **4** **5**

[1] I. A. Blair, A. Ellis, D. W. Johnson, and L. N. Mander, *Aust. J. Chem*, **31**, 405 (1978), and references cited therein.
[2] L. Lombardo, L. N. Mander, and J. V. Turner, *Am. Soc.*, **102**, 6628 (1980).
[3] K. C. Nicolaou and R. E. Zipkin, *Angew. Chem. Int. Ed.*, **20**, 785 (1981).
[4] L. M. Harwood, *J.C.S. Chem. Comm.*, 1120 (1982).
[5] W. G. Dauben and A. Chollet, *Tetrahedron Letters*, **22**, 1583 (1981).
[6] U. Widmer, J. Zsindely, H. J. Hansen, and H. Schmid, *Helv.*, **56**, 75 (1973).

Trifluoroacetyl chloride, 10, 419.

Butenolides. The final steps in a synthesis of jolkinolide E (**4**) involve formation of the butenolide ring by reaction of the α-hydroxy ketone **1** with the mixed anhydride of trichloroacetic acid and α-(diethylphosphono)propionic acid (**2**) with catalysis by DMAP. The ester **3** undergoes an intramolecular Wittig-Horner reaction in the presence of NaH to give **4**.[1]

1 + $Cl_3CCOOCCH(CH_3)P(O)(OC_2H_5)_2$ (**2**) $\xrightarrow{\text{DMAP, } C_6H_6,\ 92\%}$ **3** $\xrightarrow{\text{NaH, DME, } 70\%}$ **4**

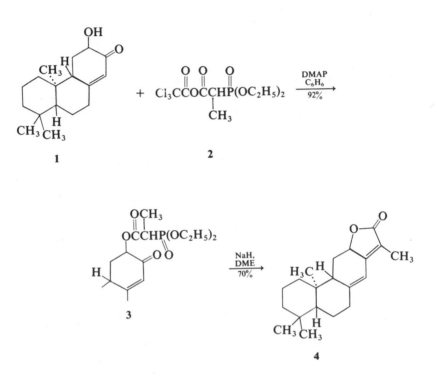

[1] S. Katsumura and S. Isoe, *Chem. Letters*, 1689 (1982).

Trifluoromethanesulfonic anhydride, 4, 533–534; **5,** 702–705; **6,** 618–620; **7,** 390; **10,** 919–920.

Keteneiminium triflates; cyclobutanones. Triflic anhydride converts tertiary amides mainly into the salt (**1**), which loses triflic acid in the presence of collidine to form the keteneiminium salt **2**. When generated in the presence of alkenes or alkynes, **2** undergoes [2 + 2] cycloaddition to form cyclobutanones or cyclobutenones, respectively.[1]

$$R^1R^2CHCON(CH_3)_2 \xrightarrow{(CF_3SO_2)_2O} R^1R^2CH{-}C(OSO_2CF_3){=}\overset{+}{N}(CH_3)_2CF_3SO_3^- \ (\mathbf{1}) \xrightarrow{\text{Collidine}}$$

$$R^1R^2C{=}C{=}\overset{+}{N}(CH_3)_2CF_3SO_3^- \ (\mathbf{2})$$

$\mathbf{2}$ + $C{=}C$ (40–80%) → cyclobutanone (R[1], R[2]); $\mathbf{2}$ + $C_6H_5C{\equiv}CC_6H_5$ (60–80%), $R^2{=}H$ → 2,3-diphenylcyclobutenone (C_6H_5, C_6H_5, O, R^1)

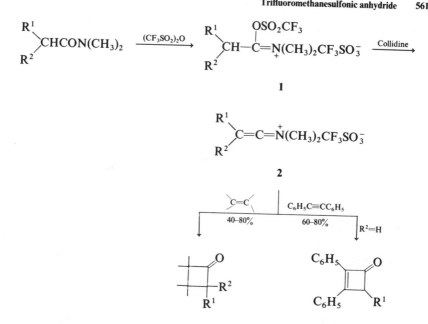

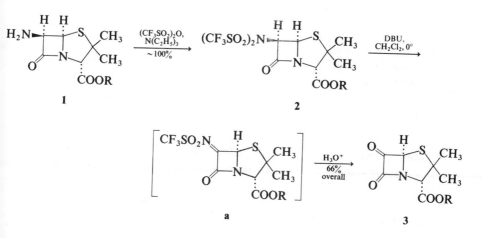

The advantage of this route to keteneiminium salts over the earlier route from α-halo enamines (**4**, 94–95; **5**, 136–138) is that R^1, and even R^1 and R^2, can be H.

***Conversion of* $>CHNH_2$ *to* $>C{=}O$.**[2] A convenient three-step conversion of alkyl 6-aminopenicillanates (**1**) into 6-oxopenicillanates (**3**) involves ditrifluoromethanesulfonation to give a ditriflamide (**2**), which in the presence of base eliminates trifluoromethanesulfinic acid to form an imine (**a**). Hydrolysis of **a** to the ketone **3** is effected with dilute HCl.

H_2N–penam (CH_3, CH_3, COOR) (**1**) $\xrightarrow[\sim 100\%]{(CF_3SO_2)_2O,\ N(C_2H_5)_3}$ $(CF_3SO_2)_2N$–penam (CH_3, CH_3, COOR) (**2**) $\xrightarrow{\text{DBU},\ CH_2Cl_2,\ 0^\circ}$

[$CF_3SO_2N{=}$penam (CH_3, CH_3, COOR)] (**a**) $\xrightarrow[66\%\ \text{overall}]{H_3O^+}$ $O{=}$penam (CH_3, CH_3, COOR) (**3**)

The same process can be used to obtain 7-ketocephalosporanic acid derivatives (**4**), but in this series triethylamine is preferable to DBU as the base.

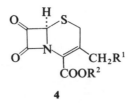

4

Aziridinones (α-lactams).[3] The hydroxamic acid (**1**) is converted in almost quantitative yield to the α-lactam **2** when treated with triflic anhydride and triethylamine in CH_2Cl_2 at − 70°. This cyclization of a hydroxamic acid was first encountered in a thebaine derivative.

$$(CH_3)_3CN(OH)-C(=O)-CH_2C_6H_5 \xrightarrow[97\%]{(CF_3SO)_2O,\ CH_2Cl_2,\ -70°} \mathbf{2}$$

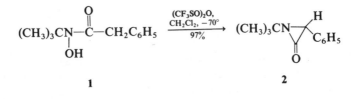

1 **2**

[1] J.-B. Falmagne, J. Escudero, S. Taleb-Sahraoui, and L. Ghosez, *Angew. Chem. Int. Ed.*, **20**, 879 (1981).
[2] D. Hagiwara, K. Sawada, T. Ohnami, M. Aratani, and M. Hashimoto, *J.C.S. Chem. Comm.*, 578 (1982).
[3] C. M. Bladon and G. W. Kirby, *ibid.*, 1402 (1982).

Trifluoroperacetic acid, 1, 821–827, **8**, 505; **10**, 421.

Sulfoxides and sulfones.[1] Sulfides are oxidized by this peracid to either sulfoxides or sulfones in high yield at 0° in TFA.

[1] C. G. Venier, T. G. Squires, Y.-Y. Chen, G. P. Hussmann, J. C. Shei, and B. F. Smith, *J. Org.*, **47**, 3773 (1982).

Triisobutylaluminum–Bis(N-methylsalicylaldimine)nickel. Ni(mesal)$_2$, m.p. 192–202°, is prepared by reaction of bis(salicylaldehyde)nickel(II) dihydrate with N-methylamine.[1]

Conjugate reductions of α,β-enones.[2] α, β-Enones react with triisobutylaluminum in pentane solution to give products of 1, 2-addition and 1, 2-reduction. The former

Examples:

$$(CH_3)_2C{=}CHCOCH_3 \xrightarrow{i\text{-}Bu_3Al,\ Ni(mesal)_2} (CH_3)_2CHCH_2COCH_3 \ (73\%) + (CH_3)_2C{=}CHCH(OH)CH_3 \ (20\%)$$

CH3 CH3 CH3 O → (i-Bu3Al, Ni(mesal)2, 93%) CH3 CH3 CH3 O

98% ↓ i-Bu3Al, (DIOP)NiCl2

CH3 CH3 CH3 OH

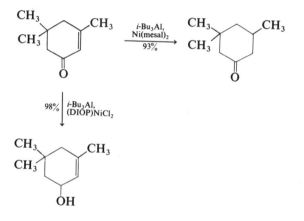

reaction is favored by an excess of the trialkylaluminum, whereas an excess of the enone favors 1,2-reduction. Addition of catalytic amounts of $Ni(mesal)_2$ favors conjugate reduction of α,β-enones. The hydrogen added to the β-position arises from the nickel species and the hydrogen added to the α-position arises from hydrolytic workup. In contrast, use of (−)-(DIOP)$NiCl_2$ as catalyst results in 1, 2-reduction to carbinols in nearly quantitative yields.

[1] L. Sacconi, P. Paoletti, and G. D. Re, *Am. Soc.*, **79**, 4062 (1957).
[2] A. M. Caporusso, G. Giacomelli, and L. Lardicci, *J. Org.*, **47**, 4640 (1982).

2,4,6-Triisopropylbenzenesulfonylhydrazine, 4, 535; **7**, 392; **9**, 486–488; **10**, 422–423.

Vinylphosphines.[1] A general route to vinylphosphines involves deprotonation of trisylhydrazones of ketones with *n*-butyllithium (or *sec*-butyllithium) followed by trapping with chlorodiphenylphosphine (equation I). Unsymmetrical trisylhydrazones are converted selectively into the less substituted vinylphosphine. The vinylphosphines are isolated conveniently as the corresponding oxides, prepared by oxidation with 30% H_2O_2 or 40% CH_3CO_3H.

$$\text{(I)}\quad R^1C(=N\text{–}NHSO_2Ar)CH_2R^2 \xrightarrow{\text{BuLi}} \left[R^1C(Li)=CHR^2\right] \xrightarrow{ClP(C_6H_5)_2} R^1C(P(C_6H_5)_2)=CHR^2 \xrightarrow[\text{65–78\% overall}]{[O]} R^1C(P(C_6H_5)_2\rightarrow O)=CHR^2$$

Allenic dianions.[2] The allenic dianions **3a** and **3b** can be generated via a modified Shapiro reaction from **1a** and **1b** by treatment with *n*-butyllithium in DME at $-78 \rightarrow 25°$.

$$R^1CH_2C(=N\text{–}NHSO_2Ar)CONHR^2 \xrightarrow[-78°]{n\text{-BuLi,}} R^1\bar{C}HC(=N\text{–}\bar{N}SO_2Ar)C(O^-)=NR^2\ \ 3Li^+ \xrightarrow{25°} R^1CH=C=C(\bar{O}Li^+)(\bar{N}R^2Li^+)$$

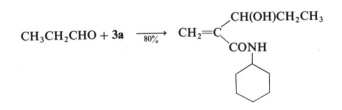

1a, $R^1=H$, $R^2=C_6H_{11}$-*c*
1b, $R^1=n$-Bu, $R^2=C_6H_{11}$-*c*

2 **3**

These carbanions react at the central carbon with aldehydes and ketones to give adducts containing an acrylamide group.

Example:

$$CH_3CH_2CHO + \mathbf{3a} \xrightarrow[80\%]{} CH_2=C(CH(OH)CH_2CH_3)(CONH\text{–}C_6H_{11})$$

Trisylamidrazones.[3] The trisylamidrazones **1** are obtained by reaction of acid chlorides with trisylhydrazine and then with PCl_5 and morpholine. On treatment

$$R^1CH_2COCl \xrightarrow{ArSO_2NHNH_2} R^1CH_2\overset{O}{\overset{\|}{C}}NHNHSO_2Ar \xrightarrow[\text{2) Morpholine}]{\text{1) PCl}_5} R^1CH_2\overset{N\sim NHSO_2Ar}{\overset{\|}{C}}-N(\text{morpholino}) \quad \mathbf{1}$$

$$\xrightarrow[-78\rightarrow 10^\circ]{t\text{-}C_4H_9Li,\ THF} \left[R^1CH{=}\overset{Li}{\overset{|}{C}}-N(\text{morpholino}) \right]_{\mathbf{a}} \xrightarrow{R^2I} R^1CH_2\overset{O}{\overset{\|}{C}}R^2$$

$$\mathbf{a} \xrightarrow{\text{cyclohexanone}} \text{1-hydroxy-1-}(COCH_2R^1)\text{cyclohexane}$$

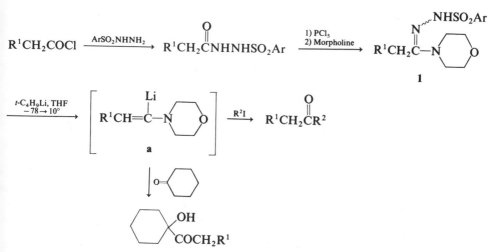

with t-butyllithium **1** undergoes the Shapiro reaction to form an α-lithioenamine **a**, which reacts as expected with various electrophiles, or even a bifunctional electrophile, as in the example.

Example:

$$C_6H_5CH_2COCl \longrightarrow C_6H_5CH{=}\overset{Li}{\overset{|}{C}}-N(\text{morpholino}) \xrightarrow[\substack{25\% \\ \text{overall}}]{\substack{\text{1) Br(CH}_2)_4\text{Br} \\ \text{2) NaI, }60^\circ}} \text{2-}C_6H_5\text{-cyclohexanone}$$

Alkyldiphenylphosphine oxides.[4] When the trisylhydrazones of aldehydes or ketones are refluxed in THF or dioxane with diphenylphosphine oxide, alkyldiphen-

$$\text{I)}\quad (C_6H_5)_2\overset{O}{\overset{\|}{P}}H + \begin{matrix}R^1\\ \\R^2\end{matrix}\!\!>C{=}NNH\overset{O}{\underset{O}{\overset{\|}{\underset{\|}{S}}}}C_6H_2(i\text{-}Pr)_3 \xrightarrow{THF,\ \Delta} \begin{matrix}R^1\\ \\R^2\end{matrix}\!\!>CH-\overset{O}{\overset{\|}{P}}(C_6H_5)_2 + N_2 + O_2$$

$$(20\text{–}90\%) \qquad + (i\text{-}Pr)_3C_6H_2SH$$

ylphosphine oxides are obtained in moderate to high yield (equation I). This new reaction is particularly successful with ketone trisylhydrazones containing a methyl group or an alkyl group with no branch at the α-position.

Aryl diazoalkanes (*cf.*, **8**, 460). Trisylhydrazones are generally superior to tosylhydrazones for the Bamford–Stevens reaction of aryl aldehydes or ketones to form aryl diazoalkanes. The products were generally identified as the corresponding 3,5-dinitrobenzoate esters. (equation I).[5]

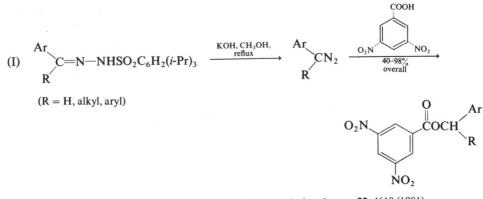

$$\text{(I)}\quad \text{Ar(R)C=N-NHSO}_2\text{C}_6\text{H}_2(i\text{-Pr})_3 \xrightarrow[\text{reflux}]{\text{KOH, CH}_3\text{OH,}} \text{Ar(R)CN}_2 \xrightarrow[\text{40-98\% overall}]{3,5\text{-(O}_2\text{N)}_2\text{C}_6\text{H}_3\text{COOH}} 3,5\text{-(O}_2\text{N)}_2\text{C}_6\text{H}_3\text{COOCH(Ar)R}$$

(R = H, alkyl, aryl)

[1] D. G. Mislanker, B. Mugrage, and S. D. Darling, *Tetrahedron Letters*, **22**, 4619 (1981).
[2] R. M. Adlington and A. G. M. Barrett, *J.C.S. Chem. Comm.*, 65 (1981); *idem*, *Tetrahedron*, **37**, 3935 (1981).
[3] J. E. Baldwin and J. C. Bottaro, *J.C.S. Chem. Comm.*, 1121 (1981).
[4] S. H. Bertz and G. Dabbagh, *Am. Soc.*, **103**, 5932 (1981).
[5] C. D. Dudman and C. B. Reese, *Synthesis*, 419 (1982).

3-Triisopropylsilylpropynyllithium, $[(CH_3)_2CH]_3SiC{\equiv}CCH_2Li$ (**1**). Mol. wt. 202.33, b.p. 100–101°/5 mm.

Preparation:

$$CH_3C{\equiv}CLi + [(CH_3)_2CH]_3SiOSO_2CF_3 \xrightarrow[87\%]{\text{ether, } -40\rightarrow 0^\circ} CH_3C{\equiv}CSi(i\text{-Pr})_3 \xrightarrow{n\text{-BuLi, THF}} \mathbf{1}.$$

Propargylation.[1] This reagent is generally superior to 3-trimethylsilylpropynyllithium (**2**, 239–241) for propargylation, since it has less tendency to form allenic products as well as the desired acetylenic products. Even aldehydes react in ether/HMPT to form only products of the type $RCH(OH)CH_2C{\equiv}CSiR_3$. The reagent also effects propargylation of oxiranes (equation I) and conjugate addition to α, β-enones (equation II).

$$\text{(I)}\quad \text{cyclohexene oxide} \xrightarrow{\mathbf{1}\text{, THF}} \text{trans-2-}(CH_2C{\equiv}CSi(i\text{-Pr})_3)\text{cyclohexanol (82\%)} + \text{allenic isomer (4.5\%)}$$

$$\text{(II)}\quad \text{cyclohexenone} \xrightarrow[81\%]{\mathbf{1}\text{, THF, HMPT}} \text{3-}(CH_2C{\equiv}CSi(i\text{-Pr})_3)\text{cyclohexanone}$$

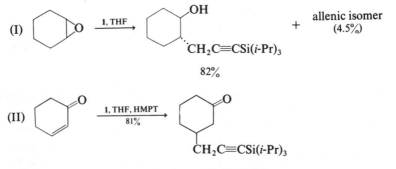

[1] E. J. Corey and C. Rücker, *Tetrahedron Letters*, **23**, 719 (1982).

Triisopropylsilyl trifluoromethanesulfonate (1). Mol. wt. 306.42, b.p. 83–87°/1.7 mm.
Preparation[1]:

$$3(CH_3)_2CHMgCl + HSiCl_3 \xrightarrow[80\%]{} [(CH_3)_2CH]_3SiH \xrightarrow[97\%]{CF_3SO_2OH} \underset{\mathbf{1}}{[(CH_3)_2CH]_3SiOSO_2CF_3}$$

This method provides a general route to trialkylsilyl triflates and has the advantage of circumventing use of the expensive silver triflate. Triisopropylsilyl triflate in combination with a base, preferably 2, 6-lutidine, is effective for silylation of alcohols.

[1] E. J. Corey, H. Cho, C. Rücker, and D. H. Hua, *Tetrahedron Letters*, **22**, 3455 (1981).

Trimethylacetic formic anhydride, $(CH_3)_3CCO_2CHO$ **(1)**. The anhydride is prepared in high yield by reaction of trimethylacetyl chloride and sodium formate (1.05 equiv.) in the presence of 18-crown-6 (10 mol %) at 0°; the yield of slightly impure material is 95–98%. The material can be stored in a freezer for several months.

Formylation.[1] The anhydride formylates even unstable amines in an alcohol-free solvent in 80–100% yield. In no cases are even traces of trimethylacetamides detectable. Substantial amounts of acetamides can be formed when acetic formic anhydride is used.

[1] E. J. Vlietstra, J. W. Zwikker, R. J. M. Nolte, and W. Drenth, *Rec. Trav. Chim.*, **101**, 460 (1982).

Trimethylaluminum–Dichlorobis(cyclopentadienyl)zirconium, 8, 506; **10**, 423.

Alkenylmercuric chlorides.[1] The reaction of alkenylalanes, available by Zr-catalyzed carboalumination of alkynes, with mercuric chloride provides a convenient synthesis of (E)-alkenylmercuric chlorides.

Example:

$$CH_3(CH_2)_5C{\equiv}CH \xrightarrow{(CH_3)_3Al,\ Cl_2ZrCp_2,\ CH_2Cl_2} \left[(CH_3(CH_2)_5)(CH_3)C{=}C(H)(Al(CH_3)_2) \right] \xrightarrow[82\%]{HgCl_2,\ THF} (CH_3(CH_2)_5)(CH_3)C{=}C(H)(HgCl)$$

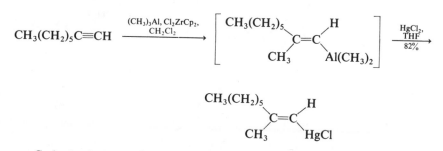

Carboalumination of propargyl and homopropargyl systems. The high *syn*-stereoselectivity of the carboalumination of alkynes with $Al(CH_3)_3$–Cl_2ZrCp_2 is retained with propargyl and homopropargyl alcohols, sulfides, and iodides. This high regioselectivity is in contrast with the Al–Ti system. The products obtained by this reaction are useful for preparation of bifunctional isoprene units.[2]

Example:

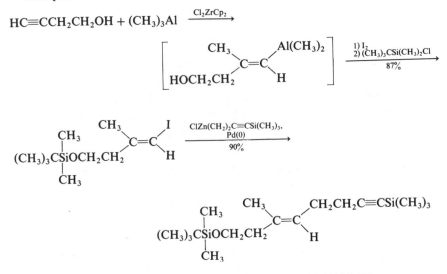

$$HC{\equiv}CCH_2CH_2OH + (CH_3)_3Al \xrightarrow{Cl_2ZrCp_2} \left[(HOCH_2CH_2)(CH_3)C{=}C(H)(Al(CH_3)_2) \right] \xrightarrow[87\%]{1)\ I_2;\ 2)\ (CH_3)_3CSi(CH_3)_2Cl}$$

$$(CH_3)_3CSi(CH_3)_2OCH_2CH_2(CH_3)C{=}C(H)I \xrightarrow[90\%]{ClZn(CH_2)_2C{\equiv}CSi(CH_3)_3,\ Pd(0)}$$

$$(CH_3)_3CSi(CH_3)_2OCH_2CH_2(CH_3)C{=}C(H)CH_2CH_2C{\equiv}CSi(CH_3)_3$$

[1] E. Negishi, K. P. Jadhav, and N. Daotien, *Tetrahedron Letters*, **23**, 2085 (1982).
[2] C. L. Rand, D. E. Van Horn, M. W. Moore, and E. Negishi, *J. Org.*, **46**, 4093 (1981).

Trimethyl orthoformate, 3, 313; **4**, 540; **5**, 714; **8**, 507–508; **10**, 425–426.

α-Dimethoxymethylation of ketones.[1] Preformed lithium enolates react regiospecifically with trimethyl orthoformate in the presence of BF_3 to give α-dimethoxymethyl ketones in moderate yield.

Examples:

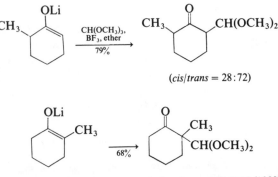

$$\text{6-methyl-1-cyclohexenyl-OLi} \xrightarrow[79\%]{CH(OCH_3)_3,\ BF_3,\ ether} \text{2-methyl-6-}CH(OCH_3)_2\text{-cyclohexanone}$$

(*cis*/*trans* = 28:72)

$$\text{2-methyl-1-cyclohexenyl-OLi} \xrightarrow[68\%]{} \text{2-methyl-2-}CH(OCH_3)_2\text{-cyclohexanone}$$

[1] S. Suzuki, A. Yanagisawa, and R. Noyori, *Tetrahedron Letters*, **23**, 3595 (1982).

N,N,P-Trimethyl-P-phenylphosphinothioic amide,

$C_6H_5P(=S)(CH_3)N(CH_3)_2$ (**1**). Mol. wt. 199.25, b.p. 110°/0.1 mm.

Preparation:

$$C_6H_5P(=S)Cl_2 \xrightarrow[72\%]{CH_3MgBr,\ ether} C_6H_5P(=S)(CH_3)\text{—}P(=S)(CH_3)C_6H_5 \xrightarrow[85\%]{SO_2Cl_2} C_6H_5P(=S)(CH_3)Cl \xrightarrow[98\%]{HN(CH_3)_2,\ N(C_2H_5)_3,\ THF} \mathbf{1}$$

Alkylidenation.[1] Methylenation of ketones can be effected in 50–99% yield by reaction with the lithium anion (**2**) of **1** with $O{=}CR^1R^2$ to give a stable adduct (**3**) that undergoes cycloelimination to alkenes (**4**) on methylation. When applied to aldehydes, the yield is low except for benzaldehydes or α,β-unsaturated enals.

$$\underset{\mathbf{2}}{C_6H_5P(=S)(CH_2Li)N(CH_3)_2} \xrightarrow{R^1R^2C{=}O} \underset{\mathbf{3}}{C_6H_5P(=S)(N(CH_3)_2)CH_2\text{—}C(OH)R^1R^2} \xrightarrow{CH_3I,\ Py} \left[C_6H_5P^+(SCH_3)(N(CH_3)_2)CH_2\text{—}C(O^-)R^1R^2\right]$$

$$\xrightarrow[\text{overall}]{50\text{–}99\%} \underset{\mathbf{4}}{CH_2{=}CR^1R^2} + C_6H_5P(=O)(N(CH_3)_2)\text{—}SCH_3$$

Alkylation of **2** yields homologs of **1**, which can be used to prepare tri- and tetrasubstituted alkenes.

Example:

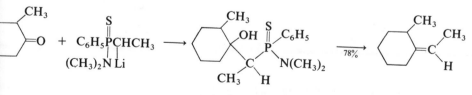

$$\text{2-methylcyclohexanone} + C_6H_5P(=S)(N(CH_3)_2)CHCH_3Li \longrightarrow \underset{\text{(major adduct, 86\%)}}{\text{1-[CH(CH}_3\text{)P(=S)(C}_6\text{H}_5\text{)N(CH}_3\text{)}_2\text{]-2-methylcyclohexan-1-ol}} \xrightarrow{78\%} \text{2-methyl-1-ethylidenecyclohexane } (C{=}C(CH_3)H)$$

Ketone methylenation can be coupled with concomitant optical resolution effected by chromatography by use of the related reagent **5**. The adduct with a prochiral ketone has three asymmetric centers.[2]

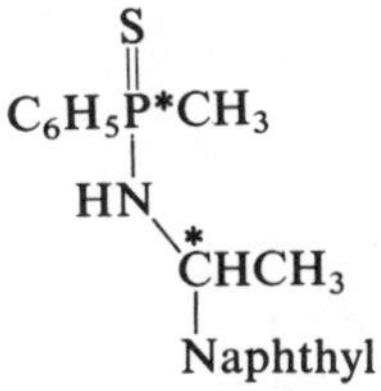

$C_6H_5P^*(=S)CH_3$—NH—$\overset{*}{C}HCH_3$—Naphthyl

5

[1] C. R. Johnson and R. C. Elliott, *Am. Soc.*, **104**, 7041 (1982).
[2] C. R. Johnson, R. C. Elliott, and N. A. Meanwell, *Tetrahedron Letters*, 5005 (1982).

Trimethyl phosphite, 1, 1233–1235; **2**, 439–441; **3**, 315–316; **4**, 591–592; **5**, 717; **7**, 393–399; **9**, 490–491.

>C=S → >C=O. The thionolactone **1** is converted in almost quantitative yield into the corresponding lactone without cleavage of the lactone ring when heated with trimethyl phosphite at 100–105° or with iron carbonyl in toluene at 100°.[1]

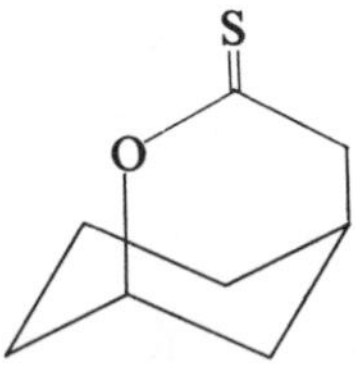

1

Allyl sulfoxide → sulfenate rearrangement.[2] A key step in a stereoselective first synthesis of withaferin A (**3**), a steroid antitumor agent, is the production of the desired A/B ring system by an allyl sulfoxide → sulfenate rearrangement in the presence of trimethyl phosphite (equation I).

(I) **1** ($O{=}SC_6H_5$) $\xrightarrow[52\%]{P(OCH_3)_3,\ 25°,\ CH_3OH\text{–}THF}$ **2** (OH) $\xrightarrow{ClC_6H_4CO_3H}$ **3** (HO)

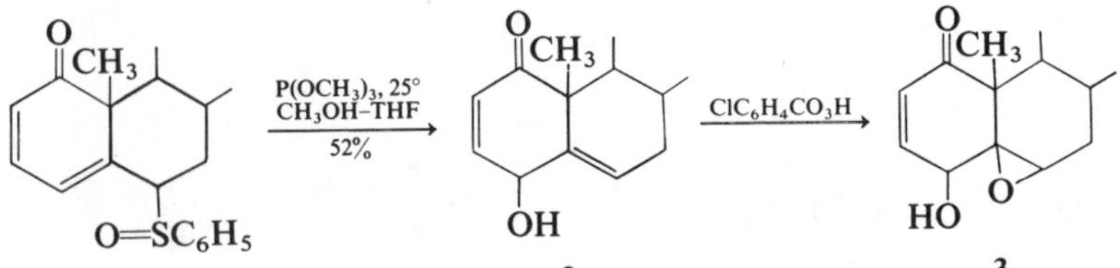

[1] S. Ayral-Kaloustian and W. C. Agosta, *Syn. Comm.*, **11**, 1011 (1981).
[2] M. Hirayama, K. Gamoh, and N. Ikekawa, *Tetrahedron Letters*, **23**, 4725 (1982).

2,4,6-Trimethylpyrylium sulfoacetate (1). Mol. wt. 259.26, m.p. 115–116°, soluble in protic solvents, insoluble in ether.

$$CH_2{=}C(CH_3)_2 + 3(CH_3CO)_2O + HO_3SCH_2COOH \xrightarrow[43\text{–}47\%]{-4CH_3COOH}$$

CH_3, CH_3, CH_3, $\overset{+}{O}$, $^{-}O_3SCH_2COOH$

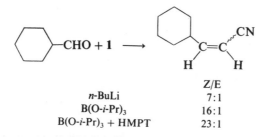

1

Pyrylium salts are useful precursors to other heterocyclic compounds such as furanes and pyridines. The perchlorates usually used are explosive, and tetrafluoroborates are formed only in low yield. Sulfoacetates are nonexplosive; if necessary for further synthetic uses, they can be converted into perchlorates.[1]

[1] A. Dinculescu and A. T. Balaban, *Org. Prep. Proc. Int.*, **14**, 39 (1982).

Trimethylsilylacetonitrile, $(CH_3)_3SiCH_2CN$ **(1)**. Mol. wt. 113.24, b.p. 65–70°/20 mm. The reagent is readily obtained by reaction of chlorotrimethylsilane with bromoacetonitrile in the presence of activated zinc (80% yield).[1]

(Z)-α,β-Unsaturated nitriles.[2] The lithio derivative of **1** reacts with aldehydes to form a mixture of (Z)- and (E)-α,β-unsaturated nitriles with some preference for the former isomer. The stereoselectivity in favor of the (Z)-isomer in reactions with aliphatic aldehydes is markedly improved by use of the boron derivative, obtained by treatment of the lithium reagent with boron isopropoxide. Addition of HMPT also favors the (Z)-isomer.

Example:

$$C_6H_{11}CHO + \mathbf{1} \longrightarrow C_6H_{11}CH{=}CHCN$$

	Z/E
n-BuLi	7:1
$B(O\text{-}i\text{-}Pr)_3$	16:1
$B(O\text{-}i\text{-}Pr)_3$ + HMPT	23:1

[1] I. Matsuda, S. Murata, and Y. Ishii, *J.C.S. Perkin I*, 26 (1979).

[2] R. Haruta, M. Ishiguro, K. Furuta, A. Mori, N. Ikeda, and H. Yamamoto, *Chem. Letters*, 1093 (1982).

(Trimethylsilyl)acetyltrimethylsilane (1). Mol. wt. 188.41, b.p. 50°/3 mm.

Preparation[1]:

$$(CH_3)_3SiC{\equiv}CSi(CH_3)_3 \xrightarrow{BH_3 \cdot S(CH_3)_2} \underset{H}{(CH_3)_3Si}\!\!>\!C{=}C\!<\!\!\overset{Si(CH_3)_3}{\underset{3}{\,}B} \xrightarrow[\text{75\% overall}]{(CH_3)_3NO} (CH_3)_3SiCH_2\overset{O}{\overset{\|}{C}}Si(CH_3)_3$$

1

(E)-α,β-Unsaturated carboxylic acids.[2] A new synthesis of these acids from **1** involves deprotonation–alkylation–aldolization (equation I). One limitation is that the alkylation is limited to methyl or ethyl iodide and allylic or benzylic bromides. The nearly exclusive formation of (E)-**3** is a result of the stereoselectivity of the aldol reaction of **2**.

$$\mathbf{1} \xrightarrow[2)\,R^1X]{1)\,LDA} \left[(CH_3)_3Si\underset{R^1}{\underset{|}{C}}H\overset{O}{\overset{\|}{C}}Si(CH_3)_3\right]_{\mathbf{2}} \xrightarrow[70\text{–}90\%]{1)\,LDA;\ 2)\,R^2CHO} \underset{R^2}{\overset{H}{>}}C{=}C\underset{R^1}{\overset{\overset{O}{\|}{C}Si(CH_3)_3}{<}}\ (\mathbf{3}) \xrightarrow[>90\%]{H_2O_2,\,OH^-} \underset{R^2}{\overset{H}{>}}C{=}C\underset{R^1}{\overset{\overset{O}{\|}{C}OH}{<}}\ (\mathbf{4})$$

[1] J. A. Miller and G. Zweifel, *Synthesis*, 288 (1981).

[2] *Idem, Am. Soc.*, **103**, 6217 (1981).

Trimethylsilylallyllithium, 8, 273–274; **10,** 430.

(Z)- and (E)-Terminal* 1,3-*dienes.[1] The pinacol boronate (**1**) derived from trimethylsilylallyllithium provides a stereoselective route to (Z)- and (E)-terminal 1,3-dienes. Deoxysilylation of **2** can be effected to give either **3** or **4** with less than 3% of the other isomer.

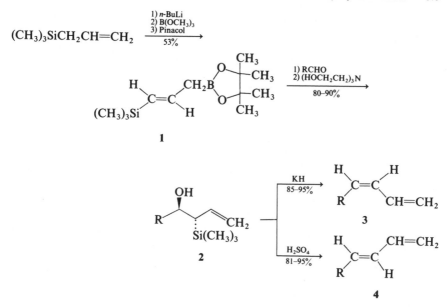

$$(CH_3)_3SiCH_2CH{=}CH_2 \xrightarrow[53\%]{1)\ n\text{-BuLi};\ 2)\ B(OCH_3)_3;\ 3)\ \text{Pinacol}}$$

(E)-$(CH_3)_3SiCH{=}CHCH_2B$(pinacolato) **1** $\xrightarrow[80\text{–}90\%]{1)\ RCHO;\ 2)\ (HOCH_2CH_2)_3N}$ $RCH(OH)CH(Si(CH_3)_3)CH{=}CH_2$ **2**

2 $\xrightarrow[85\text{–}95\%]{KH}$ (Z)-$RCH{=}CHCH{=}CH_2$ **3**

2 $\xrightarrow[81\text{–}95\%]{H_2SO_4}$ (E)-$RCH{=}CHCH{=}CH_2$ **4**

[1] D. J. S. Tsai and D. S. Matteson, *Tetrahedron Letters*, **22**, 2751 (1981).

Trimethylsilyldiazomethane, 10, 431. An efficient synthesis involves diazo transfer of $(CH_3)_3SiCH_2MgCl$ with diphenyl phosphoroazidate (55–58% yield).[1]

Tetrazoles.[2] Lithiotrimethylsilyldiazomethane (**1**), prepared with LDA or *n*-BuLi, reacts with methyl esters to give 2-substituted 5-trimethylsilyltetrazoles in 50–90% yield.

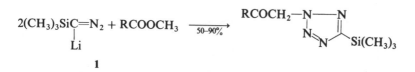

$$2(CH_3)_3SiC(Li){=}N_2\ (\mathbf{1}) + RCOOCH_3 \xrightarrow{50\text{–}90\%} \text{RCOCH}_2\text{-tetrazole-Si(CH}_3)_3$$

1,2,3-*Triazoles.*[3] Lithiotrimethylsilyldiazomethane (*n*-BuLi or LDA) reacts with a variety of nitriles to give 5-trimethylsilyl-1,2,3-triazoles in high yield (equation I).

$$\text{(I)}\quad (CH_3)_3SiC(Li){=}N_2 + RCN \longrightarrow \left[\text{4-R-5-Li-5-(CH}_3)_3\text{Si-triazoline}\right] \xrightarrow[75\text{–}95\%]{H_2O} \text{4-R-5-(CH}_3)_3\text{Si-1}H\text{-1,2,3-triazole}$$

Homologation of ketones.[4] Ketones undergo homologation when treated with trimethylsilyldiazomethane in the presence of BF_3 etherate. Yields are generally superior to those obtained with diazomethane. Another advantage is that the methylene group is inserted selectively from the less hindered side of the ketone, probably because of the bulky silyl group.

[1] S. Mori, I. Sakai, T. Aoyama, and T. Shioiri, *Chem. Pharm. Bull.*, **30**, 3380 (1982).
[2] T. Aoyama and T. Shioiri, *ibid.*, **30**, 3450 (1982).
[3] T. Aoyoma. K. Sudo, and T. Shioiri, *ibid*, **30**, 3849 (1982).
[4] N. Hashimoto, T. Aoyama, and T. Shioiri, *ibid*, **30**, 119 (1982).

2-Trimethylsilylethanol, 8, 510–511.

The reagent can be prepared in high yield by reaction of the Grignard derivative of the inexpensive chloromethyl(trimethyl)silane (Petrach Systems) with paraformaldehyde (*ca.* 95% yield).[1]

Protection of anomeric centers.[2] The hemiacetal group of carbohydrates is protected traditionally as an alkyl glycoside. An alternative is a β-trimethylsilylethyl glycoside, which can be prepared in satisfactory yield by oxymercuration ($Hg(OAc)_2$–$HOCH_2CH_2Si(CH_3)_3$, Koenig-Knorr reaction, or Fischer glycosidation. They are cleaved in > 80% yield by lithium tetrafluoroborate (**10**, 248) in dry CH_3CN (70°, 3–8 hours). They are also cleaved by $LiF/BF_3 \cdot (C_2H_5)_2O$ (1:1). Presumably, $LiBF_4$ dissociates into Li^+, F^-, and BF_3.

[1] M. L. Mancini and J. F. Honek, *Tetrahedron Letters*, **23**, 3249 (1982).
[2] B. H. Lipshutz, J. J. Pegram, and M. C. Morey, *Tetrahedron Letters*, **22**, 4603 (1981).

β-[(Trimethylsilyl)ethyl] lithium, $(CH_3)_3SiCH_2CH_2Li$ (**1**). The reagent is generated *in situ* from the corresponding bromide with *t*-butyllithium at – 78°. It decomposes on standing overnight at 25°. The corresponding Grignard reagent is stable at 0° for weeks.

Reductive vinylation of carbonyl compounds.[1] The reagent **1** or the corresponding Grignard reagent adds to aldehydes and ketones to give an alcohol, which on treatment with $BF_3 \cdot HOAc$ undergoes dehydration and protodesilylation to give a vinyl compound.

Examples:

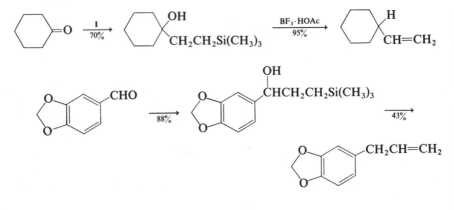

Cyclohexanone $\xrightarrow[70\%]{1}$ 1-(OH)-1-($CH_2CH_2Si(CH_3)_3$)cyclohexane $\xrightarrow[95\%]{BF_3 \cdot HOAc}$ cyclohexyl($CH{=}CH_2$) (H)

Piperonal (CHO) $\xrightarrow{88\%}$ Ar–CH(OH)$CH_2CH_2Si(CH_3)_3$ $\xrightarrow{43\%}$ Ar–$CH_2CH{=}CH_2$

The Wittig reagent $(C_6H_5)_3P{=}CHCH_2Si(CH_3)_3$ (**9**, 492) has also been used for this transformation, but this reagent does not react satisfactorily with most ketones.

[1] S. R. Wilson and A. Shedrinsky, *J. Org.*, **47**, 1983 (1982).

N-(Trimethylsilyl)imidazole, 7, 399–400.

3-Alkyl-2-cycloalkene-1-ones. The silyl enol ethers (**1**) of 1,3-cycloalkanediones, readily obtained using this imidazole, react with either methyl- or *n*-butyllithium in ether to form 3-alkyl-2-cycloalkene-1-ones (**2**). The yield can be improved and the reaction time can be shorted by addition of HMPT as catalyst.[1]

Example:

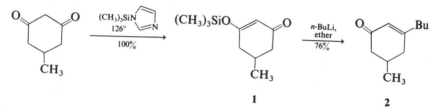

5-Methyl-1,3-cyclohexanedione (CH₃ substituent) $\xrightarrow[100\%]{(CH_3)_3SiN\text{(imidazole)},\ 126°}$ $(CH_3)_3SiO$–cyclohexenone with CH_3 (**1**) $\xrightarrow[76\%]{n\text{-BuLi, ether}}$ 3-Bu-cyclohexenone with CH_3 (**2**)

[1] Y. F. Zhou and N. Z. Huang, *Syn. Comm.*, **12**, 795 (1982).

Trimethylsilyllithium, 7, 400; **9**, 493.

Allylic silanes.[1] Trimethylsilyllithium reacts with isoprenyl chloride in THF ($-78° \rightarrow 25°$) to give isoprenyltrimethylsilane with high selectivity in 82% yield. This method is superior with respect to yield and selectivity to the reaction of allylmagnesium halides with chlorotrimethylsilane (**7**, 9).

The paper describes two other useful routes to allylsilanes, outlined in equations (I) and (II).

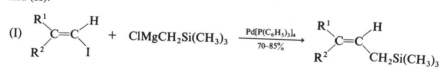

$$\text{(I)}\quad R^1R^2C{=}CHI + ClMgCH_2Si(CH_3)_3 \xrightarrow[70\text{–}85\%]{Pd[P(C_6H_5)_3]_4} R^1R^2C{=}CHCH_2Si(CH_3)_3$$

$$\text{(II)}\quad HC{\equiv}CCH_2Si(CH_3)_3 \xrightarrow{(CH_3)_3Al,\ Cl_2ZrCp_2} \left[(CH_3)_2Al(H)C{=}C(CH_3)CH_2Si(CH_3)_3\right] \xrightarrow[63\%]{I_2}$$

$$I(H)C{=}C(CH_3)CH_2Si(CH_3)_3 \xrightarrow[78\%]{n\text{-BuLi}} n\text{-Bu}(H)C{=}C(CH_3)CH_2Si(CH_3)_3$$

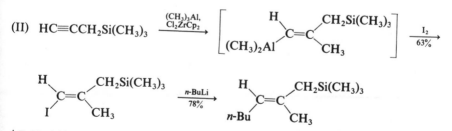

[1] E. Negishi, F.-T. Luo, and C. L. Rand, *Tetrahedron Letters*, **23**, 27 (1982).

Trimethylsilylmethanethiol, $(CH_3)_3SiCH_2SH$ **(1)**. Mol. wt. 120.29, b.p. 115°/749 mm. The thiol is prepared in two steps from bromomethyltrimethylsilane in about 20% yield.[1]

β,γ-Unsaturated aldehydes.[2] A new synthesis of β,γ-unsaturated aldehydes involves preparation of an allylic silylmethylsulfonium salt such as **4**. The derived ylide (**a**) rearranges readily to a homoallylic α-methylthiosilane (**5**). This product undergoes a sila-Pummerer reaction on oxidation to give an O-trimethylsilylhemiacetal (**6**), which is hydrolyzed to the corresponding aldehyde (**7**) by 1 equiv. of oxalic acid at 20°.

CH_2Cl **2** —(1, $NaOCH_3$)→ **3** ($CH_2Si(CH_3)_3$, S) —(CH_3OSO_2F)→ **4** ($CH_2Si(CH_3)_3$, S^+, CH_3, $\bar{O}SO_2F$)

—(n-BuLi, −78°)→ [$\bar{C}HSi(CH_3)_3$, S^+, CH_3] **a** —(−20°, 60–75%)→ **5** (CH, $Si(CH_3)_3$, SCH_3, CH_2) —($ClC_6H_4CO_3H$)→

[CH, $Si(CH_3)_3$, SCH_3 ↓ O, CH_2] **b** —(40–60%)→ **6** (CH, $OSi(CH_3)_3$, SCH_3, CH_2) —(H^+)→ **7** (CHO, CH_2)

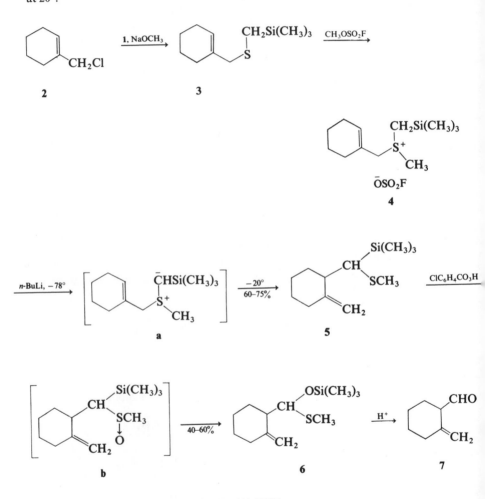

[1] D. C. Noller and H. W. Post, *J. Org.*, **17**, 1393 (1952).
[2] P. J. Kocienski, *J.C.S. Chem. Comm.*, 1096 (1980).

1-Trimethylsilyl-1-methoxyallene (1).

Preparation:

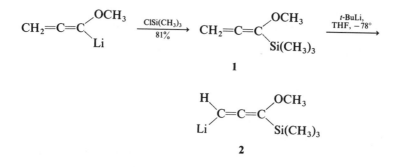

$$CH_2{=}C{=}C(OCH_3)Li \xrightarrow[81\%]{ClSi(CH_3)_3} CH_2{=}C{=}C(OCH_3)Si(CH_3)_3 \;(\mathbf{1}) \xrightarrow{t\text{-BuLi, THF, }-78^\circ} (H)(Li)C{=}C{=}C(OCH_3)Si(CH_3)_3 \;(\mathbf{2})$$

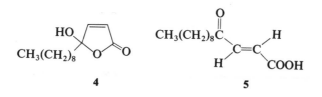

2-*Trimethylsilylfuranes.*[1] The reagent **2** reacts with aliphatic aldehydes to give an adduct that is converted to a 2-trimethylsilylfurane on mild acid treatment.

Example:

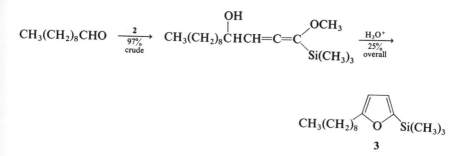

$$CH_3(CH_2)_8CHO \xrightarrow[97\%\text{ crude}]{\mathbf{2}} CH_3(CH_2)_8CH(OH)CH{=}C{=}C(OCH_3)Si(CH_3)_3 \xrightarrow[25\%\text{ overall}]{H_3O^+} \mathbf{3}$$

3: 2-$CH_3(CH_2)_8$-5-$Si(CH_3)_3$-furan

These furanes undergo ready electrophilic substitution. Thus **3** is converted by singlet oxygen into the γ-hydroxybutenolide **4** (90% yield). Attempted distillation of **3** results in the *trans*-γ-keto -α,β-unsaturated acid **5**.

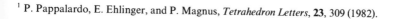

4: 5-HO-5-$CH_3(CH_2)_8$-furan-2(5H)-one

5: $CH_3(CH_2)_8C(=O)CH{=}CHCOOH$ (*trans*)

[1] P. Pappalardo, E. Ehlinger, and P. Magnus, *Tetrahedron Letters*, **23**, 309 (1982).

2-(Trimethylsilylmethyl)-3-acetoxy-1-propene.

$(CH_3)_3SiCH_2C(=CH_2)CH_2OAc$ (**1**), **9**, 454–455. Mol. wt. 186.32, b.p. 95°/mm.

Preparation:

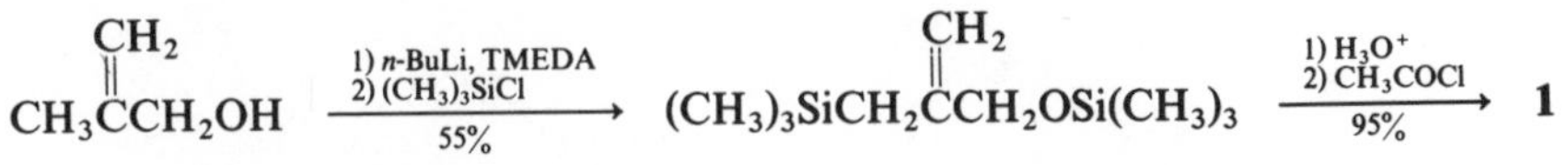

$$CH_3C(=CH_2)CH_2OH \xrightarrow[55\%]{1)\ n\text{-BuLi, TMEDA};\ 2)\ (CH_3)_3SiCl} (CH_3)_3SiCH_2C(=CH_2)CH_2OSi(CH_3)_3 \xrightarrow[95\%]{1)\ H_3O^+;\ 2)\ CH_3COCl} \mathbf{1}$$

Cyclopentene annelation by* [3 + 2] *cycloaddition.[1] A key step in a synthesis of the alkene **6** is the cycloaddition of **1** to 2,3-dicarbomethoxynorbornene (**2**) catalyzed by tetrakis(triisopropyl phosphite)palladium(0). The product (**3**) was converted to the

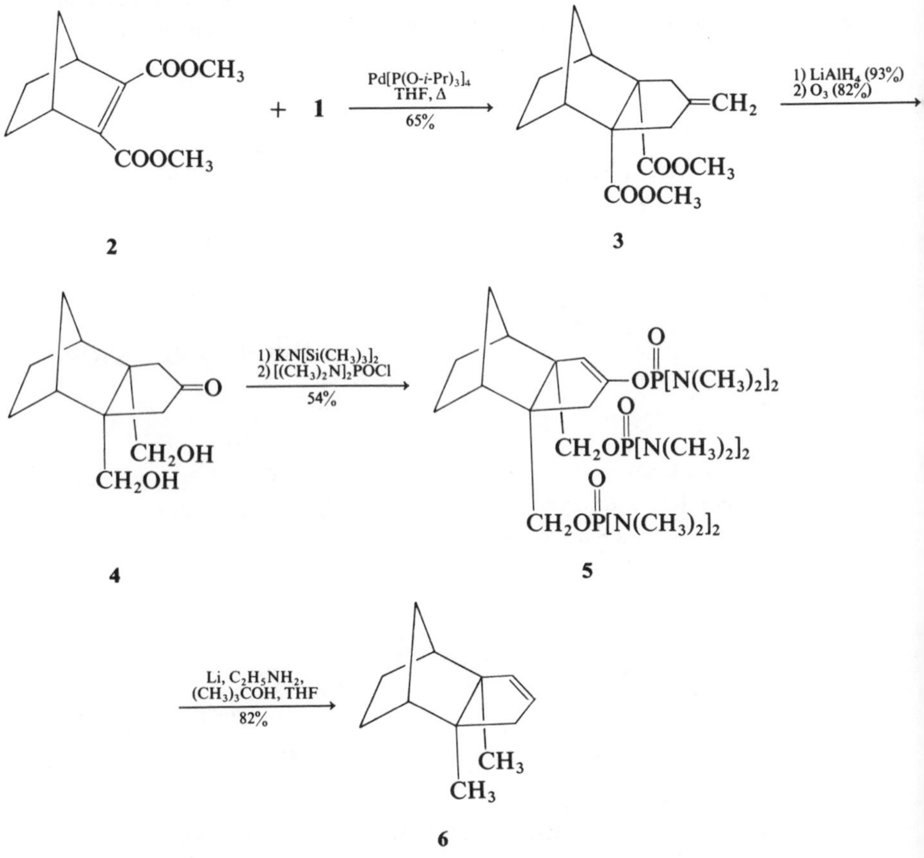

2 (COOCH₃, COOCH₃) + **1** $\xrightarrow[65\%]{Pd[P(O\text{-}i\text{-}Pr)_3]_4,\ THF,\ \Delta}$ **3** (=CH₂; COOCH₃, COOCH₃) $\xrightarrow{1)\ LiAlH_4\ (93\%);\ 2)\ O_3\ (82\%)}$

4 (=O; CH₂OH, CH₂OH) $\xrightarrow[54\%]{1)\ KN[Si(CH_3)_3]_2;\ 2)\ [(CH_3)_2N]_2POCl}$ **5** ($OP(O)[N(CH_3)_2]_2$; $CH_2OP(O)[N(CH_3)_2]_2$, $CH_2OP(O)[N(CH_3)_2]_2$)

$\xrightarrow[82\%]{Li,\ C_2H_5NH_2,\ (CH_3)_3COH,\ THF}$ **6** (CH_3, CH_3)

keto diol **4**. Deoxygenation of all three groups is effected by lithium–ethylamine reduction of the corresponding phosphoramidates. This reaction should be useful for conversion of neopentyl-type esters to methyl groups.

[1] B. M. Trost and P. Renaut, *Am. Soc.*, **104**, 6668 (1982).

2-(Trimethylsilyl)methylallyl iodide (1).

Preparation[1]:

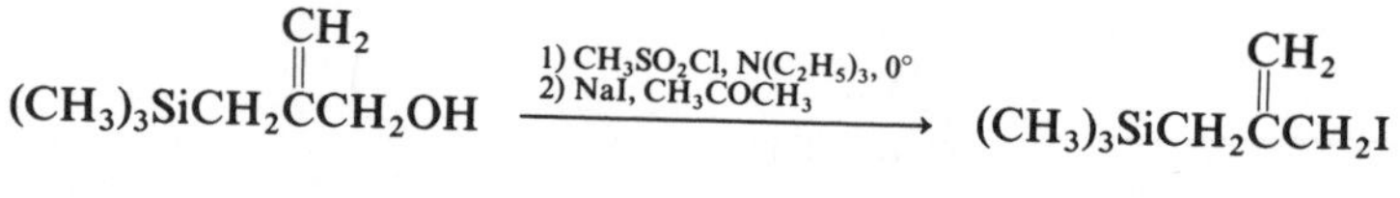

$$(CH_3)_3SiCH_2\overset{\overset{\displaystyle CH_2}{\|}}{C}CH_2OH \xrightarrow[\text{2) NaI, } CH_3COCH_3]{\text{1) } CH_3SO_2Cl,\ N(C_2H_5)_3,\ 0°} (CH_3)_3SiCH_2\overset{\overset{\displaystyle CH_2}{\|}}{C}CH_2I$$

1

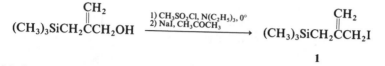

Methylenecyclopentanes.[2] Trost and Curran have used **1** as an electrophilic synthon for trimethylenemethane. Thus **1** reacts with the anion of **2** to give **3** in 72% yield. This product was oxidized to the disulfone, which was then cyclized to **4** by F^-. Various transformations are possible for the exocyclic methylene group. For the synthesis of coriolin (**6**), the group was converted to a *gem*-dimethyl group by cyclopropanation and hydrogenolysis. The product (**5**) was converted in several more steps into **6**.

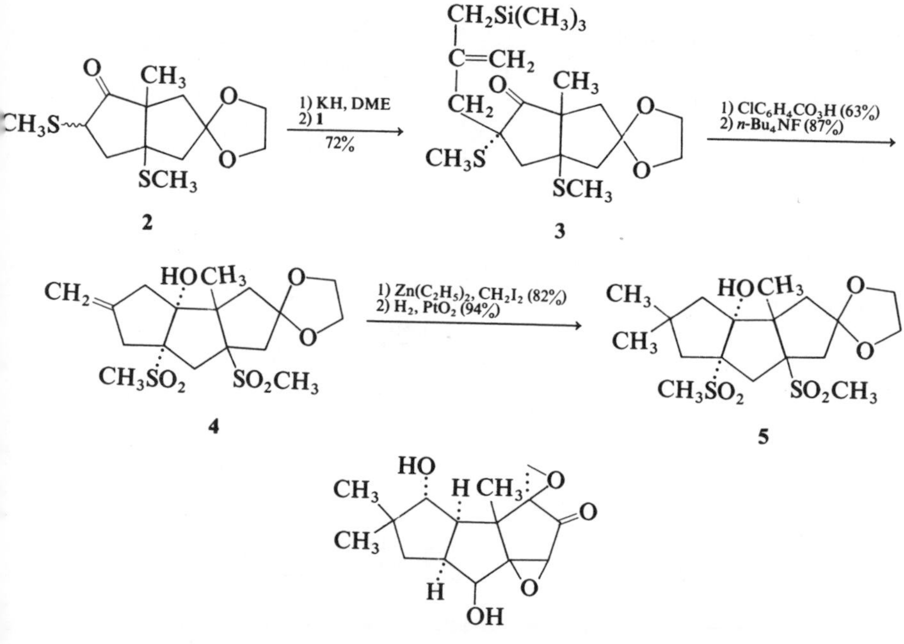

2 $\xrightarrow[\text{72\%}]{\text{1) KH, DME; 2) 1}}$ **3** $\xrightarrow{\text{1) } ClC_6H_4CO_3H\ (63\%)\text{; 2) } n\text{-}Bu_4NF\ (87\%)}$ **4** $\xrightarrow{\text{1) } Zn(C_2H_5)_2,\ CH_2I_2\ (82\%)\text{; 2) } H_2,\ PtO_2\ (94\%)}$ **5**

6

[1] B. M. Trost and J. E. Vincent, *Am. Soc.*, **102**, 5680 (1980).
[2] B. M. Trost and D. P. Curran, *ibid*, **103**, 7380 (1981).

2-Trimethylsilylmethyl-1,3-butadiene, 9, 493–494; **10**, 432–433.

Mono- and bicyclic terpenes.[1] This isoprenylsilane is useful for synthesis of mono- and bicyclic terpenes. It reacts with a variety of dienophiles under $AlCl_3$ catalysis to give mainly the so-called "para" adducts with the desired isoprenoid skeletons.

Examples:

$(CH_3)_3SiCH_2C(=CH_2)CH=CH_2$ + $CH_2=CHCOCH_3$ $\xrightarrow[64\%]{AlCl_3,\ 15-20°}$ 4-acetyl-1-(trimethylsilylmethyl)cyclohexene (COCH$_3$) $\xrightarrow[72\%]{(CH_3)_3SiCH_2MgCl}$ 4-($CCH_3=CH_2$)-1-(trimethylsilylmethyl)cyclohexene $\xrightarrow[46\%]{CsF,\ DMSO}$ 1-methyl-4-isopropenylcyclohexene (CH_3, $C(CH_3)=CH_2$)

$(CH_3)_3SiCH_2C(=CH_2)CH=CH_2$ + $CH_2=CHCOCH_2CH_2CH=C(CH_3)_2$ $\xrightarrow{85\%}$ $(CH_3)_3SiCH_2$-cyclohexenyl–$C(=O)CH_2CH_2CH=C(CH_3)_2$ $\xrightarrow{1)\ CsF,\ DMSO\ (75\%)\quad 2)\ (CH_3)_3SiCH_2MgCl\ (73\%)}$ 4-methylcyclohexenyl–$C(=CH_2)CH_2CH_2CH=C(CH_3)_2$

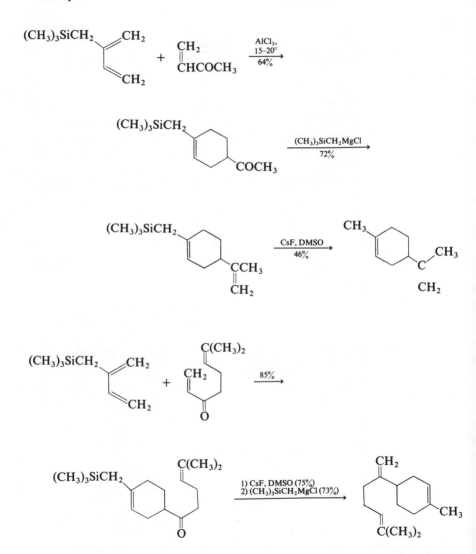

[1] A. Hosomi, H. Iguchi, J. Sasaki, and H. Sakurai, *Tetrahedron Letters*, **22**, 551 (1982).

Trimethylsilylmethyllithium, 6, 635–636; **9**, 495–496; **10**, 433.

The reagent (**1**) can be generated conveniently *in situ* by transmetallation of (trimethylsilylmethyl)tri-*n*-butyltin with *n*-butyllithium in THF/hexane at 0° (equation I).[1]

$$\text{(I)}\quad (CH_3)_3SiCH_2Cl + (n\text{-}Bu)_3SnLi \xrightarrow[92\%]{\text{THF, hexane}} (CH_3)_3SiCH_2Sn(n\text{-}Bu)_3$$

$$\xrightarrow{n\text{-}BuLi} \underset{\mathbf{1}}{(CH_3)_3SiCH_2Li} + n\text{-}Bu_4Sn$$

The *in situ* procedure can be used for methylenation of aldehydes and ketones (Peterson reaction) in comparable yields to those previously reported (**10**, 433), and for conversion of carboxylic acid esters or chlorides into α-trimethylsilyl ketones (equation II).[2]

$$\text{(II)}\quad C_6H_{11}\text{-}COCl + \mathbf{1} \xrightarrow[88\%]{\text{THF, }-78°} C_6H_{11}\text{-}\overset{O}{\overset{\|}{C}}CH_2Si(CH_3)_3$$

Allyl phenyl selenides.[3] α-Phenylseleno ketones react with trimethylsilylmethyllithium to afford an adduct that on treatment with a catalytic amount of $SnCl_2$ loses the elements of $(CH_3)_3SiOH$ accompanied with a 1,3-migration of the phenylselenyl group to give a primary allyl phenyl selenide.

Examples:

$$C_6H_5SeCH_2\overset{O}{\overset{\|}{C}}CH_3 \xrightarrow[68\%]{\text{1) }(CH_3)_3SiCH_2Li\text{; 2) }H_2O} C_6H_5SeCH_2\underset{CH_3}{\overset{OH}{C}}CH_2Si(CH_3)_3 \xrightarrow[\text{40\% overall}]{SnCl_2,\ CH_2Cl_2,\ 0°}$$

$$CH_2{=}C(CH_3)CH_2SeC_6H_5$$

$$\text{2-(phenylseleno)cyclohexanone} \xrightarrow[\text{33\% overall}]{} \text{1-}(CH_2SeC_6H_5)\text{cyclohexene}$$

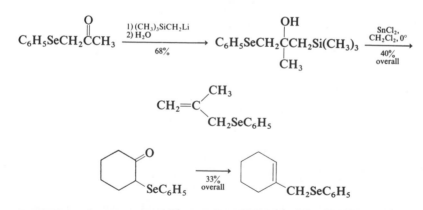

[1] D. E. Seitz and A. Zapata, *Tetrahedron Letters*, **21**, 3451 (1980).
[2] *Idem, Synthesis*, 557 (1981).
[3] N. Nishiyama, K. Itagaki, N. Osaka, and K. Itoh, *Tetrahedron Letters*, **23**, 4103 (1982).

(Trimethylsilyl)methyl trifluoromethanesulfonate, 10, 934–936.

Preparation[1]:

$$CH_3OCH_2Cl \xrightarrow[81\%]{\text{1) Li, DME; 2) ClSi(CH}_3)_3} (CH_3)_3SiCH_2OCH_3 \xrightarrow[95\%]{CH_3OSO_2CF_3} (CH_3)_3SiCH_2OSO_2CF_3$$

(b.p. 83°)

[1] R. F. Cunico and H. S. Gill, *Organometallics*, **1**, 1 (1982).

N-Trimethylsilyl-2-oxazolidinone, [2-oxazolidinone ring]N—$Si(CH_3)_3$ (**1**). Mol. wt. 159.26, b.p. 100°/6 mm. The reagent is prepared by reaction of 2-oxazolidinone with hexamethyldisilazane or with $ClSi(CH_3)_3$ and $N(C_2H_5)_3$; 90–95% yield.

Silylation of carboxylic acids.[1] The reagent converts carboxylic acids into silyl carboxylates within 15–20 minutes in CCl_4 at 65–77° in the absence of a catalyst. The by-product is 2-oxazolidinone (insoluble in CCl_4). It is particularly useful for silylation of *gem-* or 1,2-dicarboxylic acids, which are easily decarboxylated when heated or in the presence of bases.

$$RCOOH + \mathbf{1} \xrightarrow{CCl_4} RC(=O)OSi(CH_3)_3 + \text{2-oxazolidinone (NH)}$$

(>90%) (>95%)

[1] C. Palomo, *Synthesis*, 809 (1981).

(Z)-(Trimethylsilyloxy)vinyllithium, (H)(OSi($CH_3)_3$)C=C(H)(Li) (**1**). Mol. wt. 122.17, stable at − 70° for at least 20 hours. The acetaldehyde equivalent is prepared by reaction of *t*-butyllithium in ether at − 70° with (Z)-2-bromo-1-(trimethylsilyloxy)ethylene.

α,β-Unsaturated aldehydes. The reagent reacts with carbonyl compounds to form, after hydrolysis, α,β-unsaturated aldehydes.[1]

Examples:

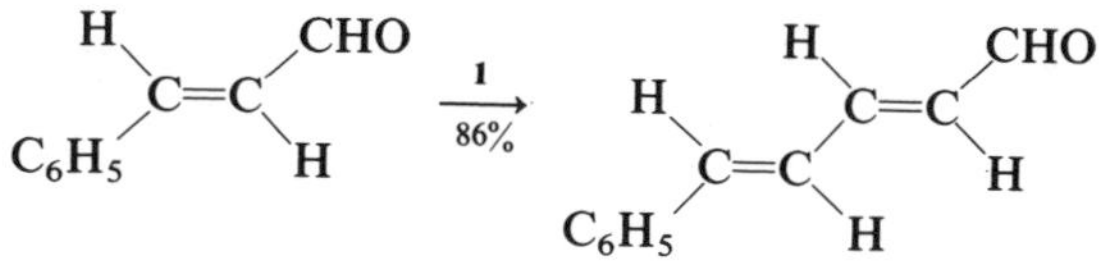

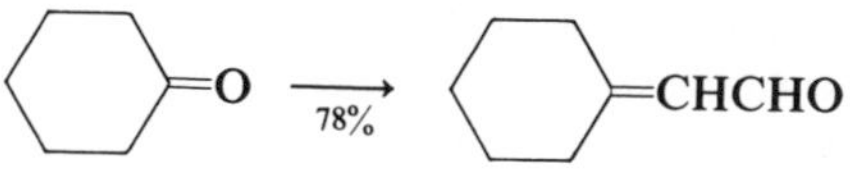

$$\text{C}_6\text{H}_5\text{CH=CHCHO} \xrightarrow[86\%]{1} \text{C}_6\text{H}_5\text{CH=CH–CH=CHCHO}$$

$$\text{Cyclohexanone} \xrightarrow[78\%]{} \text{Cyclohexylidene=CHCHO}$$

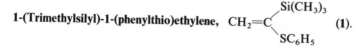

[1] L. Duhamel and F. Tombret, *J. Org.*, **46**, 3741 (1981).

1-(Trimethylsilyl)-1-(phenylthio)ethylene, $CH_2{=}C(Si(CH_3)_3)(SC_6H_5)$ **(1)**.

Mol. wt. 208.40, b.p. 45°/0.025 mm. Preparation from trimethylvinylsilane.[1]

Cyclopentenone annelation (*cf.* **10**, 444). The reagent undergoes a Nazarov-type cyclization with an α,β-unsaturated acid chloride to give an annelated 3-phenylthiocyclopentenone.[2] This reaction was used to prepare the bicyclic cyclopentenone **3**, a useful intermediate in synthesis of cyclopentenoid natural products, such as hirsutene (**4**).[3]

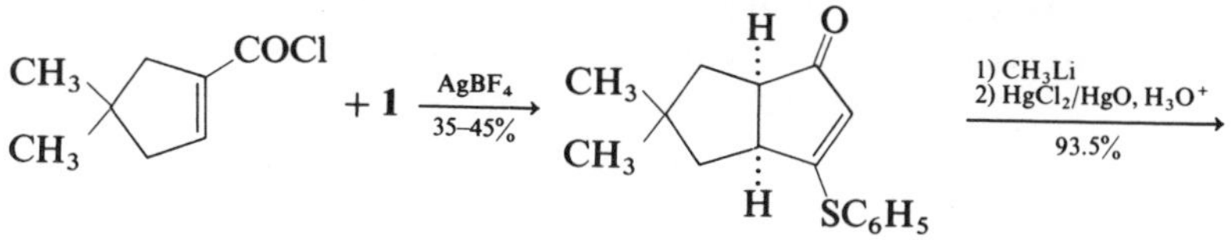

$$\text{COCl} + \mathbf{1} \xrightarrow[35\text{–}45\%]{AgBF_4} \mathbf{2} \xrightarrow[93.5\%]{1)\ CH_3Li;\ 2)\ HgCl_2/HgO,\ H_3O^+} \mathbf{3} \xrightarrow{\text{Several steps}} \mathbf{4}$$

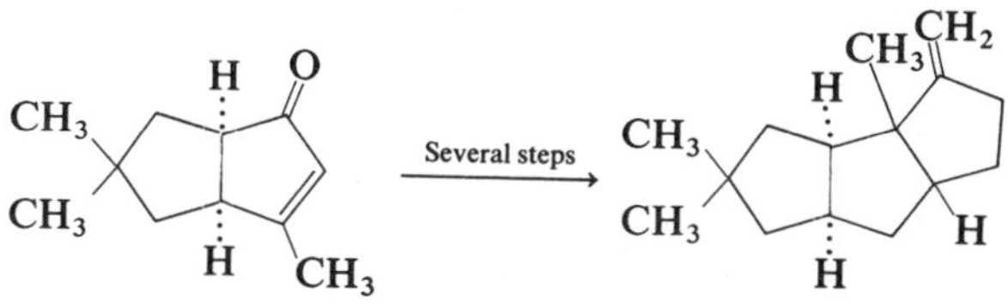

The reaction of 1-(trimethylsilyl)-2-(phenylthio)ethylene (**5**) with an α,β-unsaturated acid chloride such as **6** in the presence of $AlCl_3$ involves an unusual rearrangement to give **7**.[3]

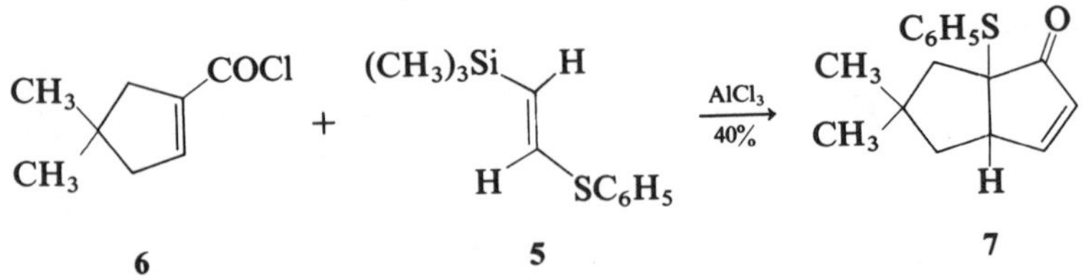

$$\mathbf{6} + (CH_3)_3SiCH{=}CHSC_6H_5\ (\mathbf{5}) \xrightarrow[40\%]{AlCl_3} \mathbf{7}$$

[1] F. Cooke, R. Moerck, J. Schwindeman, and P. Magnus, *J. Org.*, **45**, 1046 (1980).
[2] P. Magnus, D. A. Quagliato, and J. C. Huffman, *Organometallics*, **1**, 1240 (1982).
[3] P. Magnus and D. A. Quagliato, *ibid*, **1**, 1243 (1982).

Trimethylsilyl trifluoromethanesulfonate (1), 8, 514–515; **9**, 497–498; **10**, 438–441.

Two laboratories have prepared this triflate by reaction of allyltrimethylsilane with triflic acid (equation I). This method is particularly useful for *in situ* preparation of

$$\text{(I)}\quad CH_2{=}CHCH_2Si(CH_3)_3 + CF_3SO_3H \xrightarrow{CH_2Cl_2} \underset{\mathbf{1}\ (85\%)}{(CH_3)_3SiOSO_2CF_3} + CH_3CH{=}CH_2$$

the reagent. Thus, alcohols or carboxylic acids are silylated in high yield when added to allyltrimethylsilane and a catalytic amount of triflic acid.[1,2]

For a general synthesis of trialkylsilyl triflates, *see* triisopropyl triflate, this volume.

Glycoside synthesis.[3] This reagent in the presence of an acid scavenger (4 Å molecular sieve, 1,1,3,3-tetramethylurea) is a catalyst for preparation of glycosides.

Examples:

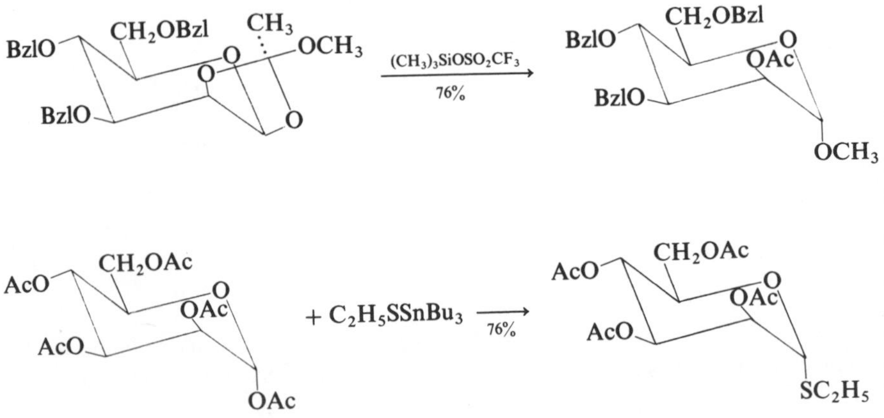

$$\xrightarrow[76\%]{(CH_3)_3SiOSO_2CF_3}$$

$$+ C_2H_5SSnBu_3 \xrightarrow[76\%]{}$$

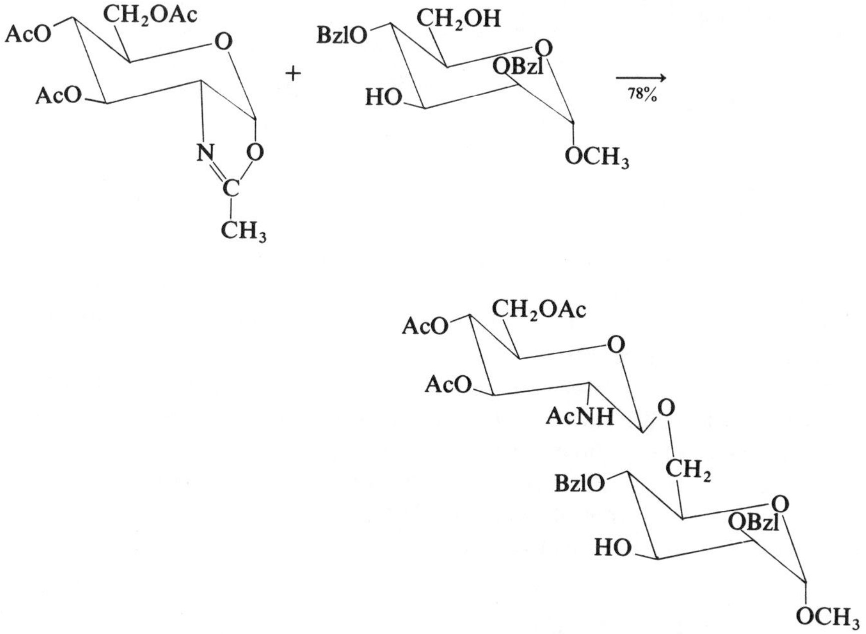

In the presence of catalytic amounts of $(CH_3)_3SiOTf$, the protected derivative (**1**) of 1-O-trimethylsilyl-β-D-glucopyranose reacts with acetals (**2**) to afford 1,1′-diacetal-β-glucosides (**3**) in yields of 75–85%.[4]

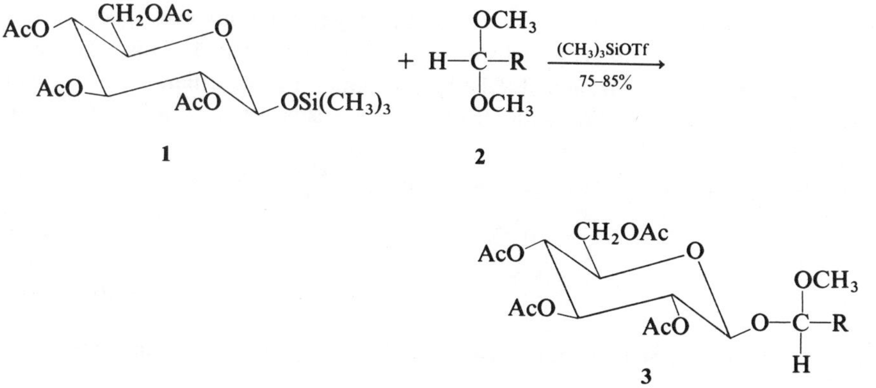

The reaction of aryl trimethylsilyl ethers catalyzed by $(CH_3)_3SiOTf$ with the D-glucoside **1a** (anomeric mixture) results in the formation of essentially only the aryl-β-glucoside.[5]

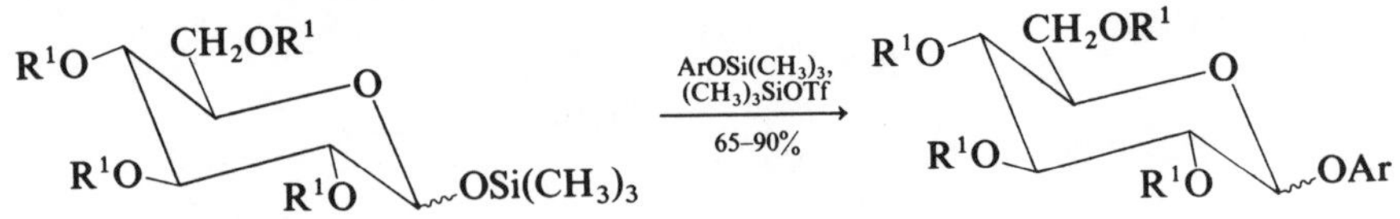

1a, $R^1 = Ac$
b, $R^1 = Bzl$

$\xrightarrow[65–90\%]{ArOSi(CH_3)_3,\ (CH_3)_3SiOTf}$

4, $\alpha/\beta \sim 1:10$
$\sim 3.5:1$

Reviews. Simchen *et al.*[6] have reviewed use of trialkylsilyl triflates for silylation and as catalysts in Friedel-Crafts acylations (191 references). The review includes a method for recovery of triflic acid from the triethylammonium triflates formed in silylation in the presence of triethylamine.

Noyori *et al.*[7] have reviewed uses of this triflate (**1**) for silylation and as a catalyst in various nucleophilic reactions. The review includes several unpublished results obtained by the Nagoya group. The Si atom in the triflate can interact with various heteroatoms, particularly with oxygen to form onium cations, which can act as supercations in aprotic solvents. For example, the reaction of **1** with the keto epoxide **2** involves ring fusion followed by cyclization to **3**.

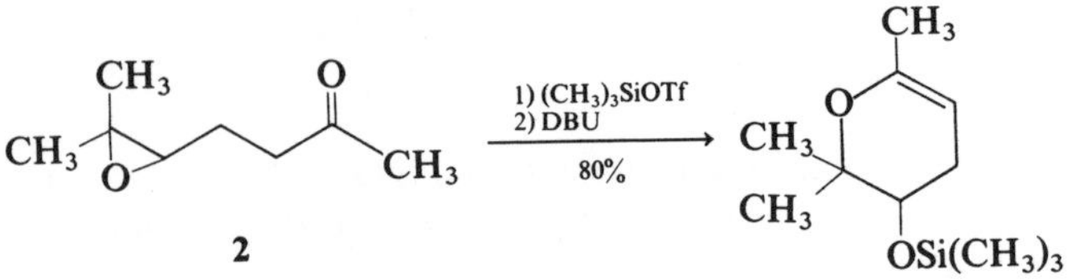

2 $\xrightarrow[80\%]{1)\ (CH_3)_3SiOTf;\ 2)\ DBU}$ **3**

A convenient method for tritylation of carboxylic acids involves reaction of a trimethylsilyl ester with trityloxytrimethylsilane catalyzed by **1**. This reaction is limited to tertiary alkyl trimethylsilyl ethers (equation I). The triflate is an excellent catalyst for thioacetalization with ethylthiotrimethylsilane (equation II). A related displacement reaction involving an N-nucleophile is formulated in equation (III).

$$\text{(I)}\quad (C_6H_5)_3COSi(CH_3)_3 + H_2C{=}C(CH_3)COOSi(CH_3)_3 \xrightarrow[73\%]{\mathbf{1},\ CH_2Cl_2} H_2C{=}C(CH_3)COOC(C_6H_5)_3 + [(CH_3)_3Si]_2O$$

$$\text{(II)}\quad \text{cyclohexanone} + 2C_2H_5SSi(CH_3)_3 \xrightarrow[84\%]{\mathbf{1},\ CH_2Cl_2} C_6H_{10}(SC_2H_5)_2 + [(CH_3)_3Si]_2O$$

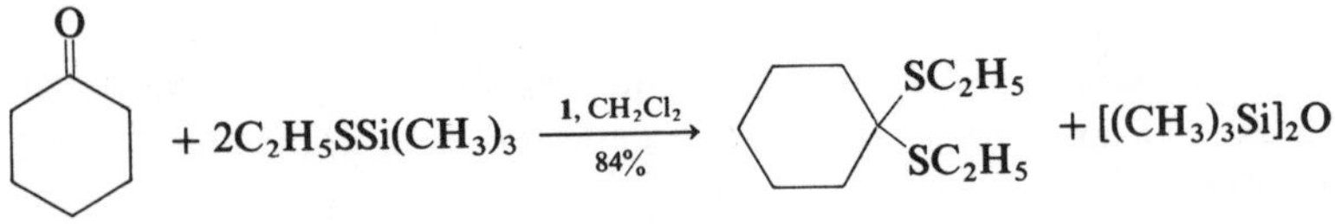

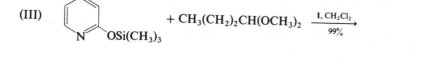

(III) [2-(trimethylsilyloxy)pyridine] $OSi(CH_3)_3$ + $CH_3(CH_2)_2CH(OCH_3)_2$ $\xrightarrow[99\%]{1,\ CH_2Cl_2}$

[N-substituted 2-pyridone: N–CH(CH_3O)($(CH_2)_2CH_3$)] + $(CH_3)_3SiOCH_3$

Nafion-Trimethylsilyl ester.[8] Immobilization of **1** on Nafion-H provides the usual advantages and, in addition, improves the stability to water. In addition to the usual silylation reactions effected with **1** itself, Nafion-TMS can be used for some catalytic reactions.

Example:

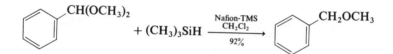

$C_6H_5CH(OCH_3)_2$ + $(CH_3)_3SiH$ $\xrightarrow[92\%]{\text{Nafion-TMS},\ CH_2Cl_2}$ $C_6H_5CH_2OCH_3$

[1] T. Morita, Y. Okamoto, and H. Sakurai, *Synthesis*, 745 (1981).
[2] G. A. Olah, A. Husain, B. G. B. Gupta, G. F. Salem, and S. C. Narang, *J. Org.*, **46**, 5212 (1981).
[3] T. Ogawa, K. Beppu, and S. Nakabayashi, *Carbohydrate Research*, **93**, C6 (1981).
[4] L.-F. Tietze and R. Fischer, *Angew. Chem. Int. Ed.*, **20**, 969 (1981).
[5] L.-F. Tietze, R. Fischer, and H.-J. Guder, *Tetrahedron Letters*, **23**, 4661 (1982).
[6] H. Emde, D. Domsch, H. Feger, U. Frick, A. Götz, H. H. Hergott, K. Hofmann, W. Kober, K. Krageloh, T. Oesterle, W. Steppan, W. West, and G. Simchen, *Synthesis*, 1 (1982).
[7] R. Noyori, S. Murata, and M. Suzuki, *Tetrahedron*, **37**, 3899 (1981).
[8] S. Murata and R. Noyori, *Tetrahedron Letters*, **21**, 767 (1980).

Triphenylbismuth carbonate, 9, 501;[1] **10,** 446.

α-Glycol cleavage.[2] In the presence of potassium carbonate and a little water, triphenylbismuth itself can be used in catalytic amounts if NBS or NBA is present to effect oxidation of Bi(III) to Bi(V). Yields are generally comparable or even superior to those obtained in the stoichiometric cleavage with triphenylbismuth carbonate.

[1] D. H. R. Barton, J. P. Kitchin, D. J. Lester, W. B. Motherwell, and M. T. B. Papoula, *Tetrahedron*, Suppl. No. 9, **37**, 73 (1981).
[2] D. H. R. Barton, W. B. Motherwell, and A. Stobie, *J.C.S. Chem. Comm.*, 1232 (1981).

Triphenylphosphine, 1, 1238–1247; **2**, 443–445; **3**, 317–320; **4**, 548–550; **6**, 643–644; **7**, 403–404.

Aza-Wittig reaction; cyclic imines.[1] Treatment of the azido ketones **1** with triphenylphosphine at room temperature results in a cyclic imine (**2**) in good yield, formed by an intramolecular aza-Wittig reaction.[2]

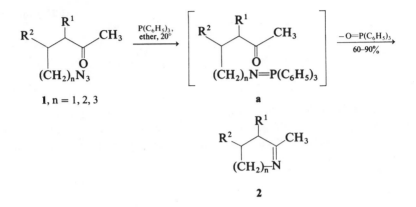

$$\underset{\mathbf{1},\ n = 1, 2, 3}{R^2CH(CH_2)_nN_3-CH(R^1)-C(=O)CH_3} \xrightarrow[\text{ether, }20^\circ]{P(C_6H_5)_3,} \left[\underset{\mathbf{a}}{R^2CH((CH_2)_nN{=}P(C_6H_5)_3)-CH(R^1)-C(=O)CH_3}\right] \xrightarrow[60\text{–}90\%]{-O{=}P(C_6H_5)_3} \underset{\mathbf{2}}{\text{cyclic imine: } R^2, R^1, CH_3, (CH_2)_n{-}N{=}}$$

[1] P. H. Lambert, M. Vaultier, and R. Carrié, *J.C.S. Chem. Comm.*, 1224 (1982).
[2] H. Staudinger and I. Meyer, *Helv.*, **2**, 635 (1919).

Triphenylphosphine–Carbon tetrachloride, 1, 1247; **2**, 445; **3**, 320; **4**, 551–552; **5**, 727; **6**, 644–645; **7**, 404; **8**, 516; **9**, 503; **10**, 447–448.

Δ^2-Oxazolines. The intramolecular cyclization of β-hydroxy hydroxamates to β-lactams by $P(C_6H_5)_3$–CCl_4 (**10**, 447–448) can be used for cyclization of a carboxylic acid and a β-amino alcohol to form a Δ^2-oxazoline (equation I).[1]

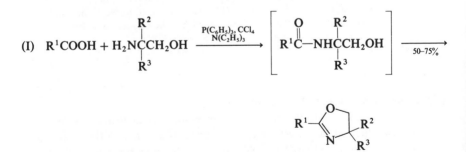

$$\text{(I)}\quad R^1COOH + H_2NC(R^2)(R^3)CH_2OH \xrightarrow[N(C_2H_5)_3]{P(C_6H_5)_3,\ CCl_4} \left[R^1C(=O){-}NHC(R^2)(R^3)CH_2OH\right] \xrightarrow[50\text{–}75\%]{} \text{2-}R^1\text{-4-}R^2\text{-4-}R^3\text{-}\Delta^2\text{-oxazoline}$$

β-Amino mercaptans or 1,3-diamines can replace the β-amino alcohol for the synthesis of Δ^2-thiazolines or Δ^2-imidazolines, respectively.

[1] H. Vorbrüggen and K. Krolikiewicz, *Tetrahedron Letters*, **22**, 4471 (1981).

Triphenylphosphine–Diethyl azodicarboxylate. **1**, 245–247; **4**, 553–555; **5**, 727–728; **6**, 645; **7**, 404–406; **8**, 517; **9**, 504–506; **10**, 448–449.

Mitsunobu reaction.[1] Two intermediates have been isolated from the Mitsunobu esterification of carboxylic acids with phenols. One is the betaine **1**, which has been generally assumed to be involved; the other is the phosphorane **2**.[2]

$$(C_6H_5)_3\overset{+}{P}-N(COOC_2H_5)-\overset{-}{N}-COOC_2H_5 \qquad C_6H_5-P(C_6H_5)_2(OAr)_2$$

1 **2**

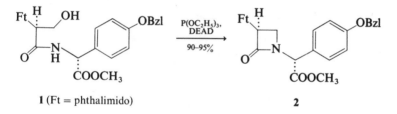

The mechanism proposed is shown in equation (I).

$$\text{(I)}\quad P(C_6H_5)_3 + \text{DEAD} \longrightarrow \mathbf{1} \xrightarrow{\text{2ROH}} \mathbf{2} \xrightarrow{R^1COOH} R^1COOR + O{=}P(C_6H_5)_3 + ROH$$

β-Lactams.[3] The Mitsunobu reaction has been used to cyclize the protected peptide **1** to the β-lactam **2** in high yield. In practice, $P(OC_2H_5)_3$ is superior to triphenylphosphine. The product is a protected form of (−)-3-aminocardicinic acid.

[Reaction scheme: **1** (Ft = phthalimido) $\xrightarrow[90\text{–}95\%]{P(OC_2H_5)_3,\ \text{DEAD}}$ **2**; structures bear labels Ft, H, OH, OBzl, NH, O, $COOCH_3$]

1 (Ft = phthalimido) **2**

Carbonates.[4] Alcohols can be converted into the carbonates, ROCOOR, by reaction with CO_2 in THF under conditions of Mitsunobu esterification. Isolated yields (five examples) are in the range 40–80%.

Thiolesters. These esters are easily obtained by addition of an alcohol and then thiolacetic acid to the preformed adduct of $P(C_6H_5)_3$ and diisopropyl azodicarboxylate in THF (equation I). The reaction is accompanied by inversion of configuration.[5]

$$\text{(I)}\quad ROH + CH_3COSH \xrightarrow[90\text{–}98\%]{P(C_6H_5)_3,\ [NCOOCH(CH_3)_2]_2} RSCOCH_3$$

[1] Review: O. Mitsunobu, *Synthesis*, 1 (1981).
[2] E. Grochowski, B. D. Hilton, R. J. Kupper, and C. J. Michejda, *Am. Soc.*, **104**, 6876 (1982).
[3] C. A. Townsend and L. T. Ngugen, *ibid.*, **103**, 4582 (1981); *idem*, *Tetrahedron Letters*, **23**, 4859 (1982).
[4] W. A. Hoffman, III, *J. Org.*, **47**, 5209 (1982).
[5] R. P. Volante, *Tetrahedron Letters*, 3119 (1981).

Tris(dimethylamino)borane, $B[N(CH_3)_2]_3$ **(1)**. Mol. wt. 143.03, b.p. 38°/9 mm. The borane is prepared in 55–70% yield by reaction of dimethylamine with boron trichloride.[1]

Macrolactams.[2] Reaction of an open-chain triamino ester such as **2** with this borane results in a macrolactam (**3**) formed via a triaminoborane intermediate (**a**).

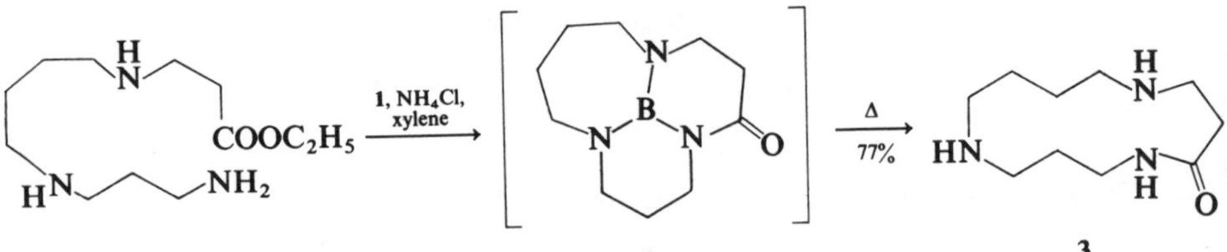

$$\mathbf{2} \xrightarrow[\text{xylene}]{\mathbf{1},\ NH_4Cl} [\mathbf{a}] \xrightarrow[77\%]{\Delta} \mathbf{3}$$

This cyclization has been used for synthesis of four spermidine alkaloids, such as celacinnine (**4**).

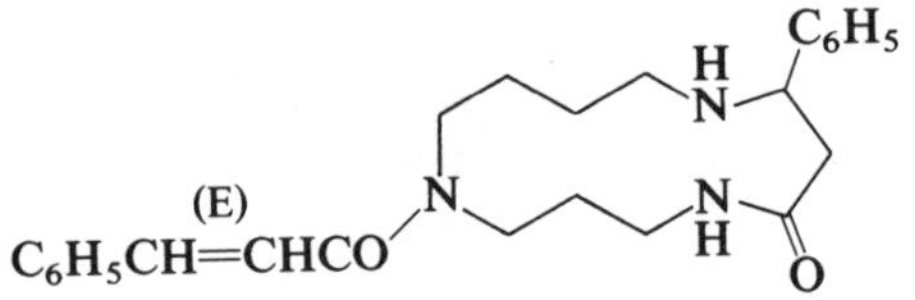

4

[1] K. Niedenzu and J. W. Dawson, *Inorg. Syn.*, **10**, 135 (1967).
[2] H. Yamamoto and K. Maruoka, *Am. Soc.*, **103**, 6133 (1981).

Tris(dimethylamino)sulfonium difluorotrimethylsilicate (1), 10, 452–453. Mol. wt. 275.49, m.p. 98–101°, hygroscopic.

Preparation:

$$3(CH_3)_3SiN(CH_3)_2 + SF_4 \xrightarrow[82\text{–}90\%]{\text{ether}} \underset{\mathbf{1}}{[(CH_3)_2N]_3\overset{+}{S}F_2Si(CH_3)_3^-} + 2FSi(CH_3)_3$$

The reagent is an anhydrous source of fluoride ion with high anionic activity and solubility in organic solvents.[1]

[1] W. J. Middleton, U. S. Patent 3, 940, 402 (1976); *Org. Syn.*, submitted (1982).

Tris(phenylthio)methyllithium, 7, 412.

α-Hydroxy esters. Some time ago Seebach[1] showed that this carbanion adds to aldehydes to form α-hydroxy trithioortho esters in 80–95% yield. These adducts can be converted into esters of α-hydroxy carboxylic acids in 60–95% yield by treatment with an alcohol, HgO, and 50% aqueous HBF_4 at 25° (equation I).[2]

$$R^1CHO + LiC(SC_6H_5)_3 \xrightarrow[80\text{–}90\%]{} R^1\underset{OH}{\underset{|}{C}}HC(SC_6H_5)_3 \xrightarrow[60\text{–}95\%]{R^2OH,\ HgO,\ HBF_4,\ H_2O} R^1\underset{OH}{\underset{|}{C}}HCOOR^2$$

[1] D. Seebach, *Ber.*, 487 (1972).
[2] D. Scholz, *Syn. Comm.*, **12**, 527 (1982).

Tris(trimethylsilyl)ketenimine, $[(CH_3)_3Si]_2C{=}C{=}NSi(CH_3)_3$. Mol. wt. 257.60, b.p. 117°/30 mm. The reagent is prepared by silylation of trimethylsilylacetonitrile (80% yield).

(Z)-α,β-Unsaturated nitriles. A highly selective synthesis of essentially pure (Z)-α,β-unsaturated nitriles from aldehydes is possible by use of **1** (equation I).[1]

$$\text{(I)}\quad RCHO + \mathbf{1} \xrightarrow{BF_3\cdot(C_2H_5)_2O,\ C_6H_6,\ 25^\circ} \left[\begin{array}{c} (CH_3)_3SiO\quad Si(CH_3)_3 \\ RCH{-}C{-}CN \\ Si(CH_3)_3 \end{array} \right] \xrightarrow[65\text{–}90\%]{\Delta,\ -[(CH_3)_3Si]_2O}$$

$$\underset{H}{\overset{R}{}}C{=}C\overset{CN}{\underset{Si(CH_3)_3}{}} \xrightarrow[65\text{–}95\%]{NaOH,\ CH_3OH} \underset{H}{\overset{R}{}}C{=}C\overset{CN}{\underset{H}{}} + (CH_3)_3SiOH$$

[1] Y. Sato and Y. Niinomi, *J.C.S. Chem. Comm.*, 56 (1982).

Tungsten(VI) oxide–Chlorosulfonic acid, WO_3–$ClSO_3H$.

1,2,4-*Trioxanes.* A general synthesis of 1,2,4-trioxanes involves treatment of an α-hydroxy hydroperoxide (**1**) and an epoxide such as **2** with WO_3 and catalytic amounts of chlorosulfonic acid; yields of 1,2,4-trioxanes are 7–63%.[1]

Examples:

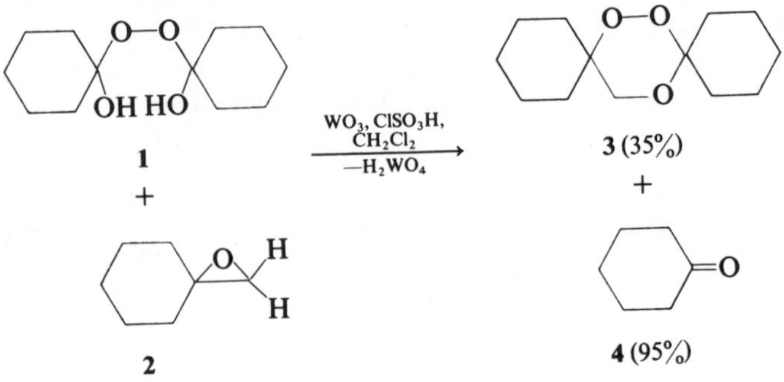

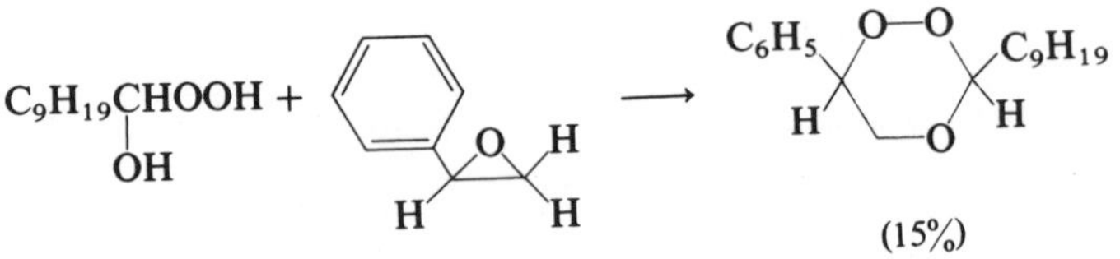

$$\mathbf{1} + \mathbf{2} \xrightarrow[-H_2WO_4]{WO_3,\ ClSO_3H,\ CH_2Cl_2} \mathbf{3}\ (35\%) + \mathbf{4}\ (95\%)$$

$$C_9H_{19}CH(OH)OOH + \text{styrene oxide} \longrightarrow (15\%)$$

[1] M. Miura, M. Nojima, and S. Kusabayashi, *J.C.S. Chem. Comm.*, 581 (1981).

V

Vanadium(II) chloride, 7, 418; **10,** 452.

Reductive hydrolysis of 2,4-dinitrophenylhydrazones.[1] Carbonyl compounds are regenerated from their 2,4-dinitrophenylhydrazones by reduction with VCl_2 in aqueous THF (reflux 1 hour). The initial step is reduction of the nitro groups to amino groups. Yields are 65–95%.

[1] G. A. Olah, Y.-L. Chao, M. Arvanaghi, and G. K. S. Prakash, *Synthesis*, 476 (1981).

Vanadyl trichloride, 3, 331–332; **5,** 744; **8** 527.

Oxidative phenolic coupling.[1] A new biomimetic approach to morphine alkaloids involves oxidative intramolecular coupling of the reticuline derivative **1** to a salutaridine derivative **2** with $VOCl_3$ in ether.[1] If the reaction is conducted in CH_2Cl_2 **2** is obtained in markedly lower yield and the undesired aporphine **3** is formed as a major product. The dienone **2** has been converted into 2-hydroxycodeine (**4**).

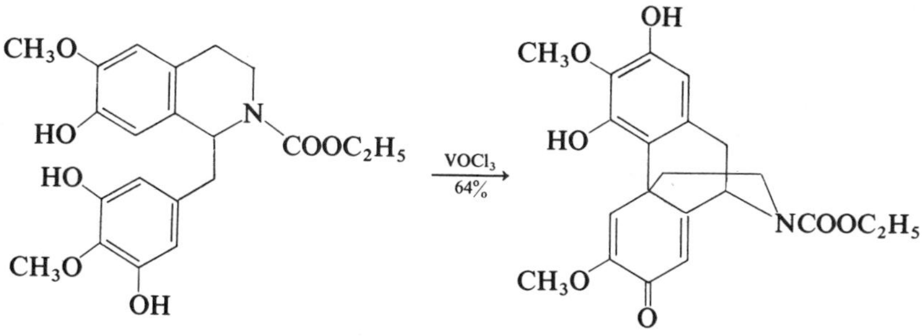

CH_3O, HO, HO, CH_3O, OH, N, $COOC_2H_5$ **1** $\xrightarrow[64\%]{VOCl_3}$ OH, CH_3O, HO, CH_3O, O, $NCOOC_2H_5$ **2**

CH_3O, HO, HO, CH_3O, OH, N, $COOC_2H_5$ **3** OH, CH_3O, O, HO, H, N, CH_3 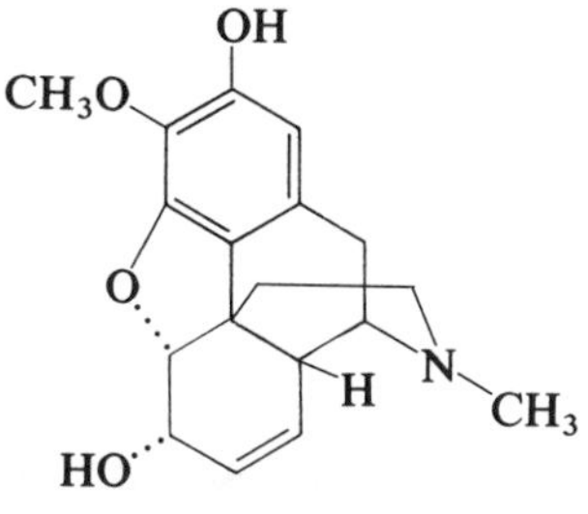 **4**

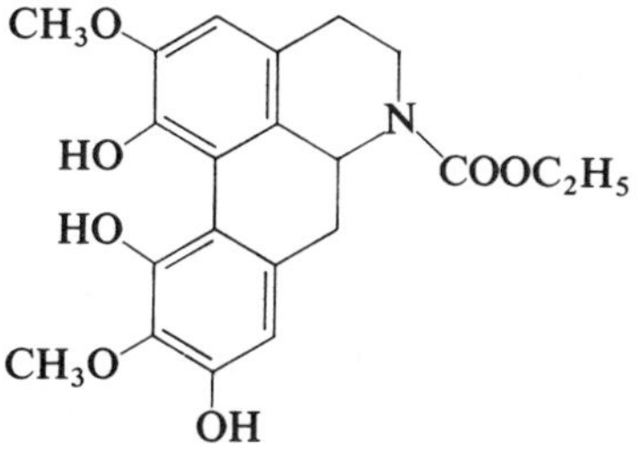

[1] M. A. Schwartz and M. F. Zoda, *J. Org.*, **46**, 4623 (1981).

β-Vinylbutenolide,

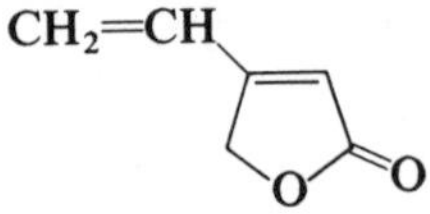

$CH_2{=}CH$ (1), **9**, 515–516.

***α-(Phenylthio)-β-vinylbutenolide* (2).** The reagent is accessible from **3** by an eliminative Pummerer rearrangement (equation I).

(I) $CH_2{=}CH$, SC_6H_5 — 1) $ClC_6H_4CO_3H$; 2) $(CF_3CO)_2O$, 20°; 71% → $CH_2{=}CH$, SC_6H_5

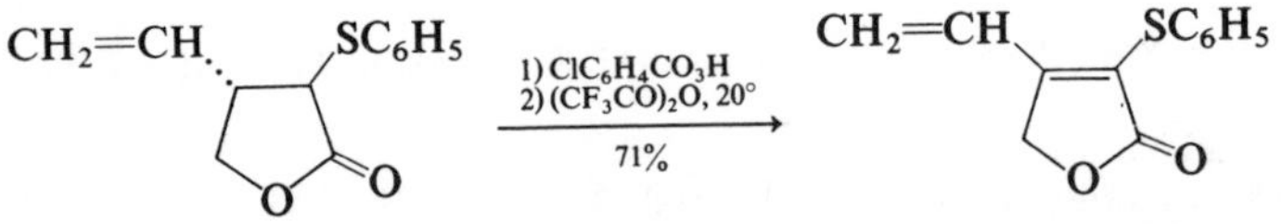

3 **2** (m.p. 70°)

The reagent converts aldehydes or ketones into oxygenated perhydrooxobenzofuranes by a 1,6-conjugate addition followed by aldol-type cyclization. Remaining steps involve dehydration, oxidation to the sulfoxide, and allylic sulfoxide-sulfenate rearrangement.[1]

Example:

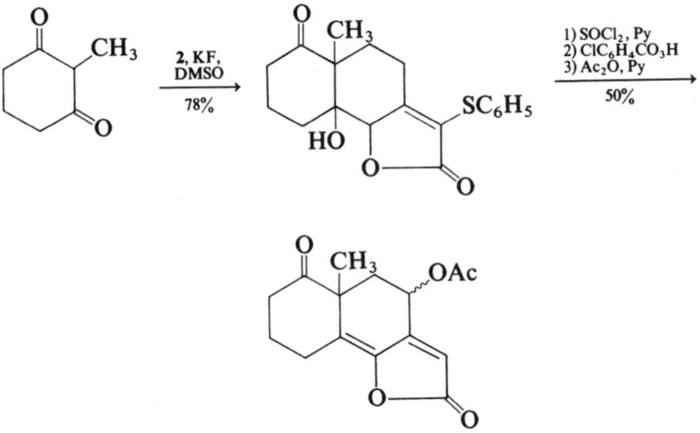

O, CH_3, O — **2**, KF, DMSO; 78% → O, CH_3, HO, SC_6H_5, O, O — 1) $SOCl_2$, Py; 2) $ClC_6H_4CO_3H$; 3) Ac_2O, Py; 50% → O, CH_3, OAc, O, O

($\alpha/\beta = 80{:}20$)

The reagent (**2**) has been used for a total synthesis of paniculide A (**4**), a highly oxygenated bisabolene, by annelation of methyl α-formylpropionate with **2** followed by dehydration.[2]

[1] F. Kido, Y. Noda, and A. Yoshikoshi, *Am. Soc.*, **104**, 5509 (1982).
[2] *Idem, J.C.S. Chem. Comm.*, 1209 (1982).

$$CH_3CHCOOCH_3\ (CHO) \xrightarrow[2)\,-H_2O]{1)\,\mathbf{2},\,KF}$$ [bicyclic lactone with CH_3, CO_2CH_3, SC_6H_5] $\xrightarrow{\text{Several steps}}$

[tricyclic lactone with CH_3] $\xrightarrow{\text{Several steps}}$ [epoxy lactone with CH_3, OH, $(CH_2)_2CH{=}C(CH_3)_2$]

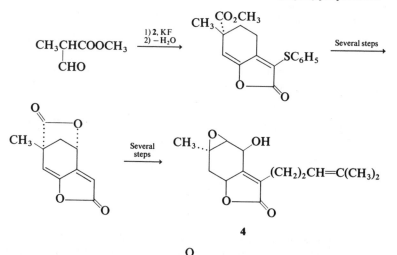

4

Vinyldiphenylphosphine oxide, $(C_6H_5)_2\overset{O}{\overset{\|}{P}}CH{=}CH_2$ (**1**). Mol. wt. 228.22, m.p. 117°. The oxide is prepared by reaction of diphenylchlorophosphine with ethylene oxide at 40 → 140° followed by isomerization (I_2) and dehydrohalogenation (NaOH); yield 78%.[1]

Allylic tertiary amines.[2] The reagent serves as a starting material for a regio- and stereospecific synthesis of allylic tertiary amines by the Wittig-Horner reaction shown in the example.

Example:

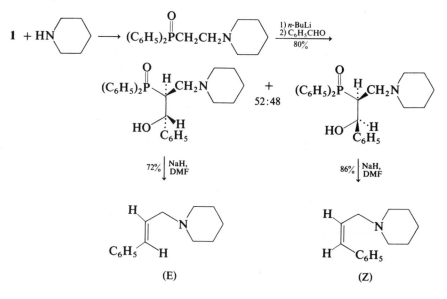

$$\mathbf{1} + HN(CH_2)_5 \longrightarrow (C_6H_5)_2\overset{O}{\overset{\|}{P}}CH_2CH_2N(CH_2)_5 \xrightarrow[80\%]{1)\,n\text{-BuLi};\ 2)\,C_6H_5CHO}$$

$(C_6H_5)_2P(O)CH(CH_2N(CH_2)_5)CH(OH)C_6H_5$ + $(C_6H_5)_2P(O)CH(CH_2N(CH_2)_5)CH(OH)C_6H_5$ (52:48)

72% ↓ NaH, DMF   86% ↓ NaH, DMF

(E) $C_6H_5CH{=}CHCH_2N(CH_2)_5$ (Z) $C_6H_5CH{=}CHCH_2N(CH_2)_5$

[1] F. J. Welch and H. J. Paxton, *J. Polymer Sci., A*, **3**, 3427 (1965).
[2] D. Cavalla and S. Warren, *Tetrahedron Letters*, **23**, 4505 (1982).

Vinylketene (1). This ketene has been generated by vacuum pyrolysis of crotonic anhydride (equation I) and identified by low-temperature ^{1}H and ^{13}C NMR.

$$\text{(I)}\quad (CH_3CH{=}CHCO)_2O \xrightarrow[0.1\text{ mm}]{550^\circ} \underset{\mathbf{1}\ (22\%)}{(CH_2{=}CH)(H)C{=}C{=}O} + CH_3CH{=}CHCOOH + \underset{(40\%)}{CH_3C{\equiv}CH}$$

Cycloaddition reactions.[1] When the above mixture is warmed to 20°, the [4 + 2] cycloadduct dimer **2** of **1** is obtained in 70% yield. The adduct is isomerized mainly to the α-pyrone sibirinone (**3**) by tosic acid and mainly to the isomeric pyrone **4** by $NaHCO_3$.

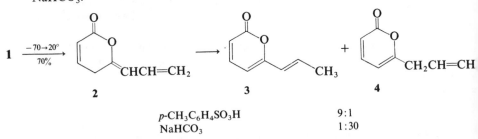

$$\mathbf{1} \xrightarrow[70\%]{-70\to20^\circ} \underset{\mathbf{2}}{\text{(dihydropyranone)}{=}CHCH{=}CH_2} \longrightarrow \underset{\mathbf{3}}{\text{(α-pyrone)}CH{=}CHCH_3} + \underset{\mathbf{4}}{\text{(α-pyrone)}CH_2CH{=}CH_2}$$

	3 : 4
p-$CH_3C_6H_4SO_3H$	9:1
$NaHCO_3$	1:30

The ketene undergoes [2 + 2] cycloaddition to cyclopentadiene to give **5** in moderate yield. When heated, **5** undergoes a Cope rearrangement to give **6**.

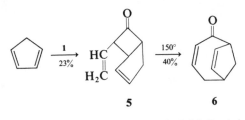

$$\text{cyclopentadiene} \xrightarrow[23\%]{\mathbf{1}} \mathbf{5}\ (\text{HC}{=}\text{CH}_2) \xrightarrow[40\%]{150^\circ} \mathbf{6}$$

[1] W. S. Trahanovsky, B. W. Surber. M. C. Wilkes, and M. M. Preckel, *Am. Soc.*, **104**, 6779 (1982).

Vinyltriphenylphosphonium bromide, 1, 1274–1275; **2,** 456–457; **3,** 333; **4,** 572; **5,** 750–751; **6,** 666–667.

(E)-Allylamines.[1] The reaction of Schweizer's reagent with sodiophthalimide and an aldehyde in the presence of LiBr as a Lewis acid mediator gives allylic phthalimides in an (E)/(Z) ratio of 2–3:1. Surprisingly, use of vinyltri-*n*-butylphosphonium bromide in this same reaction results in the (E)-allylic phthalimide (75–100% yield).

Example:

$$\text{PhthN–Na} + CH_2{=}CH\overset{+}{P}Bu_3Br^- + C_6H_5CHO \xrightarrow[83\%]{LiBr}$$

$$\text{PhthN–}CH_2C(H){=}C(H)\text{–}C_6H_5 \xrightarrow[83\%]{H_2NNH_2} (H)(C_6H_5)C{=}C(H)(CH_2NH_2)$$

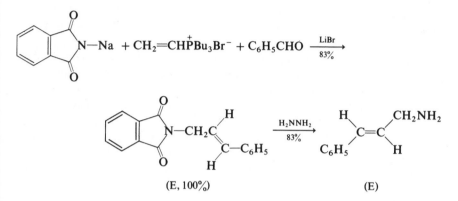

(E, 100%) (E)

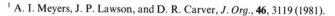

[1] A. I. Meyers, J. P. Lawson, and D. R. Carver, *J. Org.*, **46**, 3119 (1981).

Z

Zinc, 1, 1276–1284; **2**, 459–462; **3**, 334–337; **4**, 574–577; **5**, 753–756; **6**, 672–675; **7**, 426–428; **8**, 532; **10**, 459.

Highly active powder (**5**, 753; **6**, 674–675). An exceptionally reactive form can be prepared by reduction of $ZnCl_2$ with lithium and 10% naphthalene in glyme (superior to THF).[1]

Even unreactive alkenes undergo the Simmons-Smith reaction (cyclopropanation) in high yield with dibromomethane and this zinc powder. Almost quantitative yields are obtained in the Reformatsky reaction of ethyl α-bromoacetate and this zinc powder. Ethyl α-chloroacetate can also be used if the reaction is conducted at higher temperatures for a longer period.

γ-Butyrolactones (*cf.* **8**, 532). Dimethyl maleate condenses with aldehydes or ketones in the presence of high purity zinc powder to form γ-substituted-β-carbomethoxy-γ-butyrolactones in 40–85% yield. Some related reductive coupling reactions are also possible.[2]

Examples:

H, COOCH$_3$ / H, COOCH$_3$ (dimethyl maleate) + cyclopentanone $\xrightarrow[74\%]{Zn,\ CH_3CN,\ \Delta}$ spirolactone (COOCH$_3$, O, O)

$$CH_3OOCC{\equiv}CCOOCH_3 + (CH_3)_2C{=}O \xrightarrow[35\%]{} (CH_3)_2C(OH)\text{–}C(COOCH_3){=}CHCOOCH_3$$

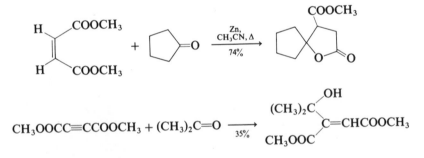

o-*Xylylene.*[3] Ultrasound induces a reaction between activated zinc[4] and α, α′-dibromo-*o*-xylene (**1**) to produce *o*-xylylene (**a**), which can be trapped by various dienophiles to form adducts such as **2** in useful yields.

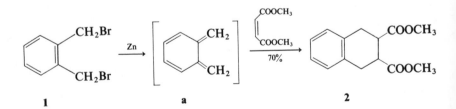

1 (CH_2Br, CH_2Br) $\xrightarrow{Zn}$ [**a** (CH_2, CH_2)] $\xrightarrow[70\%]{\text{dimethyl maleate } (COOCH_3,\ COOCH_3)}$ **2** ($COOCH_3$, $COOCH_3$)

Trifluoromethylation.[5] Ultrasonically dispersed zinc in DMF or THF reacts with trifluoromethyl iodide to form trifluoromethylzinc iodide, which adds as formed to aldehydes or ketones (equation I).

$$\text{(I)}\quad R^1R^2C{=}O + CF_3I + Zn \xrightarrow[\text{ultrasound}]{\text{DMF,}} \left[R^1{-}\underset{R^2}{\overset{CF_3}{C}}{-}OZnI \right] \xrightarrow[50\text{–}70\%]{H^+} R^1{-}\underset{R^2}{\overset{CF_3}{C}}{-}OH$$

Reformatsky reaction.[6] Ultrasonic radiation is beneficial in the Reformatsky reaction. Yields are > 90%, and the rate is enhanced. Specially activated zinc is not necessary. However, iodine and potassium iodide are effective additives, possibly by suppressing enolization. The solvent of choice in this variation is dioxane.

[1] R. D. Rieke, P. T.-J. Li, T. P. Burns, and S. T. Uhm, *J. Org.*, **46**, 4323 (1981).
[2] T. Shono, H. Hamaguchi, I. Nishiguchi, M. Sasaki, T. Miyamoto, M. Miyamoyo, and S. Fujita, *Chem. Letters*, 1217 (1981).
[3] B. H. Han and P. Boudjouk, *J. Org.*, **47**, 751 (1982).
[4] F. A. J. Kerdesky, R. J. Ardecky, M. V. Lakshimikantham, and M. P. Cava, *Am. Soc.*, **103**, 1992 (1981).
[5] T. Kitazume and N. Ishikawa, *Chem. Letters*, 1679 (1981).
[6] B.-H. Han and P. Boudjouk, *J. Org.*, **47**, 5030 (1982).

Zinc–Copper couple, 1, 1292–1293; **5**, 758–760; **7**, 428–429; **8**, 533–534; **10**, 459.

Deblocking of 2,2,2-trichloroethyl protecting group.[1] Reduction with zinc or Zn/Cu in HOAc is the classical method for cleavage of the 2,2,2-trichloroethyl group. In the case of protected oligonucleotides, the reduction does not proceed to completion. Addition of acetylacetone to chelate the zinc cation results in a faster, but still incomplete, reduction. Increasing the copper content of the couple to 16–32% results in almost complete deblocking.

[1] J. Imai and P. F. Torrence, *J. Org.*, **46**, 4015 (1981).

Zinc–Titanium(IV) chloride.

Debromination.[1] *vic*-Dibromides are converted into alkenes by reaction with zinc dust and a catalytic amount of $TiCl_4$ in THF at 0°. Isolated yields are generally in the range 80–90%.

[1] F. Sato, T. Akiyama, K. Iida, and M. Sato, *Synthesis*, 1025 (1982).

Zinc borohydride, 3, 337–338; **5**, 761–762; **10**, 460–461.

Reduction of α,β-epoxy ketones. These ketones are reduced with high stereoselectivity to *erythro*-epoxy alcohols (>90% *erythro*) by $Zn(BH_4)_2$ in ether, regardless of substitution on the epoxide ring. In fact, the stereoselectivity is higher than that obtained by Sharpless epoxidation of allylic alcohols (**9**, 81–82).[1]

This reduction was used in a highly stereoselective synthesis of *dl*-muscarine (**6**).[2] Reduction of the epoxide (**1**) of methyl vinyl ketone gives the desired alcohol **2** in high diastereoselectivity. The remaining center of asymmetry in **6** at C_2 is generated with high stereoselectivity by cyclization of **4** with iodine.[3] In contrast, cyclization of the diol corresponding to **4** shows only slight stereoselectivity.

CH_3 **1** —[$Zn(BH_4)_2$, ether; 80%]→ CH_3, OH **2** (94% de) —[1) NaH 2) $BrCH_2C_6H_3Cl_2$-2, 6; 71%]→ CH_3, H, OCH_2Ar **3**

—[CH_2=CHMgBr, CuI, THF; 85%]→ $ArCH_2O$, OH, H_3C, CH_2CH=CH_2 **4** —[I_2, CH_3CN; 91%]→ HO, CH_3, CH_2I **5** (≥97% de) —[$N(CH_3)_3$; 60%]→

HO, CH_3, $CH_2\overset{+}{N}(CH_3)_3$ I^- **6**

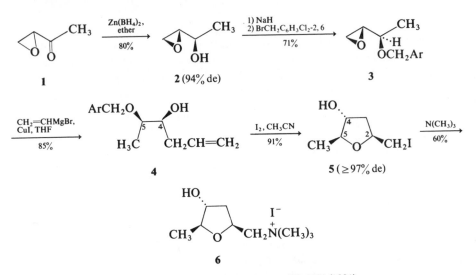

[1] T. Nakata, T. Tanaka, and T. Oishi, *Tetrahedron Letters*, **22**, 4723 (1981).
[2] R. Amouroux, G. Gerin, and M. Chastrette, *ibid.*, **23**, 4341 (1982).
[3] S. D. Rychnovsky and P. A. Bartlett, *Am. Soc.*, **103**, 3963 (1981).

Zinc bromide, 2, 463–464; **8**, 535; **9**, 520–522; **10**, 461.

Phenylthioalkylation of ketene bis(trimethylsilyl)acetals; γ- and δ-lactones. The zinc bromide-catalyzed alkylation of these acetals with α-chlorosulfides gives phenylthiolactones, which can be converted to α,β unsaturated lactones.[1]

Examples:

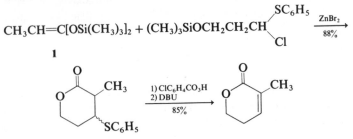

CH_3CH=$C[OSi(CH_3)_3]_2$ (**1**) + $(CH_3)_3SiOCH_2CH_2CH(SC_6H_5)Cl$ —[$ZnBr_2$; 88%]→

CH_3, SC_6H_5 —[1) $ClC_6H_4CO_3H$ 2) DBU; 85%]→ CH_3

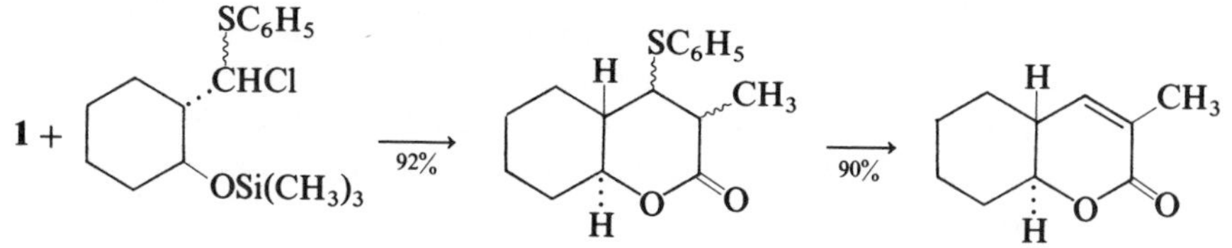

Unsaturated 1,5-dicarbonyl compounds. The phenylthioalkylation of silyl enol ethers of carbonyl compounds (**9**, 521–522) can be extended to the synthesis of unsaturated 1,5-dicarbonyl compounds. In a typical reaction the enol silyl ether of a ketone is alkylated with the unsaturated chloride **1** under $ZnBr_2$ catalysis to give a homoallyl sulfide. Ozonolysis of the methylene group is accompanied by oxidation of the phenylthio group; sulfoxide elimination results in an unsaturated 1,5-aldehydo ketone (equation I). Alkylation with **2** results in a methyl ketone (equation II).

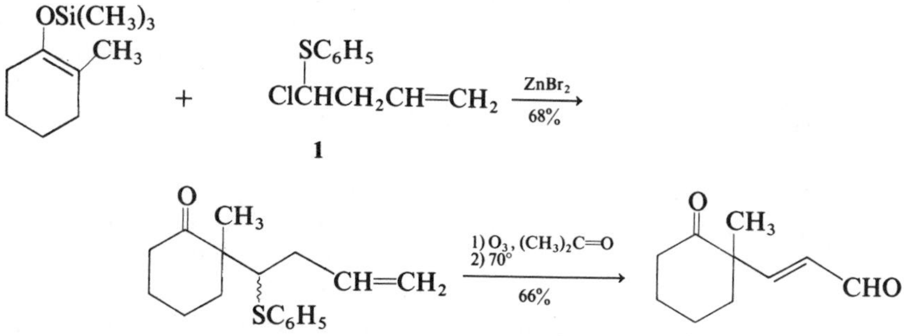

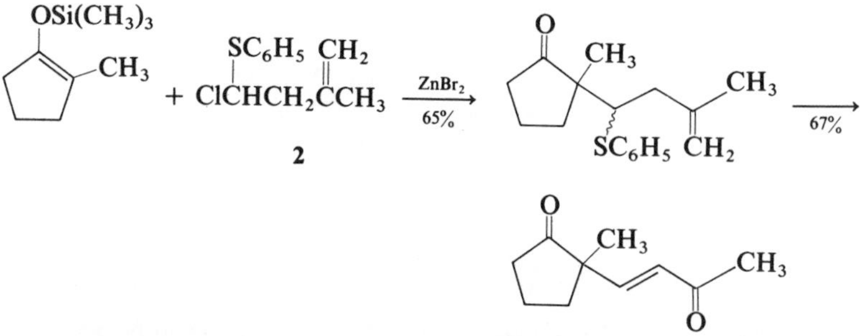

The same sequence applied to a silyl enol ether of an ester results in a 1,5-unsaturated aldehydo ester (from **1**), or a 1,5-unsaturated keto ester (from **2**).[2]

Reduction of ketoaziridines.[3] Ketoaziridines form a 2:1 complex with $ZnBr_2$, which is reduced by $NaBH_4$ in methanol stereospecifically and quantitatively to the *erythro* alcohol (equation I).

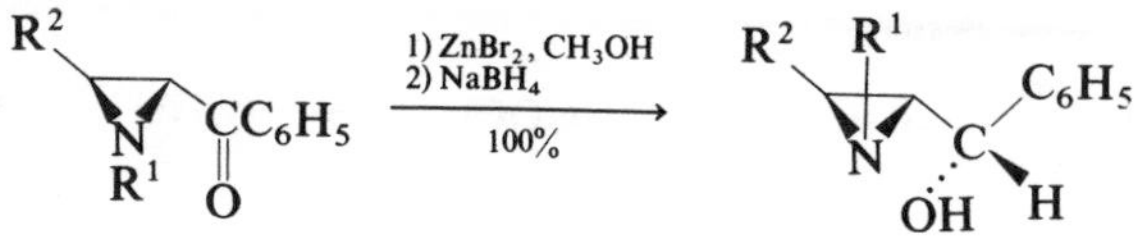

Selective N-deacylation.[4] N,O-Peracylated nucleotides are converted to the O,O-diacylated nucleotides by treatment with zinc bromide in chloroform and either methanol or ethanol ($CHCl_3$/alcohol = 1:4 v/v). Higher alcohols are less effective. However, the mixed solvent CH_2Cl_2–$(CH_3)_2CHOH$ (85:15 v/v) is recommended for detritylation with $ZnBr_2$.

Detritylation (**10**, 461).[5] A detailed study of detritylation of nucleotide derivatives with $ZnBr_2$ in CH_2Cl_2–2-propanol indicates that recombination to the trityl derivative can occur on neutralization. This reaction can be decreased by use of a large excess of $ZnBr_2$ and by neutralization of $ZnBr_2$ by addition of 4-*t*-butylpyridine, which forms a complex with $ZnBr_2$ (m.p. 205–210°) that precipitates on addition of CH_3OH.

[1] H. A. Khan and I. Patterson, *Tetrahedron Letters*, **23**, 5083 (1982).
[2] *Idem., ibid.*, **23**, 2399 (1982).
[3] R. Bartnik, S. Lesniak and A. Laurent, *ibid.*, **22**, 4811 (1981).
[4] R. Kierzek, H. Ito, R. Bhatt, and K. Itakura, *ibid.*, **22**, 3761 (1981).
[5] F. Waldmeier, S. De Bernardini, C. A. Leach, and C. Tamm, *Helv.*, **65**, 2472 (1982).

Zinc chloride, 1, 1289–1292; **2**, 464; **3**, 338; **5**, 763–764; **6**, 676; **7**, 430; **8**, 536–537; **9**, 522–523; **10**, 461–462.

erythro-*Selective aldol reaction.*[1] The lithium enolate of butyrolactone reacts with benzaldehyde to give mainly the expected *threo*-aldol (**1**), since the enolate has the (E)-configuration. Addition of $ZnCl_2$ (0.5 equiv.) reverses the diastereoselection and the *erythro*-aldol (**2**) is the main product. The reaction of the zinc enolate may involve a boat conformation in the transition state, which would result in *erythro*-selectivity.

butyrolactone + C_6H_5CHO $\xrightarrow{\text{THF, } -30°}$ **1** + **2**

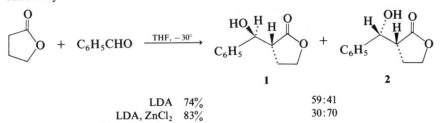

		1 : 2
LDA	74%	59:41
LDA, $ZnCl_2$	83%	30:70

1,3-Diketones.[2] Trimethylsilyl enol ethers are C-acylated by acid chlorides in the presence of zinc chloride or antimony(III) chloride. No other Lewis acids are useful. The reaction provides a convenient route to 1,3-diketones in good to excellent yield. Addition of diethyl ether before quenching of the reaction catalyzed by $ZnCl_2$ improves the yield but has no effect on the $SbCl_3$-promoted reaction.

Examples:

$$(CH_3)_2CHC(OSi(CH_3)_3){=}C(CH_3)_2 + CH_3COCl \xrightarrow{94\%} (CH_3)_2CHC(=O)C(CH_3)_2COCH_3$$

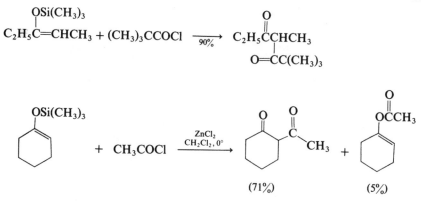

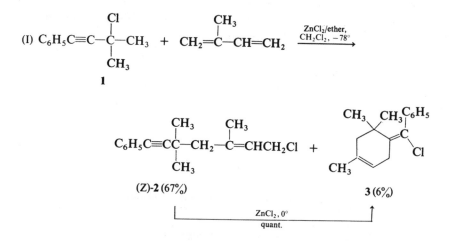

$$\underset{\mathrm{C_2H_5C{=}CHCH_3}}{\overset{\mathrm{OSi(CH_3)_3}}{|}} + (CH_3)_3CCOCl \xrightarrow[90\%]{} \mathrm{C_2H_5\overset{O}{\overset{\|}{C}}CHCH_3} \atop \mathrm{O{=}CC(CH_3)_3}$$

$$\text{1-(trimethylsilyloxy)cyclohexene} + CH_3COCl \xrightarrow{ZnCl_2,\ CH_2Cl_2,\ 0^\circ} \text{2-acetylcyclohexanone}\ (71\%) + \text{1-acetoxycyclohexene}\ (5\%)$$

Addition of propargyl chlorides to 1,3-dienes.[3] The homogeneous catalyst zinc chloride–diethyl ether effects the addition of propargyl chlorides to acyclic 1,3-dienes at low temperatures. The products undergo cyclization at higher temperatures. An example is the addition of the propargyl chloride (**1**) to isoprene (equation I). Similar reactions are obtained on reaction of **1** with 1,3-butadiene, piperylene, and 2,3-dimethyl-1,3-butadiene.

$$\text{(I)}\quad C_6H_5C{\equiv}C{-}C(Cl)(CH_3){-}CH_3\ (\mathbf{1}) + CH_2{=}C(CH_3){-}CH{=}CH_2 \xrightarrow{ZnCl_2/\text{ether},\ CH_2Cl_2,\ -78^\circ}$$

$$C_6H_5C{\equiv}CC(CH_3)_2{-}CH_2{-}C(CH_3){=}CHCH_2Cl\ \ (Z)\text{-}\mathbf{2}\ (67\%) + \mathbf{3}\ (6\%)$$

(3: 1,5,5-trimethyl-4-[C(Cl)C$_6$H$_5$ ylidene]cyclohexene; (Z)-**2** $\xrightarrow[\text{quant.}]{ZnCl_2,\ 0^\circ}$ **3**)

Allyl cations (*cf.*, **10**, 461–462); ***cyclopentenes.*** Allyl chlorides are converted into allyl cations by $ZnCl_2$ complexed with ether. These cations undergo [3+2] cycloaddition to C=C bonds to form cyclopentenes. Unsymmetrical cations are preferentially attacked at the less substituted terminus; unsymmetrical alkenes react in the Markovnikow sense. The most thermodynamically stable cyclopentene is formed.[4]

Examples:

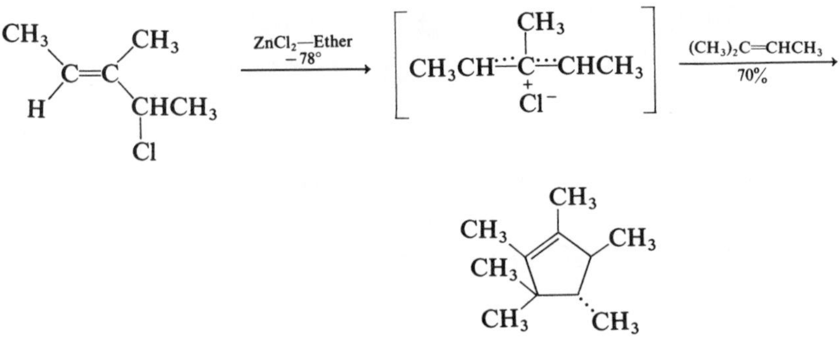

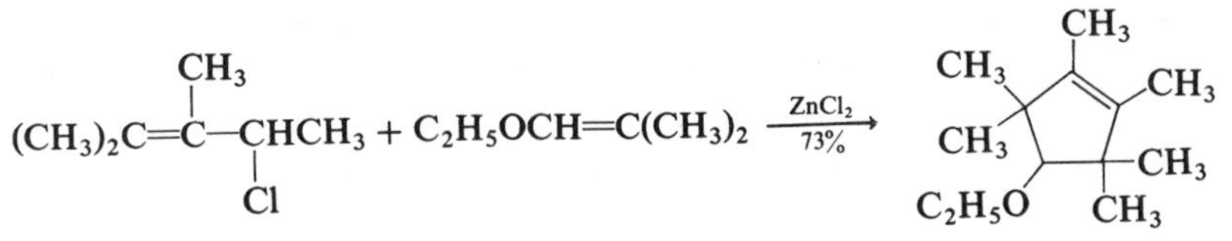

$$\underset{\text{H}}{\overset{\text{CH}_3}{}}\text{C}=\text{C}(\text{CH}_3)\text{CHClCH}_3 \xrightarrow[-78^\circ]{\text{ZnCl}_2\text{—Ether}} \left[\text{CH}_3\text{CH}\cdots\overset{\text{CH}_3}{\underset{+}{\text{C}}}\cdots\text{CHCH}_3 \ \text{Cl}^-\right] \xrightarrow[70\%]{(\text{CH}_3)_2\text{C}=\text{CHCH}_3}$$

(1,2,3,3,4,5-hexamethylcyclopentene: ring with CH_3, CH_3, CH_3 on C=C and C-5; gem-$(CH_3)_2$; CH_3)

$$(\text{CH}_3)_2\text{C}=\overset{\text{CH}_3}{\text{C}}-\underset{\text{Cl}}{\text{CHCH}_3} + \text{C}_2\text{H}_5\text{OCH}=\text{C}(\text{CH}_3)_2 \xrightarrow[73\%]{\text{ZnCl}_2}$$

(cyclopentene bearing CH_3, CH_3, CH_3, CH_3, CH_3, CH_3 and C_2H_5O)

Protected* 1,3-*dicarbonyl compounds.[5] Use of $ZnCl_2$ as catalyst promotes the reaction of enol silyl ethers with 2-ethoxy-1,3-dithiolane (**1**) to give the protected derivative (**2**) of 1,3-dicarbonyl compounds.

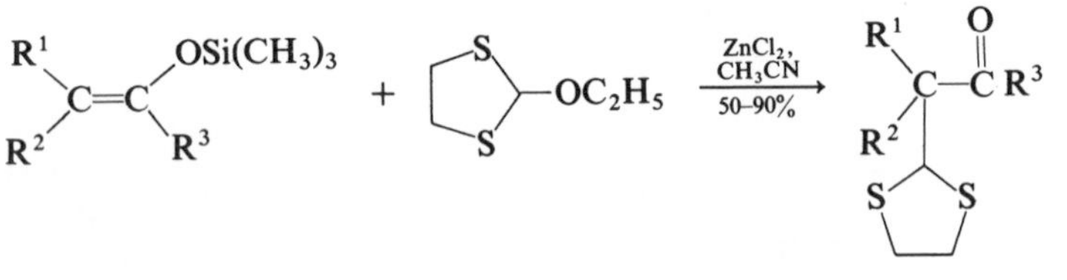

$$\text{R}^1\text{R}^2\text{C}=\text{C}(\text{R}^3)\text{OSi}(\text{CH}_3)_3 + \underset{\mathbf{1}}{\text{2-ethoxy-1,3-dithiolane (OC}_2\text{H}_5)} \xrightarrow[50\text{–}90\%]{\text{ZnCl}_2,\ \text{CH}_3\text{CN}} \underset{\mathbf{2}}{\text{R}^1\text{R}^2\text{C}(\text{1,3-dithiolan-2-yl})-\overset{\text{O}}{\overset{\|}{\text{C}}}\text{R}^3}$$

[1] D. A. Widdowson, G. H. Wiebecke, and D. J. Williams, *Tetrahedron Letters*, **23**, 4285 (1982).
[2] R. E. Tirpak and M. W. Rathke, *J. Org.*, **47**, 5099 (1982).
[3] H. Mayr and H. Klein, *ibid.*, **46**, 4097 (1981).
[4] H. Klein and H. Mayr, *Angew. Chem., Int. Ed.*, **20**, 1027 (1981).
[5] K. Hatanaka, S. Tanimoto, T. Sugimoto, and M. Okano, *Tetrahedron Letters*, **22**, 3243 (1981).

Zinc iodide, 1, 1299; **10,** 462–463.

Diels-Alder catalyst. ZnI_2 greatly accelerates Diels-Alder reactions of furane and some dienophiles. The products can be converted into cyclohexadienols by β-elimination of the oxa bridge.[1]

Example:

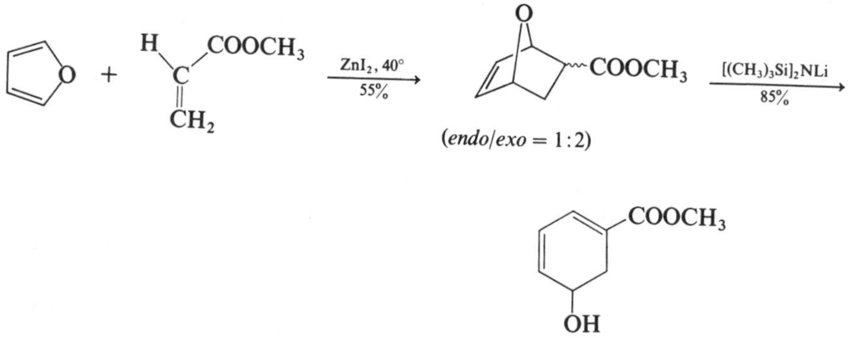

Furan + $H(COOCH_3)C{=}CH_2$ $\xrightarrow[55\%]{ZnI_2,\ 40°}$ 7-oxabicyclo[2.2.1]heptene-$COOCH_3$ (*endo*/*exo* = 1:2) $\xrightarrow[85\%]{[(CH_3)_3Si]_2NLi}$ cyclohexadiene-$COOCH_3$, OH

Other chemists have also reported that high pressures are not essential for Diels-Alder reactions of furane as originally believed (**7**, 161; **10**, 187).[2]

1 F. Brion, *Tetrahedron Letters*, **23**, 5239 (1982).
2 P. F. Schuda and J. M. Bennett, *ibid.*, **23**, 5525 (1982), and references cited therein.

Zinc tosylate, $(CH_3C_6H_4SO_3)_2Zn$. The salt is prepared by reaction of tosic acid monohydrate in water with zinc chloride (J. T. Baker).

Tosylation with inversion.[1] The reaction of a secondary alcohol with zinc tosylate, diethyl azodicarboxylate, and triphenylphosphine (Mitsunobu inversion, **5**, 728) leads to the inverted tosylate in about 80–95% yield. Lithium tosylate is less effective. The reaction is sensitive to steric hindrance.

1 I. Galynker and W. C. Still, *Tetrahedron Letters*, **23**, 4461 (1982).

Zirconium(IV) *n*-propoxide, $(C_3H_7O)_4Zr \cdot 2C_3H_7OH$. Supplier: Alfa.

trans-*Hydrindanes.* Stork's group[1] has shown that an internal Michael addition is useful for construction of the *trans*-hydrindenone **2** (equation I). Thus, treatment of the aldehyde enone **1** with sodium methoxide in methanol gives a mixture of ketols,

(I) 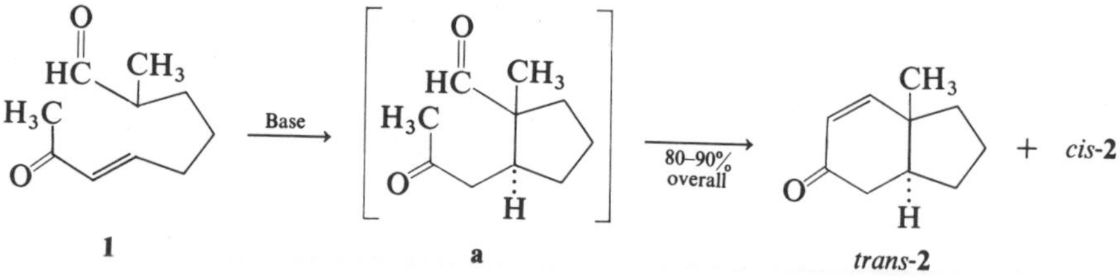 **1** $\xrightarrow{\text{Base}}$ [**a**] $\xrightarrow[\text{overall}]{80\text{–}90\%}$ *trans*-**2** + *cis*-**2**

1) $NaOCH_3$, CH_3OH 2) $-H_2O$	3:1
1) LiOH, CH_3OH 2) $-H_2O$	4:1
1) $Zr(O\text{-}n\text{-}Pr)_4$, LiOH 2) $(CF_3CO)_2O$, DBU	40:1

which is dehydrated to *trans*-**2**/*cis*-**2** in the ratio of 3:1. The stereoselectivity is improved to 4:1 in favor of the *trans*-isomer by use of LiOH as base. But the stereoselectivity markedly changes in favor of *trans*-**2** by cyclization with zirconium *n*-propoxide. Other cations such as $Mg(OCH_3)_2$ and $Ca(OCH_3)_2$ show similar but less pronounced selectivity. The *trans*-selectivity of these cations can be correlated with increasing strength of the metal–oxygen bond. The geometry of the aldehyde enolate may also play a role.

This intramolecular Michael addition has been used to obtain the C/D ring in a short synthesis of adrenosterone (**6**).[2] Thus cyclization of **3** with zirconium tetra-*n*-propoxide results in the desired **4** and the corresponding *cis*-isomer in a 25:1 ratio. The B ring was added to give **5** in 62% yield by reaction of the anion of **4** with 2-(trimethylsilyl)-1-pentene-3-one (**5**, 462). The last ring was then added to obtain **6** by a method developed previously by Stork and Logusch (**10**, 234–235).

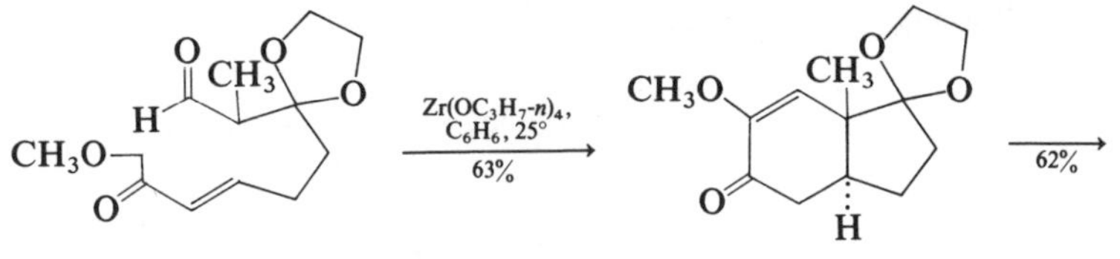

3 → $Zr(OC_3H_7\text{-}n)_4$, C_6H_6, 25°, 63% → **4** → 62% → **5** → 56% → **6**

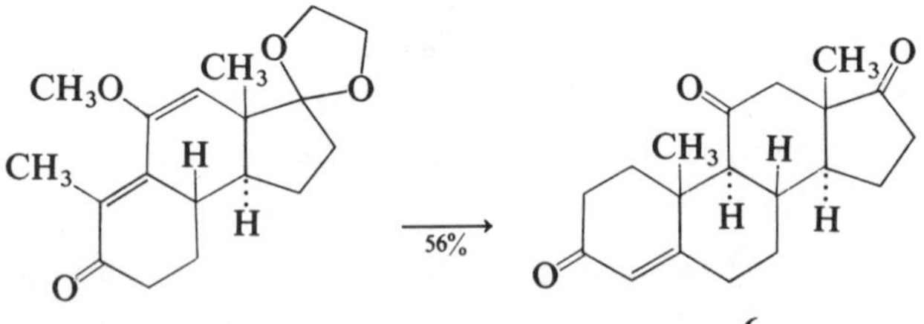

[1] G. Stork, C. S. Shiner, and J. D. Winkler, *Am. Soc.*, **104**, 310 (1982).
[2] G. Stork, J. D. Winkler, and C. S. Shiner, *ibid.*, **104**, 3767 (1982).

INDEX OF REAGENTS ACCORDING TO TYPES

AUTHOR INDEX

SUBJECT INDEX

SUBJECT INDEX